Fred Nordai
471 A Weil Hall

COMMUNICATION, TRANSMISSION, AND TRANSPORTATION NETWORKS

HOWARD FRANK · IVAN T. FRISCH

Network Analysis Corporation,
formerly, University of California, Berkeley

ADDISON-WESLEY PUBLISHING COMPANY
Reading, Massachusetts · Menlo Park, California · London · Don Mills, Ontario

This book is in the
ADDISON-WESLEY SERIES IN ELECTRICAL ENGINEERING

Consulting Editors
David K. Cheng, Leonard A. Gould, Fred K. Manasse

 Printed in the United States of America. Published simultaneously in Canada. Library of Congress Catalog Card No. 78-119666.

To

Our Parents
Herman and Tina Frank
Laszlo and Rose Frisch

Our Wives
Jane and Vivian

PREFACE

Modern society is dominated by a complex of networks for the transmission of information, the transportation of people, and the distribution of goods and energy. This complex includes such diverse systems as the telephone network, gas and oil pipelines, airline networks, and networks of computers serving as data banks and remote processing units. The enormous cost of these networks demands that existing ones be rationally used and new ones intelligently planned.

The purpose of this book is to develop a unified treatment of the fundamental theory of networks. We present the underlying problems and properties common to many classes, and techniques for their solution. We have intended the treatment to be rigorous and include many problems concerning new mathematical developments relevant to the study of physical networks. In all cases, we present algorithms which are computationally efficient. These algorithms usually require the use of a computer for implementation. In a few cases, where a rigorous theory is not available, we give heuristic methods whose merits have been substantiated by successful, documented computer programs.

The material is organized with respect to classes of physical problems rather than the techniques used to solve them. For example, a unified treatment of network vulnerability is presented even though significantly different mathematical techniques are required to treat various aspects of the problem. In spite of this problem orientation, many new theoretical results appear throughout the book. Problems or techniques which do not seem to be presently applicable to the study of physical systems are omitted or mentioned only in problems at the ends of chapters.

All basic concepts are defined and many exercises are included to illustrate them. Hence no prior background in networks is required. However, in almost all cases, the material is developed to the level of present research. Indeed, we believe that much of the material in the book is new or novel. As far as we know, this is the first book to treat the theory of probabilistic graphs in depth. The material we present on the graph theoretic formulation of vulnerability has never appeared in any book; nor has the general solution of the multiterminal synthesis problem. Numerous other algorithms are presented in book form for the first time. Many of these are based on our own research and that of our graduate students.

The book contains parallel treatments of deterministic and probabilistic networks. A knowledge of the fundamental concepts of probability theory is required for the latter. The entire book can be covered in a one-year sequence of courses for students with no previous background in graph theory or networks. Several possible one-semester or quarter courses can also be taught from it. A course treating only deterministic problems could be based on Chapters 1, 2, 3, 5, Sections 1 through 7, 9, 10, and 13 of Chapter 6, and Chapter 7. A course emphasizing probabilistic concepts could be built on Chapters 1, 2, Sections 1, 2, 3, and 6 of Chapter 3, Chapter 4, Sections 1, 2, 3, 5 through 8, and 10 through 12 of Chapter 6, and Chapters 8 and 9. A course combining both deterministic and probabilistic treatments would consist of Chapters 1, 2, Sections 1 through 8 of Chapter 3, Sections 1 through 6 of Chapter 4, Sections 1 through 6 of Chapter 5, and either Sections 1 through 10 of Chapter 6 or Chapter 7 and either Sections 2, 3, and 7 of Chapter 8 or Chapter 9. We have taught courses using a number of these options.

Each of the above sequences requires approximately forty hours of lectures. For students who have already had a course in network flows the following option might be desirable: Chapter 4, Sections 1 through 6 of Chapter 5, Sections 5 through 13 of Chapter 6, Chapter 7, and Chapter 8. A shorter course on vulnerability can be taught using Chapters 1, 2, Sections 1, 2, 3, and 6 of Chapter 3, and Chapters 7, 8, and 9. For students with a background in network flow theory, Chapters 7 and 8 can form the basis of a twenty-hour course on vulnerability. We have taught several such variations at Berkeley.

ACKNOWLEDGMENTS

We are indebted to many people for their generous help. We both were initiated into the study of flow networks as doctoral students—Howard Frank, while working with S. L. Hakimi at Northwestern University, and Ivan T. Frisch, while working with W. H. Kim at Columbia University. At Berkeley these studies were supported in part by an Army Research Office Contract on "Information Processing" and in part by grants from the Joint Services Electronics Program and the National Science Foundation. In this regard we are grateful to J. E. Norman of the Army Research Office, Durham, North Carolina, for his continued interest and encouragement. Some material was developed while one of us (I. T. F.) was on a leave of absence at the Bell Telephone Laboratories, Whippany, New Jersey, as a Ford Foundation Resident in Engineering Practice. The original motivation for some of the work discussed in Chapter 7 is due to D. Gillette, Executive Director, Transmission Systems, at the Bell Laboratories in Holmdel, and to A. R. Eckler and F. D. Benedict, who headed groups at the Bell Laboratories in Whippany, working on the vulnerability problem. Some of the results presented in the book were developed while one of us (H. F.) was on a leave of absence at the Office of Emergency Preparedness (O.E.P.), Executive Office of the President, Washington, D.C. The enthusiastic support of R. Truppner, Director of the O.E.P.'s National Resource Analysis Center (N.R.A.C.), and R. H. Kupperman, Chief of the N.R.A.C.'s Systems Evaluation Division, enabled us to apply many of our results to problems of national importance. Much of the work at the O.E.P. was strongly influenced by the far-sighted recommendations of David Rosenbaum, presently with the MITRE Corporation, Maclean, Virginia.

We are grateful to L. A. Zadeh and E. S. Kuh, who encouraged us to develop our network courses at Berkeley, and to L. Farbar, Director of Engineering Extension at the University of California at Berkeley, for his aid in presenting a two-week extension course. Helen Barry was most helpful in administering this course. In addition, portions of the book were presented in the UCLA extension courses on Large Scale Systems, organized by C. T. Leondes, and Queueing Systems, organized by L. Kleinrock, and in short courses sponsored by the Network Analysis Corporation.

We have incorporated many suggestions from our former graduate students: J. Ayoub, S. Chaubey, W. S. Chou, M. El-Bardai, M. Malek-Zaverei, S. Sankaran, D. K. Sen, N. P. Shein, and especially P. Jabedar-Maralani. B. Rothfarb made numerous valuable suggestions about several versions of the manuscript and was one of the reviewers of the final manuscript. A number of our other colleagues have read portions of the manuscript and offered criticisms. Among them are F. T. Boesch, of the Bell Telephone Laboratories, S. L. Hakimi, of Northwestern University, T. C. Hu, of the University of Wisconsin, W. S. Jewell, of the University of California, Berkeley, L. Kleinrock, of the University of California, Los Angeles, D. J. Kleitman, of the Massachusetts Institute of Technology, E. Lawler, of the University of California, Berkeley, W. Mayeda, of the University of Illinois, and K. Steiglitz, of Princeton University.

Jane Frank and Vivian Frisch typed many pages of original notes; helped with all sorts of clerical details; strove mightily to suppress laughter while we argued bitterly over whether ρ should represent connectivity or number of components; and acted as witnesses while the order of our names on the title page was decided by the flip of a quarter.

Glen Cove, New York
September 1970

H. F.
I. T. F.

NOTES TO THE READER

1. Sections marked with an asterisk contain advanced topics not essential to the continuity of the main development and may be omitted on first reading.
2. Unless otherwise indicated, a reference to a section, figure, formula, etc., is to an item in the same chapter as the reference.
3. References in the text are indicated by the first two letters of the author's surname followed by an indexing number. The references are listed alphabetically at the end of each chapter.
4. Many of the problems at the ends of chapters are based on results from published papers. In such cases, a reference to the appropriate paper is given in the problem.

CONTENTS

Chapter 5. Multiterminal Analysis and Synthesis

Chapter 6. Minimum Cost Problems

Chapter 7. Connectivity and Vulnerability of Deterministic Graphs

Chapter 8. Connectivity and Vulnerability of Probabilistic Graphs

CHAPTER 1

GRAPHS AND PHYSICAL MODELS

1.1 GRAPHS AS STRUCTURAL MODELS

Many systems involve the communication, transmission, transportation, flow, or movement of commodities. The commodity under consideration may be a tangible item, such as railway cars, automobiles, oil drums or water, or an intangible item, such as information, disease, "friendship," or heredity. Thus, a highway system, a telephone network, an interconnection of warehouses and retail outlets, a power grid, or an airline network all involve the flow of commodities through a network. Often these networks can be modeled by a mathematical entity called a *graph*.

A graph may be considered to be a collection of points called vertices connected by lines called branches. The modeling of some physical systems by graphs is quite natural, while for others the relationship between the graph theoretic model and the original system is extremely subtle. In the former group are communication or transportation systems. The branches of the graph can represent roads, telegraph wires, railroad tracks, airline routes, water pipes—in general, channels through which the commodities are transmitted. The vertices of the graph can represent communities, highway intersections, telegraph stations, railroad yards, airline terminals, water reservoirs and outlets—in general, points where flow originates, is relayed, or terminates.

Two physical networks may be structurally similar, but have significantly different characteristics. For example, the interconnections in both an electrical network and a telephone system can be specified by a graph. However, the branches of the electrical network model are characterized by parameters such as resistance, inductance, or capacitance whereas the branches of the telephone network model are characterized by parameters such as the number of wires per trunk, the maximum transmission rate, and the cost per unit length. We must account for these parameters as part of the model, to obtain meaningful results.

With each branch and vertex of a graph, we can associate a number of parameters that represent the natural limitations and capabilities of the branches and vertices. For example, a power system might be modeled by a graph in which the branches represent power lines and the vertices represent power generation stations, substations, and customers. The important parameters of the system are incorporated into the model as numbers, or *weights*, on the branches and vertices of the graph. These weights may be either fixed or random. Thus, for the power system, a typical vertex

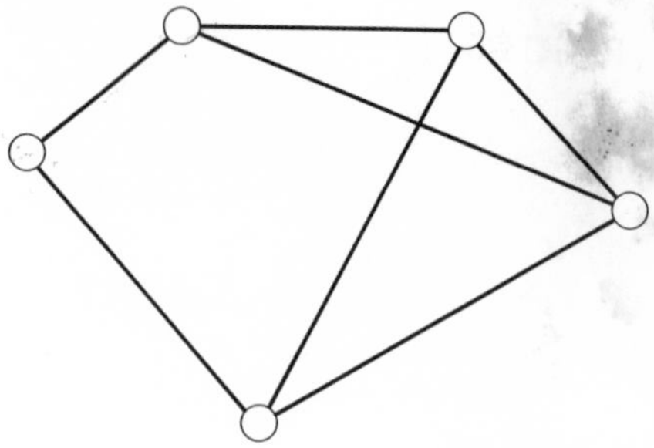

Fig. 1.1.1 Pictorial representation of a graph.

representing a power generator might have the following weights: maximum power output, number of generators at the station, reliability of each generator, and cost per kilowatt-hour. A typical branch might have three weights corresponding to the maximum power-handling capacity, reliability, and cost.

The purpose of the branch and vertex weights is to include nonstructural information into the graph theoretic model of a system. The following examples further illustrate this point.

Example 1.1.1 A traffic network. Let each vertex of a graph represent a city. Two vertices are connected by a branch if there is a highway between the corresponding cities. A number is associated with each branch to indicate the length of the corresponding highway. A second weight represents the maximum number of cars that can be accommodated per unit length per unit time, and a third branch weight could be the speed limit.

Example 1.1.2 An airline system. Let each vertex of a graph represent an airline terminal. Two vertices are connected by a branch if there is a direct air link between the terminals. Each vertex of the graph has a weight indicating the number of airplanes that the terminal can handle in a given interval of time. This vertex weight could be a fixed number if the traffic-handling capability of the terminal is assumed to be constant, it could be time-varying if the traffic-handling capability fluctuates, or it could be a random variable if it depends on unpredictable elements such as weather.

Example 1.1.3 A telegraph system. Let each vertex of a graph represent a telegraph station. Two vertices are connected if there is a telegraph wire between the corresponding telegraph stations. That is, there will be a branch between two vertices if the stations can communicate directly without any intermediate relay station. The number of telegraph operators at each station is limited, so there is a maximum number of messages which can be simultaneously transmitted and received. This can be included in the model by an appropriate vertex weight. The maximum number of messages on a branch is determined by the number of telegraph wires. We weight each branch by the maximum number of simultaneous messages that can be handled. Another possible consideration is the time delay required to send a message through the network. At each station there will be a "waiting" time for an available operator or channel, as well as the time required to transmit the message. The total time delay at a given station will usually be a random variable and can be represented by an additional vertex weight.

Example 1.1.4 An economic model. Suppose we are given a system of factories, warehouses, and outlets connected by a set of highways, railroads, and waterways. This system can be structurally modeled by a graph with the branches representing transportation channels and the vertices representing factories, warehouses, and outlets. In the graph, the factories are source vertices, the outlets are terminal vertices, and the warehouses are relay vertices. We could further distinguish between the vertices. For example, some source vertices might produce one type of commodity while others produce a different type. Among the many possible vertex weights are the rate of production of the ith commodity, the production cost per unit of the ith commodity, and the time required to produce a unit of the ith commodity. For the relay vertices, a single weight representing the storage space might suffice. A terminal vertex could be weighted with numbers which indicate the types of commodity which are sold at that vertex, the price of each commodity, the amount of local storage available, and the demand (usually a random variable) for each commodity. Typical branch weights could be maximum volume per unit time, cost of transportation per unit of ith commodity, and transmission time per unit of ith commodity.

1.2 TYPICAL PROBLEMS

The utility of graphs as models depends on the nature of the physical problem to be solved. The type of problem for which a graph is most obviously useful is that of connectivity. Given a system and its graph, we might be interested in determining whether or not a particular commodity can be transported from one given location to another. We are interested in finding a "path" between two given vertices over which the commodity can be sent. A more general, but similar, problem is to establish whether or not a given commodity can be sent from *any* point to *any other* point. Here, we must determine whether there is at least one path from any vertex to any other vertex.

The connectivity problems mentioned above are structural problems. The existence of a path between a pair of vertices implies that some amount of flow can be transmitted between these two vertices. There is no information about the quantity of flow which can be sent. To include this information, we must consider weighted graphs. Suppose we weight each branch and vertex by a number which indicates the maximum amount of flow that it can accommodate. These weights represent the capacities of the channel, sources, terminals, and relay points in the original system. We might then ask: What is the maximum amount of flow which can be sent between a given pair of vertices? In a power system, this number might correspond to the maximum power that a particular generating station could supply to a particular user; in a telegraph system, it might represent the maximum attainable rate of information transmission between two telegraph stations.

The problem of finding the maximum amount of a given quantity which can be transmitted between two points is known as the Maximum Flow Problem. A generalization of this problem is to find the maximum amount of several commodities which can be *simultaneously* sent between several pairs of points. This is known as the Multicommodity Maximum Flow Problem. Both of these are analysis problems. Given a system and its graph theoretic model, we can attempt to analyze the graph and find the maximum flow rates. We can also formulate an analogous synthesis problem. Suppose we are given a set of stations and maximum flow-rate requirements. We would like to design a system which satisfies these requirements. Furthermore, since there may be many such systems, we would like to select a system which is in some sense "optimal." One possible optimality criterion could be minimum cost. This synthesis problem has many variations. For example, we may assume that the behavior of the network to be designed can be precisely predicted. In this case, flow-rate requirements can be exactly satisfied. In other cases, there may be random elements in the design or behavior of the system and so the meaning of the phrase "satisfy flow rate requirement" must be interpreted probabilistically.

The preceding connectivity and maximum-flow problems are closely related to a group of problems which may be termed problems of "vulnerability" and "reliability." Here, we are given a system operating in a "hostile" environment. This hostility may be the result of natural disturbances, equipment failure, or enemy attack. The effect of the hostility is to disrupt communications. Given an existing system, we must analyze it to determine the system degradation which could occur. Given a set of performance criteria, we must design a system which minimizes the possible system degradation. Again, both the analysis and synthesis problems can be posed in either deterministic or probabilistic terms.

It is usually possible to route a given commodity over many different paths, with one routing possibly better than another. If a poor route is selected, it may block an additional flow which might otherwise have been established. Among the problems we must therefore consider are finding the shortest, the least expensive, or the most reliable route for a given commodity.

In many physical systems network traffic is a function of time. All available routes between a pair of stations may be occupied and so it is impossible to send additional flow between these stations. Thus, a subscriber at one of these stations must wait until a channel is available. The expected value of the time he must wait is an important parameter of the system. Typical analysis problems are to find the average waiting time and to investigate the effect of network structure and various routing procedures on this waiting time.

The examples given above should suffice to point out the wide range and applicability of the model and the nature of the problems that can be posed. To solve these problems, we require some concepts and results from the theory of graphs. These are the subject of the next chapter.

1.3 RELATED READING

The material relevant to the study of networks is scattered throughout the journals of a number of different disciplines, because, historically, problems modeled by graphs have arisen in varied and seemingly unrelated situations. Furthermore, solutions to these problems have been achieved by the use of several different branches of mathematics.

As we have seen, one motivation for the study of networks is the investigation of traffic in communication systems. A. K. Erlang, whose work is summarized by Brockmeyer *et al.* [BR1], was one of the earliest and most significant innovators in this area. Since his work, a tremendous body of literature has developed. References to this literature can be found in the recent books by Benes [BE2] and Kleinrock [KL1]. A good summary of the physical bases of these problems is given by Rubin and Haller [RU1]. The results of extensive computer simulation are given by Weber [WE1].

The second major impetus for the development of the theory of networks was the formulation of mathematical models for economic and distribution systems, both of which are included under the generic title of operations research. Probably the earliest link between problems in this area and communication networks was Hitchcock's solution to the "transportation problem" [HI1]. A number of the major results and references in this area are given by Ford and Fulkerson [FO1], who have been among the most original and prolific contributors to the development of the theory of flows in networks.

Another stimulus to the study of networks has been in the study of steady-state flow of information through a communication system. This approach seems to have been first adopted by Elias, Feinstein and Shannon [EL1], although Mayeda [MA1] was the first to formulate and solve significantly different problems arising primarily from the new viewpoint. This new stimulus came at a time when many electrical engineers were prepared to apply their knowledge of graph theory acquired through the study of electrical networks. Hence, a new body of literature was developed by this group. Some of the early work in this area is given by Kim and Chien [KI1].

The primary mathematical disciplines relevant to the study of networks in this book are graph theory, combinatorics, probability theory, and statistics. Secondary use is made of mathematical programming and queuing theory. All the necessary results from the theory of graphs are derived in this book and hence no outside references are required. However, for those readers wishing to delve more deeply into this subject, a number of excellent books and bibliographies are now available [BE3, BE4, BU1, HA3, HA4, KI1, KO1, OR1, DE1, SE1, TU1, TU2, ZY1].

The book is also self-contained with regard to theorems from combinatorics. Indeed, much of the material on flows in networks actually comprises a significant new branch of combinatorics. We do not emphasize this viewpoint. However, as an example, Ford and Fulkerson [FO1] use the theory of flows to solve many purely combinatorial problems such as the assignment of entries to matrices of zeros and ones to satisfy various constraints. The classical results of combinatorics can be found in several references [BE1, RI1, RY1].

Probability theory and statistics are the only essential disciplines in which we must assume some background on the part of the reader. However, except for a few sections, only elementary knowledge is required. Furthermore, the book is organized so that the material on random networks appears in separate chapters or sections parallel to corresponding sections on deterministic networks. Hence, it is possible to read those sections which do not require probabilistic theory independently of the others. The required material on probability and statistics can be found in references [CR1, FI1, LE1, MO1]. Reference [CR1] is a classic work in probability theory and statistics, which contains all of the necessary foundations in probability theory. Reference [KO1] is an elementary text on statistics while references [FI1] and [LE1] are intermediate and advanced level texts, respectively. Fisz's book [FI1] may be considered as the primary reference for both probability theory and statistics. Queuing theory is required in only one chapter of the book and only the most elementary concepts are needed. For a more extensive treatment and a guide to the literature in queueing theory as applied to networks one can refer to the work of Kleinrock [KL1].

Mathematical programming is useful in the study of communication networks, first because it is an efficient computational tool and second because it enables one to place many algorithms and results in a general framework. We regard mathematical programming as a secondary discipline for this book, since we can derive most of our results without resorting to it. Furthermore, where flow techniques are applicable, they are usually more efficient than general programming methods. The few theorems we do need are given in the next chapter without their proofs. The proofs of the theorems and the general theory of mathematical programming are expounded in a number of excellent books among which are those of Dantzig [DA1], Hadley [HA1, HA2], Berge and Ghouila-Houri [BE4], and Simonard [SI1]. The relationship between programming techniques and flow problems is developed in the books by Kaufmann [KAI] and Hu [HU1]. Further generalizations of graph theory and programming techniques using matroids appear in a book being written by E. Lawler.

REFERENCES

BE1 E. F. Beckenbach (ed.), *Applied Combinatorial Mathematics*, Wiley, New York, 1964.

BE2 V. R. Benes, *Mathematical Theory of Connecting Networks and Telephone Traffic*, Academic Press, New York, 1965.

BE3 C. Berge, *The Theory of Graphs and Its Applications*, Wiley, New York, 1962.

BE4 C. Berge and A. Ghouila-Houri, *Programming, Games and Transportation Networks*, Wiley, New York, 1962.

BR1 E. Brockmeyer, H. L. Helstrom, and A. Jensen, *The Life and Works of A. K. Erlang*, Danish Academy of Technical Sciences, No. 2, Copenhagen, 1948.

BU1 R. G. Busacker and T. L. Saaty, *Finite Graphs and Networks: An Introduction with Applications*, McGraw-Hill, New York, 1965.

CR1 H. Cramer, *Mathematical Models of Statistics*, Princeton University Press, Princeton, 1963.

DA1 G. B. Dantzig, *Introduction to Linear Programming*, Princeton University Press, Princeton, 1963.

DE1 N. Deo, *An Extensive English Language Bibliography on Graph Theory and Its Applications*, National Aeronautics and Space Administration, Jet Propulsion Laboratory, California Institute of Technology, Pasadena, Cal., 1969.

EL1 P. Elias, A. Feinstein, and C. E. Shannon, "A Note on the Maximum Flow through a Network," *IRE Trans. Inform. Theory* **IT-2,** 117–119 (1956).

FI1 M. Fisz, *Probability Theory and Mathematical Statistics*, Wiley, New York, 1963.

FO1 L. R. Ford, Jr., and D. R. Fulkerson, *Flows in Networks*, Princeton University Press, Princeton, 1962.

HA1 G. Hadley, *Linear Programming*, Addison-Wesley, Reading, Mass., 1962.

HA2 G. Hadley, *Nonlinear and Dynamic Programming*, Addison-Wesley, Reading, Mass., 1964.

HA3 F. Harary, *Graph Theory*, Addison-Wesley, Reading, Mass., 1969.

HA4 F. Harary, R. Z. Norman, and D. Cartwright, *Structural Models: An Introduction to the Theory of Directed Graphs*, Wiley, New York, 1965.

HI1 F. L. Hitchcock, "The Distribution of a Product from Several Sources to Numerous Localities," *J. Math. Phys.* **20,** 224–230 (1941).

HU1 T. C. Hu, *Integer Programming and Network Flows*, Addison-Wesley, Reading, Mass., 1969.

KA1 A. Kaufmann, *Graphs, Dynamic Programming, and Finite Games*, Academic Press, New York, 1967.

KI1 W. H. Kim and R. T. Chien, *Topological Analysis and Synthesis of Communication Networks*, Columbia University Press, New York, 1962.

KL1 L. Kleinrock, *Communication Nets: Stochastic Message Flow and Delay*, McGraw-Hill, New York, 1964.

KO1 D. König, *Theorie der Endlichen und Unendlichen Graphen*, Chelsea, New York, 1950.

LE1 E. L. Lehman, *Testing Statistical Hypotheses*, Wiley, New York, 1959.

MA1 W. Mayeda, "Terminal and Branch Capacity Matrices of a Communication Net," *IRE Trans. Circuit Theory* **CT-7.** 261–269 (1960).

MO1 A. M. Mood and F. A. Graybill, *Introduction to the Theory of Statistics*, McGraw-Hill, New York, 1963.

OR1 O. Ore, *Theory of Graphs*, American Mathematical Society, Vol. XXXVIII, Colloquium Publications, Providence, R.I., 1962.

RI1 J. Riordan, *An Introduction to Combinatorial Analysis*, Wiley, New York, 1958.

RU1 M. Rubin and C. E. Haller, *Communication Switching Systems*, Reinhold, New York, 1966.

RY1 H. J. Ryser, *Combinatorial Mathematics*, Carus Mathematical Monograph 14, Mathematical Association of America, Wiley, New York, 1963.

SE1 S. Seshu and M. B. Reed, *Linear Graphs and Electrical Networks*, Addison-Wesley, Reading, Mass., 1961.

SI1 M. Simonard, *Linear Programming*, Prentice-Hall, Englewood Cliffs, N.J., 1966.

TU1 J. Turner, *Key Word Indexed Bibliography on Graph Theory*, Stanford Research Institute Report, SRI Project 145591-W.O.A14, February 1964.

TU2 W. Tutte, *Connectivity in Graphs*, University of Toronto Press, Toronto, 1968.

WE1 J. H. Weber, "A Simulation Study of Routing and Control in Communication Networks," *Bell System Tech. J.* **43.** 2639–2676 (1964).

ZY1 A. A. Zykov, "Bibliography on Graph Theory," in *Theory of Graphs and Its Applications: Proceedings of the Symposium Held in Smolenice in June 1963*, Academic Press, New York, 1964.

CHAPTER 2

DEFINITIONS AND FUNDAMENTAL PRINCIPLES

2.1 DIRECTED AND UNDIRECTED GRAPHS

We have already given an informal description of a graph. It is necessary to distinguish between two types of graphs, directed and undirected. In a directed graph, some of the branches may not be able to accommodate "flow" in both directions. In an undirected graph, all branches are "two-way" branches. We can make these concepts precise by the following definitions. A *directed graph*, G, consists of a set of elements called *vertices* and a set of ordered pairs of vertices called *directed branches*. The set of vertices is denoted by the symbol V and the set of directed branches by the symbol Γ. To denote G, we write $G = (V,\Gamma)$. The graph G is *finite* if both V and Γ are finite sets. Unless otherwise stated, all graphs considered are finite.

The kth vertex in V is denoted by v_k, where k is any lower-case letter or a numeral. Thus, if V has n elements, we can write V as $V = \{v_1,v_2,\ldots,v_n\}$. Pictorially, the elements of V are represented by circles or dots. The kth branch in Γ is denoted by $b_k(i,j)$ and is said to be directed from v_i to v_j. The branch may be represented by a line connecting vertices v_i and v_j, with an arrowhead pointing from v_i to v_j. Note that as we have defined a graph, there may be at most one branch from any vertex v_i to any vertex v_j. Vertex v_i is called the *initial* vertex of $b_k(i,j)$ and vertex v_j is called the *terminal* vertex. Branch $b_k(i,j)$ is said to be *incident out of* its initial vertex v_i and *incident into* its terminal vertex v_j or simply *incident* at v_i and v_j; v_i and v_j are said to be *adjacent*.

The symbol $b_k(i,j)$ is mnemonically convenient but rather awkward to manipulate. Often, when the meaning is clear, branch $b_k(i,j)$ will be indicated by either the ordered pair (i,j) or by the symbol b_k. The symbol (i,j) will be used when the only fact of interest about the branch is its incidence relation. If for all i and j we allow more than one branch from v_i to v_j, but no more than s branches from v_i to v_j, we will designate the branches by $(i,j)_1,(i,j)_2,\ldots,(i,j)_r$ $(r \leq s)$ and call the resulting structure an *s-graph*. On the other hand, the incidence relations of the branches may not be important and only their indexing may be of interest. It will sometimes be more convenient to denote the set of branches as $\Gamma = \{b_1,b_2,\ldots,b_m\}$, where m is the total number of branches in the graph. As an example of our notation, if $G = (\{v_1,v_2,v_3,v_4\}, \{(1,1), (1,2)_1, (1,2)_2, (2,3), (3,4), 4,2)\})$, the pictorial representation of the 2-graph is as given in Fig. 2.1.1. Henceforth, we will not distinguish between graph G and its pictorial representation.

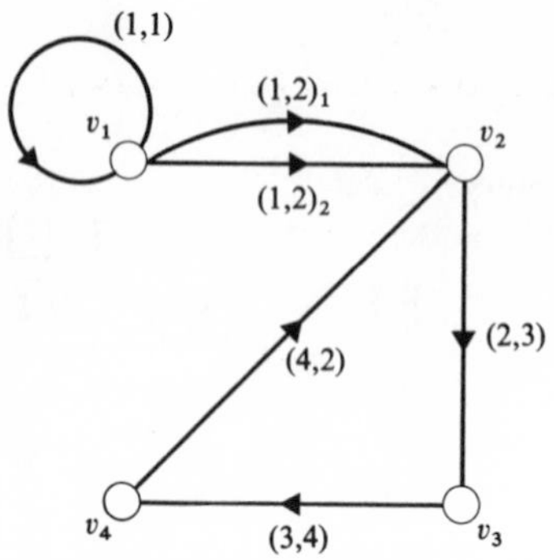

Fig. 2.1.1 Pictorial representation of a 2-graph.

Often, the existence of a branch from v_i to v_j may imply the existence of a branch from v_j to v_i. In this case, we call the graph *symmetric*. (Such a graph corresponds to a symmetric relation.[1]) For some problems, if $(i,j) \in \Gamma$ implies that $(j,i) \in \Gamma$ for all i and j, then it will be useful to replace the two ordered pairs (i,j) and (j,i) by a single *unordered* pair denoted by $[i,j]$. In the pictorial representation of the graph, we replace the two directed branches (i,j) and (j,i) by an undirected branch $[i,j]$ represented by a line with no arrowhead. Thus, the graph in Fig. 2.1.2(a) can be replaced by the graph in Fig. 2.1.2(b). This idea leads to the following definition. An *undirected graph* consists of a set of vertices and a set of unordered pairs of these vertices, called branches. In the remainder of the book, an *undirected* branch between v_i and v_j will be denoted by $[i,j]$ while a *directed* branch from v_i to v_j will be denoted by (i,j).

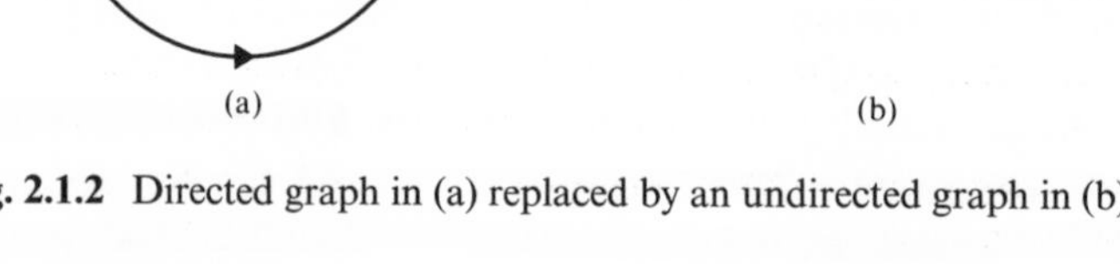

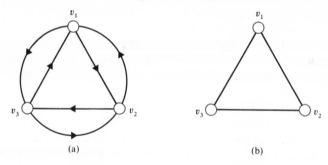

Fig. 2.1.2 Directed graph in (a) replaced by an undirected graph in (b).

The class of systems that can be modeled by graphs is immense. From the definition of the graph G, any system that can be modeled by a relation $\mathcal{R}$ on a set X of elements can be modeled by a graph. The relation is easily represented by a graph G by letting $X = V$ and $\mathcal{R} = \Gamma$.

---

[1] A *relation* $\mathcal{R}$ on a set X is a set of ordered pairs of (x,y) such that $x \in X$ and $y \in X$.

Example 2.1.1 Let the elements of X represent power generators and users. Let (x,y) be a member of $\mathscr{R}$ if and only if user y can receive power directly from generator x. Let $X = V$ and $\mathscr{R} = \Gamma$. The resulting graph is a possible model for the power system. The same model will obviously result if X represents information sources and terminals (such as transmitters and receivers) and $(x,y) \in \mathscr{R}$ if and only if x can transmit information directly to y. If the elements of X are factories and outlets, then $(x,y) \in \mathscr{R}$ if and only if factory x can ship goods directly to outlet y. In all these cases, relay stations, switching centers, or warehouses can be included by considering them as both generators and users, receivers and transmitters, or factories and outlets, respectively.

Example 2.1.2 Let X be the set of integers from 1 to 10 and let $\mathscr{R}$ be the relation "divisible by." In other words, $(x,y) \in \mathscr{R}$ if and only if x/y is an integer. The graph G with $V = \{v_1,\ldots,v_{10}\}$ and $\Gamma = \mathscr{R}$ is given in Fig. 2.1.3.

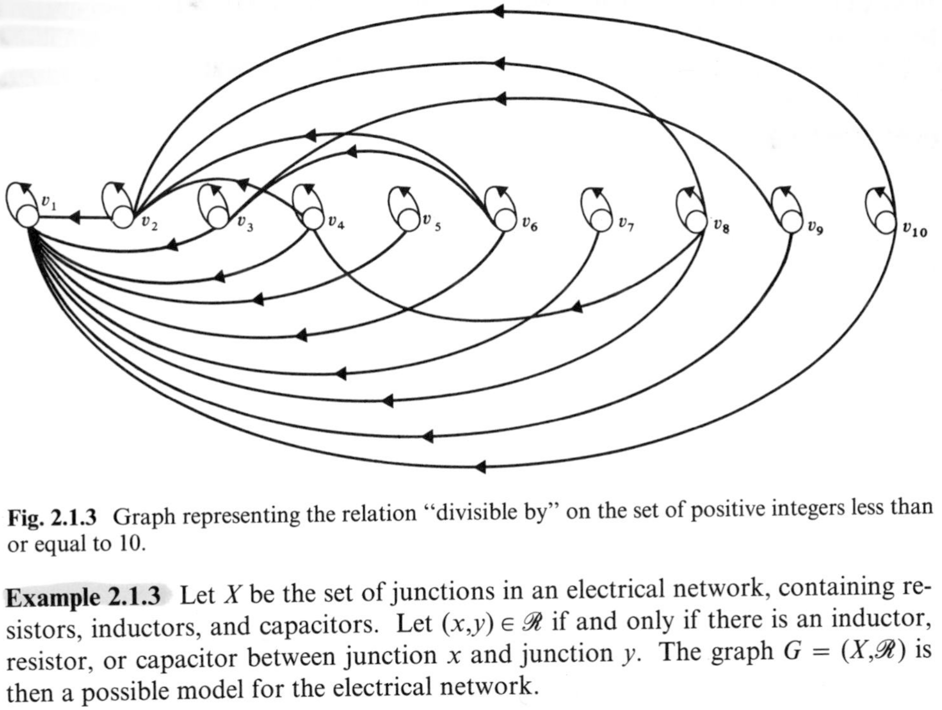

Fig. 2.1.3 Graph representing the relation "divisible by" on the set of positive integers less than or equal to 10.

Example 2.1.3 Let X be the set of junctions in an electrical network, containing resistors, inductors, and capacitors. Let $(x,y) \in \mathscr{R}$ if and only if there is an inductor, resistor, or capacitor between junction x and junction y. The graph $G = (X,\mathscr{R})$ is then a possible model for the electrical network.

As we shall see in later chapters, a graph theoretic model G of a system can yield many significant properties of the system. Thus, if the graph represents a power system, we can use it to determine such factors as the possible routes over which power can be sent and the number of relay stations that must be destroyed before power transmission is interrupted for some users. However, to find other properties of the system we introduce additional information in the form of branch and vertex weights. Consequently, we consider weighted graphs in the next section.

2.2 WEIGHTED GRAPHS

A *weighted graph* is a graph in which numbers, called weights, are associated with the branches or vertices. We will allow any number of weights to be associated with each vertex or branch. The set of weights associated with vertex v_k will be denoted by $\{w_i(k)\}$. The set of weights associated with $b_k(i,j)$ will be denoted by $\{w_r[b_k(i,j)]\}$ and will usually be abbreviated as $\{w_r(i,j)\}$ or $\{w_{k_r}\}$. We shall say that a weighted graph is *symmetric* if $w_k(i,j) = w_k(j,i)$ for all i,j, and k and *pseudo-symmetric* if

$$\sum_{i \in V} w_k(i,j) = \sum_{i \in V} w_k(j,i)$$

for all j and k.

Some symbols will be reserved to denote weights of a specific nature throughout the book. Thus, the weight $c(i,j)$ will represent the capacity for transmission through the medium (a wire, channel, or road) represented by branch (i,j); the symbol $f(i,j)$ will represent a flow (information, goods, water, etc.) through a branch (i,j); and the symbol $p(i,j)$ will be a probabilistic weight associated with (i,j). For example, $p(i,j)$ could be the probability that $(i,j) \in \Gamma$.

Example 2.2.1 Suppose we are given a telegraphic communication system in which there are three stations, s_1,s_2,s_3, with wire connections between them. Furthermore, assume that the wire between s_i and s_j is of length $l(i,j)$, has a cost $\alpha l(i,j)$, an information capacity $c(i,j)$ in bits per second, and a probability $p(i,j)$ of normal operation. Also assume that each station s_j has capacity $c(j)$ and probability $p(j)$ of normal operation. The system can be represented by the weighted graph G shown in Fig. 2.2.1, where vertex v_k corresponds to station s_k and branch (i,j) corresponds to the wire between s_i and s_j.

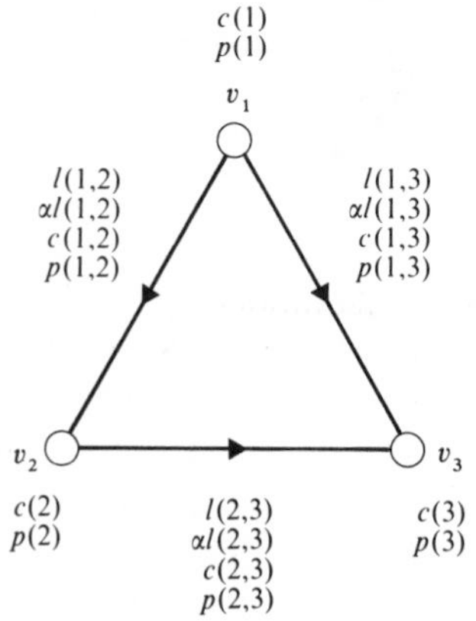

Fig. 2.2.1 A weighted graph.

Example 2.2.2 A flow problem. Suppose we associate real, nonnegative weights $f(i,j)$ and $c(i,j)$ with each branch (i,j) in the graph such that $f(i,j) \leq c(i,j)$. We

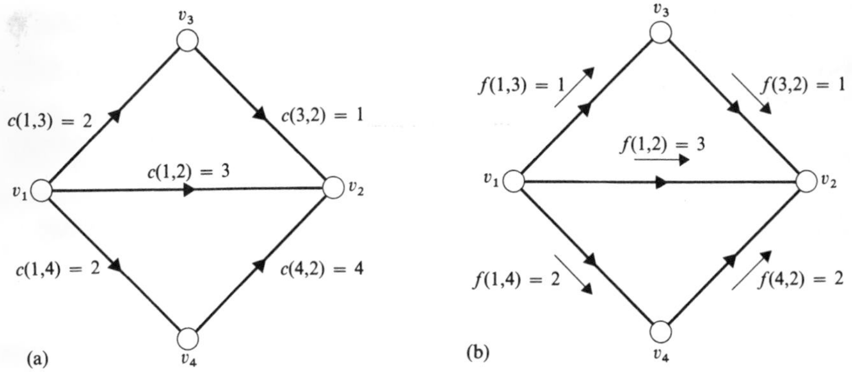

Fig. 2.2.2 (a) Graph with branch capacities. (b) Flows in graph shown in (a).

further require that at every vertex of G, except for a possible "source" vertex v_s and a possible "terminal" vertex v_t, the sum of the incoming flows is equal to the sum of the outgoing flows. This condition is a conservation of flow condition and may be written as

$$\sum_{\substack{v_i \in V \\ (j,i) \in \Gamma}} f(j,i) - \sum_{\substack{v_j \in V \\ (i,j) \in \Gamma}} f(i,j) = 0 \qquad \text{for } i \neq s,t. \tag{2.2.1}$$

If we now attempt to find a set of $f(i,j)$ subject to the above constraints, we are said to be seeking a feasible flow pattern $\mathscr{F}$ or to be solving a "flow problem." Again, both $f(i,j)$ and $c(i,j)$ may be random variables. Thus, for the graph shown in Fig. 2.2.2(a), a feasible flow $\mathscr{F}$ from source v_1 to terminal v_2 is

$$f(1,3) = f(3,2) = 1, \tag{2.2.2a}$$

$$f(1,2) = 3, \tag{2.2.2b}$$

$$f(1,4) = f(4,2) = 2. \tag{2.2.2c}$$

2.3 PATHS AND CIRCUITS

Since our primary concern is systems modeled by graphs, we must first develop a framework and language for graph theory. We present only those concepts which are specifically used in the sequel.

Let $G_1 = (V_1,\Gamma_1)$ and $G_2 = (V_2,\Gamma_2)$ be graphs. The *union* $G_3 = G_1 \cup G_2$ of G_1 and G_2 is a graph $G_3 = (V_3,\Gamma_3)$ for which $V_3 = V_1 \cup V_2$ and $\Gamma_3 = \Gamma_1 \cup \Gamma_2$. A *subgraph* G_1 of the graph $G = (V,\Gamma)$ is a graph $G_1 = (V_1,\Gamma_1)$ such that $V_1 \subseteq V$ and $\Gamma_1 \subseteq \Gamma$. Thus, the graph $G_1 = (\{v_1,v_2,v_3\}, \{(1,1), (2,2), (3,3), (2,1), (3,1)\})$ shown in Fig. 2.3.1 is a subgraph of the graph shown in Fig. 2.1.3.

The set of vertices which are connected to a vertex v_j by branches originating at v_j will be denoted by the symbols $\Gamma(j)$. In other words,

$$\Gamma(j) = \{v_k \in V | (j,k) \in \Gamma\}. \tag{2.3.1}$$

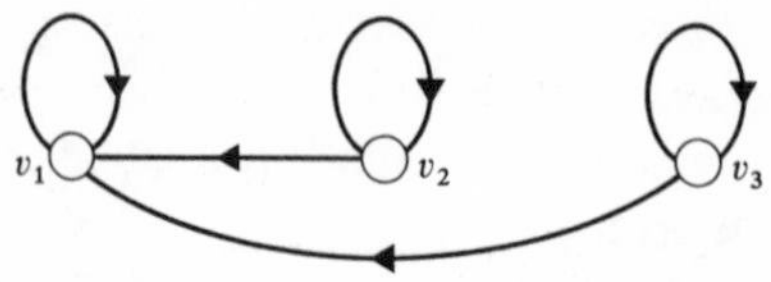

Fig. 2.3.1 Subgraph of graph in Fig. 2.1.3.

For G_1 in Fig. 2.3.1,

$$\Gamma(1) = \{v_1\}, \tag{2.3.2a}$$
$$\Gamma(2) = \{v_1,v_2\}, \tag{2.3.2b}$$
$$\Gamma(3) = \{v_1,v_3\}. \tag{2.3.2c}$$

The set of vertices connected to vertex v_j by a branch terminating at v_j will be denoted by $\Gamma^{-1}(j)$. Thus

$$\Gamma^{-1}(j) = \{v_k \in V | (k,j) \in \Gamma\}. \tag{2.3.3}$$

For G_1 in Fig. 2.3.1,

$$\Gamma^{-1}(1) = \{v_1,v_2,v_3\}, \tag{2.3.4a}$$
$$\Gamma^{-1}(2) = \{v_2\}, \tag{2.3.4b}$$
$$\Gamma^{-1}(3) = \{v_3\}. \tag{2.3.4c}$$

In an undirected graph, $\Gamma(j) = \Gamma^{-1}(j)$ for all j. If X is the set of vertices $\{v_{i_1},v_{i_2},\ldots,v_{i_k}\}$, $\Gamma(X)$ and $\Gamma^{-1}(X)$ are respectively defined by

$$\Gamma(X) = \bigcup_{j=1}^{k} \Gamma(i_j) \qquad \text{and} \qquad \Gamma^{-1}(X) = \bigcup_{j=1}^{k} \Gamma^{-1}(i_j).$$

For example, if $X = \{v_1,v_2\}$ in the graph in Fig. 2.3.1, $\Gamma(X) = \{v_1,v_2\}$.

The *degree* $d(j)$ of a vertex v_j is the total number of branches incident at that vertex. For a directed graph, the *inward demidegree* $d^-(j)$ of a vertex v_j is the total number of branches with v_j as a terminal vertex. The *outward demidegree* $d^+(j)$ of a vertex v_j is the total number of branches with v_j as an initial vertex. Thus, if we denote the number of elements in an arbitrary set X by $|X|$,

$$d^+(j) = |\Gamma(j)|, \tag{2.3.5a}$$
$$d^-(j) = |\Gamma^{-1}(j)|. \tag{2.3.5b}$$

Clearly, $d(j) = d^-(j) + d^+(j)$ and furthermore, since each branch has exactly one initial and one terminal vertex,

$$\frac{1}{2}\sum_{v_k \in V} d(k) = \sum_{v_i \in V} d^-(i) = \sum_{v_j \in V} d^+(j) = |\Gamma|. \tag{2.3.6}$$

A *self-loop* is a branch of the form (i,i). Branches (1,1), (2,2), and (3,3) in Fig. 2.3.1 are all self-loops. For self-loops the initial and terminal vertices are identical, and hence contribute to both $d^-(i)$ and $d^+(i)$. In the remainder of the book, we will specifically *exclude* self-loops unless otherwise stated. This means that the relation $\mathscr{R}$ which defines the graph is irreflexive.[2] A graph is said to be *homogeneous of degree* α if for all $v_i \in V, d(i) = \alpha$.

A *directed path* is a subgraph of G, specified by the sequence of vertices and branches $v_{i_1}(i_1,i_2)v_{i_2}(i_2,i_3)\ldots v_{i_{k-1}}(i_{k-1},i_k)v_{i_k}$ such that in the subgraph, $d^-(i_1) = 0$, $d^+(i_1) = 1$, $d^-(i_k) = 1$, $d^+(i_k) = 0$ and for $j = 2,3,\ldots,k-1$, $d^-(i_j) = d^+(i_j) = 1$ (i.e., all vertices and branches are distinct). An *undirected path* is a subgraph of G, specified by the sequence of vertices and undirected branches[3] $v_{i_1}[i_1,i_2]v_{i_2}[i_2,i_3]\ldots v_{i_{k-1}}[i_{k-1},i_k]v_{i_k}$ such that in the subgraph, $d(i_1) = d(i_k) = 1$ and for $j = 2,3,\ldots, k-1$, $d(i_j) = 2$. A *path* is a subgraph of G such that when branch directions are removed, it is an undirected path. When branch directions are considered, a path is not necessarily a directed path.

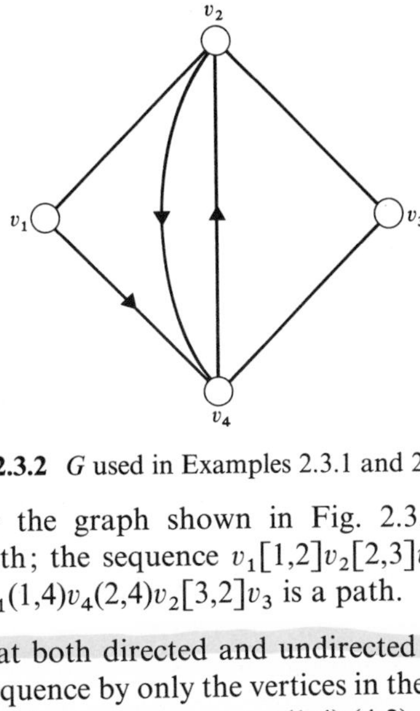

Fig. 2.3.2 G used in Examples 2.3.1 and 2.3.2.

Example 2.3.1 Consider the graph shown in Fig. 2.3.2. The sequence $v_1(1,4)$ $v_4(4,2)v_2$ is a directed path; the sequence $v_1[1,2]v_2[2,3]v_3[4,3]v_4$ is an undirected path; and the sequence $v_1(1,4)v_4(2,4)v_2[3,2]v_3$ is a path.

It should be clear that both directed and undirected paths are also paths. We sometimes describe the sequence by only the vertices in the sequence or the branches. Thus, $v_1(1,4)v_4(4,2)v_2$ may become v_1,v_4,v_2 or $(1,4),(4,2)$. Hereafter we will usually refer to all three types of path as simply paths if the type is clear from the context.

We can refer to a path as an i_1-i_k path, denoted by π_{i_1,i_k}, indicating that the first vertex in the sequence (the first vertex in the path) is v_{i_1} and the last vertex in the sequence (the last vertex in the path) is v_{i_k}. Moreover, if we refer to the jth vertex or branch in the path, we mean that jth vertex v_{i_j} in the sequence or the jth branch (i_j,i_{j+1}). Finally, if there is a directed i_1-i_k path in G, then vertex v_{i_k} will be called a *descendant* of vertex v_{i_1} and vertex v_{i_1} will be called an *ancestor* of vertex v_{i_k}. If G is undirected and there is an i-j path then v_i is both a descendant and an ancestor of v_j.

---

[2] A relation $\mathscr{R}$ is reflexive if $(x,y) \in \mathscr{R}$ for all x and is irreflexive if $(x,x) \notin \mathscr{R}$ for any x.

[3] Note that $[i,j] = [j,i]$.

The graph G is said to be *connected* if for any pair of vertices v_i and v_j there is an i-j path in G. Otherwise it is *disconnected.* G is *strongly connected* if for any pair of vertices v_i and v_j there is a directed i-j path and a directed j-i path in G. A maximal[4] connected subgraph of G is called a *component.* The components of G define a unique partition of the vertices of G. If ρ is the number of components of G, then G is connected if and only if $\rho = 1$.

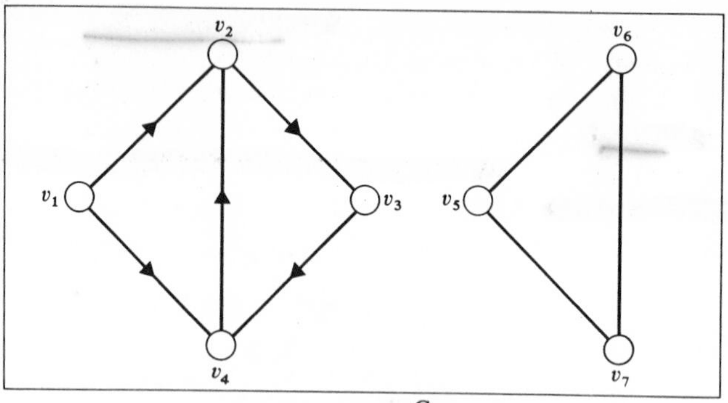

$$V_1 = \{v_1,v_2,v_3,v_4\},\ V_2 = \{v_5,v_6,v_7\}$$
$$\Gamma_1 = \{(1,2),(1,4),(4,2),(3,4),(2,3)\}$$
$$\Gamma_2 = \{[5,6],[5,7],[6,7]\}$$
$$G = (V_1 \cup V_2, \Gamma_1 \cup \Gamma_2)$$

Fig. 2.3.3 Graph used in Example 2.3.2.

Example 2.3.2 For the graph $G = (V_1 \cup V_2, \Gamma_1 \cup \Gamma_2)$ in Fig. 2.3.3, the subgraphs $G_1 = (V_1,\Gamma_1)$ and $G_2 = (V_2,\Gamma_2)$ are the unique components of G, $\rho = 2$, and hence G is not connected. (There is no path from any vertex in V_1 to any vertex in V_2.)

Suppose in the sequence of vertices describing a path, vertex v_i is immediately followed by vertex v_j. Then, either (i,j) or (j,i) is in the path. If (i,j) is in the path, then we will say that (i,j) is a *forward* branch with respect to the path. If (j,i) is in the path, then we will say that (j,i) is a *backward* branch with respect to the path. For example, in Fig. 2.3.2 the path $v_1(1,4)v_4(2,4)v_2$ has the backward branch (2,4) and forward branch (1,4).

A *directed circuit* is a subgraph of G specified by the sequence of vertices and branches $v_{i_1}(i_1,i_2)v_{i_2}\cdots v_{i_{k-1}}(i_{k-1},i_k)v_{i_k}$ such that in the subgraph, $d^-(i_j) = d^+(i_j) = 1$ for $j = 1,2,\ldots,k$. Thus, a directed circuit is essentially a directed path in which the first and last vertices in the path are not required to be distinct but are required instead to be identical. We can define undirected circuits and circuits in an analogous fashion. An *undirected circuit* is a sequence of vertices which satisfies all the require-

[4] A subgraph is maximal with respect to a property P if it loses property P if additional branches or vertices are added.

ments of an undirected path except that the first and last vertices are identical. A *circuit* is a sequence which satisfies all the requirements of a path except that the first and last vertices are identical. More rigorously, an undirected circuit is a subgraph of G specified by the sequence of vertices and branches $v_{i_1}[i_1,i_2]v_{i_2}[i_2,i_3]\ldots v_{i_{k-1}}[i_{k-1},i_k]v_{i_k}$ such that in the subgraph, $d(i_j) = 2$ for $j = 1,2,\ldots,k$. A *circuit* is a subgraph of G such that, when branch directions are removed, it is an undirected circuit.

Example 2.3.3 In Fig. 2.3.3, $v_2(2,3)v_3(3,4)v_4(4,2)v_2$ is a directed circuit; $v_1(1,2)v_2(2,3)v_3(3,4)v_4(1,4)v_1$ is a circuit; and $v_5[5,6]v_6[6,7]v_7[7,5]v_5$ is an undirected circuit.

A *Hamilton circuit* for a graph G is a circuit containing every vertex in G.

Example 2.3.4 The graph in Fig. 2.3.3 does not contain a Hamilton circuit since it is disconnected. For the graph in Fig. 2.3.2, [1,2],[2,3],[3,4],[4,1] is a Hamilton circuit.

2.4 CUT-SETS, CUTS, AND TREES

Suppose we are given a graph G through which flow is to be transmitted from one vertex to another. It may be that the branches of G are unreliable. If "enough" branches fail, we will not be able to transmit the desired flow. The graph theoretic quantities which are useful to describe this type of phenomenon are cut-sets, cuts, and trees.

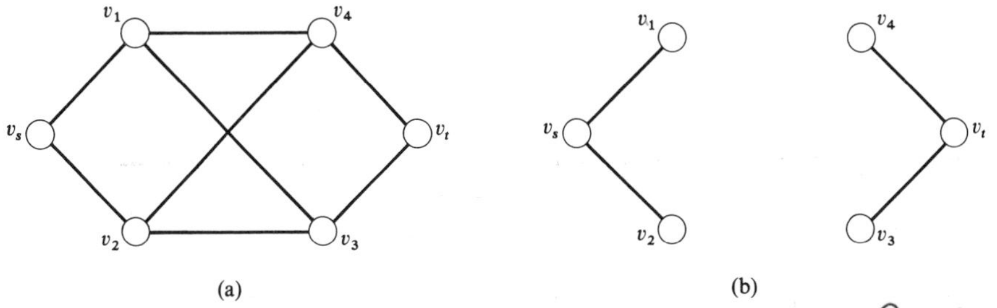

Fig. 2.4.1 Graphs illustrating the concept of a cut-set.

We will first consider undirected, connected graphs and then generalize our definitions. An *undirected branch cut-set* of an undirected, connected graph is a *minimal*[5] set of branches, the removal of which yields a graph with two or more[6] components. As an example, consider the graph in Fig. 2.4.1(a). The removal of the set $\{[1,4],[1,3],[2,4],[2,3]\}$ yields a graph with the two components shown in Fig. 2.4.1(b), while there is no proper subset of this set whose removal yields a disconnected graph.

[5] A set X is said to be minimal with respect to property P if no proper subset of X has property P. *Note:* The concepts of *minimum* and *minimal* are different.

[6] We will show in Theorem 2.4.1 that removal of a branch cut-set yields a graph with *exactly* two components.

It is often necessary to distinguish between the vertices in the components which result from the removal of a cut-set. Thus, we will say that an undirected branch cut-set of an undirected, connected graph G is an *i-j* cut-set if, after the removal of the cut-set from G, v_i and v_j are not in the same component. Again referring to Fig. 2.4.1, we see that $\{[1,4],[1,3],[2,4],[2,3]\}$ is an s-4, s-t, s-3, 1-4, 1-t, 1-3, 2-4, 2-t, and 2-3 cut-set.

If the original graph G is not connected, we will say that a set of branches is an undirected branch cut-set if it is an undirected branch cut-set of some component of G. Now suppose that G is directed. A set of branches of a directed graph G is a *directed branch cut-set* if its removal from G breaks all *directed* paths from at least one vertex of G to at least one other vertex of G and no proper subset breaks all directed paths between the same vertices. For example, consider the graph $\hat{G}$ of Fig. 2.4.2, obtained from Fig. 2.4.1(a) by adding directions to the branch of G. The set $\{(1,4),(1,3),(2,3)\}$ is a directed branch cut-set since its removal destroys all directed s-t paths and no proper subset does so. Again, to indicate the vertices which are affected by the removal of a cut-set, we could call the cut-set a directed s-t branch cut-set. More formally, a *directed s-t branch cut-set* is a minimal set of branches of G whose removal from G breaks all directed s-t paths. Henceforth, we shall refer to directed branch cut-sets and undirected branch cut-sets simply as branch cut-sets or cut-sets when the exact type is clear from the context.

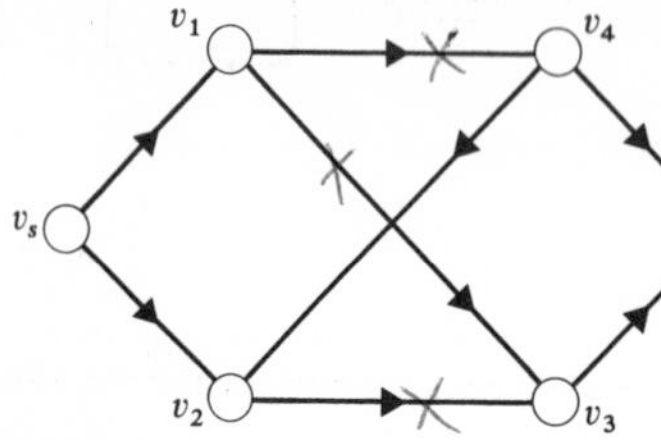

Fig. 2.4.2 Graph $\hat{G}$ illustrating the concept of a directed cut.

In many applications, the vertices of a graph may be the unreliable elements. For example, in an airline network, the vertices represent airports and the branches represent airlinks. Suppose the network is the subject of an attack aimed at disrupting service between various terminals. It will often be much easier to destroy airports than to achieve enough air superiority to close airlinks.

If a vertex v_s is removed from G, all branches incident at v_s are also removed from G. An *undirected vertex cut-set* of an undirected, *connected graph* G is a minimal set of vertices whose removal from G separates the graph into two or more components. A set of vertices of an undirected, *disconnected graph* G is an undirected vertex cut-set if it is an undirected vertex cut-set of some component of G. In other words, an undirected vertex cut-set of an undirected graph is a minimal set of vertices whose removal breaks all paths between at least one pair of *remaining* vertices.[7] As an

[7] Obviously, the removal of any single vertex breaks all paths between that vertex and any other vertex in G. By requiring that all paths between at least one pair of *remaining* vertices be broken, we specifically exclude the former type of path destruction from consideration.

example, again consider the graph in Fig. 2.4.1. The sets $\{v_1,v_2\}$ and $\{v_3,v_4\}$ are both undirected vertex cut-sets since all paths between v_s and v_t are destroyed by their removal. To further illustrate this concept, the only undirected vertex cut-set in the graph in Fig. 2.4.3 is $\{v_1\}$. In the above examples the removal of the undirected vertex cut-sets yields a graph in which v_s and v_t are in separate components. We therefore call the cut-set in question an *undirected s-t vertex cut-set.* The removal of an undirected *s-t* vertex cut-set from a graph breaks all undirected *s-t* paths.

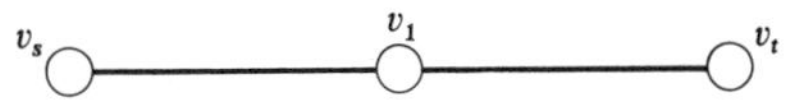

Fig. 2.4.3 Graph illustrating a vertex cut-set.

By analogy with the preceding development, we can define *directed vertex cut-sets* and *directed s-t vertex cut-sets.* A set of vertices of a directed graph G is a *directed vertex cut-set* if its removal from G destroys all directed paths from at least one of the remaining vertices to at least one other remaining vertex and no proper subset breaks all directed paths between these vertices. A *directed s-t vertex cut-set* of G is a minimal set of vertices (not containing v_s or v_t) whose removal from G breaks all directed paths from vertex v_s to vertex v_t. We usually speak of vertex cut-sets leaving the context to differentiate between directed and undirected.

In subsequent chapters, it will usually be clear whether the graphs we are considering are directed or undirected and whether the cut-sets of interest are branch or vertex cut-sets. Therefore, in such situations, to avoid a cumbersome set of modifiers, we will simply use the term cut-set to denote any of the above types of cut-sets. The set of *s-t* cut-sets of a graph G will usually be denoted $\{A^k_{s,t} | k = 1,\ldots,q\}$, where q is the number of such cut-sets.

Closely related to the concept of a branch cut-sets is the concept of a cut. If $A \subseteq V$ and $B \subseteq V$ (A and B not necessarily disjoint), let (A,B) denote the set of all branches which are incident out of an element of A and incident into an element of B. That is, $(A,B) = \{(i,j) \in \Gamma | v_i \in A \text{ and } v_j \in B\}$. Given a subset V_1 of vertices of the graph $G = (V,\Gamma)$, the set theoretic complement of V_1 in V will be denoted by $\bar{V}_1$. We will say that for any $V_1 \subseteq V$, the set of branches $(V_1,\bar{V}_1)$ is a *cut.* For example, in Fig. 2.4.2, if $V_1 = \{v_s\}$, then $(\{v_s\},\{v_1,v_2,v_3,v_4,v_t\}) = \{(s,1),(s,2)\}$ is a cut, while if $V_1 = \{v_s,v_1,v_2\}$, the set of branches $(\{v_s,v_1,v_2\},\{v_3,v_4,v_t\}) = \{(1,4),(1,3),(2,3)\}$ is another cut. Furthermore, either of these may be called *s-t* cuts since $v_s \in V_1$ and $v_t \in \bar{V}_1$. In general the notation $(X,\bar{X})_{s,t}$, where $X \subseteq V$ will represent any cut for which $v_s \in X$ and $v_t \in \bar{X}$. In an analogous fashion, we will say that in an undirected graph $[A,B] = \{[i,j] \in \Gamma | v_i \in A, v_j \in B\}$ and $[V_1,\bar{V}_1]$ is a cut. A graph $G = (V,\Gamma)$ is said to be *bipartite* if for some $V_1 \subseteq V$, $\Gamma = (V_1,\bar{V}_1) \cup (\bar{V}_1,V_1)$. The graph in Fig. 2.4.4 is bipartite since $\Gamma = (\{v_1,v_3\},\{v_2,v_4\})$.

We have seen that in Fig. 2.4.2 $\{(1,4),(1,3),(2,3)\}$ is a cut. We have also seen previously that this set of branches is also a directed cut-set. Consequently, we might conjecture that the concept of cut is no more general than the concept of directed cut-set. This conjecture is false as is shown by the following theorem.

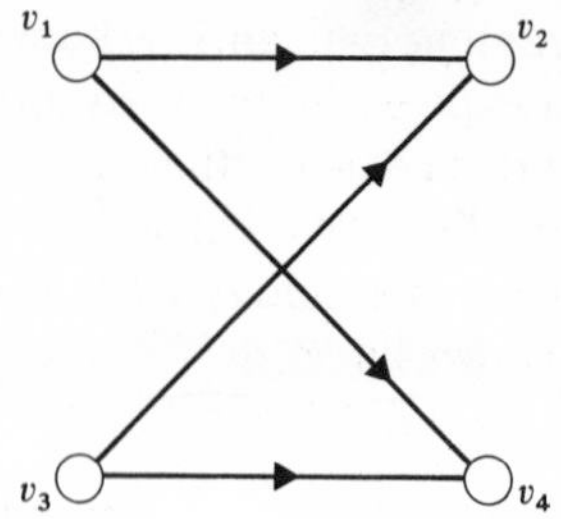

Fig. 2.4.4 Graph used in the proof of Theorem 2.4.1.

Theorem 2.4.1

a) Not every cut is a directed branch cut-set.
b) Every directed s-t branch cut-set is an s-t cut.

Proof. To show (a), we need simply provide a graph in which some cut is not a directed cut-set. Consider the graph shown in Fig. 2.4.4. The set of branches $(\{v_1,v_3\},\{v_2,v_4\}) = \{(1,2),(1,4),(3,2),(3,4)\}$ is by definition a 1-4 cut. Furthermore, the removal of these branches from G certainly breaks all directed paths from v_1 to v_4. However, this set of branches is not a directed 1-4 cut-set because it is not minimal. This follows since if the subset $\{(1,4)\}$ of the cut is removed from the graph, all directed paths from v_1 to v_4 are broken. Therefore not all cuts are directed cut-sets.

To show (b), let $A^k_{s,t}$ be a directed cut-set which breaks all directed s-t paths. We will partition the vertices of G into two disjoint sets V_1 and $\bar{V}_1$ so that $A^k_{s,t} = (V_1,\bar{V}_1)$. To do this remove the branches in $A^k_{s,t}$ from G and let V_1 be the set of all vertices to which there is a directed path from v_s. $\bar{V}_1$ is therefore the set of all vertices which cannot be reached via a directed path from v_s. Now, in G, the only branches from the vertices in V_1 to the vertices in $\bar{V}_1$ are the branches in $A^k_{s,t}$. Furthermore, if $v_i \in V_1$, $v_j \in \bar{V}_1$ and $(i,j) \in \Gamma$, then $(i,j) \in A^k_{s,t}$, since if $(i,j) \notin A^k_{s,t}$, in the graph formed by removing the $A^k_{s,t}$ from G, there is a directed path from v_s to v_j. This directed path is formed by a directed path from v_s to v_i followed by branch (i,j). However, by definition this implies that $v_j \in V_1$ which is a contradiction. Consequently, $(i,j) \in (V_1,\bar{V}_1)$ if and only if $(i,j) \in A^k_{s,t}$. Therefore $A^k_{s,t}$ is a cut and the theorem is proved. //

Let Γ_1 be a subset of branches of G and let c_k be the *capacity* of branch b_k of G for $k = 1,2,\ldots,m$ (where m is the number of branches of G). We will often consider the sum of the branch capacities of branches in an arbitrary set Γ_1. We will denote this sum as $c(\Gamma_1)$. That is,

$$c(\Gamma_1) = \sum_{b_k \in \Gamma_1} c_k. \tag{2.4.1}$$

If the capacity of each branch is unity, then $c(\Gamma_1) = |\Gamma_1|$ is the number of branches in Γ_1.

Using the above notation, we will say that the *capacity of a cut-set* A^k is $c(A^k)$ and the *capacity of a cut* $(V_1,\bar{V}_1)$ is $c(V_1,\bar{V}_1)$. (Strictly speaking, the latter capacity should be written as $c((V_1,\bar{V}_1))$. However, this simplification in notation should cause no confusion.) The capacity of an undirected cut $[V_1,\bar{V}_1]$ is denoted by $c[V_1,\bar{V}_1]$. For many applications, the capacities of the cuts or cut-sets are of interest. In particular, the cuts and cut-sets with smallest capacities are critical. Consequently, we will prove a theorem which relates these cuts and cut-sets. The theorem is useful since it is generally easier to find cuts than to find cut-sets. An *s-t* cut of minimum capacity is called a *minimum s-t cut*. Similarly an *s-t* cut-set of minimum capacity is called a *minimum s-t cut-set*.

Theorem 2.4.2 *Let $\{(V_i,\bar{V}_i)_{s,t}\}$ be the set of all s-t cuts of a directed graph G and let $\{A^k_{s,t}\}$ be the set of all directed s-t branch cut-sets of G. Then if the capacities of all branches are positive,*

$$\operatorname*{Min}_i \{c(V_i,\bar{V}_i)_{s,t}\} = \operatorname*{Min}_k \{c(A^k_{s,t})\}. \tag{2.4.2}$$

Proof. By Theorem 2.4.1, every directed *s-t* cut-set is an *s-t* cut. Therefore,

$$\{A^k_{s,t}\} \subseteq \{(V_i,\bar{V}_i)_{s,t}\} \tag{2.4.3}$$

and clearly

$$\operatorname*{Min}_i \{c(V_i,\bar{V}_i)_{s,t}\} \le \operatorname*{Min}_k \{c(A^k_{s,t})\}. \tag{2.4.4}$$

We will now show that the reverse inequality is also true and hence equality must hold. Let $(V_0,\bar{V}_0)_{s,t}$ be an *s-t* cut with minimum capacity. Since $(V_0,\bar{V}_0)_{s,t}$ is a minimum *s-t* cut and the capacities of all branches are positive, no proper subset of $(V_0,\bar{V}_0)_{s,t}$ can be an *s-t* cut; if there were a proper subset which was an *s-t* cut, it would have a smaller capacity than $(V_0,\bar{V}_0)_{s,t}$. This means that no proper subset of branches of $(V_0,\bar{V}_0)_{s,t}$ breaks all directed *s-t* paths. Consequently $(V_0,\bar{V}_0)_{s,t}$ is a minimal set of branches whose removal breaks all directed *s-t* paths. Hence, by definition, $(V_0,\bar{V}_0)_{s,t}$ is a directed *s-t* cut-set which implies

$$\operatorname*{Min}_i \{c(V_i,\bar{V}_i)_{s,t}\} \ge \operatorname*{Min}_k \{c(A^k_{s,t})\}. \tag{2.4.5}$$

The only way for both (2.4.4) and (2.4.5) to be simultaneously true is for equality to hold in both expressions. This proves the theorem. //

To see that the hypothesis that all branch capacities are positive is necessary for the theorem to hold, consider the graph shown in Fig. 2.4.5. The cut $(\{v_s,v_1\}, \{v_2,v_t\}) = \{(s,t),(s,2),(1,t),(1,2)\}$ has capacity $c(X,\bar{X}) = 0$ and is the minimum *s-t* cut. However, it is not a directed cut-set. The value of the smallest *s-t* cut-set is $c(s,t) = 1$.

We now prove a relationship between cuts and directed circuits which will be of use in the next chapter.

Theorem 2.4.3 *$(X,\bar{X})$ and $(\bar{X},X)$ contain an equal number of branches in common with any directed circuit.*

Proof. Select a vertex v_1 in a directed circuit and sequentially list the branches and vertices in the circuit. Compare the number of branches in $(X,\bar{X})$ and $(\bar{X},X)$.

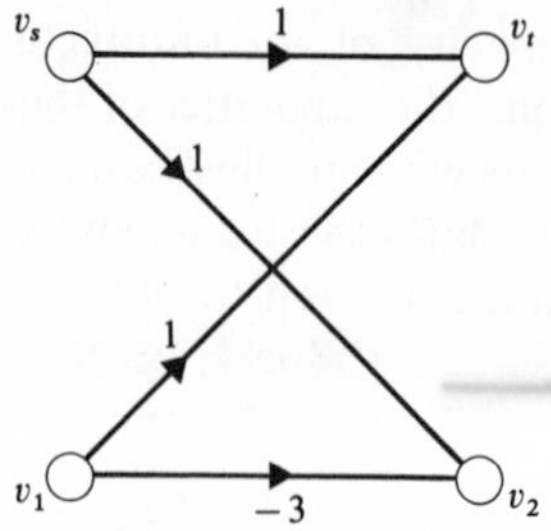

Fig. 2.4.5 Graph in which the minimum *s-t* cut does not have the same value as the minimum *s-t* cut-set.

If none of the branches in the circuit is in $(X,\bar{X})$ or $(\bar{X},X)$, the theorem is trivially true. Without loss of generality, suppose the first branch in the circuit in common with $(X,\bar{X})$ or $(\bar{X},X)$ is in $(X,\bar{X})$, that is, $v_1 \in X$. In listing the branches we must eventually reach a branch in $(\bar{X},X)$, since otherwise we could not reach v_1 again. If there are no more branches in $(X,\bar{X})$, the theorem is proved. If there is another branch in $(X,\bar{X})$, continue the above procedure until the last branch in the circuit and $(X,\bar{X})$ has been considered. Each time, an equal number of branches in $(X,\bar{X})$ and $(\bar{X},X)$ are found and hence the theorem is proved. //

If we remove an undirected cut-set from a connected graph, we disconnect the graph, while if we remove any proper subset of an undirected cut-set, the graph remains connected. Different cut-sets have different numbers of branches so we cannot readily determine the minimum number of branches that must be removed to disconnect the graph or, if branches are judiciously selected, the maximum number of branches which can be removed without disconnecting the graph. The latter problem can be completely solved by introducing the concept of a tree. A *tree* $T = (V,U)$ of a connected graph $G = (V,\Gamma)$ is a connected subgraph of G containing all of the vertices of G and no circuits. For G shown in Fig. 2.4.6(a), one possible tree is shown in

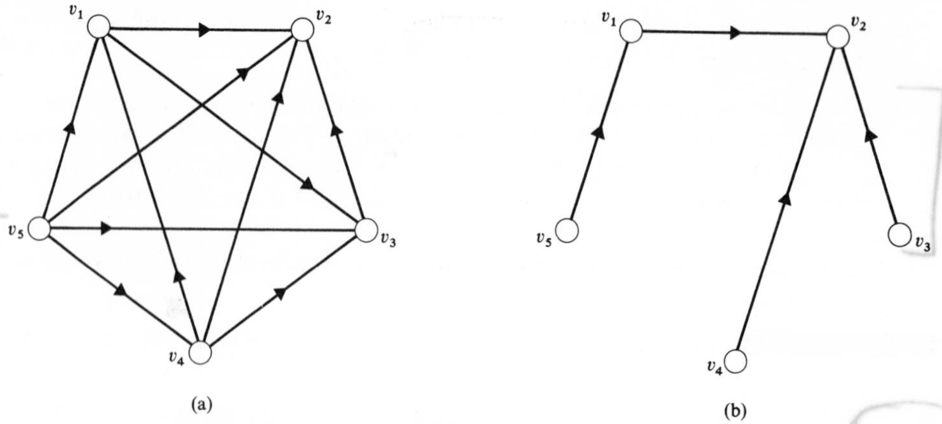

Fig. 2.4.6 Graphs illustrating the concept of a tree.

Fig. 2.4.6(b). A number of properties of a tree follow easily from this elementary definition and are left as exercises. Among these properties are:

1. U contains $n - 1$ branches.
2. $T = (V,U)$ is connected but becomes disconnected if any branch of T is removed.
3. Every pair of vertices v_i and v_j of the graph $T = (V,U)$ is joined by one and only one i-j path.
4. Any subgraph of G, which contains T as a proper subgraph, has at least one circuit.
5. Every undirected branch cut-set has at least one branch in common with every tree.

The concept of a tree is one of the most important in the theory of graphs.

2.5 MATRICES ASSOCIATED WITH GRAPHS

For the purposes of computer calculation it is necessary to describe graphs without resorting to their pictorial representation. One way to describe the incidence relations of G is by means of the "incidence matrix."

Let $G = (V,\Gamma)$ be an n vertex, m branch directed graph. Let $\mathscr{U} = [u_{i,j}]$ be an $n \times m$ matrix whose i-jth entry is $u_{i,j}$. Then $\mathscr{U}$ is the *incidence matrix* of G if for $i = 1, \ldots, n$ and $j = 1, \ldots, m$,

$$u_{i,j} = \begin{cases} +1 & \text{if } b_j \text{ is directed away from } v_i \\ -1 & \text{if } b_j \text{ is directed toward } v_i \\ 0 & \text{otherwise.} \end{cases} \tag{2.5.1}$$

If G is an undirected graph, $\mathscr{U} = [u_{i,j}]$ is the incidence matrix of G if for $i = 1, \ldots, n$ and $j = 1, \ldots, m$,

$$u_{i,j} = \begin{cases} 1 & \text{if } b_j \text{ is incident at } v_i \\ 0 & \text{otherwise.} \end{cases} \tag{2.5.2}$$

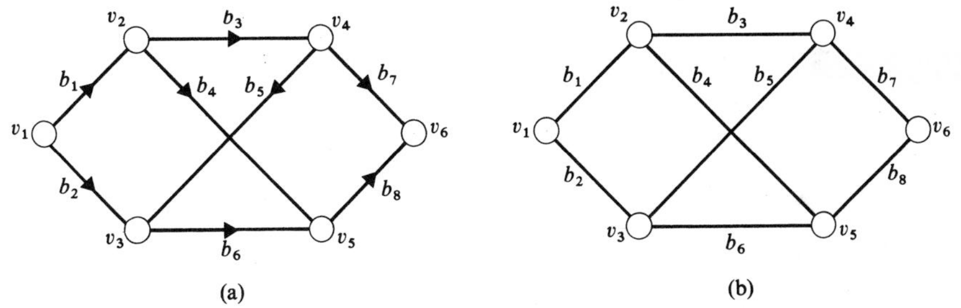

Fig. 2.5.1 Graphs for which matrices are given in (2.5.3), (2.5.4), (2.5.7), (2.5.8), (2.5.10), and (2.5.12).

Example 2.5.1 Consider the graph shown in Fig. 2.5.1(a). The incidence matrix of G is

$$\mathscr{U} = \begin{bmatrix} 1 & 1 & 0 & 0 & 0 & 0 & 0 & 0 \\ -1 & 0 & 1 & 1 & 0 & 0 & 0 & 0 \\ 0 & -1 & 0 & 0 & -1 & 1 & 0 & 0 \\ 0 & 0 & -1 & 0 & 1 & 0 & 1 & 0 \\ 0 & 0 & 0 & -1 & 0 & -1 & 0 & 1 \\ 0 & 0 & 0 & 0 & 0 & 0 & -1 & -1 \end{bmatrix} \tag{2.5.3}$$

When we remove the arrowheads from the branches of G, we obtain the undirected graph shown in Fig. 2.5.1(b). The incidence matrix of this graph can be found directly from the graph or by removing the minus signs from the incidence matrix of the directed graph. Thus, for the undirected graph,

$$\mathscr{U} = \begin{bmatrix} 1 & 1 & 0 & 0 & 0 & 0 & 0 & 0 \\ 1 & 0 & 1 & 1 & 0 & 0 & 0 & 0 \\ 0 & 1 & 0 & 0 & 1 & 1 & 0 & 0 \\ 0 & 0 & 1 & 0 & 1 & 0 & 1 & 0 \\ 0 & 0 & 0 & 1 & 0 & 1 & 0 & 1 \\ 0 & 0 & 0 & 0 & 0 & 0 & 1 & 1 \end{bmatrix} \tag{2.5.4}$$

Other important matrices associated with a graph are the "circuit matrix" and the "cut-set matrix." Both of these matrices can be defined for both directed and undirected graphs. We will only need the undirected versions of these definitions.

Let the undirected circuits of the graph G be labeled from 1 to β. Then, the $\beta \times m$ matrix $\mathscr{B} = [b_{i,j}]$ is the *circuit matrix* of G if and only if for $i = 1,\ldots,\beta$ and $j = 1,\ldots,m$,

$$b_{i,j} = \begin{cases} 1 & \text{if } b_j \text{ is in circuit } i \\ 0 & \text{otherwise.} \end{cases} \tag{2.5.5}$$

Let the ith undirected cut-set of G be denoted by A^i and let i range from 1 to $\hat{q}$. Then the $\hat{q} \times m$ matrix $\mathscr{A} = [a_{i,j}]$ is the *branch cut-set matrix* of G if and only if for $i = 1,\ldots,\hat{q}$ and $j = 1,\ldots,m$,

$$a_{i,j} = \begin{cases} 1 & \text{if } b_j \text{ is in cut-set } A^i \\ 0 & \text{otherwise.} \end{cases} \tag{2.5.6}$$

Example 2.5.2 Again consider the graph shown in Fig. 2.5.1(b). The circuit matrix of G is

$$\mathscr{B} = \begin{bmatrix} 1 & 1 & 0 & 1 & 0 & 1 & 0 & 0 \\ 0 & 0 & 1 & 1 & 0 & 0 & 1 & 1 \\ 1 & 1 & 1 & 0 & 0 & 1 & 1 & 1 \\ 1 & 1 & 1 & 0 & 1 & 0 & 0 & 0 \\ 0 & 0 & 0 & 0 & 1 & 1 & 1 & 1 \\ 0 & 0 & 1 & 1 & 1 & 1 & 0 & 0 \\ 1 & 1 & 0 & 1 & 1 & 0 & 1 & 1 \end{bmatrix} \tag{2.5.7}$$

and the cut-set matrix of G is

$$\mathscr{A} = \begin{bmatrix} 1 & 1 & 0 & 0 & 0 & 0 & 0 & 0 \\ 1 & 0 & 1 & 1 & 0 & 0 & 0 & 0 \\ 0 & 1 & 0 & 0 & 1 & 1 & 0 & 0 \\ 0 & 0 & 0 & 1 & 0 & 1 & 0 & 1 \\ 0 & 0 & 1 & 0 & 1 & 0 & 1 & 0 \\ 0 & 0 & 0 & 0 & 0 & 0 & 1 & 1 \\ 1 & 0 & 0 & 0 & 1 & 1 & 0 & 0 \\ 0 & 1 & 1 & 1 & 0 & 0 & 0 & 0 \\ 1 & 0 & 0 & 1 & 1 & 0 & 0 & 1 \\ 1 & 0 & 0 & 1 & 1 & 0 & 1 & 0 \\ 0 & 0 & 0 & 1 & 0 & 1 & 1 & 0 \\ 0 & 0 & 1 & 0 & 1 & 0 & 0 & 1 \\ 0 & 1 & 0 & 1 & 1 & 0 & 1 & 0 \\ 0 & 1 & 0 & 1 & 1 & 0 & 0 & 1 \\ 0 & 0 & 1 & 1 & 1 & 1 & 0 & 0 \end{bmatrix} \tag{2.5.8}$$

Finally, it may be necessary to enumerate the cut-sets of G which separate a given pair of vertices, v_s and v_t. If the set of s-t cut-sets is $\{A^1_{s,t}, \ldots, A^q_{s,t}\}$, the $q \times m$ matrix $\mathscr{A}_{s,t} = [a_{i,j}]$ is an *s-t branch cut-set matrix*[8] if and only if for $i = 1, \ldots, q$ and $j = 1, \ldots, m$,

$$a_{i,j} = \begin{cases} 1 & \text{if } b_j \text{ is in } A^i_{s,t} \\ 0 & \text{otherwise.} \end{cases} \tag{2.5.9}$$

Example 2.5.3 For the graph in Fig. 2.5.1(b), $\mathscr{A}_{1,6}$ is given by

$$\mathscr{A}_{1,6} = \begin{bmatrix} 1 & 1 & 0 & 0 & 0 & 0 & 0 & 0 \\ 0 & 1 & 1 & 1 & 0 & 0 & 0 & 0 \\ 1 & 0 & 0 & 0 & 1 & 1 & 0 & 0 \\ 0 & 0 & 1 & 1 & 1 & 1 & 0 & 0 \\ 0 & 0 & 0 & 1 & 0 & 1 & 1 & 0 \\ 0 & 0 & 1 & 0 & 1 & 0 & 0 & 1 \\ 0 & 0 & 0 & 0 & 0 & 0 & 1 & 1 \\ 1 & 0 & 0 & 1 & 1 & 0 & 0 & 1 \\ 0 & 1 & 0 & 1 & 1 & 0 & 1 & 0 \end{bmatrix} \tag{2.5.10}$$

As an alternative to the incidence matrix it is often more convenient to represent a graph by its *connection matrix, $\mathscr{K}$*. For an undirected graph G, $\mathscr{K}$ is an $n \times n$ matrix whose i-jth entry is $k_{i,j}$ where

$$k_{i,j} = \begin{cases} 1 & \text{if } [i,j] \in \Gamma \\ 0 & \text{otherwise.} \end{cases} \tag{2.5.11}$$

[8] Whenever there is no possibility of confusion we will call the matrix $\mathscr{A}_{s,t}$ simply the cut-set matrix since the subscripts s,t clearly indicate which vertices are separated.

Example 2.5.4 For the graph G in Fig. 2.5.1 the connection matrix is

$$\mathscr{K} = \begin{bmatrix} 0 & 1 & 1 & 0 & 0 & 0 \\ 1 & 0 & 0 & 1 & 1 & 0 \\ 1 & 0 & 0 & 1 & 1 & 0 \\ 0 & 1 & 1 & 0 & 0 & 1 \\ 0 & 1 & 1 & 0 & 0 & 1 \\ 0 & 0 & 0 & 1 & 1 & 0 \end{bmatrix} \tag{2.5.12}$$

Several interesting properties of the above matrices are the subjects of Problems 17 to 22.

2.6 MATHEMATICAL PROGRAMMING

In solving network problems, one of our tasks will be to devise algorithms which can yield answers for problems of a reasonable size in a small amount of computer time. Because of the computational efficiency that has been achieved for mathematical programs, our effort in some cases will be to formulate the problem as a mathematical programming problem (preferably a quadratic or linear programming problem) with a small number of constraints.

For this reason we now give a very brief review of the exact formulations of mathematical programming problems for which efficient computational techniques are available. For the actual techniques of solution the reader is referred to the book by Dantzig [DA1] and the two books by Hadley [HA1, HA2], listed in Chapter 1. For the case of linear programs we present some structural properties which will be of use in several proofs in the book.

In the *convex programming problem*, we seek to maximize the objective function of n variables $x_1, \ldots, x_n$,

$$z = h(x_1, \ldots, x_n), \tag{2.6.1}$$

subject to a set of m constraints:

$$g_i(x_1, \ldots, x_n) = b_i \quad \text{for } i = 1, \ldots, m, \tag{2.6.2a}$$

$$x_i \geq 0 \qquad \text{for } i = 1, \ldots, n, \tag{2.6.2b}$$

where for $i = 1, \ldots, m$ the b_i are known constants. We place the following requirements on h and g_i for $i = 1, \ldots, m$.

a) h is concave. That is, the surface z lies below any of its tangent planes. For example, the surface in Fig. 2.6.1 is concave.
b) For $i = 1, \ldots, m$, $g_i(x_1, \ldots, x_n)$ is convex. A convex function is the negative of a concave function.

Properties (a) and (b) guarantee that any local maximum of z is a global maximum.

If the constraints are presented in terms of inequalities they can be converted to equalities as in (2.6.2) by introducing an additional variable, x_{n+i}, called a "slack" or "surplus" variable into the ith constraint, for $i = 1, \ldots, m$. If a variable x_i is unrestricted in sign it can be replaced by variables x_i' and x_i'' so $x_i = x_i' - x_i''$, with

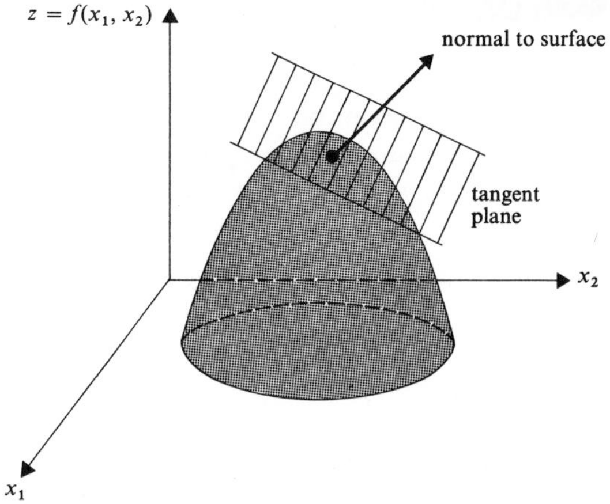

Fig. 2.6.1 Concave function of two variables.

$x_i' \geq 0$ and $x_i'' \geq 0$ (2.6.2b is then satisfied). If $x_i \leq 0$ we can replace it with $x_i' = -x_i \geq 0$. If the objective function z is to be minimized it can be replaced by $-z$ which is then to be maximized. Hence the problem formulated in (2.6.1) and (2.6.2) is very general.

A *quadratic program* is a special case of the convex program in which the objective function is a quadratic and the constraints are linear. The problem is to maximize

$$z = h(x_1, \ldots, x_n) = \sum_{j=1}^{n} c_j x_j + \sum_{i=1}^{n} \sum_{j=1}^{n} h_{i,j} x_i x_j \tag{2.6.3}$$

subject to a set of constraints

$$\sum_{j=1}^{n} a_{i,j} x_j = b_i, \quad \text{for } i = 1, \ldots, m, \tag{2.6.4a}$$

$$x_i \geq 0 \quad \text{for } i = 1, \ldots, n. \tag{2.6.4b}$$

We require that the quadratic form

$$\sum_{i=1}^{n} \sum_{j=1}^{n} h_{i,j} x_i x_j$$

be positive semidefinite.

If in (2.6.3) we let $h_{i,j} = 0$ for all i and j, the resulting problem is called a *linear program*.

The three types of program have been presented in descending order of difficulty with the most difficult first. For all nonlinear problems, techniques are available with suggestive names like "gradient methods," "simplicial methods," and "cutting plane methods." For linear and quadratic programs the simplex and primal-dual algorithms are the most useful. If in any of the above problems we constrain some of the variables to be integers, the resulting problem is called an *integer programming* problem. Several techniques are available for the solution of quadratic and linear programs with integer variables but the solutions are difficult and time-consuming. For the quadratic and linear programming problem, exact solutions can be obtained in a finite number of steps, whereas for the general convex programming problem only an approximate maximum can be obtained.

We now examine in more detail the structure of the linear programming problem. The linear programming problem as defined above is specified in matrix form as follows.

Maximize the objective function

$$z = \mathbf{c}'\mathbf{X} \tag{2.6.5}$$

subject to constraints

$$A\mathbf{X} = \mathbf{b} \tag{2.6.6a}$$

$$\mathbf{X} \geq \mathbf{0} \tag{2.6.6b}$$

where:

$\mathbf{X} = (x_1,x_2,\ldots,x_n)'$ is an n vector of variables;
$\mathbf{c}$ is an n vector of constants ($'$ is the notation for transpose);
$\mathbf{X} \geq \mathbf{0}$ means $x_i \geq 0$ for $i = 1,\ldots,n$;
A is an $m \times n$ matrix of constants;
$\mathbf{b}$ is an m vector of constants.

Another linear program is associated with the one specified by (2.6.5) and (2.6.6). It is called the dual linear program and is specified as follows.

Minimize the objective function

$$q = \mathbf{b}'\boldsymbol{\lambda}, \tag{2.6.7}$$

subject to constraints

$$A'\boldsymbol{\lambda} \geq \mathbf{c}, \tag{2.6.8}$$

where $\boldsymbol{\lambda}$ is an m vector of constants, whose entries are unrestricted in sign. The problem in (2.6.5) and (2.6.6) is called the *primal problem* for the *dual problem* in (2.6.7) and (2.6.8).

Let us assume the matrix A has rank r. A set of r linearly independent columns of A (or A') is said to form a *basis* for the primal (or dual) problem. The set of variables corresponding to the columns in a basis of A (or A') is said to be a set of *basic variables* for the primal (or dual) problem. If all variables which are not in a basis are set equal to zero and (2.6.7) and (2.6.8) are solved for the basic variables, the resulting solution is called a *basic solution* for the primal or dual problem respectively. A solution is *feasible* if it satisfies all constraints. A basic solution satisfying the sign constraints on the variables is called a *basic feasible solution.* A feasible solution for which a finite optimum value of the objective function is attained is said to be an *optimum solution.* There exists at least one optimum solution which is a basic solution.

From the theory of linear programming we have the following theorems.

Complementary slackness theorem

a) *If a slack or surplus variable x_{n+i} which has been added to the primal constraint appears in an optimal basic solution, then in the dual optimal solution the corresponding dual variable λ_i is zero.*
b) *If the variable x_j appears in an optimal solution to the primal problem, then in the dual optimal solution the jth dual constraint is a strict equality.*

Duality theorem

a) *If both the primal and dual problems possess a feasible solution then both have optimal solutions. Furthermore, the objective functions are equal when optimal solutions are obtained.*
b) *If either the primal or dual problem does not possess a feasible solution then neither has an optimal solution.*

PROBLEMS

1. Construct a directed graph $G = (V,\Gamma)$ where $\Gamma = \mathscr{R}$ and $(i,j) \in \mathscr{R}$ if and only if $i+j$ is even.
2. For the graphs in Fig. P2.1 find all:

   a) 3-2 paths
   b) directed 3-2 paths
   c) undirected 3-2 paths
   d) directed 2-3 paths
   e) circuits
   f) directed circuits
   g) undirected circuits
   h) 4-2 branch cuts
   i) undirected 4-2 branch cut-sets
   j) directed 4-2 branch cut-sets
   k) 4-2 vertex cut-sets
   l) trees

3. For the graphs in Fig. P2.1 find

   a) $\Gamma(1)$
   b) $\Gamma^{-1}(1)$
   c) $\Gamma(\{v_1,v_2\})$
   d) $\Gamma^{-1}(\{v_1,v_2\})$
   e) the incidence matrix $\mathscr{U}$
   f) the circuit matrix $\mathscr{B}$
   g) the cut-set matrix $\mathscr{A}$
   h) the connection matrix $\mathscr{K}$

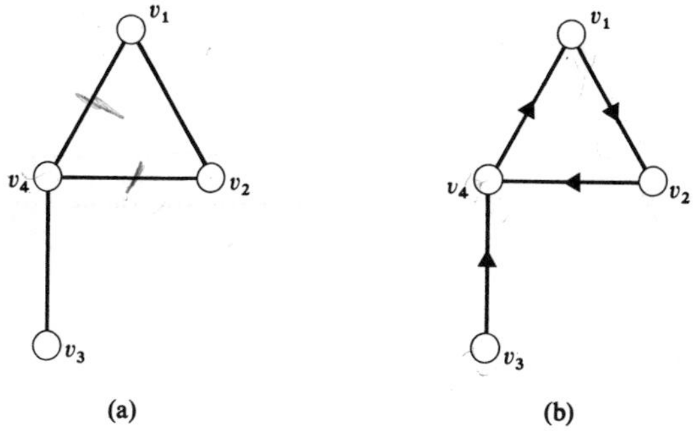

Figure P2.1

4. In Fig. P2.2 show that the value of the minimum 1-2 cut is not the same as the value of the minimum 1-2 cut-set.

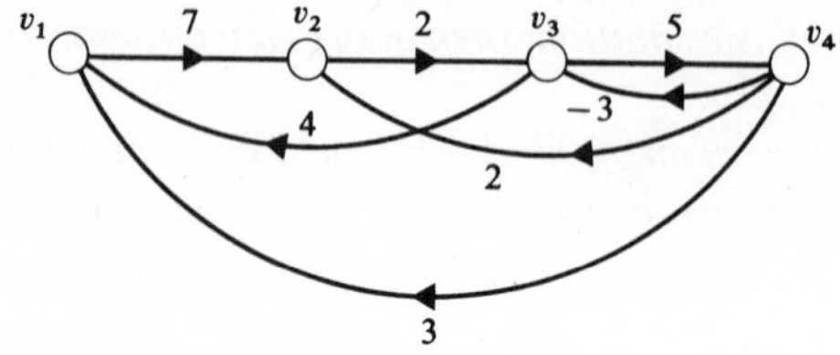

Figure P2.2

5. Let $T = (V,U)$ be a tree of an undirected graph G with n vertices. Prove
   a) U contains $(n\text{-}1)$ branches.
   b) T is connected but becomes unconnected if a single branch of T is removed.
   c) Every pair of vertices in G is joined by a single path in T.
   d) T has no circuits and any subgraph of G which contains T as a proper subgraph has at least one circuit.
   e) Every undirected branch cut-set of G has at least one branch in common with every tree.
6. Prove that any branch of a connected graph is contained in at least one tree.
7. A directed graph $G = (V,\Gamma)$ with n vertices is said to be "descending" if its vertices can be numbered from v_1 to v_n so that for $i > j$, $(i,j) \notin \Gamma$. Prove that a graph is descending if and only if it contains no directed circuits.
8. Show that in an undirected graph with exactly two vertices of odd degree, there is a path joining these vertices; i.e., the two vertices of odd degree cannot be in separate components.
9. Prove that in an undirected graph there are an even number of vertices of odd degree.
10. An undirected graph is called an Euler graph if every vertex has even degree. Prove that a graph G is an Euler graph if and only if G is a union of circuits, no two of which have a branch in common.
11. An Euler decomposition of an undirected graph G is a branch disjoint union of one path and an arbitrary number of circuits which contains every branch of G exactly once. Prove that G has an Euler decomposition if and only if G is connected and the number of vertices of odd degree is zero or two.
12. Prove that a set of integers $d_1 \leq d_2 \leq \cdots \leq d_n$ is realizable as the degrees of the vertices of an undirected s-graph for some s (i.e., a graph with parallel branches) if and only if

    i) $\sum_{i=1}^{n} d_i = 2e$, where e is an integer,

    ii) $\sum_{i=1}^{n-1} d_i \geq d_n$.

    [*Hint:* Prove sufficiency by induction on the number of vertices. Begin by proving that the statement is true for two and three vertex s-graphs.]
13. Let $G = (V,\Gamma)$ be a graph such that for all $l \neq k$ if $(l,k) \notin \Gamma$ then $(k,l) \in \Gamma$. Show that there exists a vertex v_i such that for all $j \neq i$ there is a directed i-j path containing at most two

branches. Moreover show that a sufficient condition for v_i to be such a vertex is that

$$|\Gamma(i) - \{v_i\}| = \operatorname*{Max}_{v_j \in V} |\Gamma(j) - \{v_j\}|.$$

[*Hint:* Prove by contradiction.]

14. Let $G = (V,\Gamma)$ and let v_u,v_w,v_x,v_y be elements of V such that

$$(x,y) \in \Gamma, \qquad (u,w) \in \Gamma,$$
$$(x,w) \notin \Gamma, \qquad (u,y) \notin \Gamma.$$

Construct the graph $\hat{G} = (V,\hat{\Gamma})$ from G by removing (x,y) and (u,w) and adding (x,w) and (u,y). The process of obtaining $\hat{G}$ from G is called a *branch exchange*. Suppose $G_1 = (V,\Gamma_1)$ and $G_2 = (V,\Gamma_2)$ and that, for all $v_i \in V$, $d^+(i)$ and $d^-(i)$ have the same values in G_1 and G_2. Show that G_2 can be derived from G_1 by means of a finite number of branch exchanges.

15. Prove that an undirected graph is bipartite if and only if every circuit contains an even number of branches. — This gives way to see if graph is bipartite — label it with all a's & b's

16. Prove that a complete undirected graph contains a Hamilton circuit.

17. A matrix is said to be "totally unimodular" if the determinant of every square submatrix of any order is 0, +1, or −1. Prove that the incidence matrix of any directed graph is totally unimodular. [*Hint:* Use induction on the order of the square submatrices.]

18. Prove that the rank of the incidence matrix of a connected directed graph with n vertices is $n - 1$. If a directed graph with n vertices has ρ components, what is the rank of its incidence matrix?

19. Let $\mathscr{U}$ be the incidence matrix of a connected directed graph with n vertices and m branches. Let the solution space $\mathbf{x} = (x_1, \ldots, x_m)'$ of the equation

$$\mathscr{U}\mathbf{x} = \mathbf{0}$$

have a basis $\mathbf{S}_1,\mathbf{S}_2,\ldots,\mathbf{S}_{m-n+1}$ with $\mathbf{S}_i = (S_{i_1},\ldots,S_{i_m})'$. Show that there exists a set of $\mathbf{S}_i$ such that $S_{i_j} = 0, +1$, or -1 for $i = 1,\ldots,m - n + 1$ and $j = 1,\ldots,m$.

20. Let $\mathscr{U}$ be the incidence matrix of a directed graph G.

a) Let β^* be a circuit $v_{i_1}[i_1,i_2]v_{i_2}\cdots v_{i_{k-1}}[i_{k-1},i_k]v_{i_k}$ with $v_{i_k} = v_{i_1}$. Assign a direction to the circuit corresponding to the order with which the vertices are enumerated above. Define a vector $\boldsymbol{\beta} = (\beta_1,\ldots,\beta_m)'$ such that for all j: $\beta_j = 0$ if branch b_j is not in the circuit; $\beta_j = +1$ if branch b_j is in the circuit and has the same direction as the circuit; $\beta_j = -1$ if branch b_j is in the circuit and has the opposite direction to the circuit. Show that

$$\mathscr{U}\boldsymbol{\beta} = 0.$$

b) Let $\boldsymbol{\beta} = (\beta_1,\ldots,\beta_m)'$ be any $m \times 1$ vector all of whose entries are 0, +1, or −1, such that $\mathscr{U}\boldsymbol{\beta} = 0$. Let $G^* = (V^*,\Gamma^*)$, be a subgraph of G such that $b_i \in \Gamma^*$ if and only if $\beta_i \neq 0$ and such that V^* contains the endpoints of the branches in Γ^*. Show that G^* is a circuit or a branch disjoint union of circuits.

c) Suppose the circuits of G are numbered from 1 to μ and an arbitrary direction is assigned to each. Define the directed circuit matrix $\mathscr{B}^* = [b^*_{i,j}]$ such that $b^*_{i,j} = 0$ if branch b_j is not in the ith circuit, $b^*_{i,j} = 1$ if b_j is in the ith circuit and has the same direction, and $b^*_{i,j} = -1$ if b_j is in the ith circuit and has the opposite direction. What can you say about $\mathscr{U}'\mathscr{B}^*$ and $\mathscr{B}^{*\prime}\mathscr{U}$?

d) What is the rank of $\mathscr{B}^*$ if G is connected?

21. Let $A^1, \ldots, A^q$ be the directed cut-sets of a connected directed graph G. The removal of each directed cut-set A^i breaks all directed paths from a set of vertices X_i to its complement $\bar{X}_i$. Suppose that an arbitrary direction for each cut-set is assigned (e.g., from X_i to $\bar{X}_i$). Define the matrix $\mathscr{A}^* = [a^*_{i,j}]$ such that: $a^*_{i,j} = 0$ if b_j is not in A^i; $a^*_{i,j} = +1$ if b_j is in A^i and has the same direction as A^i; $a^*_{i,j} = -1$ if b_j is in A^i and has the opposite direction to A^i.

   a) Let $\mathscr{U}^*$ be any $n - 1$ row submatrix of the incidence matrix of G. Show that there exists a nonsingular $n - 1 \times n - 1$ matrix M such that if $\hat{\mathscr{A}}^*$ is any $n-1$ row submatrix of $\mathscr{A}^*$, then

$$\hat{\mathscr{A}}^* = M\mathscr{U}^*.$$

   b) What is the rank of $\mathscr{A}^*$? Show that $\mathscr{A}^{*\prime}\mathscr{B}^* = 0$ where $\mathscr{B}^*$ is defined in part (c) of the preceding problem.

22. Let $\mathscr{A}^*$ and $\mathscr{B}^*$ be the directed cut-set matrix and directed circuit matrix of a connected graph G, as defined in Problems 20 and 21, respectively.

   a) Show that any $n - 1 \times n - 1$ submatrix of $\mathscr{A}^*$ is nonsingular if and only if the columns of this submatrix correspond to the branches of a tree.
   b) Show that any $m - n + 1 \times m - n + 1$ submatrix of $\mathscr{B}^*$ is nonsingular if and only if its columns correspond to the branches in the complement of a tree (i.e., the branches not in the tree).

23. Let G be a directed connected graph and let T be any tree of G. Suppose the branches of T are labeled $b_1, \ldots, b_{n-1}$ and the branches in the complement are labeled $b_n, \ldots, b_m$. Let $\mathbf{F} = (f_1, \ldots, f_m)'$ be a vector such that f_i represents the "flow" in b_i. Show that any solution of the equation $\mathscr{U}^*\mathbf{F} = \mathbf{0}$ can be generated by specifying the flows $f_n, \ldots, f_m$ in the branches in the complement of T. (Assume that at any vertex v_i,

$$\sum_{(i,j)\in\Gamma} f(i,j) - \sum_{(j,i)\in\Gamma} f(j,i) = 0.)$$

24. Consider m mills delivering flour to n bakeries. The cost of shipping flour from mill i to bakery j is $h(i,j)$ cents per pound. Mill i can produce at most a_i pounds of flour and bakery j requires b_j pounds. In this case the transportation problem consists of finding the minimum cost routing of the flour to satisfy the requirements of all bakeries.

   a) Formulate the transportation problem as a linear programming problem.
   b) Find the linear program dual to that formulated in (a).

25. Consider an undirected complete graph $G = (V,\Gamma)$ with n vertices. Associated with each branch (i,j) is a nonnegative length $l(i,j)$. The traveling salesman problem consists of finding the shortest Hamilton circuit in G, where the length of a circuit is the sum of the lengths of the branches in the circuit. Formulate the traveling salesman problem as an integer programming problem.

CHAPTER 3

MAXIMUM FLOW IN DETERMINISTIC GRAPHS

3.1 FLOWS IN GRAPHS

The objective of many of the flow problems used to model physical systems is to maximize specified flows subject to a set of constraints. These maximizations are the subject of this chapter. The techniques introduced are basic and are used throughout the book.

Before proceeding to the simplest maximization problem, we formalize the material on flows presented in Chapter 2. Assign to each branch (i,j) or $[i,j]$ a weight $f(i,j)$ or $f[i,j]$ called the flow in the branch from v_i to v_j. We require $f[i,j] = -f[j,i]$ but $f(i,j)$ is not necessarily equal to $-f(j,i)$. Also, assign to each branch (i,j) or $[i,j]$ a weight $c(i,j)$ or $c[i,j]$ called the capacity of the branch. We have $c[i,j] = c[j,i]$ but $c(i,j)$ is not necessarily equal to $c(j,i)$. If A and B are arbitrary sets of vertices, define $f(A,B)$ and $f[A,B]$ by the relationships

$$f(A,B) = \sum_{(i,j) \in (A,B)} f(i,j) \tag{3.1.1a}$$

and

$$f[A,B] = \sum_{[i,j] \in [A,B]} f[i,j]. \tag{3.1.1b}$$

The set of flows associated with the branches in a graph is denoted by $\mathscr{F}$ and is called a flow pattern. $\mathscr{F}$ is said to be *feasible* in a directed graph if it satisfies (3.1.2) and (3.1.3) for some nonnegative number $f_{s,t}$; it is feasible in an undirected graph if it satisfies (3.1.4) and (3.1.5) for some nonnegative constant $f_{s,t}$. v_s and v_t are called the source and terminal vertices respectively, and $f_{s,t}$ is called the *value* of $\mathscr{F}$.

For all $v_i \in V$,

$$f(i,V) - f(V,i) = \begin{cases} f_{s,t} & \text{if } i = s, \\ 0 & \text{if } i \neq s,t, \\ -f_{s,t} & \text{if } i = t; \end{cases} \tag{3.1.2}$$

$$c(i,j) \geq f(i,j) \geq 0 \qquad \text{for all } i \text{ and } j. \tag{3.1.3}$$

For all $v_i \in V$

$$f[i,V] = \begin{cases} f_{s,t} & \text{if } i = s, \\ 0 & \text{if } i \neq s,t, \\ -f_{s,t} & \text{if } i = t; \end{cases} \tag{3.1.4}$$

$$c[i,j] \geq f[i,j] \geq -c[i,j]. \tag{3.1.5}$$

Equations (3.1.2) and (3.1.4) are simply expressions of the fact that flow is conserved at all vertices except v_s and v_t. Equations (3.1.3) and (3.1.5) indicate that flows in branches are bounded by the capacitites associated with the branches.

Fig. 3.1.1 Directed branches representing an undirected branch.

In this chapter we consider directed graphs whenever possible. The reason is that an undirected branch $[i,j]$ with capacity $c[i,j]$ can be represented by two directed branches (i,j) and (j,i), as shown in Fig. 3.1.1 where $c(i,j) = c(j,i) = c[i,j]$. Then a set of flows $\{f(i,j)\} = \mathscr{F}$ satisfying (3.1.2) and (3.1.3) can be replaced by a set of flows $\{f[i,j]\} = \mathscr{F}$ satisfying (3.1.3) and (3.1.4) by defining

$$f[i,j] = f(i,j) - f(j,i). \tag{3.1.6}$$

Our main concern in the next three sections will be to relate the maximum value of $f_{s,t}$ to the values of the s-t cuts, and in particular, the value of the minimum capacity s-t cut and to develop algorithms which maximize $f_{s,t}$. A number of algorithms used to maximize flow are based on fundamental relationships between paths and flows. With respect to an s-t path $\pi_{s,t}$ of a *directed* graph, $f(i,j)$ is said to be a *forward flow* if (i,j) is a forward branch in the path; $f(i,j)$ is said to be a *backward flow* if (i,j) is a backward branch in the path. Suppose that in an *undirected* graph $[i,j]$ is replaced by (i,j) resulting in a graph $\hat{G}$. Then $f[i,j]$ is a *forward flow* with respect to a path $\pi_{s,t}$ if (i,j) is a *forward branch* in $\pi_{s,t}$ in the directed graph $\hat{G}$; it is a *backward flow* if (i,j) is a backward branch in $\pi_{s,t}$. The graphs in Fig. 3.1.2 illustrate these concepts. In the s-t path shown in Fig. 3.1.2(a) the forward flows are $f(s,1) = 1$ and $f(2,t) = 1$. The backward flow is $f(2,1) = 2$. In the s-t path in Fig. 3.1.2(b), the forward flows are $f[s,1] = 1, f[1,2] = -2$ and $f[2,t] = 1$. The backward flows are $f[1,s] = -1$, $f[2,1] = 2$ and $f[t,2] = -1$.

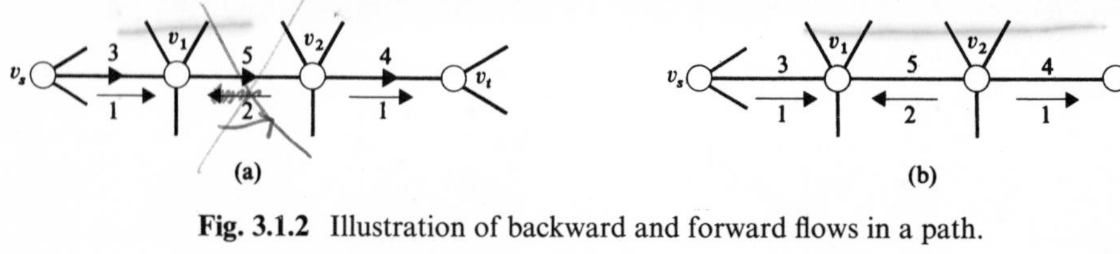

Fig. 3.1.2 Illustration of backward and forward flows in a path.

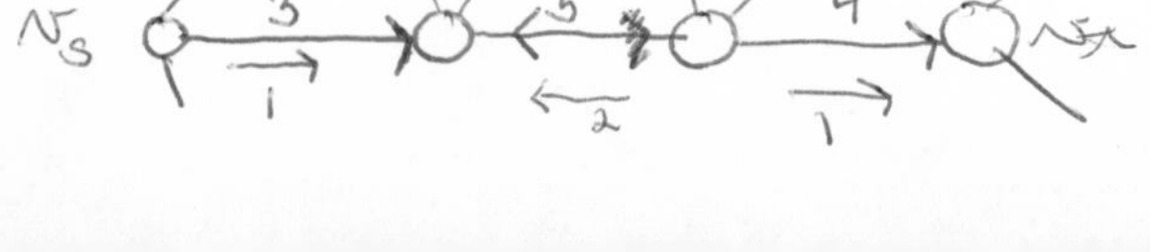

The *residual capacity* $r(\pi_{s,t})$ of an *s-t* path in a *directed* graph is defined by

$$r(\pi_{s,t}) = \text{Min}\left[\underset{\substack{\text{backward branches}\\(i,j)\in\pi_{s,t}}}{\text{Min}}[f(i,j)],\quad \underset{\substack{\text{forward branches}\\(a,b)\in\pi_{s,t}}}{\text{Min}}[c(a,b)-f(a,b)]\right] \tag{3.1.7}$$

The *residual capacity* of an *s-t* path in an *undirected* graph is defined by

$$r(\pi_{s,t}) = \text{Min}\left[\underset{\substack{\text{branches } [i,j]\in\pi_{s,t}\\ \text{with positive}\\ \text{backward flow}\\ f[i,j]}}{\text{Min}}[f[i,j]],\quad \underset{\substack{\text{branches}\\ [a,b]\in\pi_{s,t}\text{ with}\\ \text{nonnegative}\\ \text{forward flow}\\ f[i,j]}}{\text{Min}}[c[a,b]-f[a,b]]\right] \tag{3.1.8}$$

Thus, for $\pi_{s,t}$ in Fig. 3.1.2(a)

$$r(\pi_{s,t}) = \text{Min}\,[\text{Min}\,[2], \text{Min}\,[(3-1), (4-1)]] = 2, \tag{3.1.9}$$

and for $\pi_{s,t}$ in Fig. 3.1.2(b)

$$r(\pi_{s,t}) = \text{Min}\,[\text{Min}\,[2], \text{Min}\,[(3-1), (4-1)]] = 2. \tag{3.1.10}$$

A branch (i,j) is said to be *saturated* if $f(i,j) = c(i,j)$ and to be *unsaturated* otherwise; it is said to be *empty* if $f(i,j) = 0$. Similarly, $[i,j]$ is saturated if $|f[i,j]| = c[i,j]$ and unsaturated otherwise; it is empty if $|f[i,j]| = 0$. A path $\pi_{s,t}$ is said to be *saturated* if $r(\pi_{s,t}) = 0$ and to be *unsaturated* otherwise. If the path is unsaturated then we can add $r(\pi_{s,t})$ to all forward flows in the path and subtract $r(\pi_{s,t})$ from all backward flows, thereby obtaining a new flow pattern for which flow is still conserved and branch capacities are not exceeded. Similarly, we say that a directed circuit is saturated if every branch in the circuit is saturated.

It is sometimes desirable to represent a given flow pattern as a union of flows along directed paths. However, this is not always possible. In Fig. 3.1.3 we cannot

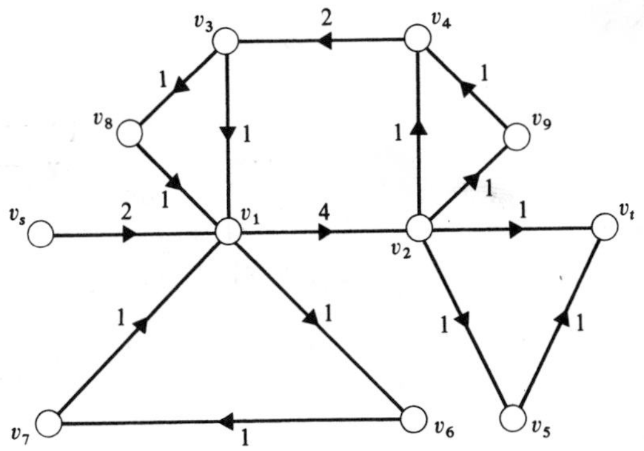

Fig. 3.1.3 Graph illustrating directed semipaths; branch weights are flows.

account for all the flow by flow along paths. We therefore define a *directed s-t semipath* as a branch disjoint union of a directed *s-t* path and directed circuits. Similarly, a directed *semicircuit* is a branch disjoint union of directed circuits.

To state our goals more precisely, we must introduce the concepts of a semipath of flow and a semicircuit of flow. Let S be a directed *s-t* semipath or a semicircuit. Suppose a flow of x units is assigned to each branch of S, while zero flow is assigned to all other branches. The resulting flow pattern, denoted by $x(S)$, is called a *semicircuit* or *semipath of flow* of x units. If $x = 1$, then $1(S)$ is called a *unit semipath* or *semicircuit of flow*. If S is a path then we write $x(S)$ as $x(\pi)$.

If $\mathscr{F}$ and $\mathscr{G}$ are two flow patterns in a directed graph, then $\mathscr{F} \oplus \mathscr{G}$ denotes a flow pattern $\mathscr{L}$ for which $l(i,j) = f(i,j) + g(i,j)$ for each (i,j). Similarly $\mathscr{F} \ominus \mathscr{G}$ denotes a flow pattern $\mathscr{L}$ for which $l(i,j) = f(i,j) - g(i,j)$ for all (i,j). Note that $\mathscr{F} \ominus \mathscr{G}$ may not be a feasible flow pattern since capacity constraints may be violated. If $\mathscr{F}_1, \mathscr{F}_2, \ldots, \mathscr{F}_k$ are semicircuits of flow or flow patterns in which all flows are zero then the value of $\mathscr{F}_1 \oplus \mathscr{F}_2 \oplus \cdots \oplus \mathscr{F}_k$ is zero and we call the flow pattern a *zero flow pattern* denoted by $\mathscr{Z}$.

A *semipath decomposition* of a flow pattern $\mathscr{F}$ with value $f_{s,t}$ is said to be achieved when $\mathscr{F}$ can be written as

$$\mathscr{F} = x_1(\pi_1) \oplus x_2(\pi_2) \oplus \cdots \oplus x_k(\pi_k) \oplus \mathscr{Z} \tag{3.1.11}$$

It is seldom necessary in practice actually to perform a semipath decomposition of a flow pattern. However, a number of proofs depend upon a concept of a flow in semipath decomposed form. For this reason we will not be interested in the efficiency of proposed algorithms to perform this decomposition but only in their existence. If, in (3.1.11), $x_i = 1$ for $i = 1, \ldots, k$ then the resulting decomposition is called a unit semipath decomposition. The basis for a semipath decomposition algorithm is presented in Lemmas 3.1.1 and 3.1.2 below.

Lemma 3.1.1 *If S is a directed s-t semipath then $1(S) = 1(\pi) \oplus \mathscr{Z}$ for some directed s-t path, π.*

Proof. If S is a directed *s-t* path then $1(S) = 1(\pi) \oplus \mathscr{Z}$ where $\mathscr{Z}$ is the flow pattern consisting of zero flow in every branch. If S is not a directed *s-t* path then S can be identified by the sequence of vertices $v_s, v_{j_1}, v_{j_2}, \ldots, v_{j_l}, v_t$ where some of the vertices are repeated. Let v_i be the first vertex which is repeated. Then $1(S) = 1(S_1) \oplus \mathscr{Z}_1$ where $\mathscr{Z}_1$ is determined by the sequence of vertices between the first and second appearances of v_i, and S_1 is an *s-t* semipath. If S_1 is a directed *s-t* path then the proof is completed by setting $\pi = S_1$ and $\mathscr{Z} = \mathscr{Z}_1$. If not, then the above procedure is repeated, expressing $1(S_1)$ as $1(S_2) \oplus \mathscr{Z}_2$ so that $1(S) = 1(S_2) \oplus \mathscr{Z}_1 \oplus \mathscr{Z}_2$. As long as each successive semipath S_l is not a path, this procedure is repeated. Eventually, $1(S) = 1(S_k) \oplus \mathscr{Z}_1 \oplus \mathscr{Z}_2 \oplus \cdots \oplus \mathscr{Z}_k$ where S_k is a directed *s-t* path. Then $\pi = S_k$ and $\mathscr{Z} = \mathscr{Z}_1 \oplus \mathscr{Z}_2 \oplus \cdots \oplus \mathscr{Z}_k$.

Lemma 3.1.2 *If $\mathscr{F}$ is an s-t flow pattern of integer value $f_{s,t} > 0$, then $\mathscr{F} = 1(\pi_1) \oplus 1(\pi_2) \oplus \cdots \oplus 1(\pi_k) \oplus \mathscr{Z}$ for some directed s-t paths, $\pi_1, \ldots, \pi_k$.*

Proof. Since $f_{s,t} > 0$, there is a branch (s,j_1) with nonzero flow. Because flow must be conserved at v_1, there must be another branch (j_1,j_2) with nonzero flow. We can continue to identify successive vertices $v_s, v_{j_1}, \ldots, v_{j_l}$ until eventually v_t or a vertex that has previously been encountered is reached. In the latter case, a circuit of nonzero flow has been traced. The branches in this circuit may be removed and the tracing continue. Eventually v_t must be reached.

Consider the flow pattern obtained by assigning one unit of flow to each branch thus identified. The resulting flow pattern is a unit flow semipath $1(S)$ and $\mathscr{F}$ can then be written as $\mathscr{F} = \mathscr{F}_1 \oplus 1(S_1)$ where $\mathscr{F}_1$ is a flow pattern of value $f_{s,t} - 1$. If $f_{s,t} - 1 > 0$ the same procedure can be applied to $\mathscr{F}_1$ and every resulting flow pattern $\mathscr{F}_i$ until eventually $\mathscr{F} = \mathscr{Z}' \oplus 1(S_1) \oplus 1(S_2) \oplus \cdots \oplus 1(S_k)$. By Lemma 3.1.1, every $1(S_i)$ can be decomposed into $1(S_i) = 1(\pi_i) \oplus \mathscr{Z}_i$. Hence, $\mathscr{F} = \mathscr{Z} \oplus 1(\pi_1) \oplus \cdots \oplus 1(\pi_k)$ where $\mathscr{Z} = \mathscr{Z}' \oplus \mathscr{Z}_1 \oplus \mathscr{Z}_2 \oplus \cdots \oplus \mathscr{Z}_k$.

The methods used to prove Lemmas 3.1.1 and 3.1.2 yield the following algorithm to find a semipath decomposition of an *s-t* flow of integer value *k* in a directed graph.

Semipath Decomposition Algorithm

Step 0. Let $i = 0$.
Step 1. Find a unit *s-t* semipath of flow, $1(S_i)$.
Step 2. Let $\mathscr{F}_{i+1} = \mathscr{F}_i \ominus 1(S_i)$ where $\mathscr{F}_i$ is the remaining *s-t* flow in the graph.
Step 3. Decompose S_i as $1(S_i) = 1(\pi_i) \oplus \mathscr{Z}_i$.
Step 4. If the value of $\mathscr{F}_{i+1}$ is greater than zero, increase i by 1 and go to 2. Otherwise find the remaining zero flow pattern $\mathscr{Z}'$.
Step 5. Let $\mathscr{Z} = \mathscr{Z}' \oplus \mathscr{Z}_1 \oplus \mathscr{Z}_2 \oplus \cdots \oplus \mathscr{Z}_k$.
Step 6. $\mathscr{F} = \mathscr{Z} \oplus 1(\pi_1) \oplus 1(\pi_2) \oplus \cdots \oplus 1(\pi_k)$.

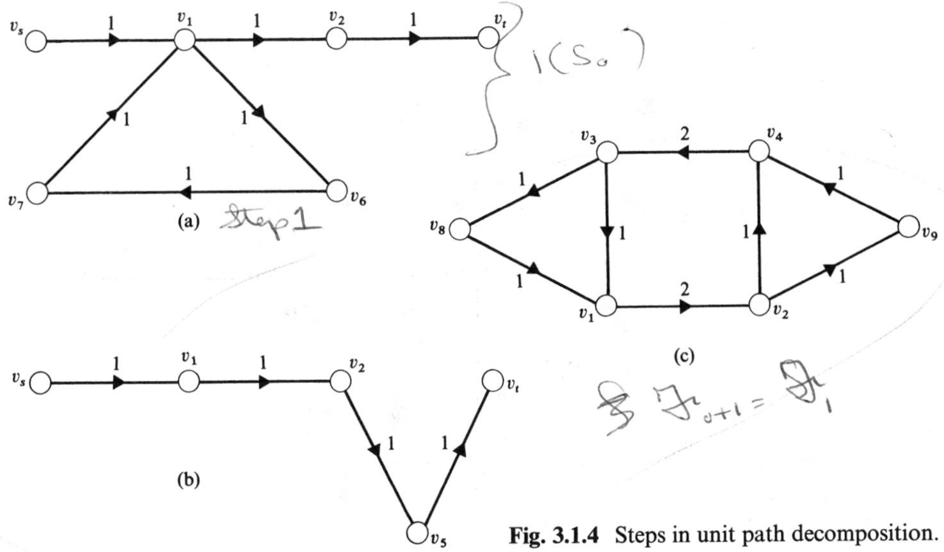

Fig. 3.1.4 Steps in unit path decomposition.

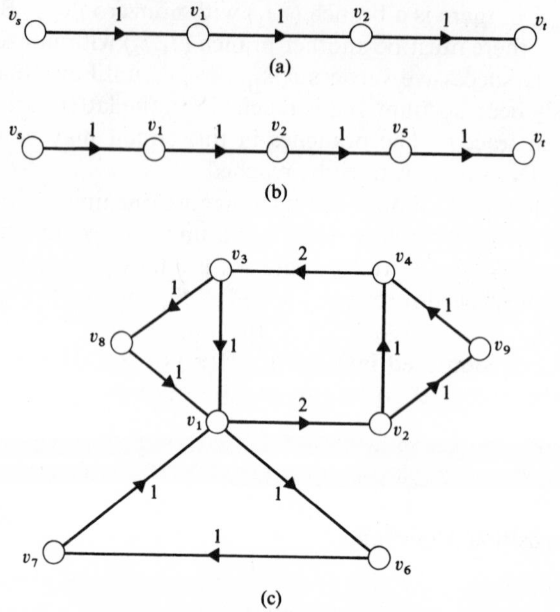

Fig. 3.1.5 Unit path decomposition.

Example 3.1.1 Consider the flow pattern in Fig. 3.1.3. By applying Steps 1, 2, and 3 the semipaths of flow in Figs. 3.1.4(a) and 3.1.4(b) can be found, leaving the zero flow pattern in Fig. 3.1.4(c). Using Steps 4, 5, and 6, $1(\pi_1)$ is in Fig. 3.1.5(a), $1(\pi_2)$ is in Fig. 3.1.5(b), and $\mathscr{L}$ is in Fig. 3.1.5(c). If a decomposition with nonunit flow paths

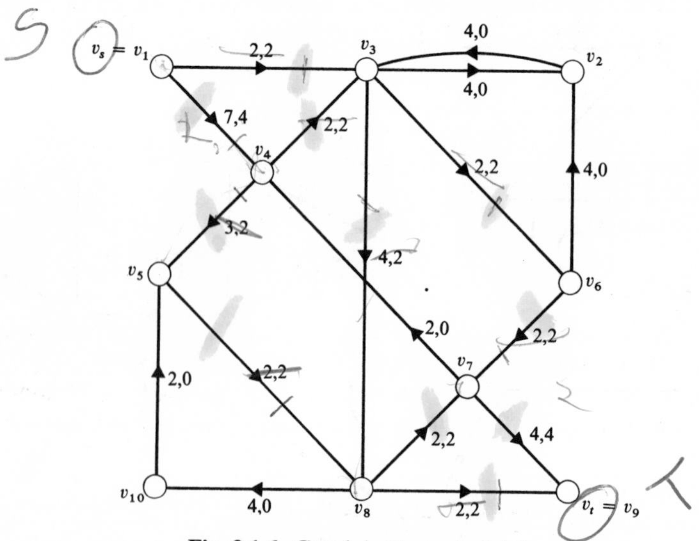

Fig. 3.1.6 Graph in Example 3.1.2.

is desired then $1(\pi_i) \oplus 1(\pi_i) \oplus \cdots \oplus 1(\pi_i)$ can be combined into $x(\pi_i)$ where x is the number of unit flow paths using π_i.

Example 3.1.2 The s-t flow pattern in Fig. 3.1.6 can be decomposed in several ways, three of which are indicated in (3.1.12), (3.1.13), and (3.1.14) below.

$$\mathscr{F} = 1(\pi_1) \oplus 1(\pi_1) \oplus 1(\pi_2) \oplus 1(\pi_2) \oplus 1(\pi_3) \oplus 1(\pi_3) \oplus \mathscr{Z}, \tag{3.1.12}$$

where

$$\pi_1 = v_s, v_3, v_6, v_7, v_t,$$
$$\pi_2 = v_s, v_4, v_3, v_8, v_t,$$
$$\pi_3 = v_s, v_4, v_5, v_8, v_t,$$

and $\mathscr{Z}$ is the flow pattern in which all flows are zero.

$$\mathscr{F} = 2(\pi_1) \oplus 2(\pi_2) \oplus 2(\pi_3), \tag{3.1.13}$$

where π_1, π_2, π_3, and $\mathscr{Z}$ are the same as in (3.1.12).

$$\mathscr{F} = 1(\pi_1) \oplus 1(\pi_2) \oplus 1(\pi_3) \oplus 2(\pi_4) \oplus 1(\pi_5) \oplus \mathscr{Z}, \tag{3.1.14}$$

where

$$\pi_1 = v_s, v_3, v_6, v_7, v_t,$$
$$\pi_2 = v_s, v_4, v_3, v_6, v_7, v_t,$$
$$\pi_3 = v_s, v_4, v_3, v_8, v_t,$$
$$\pi_4 = v_s, v_4, v_5, v_8, v_t,$$
$$\pi_5 = v_s, v_4, v_3, v_6, v_7, v_t,$$

and $\mathscr{Z}$ is the flow pattern in which all flows are zero.

3.2 THE MAX-FLOW MIN-CUT THEOREM

The problem which arises most often in flow problems is to find a feasible flow pattern which maximizes $f_{s,t}$ in a given graph. The answer to this problem is therefore one of the most important and basic results in this book. Let $\tau_{s,t}$ be the value of the minimum s-t cut, that is,

$$\tau_{s,t} = \operatorname*{Min}_k \; [c(A^k_{s,t})]. \tag{3.2.1}$$

Max-Flow Min-Cut Theorem *For all feasible flow patterns $\mathscr{F}$,*

$$\operatorname*{Max}_{\mathscr{F}} \; [f_{s,t}] = \operatorname*{Min}_k \; [c(A^k_{s,t})] = \tau_{s,t}. \tag{3.2.2}$$

In 1956 three independent proofs of this theorem were given [DA1, EL1, FO1]. The proof by Ford and Fulkerson [FO1] is combinatoric in that it relies on the structure of flow patterns which maximize $f_{s,t}$. The proof leads to a very efficient method of constructing flows to maximize $f_{s,t}$, which will serve as the foundation for a number of other algorithms in this book. The proof of Elias, Feinstein, and Shannon [EL1] gives an ingenious graph theoretic method of breaking a graph into simpler graphs until the solution becomes obvious for the simpler graphs and it is also apparent that a solution for the simpler graphs implies a solution for the original graph. The proof by Dantzig and Fulkerson [DA1] uses the Duality Theorem of linear programming and hence places the problem in a broad framework that emphasizes some of the fundamental properties of the solution. The following lemma is proved similarly in all three papers.

Lemma 3.2.1 $f_{s,t} \le \tau_{s,t}$. (3.2.3)

Proof. From (3.1.2),

$$f_{s,t} = f(s,V) - f(V,s) \tag{3.2.4a}$$

and

$$0 = f(i,V) - f(V,i) \qquad \text{for } i \ne s,t. \tag{3.2.4b}$$

Thus, for any X such that $v_s \in X$ and $v_t \in \bar{X}$,

$$\sum_{v_i \in X} [f(i,V) - f(V,i)] = f_{s,t}, \tag{3.2.5}$$

or

$$f(X,V) - f(V,X) = f_{s,t}. \tag{3.2.6}$$

Substituting, $V = X \cup \bar{X}$ in (3.2.6) and using the fact that $X \cap \bar{X} = \phi$ we have

$$f(X,\bar{X}) - f(\bar{X},X) + f(X,X) - f(X,X) = f_{s,t}$$

or

$$f(X,\bar{X}) - f(\bar{X},X) = f_{s,t}. \tag{3.2.7}$$

However,

$$f(\bar{X},X) \ge 0,$$

and therefore

$$f_{s,t} \le f(X,\bar{X}) = \sum_{\substack{v_i \in X \\ v_j \in \bar{X}}} f(i,j). \tag{3.2.8}$$

Combining (3.2.8) with the fact that

$$f(i,j) \le c(i,j), \tag{3.2.9}$$

we have for any $f_{s,t}$

$$f_{s,t} \le \sum_{\substack{v_i \in X \\ v_j \in \bar{X}}} c(i,j) = c(X,\bar{X}). \tag{3.2.10}$$

Since (3.2.10) is true for all X, the proof is completed.

Because of the central importance of the Max-Flow Min-Cut Theorem we present all three proofs in the next three sections. The theorem does not require that the branch capacities be integers. However, the proof by Ford and Fulkerson does require integrality of branch capacities and yields integer branch flows. Finally it will be obvious from the proof of the Max-Flow Min-Cut Theorem, that a specific flow requirement $r_{s,t} \ge 0$ can be satisfied if and only if $r_{s,t} \le \tau_{s,t}$.

3.3 COMBINATORIC PROOF OF THE MAX-FLOW MIN-CUT THEOREM

In this section we prove the Max-Flow Min-Cut Theorem by using the Ford–Fulkerson Augmentation Algorithm. This algorithm provides a method to increase $f_{s,t}$ systematically from an arbitrary value less than $\tau_{s,t}$. We then show that the algorithm terminates if and only if $f_{s,t} = \tau_{s,t}$. In order to prove that $f_{s,t} = \tau_{s,t}$ can be attained by the Augmentation Algorithm, we must assume that for all i and j, $c(i,j)$ is an integer. Of course, if branch capacities are rational numbers, they can be converted to integers before flow is maximized and then reconverted to rationals. The integrality restriction is not needed in the graph theoretic proof given in the next section.

Augmentation Algorithm

Step 1. Find any feasible flow pattern ($f(i,j) = 0$ for all $(i,j) \in \Gamma$ is always a possible choice). If $f_{s,t} = \tau_{s,t}$ then the algorithm terminates. Otherwise proceed to Step 2.
Step 2. Find an unsaturated s-t path $\pi_{s,t}$ with residual capacity $r(\pi_{s,t}) > 0$. Such a path will be called an *augmentation path.*
Step 3. Increase the flow in the augmentation path by $r(\pi_{s,t})$.
Step 4. Repeat Steps 2 and 3 until no more augmentation paths can be found.

It is clear that the Augmentation Algorithm will yield integer feasible flows and that $f_{s,t}$ is increased by an integer each time an augmentation path is found. The following theorem in conjunction with Lemma 3.2.1 proves the Max-Flow Min-Cut Theorem.

Theorem 3.3.1 *The Augmentation Algorithm terminates if and only if $f_{s,t} \geq \tau_{s,t}$.*

Proof. If the Augmentation Algorithm is applied it must terminate since $f_{s,t}$ is increased by an integer at each application of Step 3 and $f_{s,t}$ is bounded above by the integer $\tau_{s,t}$. Therefore, we assume the algorithm has terminated. We show that $f_{s,t} \geq \tau_{s,t}$. Define a set of vertices L such that

$$v_s \in L. \tag{3.3.1}$$

$$\text{If } v_i \in L \text{ and } c(i,j) > f(i,j) \text{ or } f(j,i) > 0 \text{ then } v_j \in L. \tag{3.3.2}$$

We now show that $f_{s,t} = c(L,\bar{L}) \geq \tau_{s,t}$.

We know that $v_t \in \bar{L}$ because otherwise there would be an s-t augmentation path contradicting the assumption that the Augmentation Algorithm has terminated. Hence $\bar{L}$ is not empty.

From (3.2.7),

$$f_{s,t} = f(L,\bar{L}) - f(\bar{L},L). \tag{3.3.3}$$

But if $(i,j) \in (L,\bar{L})$ then $f(i,j) = c(i,j)$ and if $(i,j) \in (\bar{L},L)$ then $f(i,j) = 0$ from the definition of L. Hence,

$$f(L,\bar{L}) = c(L,\bar{L}) \tag{3.3.4}$$

and

$$f(\bar{L},L) = 0. \tag{3.3.5}$$

Combining (3.3.3), (3.3.4), and (3.3.5) yields

$$f_{s,t} = c(L,\bar{L}) \geq \tau_{s,t}, \tag{3.3.6}$$

which proves the theorem. //

The Augmentation Algorithm does not provide a practical algorithm for maximizing $f_{s,t}$ since it does not contain a method for finding an augmentation path. However, the Labeling Algorithm given below is an extremely efficient one for maximizing $f_{s,t}$. It contains a Labeling Routine for finding an augmentation path, as well as a systematic procedure for determining L as defined in the proof of Theorem 3.3.1. The efficiency of the algorithm is due to the fact that in finding an augmentation path each vertex must be examined at most once.

The algorithm assigns ordered triples, called labels, to the vertices. In the label of a typical vertex v_j, the first term in the triple is the index i, $v_i \in L$, which is being examined to see whether (3.3.2) is satisfied for some j. The second term is a plus if $c(i,j) - f(i,j) > 0$ and a minus if $f(j,i) > 0$. The third term is the residual capacity of the section of augmentation path already found. Once the Labeling Routine finds an augmentation path $\pi_{s,t}$ the Augmentation Routine increases the flow along the path by $r(\pi_{s,t})$. The proof of the convergence of the algorithm is identical to the proof of Theorem 3.3.1.

Labeling Algorithm

Labeling Routine

Step 1. Label v_s by $(s, +, \varepsilon(s) = \infty)$. v_s is now labeled and unscanned and all other vertices are unlabeled and unscanned.

Step 2. Select any labeled and unscanned vertex v_x.

a) For all v_y such that $(y,x) \in \Gamma$, v_y is unlabeled, and $f(y,x) > 0$, label v_y by $(x, -, \varepsilon(y))$ where $\varepsilon(y) = \text{Min}\,[\varepsilon(x), f(y,x)]$. Then, v_y is labeled and unscanned.

b) For all v_y such that $(x,y) \in \Gamma$, v_y is unlabeled and $c(x,y) > f(x,y)$, label v_y by $(x, +, \varepsilon(y))$ where $\varepsilon(y) = \text{Min}\,[\varepsilon(x), c(x,y) - f(x,y)]$. Then, v_y is labeled and unscanned.

Change the label on v_x by circling the $+$ or $-$ entry. Then, v_x is now labeled and scanned.

Step 3. Repeat Step 2 until v_t is labeled or until no more labels can be assigned. In the latter case the algorithm terminates. In the former case proceed to the Augmentation Routine.

Augmentation Routine

Step 1. Let $z = t$ and go to Step 2 of the Augmentation Routine.

Step 2. If the label on v_z is $(q, +, \varepsilon)$, increase $f(q,z)$ by $\varepsilon(t)$. If the label on v_z is $(q, -, \varepsilon)$, decrease $f(z,q)$ by $\varepsilon(t)$.

Step 3. If $q = s$, erase all labels and return to Step 1 of the Labeling Routine. Otherwise let $z = q$ and return to Step 2 of the Augmentation Routine.

The use of the Labeling Algorithm is illustrated in the example below.

Example 3.3.1 Consider the graph in Fig. 3.3.1(a). An undirected branch $[i,j]$ with capacity $c[i,j]$ has been replaced by two directed branches (i,j) and (j,i) with $c(i,j) = c(j,i) = c[i,j]$. In the ordered pair associated with each branch the first entry is the capacity of the branch and the second entry is the flow. An initial flow pattern is assumed as indicated. The Labeling Algorithm then assigns the labels shown. Since v_t is labeled, an augmentation path exists and the Augmentation Routine gives the path as $\pi_{s,t} = (1,3), (2,3), (6,2), (6,7), (7,9)$ with $r(\pi_{s,t}) = \varepsilon(9) = 2$. Hence $f(\pi_{s,t})$ can be increased by 2 to yield the flow pattern shown in Fig. 3.3.1(b). The labels are erased and the Labeling Routine assigns the new labels as in Fig. 3.3.1(b). Since v_t cannot be labeled, $f_{s,t}$ cannot be increased. L is given by the set of labeled vertices; hence $L = \{v_1,v_4,v_5\}$ and $\bar{L} = \{v_2,v_3,v_6,v_7,v_8,v_9,v_{10}\}$. As predicted, $f(L,\bar{L}) = c(L,\bar{L})$ and $f(\bar{L},L) = 0$. Note that each vertex need be scanned at most once when finding an augmentation path.

*3.4 GRAPH THEORETIC PROOF OF THE MAX-FLOW MIN-CUT THEOREM

We next present the proof of the Max-Flow Min-Cut Theorem developed by Elias, Feinstein, and Shannon. As mentioned previously, this proof does not require that the capacities be rational. We first introduce three operations on a graph—reduction, splitting, and grouping—which result in a new "flow equivalent graph." For the new graph, the validity of the Max-Flow Min-Cut Theorem is readily established.

1. *Reduction.* If there is a branch which is not in at least one s-t cut of value $\tau_{s,t}$, reduce its capacity until it is in such a cut or until its capacity is zero. Continue this process, each time using the new resulting graph, until all branches have been examined. The final graph is said to be *reduced.* In a reduced graph every branch of nonzero capacity is in at least one s-t cut of capacity $\tau_{s,t}$ and the capacity of the minimum s-t cut is still $\tau_{s,t}$.

2. *Splitting.* If there exist single vertices $v_1,v_2,\ldots,v_j$ which are s-t vertex cut-sets partition the graph such that all vertices and branches are included in the subgraphs, $G_1,G_2,\ldots,G_{j+1}$ so that G_1 contains v_s and has only v_1 in common with G_2; for $l = 2,\ldots,j$, G_l contains only v_l in common with G_{l+1}; finally G_{j+1} contains v_t. $G_1,\ldots,G_{j+1}$ are called *series subgraphs.* If there are no s-t cuts consisting of single vertices then G is the only series subgraph. If there is an s-t path containing more than two branches in a series subgraph G_i, find the second branch in the path in the subgraph and find a minimum s-t cut $(X,\bar{X})$ containing that branch. Add a vertex v_{c_i} to G_i, replace each branch, (a,b), in $(X,\bar{X})$ by two branches (a,c_i) and (c_i,b), and if $(b,a) \in \Gamma$, replace it by (b,c_i) and (c_i,a). The operation of splitting is repeated until the longest path in every resulting series subgraph has at most two branches. With repetition of splitting, the number of series subgraphs ultimately is increased and each new series subgraph has a shorter maximum length path than the original series subgraph from which it was obtained. Note that the resulting structure may contain parallel branches and is therefore an s-graph (with $s \geq 1$).

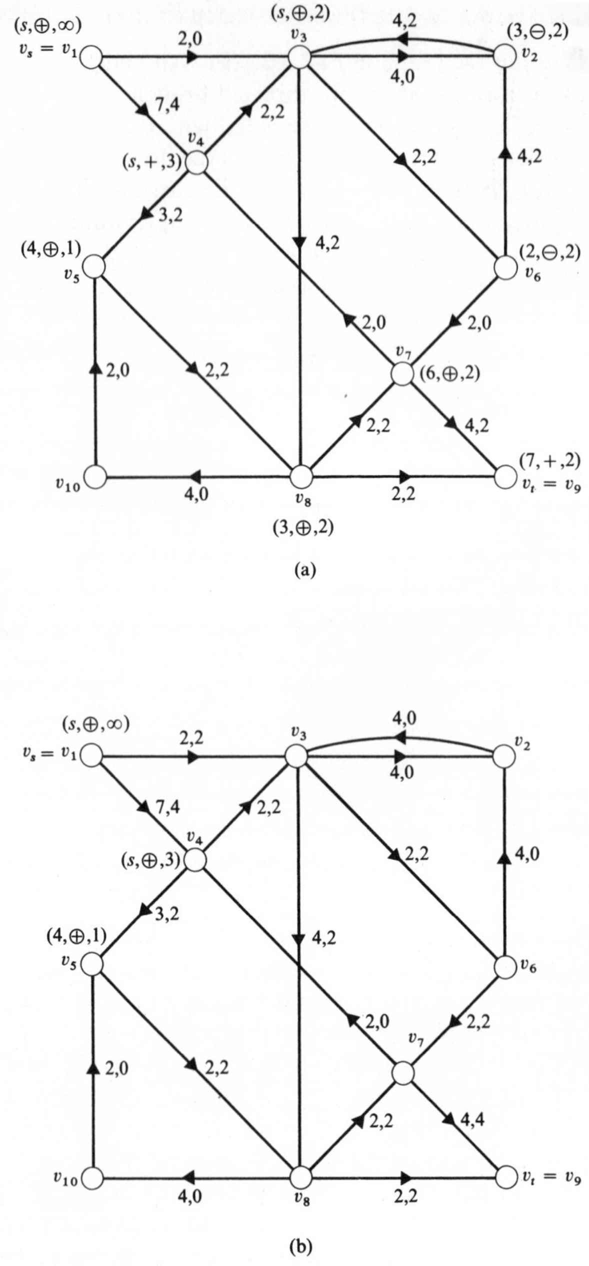

Fig. 3.3.1 (a) Initial flow in G; (b) maximum flow in G.

3. *Grouping.* If there are $k > 1$ branches $(i,j)_1,(i,j)_2,\ldots,(i,j)_k$ directed from v_i to v_j, replace them by a single branch (i,j) with

$$c(i,j) = \sum_{l=1}^{k} c(i,j)_l.$$

Repeat this operation for all (i,j).

The three operations and their properties are illustrated in the following example.

Example 3.4.1 Consider the graph shown in Fig. 3.4.1.

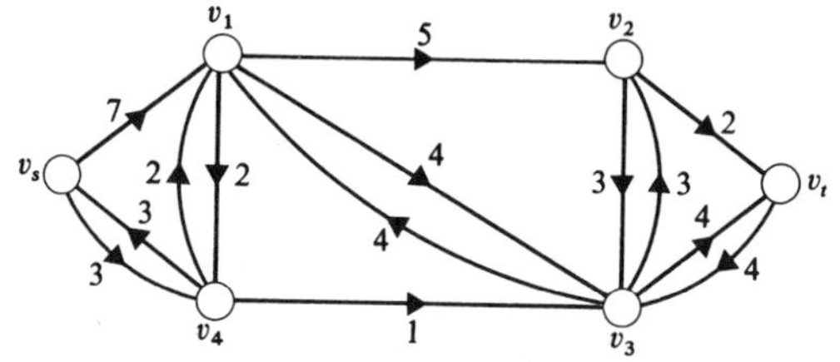

Fig. 3.4.1 Graph to which reduction is applied in Example 3.4.1.

1. *Reduction.* The minimum s-t cut consists of branches $(2,t)$ and $(3,t)$ and has capacity 6. We first examine branch (4,3). The smallest s-t cut containing (4,3) is $(\{v_1,v_4,v_s\},\{v_2,v_3,v_t\})$ and has value 10. We therefore reduce $c(4,3)$ to zero, and thus

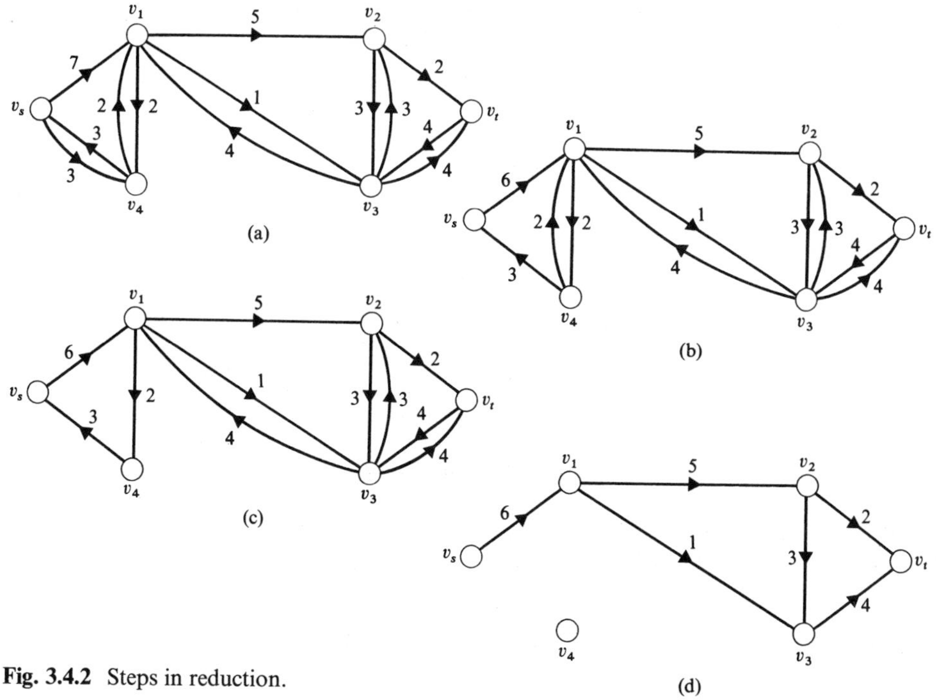

Fig. 3.4.2 Steps in reduction.

eliminate (4,3). We next consider branch (1,3). The smallest s-t cut containing (1,3) is $(\{v_1,v_4,v_s\},\{v_2,v_3,v_t\})$ with value 9 (since (4,3) has been removed). We therefore reduce $c(1,3)$ to 1. We next consider (1,2). It is now in the cut $(\{v_1,v_4 v_s\},\{v_2,v_3,v_t\})$ with value 6 and hence we do not change its capacity. The resulting graph is shown in Fig. 3.4.2(a). After considering $(s,4)$ and $(s,1)$ the resulting graph is shown in Fig. 3.4.2(b). After considering (4,1) the resulting graph is shown in Fig. 3.4.2(c). Finally $c(4,s)$, $c(1,4)$, $c(3,1)$, $c(3,2)$ and $c(t,3)$ are reduced to zero by the same process and the reduced graph is given in Fig. 3.4.2(d). Each branch is now in at least one s-t cut of value $\tau_{s,t} = 6$ and the value of the minimum s-t cut has not been reduced. Note that the reduced graph is not unique since the order in which we considered the branches was arbitrary.

2. *Splitting*. The series subgraphs of G in Fig. 3.4.2(d) are

$$G_1 = (\{v_s,v_1,v_4\},\{(s,1)\}) = (V_1,\Gamma_1)$$

and

$$G_2 = (\{v_1,v_2,v_3,v_t\},\{(1,2),(2,3),(2,t),(3,t),(1,3)\}) = (V_2,\Gamma_2).$$

G_2 contains a path with three branches (1,3), (2,3), and $(2,t)$. The second branch in the path is (2,3). We therefore let $(X,\bar{X}) = (\{v_1,v_2\},\{v_3,v_t\})$ and apply splitting. The resulting 2-graph is shown in Fig. 3.4.3, in which the longest path in every series subgraph has at most two branches.

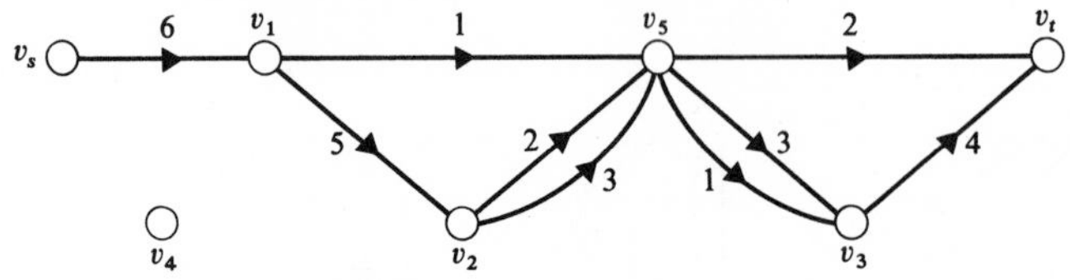

Fig. 3.4.3 A 2-graph after splitting.

3. *Grouping*. The graph that results after grouping is shown in Fig. 3.4.4.

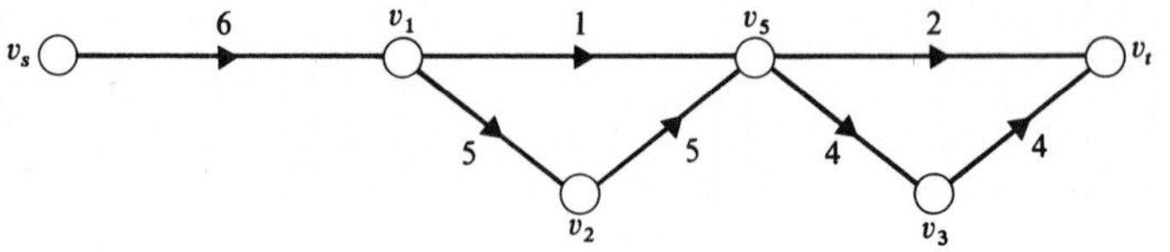

Fig. 3.4.4 Graph after grouping.

We now show that the three operations yield a graph $\hat{G}$ in which the maximum s-t flow equals $\tau_{s,t}$. Moreover, once we have such a flow pattern in $\hat{G}$ we will be able to find a flow pattern in G with $f_{s,t} = \tau_{s,t}$.

Lemma 3.4.1 *The value of $\tau_{s,t}$ is invariant under the operations of reduction, splitting, and grouping.*

Proof. In the reduction operation the values of *s-t* cuts may be decreased, but never below the value of the minimum *s-t* cut in G. No cuts are increased in value. Splitting replaces a minimum *s-t* cut by two minimum *s-t* cuts of the same value but does not create any new cuts. Grouping leaves the values of all *s-t* cuts unchanged.

When the operations of reduction, splitting, and grouping are completed the resulting graph has the appearance of the graph in Fig. 3.4.5; namely, it is a number of series subgraphs each consisting of at most one directed path of length one and any number of lengths two. Furthermore all branches are forward branches in *s-t* paths since backward branches would not be in minimum *s-t* cuts and hence would have been eliminated by reduction. Hence we can use Lemma 3.4.1 to prove the following lemma. Let $\hat{G}$ be the graph resulting from reduction, splitting and grouping.

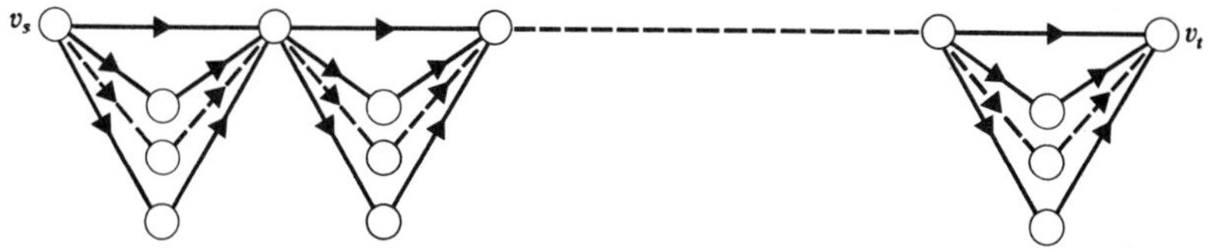

Fig. 3.4.5 Representation of graph $\hat{G}$ after reduction, splitting, and grouping.

Lemma 3.4.2

a) In each series subgraph of $\hat{G}$, *there is an s-t cut value* $\tau_{s,t}$.
b) If two branches are in series in a series subgraph of $\hat{G}$ *they have the same capacity.*

Proof. Every branch must be in an *s-t* cut of value $\tau_{s,t}$ and hence (a) follows. Suppose (i,j) and (j,k) are in a series subgraph and without loss of generality $c(i,j) > c(j,k)$. There is an *s-t* cut $(Y,\bar{Y})$ containing (i,j) in the series subgraph, such that $c(Y,\bar{Y}) = \tau_{s,t}$. But then the cut $(Z,\bar{Z})$ formed by replacing (i,j) by (j,k) in $(Y,\bar{Y})$ is an *s-t* cut with $c(Z,\bar{Z}) < \tau_{s,t}$. This is a contradiction since reduction, splitting, and grouping do not reduce the value of the minimum *s-t* cut. Hence the lemma is proved.

From Lemma 3.4.2 we see that we can obtain a maximum *s-t* flow of value $\tau_{s,t}$ in $\hat{G}$ by saturating each branch. Hence we can achieve an *s-t* flow in G of value $\tau_{s,t}$ by assigning the same flows as in $\hat{G}$; that is, we first assign the flows in $\hat{G}$ to the graph G' which we had before grouping. Next, we assign to (i,j) in G the flow in (l,k) in G' if (l,k) was obtained from (i,j) by reduction and splitting. Therefore $\max_{\mathcal{F}} [f_{s,t}] \geq \tau_{s,t}$.

*3.5 LINEAR PROGRAMMING PROOF OF THE MAX-FLOW MIN-CUT THEOREM

In this section we show that the Max-Flow Min-Cut Theorem is a consequence of the Duality Theorem of linear programming. To show this we first formulate the maximum flow problem as a linear program. By investigating the properties of this linear program we develop a method to show that an integer solution exists if the branch capacities are integers. Using this method, we show that the dual problem has integer solutions in which certain variables have value 0 or 1 and then, using the Complementary Slackness Theorem and the Duality Theorem, we show that the value of the maximum s-t flow is equal to the value of the minimum s-t cut.

Consider a directed graph $G = (V,\Gamma)$ with n vertices, $v_s,v_t,v_1,\ldots,v_N$, where $N = (n - 2)$. Assume for convenience that there is no branch between v_s and v_t. There is no loss of generality since if there were such a branch, the maximum value of $f_{s,t}$ would be increased by $c(s,t)$. For the sake of symmetry in the equations to be considered call either v_s or v_t an end vertex, and denote it by v_e. It is also clear that if we delete all branches directed to v_s or from v_t the maximum flow is not changed. Therefore delete these branches, let $f(e,i)$ denote a flow $f(s,i)$, and let $f(i,e)$ denote a flow $f(i,t)$.

Let $f(i,i)$ for $i = e,1,\ldots,N$ denote the total flow "through" a vertex. For $i = e$, $f(i,i)$ is the total flow into v_t or the total flow out of v_s. For all other i, it represents the total flow into v_i and the total flow out of v_i, because of conservation of flow. Let $f(i,i) \leq c(i,i)$ for $i = e,1,\ldots,N$, where

$$c(i,i) = \text{Min}\left[\sum_{j,j\neq i} c(i,j), \sum_{j,j\neq i} c(j,i)\right]. \tag{3.5.1}$$

In terms of these variables, the conservation constraints, the capacity constraints, and the sign requirements on flows are, respectively,

$$-f(i,i) + \sum_{j,j\neq i} f(i,j) = 0 \qquad \text{for } j = e,1,\ldots,N; \tag{3.5.2a}$$

$$-f(j,j) + \sum_{i,i\neq j} f(i,j) = 0 \qquad \text{for } j = e,1,\ldots,N; \tag{3.5.2b}$$

$$f(i,j) + y(i,j) = c(i,j) \qquad \text{for } i,j = e,1,\ldots,N; \tag{3.5.3}$$

$$f(i,j) \geq 0 \qquad \text{for } i,j = e,1,\ldots,N; \tag{3.5.4a}$$

$$y(i,j) \geq 0 \qquad \text{for } i,j = e,1,\ldots,N. \tag{3.5.4b}$$

For $(i,j) \notin \Gamma$, $c(i,j)$ is assumed to be zero. In (3.5.3) and (3.5.4), the $y(i,j)$ for $i,j = e,1,\ldots,N$ are the dummy variables, called slack variables, introduced to form equalities in (3.5.3). The linear program whose solution maximizes $f_{s,t}$ is equivalent to the maximization of $f(e,e)$ subject to (3.5.2) to (3.5.4). We consider this linear program for the remainder of the section. Let 0 be the zero matrix. Let I be the identity matrix and let $\mathbf{F}$ be a column matrix of flows lexicographically ordered, i.e.

$$\mathbf{F} = [f(e,e),f(e,1),\ldots,f(e,N),f(1,e),f(1,1),\ldots,f(1,N),\ldots,f(N,e),\ldots,f(N,N)]'.$$

Let $\mathbf{Y}$ and $\mathbf{c}$ be column vectors of $y(i,j)$ and $c(i,j)$, again lexicographically ordered. The equations in (3.5.2) and (3.5.3) can be written in matrix form as

$$\left[\begin{array}{c|c} M & 0 \\ \hline I & I \end{array}\right] \left[\begin{array}{c} \mathbf{F} \\ \hline \mathbf{Y} \end{array}\right] = \left[\begin{array}{c} \mathbf{0} \\ \hline \mathbf{c} \end{array}\right] \tag{3.5.5}$$

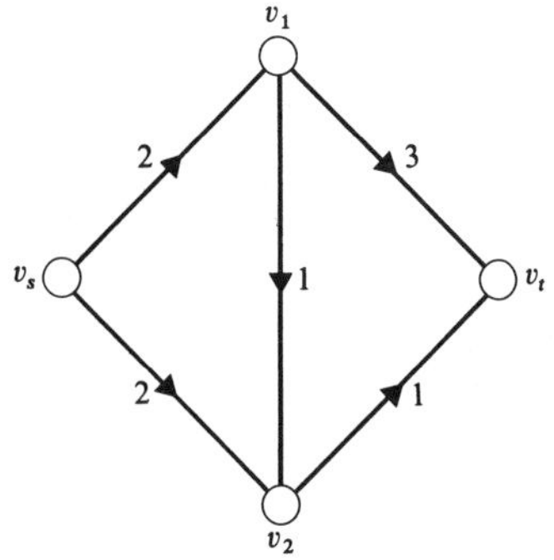

Fig. 3.5.1 Graph in which $f_{s,t}$ is to be maximized.

where

$$M = \left[\begin{array}{c|c|c|c} R_1 & R_2 & \cdots & R_{N+1} \\ \hline U_1 & U_2 & \cdots & U_{N+1} \end{array}\right]. \tag{3.5.6}$$

For $k = 1, \ldots, (N + 1)$, U_k is a diagonal matrix of order $(N + 1)$ in which the kth diagonal entry is -1 and all other diagonal entries are $+1$; and R_k is a square matrix of order $(N + 1)$ in which all entries are zero except in the kth row. In the kth row the kth element is -1 and all others are $+1$. Clearly M is a $2(N + 1) \times 2(N + 1)$ matrix and I is a $(N + 1)^2 \times (N + 1)^2$ matrix. There are therefore $(N + 1)^2 + 2(N + 1)$ equations in (3.5.5). The first $2(N + 1)$ of them are in (3.5.2) and the last $(N + 1)^2$ of them are in (3.5.3).

Example 3.5.1 As an illustration, the constraints (3.5.5) are written in (3.5.7) on p. 50, for the graph in Fig. 3.5.1.

The equations in (3.5.2) can be represented by the matrix in (3.5.8). The equations are obtained by setting the sum of the entries in any row or column equal to zero:

$$\begin{bmatrix} -f(e,e) & +f(e,1) & +f(e,2) & +f(e,3) + \cdots + f(e,N) \\ +f(1,e) & -f(1,1) & +f(1,2) & +f(1,3) + \cdots + f(1,N) \\ +f(2,e) & +f(2,1) & -f(2,2) & +f(2,3) + \cdots + f(2,N) \\ \vdots & & & \vdots \\ +f(N,e) & +f(N,1) & +f(N,2) & +f(N,3) + \cdots - f(N,N) \end{bmatrix}. \tag{3.5.8}$$

There is exactly one equation resulting from a column summation and one equation resulting from a row summation. Hence by negating all equations resulting from column summations and adding to the equations resulting from row summations we obtain a zero constraint matrix. Therefore the equations are linearly dependent.

It is left as Problem 4 to show that an independent set of equations is obtained by eliminating the first equation in (3.5.2a). There are therefore $(N + 1)^2 + 2(N + 1) - 1$ linearly independent equations in (3.5.5) and $(N + 1)^2 + 2(N + 1) - 1$ linearly independent columns in a basis of the proposed linear program.

f(e,e)	f(e,1)	f(e,2)	f(1,e)	f(1,1)	f(1,2)	f(2,e)	f(2,1)	f(2,2)	y(e,e)	y(e,1)	y(e,2)	y(1,e)	y(1,1)	y(1,2)	y(2,e)	y(2,1)	y(2,2)
−1	+1	+1	0	0	0	0	0	0	0	0	0	0	0	0	0	0	0
0	0	0	+1	−1	+1	0	0	0	0	0	0	0	0	0	0	0	0
0	0	0	0	0	0	+1	+1	−1	0	0	0	0	0	0	0	0	0
−1	0	0	+1	0	0	+1	0	0	0	0	0	0	0	0	0	0	0
0	+1	0	0	−1	0	0	+1	0	0	0	0	0	0	0	0	0	0
0	0	+1	0	0	+1	0	0	−1	0	0	0	0	0	0	0	0	0
+1	0	0	0	0	0	0	0	0	+1	0	0	0	0	0	0	0	0
0	+1	0	0	0	0	0	0	0	0	+1	0	0	0	0	0	0	0
0	0	+1	0	0	0	0	0	0	0	0	+1	0	0	0	0	0	0
0	0	0	+1	0	0	0	0	0	0	0	0	+1	0	0	0	0	0
0	0	0	0	+1	0	0	0	0	0	0	0	0	+1	0	0	0	0
0	0	0	0	0	+1	0	0	0	0	0	0	0	0	+1	0	0	0
0	0	0	0	0	0	+1	0	0	0	0	0	0	0	0	+1	0	0
0	0	0	0	0	0	0	+1	0	0	0	0	0	0	0	0	+1	0
0	0	0	0	0	0	0	0	+1	0	0	0	0	0	0	0	0	+1

$$\begin{bmatrix} f(e,e) \\ f(e,1) \\ f(e,2) \\ f(1,e) \\ f(1,1) \\ f(1,2) \\ f(2,e) \\ f(2,1) \\ f(2,2) \\ y(e,e) \\ y(e,1) \\ y(e,2) \\ y(1,e) \\ y(1,1) \\ y(1,2) \\ y(2,e) \\ y(2,1) \\ y(2,2) \end{bmatrix} \qquad \begin{matrix} = 0 \\ = 0 \\ = 0 \\ = 0 \\ = 0 \\ = 0 \\ = 4 \\ = 2 \\ = 2 \\ = 3 \\ = 2 \\ = 1 \\ = 1 \\ = 1 \\ = 1 \end{matrix} \qquad (3.5.7)$$

Let a column in the constraint matrix in (3.5.5) be denoted by $\mathbf{Y}(i,j)$ or $\mathbf{F}(i,j)$, if that column is multiplied by $y(i,j)$ or $f(i,j)$, respectively. Given a basis B, we classify the ordered pair (i,j) for $i,j = e,1,\ldots,N$ as being of type α, β, or γ as follows:

$$(i,j) \text{ is type } \alpha \quad \text{if} \quad \mathbf{F}(i,j) \in B \quad \text{and} \quad \mathbf{Y}(i,j) \in B; \tag{3.5.9a}$$

$$(i,j) \text{ is type } \beta \quad \text{if} \quad \mathbf{F}(i,j) \in B \quad \text{and} \quad \mathbf{Y}(i,j) \notin B; \tag{3.5.9b}$$

$$(i,j) \text{ is type } \gamma \quad \text{if} \quad \mathbf{F}(i,j) \notin B \quad \text{and} \quad \mathbf{Y}(i,j) \in B. \tag{3.5.9c}$$

Any (i,j) can be classified as type α, β, or γ since B must span the space generated by the columns of the constraint matrix in (3.5.5), which would be impossible if both $\mathbf{F}(i,j) \notin B$ and $\mathbf{Y}(i,j) \notin B$.

We now examine the properties of a basis more closely.

Lemma 3.5.1 *The number of* (i,j) *of type* α *corresponding to a basis* B *for* (3.5.5) *is* $2N + 1$.

Proof. Assume there are k branches (i,j) of type α corresponding to a basis. There must therefore be

$$(N + 1)^2 - k \tag{3.5.10}$$

branches (i,j) of type β or γ.

Corresponding to the k ordered pairs (i,j) of type α there are $2k$ columns in the basis, that is, $\mathbf{F}(i,j)$ and $\mathbf{Y}(i,j)$ for each (i,j). Corresponding to each $(N + 1)^2 - k$ ordered pairs (i,j) of type β or γ there is one column in the basis. The total number of columns is therefore $2k + (N + 1)^2 - k$ columns in the basis. However, the number of columns in the basis is $(N + 1)^2 + 2(N + 1) - 1$. Therefore

$$2k + (N + 1)^2 - k = (N + 1)^2 + 2(N + 1) - 1. \tag{3.5.11}$$

Solving (3.5.11), we obtain

$$k = 2N + 1 \tag{3.5.12}$$

and the lemma is proved.

Form an undirected graph $G_B = (V_B,\Gamma_B)$ to represent the basis, using the following rules.

1) There are $(N + 1)$ vertices $v_{a_e},v_{a_1},\ldots,v_{a_N}$ and $(N + 1)$ vertices $v_{b_e},v_{b_1},\ldots,v_{b_N}$.
2) $[a_i,b_j] \in \Gamma_B$ if and only if (i,j) is of type α.

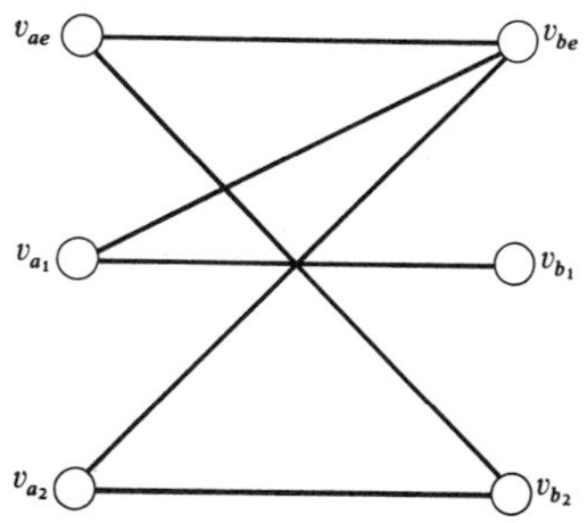

Figure 3.5.2 G_B.

Example 3.5.2 A set of basic variables for the constraints in Example 3.5.1 is given by selecting all variables except $y(s,1), y(1,2), f(2,1)$, and $y(2,2)$. Hence $y(s,1) = y(1,2) = f(2,1) = y(2,2) = 0$ in the corresponding basic solution.

The graph G_B corresponding to this set of basic variables is given in Fig. 3.5.2.

***Lemma* 3.5.2** *G_B is a tree.*

Proof. From Lemma 3.5.1, the number of branches in G_B is $2N + 1$, and the number of vertices in G_B is $2N + 2$. All that remains to be proved is that G_B contains no circuits. However, this is true, for otherwise it is seen (Problem 5) by examining (3.5.5) that the set of the columns corresponding to $f(i,j)$ and $y(i,j)$ for the (i,j) in the circuit, is linearly independent. Hence the columns in the basis are not independent and we have a contradiction.

Using G_B we can now prove a key result.

***Theorem* 3.5.1** *All basic solutions to* (3.5.5) *consist of integers if the $c(i,j)$ are integers for all i and j.*

Proof. In G_B associate with each v_{a_i} the equation

$$-f(i,i) + \sum_{j, j \neq i} f(i,j) = 0 \tag{3.5.13a}$$

and with each v_{b_j} the equation

$$-f(j,j) + \sum_{i, i \neq j} f(i,j) = 0. \tag{3.5.13b}$$

Since G_B is a tree, there are at least two vertices in G_B of degree one. Select one of them, say v_{a_k}, and suppose the single branch incident with v_{a_k} is $[a_k,b_l]$. For all (i,j) of type β, $y(i,j) = 0$ and hence $f(i,j) = c(i,j)$. For all (i,j) of type γ, $f(i,j) = 0$ and hence $y(i,j) = c(i,j)$. Thus in the equation

$$-f(k,k) + \sum_{j, j \neq k} f(k,j) = 0 \tag{3.5.14}$$

all variables are known except $f(k,l)$ and its value may be found. $y(k,l)$ can be found from $y(k,l) = c(k,l) - f(k,l)$. We can now delete v_{a_k} and $[a_k,b_l]$ from G_B. The remaining graph is still a tree. We can therefore repeat the above operations $(2N + 1)$ times to obtain the values of all variables. Since only additions and subtractions are involved, integrality is preserved and the theorem is proved.

Examples 3.5.3 From the basis chosen in Example 3.5.2 we have the graph shown in Fig. 3.5.2. From the graph

$$y(s,1) = 0 \quad \text{and hence} \quad f(s,1) = 2, \tag{3.5.15a}$$

$$y(1,2) = 0 \quad \text{and hence} \quad f(1,2) = 1, \tag{3.5.15b}$$

$$f(2,1) = 0 \quad \text{and hence} \quad y(2,1) = c(2,1) = 0, \tag{3.5.15c}$$

$$y(2,2) = 0 \quad \text{and hence} \quad f(2,2) = 1. \tag{3.5.15d}$$

For the end vertex v_{a_2}

$$-f(2,2) + f(2,1) + f(2,t) = 0 \tag{3.5.16}$$

and from (3.5.15c) and (3.5.15d) we know $f(2,2)$ and $f(2,1)$. Therefore $f(2,t) = 1$. For the end vertex v_{b_2}

$$-f(2,2) + f(s,2) + f(1,2) = 0. \tag{3.5.17}$$

Combining (3.5.17) with (3.5.15b) and (3.5.15d), we get $f(s,2) = 0$. For the end vertex v_{b_1}

$$-f(1,1) + f(s,1) + f(2,1) = 0. \tag{3.5.18}$$

Combining (3.5.18) with (3.5.15a) and (3.5.15c), $f(1,1) = 2$. Removing branches $[a_2,b_0]$, $[b_1,a_1]$, and $[b_2,a_0]$ we next examine the end vertex v_{a_1} to obtain

$$-f(1,1) + f(1,2) + f(1,t) = 0. \tag{3.5.19}$$

Combining (3.5.19) with (3.5.15b) yields $f(1,t) = 1$. Next, examining end vertex v_{a_0} gives

$$-f_{s,t} + f(s,1) + f(s,2) = 0. \tag{3.5.20}$$

Combining (3.5.20) with (3.5.15a) yields $f_{s,t} = 2$. Finally, removing branch $[a_0,b_0]$, we have for v_{b_0}

$$-f_{s,t} + f(1,t) + f(2,t) = 0. \tag{3.5.21}$$

Therefore $f(2,t) = 1$ and the last step yields no new information.

Having obtained all the required information about the primal problem, we proceed to the dual problem, to minimize

$$\sum_{i,j} c(i,j)w(i,j)$$

subject to the constraints

$$-u_e - x_e + w(e,e) \geq 1, \tag{3.5.22a}$$

$$-u_i - x_i + w(i,i) \geq 0 \qquad \text{for } i = 1,\ldots,N, \tag{3.5.22b}$$

$$u_i + x_j + w(i,j) \geq 0 \qquad \text{for } i \neq j, \tag{3.5.22c}$$

$$w(i,j) \geq 0 \qquad \text{for all } i,j, \tag{3.5.22d}$$

where u_i and x_j are unrestricted in sign. The u_i,x_i, and $w(i,j)$ are the dual variables corresponding to (3.5.2a), (3.5.2b), and (3.5.3) respectively.

Let B^* be an optimum basis for the primal problem. From the Complementary Slackness Theorem, if $\mathbf{F}(i,j) \in B^*$, the corresponding equation in (3.5.22a, b, and c) holds with equality and if $\mathbf{Y}(i,j) \in B^*$, the corresponding equation in (3.5.22d) holds with equality. Using properties of the primal we will prove the following results on the dual problem.

Lemma 3.5.3 *If* $\mathbf{Y}(e,e) \notin B^*$, *the maximum s-t flow is equal to the value of an s-t cut.*

Proof. If $\mathbf{Y}(e,e) \notin B^*$, from (3.5.3) and (3.5.1)

$$f(e,e) = f_{s,t} = c(i,i) = \text{Min}\,[c(s,V),c(V,t)] \tag{3.5.23}$$

and the proof is completed.

We now need only examine the case in which $\mathbf{Y}(e,e) \in B^*$. We exclude the case in which the maximum value of $f_{s,t}$ is zero since then there is no directed s-t path in the graph and the Max-Flow Min-Cut Theorem is clearly true.

Lemma 3.5.4 *If* $\mathbf{Y}(e,e) \in B^*$ *and* $\text{Max}\,[f_{s,t}] > 0$ *then* $w(i,j) = 0$ *or* 1 *for all* (i,j) *such that* $Y(i,j) \notin B^*$.

Proof. Ignoring the redundant equation in (3.5.5) is equivalent to setting $u_e = 0$. By assumption $\mathbf{Y}(e,e) \in B^*$ and since $f(e,e) > 0$ we have $\mathbf{F}(e,e) \in B^*$. From part (b) of the Complementary Slackness Theorem, $w(e,e) = 0$. Therefore, $x_e = -1$. For all (i,j) of type α, other than (e,e) we know from (3.5.22d) and part (b) of the Complementary Slackness Theorem that $w(i,j) = 0$. Hence, from (3.5.22b) and part (b) of the Complementary Slackness Theorem,

$$u_i + x_j = 0. \tag{3.5.24}$$

We can then easily see from (3.5.24) that for all $i \neq e$

$$u_i = 0 \text{ or } 1 \tag{3.5.25}$$

and

$$x_j = 0 \text{ or } -1. \tag{3.5.26}$$

In particular, this is shown by considering G_B^* the graph corresponding to B^* and associating u_i with v_{a_i} and x_i with v_{b_i} and the equation $u_i + x_j = 0$ with branch $[a_i,b_j]$.

Substituting $u_e = 0$, $x_e = -1$, $u_i = 0$ or 1, and $x_j = 0$ or -1 into (3.5.22a, b, and c) and noting (3.5.22d), we see for the cases in which $\mathbf{Y}(i,j) \notin B^*$ that $w(i,j) = 0$ or 1. This is the desired result.

We now use Lemma 3.5.4 to prove the sufficiency part of the Max-Flow Min-Cut Theorem. From the Duality Theorem,

$$\text{Max}\,[f_{s,t}] = \text{Min}\left[\sum_{i,j} w(i,j)c(i,j)\right]. \tag{3.5.27}$$

But $w(e,e) = 0$. Hence

$$\text{Max}\,[f_{s,t}] = \text{Min}\left[\sum_{i,i \neq j} w(i,j)c(i,j) + \sum_{i=1,\ldots,N} w(i,i)c(i,i)\right]. \tag{3.5.28}$$

As we have seen from the Complementary Slackness Theorem, $w(i,j) = 0$ if $\mathbf{Y}(i,j) \in B^*$ and from Lemma 3.5.4 we know $w(i,j) = 0$ or 1 otherwise. Therefore,

$$\text{Max}\,[f_{s,t}] = \text{Min}\left[\sum_{\substack{W\\ i,i\neq j}} c(i,j) + \sum_{\substack{W\\ i=1,\ldots,N}} c(i,i)\right]. \tag{3.5.29}$$

Here W is the subset of branches (i,j) for which $\mathbf{Y}(i,j) \notin B^*$ and $w(i,j) = 1$ and the set of vertices $v_i \neq v_e$ for which $\mathbf{Y}(i,i) \in B^*$ and $w(i,i) = 1$. We now claim that W constitutes a set of branches and vertices whose removal from the graph breaks all s-t paths. Suppose we increase $c(i,j)$ by $\varepsilon > 0$ for any $(i,j) \notin W$. This does not affect (3.5.29), but it does mean that among the branches and vertices not in W there is no directed s-t path. Hence the claim is substantiated. Furthermore, because of (3.5.29) the set W must be minimal.

If W contains no vertices, the proof of the Max-Flow Min-Cut Theorem is completed. If W does contain vertices, we can replace a vertex v_i in W by either the branches directed from v_i or the branches directed to v_i, depending on which set has the lower total capacity. Since $\mathbf{Y}(i,j) \notin B^*$,

$$f(i,i) = c(i,i) = \text{Min}\left[\sum_{j,j\neq i} c(i,j), \sum_{j,j\neq i} c(j,i)\right] \tag{3.5.30}$$

and the smallest of the two branch sets is saturated. Hence we can again form a cut which is saturated and the proof is completed.

3.6 EXTENSIONS OF THE MAX-FLOW MIN-CUT THEOREM

A number of problems with apparently different constraints from the maximum flow problem can be reduced to it by only slight modifications. For example, consider the problem of finding the maximum s-t flow in a directed graph in which both branches and vertices have capacities. If the capacity of vertex v_i is $c(i)$, we require the flow out of vertex v_i to be less than or equal to $c(i)$. We convert the graph with vertex capacities to a graph with only branch capacities by using the transformation indicated in Fig. 3.6.1. Each vertex v_i is replaced by two vertices $v_{i'}$ and $v_{i''}$ and a branch (i',i''). All branches directed to v_i are now directed to $v_{i'}$ and all branches directed from v_i are now directed from $v_{i''}$.

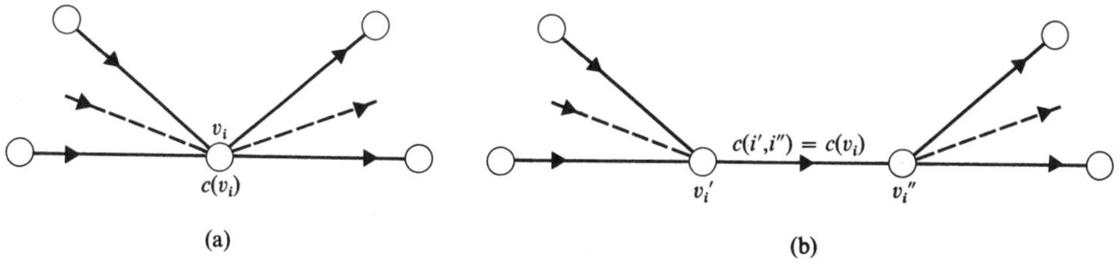

Fig. 3.6.1 Vertex weights replaced by branch weights.

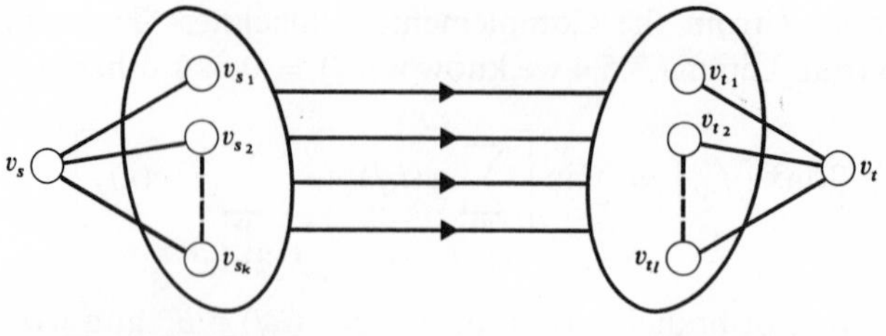

Fig. 3.6.2 Graph with multiple sources and terminals.

Another problem equivalent to the maximum flow problem is that of attaining a given flow or maximizing the total flow from a set of sources v_{s_i} for $i = 1, \ldots, k$ to a disjoint set of terminals v_{t_i} for $i = 1, \ldots, l$. The flow in this problem is equal to the s-t flow in the graph shown in Fig. 3.6.2, formed by creating a new source vertex v_s with branches (s,s_i), and a new terminal vertex v_t with branches (t_i,t) for $i = 1, \ldots, l$. Each new branch is assigned infinite capacity. In the new graph, the total flow will be equal to the value of a minimum cut separating $v_{s_i}, \ldots, v_{s_k}$ from $v_{t_1}, \ldots, v_{t_l}$.

3.7 THE MULTICOMMODITY FLOW PROBLEM

In a communication or transportation system it is likely that many commodities are to be transmitted simultaneously between various pairs of sources and terminals. If we seek the maximum amount of flow that can be simultaneously sent between various source-terminal pairs we arrive at what is known as the Multicommodity Flow Problem. Assume there are k commodities and that the flow of the lth commodity in branch (i,j) is denoted by $f^l(i,j)$. A set of $f^l(i,j)$ satisfying the appropriate conservation constraints, called a flow pattern of the lth commodity, is denoted by $\mathscr{F}^l$. We will assume that the source and terminal for each flow are represented by distinct vertices, such that the lth commodity originates at source v_{s_l} and terminates at terminal v_{t_l}. The flow value of the lth commodity from its source v_{s_l} to its terminal v_{t_l} is denoted by $f^l_{s_l,t_l}$. For directed graphs no generality is lost in this model since, if a vertex serves as source and terminal for several commodities (Fig. 3.7.1a), branches and vertices can be added as in Fig. 3.7.1(b).

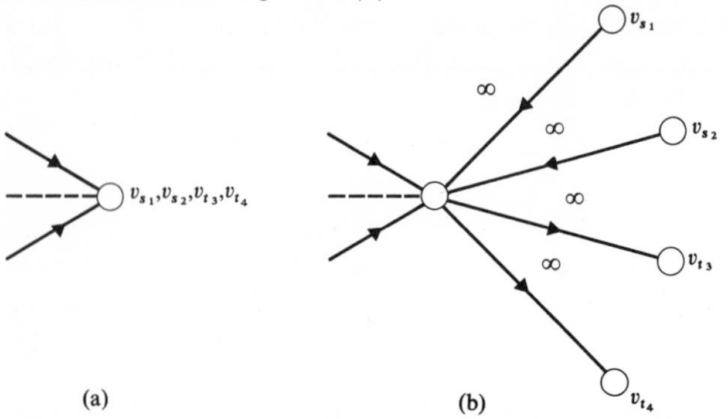

Fig. 3.7.1 Representation of sources and terminals for multicommodity problems.

By analogy with the single commodity case, we have the constraints in (3.7.1) for directed graphs and (3.7.2) for undirected graphs.

for all i and l,
$$f^l(i,V) - f^l(V,i) = \begin{cases} f^l_{s_l,t_l} & \text{if } i = s_l, \\ 0 & \text{if } i \neq s_l,t_l, \\ -f^l_{s_l,t_l} & \text{if } i = t_l; \end{cases} \tag{3.7.1a}$$

for all (i,j),
$$c(i,j) \geq \sum_{l=1}^{k} f^l(i,j); \tag{3.7.1b}$$

for all (i,j) and l,
$$f^l(i,j) \geq 0; \tag{3.7.1c}$$

for all i and l,
$$f^l[i,V] = \begin{cases} f^l_{s_l,t_l} & \text{if } i = s_l, \\ 0 & \text{if } i \neq s_l,t_l, \\ -f^l_{s_l,t_l} & \text{if } i = t_l; \end{cases} \tag{3.7.2a}$$

for all $[i,j]$,
$$c[i,j] \geq \sum_{l=1}^{k} |f^l[i,j]|\,. \tag{3.7.2b}$$

We denote by r_{s_l,t_l}, or simply by r_l, the requirement of flow of commodity l from source v_{s_l} to terminal v_{t_l}. We can then state two multicommodity flow problems.

Problem 1 Determine necessary and sufficient conditions on a set of r_l for $l = 1, \ldots, k$ to be satisfied by a set of flows satisfying (3.7.1) or (3.7.2) in a graph G.

Problem 2 Determine the maximum value of

$$\sum_{l=1}^{k} f^l_{s_l,t_l}$$

in a graph G.

It may be that

$$\sum_{l=1}^{k} r_l$$

is less than the maximum determined in Problem 2 and still is unfeasible. To illustrate this point consider the graph in Fig. 3.7.2. In the triple associated with each directed branch, the first entry is the branch capacity, the second is the flow of commodity 1, and the third is the flow of commodity 2. In Fig. 3.7.2(b) a flow pattern is shown which maximizes $f^1_{s_1,t_1} + f^2_{s_2,t_2}$ by achieving $f^1_{s_1,t_1} = 0$ and $f^2_{s_2,t_2} = 2$. On the other hand, we cannot achieve $r_1 = 1$ and $r_2 = 1$, since if $r_1 = 1$ no more flow can be sent.

In the remainder of this chapter we consider several special cases in which the solutions to Problems 1 and 2 are known in terms of explicit graph theoretic quantities. The general solution is not yet known. We first introduce several basic definitions and theorems which will be useful throughout the remainder of the chapter.

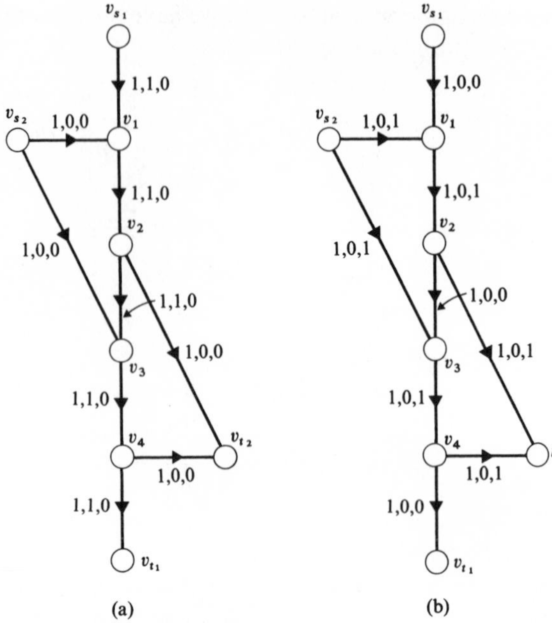

Fig. 3.7.2 Graphs for which $f^1_{s_1,t_1}$ and $f^2_{s_2,t_2}$ cannot be independently maximized.

A *separating set* for a multicommodity problem is a set of branches whose removal from G breaks all directed s_l-t_l paths for $l = 1, \ldots, q \le k$. The value of the separating set is the sum of the capacities of the branches in the set. $\tau_{s_1,\ldots,s_q;t_1,\ldots,t_q}$ is taken to be the value of the smallest such separating set. Let $\tau_{s_1\text{-}s_2;t_1\text{-}t_2}$ be the value of the smallest separating set in the graph formed by superimposing v_{s_1} with v_{s_2} and v_{t_1} with v_{t_2}. Moreover, let $\tau_{s_1\text{-}t_2;s_2\text{-}t_1}$ be the value of the smallest separating set in the graph formed by superimposing the vertices v_{s_1} and v_{t_2} and the vertices v_{s_2} and v_{t_1}. In terms of these definitions we can derive several properties of maximum multicommodity flows.

Theorem 3.7.1 Max $\displaystyle\sum_{l=1}^{q} f^{a_l}_{s_{a_l},t_{a_l}} \le \tau_{s_{a_1},\ldots,s_{a_q};t_{a_1},\ldots,t_{a_q}}$ (3.7.3)

for all combinations $(a_1, \ldots, a_q)$ *from the set* $\{1, \ldots, q\}$ *with* $q = 1, \ldots, k$.

Proof. For any feasible flow pattern, let $\mathscr{F}^j$ be the flow pattern of commodity j. We can find an s_j-t_j semipath decomposition for $\mathscr{F}^j$ for $j = 1, \ldots, q$. Every semipath in these decompositions must have at least one branch in common with the separating set corresponding to $\tau_{s_1,\ldots,s_q;t_1,\ldots,t_q}$. Therefore, the capacity of the separating set must be at least

$$\sum_{l=1}^{q} f^l_{s_l,t_l}.$$

Example 3.7.1 As an example of Theorem 3.7.1 for $k = 3$ and $q = 1$,

$$f^1_{s_1,t_1} \leq \tau_{s_1,t_1}, \tag{3.7.4a}$$

$$f^2_{s_2,t_2} \leq \tau_{s_2,t_2}, \tag{3.7.4b}$$

$$f^3_{s_3,t_3} \leq \tau_{s_3,t_3}. \tag{3.7.4c}$$

For $q = 2$,

$$f^1_{s_1,t_1} + f^2_{s_2,t_2} \leq \tau_{s_1,s_2;t_1,t_2}, \tag{3.7.5a}$$

$$f^2_{s_2,t_2} + f^3_{s_3,t_3} \leq \tau_{s_2,s_3;t_2,t_3}, \tag{3.7.5b}$$

$$f^1_{s_1,t_1} + f^3_{s_3,t_3} \leq \tau_{s_1,s_3;t_1,t_3}. \tag{3.7.5c}$$

For $q = 3$,

$$f^1_{s_1,t_1} + f^2_{s_2,t_2} + f^3_{s_3,t_3} \leq \tau_{s_1,s_2,s_3;t_1,t_2,t_3}. \tag{3.7.6}$$

The main question is whether the maximum q commodity flow is equal to $\tau_{s_1,\ldots,s_q;t_1,\ldots,t_q}$. In order to simplify further discussion we will say that, for a given set of sources and terminals, a graph for which this equality holds is *gapless* (since there is no gap between the upper bound and the maximum flow value). For a single commodity, all graphs are gapless. As we shall see, for two commodities undirected graphs with even branch capacities are also gapless. However, the counter-example below shows that there are graphs with as few as three commodities which are not gapless.

Consider the undirected graph in Fig. 3.7.3 in which all branches have capacity 2. For $l = 1,2,3$, each s_l-t_l path contains at least three branches. Hence each unit of f_{s_l,t_l} requires at least three units of branch capacity. The sum of all the branch

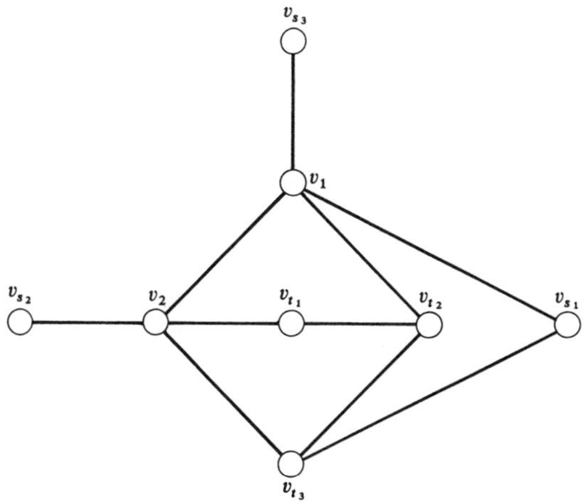

Fig. 3.7.3 Graph which is not gapless.

capacities is 20. Therefore

$$3 \operatorname*{Max}_{\{\mathscr{F}^l\}} \left[\sum_{l=1}^{3} f_{s_l,t_l}^l \right] \leq 20.$$

However, $\tau_{s_1,s_2,s_3;t_1,t_2,t_3} = 8$ since branches $[s_2,2]$, $[s_3,1]$, $[2,t_1]$, and $[t_1,t_2]$ form a separating set. The graph is therefore not gapless since

$$\operatorname*{Max}_{\{\mathscr{F}^l\}} \left[\sum_{l=1}^{3} f_{s_l,t_l}^l \right] \leq 20/3 < 8.$$

It would also be desirable if the solution consisted of integer flows when the capacities are integers. Otherwise, the convergence of a proposed algorithm often becomes doubtful. However, the solution of the general multicommodity problem does not consist of integer flows even if the graph is gapless and the capacities are integers. A solution which consists of integer flows will be called an *integer solution*. The solution of the single commodity flow problem is therefore integer if the branch capacities are integer.

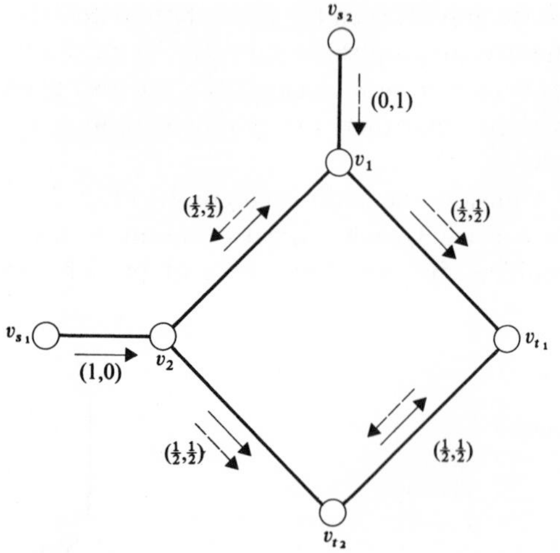

Fig. 3.7.4 Gapless graph which is noninteger.

Consider the undirected graph in Fig. 3.7.4 in which all capacities are unity. In the graph, the first number associated with each branch is the flow of commodity 1 and the second number is the flow of commodity 2. Solid arrows indicate directions of flow of commodity 1 and dotted arrows, the flow of commodity 2. The graph is gapless since the branches $(s_1,2)$ and $(s_2,1)$ form a saturated separating set. However, the total flow of value 2 cannot be attained with integer flows.

Before attempting the multicommodity problem several additional concepts are presented to facilitate the subsequent development. We first show that for an undirected graph (3.7.3) can be replaced by the following set of inequalities which involve only separating sets

$$\sum_{\substack{v_{s_l} \in X \\ v_{t_l} \in \bar{X}}} f^l_{s_l,t_l} \le \sum_{\substack{v_{s_l} \in X \\ v_{t_l} \in \bar{X}}} c(X,\bar{X}) \qquad \text{for all } X \subset V. \tag{3.7.7}$$

Theorem 3.7.2 *For an undirected graph (3.7.3) and (3.7.7) are equivalent.*

Proof. Clearly (3.7.3) implies (3.7.7) and we must show that (3.7.7) implies (3.7.3). Consider an arbitrary inequality in (3.7.3), namely,

$$\sum_{l=1}^{q} f^l_{s_l,t_l} \le \tau_{s_1,\ldots,s_q;t_1,\ldots,t_q}. \tag{3.7.8}$$

Assume that $\tau_{s_1,s_2,\ldots,s_q;t_1,\ldots,t_q}$ corresponds to a separating set whose removal from G breaks the graph into α disjoint connected subgraphs $G_1 = (V_1,\Gamma_1),\ldots,G_\alpha = (V_\alpha,\Gamma_\alpha)$. For any G_i, assume the cut $[V_i,\bar{V}_i]$ in G breaks all s_{a_j}-t_{a_j} paths for $j = 1,\ldots, p_i \le q$. From (3.7.7),

$$\sum_{j=1}^{p_i} f^{a_j}_{s_{a_j},t_{a_j}} \le c[V_i,\bar{V}_i]. \tag{3.7.9}$$

We now show that (3.7.8) can be obtained from (3.7.9). The a_jth source-terminal pair is separated by exactly two cuts of the form $[V_i,\bar{V}_i]$. Therefore

$$2\sum_{i=1}^{q} f^i_{s_i,t_i} = \sum_{j=1}^{p_i} f^{a_j}_{s_{a_j},s_{a_j}} \le \sum_{i=1}^{\alpha} c[V_i,\bar{V}_i]$$

$$= 2\tau_{s_1,\ldots,s_q;t_1,\ldots,t_q}, \tag{3.7.10}$$

so each inequality of the type (3.7.8) is contained among the inequalities of the type (3.7.7). This proves the theorem.

To illustrate Theorem 3.7.2 consider the following example.

Example 3.7.2 For the graph in Fig. 3.7.5,

$$f^1_{s_1,t_1} + f^2_{s_2,t_2} + f^3_{s_3,t_3} \le c[s_1,s_2] + c[s_3,t_1] + c[t_2,t_3] = 3. \tag{3.7.11}$$

Branches $[s_1,s_2]$, $[s_3,t_1]$, and $[t_2,t_3]$ do not constitute a cut. We also find by examin-

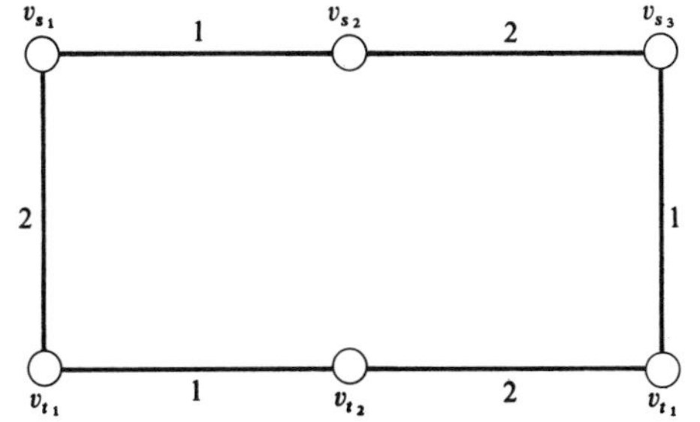

Fig. 3.7.5 Graph for Example 3.7.2.

ing cuts that

$$f^1_{s_1,t_1} + f^2_{s_2,t_2} \le 2, \qquad (3.7.12)$$

$$f^2_{s_2,t_2} + f^3_{s_3,t_3} \le 2, \qquad (3.7.13)$$

$$f^1_{s_1,t_1} + f^3_{s_3,t_3} \le 2. \qquad (3.7.14)$$

Adding (3.7.12), (3.7.13), and (3.7.14) and dividing by two yields (3.7.11).

Theorem 3.7.2 does not imply that a particular $\tau_{s_1,s_2,\ldots,s_q;t_1,t_2,\ldots,t_q}$ corresponds to a cut but only that the value of this upper bound on the

$$\sum_{i=1}^{q} f^i_{s_i,t_i}$$

given in (3.7.3) can be determined by examining only cuts. In the undirected two-commodity problem we can even be more specific; we will show that $\tau_{s_1,s_2;t_1,t_2}$ corresponds to a cut.

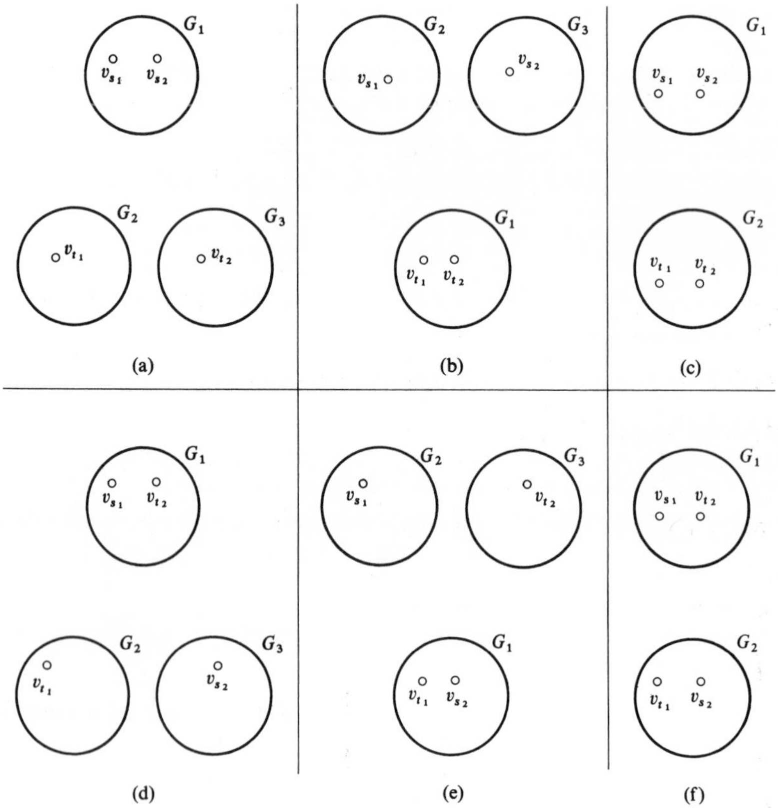

Fig. 3.7.6 Graphs used in proof of Theorem 3.7.3.

We first prove Lemma 3.7.1.

Lemma 3.7.1 *The removal of a minimum separating set for k sources and k corresponding terminals will break an undirected connected graph into at most* $(k + 1)$ *components.*

Proof. Remove the branches in the minimum separating set one at a time. Each time we remove a branch the number of components is either unchanged or increased by one. If it is increased by one, at least one new pair of corresponding sources and terminals must be separated. Otherwise the separating set would not be minimum. Hence, when the number of components reaches $(k + 1)$ all corresponding sources and terminals must be separated.

We now show

Theorem 3.7.3 *In an undirected graph*

$$\tau_{s_1,s_2;t_1,t_2} = \text{Min}\left[\tau_{s_1\text{-}s_2;t_1\text{-}t_2}, \tau_{s_1\text{-}t_2;s_2\text{-}t_1}\right].$$

Proof. From Lemma 3.7.1 there are at most three components after removing a separating set of value $\tau_{s_1,s_2;t_1,t_2}$. The only possible arrangements of vertices is then shown in Fig. 3.7.6, where $G_i = (V_i,\Gamma_i)$ for $i = 1,2,3$ represent components of the resulting graph. In each case,

$$\tau_{s_1,s_2;t_1,t_2} = c[V_1,\bar{V}_1]. \tag{3.7.15}$$

In Fig. 3.7.6(a), (b), and (c)

$$c[V_1,\bar{V}_1] = \tau_{s_1\text{-}s_2;t_1\text{-}t_2} \leq \tau_{s_1\text{-}t_2;s_2\text{-}t_1}, \tag{3.7.16}$$

and in Fig. 3.7.6(d), (e), and (f)

$$c[V_1,\bar{V}_1] = \tau_{s_1\text{-}t_2;s_2\text{-}t_1} \leq \tau_{s_1\text{-}s_2;t_1\text{-}t_2}. \tag{3.7.17}$$

Combining (3.7.15), (3.7.16), and (3.7.17) gives the theorem.

3.8 MAXIMUM TWO-COMMODITY FLOW

In this section we consider undirected graphs in which all branches have even integer capacities. The major result of the section is given by Theorem 3.8.1. The necessity part of the theorem was established in the previous section. We first state the theorem and then devote the remainder of the section to its illustration and proof.

Theorem 3.8.1

a) *Requirements* r_1 *and* r_2 *are feasible in an undirected graph with even integer capacities if and only if*

$$r_1 \leq \tau_{s_1,t_1}, \tag{3.8.1}$$

$$r_2 \leq \tau_{s_2,t_2}, \tag{3.8.2}$$

$$r_1 + r_2 \leq \tau_{s_1,s_2;t_1,t_2}. \tag{3.8.3}$$

b) *The maximum value of* $f^1_{s_1,t_1} + f^2_{s_2,t_2}$ *is* $\tau_{s_1,s_2;t_1,t_2}$.

The method of proof in this section is similar to the proof of the Max-Flow Min-Cut Theorem using the Augmentation Algorithm. We provide a method for first obtaining a maximum $f^1_{s_1,t_1}$ and then increasing $f^2_{s_2,t_2}$. For this problem we use the Two-Commodity Augmentation Algorithm, one step of which recirculates flow of commodity 1 in order to "unblock" paths along which flow of commodity 2 can be increased.

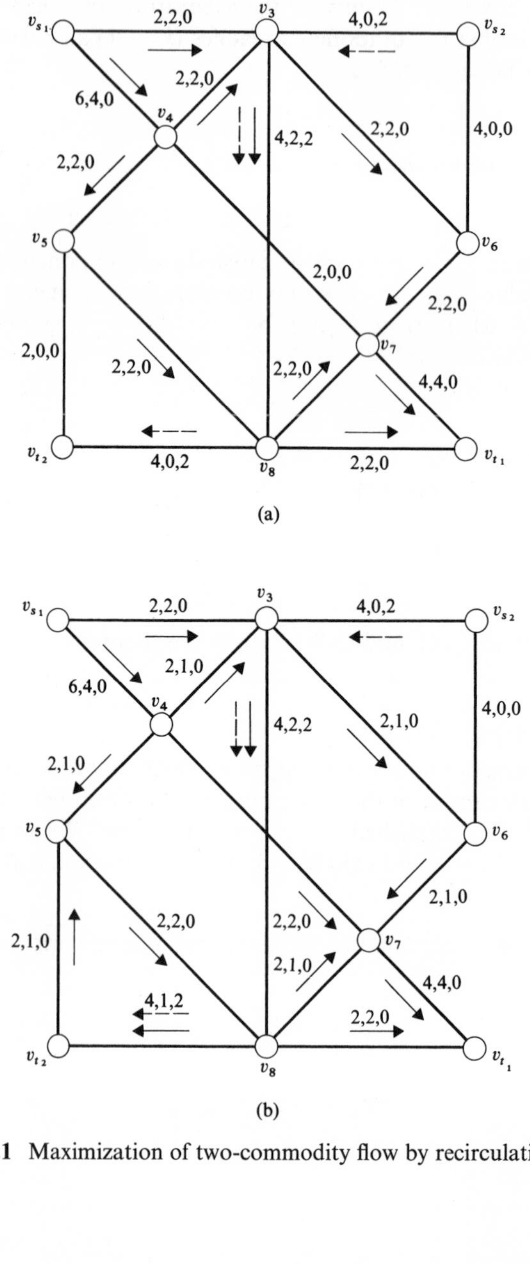

Fig. 3.8.1 Maximization of two-commodity flow by recirculating flow.

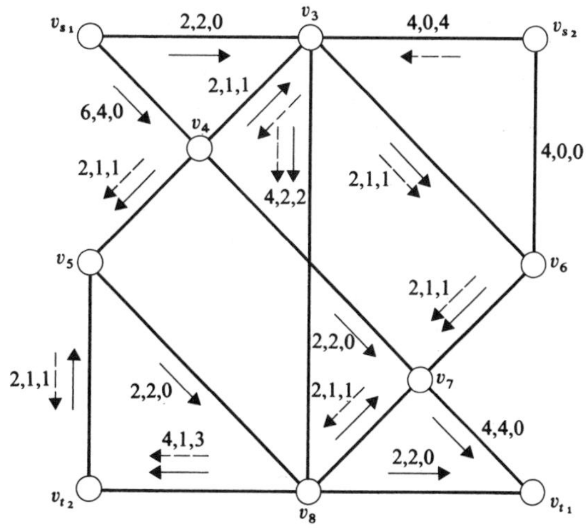

(c)

Figure 3.8.1 (continued)

Example 3.8.1 Consider the graph in Fig. 3.8.1(a). Again, associated with each branch $[i,j]$ is an ordered triple $(c[i,j], f^1[i,j], f^2[i,j])$. The direction of $f^1[i,j]$ is indicated by a solid arrow and the direction of $f^2[i,j]$ by a dashed arrow. A flow pattern which maximizes $f^1_{s_1,t_1}$ is given in Fig. 3.8.1(a) by solid arrows. A flow pattern which maximizes $f^2_{s_2,t_2}$ in the graph obtained by converting the capacity of branch $[i,j]$ to $c[i,j] - |f^1[i,j]|$ is given by dashed arrows. We now recirculate some of the flow of commodity 1 without changing $f^1_{s_1,t_1}$ in order to create an unsaturated augmentation path for commodity 2. If we increase $f^1_{s_2,t_2}$ by 1 along $\pi_{s_2,t_2} = [s_2,3],[3,4],[4,7],[7,8],[8,t_2]$ and if we increase $f^1_{t_2,s_2}$ by 1 along $\pi_{t_2,s_2} = [t_2,5],[5,4],[4,7],[7,6],[6,3],[3,s_2]$, then flow 1 is still conserved at all vertices. In addition, we have recirculated flow 1 so that $f^2_{s_2,t_2}$ can now be increased. The resulting flow pattern after the recirculation is given in Fig. 3.8.1(b). We can now increase $f^2_{s_2,t_2}$ by sending one unit of flow along π_{s_2,t_2} from v_{s_2} to v_{t_2} and one unit of flow along π_{t_2,s_2} from v_{t_2} to v_{s_2}. The resulting flow pattern is given in Fig. 3.8.1(c). $f^1_{s_1,t_1} + f^2_{s_2,t_2}$ has now been maximized since in Fig. 3.8.1(c) the cut $[\{v_{s_1}, v_{s_2}, v_3, v_4, v_6\}, \{v_{t_1}, v_{t_2}, v_5, v_7, v_8\}]$ is saturated.

We now formalize the above procedure as an algorithm which always maximizes $f^1_{s_1,t_1} + f^2_{s_2,t_2}$ or satisfies a given r_1 and r_2. The algorithm will determine a "permissible" s_2-t_2 path and a "permissible" t_2-s_2 path such that a flow of α units of commodity 1 can be sent from v_{s_2} to v_{t_2} along the s_2-t_2 path and from v_{t_2} to v_{s_2} along the t_2-s_2 path in such a way that f_{s_2,t_2} can be increased.

Definition 3.8.1 *An s_2-t_2 path is said to be permissible if for each branch, say $[i,j]$, in the path*

a) there is a positive backward flow of commodity 1, or

b) there is a positive backward flow of commodity 2, or

c) $c[i,j] - |f^1[i,j]| - |f^2[i,j]| > 0$.

Definition 3.8.2 *A t_2-s_2 path is said to be permissible if for each branch, say $[i,j]$, in the path*

a) there is a positive backward flow of commodity 1, or

b) there is a positive forward flow of commodity 2, or

c) $c[i,j] - |f^1[i,j]| - |f^2[i,j]| > 0$.

Definition 3.8.3 *A permissible pair of paths consists of a permissible s_2-t_2 path and a permissible t_2-s_2 path.*

Note that the permissible pair of paths do not necessarily form a circuit since the permissible s_2-t_2 path and the permissible t_2-s_2 path may contain branches in

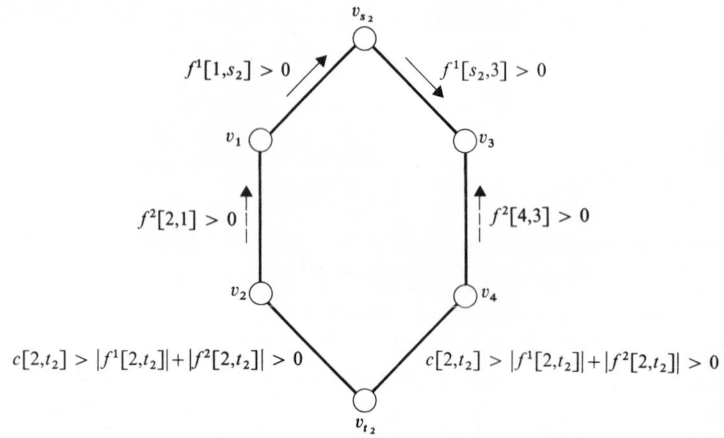

(a)

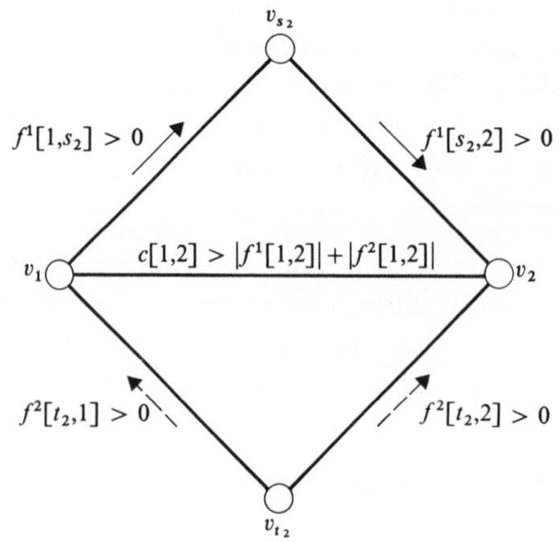

(b)

Fig. 3.8.2 Permissible pairs of paths.

common. The circuit in Fig. 3.8.2(a) comprises a permissible pair of paths. The permissible s_2-t_2 path is given by $[s_2,1],[1,2],[2,t_2]$ and the permissible t_2-s_2 path is $[t_2,4],[4,3],[3,s_2]$. The graph in Fig. 3.8.2(b) also comprises a permissible pair of paths. The permissible s_2-t_2 path is $[s_2,1],[1,2],[2,t_2]$ and the permissible t_2-s_2 path is $[t_2,1],[1,2],[2,s_2]$.

Finally, we define a quantity α, which is the amount by which flow may be increased along the permissible paths.

Definition 3.8.4 *For $[i,j]$ in a permissible s_2-t_2 path*

$$\alpha[i,j] = \begin{cases} 1/2(c[i,j] - f^1[i,j] - f^2[i,j]), \text{ if } f^1[i,j] \text{ and } f^2[i,j] \text{ are positive forward flows.} \\ \text{Max}\,[f^1[j,i],f^2[j,i]], \text{ otherwise.} \end{cases} \tag{3.8.4a}$$

For $[i,j]$ in a permissible t_2-s_2 path

$$\alpha[i,j] = \begin{cases} 1/2(c[i,j] - f^1[i,j] - f^2[j,i], \text{ if } f^1[i,j] \text{ is a positive forward flow and } f^2[j,i] \text{ is a positive backward flow.} \\ \text{Max}\,[f^1[j,i],f^2[i,j]], \text{ otherwise.} \end{cases} \tag{3.8.4b}$$

Then for a permissible pair of paths π_{s_2,t_2} and π_{t_2,s_2}, let

$$\alpha = \min_{[i,j] \in \pi_{s_2,t_2} \cup \pi_{t_2,s_2}} [\alpha[i,j]]. \tag{3.8.5}$$

Example 3.8.2 To illustrate Definition 3.8.4, reconsider the permissible pair of paths in Example 3.8.1. For the permissible s_2-t_2 path we have $\alpha[s_2,3] = 1$, $\alpha[3,4] = 2$, $\alpha[4,7] = 1$, $\alpha[7,8] = 2$, and $\alpha[8,t_2] = 1$. For the permissible t_2-s_2 path we have $\alpha[t_2,5] = 1$, $\alpha[5,4] = 2$, $\alpha[4,7] = 1$, $\alpha[7,6] = 2$, $\alpha[6,3] = 2$, and $\alpha[3,s_2] = 1$. Hence, $\alpha = 1$.

We now give the algorithm either to satisfy a given set of requirements r_1 and r_2 or to maximize $f^1_{s_1,t_1} + f^2_{s_2,t_2}$.

Two-Commodity Augmentation Algorithm

Step 1. Find an even integer flow pattern[1] with $f^1_{s_1,t_1} = r_1$ (or maximize $f^1_{s_1,t_1}$ if $f^1_{s_1,t_1} + f^2_{s_2,t_2}$ is to be maximized) by any method such as the Labeling Algorithm. If r_1 cannot be satisfied the algorithm terminates.

Step 2. Find an even integer flow pattern with maximum $f^2_{s_2,t_2}$ in the graph in which the capacity of branch $[i,j]$ is $c[i,j] - |f^1[i,j]|$ for all $[i,j]$. If $f^2_{s_2,t_2} \geq r_2$ the algorithm terminates.

Step 3. Locate a permissible pair of paths. If such a pair cannot be found, the algorithm terminates. Increase the flow of commodity 1 by α units from v_{s_2} to v_{t_2}

[1] It is left as an exercise to show that if the initial flow in the Labeling Algorithm consists of even integer flows in a graph with even integer branch capacities then all flows remain even integers.

along the permissible s_2-t_2 path and increase the flow of commodity 1 by α units from v_{t_2} to v_{s_2} along the permissible t_2-s_2 path.

Step 4. Increase $f^2_{s_2,t_2}$ by 2α units by increasing the flow of commodity 2 by α units from v_{s_2} to v_{t_2} along each permissible path.

Step 5. Repeat Steps 2, 3, and 4 until the algorithm terminates.

We now give four lemmas and defer their proofs until all four are stated. Theorem 3.8.1 follows directly from these lemmas.

Lemma 3.8.1 *After each application of Steps 1 to 4 of the Two-Commodity Augmentation Algorithm, the resulting flows are feasible (although they may not be feasible after only Steps 1, 2, and 3 are applied).*

Lemma 3.8.2 *If r_1, r_2, and $c[i,j]$ are even integers for all i and j, the Two-Commodity Augmentation Algorithm yields integer flows. (If we wish to maximize $f^1_{s_1,t_1} + f^2_{s_2,t_2}$ then only the integer $c[i,j]$ are required.)*

Using Lemmas 3.8.1 and 3.8.2, we prove:

Lemma 3.8.3 *If r_1, r_2, and $c[i,j]$ are even integers for all i and j, the Two-Commodity Augmentation Algorithm terminates after a finite number of steps. (If we wish to maximize $f^1_{s_1,t_1} + f^2_{s_2,t_2}$ then only even integer $c[i,j]$ are required.)*

Using Lemma 3.8.3 we next prove:

Lemma 3.8.4 *For even integer r_1, r_2, and $c[i,j]$, the Two-Commodity Augmentation Algorithm terminates if and only if (a), (b), or (c) below is true:*
a) r_1 and r_2 are satisfied,
b) $f^1_{s_1,t_1} + f^2_{s_2,t_2}$ is maximized,
c) r_1 and/or r_2 and/or $r_1 + r_2$ are not feasible.

Proof of Lemma 3.8.1. Flow is clearly conserved at each vertex at the completion of each step. Therefore, we must only show that branch capacities are not exceeded. After completion of Steps 1 and 2 this is clearly true. Now consider the flow in a branch $[i,j]$. Proceeding by induction, assume that before a particular application of Step 3 and Step 4 the flows in $[i,j]$ are $f^1[i,j]$ and $f^2[i,j]$, where

$$|f^1[i,j]| + |f^2[i,j]| \leq c[i,j]. \tag{3.8.6}$$

There are now three cases to consider depending upon the exact nature of the flows in $[i,j]$.

CASE 1. Assume $[i,j]$ is not in both permissible paths in a permissible pair of paths and suppose that $f^1[i,j]$ is a positive backward flow in either a permissible s_2-t_2 path or a permissible t_2-s_2 path. Then after Step 3 is applied, the magnitude of the new flow of commodity 1 is

$$|f^1[i,j]| - \alpha. \tag{3.8.7}$$

After application of Step 4, the magnitude of the flow of commodity 2 is

$$|f^2[i,j]| + \alpha \quad \text{or} \quad |f^2[i,j]| - \alpha. \tag{3.8.8}$$

Combining (3.8.7) and (3.8.8), we see that the magnitude of the new flows is at most

$$[|f^1[i,j]| - \alpha] + [|f^2[i,j]| + \alpha] = |f^1[i,j]| + |f^2[i,j]| \leq c[i,j]. \quad (3.8.9)$$

Hence the proof is completed for Case 1.

CASE 2. Assume again that $[i,j]$ is not in both permissible paths in a permissible pair of paths and suppose $f^1[i,j]$ is a nonnegative forward flow in a permissible s_2-t_2 path or a permissible t_2-s_2 path. Then, after the application of Step 3, the magnitude of the flow of commodity 1 is

$$|f^1[i,j]| + \alpha. \quad (3.8.10)$$

After the application of Step 4, the magnitude of the value of flow 2 is either

$$|f^2[i,j]| - \alpha \quad (3.8.11)$$

or

$$|f^2[i,j]| + \alpha. \quad (3.8.12)$$

In the former case, the magnitude of the flow in $[i,j]$ satisfies

$$[|f^1[i,j]| + \alpha] + [|f^2[i,j]| - \alpha] = |f^1[i,j]| + |f^2[i,j]| \leq c[i,j]. \quad (3.8.13)$$

In the latter case the magnitude of the flow is

$$\begin{aligned} &|f^1[i,j]| + |f^2[i,j]| + 2\alpha \\ &\leq |f^1[i,j]| + |f^2[i,j]| + 2 \cdot 1/2 \cdot (c[i,j] - |f^1[i,j]| - |f^2[i,j]|) \\ &= c[i,j]. \end{aligned} \quad (3.8.14)$$

Hence, we have also proved the lemma for Case 2. (Note that after Step 3 there is no guarantee that the branch capacity is not exceeded with the flow of commodity 1 given in (3.8.10).)

CASE 3. Suppose that $[i,j]$ is in both the permissible s_2-t_2 path and the permissible t_2-s_2 path. Then the flow of commodity 1 may be unchanged by the application of Step 3 and $|f^2[i,j]|$ is increased by 2α or $|f^2[i,j]|$ may be unchanged and $|f^1[i,j]|$ increased by 2α. Hence, the value of the new flow is

$$|f^1[i,j]| + |f^2[i,j]| + 2\alpha \quad (3.8.15)$$

and the remainder of the proof follows as in (3.8.14).

Proof of Lemma 3.8.2. By hypothesis, r_1, r_2, and $c[i,j]$ are all even integers. Furthermore, in Steps 1 and 2 we require that $f^1[i,j]$ and $f^2[i,j]$ be even integers for all i and j. Hence, after Steps 1 and 2, all flows are even integers and α is an integer.

Assume that after k or fewer applications of Steps 3 and 4 that α is again an integer. As shown in the proof of Lemma 3.8.1, after application of Steps 3 and 4, $|f^1[i,j]| + |f^2[i,j]|$ is either unchanged or changed by 2α. Hence, after the $(k + 1)$st application of Steps 3 and 4, α is still an integer. Therefore, α is always an integer since the flows of commodities 1 and 2 are changed by α in each step or are unchanged. The flows are therefore integers throughout the algorithm.

Proof of Lemma 3.8.3. After the application of Step 1 the even integer value of $f^1_{s_1,t_1}$ is unchanged throughout the algorithm. From Lemma 3.8.2 we see that the value of $f^2_{s_2,t_2}$ is increased by an even integer with each application of Steps 2, 3, and 4. Therefore, either r_1 and r_2 (which are even integers) are satisfied in a finite number of steps or $f^1_{s_1,t_1} + f^2_{s_2,t_2}$ reaches its even integer maximum, $\tau_{s_1,s_2;t_1,t_2}$, in a finite number of steps. Hence the lemma is proved.

Proof of Lemma 3.8.4. From Lemma 3.8.3, we know that the algorithm terminates after a finite number of iterations, following Step 1, 2, or 3. Clearly, if r_1 and r_2 are satisfied the algorithm terminates after Step 2; if $f^1_{s_1,t_1} + f^2_{s_2,t_2}$ is maximized the algorithm terminates after Step 3; if r_1 is not feasible, the algorithm terminates after Step 1; if r_2 or $r_1 + r_2$ is not feasible, the algorithm terminates after Step 3. We must now show that these are the only conditions under which the algorithm terminates. To do this, we show that if the algorithm terminates, (a), (b), or (c) of Lemma 3.8.4 is true.

CASE 1. If the algorithm terminates after Step 1, by the Max-Flow Min-Cut Theorem, r_1 is not feasible and (a) is true.

CASE 2. If the algorithm terminates after Step 2, r_1 and r_2 are satisfied and (a) is true.

CASE 3. If the algorithm terminates after Step 3, either a permissible s_2-t_2 path cannot be found or a permissible t_2-s_2 path cannot be found or neither can be found.

a) Suppose a permissible s_2-t_2 path cannot be found. Recursively define a set of vertices X as follows:
   i) $v_{s_2} \in X$.
   ii) If $v_j \in X$ then $v_k \in X$ provided
      1) $f^1[k,j]$ is a positive flow from v_k to v_j, or
      2) $f^2[k,j]$ is a positive flow from v_k to v_j, or
      3) $c[k,j] - |f^1[k,j]| - |f^2[k,j]| > 0.$

The cut $[X,\bar{X}]$ is not empty since $v_{t_2} \in \bar{X}$. If $v_{t_2} \in X$ then there would be a permissible s_2-t_2 path which contradicts the hypothesis of Case 3(a). For all $[i,j] \in [X,\bar{X}]$ with $v_i \in X$,

$$c[i,j] - |f^1[i,j]| - |f^2[i,j]| = 0, \tag{3.8.16}$$

and

$$f^1[i,j] \geq 0, \tag{3.8.17}$$

and

$$f^2[i,j] \geq 0. \tag{3.8.18}$$

If $f^1[i,j] = 0$ for all $[i,j] \in [X,\bar{X}]$ with $v_i \in X$, then $v_{s_2} \in X$, $v_{t_2} \in \bar{X}$, and $c[X,\bar{X}] = f^2[X,\bar{X}]$ with all flow positive in the direction from X to $\bar{X}$. Hence, $f^2_{s_2,t_2}$ is maximum and either $f^1_{s_1,t_1} + f^2_{s_2,t_2}$ has been maximized or $r_1 + r_2$ is not feasible. If for some $[i,j] \in [X,\bar{X}]$ with $v_i \in X$, $f^1[i,j]$ is a positive flow from v_i to v_j with $v_i \in X$, then $v_{s_1} \in X$ and $v_{t_1} \in \bar{X}$. Otherwise, there would be a branch $[l,k] \in [X,\bar{X}]$ with positive flow from v_k to v_l with $v_l \in X$. We therefore have $v_{s_1}, v_{s_2} \in X$, $v_{t_1}, v_{t_2} \in \bar{X}$ and $r_1 + r_2 \geq c[X,\bar{X}] \geq \tau_{s_1,s_2;t_1,t_2}$. Hence $r_1 + r_2$ is not feasible or $f^1_{s_1,t_1} + f^2_{s_2,t_2}$ is maximized.

b) If a permissible t_2-s_2 path cannot be found the proof is similar to the proof in Case 3(a) and is left as Problem 13.

Based upon the definition of X in (i) and (ii) a labeling algorithm can be devised for finding permissible s_2-t_2 paths. This and the formulation of a corresponding labeling algorithm to find a permissible t_2-s_2 path are straightforward and are left as Problem 14.

3.9 COMMON TERMINAL MULTICOMMODITY FLOW

In this section we consider the special case of the multicommodity flow problem in which a directed graph contains a single terminal vertex v_t and an arbitrary number N of source vertices $v_{s_1}, \ldots, v_{s_N}$. Without loss of generality, each source is assumed to supply a single distinct commodity. For this case we show that the maximum flow is limited only by the value of a minimum separating set. Moreover, for a given set of nonnegative constants $\alpha_1 > \alpha_2 > \cdots > \alpha_N$ we give an algorithm to maximize

$$\sum_{i=1}^{N} \alpha_i f^i_{s_i,t}.$$

In practice this function may represent a performance criterion such as profit.

Common-Terminal Multicommodity Flow Algorithm

Step 1. Create a new source vertex v_s.

Step 2. Create an infinite capacity branch (s,s_i), where initially $i = 1$.

Step 3. Treating all flows as a single commodity, maximize the s-t flow using the Labeling Algorithm.

Step 4. If $i < N$, increase i by one and return to Step 2. If $i = N$, go to Step 5.

Step 5. Perform a semipath decomposition on the flow pattern. If a flow semipath uses (s,s_i) assign that flow to commodity i.

Theorem 3.9.1 *The flow values obtained from the Common-Terminal Multicommodity Flow Algorithm maximize the objective function*

$$\sum_{i=1}^{N} \alpha_i f^i_{s_i,t}.$$

Proof. Define

$$f_k = \sum_{i=1}^{k} f^i_{s_i,t} \quad \text{and} \quad C_k = \sum_{i=1}^{k} \alpha_i f^i_{s_i,t}$$

and denote by $\tilde{f}_k$ and $\tilde{C}_k$ the respective maximum values of these sums. The subscript o will be used to denote values obtained from the algorithm. In successive iterations of Step 3 of the algorithm, f_k is maximized in the order $k = 1,2,\ldots,N$. Thus,

$$f^k_{s_k,to} = \tilde{f}_k - \tilde{f}_{k-1} \qquad \text{for } k = 2,3,\ldots,N \tag{3.9.1a}$$

and

$$f^i_{s_i,to} = \tilde{f}_1. \tag{3.9.1b}$$

We now show by induction on k that the algorithm maximizes C_1. For $k = 1$, $C_1 = \alpha_1 f^1_{s_1,t}$. Therefore to maximize C_1 we need only maximize $f^1_{s_1,t}$. Thus, from (3.9.1b) $C_{1_o} = \tilde{C}_1$ and the statement is true for $k = 1$.

We now assume

$$C_{ko} = \tilde{C}_k \tag{3.9.2}$$

and show that $C_{(k+1)o} = \tilde{C}_{k+1}$. From the definition of $C_{(k+1)o}$,

$$C_{(k+1)o} = C_{ko} + \alpha_{k+1} f^{k+1}_{s_{k+1},to}. \tag{3.9.3}$$

Combining (3.9.1a), (3.9.2), and (3.9.3) gives

$$C_{(k+1)o} = \tilde{C}_k + \alpha_{k+1}\tilde{f}_{k+1} - \alpha_{k+1}\tilde{f}_k. \tag{3.9.4}$$

We prove that $C_{(k+1)o} = \tilde{C}_{k+1}$ by showing that, if $f^{k+1}_{s_{k+1},t}$ is either smaller or larger than $f^{k+1}_{s_{k+1},to}$, the resulting C_{k+1} is less than $C_{(k+1)o}$.

CASE A. In (3.9.3), $C_{ko} = \tilde{C}_k$ by the inductive hypothesis (3.9.2). Thus, $f^{k+1}_{s_{k+1},t} < f^{k+1}_{s_{k+1},to}$ results in $C_{k+1} < C_{(k+1)o}$.

CASE B. Consider a flow pattern for which $f^{k+1}_{s_{k+1},t} > f^{k+1}_{s_{k+1},to}$, that is,

$$f^{k+1}_{s_{k+1},t} = \tilde{f}_{k+1} - \tilde{f}_k + \beta, \text{ for } \beta > 0. \tag{3.9.5}$$

By definition $f_k = f_{k+1} - f^{k+1}_{s_{k+1},t}$. Together with (3.9.5) this yields

$$f_k = (f_{k+1} - \tilde{f}_{k+1}) + \tilde{f}_k - \beta. \tag{3.9.6}$$

Since $f_{k+1} \le \tilde{f}_{k+1}$, (3.9.6) gives

$$f_k \le \tilde{f}_k - \beta. \tag{3.9.7}$$

As an intermediate step we now use (3.9.7) to show that the corresponding C_k satisfy

$$C_k \le \tilde{C}_k - \alpha_k\beta. \tag{3.9.8}$$

Suppose that $C_k > \tilde{C}_k - \alpha_k\beta$. By (3.9.7), we can then increase f_k by β. Hence we can increase C_k by at least $\alpha_k\beta$ since $\alpha_j > \alpha_k$ for $j < k$. The new value of C_k would then be greater than $\tilde{C}_k$. This contradiction establishes (3.9.8).

Combining the definition $C_{k+1} = C_k + \alpha_{k+1} f^{k+1}_{s_{k+1},t}$ with (3.9.8) gives

$$C_{k+1} \le \tilde{C}_k + \alpha_{k+1} f^{k+1}_{s_{k+1},t} - \alpha_k\beta. \tag{3.9.9}$$

Substituting (3.9.5) into (3.9.9) gives

$$C_{k+1} \le [\tilde{C}_k + \alpha_{k+1}\tilde{f}_{k+1} - \alpha_{k+1}\tilde{f}_k] - (\alpha_k - \alpha_{k+1})\beta. \tag{3.9.10}$$

Comparing (3.9.4) and (3.9.10) and noting that $\alpha_k > \alpha_{k+1}$ gives $C_{k+1} < C_{(k+1)o}$.

Cases A and B establish that the algorithm yields the optimum value of $f^{k+1}_{s_{k+1},t}$. Since C_{ko} was assumed maximum, $C_{(k+1)o} = \tilde{C}_{k+1}$ and in particular, $C_{no} = \tilde{C}_n$. Since

$$C_n = \sum_{i=1}^{n} \alpha_i f^i_{si,t},$$

it follows that the algorithm maximizes the given objective function.

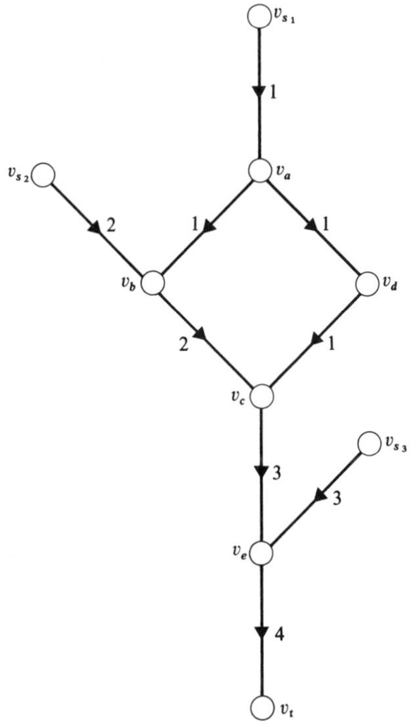

Fig. 3.9.1 Illustration of Step 3 of the Common-Terminal Multicommodity Algorithm.

Example 3.9.1 We illustrate the algorithm by maximizing the objective function

$$\sum_{i=1}^{3} \alpha_i f^i_{i,t}$$

for the graph shown in Fig. 3.9.1. Suppose the first iteration of the algorithm results in one unit of flow along the path $v_{s_1},v_a,v_b,v_c,v_e,v_t$. The second iteration sends one unit of flow along the path v_{s_2}, v_b,v_a,v_d,v_c,v_e,v_t and one unit of flow along the path v_{s_2},v_b,v_c,v_e,v_t. The final iteration sends one unit of flow along the path v_{s_3},v_e,v_t. The individual commodities are then easily identified by a semipath decomposition.

It is important to note that the validity of the algorithm depends upon waiting until the final step to identify the flows of individual commodities. Simply sending maximum amounts of each commodity in priority order will not in general yield an optimum solution.

When some of the coefficients in the objective function are equal, we can still apply the algorithm if we treat as a single source those sources which are associated with equal α's. At the completion of the algorithm, flows of the individual commodities are identified by a semipath decomposition.

3.10 FURTHER REMARKS

The three independent proofs of the Max-Flow Min-Cut Theorem presented in this chapter all appeared in 1956. The combinatoric proof is due to Ford and Fulkerson [FO1]. The graph theoretic proof is due to Elias, Feinstein, and Shannon [EL1] and the programming proof to Dantzig and Fulkerson [DA1]. Other proofs of the theorem are also available, such as Ore [OR1] and, surprisingly, the proofs are quite different in nature. An earlier algorithm by Boldyreff [BO1] is a heuristic procedure for maximizing flow, which is effective computationally but does not yield a Max-Flow Min-Cut Theorem. This procedure is the subject of Problem 10. Theorem 3.7.2 is due to Tang [TA2]. Lemma 3.7.1 and Theorem 3.7.3 are due to Hu [HU2]. The treatment for the two-commodity flow problem based on recirculation of flow is modeled after that given by Hu [HU1]. Hakimi was the first to use the idea of recirculation of flow for solving the two-commodity flow problem [HA2]. An alternative approach based on the properties of Euler graphs has been given by Rothschild and Whinston [RO5, RO6] for the case in which the sum of the capacities of the branches incident at any vertex is even. In this case integer flow patterns can still be obtained. The results on the multicommodity problem for the single-terminal case are due to Rothfarb, Shein, and Frisch [RO4] and have been extended by Rothfarb and Frisch [RO3] to include the case of concave profit functions.

Several other results are available on the multicommodity flow problem which have not been discussed in this chapter. Tang [TA1, TA2, TA4] has shown that conditions (3.7.7) are necessary and sufficient for a set of multicommodity requirements to be satisfied in a graph which is a tree and has used this fact to synthesize trees which satisfy time-varying requirements and minimize a cost function. Tang [TA3] has also shown that the same conditions are necessary and sufficient for the class of "bipath graphs," defined as the class of graphs in which there are at most two vertex disjoint paths between any two vertices in the graph. Rothfarb and Frisch [RO2] have shown that conditions (3.7.7) are necessary and sufficient for a three-commodity flow in a graph with only source and terminal vertices. These results have been extended to a class of graphs which are not gapless and contain internal vertices, by means of simple transformations used by Rothfarb [RO1] (see Problem 19). The results have also been extended to show sufficiency of the conditions for a class of n-commodity graphs. Sakarovitch [SA1] (see Problem 15) has proved that conditions (3.7.7) are necessary and sufficient for graphs which are "completely planar."

The n-commodity flow problem can, of course, be formulated as a linear programming problem. As we have seen in Section 3.5, for one commodity the maximum flow problem has a special form which guarantees that integer capacities yield integer flows. This simplification is characteristic of a larger class of linear programming problems. Hoffman and Kruskal [HO1] have shown that a linear programming problem Max $\mathbf{cx}$ with integer $\mathbf{c}$ subject to $A\mathbf{x} \geq \mathbf{b}$ and $\mathbf{x} \geq 0$ has integer solutions for *all* integer $\mathbf{b}$ if and only if A is "totally unimodular"; i.e. the determinant of any square submatrix of A has value $+1$, -1, or 0. The maximum flow problem as well as the "transportation problem" and "assignment problem" (see Problem 7) are characterized by totally unimodular constraint matrices.

However, the n-commodity flow problem does not fit in this category and in general, as we have seen, integer solutions cannot be guaranteed. General linear programming techniques for n-commodity flow problems have been developed by Ford and Fulkerson [FO2] and Gomory and Hu [GO1].

PROBLEMS

1. In Fig. P3.1, maximize the s-t flow, $f_{s,t}$ by using the Labeling Algorithm. Find $\tau_{s,t}$ and an s-t cut with value $\tau_{s,t}$.

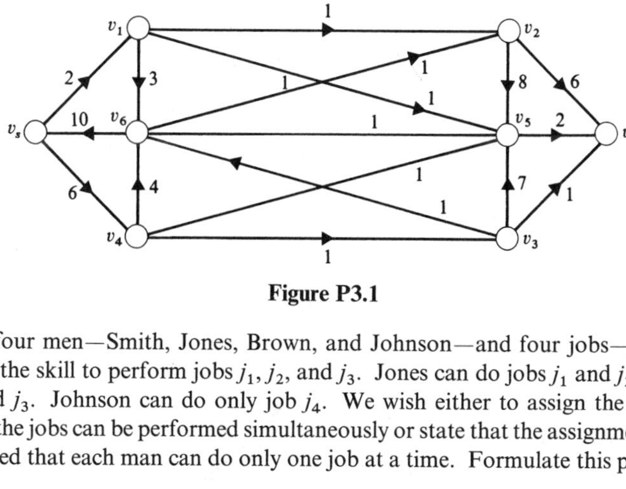

Figure P3.1

2. Consider four men—Smith, Jones, Brown, and Johnson—and four jobs—j_1, j_2, j_3, and j_4. Smith has the skill to perform jobs j_1, j_2, and j_3. Jones can do jobs j_1 and j_2. Brown can do jobs j_2 and j_3. Johnson can do only job j_4. We wish either to assign the men to the jobs so that all the jobs can be performed simultaneously or state that the assignment is impossible. It is assumed that each man can do only one job at a time. Formulate this problem as a flow problem.
3. Prove that the Labeling Algorithm yields *even* integer flows if the branch capacities and the initial flows are *even* integer flows.
4. Show that an independent set of equations is obtained by eliminating the first equation in (3.5.2a).
5. Consider the constraint matrix in (3.5.5). Prove that the set of columns corresponding to $f(i,j)$ in a circuit of G_B are linearly dependent.
6. Consider a directed graph $G = (V,\Gamma)$ with an associated flow pattern which satisfies conservation at vertices and the constraint
$$f(i,j) \geq l(i,j) \quad \text{for all } (i,j) \in \Gamma,$$
where $l(i,j)$ is a nonnegative integer.
a) Give a Labeling Algorithm to find the value $f_{s,t}$ for the minimum s-t flow.
b) Prove $\underset{\mathscr{F}}{\text{Min}}\,[f_{s,t}] = \underset{X}{\text{Max}}\,[c(X,\bar{X})_{s,t}]$.

7. Let $G = (V,\Gamma)$ be a graph with integer branch capacities. Let $S = \{v_{s_1}, \ldots, v_{s_g}\}$ and $T = \{v_{t_1}, \ldots, v_{t_h}\}$ be two distinct subsets of V. We wish to determine whether we can satisfy a set of integer requirements $\{r_1, \ldots, r_h\}$ at T from a set of available integer supplies $\{a_1, \ldots, a_g\}$ at S. That is, the constraints are

$$\sum_l f_{s_k,t_l} \leq a_k \qquad \text{for all } k,$$

and

$$\sum_k f_{s_k,t_l} \geq r_l \qquad \text{for all } l,$$

where

$$\sum_k a_k \geq \sum_l r_l.$$

We also require

$$f(i,j) \leq c(i,j) \qquad \text{for all } (i,j),$$

and

$$f(i,V) - f(V,i) = \begin{cases} \sum_l f_{s_k,t_l} & \text{if } i = s_k \\ 0 & \text{for } i \neq s_k,t_k \text{ for any } k \\ -\sum_k f_{s_k,t_l} & \text{if } i = t_l. \end{cases}$$

This is known as the "transportation problem." For $a_k = 1$ for all k and $v_l = 1$ for all l it is called the "assignment problem." Problem 2 is an example of the latter.

a) Formulate the above problem as the maximization of flow in a graph with a single source and a single terminal.

b) Prove that a flow pattern satisfying the given constraints can be found if and only if for every subset T_q of T, any cut $(X,\bar{X})$, and any subset S_m of S

$$\sum_{v_l \in (T_q \cap \bar{X})} r_l - \sum_{v_k \in (S_q \cap X)} a_k \leq c(X,\bar{X}).$$

8. Consider a graph $G = (V,\Gamma)$, with n vertices, in which branch flows are constrained by both upper and lower bounds. We seek a flow pattern satisfying conservation at all vertices (i.e., there are no sources or terminals) and the constraint $l(i,j) \leq f(i,j) \leq c(i,j)$, where for all (i,j), $l(i,j)$ is an integer in the range $0 \leq l(i,j) \leq c(i,j)$.

For $X \subseteq V$, $Y \subseteq V$ define

$$l(X,Y) = \sum l(i,j).$$

Then form a graph $G^* = (V^*,\Gamma^*)$ from G by adding vertices v_s and v_t and branches $(\{v_s\},V)$ and $(\{v_t\},V)$. The capacities of the branches in G^* are

$$\begin{aligned} c^*(i,j) &= c(i,j) - l(i,j) && \text{for } (i,j) \in \Gamma, \\ c^*(s,i) &= l(V,\{v_i\}) && \text{for } v_i \in V, \\ c^*(i,t) &= l(\{v_i\},V) && \text{for } v_i \in V. \end{aligned}$$

Using G^* prove that a feasible flow pattern can be found in G if and only if $c(X,\bar{X}) \geq l(X,\bar{X})$ for all s-t cuts $(X,\bar{X})$.

9. Consider a directed graph G in which there are no directed circuits. We wish to find the minimum number of paths required to "cover" a given subset of branches such that each branch is contained in at least one of the chosen paths. Formulate this problem as one of finding a minimum flow in a graph.

10. The following "flooding technique" is suggested as a heuristic method for maximizing $f_{s,t}$, in a directed graph $G = (V,\Gamma)$.

    a) Starting at the source vertex v_s, assign flows $f(i,j)$ to the branches whose initial vertex is v_s.
    b) Find the vertices v_j such that $f(V,j) > 0, f(j,V) = 0$ and find one of these vertices, say v_i, for which $c(i,V)$ is minimum. Assign flows to each branch whose initial vertex is v_i in the following order: (1) advancing—a branch whose terminal vertex has zero flow through it; (2) lateral—a branch whose final vertex has nonzero flow through it. Note that after this step the branch capacity constraints must be satisfied but the conservation restriction at the vertices need not be satisfied.
    c) Repeat (b) until every vertex except v_t has been considered as an initial vertex.
    d) Remove "bottlenecks" (i.e., vertices $v_j \neq v_s,v_t$ for which $f(j,V) < f(V,j)$) by reducing the flow into v_j. After this step, flow is conserved at each vertex. Reduce each branch capacity by the amount of flow in that branch.
    e) Repeat Steps (a), (b), (c), and (d) until there is an s-t cut $(X,\bar{X})$ for which $f_{s,t} = c(X,\bar{X})$ or until no more flows can be assigned. In the former case we have maximized $f_{s,t}$. In the latter case the algorithm has failed and we must apply it again!

    A. Apply the Flooding Technique to the graph in Fig. P3.1.
    B. What are the advantages and disadvantages of the Flooding Technique as compared with the Labeling Algorithm? [BO1]

11. In an undirected graph (Fig. P3.2), the number associated with each branch is its capacity and the number associated with each arrow is the flow of commodity 1 in that direction. Use the Two-Commodity Flow Algorithm to rearrange the flows of commodity 1 in order to determine $\mathscr{F}^2$ such that $f^1_{s1,t1} = 2$ and $f^2_{s2,t2}$ is maximized.

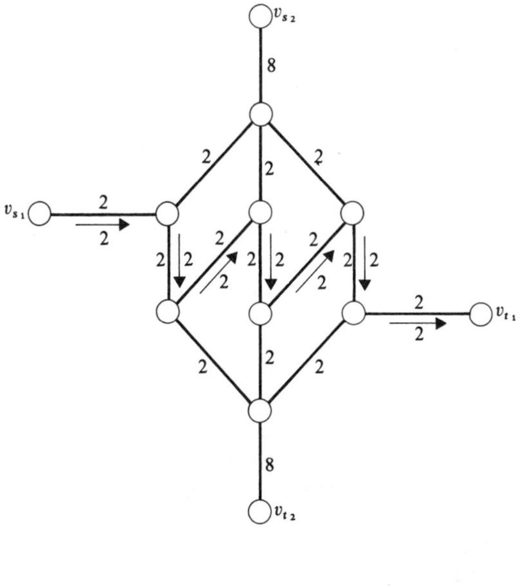

Figure P3.2

12. For a directed graph let r_1 be a requirement of commodity 1 from v_{s_1} to v_{t_1} and let r_2 be a requirement of commodity 2 from v_{s_2} to v_{t_2}. Prove

$$r_1 + r_2 \leq \tau_{s_1,s_2;t_1,t_2}.$$

13. Prove Case 3(b) of Lemma 3.8.4.
14. Devise a Labeling Algorithm to find a permissible s_2-t_2 path in an undirected graph with two-commodity requirements.
15. A graph is "planar" if it can be drawn on a plane such that no branches intersect except at vertices. For a graph G representing a multicommodity flow problem let $\hat{G}$ be the graph formed by adding a branch between the source of commodity i and the terminal of commodity i, for all i. Assume each commodity has a single source and a single terminal. G is said to be completely planar if $\hat{G}$ can be drawn with branches crossing only at vertices when the sources are arranged in a vertical column and the terminals, in another vertical column in the same order as the corresponding sources. Prove that for a completely planar graph G representing an n-commodity flow problem

$$\text{Max}\left[\sum_{i=1}^{n} f^i_{s_i,t_i}\right] = \tau_{s_1,s_2,\ldots,s_n;t_1,t_2,\ldots,t_n}. \quad \text{[SA1]}$$

16. An undirected graph is said to be bipath if between any two vertices, say v_s and v_t, there are no more than two paths each containing three or more branches, such that the only vertices common to any two paths are v_s and v_t. Prove
    a) A bipath graph is planar.
    b) Let G be a bipath graph. Suppose G contains at least three vertices none of which is a vertex cut-set. Prove that G contains a Hamilton circuit [TA3].
17. Prove that in an undirected tree a set of flow requirements $r_1, \ldots, r_k$ can be simultaneously satisfied by a set of flows with $f^l_{s_l,t_l} = r_l$, for $l = 1, \ldots, k$, if and only if for each branch $[i,j]$

$$\sum_{l} r_l \leq c[i,j],$$

where the summation is over those l for which $[i,j]$ is an s_l-t_l cut [TA4].

18. Prove that the Algorithm P1 below will yield a tree which satisfies a set of requirement $r_{i,j}$ for $i \neq j$ with $i = 1, \ldots, n$ and $j = 1, \ldots, n$, so that each branch is saturated.

Algorithm P1

Step 1. Form a graph G in which $c[i,j] = r_{i,j}$ for $i,j = 1, \ldots, n$.

Step 2. Find a tree T of G.

Step 3. For each branch $[i,j]$ of the tree replace $c[i,j]$ by the capacity of the cut-set which contains only one branch $[i,j]$ of T.

Step 4. Set the capacities of the nontree branches equal to zero [TA4].

19. Assume that the graph G in Fig. P3.3(a) is a subgraph of a graph for which there is a feasible n-commodity flow pattern with values f_{s_i,t_i} for $i = 1, \ldots, n$. Assume that only three branches shown are incident at v_d and there are no flow requirements at v_d. Show that, if the subgraph is replaced by the subgraph G^* (Fig. P3.3b), then the original flow values are still feasible. Similarly show that any flow values feasible in the second graph are feasible in the first graph.

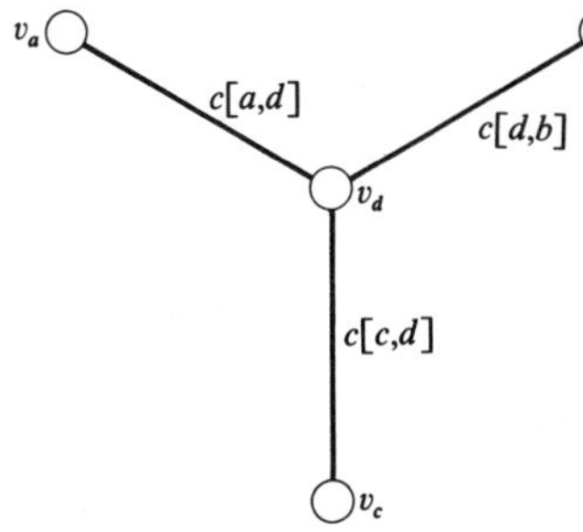

(a)

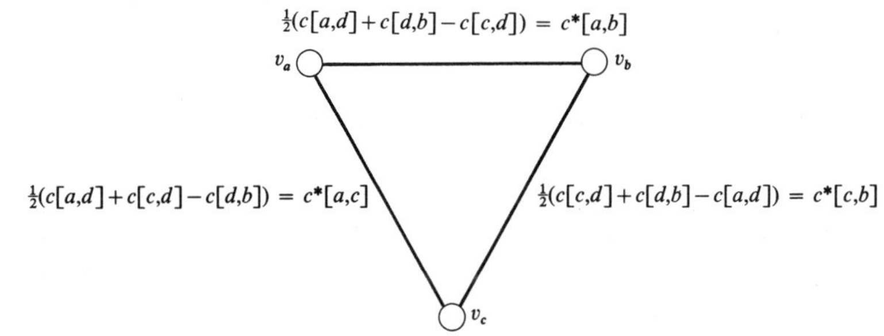

(b)

Figure P3.3

20. In the directed graph shown in Fig. P3.4 the branch weights are capacities, v_{s_i} is the source for commodity i for $i = 1, 2, 3$, and v_t is the terminal for all commodities. Find a flow pattern which maximizes the quantity β where

$$\beta = f_{s_1,t} + 2f_{s_2,t} + 3f_{s_3,t}.$$

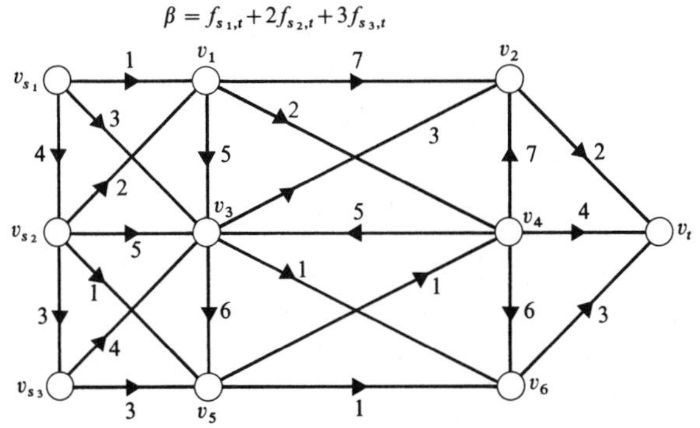

Figure P3.4

REFERENCES

BO1 A. W. Boldyreff, "Determination of the Maximal Steady Flow of Traffic through a Railroad Network," *Operations Res.* **3,** 443–465 (1955).

DA1 G. B. Dantzig and D. R. Fulkerson, "On the Max-Flow Min-Cut Theorem of Networks," in *Linear Inequalities and Related Systems*, *Ann. Math. Studies* **38,** 215–221 (1956).

EL1 P. Elias, A. Feinstein, and C. E. Shannon, "A Note on the Maximum Flow through a Network," *IRE Trans. Inform. Theory* **IT-2,** 117–119 (1956).

FO1 L. R. Ford and D. R. Fulkerson, "Maximal Flow through a Network," *Can. J. Math.* **18,** 399–404 (1956).

FO2 L. R. Ford and D. R. Fulkerson, "Suggested Computation for Maximal Multi-Commodity Network Flows," *Management Sci.* **5,** 97–101 (1958).

GO1 R. E. Gomory and T. C. Hu, "Synthesis of a Communication Network," *J. Soc. Ind. Appl. Math.* **12,** 348–368 (1964).

HA1 G. Hadley, *Linear Programming*, Addison-Wesley, Reading, Mass., 1962.

HA2 S. L. Hakimi, "Simultaneous Flows through Communication Networks," *IRE Trans. Circuit Theory* **CT-9,** 169–175 (1962).

HO1 A. J. Hoffman and J. G. Kruskal, "Integral Boundary Points of Convex Polyhedra," in *Linear Inequalities and Related Systems*, *Ann. Math. Studies* **38,** 215–222 (1956).

HU1 T. C. Hu, "Multicommodity Network Flows," *Operations Res.* **11,** 344–360 (1963).

HU2 T. C. Hu, "On the Feasibility of Simultaneous Flows in a Network," *Operations Res.* **12,** 359–360 (1964).

JE1 W. S. Jewell, *Multicommodity Network Solutions*, ORC66-23, Operations Research Center, University of California, Berkeley, 1966.

OR1 O. Ore, *Theory of Graphs*, American Mathematical Society, Vol. XXXVIII, Colloquium Publications, Providence, R.I., 1962.

RO1 B. Rothfarb, "Combinatoric Methods for Multicommodity Flows," Ph.D. Dissertation, University of California, Berkeley, 1968.

RO2 B. Rothfarb and I. T. Frisch, "On the 3-Commodity Flow Problem," *J. Soc. Ind. Appl. Math.* **16,** 202–205 (1968).

RO3 B. Rothfarb and I. T. Frisch, "Common Terminal Multicommodity Flow with a Concave Cost Function," *J. Soc. Ind. Appl. Math.* **18,** 489–502 (1970).

RO4 B. Rothfarb, N. P. Shein, and I. T. Frisch, "Common Terminal Multicommodity Flow," *Operations Res.* **16,** 202–205 (1968).

RO5 B. Rothschild and A. Whinston, "On Two Commodity Networks Flows," *Operation Res.* **14,** 377–387 (1966).

RO6 B. Rothschild and A. Whinston, "Feasibility of Two Commodity Network Flows," *Operations Res.* **14,** 1121–1129 (1966).

SA1 M. Sakarovitch, "The Multicommodity Flow Problem," Ph.D. Dissertation, Operations Research Center, University of California, Berkeley, 1966.

TA1 D. T. Tang, "Comments on Feasibility Conditions of Simultaneous Flow in a Network," *Operations Res.* **13,** 143–146 (1965).

TA2 D. T. Tang, "Communication Networks with Simultaneous Flow Requirements," *IRE Trans. Circuit Theory* **CT-9,** 176–182 (1962).

TA3 D. T. Tang, "Bipath Networks and Multicommodity Flows," *IEEE Trans. Circuit Theory* **CT-11,** 468–474 (1964).

TA4 D. T. Tang, "Optimal Trees for Simultaneous Flow Requirements," *Proc. Natl. Electron. Conf.* **19,** 28–32 (1963).

CHAPTER 4

MAXIMUM FLOW IN PROBABILISTIC GRAPHS

4.1 INTRODUCTION

In deterministic network problems, one usually assumes that complete knowledge about the system is available. For example, in a given network, we may want to find the maximum flow between a particular pair of stations. If the system is static we can simply apply the Max-Flow Min-Cut Theorem to find the value of this flow. In most practical cases a *random* number of subscribers are using the system at each moment. Thus, not all of the facilities are available for each user.

In this chapter, we assume that the existing flow or traffic within a network is probabilistic. In other words, the present flow within each branch of the network is not a known number but rather a random number with a possibly known probability distribution. The analysis problem to be considered is a generalization of the analysis problem discussed in Chapter 3: given a network, determine the amount of flow that can be established between a specified pair of stations. Now, however, the existing flows in the channels are not deterministic. It is necessary to find the *probability* that a particular flow can be attained between the two stations. Therefore, if we represent the network by a graph, and the points of interest are denoted by v_s and v_t, our problem may be stated as: find the probability that a flow of at least $r_{s,t}$ units can be attained between v_s and v_t (i.e., find Prob $\{\tau_{s,t} \geq r_{s,t}\}$). A variation of this problem is also of interest. It may be that we are given a critical probability level p_0, and are asked to determine whether Prob $\{\tau_{s,t} \geq r_{s,t}\} \geq p_0$.

The analysis problem formulated above also applies to the situation in which the existing flows are not necessarily random but the branches or vertices are unreliable. In this case, the capacity of a given branch may be reduced because of equipment failure, and so the maximum flow $\tau_{s,t}$ is again a random variable. Equipment failure could result from normal wear or from an attack on the system.

Other analysis problems can also be formulated for the model we have described. The most obvious and important one is the queuing problem associated with the time required to transmit a specified flow between vertices v_s and v_t. At a given instance of time, a flow rate $r_{s,t}$ may not be attainable between v_s and v_t. However, it may be possible partially to transmit the flow through the graph and store it at some intermediate vertex until sufficient capacity is available to complete transmission. For such a transmission system, a major analysis problem is to determine the length of time a message will spend in the graph. Problems of this nature have been considered in depth by L. Kleinrock [KL1] and will be briefly discussed in Chapter 9. The relation between the present problem and the queuing problem is that Prob $\{\tau_{s,t} \geq r_{s,t}\}$ is the probability that we can transmit $r_{s,t}$ units of flow from v_s to v_t without having to store the flow at an intermediate vertex.

Any approach to the analysis problem is strongly dependent on the statistical information available about the random flows. For example, we may have no information about the probability distributions of these flows, but we may have a record of their values at various instances of time. On the other hand, we may be given their probability distributions or we may be allowed to assume appropriate asymptotic distributions. In this chapter we will investigate these above problems under a variety of different assumptions about the available statistical information. The results do not depend on whether the graphs are directed or undirected.

We begin in Section 2 with a brief review of some elementary statistical concepts. In Section 3, we examine the problem of testing hypotheses about the magnitude of Prob $\{\tau_{s,t} \geq r_{s,t}\}$ when samples of $\tau_{s,t}$ are available but the probability distributions of branch flows and capacities are unknown. In Section 4, we examine the computational difficulties of finding Prob $\{\tau_{s,t} \geq r_{s,t}\}$ when branch flow variables have known probability distributions. In Section 5, we investigate the consequences of assuming that branch flows have normal distributions. Monte Carlo techniques are applied in Section 6 to the problem of computing Prob $\{\tau_{s,t} \geq r_{s,t}\}$. It is seen that normal approximations can often be used for this probability. Finally, in Section 7, the implications of normal approximations for the maximum flow are studied using statistical techniques. It is shown that a test with excellent statistical properties can be developed if the maximum flow is approximately normally distributed.

4.2 ELEMENTARY HYPOTHESIS TESTING

Consider a system characterized by a vector random variable $\mathbf{X} = (X_1, \ldots, X_K)$. Since $\mathbf{X}$ is random, the behavior of the system cannot be precisely predicted. However, if the probability distribution of $\mathbf{X}$ is known, it is often possible to obtain quantitative estimates of the expected performance of the system. For example, $\mathbf{X}$ may be the outcome of a battery of aptitude tests, X_1 the verbal score, X_2 the mathematical score, and so on. If the probability distribution of each X_i is known, we can predict the percentage of people who will score in a certain numerical range on each test. Furthermore, if the joint distribution of the X_i is known, then we can make an overall prediction about the test outcome.

The probability distributions of the X_i are not usually completely known. In the aptitude test, we may know that for most populations the test scores have a normal distribution, but for particular groups the mean and variance will vary. Therefore, before the test scores can be statistically valid, we must test a large, representative, group and obtain an estimate of the moments of the distribution.

The statistical problem that we must consider can be stated as follows: We are given a random variable $\mathbf{X}$ which has a probability distribution known to be a member of a class of probability distributions $P = \{P_\theta; \theta \in \Omega\}$ which is indexed by the vector parameter θ which ranges over a set Ω. Then, we must perform a statistical test to determine the true value of θ. For example, suppose X is a normal random variable with probability density

$$p_\theta(X) = \frac{1}{\sqrt{2\pi\theta_1}} e^{-\frac{1}{2}(x-\theta_2)^2/\theta_2} \tag{4.2.1}$$

and $\theta = (\theta_1, \theta_2)$ is the unknown parameter. Then θ_2 is the mean and θ_1 is the variance of X. Quite often, it is not necessary to determine θ exactly, since placing θ into a subclass, say Ω_0 of Ω, may yield reasonably accurate predictions. For example, if X is normally distributed and we know that $100 \leq \theta_1 \leq 110$ and $1 \leq \theta_2 \leq 2$, then we could easily determine the group of people who performed well on the test with scores distributed as $P_\theta(x)$.

Suppose we have the *hypothesis* $H_0 : \theta \in \Omega_0$ and we want to test H_0 against the *alternative* $\theta \notin \Omega_0$. Denote this alternative by H_1 and let $\Omega_1 = \Omega - \Omega_0$. The statistical problem we formulate is: Test the hypothesis

$$H_0 : \theta \in \Omega_0 \tag{4.2.2a}$$

against the alternative

$$H_1 : \theta \in \Omega_1 \tag{4.2.2b}$$

Naturally, the statistical test is based on a set of possibly vector observations or samples, $\mathbf{x}_1, \mathbf{x}_2, \ldots, \mathbf{x}_K$ of $\mathbf{X}$. The outcome of any test will depend upon how we sample the random variable $\mathbf{X}$. Let $\mathbf{X}_i$ be the random variable corresponding to the outcome of the ith sample. Clearly, a "good" sampling scheme must have the same probability distributions for $\mathbf{X}$ and $\mathbf{X}_i$ for each i. Therefore, we want $\mathbf{X}_1, \mathbf{X}_2, \ldots, \mathbf{X}_K$ to be identically distributed with the distribution of $\mathbf{X}$. Furthermore, we want each sample outcome $\mathbf{X}_i$ to yield as much information as possible. Consequently, $\mathbf{X}_1, \mathbf{X}_2, \ldots, \mathbf{X}_K$ should be independent random variables.

In making a decision about the truth of hypothesis H_0 on the basis of a finite number of samples, there are two types of possible error. We can *reject* H_0 (i.e., conclude that $\theta \in \Omega_1$) when H_0 is actually true, or we can accept H_0 (i.e., conclude that $\theta \in \Omega_0$) when H_0 is false (i.e., H_1 is true). The former error is called Type I error and the latter error is called Type II error. An ideal test would have

$$\text{Prob}\,\{\text{Type I error}\} = \text{Prob}\,\{\text{reject } H_0 | H_0 \text{ is true}\} = 0; \tag{4.2.3a}$$

$$\text{Prob}\,\{\text{Type II error}\} = \text{Prob}\,\{\text{accept } H_0 | H_0 \text{ is false}\} = 0. \tag{4.2.3b}$$

Ideal tests do not exist for nondegenerate random variables. Therefore, we would like to bound the probabilities of both types of error. That is, we wish to find a test such that

$$\text{Prob}\,\{\text{Type I error}\} \leq \alpha_1; \tag{4.2.4a}$$

$$\text{Prob}\,\{\text{Type II error}\} \leq \alpha_2; \tag{4.2.4b}$$

where α_1 and α_2 are nonzero but arbitrarily small probabilities.

In general, tests which exhibit the characteristics given in (4.2.4) cannot be found. However, it is often possible to find a test which will bound the probability of one testing error while simultaneously *minimizing* the probability of the other error. Thus, tests often exist for which, given an arbitrarily small constant α,

$$\text{Prob}\,\{\text{Type I error}\} \leq \alpha \tag{4.2.5a}$$

and

$$\text{Prob}\,\{\text{Type II error}\} \text{ is minimum.} \tag{4.2.5b}$$

Note that although the probability of Type II error is reduced to its smallest possible value, the minimum may still be a number close to one.

The fundamental result in the theory of hypothesis testing is the Neyman-Pearson Lemma [LE1,FI1]. This result states that if Ω_0 and Ω_1 consist of single points θ_0 and θ_1, respectively, then a test with the properties specified by (4.2.5) to test $H_0 : \theta = \theta_0$ against $H_1 : \theta = \theta_1$ always exists and is given by the following instructions:

reject H_0 if

$$P_{\theta_1}(\mathbf{x}) > \hat{K}P_{\theta_0}(\mathbf{x}); \tag{4.2.6a}$$

accept H_0 if

$$P_{\theta_1}(\mathbf{x}) < \hat{K}P_{\theta_0}(\mathbf{x}); \tag{4.2.6b}$$

reject H_0 with probability γ if

$$P_{\theta_1}(\mathbf{x}) = \hat{K}P_{\theta_0}(\mathbf{x}), \tag{4.2.6c}$$

where $P_{\theta_0}(\mathbf{x})$ is the probability density of $\mathbf{X}$ evaluated at the sample point $\mathbf{x}$ under the condition that $\theta = \theta_0; P_{\theta_1}(\mathbf{x})$ is the same probability density evaluated under the condition that $\theta = \theta_1$ and $\hat{K}$ and γ are constants obtainable from the equation

$$\text{Prob}\,\{\text{rejecting } H_0 | H_0 \text{ is true}\} = \alpha. \tag{4.2.6d}$$

Such a test is called a *Most Powerful Level α Test*. Note that if X is a continuous random variable, (4.2.6c) occurs with zero probability and so we can set γ equal to any value including zero or one.

Example 4.2.1 Let X be a normal random variable with mean θ and unity variance. Let H_0 be the hypothesis $\theta = 0$ and H_1 be the alternate $\theta = 1$. Let $X_1, \ldots, X_K$ be the independent random variables with the same probability distribution as X, such that X_i represents the random outcome of the ith sample. Let x_i denote a particular value of X_i, for $i = 1, \ldots, K$, and let $\mathbf{x} = (x_1, \ldots, x_K)$:

$$P_{\theta_0}(\mathbf{x}) = \prod_{i=1}^{K} \frac{1}{\sqrt{2\pi}} e^{-\frac{1}{2}(x_i - 0)^2} \tag{4.2.7a}$$

and

$$P_{\theta_1}(\mathbf{x}) = \prod_{i=1}^{K} \frac{1}{\sqrt{2\pi}} e^{-\frac{1}{2}(x_i - 1)^2}. \tag{4.2.7b}$$

According to the Neyman-Pearson Lemma, the statistical test with the properties specified by (4.2.5a) and (4.2.5b) rejects H_0 if

$$\prod_{i=1}^{K} \frac{1}{\sqrt{2\pi}} e^{-\frac{1}{2}(x_i - 1)^2} > \hat{K} \prod_{i=1}^{K} \frac{1}{\sqrt{2\pi}} e^{-\frac{1}{2}x_i^2}. \tag{4.2.8}$$

Simplifying (4.2.8) gives

reject H_0 if

$$\exp\left(-\frac{1}{2}\left[\sum_{i=1}^{K} (x_i - 1)^2 - x_i^2\right]\right) > \hat{K}; \tag{4.2.9}$$

or

$$\exp\left(-\frac{1}{2}\left[\sum_{i=1}^{K} x_i^2 - 2x_i + 1 - x_i^2\right]\right) = \exp\left(-\frac{1}{2}\left[\left(2\sum_{i=1}^{K} - x_i\right) + K\right]\right) > \hat{K}. \tag{4.2.10}$$

Further simplification can be obtained by taking the natural logarithm of both sides of (4.2.10). Thus we reject H_0 if

$$\sum_{i=1}^{K} x_i - \tfrac{1}{2}K > \ln \hat{K}. \tag{4.2.11}$$

Let $K^* = \ln \hat{K} + K/2$. Then the Neyman-Pearson Lemma results in the following test:

reject H_0 if

$$\sum_{i=1}^{K} x_i > K^*; \tag{4.2.12a}$$

accept H_0 if

$$\sum_{i=1}^{K} x_i < K^*, \tag{4.2.12b}$$

and by (4.2.6d), K^* is found from the equation

$$\int_D P_{\theta=0}(\mathbf{x})\,\mathbf{dx} = \alpha, \tag{4.2.12c}$$

where $\mathbf{dx} = dx_1\, dx_2 \ldots dx_K$ and D is the region in the K dimensional space of $(x_1, \ldots, x_K)$ defined by

$$\sum_{i=1}^{K} x_i > K^*.$$

Finding K^* by means of (4.2.12c) is complicated. Fortunately, there is another procedure which we can use. Note that if $\theta = 0$, the random variable

$$\sum_{i=1}^{K} X_i$$

is the sum of K independent normal random variables with zero means and unity variance. Thus,

$$\sum_{i=1}^{K} X_i$$

is a normal random variable with zero mean and variance K. Therefore the integral on the left-hand side of (4.2.12c) can be written as

$$\int_{K*}^{\infty} \frac{1}{\sqrt{2\pi K}} e^{-\frac{1}{2}z^2/K}\, dz \tag{4.2.13}$$

and so K^* is obtained from the tables of the normal variable from the equation

$$\int_{K*}^{\infty} \frac{1}{\sqrt{2\pi K}} e^{-\frac{1}{2}z^2/K}\, dz = \alpha. \tag{4.2.14}$$

The main result of the theory of hypothesis testing is the fundamental lemma of Neyman and Pearson, which states that if both the hypothesis and alternative are *simple* (Ω_0 and Ω_1 contain only single points), a "good" statistical test can be found. In certain cases, "good" tests can be found when H_0 and H_1 are *composite*, that is, Ω_0 and Ω_1 each contain more than a single point.

We will postpone discussion of composite hypotheses until we encounter specific testing problems related to the maximum flow problem.

4.3 PSEUDOPARAMETRIC TERMINAL CAPACITY ANALYSIS

Let G be a graph with n vertices $v_1, \ldots, v_n$ and m branches $b_1, \ldots, b_m$. Furthermore, let c_i represent the capacity of branch b_i for $i = 1, \ldots, m$. If f_i is the present flow in branch b_i, the amount of additional flow that the branch can accommodate is clearly $c_i - f_i$. Consequently, if $f_1, \ldots, f_m$ are fixed numbers, the problem of finding the maximum flow between v_s and v_t is equivalent to finding the maximum flow between v_s and v_t in the graph G^* where G^* has the same structure as G, and the capacity of branch b_i^* of G^* is $c_i - f_i$ for $i = 1, \ldots, m$. If the numbers $f_1, f_2, \ldots, f_m$ are actually realizations of the random variables $F_1, F_2, \ldots, F_m$ the graph G^* consists of branches with random capacities $C_1 = c_1 - F_1,\ C_2 = c_2 - F_2, \ldots, C_m = c_m - F_m$. Let $\mathbf{c}$, $\mathbf{F}$, and $\mathbf{C}$ be defined by

$$\mathbf{c} = \begin{bmatrix} c_1 \\ c_2 \\ \cdot \\ \cdot \\ \cdot \\ c_m \end{bmatrix} \qquad \mathbf{F} = \begin{bmatrix} F_1 \\ F_2 \\ \cdot \\ \cdot \\ \cdot \\ F_m \end{bmatrix} \qquad \mathbf{C} = \begin{bmatrix} C_1 \\ C_2 \\ \cdot \\ \cdot \\ \cdot \\ C_m \end{bmatrix} \tag{4.3.1}$$

Clearly, $\mathbf{C} = \mathbf{c} - \mathbf{F}$.

Assume that the branch flows are random variables with *unknown* probability distributions but that a set of K samples of these flows is available. Thus, $\mathbf{F}(k) = (f_1(k), f_2(k), \ldots, f_m(k))'$ is a measurement of the existing flows in branches $b_1, b_2, \ldots, b_m$ at time k, where $k = 1,2,\ldots,K$. Although the distributions of the $\mathbf{F}(k)$ are unknown, we will require that the $\mathbf{F}(k)$ be identically and independently distributed variables. The independence assumption is assured if the samples are taken at sufficiently large intervals of time, while the common distribution assumption follows if the flows are stationary.[1] From the above notions, we may compute the random capacity vector $\mathbf{C}(k)$ at time k by

$$\mathbf{C}(k) = \mathbf{c} - \mathbf{F}(k) = \begin{bmatrix} c_1 \\ c_2 \\ \cdot \\ \cdot \\ \cdot \\ c_m \end{bmatrix} - \begin{bmatrix} f_1(k) \\ f_2(k) \\ \cdot \\ \cdot \\ \cdot \\ f_m(k) \end{bmatrix}. \tag{4.3.2}$$

We would like to find the probability that the maximum flow between terminals v_s and v_t is at least $r_{s,t}$.

From the Max-Flow Min-Cut Theorem, the maximum flow at time k is equal to the value of the minimum cut $\tau_{s,t}(k)$ at the time k. By definition,

$$\begin{aligned} \tau_{s,t}(k) &= \text{Min}\,[c(A^1_{s,t}(k)), \ldots, c(A^q_{s,t}(k))] \\ &= \underset{\text{over rows}}{\text{Min}} \{\mathscr{A}_{s,t}\mathbf{C}(k)\} \\ &= \underset{\text{over rows}}{\text{Min}} \{\mathscr{A}_{s,t}(\mathbf{c} - \mathbf{F}(k))\}, \end{aligned} \tag{4.3.3}$$

where $A^1_{s,t}, \ldots, A^q_{s,t}$ are the cut-sets which separate v_s and v_t, $c(A^1_{s,t}(k)), \ldots, c(A^q_{s,t}(k))$ are their values at time k, and $\mathscr{A}_{s,t} = [a_{ij}]$ is the s-t cut-set matrix.

Since only a set of samples is available, we use a statistical testing procedure. Usually, we wish to determine if Prob $\{\tau_{s,t} \geq r_{s,t}\}$ is at least some quantity p_0 for $0 < p_0 < 1$. Consequently, we pose the following statistical problem: test the hypothesis

$$H_0 : p = \text{Prob}\,\{\tau_{s,t} \geq r_{s,t}\} \geq p_0 \tag{4.3.4a}$$

[1] The independence assumption is not severe. However, for many systems, the flows might not be stationary. For example, in a telephone system there are abrupt price rate changes. Clearly, the probability distributions of flow before and after these rate changes are not the same.

against the alternative

$$H_1 : p < p_0 \tag{4.3.4b}$$

at *confidence level* α, where α is the probability of Type I error.

Since $\mathbf{F}(1), \ldots, \mathbf{F}(K)$ are assumed to be identically and independently distributed, it is easy to see that the random variables which generate the observations $\tau_{s,t}(1)$, $\tau_{s,t}(2), \ldots, \tau_{s,t}(K)$ are also identically and independently distributed.

Both the hypothesis H_0 and the alternative H_1 are *composite*. Our goal now is to create a new hypothesis H_0' and alternative H_1' which are *simple* but closely related to H_0 and H_1. We will then give a most powerful test for H_0' and H_1'. This test will be structured so that we can return to H_0 and H_1 without disturbing the structure of the test.

The distribution of $\tau_{s,t}(k)$ is defined over the positive real line, and may be decomposed into two distributions by using the probability $p = \text{Prob}\,\{\tau_{s,t}(k) \geq r_{s,t}\}$ and the conditional probability densities $p_-(\tau_{s,t}(k))$ and $p_+(\tau_{s,t}(k))$ where $p_+(x)$ and $p_-(x)$ are defined by

$$p_+(x)\,dx = \text{Prob}\,\{x \leq \tau_{s,t}(k) < x + dx | \tau_{s,t} \geq r_{s,t}\}; \tag{4.3.5a}$$

$$p_-(x)\,dx = \text{Prob}\,\{x \leq \tau_{s,t}(k) < x + dx | \tau_{s,t} < r_{s,t}\}. \tag{4.3.5b}$$

Thus the probability that $x \leq \tau_{s,t}(k) \leq x + dx$ can be written in terms of conditional probability distributions as

$$\text{Prob}\,\{x \leq \tau_{s,t}(k) < x + dx\} = p_+(x)\,p dx + p_-(x)(1 - p)\,dx. \tag{4.3.6}$$

In terms of these conditional densities, the joint density of the maximum flows evaluated as a point $(\tau_{s,t}(1), \ldots, \tau_{s,t}(K))$ satisfying

$$\tau_{s,t}(i_1), \ldots, \tau_{s,t}(i_N) < r_{s,t} \leq \tau_{s,t}(j_1), \ldots, \tau_{s,t}(j_{K-N})$$

is

$$p^{K-N}(1 - p)^N p_-(\tau_{s,t}(i_1)) \ldots p_-(\tau_{s,t}(i_N)) p_+(\tau_{s,t}(j_1)) \ldots p_+(\tau_{s,t}(j_{K-N})), \tag{4.3.7}$$

since the sample variables are independent.

We now replace the composite hypotheses by simple hypotheses H_0' and H_1'. The exact justification for this replacement stems from the theory of *least favorable* distributions [LE1] and will not be discussed here.

Consider any fixed alternative, H_1', with $p = p_1$ and replace H_0 by H_0'; $p = p_0$. Let the density of $\tau_{s,t}$ under the hypothesis H_0' be denoted (p_0, p_+, p_-). The fixed alternative with $p = p_1$ which appears *most like* the hypothesis H_0' is H_1': the density of $\tau_{s,t}$ is (p_1, p_+, p_-) where p_1 is a nonzero probability less than p_0. Under these conditions the statistical hypotheses to be tested are H_0': the density of $\tau_{s,t}$ is characterized by (p_0, p_+, p_-) against H_1': the density of $\tau_{s,t}$ is characterized by (p_1, p_+, p_-). Thus, we are testing a *simple* hypothesis against a *simple* alternative and the Neyman-Pearson Lemma may be applied.

According to the Neyman-Pearson Lemma we take the ratio of the quantity given in (4.3.7) with $p = p_1$ to the same quantity with $p = p_0$. Note that the p_+ and p_- factors cancel in this ratio. Then, a *Most Powerful* test for testing H_0' against H_1'

at *level* α is

reject the hypothesis H_0' if

$$\left(\frac{p_1}{p_0}\right)^{K-N}\left(\frac{1-p_1}{1-p_0}\right)^N > \hat{K}, \tag{4.3.8a}$$

reject the hypothesis H_0' with probability γ if

$$\left(\frac{p_1}{p_0}\right)^{K-N}\left(\frac{1-p_1}{1-p_0}\right)^N = \hat{K}, \tag{4.3.8b}$$

accept the hypothesis H_0' if

$$\left(\frac{p_1}{p_0}\right)^{K-N}\left(\frac{1-p_1}{1-p_0}\right)^N < \hat{K}, \tag{4.3.8c}$$

where $\hat{K}$ and γ are constants determined by

$$\text{Prob}\left\{\left(\frac{p_1}{p_0}\right)^{K-N}\left(\frac{1-p_1}{1-p_0}\right)^N > \hat{K}\right\} + \gamma\,\text{Prob}\left\{\left(\frac{p_1}{p_0}\right)^{K-N}\left(\frac{1-p_1}{1-p_0}\right)^N = \hat{K}\right\} = \alpha \tag{4.3.8d}$$

and these probabilities are taken under the assumption that $p = p_0$ and α is the prespecified probability of Type I error. In expression (4.3.8d), N, the number of the total of K observations for which $\tau_{s,t} < r_{s,t}$, is the random variable.

Note that the test can be expressed in another form by taking the logarithm of both sides of (4.3.8a) through (4.3.8c), as illustrated below with (4.3.8a):

$$(K - N)\ln\left(\frac{p_1}{p_0}\right) + \left[\ln\left(\frac{1-p_1}{1-p_0}\right)\right]N > \ln\hat{K} \tag{4.3.9}$$

$$N\left[\ln\left(\frac{p_0}{p_1}\right) + \ln\left(\frac{1-p_1}{1-p_0}\right)\right] > \ln\hat{K} + K\ln\left(\frac{p_0}{p_1}\right) = K'. \tag{4.3.10}$$

Since the coefficient of N on the left side of (4.3.10) is positive, we obtain the following equivalent test:[2]

reject the hypothesis H_0' if $N > \hat{K}$, (4.3.11a)

reject the hypothesis H_0' with probability γ if $N = \hat{K}$, (4.3.11b)

accept the hypothesis H_0' if $N < \hat{K}$. (4.3.11c)

The constants $\hat{K}$ and γ are determined from the equation

$$\text{Prob}\,\{N > \hat{K}\} + \gamma\,\text{Prob}\,\{N = \hat{K}\} = \alpha \text{ for } p = p_0. \tag{4.3.12}$$

For the purposes of finding $\hat{K}$ and γ from (4.3.12), we assume that H_0' is true. It can be seen that N, under this assumption, is *binomially distributed* with parameters

[2] Here we have let $\hat{K} = K'[\ln(p_0/p_1) + \ln((1 - p_1)/(1 - p_0))]^{-1}$.

$1 - p_0$ and K, and does not depend on the true value of p. Thus, $\hat{K}$ and γ can be easily obtained from the tables of the binomial distribution. A test is *Uniformly Most Powerful* if it minimizes the probability of Type II error for all values of composite alternative. Since the test in (4.3.11) does not depend on the alternative H_1', it is not only Most Powerful for testing $p = p_0$ versus $p = p_1 < p_0$ but also *Uniformly* Most Powerful for testing $p = p_0$ against $p < p_0$.

Suppose now that the true value of the parameter is $p = p_2$, where $p_2 > p_0$. The probability of Type I error will still be less than α since the error of the test is a decreasing function of p and under the original hypothesis $H_0 : p \geq p_0$, it assumes its greatest value at p_0. This is a sufficient condition for the test to be *Uniformly Most Powerful* for testing the hypothesis $H_0 : p \geq p_0$ against $H_1 : p < p_0$ [LE1].

Thus, we have obtained a statistical test for testing the hypothesis that a flow rate of at least $r_{s,t}$ units can be attained in the network with at least probability p_0. The test is entirely consistent with physical reasoning and involves examining the number of times that the flow $r_{s,t}$ cannot be achieved in the statistical sample. If this number is "too large," it may be concluded that the probability p_0 is not a realistic estimate of the actual probability. The concept of "too large" is made exact by the nature of the test, which is extremely simple to apply. The test is optimum in the sense that the probability of accepting the hypothesis when it is false is minimized.

Example 4.3.1 Consider a graph G with fixed branch capacities but random branch flows. We want to determine whether a flow rate of one unit can be attained between the pair of vertices v_s and v_t at least 80 percent of the time. Thus, if we have integer capacities we want to find the probability that there is at least one unsaturated s-t path. Suppose that we have no knowledge of the probability distributions of the random branch flows, but that a record of past flows is available. Then, we can use the statistical procedure to test whether Prob $\{\tau_{s,t} \geq 1\} \geq 0.8$. Suppose that the consequences of deciding that the probability is small when it is actually large are more costly than the consequences of the opposite error (deciding that the probability is large when it actually is small). For example, the former decision may result in a major modification of the system while the latter decision will result in no changes. We must then bound the probability of the first error, and hence test

$$H_0 : p = \text{Prob}\,\{\tau_{s,t} \geq 1\} \geq 0.8 \tag{4.3.13a}$$

against

$$H_1 : p < 0.8 \tag{4.3.13b}$$

at level α. Let $\alpha = 0.1$. Then, the best test is given by (4.3.11a) to (4.3.11c), and we must find $\hat{K}$ and γ.

Let $N = X_1 + X_2 + \cdots + X_K$ where $X_i = 0$ if $\tau_{s,t}(i) \geq 1$ and $X_i = 1$ if $\tau_{s,t}(i) < 1$. Clearly, for each i, $X_i = 0$ with probability p and $X_i = 1$ with probability $1 - p$. Therefore N is a binomial variable with parameters $1 - p$ and K. Hence

$$\text{Prob}\,\{N > \hat{K}\} = \sum_{k=\hat{K}+1}^{K} \binom{K}{k} (1 - p)^k p^{K-k}, \tag{4.3.14}$$

and

$$\text{Prob}\,\{N = \hat{K}\} = \binom{K}{\hat{K}}(1 - p)^{\hat{K}}p^{K-\hat{K}}. \tag{4.3.15}$$

We must now find $\hat{K}$ and γ such that

$$\sum_{k=\hat{K}+1}^{K}\binom{K}{k}(1 - p)^{k}p^{K-k} + \gamma\binom{K}{\hat{K}}(1 - p)^{\hat{K}}p^{K-\hat{K}} = 0.1 \qquad \text{for } p = 0.8. \tag{4.3.16}$$

To solve this equation, we must specify the sample size K. Let $K = 25$. We can solve (4.3.16) by using the Poisson approximation to the binomial distribution [FI1]. In other words, we let

$$\text{Prob}\,\{N = k\} = \frac{[25(1 - 0.8)]^k}{k!}e^{-25(1-0.8)} = \frac{5^k}{k!}e^{-5} \tag{4.3.17}$$

and from a standard table of the Poisson variable we find that

$$\text{Prob}\,\{N \geq 7\} = 0.14 \tag{4.3.18a}$$

and

$$\text{Prob}\,\{N \geq 8\} = 0.07. \tag{4.3.18b}$$

Therefore, $\hat{K} = 7$. To find γ solve the equation

$$0.07 + \gamma\,\text{Prob}\,\{N = 7\} = 0.1. \tag{4.3.19}$$

Since Prob $\{N = 7\}$ = Prob $\{N \geq 7\}$ − Prob $\{N \geq 8\} = 0.07$, γ is easily found to be 0.43. Thus, the Uniformly Most Powerful Test at level 0.1 is:

reject H_0 if $N > 7$; (4.3.20a)

reject H_0 with probability 0.43 if $N = 7$; (4.3.20b)

accept H_0 if $N < 7$. (4.3.20c)

It should be noted that although the probability of making a Type II error is minimum, it still may be large. The magnitude of this probability is a function of both the sample size and the true value of the parameter, and will decrease as the sample size becomes larger and the true value of p becomes smaller. If the cost of making a Type II error is greater than the cost of making a Type I error, it would be advisable to formulate the dual statistical testing problem; $H_0 : p \leq p_0$ versus $H_1 : p > p_0$. The details of the solution of the dual problem are similar to the test discussed here and will not be considered further.

4.4 MAXIMUM FLOW RATE PROBABILITY FOR ARBITRARILY DISTRIBUTED BRANCH CAPACITIES

Let G be an n-vertex, m-branch graph. As before, in each branch b_i of G there is a random flow F_i. We now assume that F_i is a continuous random variable with *known* probability density $p_i^*(f_i)$. Thus

$$\text{Prob}\,\{F_i \geq x\} = 1 - \int_0^x p_i^*(y)\,dy. \tag{4.4.1}$$

With each branch of the graph G^* we can associate a random capacity $C_i = c_i - F_i$. Let the probability density function of C_i be $p_i(x_i)$. Given p_i^*, we can easily calculate $p_i(x)$ by means of the equation

$$p_i(x) = p_i^*(c_i - x) \qquad \text{for } i = 1,\ldots,m. \tag{4.4.2}$$

Then, the probability of an attainable flow rate x_i in branch b_i is

$$\text{Prob}\,\{C_i \geq x_i\} = 1 - \int_0^{x_i} p_i(y)\,dy = \int_0^{c_i - x_i} p_i^*(y)\,dy. \tag{4.4.3}$$

We now give a procedure to find the probability that a flow rate $\tau_{s,t} \geq r_{s,t}$ between v_s and v_t can be attained in G. We do not propose this procedure as a practical computational technique. It is valuable, however, since it indicates some of the difficulties associated with the computation of exact probabilities of maximum flow rate.

From the Max-Flow Min-Cut Theorem, a flow of value $r_{s,t}$ is attainable in G if and only if

$$\underset{k}{\text{Min}}\,[c(A_{s,t}^k)] \geq r_{s,t}, \tag{4.4.4}$$

where $c(A_{s,t}^k)$ is the value of cut-set $A_{s,t}^k$ obtained by adding the capacities of the branches in $A_{s,t}^k$. Since capacities are random variables, $c(A_{s,t}^k)$ as well as $\underset{k}{\text{Min}}\,[c(A_{s,t}^k)]$ is a random variable. The problem is to compute

$$\text{Prob}\,\{\text{Min}[c(A_{s,t}^k)] \geq r_{s,t}\} = \text{Prob}\,\{c(A_{s,t}^1) \geq r_{s,t}, c(A_{s,t}^2) \geq r_{s,t}, \ldots, c(A_{s,t}^q) \geq r_{s,t}\}. \tag{4.4.5}$$

To find this probability, we need the probability density of the random variable $\underset{k}{\text{Min}}\,[c(A_{s,t}^k)]$. Since two s-t cut-sets may have branches in common, the $c(A_{s,t}^k)$ are not independent random variables. This is the main difficulty that arises, and here it will be resolved by the following method:

Let $p(a_1,a_2,\ldots,a_q)$ be the joint probability density of variables $c(A_{s,t}^1)$, $c(A_{s,t}^2),\ldots,c(A_{s,t}^q)$. In other words,

$$p(a_1,a_2,\ldots,a_q)\,da_1\,da_2\ldots da_q = \text{Prob}\,\{a_1 \leq c(A_{s,t}^1) < a_1 + da_1,\, a_2 \leq c(A_{s,t}^2) < a_2 + da_2, \ldots, a_q \leq c(A_{s,t}^q) < a_q + da_q\}. \tag{4.4.6}$$

To find the density of $\underset{k}{\text{Min}}\,[c(A_{s,t}^k)]$, we first compute $p(a_1,a_2,\ldots,a_q)$. Let $|\mathbf{A}| = [c(A_{s,t}^1),c(A_{s,t}^2),\ldots,c(A_{s,t}^q)]'$ be a random cut-set vector and let $\mathbf{a} = [a_1,a_2,\ldots,a_q]'$ be a particular value of $|\mathbf{A}|$. Similarly let $\mathbf{C} = [C_1,C_2,\ldots,C_m]'$ be a random capacity vector and $\tilde{\mathbf{c}} = [\tilde{c}_1,\tilde{c}_2,\ldots,\tilde{c}_m]'$ be a particular value of $\mathbf{C}$. Given $\tilde{\mathbf{c}}$ (or $\mathbf{C}$), we can calculate **a** (or $|\mathbf{A}|$) as follows:

$$\mathbf{a} = \mathscr{A}_{s,t}\tilde{\mathbf{c}} \text{ (or } |\mathbf{A}| = \mathscr{A}_{s,t}\mathbf{C}), \tag{4.4.7}$$

where $\mathscr{A}_{s,t} = [a_{i,j}]$ is the $q \times m$ s-t cut-set matrix. To find the probability density of $|\mathbf{A}|$, we first compute the characteristic function[3] of the random vector **C**. Let $p(\tilde{c}_1,\tilde{c}_2,\ldots,\tilde{c}_m)$ be the joint probability density of **C**. For $\boldsymbol{\alpha} = (\alpha_1,\alpha_2,\ldots,\alpha_m)'$ and $\mathbf{d\tilde{c}} = d\tilde{c}_1 \ldots d\tilde{c}_m$, let $H(\alpha_1,\alpha_2,\ldots,\alpha_m)$ be the characteristic function of **C** defined by

$$H(\alpha_1,\alpha_2,\ldots,\alpha_m) = \int_0^\infty \int_0^\infty \cdots \int_0^\infty \exp\left(\sum_{i=1}^m \sqrt{-1}\,\alpha_i\tilde{c}_i\right) p(\tilde{c}_1,\tilde{c}_2,\ldots,\tilde{c}_m)\,\mathbf{d\tilde{c}}$$

$$= E\{\exp(\sqrt{-1}\,\boldsymbol{\alpha}'\mathbf{C})\} = H(\boldsymbol{\alpha}'). \tag{4.4.8}$$

If $C_1,C_2,\ldots,C_m$ are independent random variables, then $p(\tilde{c}_1,\tilde{c}_2,\ldots,\tilde{c}_m) = p_1(\tilde{c}_1)\cdot p_2(\tilde{c}_2)\ldots p_m(\tilde{c}_m)$, and computational difficulties are reduced since then $H(\boldsymbol{\alpha}')$ is

$$H(\alpha_1,\alpha_2,\ldots,\alpha_m) = E\{\exp(\sqrt{-1}\,\alpha_1 C_1)\}E\{\exp(\sqrt{-1}\,\alpha_2 C_2)\}\cdots E\{\exp(\sqrt{-1}\,\alpha_m C_m)\}. \tag{4.4.9}$$

Let the characteristic function of the random vector $|\mathbf{A}|$ be denoted by $\Psi(\boldsymbol{\beta}')$, where $\boldsymbol{\beta} = (\beta_1,\beta_2,\beta_3,\ldots,\beta_q)'$. Then

$$\Psi(\boldsymbol{\beta}') = E\{\exp(\sqrt{-1}\,\boldsymbol{\beta}'|\mathbf{A}|)\}. \tag{4.4.10}$$

The operators E in (4.4.8) and (4.4.10) are expectation operators with respect to different probability densities. The expectation of a function of a random variable can be computed as the integral of the product of that function and the density of the *original variable* [CR1]. Hence, using (4.4.7) we obtain

$$\Psi(\boldsymbol{\beta}') = E\{\exp(\sqrt{-1}\,\boldsymbol{\beta}'\mathscr{A}_{s,t}\mathbf{C})\}$$

$$= \int_0^\infty \cdots \int_0^\infty \exp(\sqrt{-1}\,\boldsymbol{\beta}'\mathscr{A}_{s,t}\tilde{\mathbf{c}})p(\tilde{c}_1,\tilde{c}_2,\ldots,\tilde{c}_m)\,\mathbf{d\tilde{c}}. \tag{4.4.11}$$

Therefore, according to (4.4.11), $\Psi(\boldsymbol{\beta}')$ can be calculated by

$$\Psi(\boldsymbol{\beta}') = H(\boldsymbol{\beta}'\mathscr{A}_{s,t}). \tag{4.4.12}$$

Given the characteristic function of $|\mathbf{A}|$, we can find the probability density $p(\mathbf{a}')$ of $|A|$ by inverting $\Psi(\boldsymbol{\beta}')$. At this point we consider a simple example.

---

[3] The characteristic function of the random vector $\mathbf{X} = (X_1,\ldots,X_k)$ is defined as $\Psi(u_1,\ldots,u_k) = E\{\exp(\sqrt{-1}(u_1X_1 + u_2X_2 + \cdots + u_kX_k))\}$ where E is the expectation operator. Among the many important properties of characteristic functions are: (1) if $X_1,\ldots,X_k$ are independent, then $\Psi(u_1,\ldots,u_k) = (E\{\exp(\sqrt{-1}u_1X_1)\})\,(E\{\exp(\sqrt{-1}u_2X_2)\})\cdots(E\{\exp(\sqrt{-1}u_kX_k)\})$; and (2) a characteristic function uniquely determines its distribution. That is, given a characteristic function, we can uniquely determine the probability density of the random variable which generated it. For an excellent discussion of characteristic functions see Gnedenko [GN1].

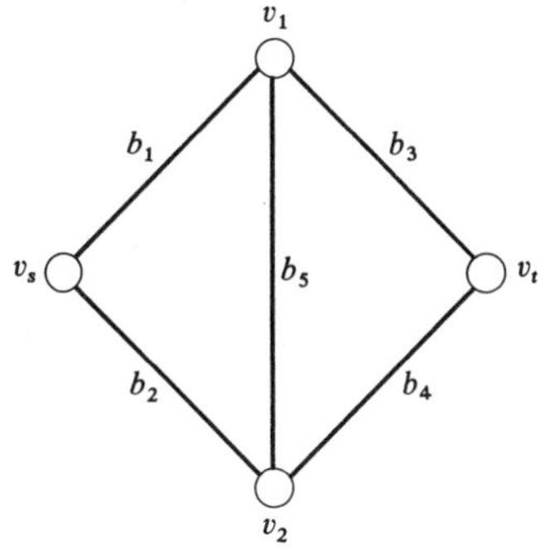

Fig. 4.4.1 Graph for Example 4.4.1.

Example 4.4.1 Consider the graph G shown in Fig. 4.4.1. Let the branch capacities of b_1,b_2,b_3,b_4,b_5 be identically and independently distributed random variables with a uniform distribution over the interval $[0,c_0]$ on the real line. In other words,

$$p_j(\tilde{c}) = \begin{cases} \dfrac{1}{c_0} & \text{for } 0 \leq \tilde{c} \leq c_0 \\ 0 & \text{otherwise} \end{cases} \qquad j = 1,\ldots,5. \tag{4.4.13}$$

The characteristic function of the random variable C_j is, for $c_0 = 1$,

$$\begin{aligned} E\{\exp\sqrt{-1}\,\alpha_j C_j\} &= \int_0^1 \exp(\sqrt{-1}\,\alpha_j\tilde{c}_j)p_j(\tilde{c}_j)\,d\tilde{c}_j \\ &= \frac{1}{\sqrt{-\alpha_j}}[\exp(\sqrt{-1}\,\alpha_j) - 1] \qquad \text{for } j = 1,\ldots,5. \end{aligned} \tag{4.4.14}$$

Therefore, the characteristic function of $\mathbf{C}$ is

$$H(\alpha_1,\alpha_2,\alpha_3,\alpha_4,\alpha_5) = \frac{1}{(\sqrt{-1})^5\alpha_1\alpha_2\alpha_3\alpha_4\alpha_5}\prod_{j=1}^{5}[\exp(\sqrt{-1}\,\alpha_j) - 1]. \tag{4.4.15}$$

To find the characteristic function of the cut-sets we must first find the s-t cut-set matrix $\mathscr{A}_{s,t}$. Since G is small, it is easy to see that if column i corresponds to b_i

$$\mathscr{A}_{s,t} = \begin{bmatrix} 1 & 1 & 0 & 0 & 0 \\ 0 & 0 & 1 & 1 & 0 \\ 1 & 0 & 0 & 1 & 1 \\ 0 & 1 & 1 & 0 & 1 \end{bmatrix}. \tag{4.4.16}$$

By (4.4.12) we let $\boldsymbol{\alpha}' = \boldsymbol{\beta}'\mathscr{A}_{s,t}$ and so $\boldsymbol{\alpha} = \mathscr{A}'_{s,t}\boldsymbol{\beta}$ or

$$\begin{bmatrix} \alpha_1 \\ \alpha_2 \\ \alpha_3 \\ \alpha_4 \\ \alpha_5 \end{bmatrix} = \begin{bmatrix} 1 & 0 & 1 & 0 \\ 1 & 0 & 0 & 1 \\ 0 & 1 & 0 & 1 \\ 0 & 1 & 1 & 0 \\ 0 & 0 & 1 & 1 \end{bmatrix} \begin{bmatrix} \beta_1 \\ \beta_2 \\ \beta_3 \\ \beta_4 \end{bmatrix} = \begin{bmatrix} \beta_1 + \beta_3 \\ \beta_1 + \beta_4 \\ \beta_2 + \beta_4 \\ \beta_2 + \beta_3 \\ \beta_3 + \beta_4 \end{bmatrix}. \tag{4.4.17}$$

The characteristic function of the cut-sets is thus

$$\Psi(\beta_1,\beta_2,\beta_3,\beta_4) = \frac{-\sqrt{-1}}{(\beta_1+\beta_3)(\beta_1+\beta_4)(\beta_2+\beta_4)(\beta_2+\beta_3)(\beta_3+\beta_4)} \cdot [\exp(\sqrt{-1}(\beta_1+\beta_3)) - 1][\exp(\sqrt{-1}(\beta_1+\beta_4)) - 1] \cdot [\exp(\sqrt{-1}(\beta_2+\beta_4)) - 1][\exp(\sqrt{-1}(\beta_2+\beta_3)) - 1] \cdot [\exp(\sqrt{-1}(\beta_3+\beta_4)) - 1]. \tag{4.4.18}$$

Now that we have found $\Psi(\boldsymbol{\beta}')$, the density function of the cut-sets has, in principle, been found. Specifically, $p(a_1,a_2,a_3,a_4)$ is given by

$$p(a_1,a_2,a_3,a_4) = \frac{1}{(2\pi)^4}\int_{-\infty}^{\infty}\int_{-\infty}^{\infty}\int_{-\infty}^{\infty}\int_{-\infty}^{\infty} \Psi(\beta_1,\beta_2,\beta_3,\beta_4)\exp\left(-\sqrt{-1}\sum_{i=1}^{4}\beta_i a_i\right) \mathbf{d\beta} \tag{4.4.19}$$

and Prob $\{\tau_{s,t} \geq r_{s,t}\}$ is

$$\text{Prob}\,\{\tau_{s,t} \geq r_{s,t}\} = \int_{r_{s,t}}^{\infty}\int_{r_{s,t}}^{\infty}\int_{r_{s,t}}^{\infty}\int_{r_{s,t}}^{\infty} p(a_1,a_2,a_3,a_4)\, da_1\, da_2\, da_3\, da_4. \tag{4.4.20}$$

Now, we return to the original problem, which was to calculate Prob $\{\text{Min}\,[c(A_{s,t}^k)] \geq r_{s,t}\}$. Since Prob $\{\tau_{s,t} \geq r_{s,t}\}$ = Prob $\{c(A_{s,t}^1) \geq r_{s,t},\ c(A_{s,t}^2) \geq r_{s,t},\ldots,$ $c(A_{s,t}^q) \geq r_{s,t}\}$ we can compute this probability by means of the integrals

$$p(a_1,a_2,\ldots,a_q) = \frac{1}{(2\pi)^q}\int_{-\infty}^{\infty}\int_{-\infty}^{\infty}\cdots \int_{-\infty}^{\infty} \Psi(\boldsymbol{\beta}')\exp(-\sqrt{-1}\,\boldsymbol{\beta}'\mathbf{a})\, d\beta_1\, d\beta_2 \cdots d\beta_q \tag{4.4.21a}$$

and

$$\text{Prob}\,\{\tau_{s,t} \geq r_{s,t}\} = \int_{r_{s,t}}^{\infty}\int_{r_{s,t}}^{\infty}\cdots\int_{r_{s,t}}^{\infty} p(a_1,a_2,\ldots,a_q)\, da_1\, da_2 \cdots da_q. \tag{4.4.21b}$$

However, (4.4.21a) and (4.4.21b) may still be inadequate. Although the characteristic function of $(c(A_{s,t}^1),\ldots,c(A_{s,t}^q))$ is well defined and uniquely defines a density function, the corresponding density function may be degenerate. For example, if X_1 and X_2 are two random variables such that $X_2 = cX_1$ where $c \neq 0$ is a constant, $p(x_1,x_2) = 0$ if $x_2 \neq cx_1$ and $p(x_1,x_2) = p(x_1)$ if $x_2 = cx_1$. In other words, the entire "mass" of the probability density is concentrated on the line $x_2 = cx_1$ of the two-dimensional (x_1,x_2) plane. In general, if $X_1,\ldots,X_q$ are the random variables in question, the characteristic function will define a degenerate or *singular* distribution if and only if with probability 1 there exists linear relationships among the X_i. That is, if and only if, with probability 1, at least one X_i can be written as a linear combination of the others. To see that the characteristic function of the cut-set vector could generate a singular distribution, consider the following example.

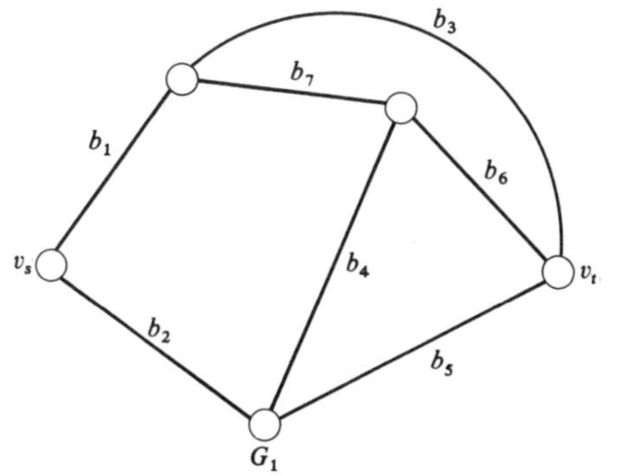

Fig. 4.4.2 Graph for Example 4.4.2.

Example 4.4.2 Consider the graph G shown in Fig. 4.4.2. The cut-set matrix of G with respect to vertices v_s and v_t is

$$\mathscr{A}_{s,t} = \begin{bmatrix} 1 & 1 & 0 & 0 & 0 & 0 & 0 \\ 1 & 0 & 0 & 1 & 1 & 0 & 0 \\ 0 & 1 & 1 & 0 & 0 & 0 & 1 \\ 1 & 0 & 0 & 0 & 1 & 1 & 1 \\ 0 & 1 & 1 & 1 & 0 & 1 & 0 \\ 0 & 0 & 1 & 1 & 1 & 0 & 1 \\ 0 & 0 & 1 & 0 & 1 & 1 & 0 \end{bmatrix}. \tag{4.4.22}$$

The value of cut-set $A_{s,t}^6$ can be written as

$$\begin{aligned} c(A_{s,t}^6) &= C_3 + C_4 + C_5 + C_7 \\ &= C_1 + C_4 + C_5 + C_2 + C_3 + C_7 - (C_1 + C_2) \end{aligned} \tag{4.4.23a}$$

or

$$c(A_{s,t}^6) = c(A_{s,t}^2) + c(A_{s,t}^3) - c(A_{s,t}^1), \tag{4.4.23b}$$

since the sixth row of $\mathscr{A}_{s,t}$ can be obtained from the first three rows of $\mathscr{A}_{s,t}$ by adding rows 2 and 3 and then subtracting the first row. Thus, if any of the $c(A_{s,t}^i)$ are linear combinations of other cut-set values, the corresponding rows in $\mathscr{A}_{s,t}$ will be linear combinations of the other rows. In other words, if $|\mathbf{A}|$ has a singular probability distribution, the matrix $\mathscr{A}_{s,t}$ does not have full rank and vice versa.

Let the rank of $\mathscr{A}_{s,t}$ over the real field be q'. If $q' < q$, instead of finding the characteristic function of $c(A_{s,t}^1), \ldots, c(A_{s,t}^q)$, we can find the characteristic function of $c(A_{s,t}^{i_1}), \ldots, c(A_{s,t}^{i_q})$, where $\{c(A_{s,t}^{i_1}), \ldots, c(A_{s,t}^{i_q})\}$ is a maximal subset of $c(A_{s.t}^i)$ which are not linearly dependent. We may then write [GN1]

$$c(A_{s,t}^j) = \sum_{k=1}^{q'} \xi_{j,i_k} c(A_{s,t}^{i_k}) \tag{4.4.24}$$

where the ξ_{j,i_k} are constants found by Gauss reducing $\mathscr{A}_{s,t}$. The characteristic function of $c(A_{s,t}^{i_1}), \ldots, c(A_{s,t}^{i_q})$ is found by a trivial modification of the above procedure. We first find a maximal set of linearly independent $c(A_{s,t}^i)$, by finding a maximal set of independent rows of $\mathscr{A}_{s,t}$. To find the characteristic function of the $c(A_{s,t}^i)$, we then replace the matrix $\mathscr{A}_{s,t}$ in the above procedure by the matrix $\hat{\mathscr{A}}_{s,t}$ whose rows are the maximal set of independent rows of $\mathscr{A}_{s,t}$ already found. Prob $\{\tau_{s,t} \geq r_{s,t}\}$ is then given by the equation

$$\text{Prob}\,\{\tau_{s,t} \geq r_{s,t}\} = \int \cdots_{\Omega} \int p(a_{i_1}, a_{i_2}, \ldots, a_{i_{q,q'}})\, da_{i_1} \ldots da_{i_{q,q'}}, \tag{4.4.25}$$

where Ω is the region defined by the set of inequalities:

$$\begin{aligned} a_{i_j} &\geq r_{s,t} \qquad \text{for } j = 1,2,\ldots,q', \\ \sum_{k=1}^{q'} \zeta_{j,i_k} a_{i_k} &\geq r_{s,t} \qquad \text{for } j = 1,\ldots,q, \quad j \neq i_k, \quad k = 1,\ldots,q'. \end{aligned} \tag{4.4.26}$$

Note that since G has m branches and q s-t cut-sets, whenever there are more cut-sets than branches $\mathscr{A}_{s,t}$ cannot have full rank. For most graphs, $q >> m$ and so in (4.4.25) we must evaluate far fewer integrals than in (4.4.21). This computational saving is achieved at the expense of an exceedingly complicated domain of integration. Clearly, the generality of the assumptions in this section limits the possibility of simplifying the computational details needed to find practical solutions to the probabilistic flow problem. Thus, we have exhibited the difficulties that we encounter when we attempt to give *exact* answers to this analysis problem. In the following sections, we will limit the analysis to classes of networks about which more detailed information is assumed. In this way, we will determine assumptions which can be made about branch flows that will lead to reasonable techniques for computing Prob $\{\tau_{s,t} \geq r_{s,t}\}$.

4.5 PARAMETRIC TERMINAL CAPACITY ANALYSIS WITH NORMALLY DISTRIBUTED BRANCH FLOWS

If the probability distribution of the branch capacity vector is known, we have seen that although we can write the solution in closed form, the task of performing the necessary operations is formidable. Therefore, it is desirable to investigate any assumptions which might simplify the computational procedure.

One of the most common assumptions in probabilistic approaches is the normality or Gaussian assumption which is usually supported by physical observations or the Central Limit Theorem. In our problem, such a normality assumption is justifiable on the following grounds: (1) in the case of many systems including telephone networks, branch flow has been observed to be Poisson-distributed [FE1] which may be adequately approximated by a normal distribution; (2) branch flow is actually the sum of a large number of independent random variables since each branch flow is composed of contributions from a number of subscribers. These independent variables may be considered to assume two values, 0 and 1. It can be shown (Theorem 6.9.3, Ref. [FI1]) that the limit distribution of the standardized sum of the variables is normal and thus, if a large number of users have access to the same line, the flow distribution is approximately normal.

Before proceeding, we must realize the limitations of the normality assumption. If F_k is normally distributed with mean μ_k and variance σ_k^2, the probability that an additional flow x may be established through b_k is

$$\text{Prob}\,\{c_k - F_k \geq x\} = 1 - \int_0^x \frac{1}{\sqrt{2\pi}\,\sigma_k} \exp\left(-\frac{1}{2}\frac{[y - (c_k - \mu_k)]^2}{\sigma_k^2}\right) dy. \quad (4.5.1)$$

The physical constraint that flow F_k must be in the interval $[0,c_k]$ requires that the mean and variance of F_k be such that the "tails" of the approximation are negligible. In other words

$$\int_0^{c_k} \frac{1}{\sqrt{2\pi}\,\sigma_k} \exp\left\{-\frac{1}{2}\frac{(y - \mu_k)^2}{\sigma_k^2}\right\} dy \approx 1. \quad (4.5.2)$$

If the above equation does not hold, we must deal with *truncated* distributions (section 19.3 of reference [CR1]). The implications of this statement are that even though the *demand* on the line is normally distributed, the actual distribution of a branch flow may not be. This is the case when a large number of people attempt to use a line of small capacity. Although the number of attempts is normally distributed, we can expect the flow F_k always to be close to the capacity of the line.

Equation (4.5.2) requires that each branch of G have a high enough capacity to accommodate the existing branch flows with high probability. If this equation does not hold for some branch, it is impractical to attempt to route additional flow through that branch. Consequently, we can assume that all branches for which (4.5.2) does not hold are not in the graph G. This assumption also differentiates between the queuing problem mentioned in Section 1 and the problem we are considering here since in that case, if we had saturated branches, we would "wait" for them to become available. For graphs whose branches are characterized by (4.5.2), meaningful parameters are the "instantaneous flow rate" and the probability of attaining a given flow rate. In other words, Prob $\{\tau_{s,t} \geq r_{s,t}\}$ gives an indication of the capability of the graph to accommodate additional flow between vertices v_s and v_t.

To specify completely the probabilistic structure of the model, we must specify the relationships among $F_1,F_2,\ldots,F_m$. In this treatment, we assume the joint distribution of the multidimensional random variable $\mathbf{F} = (F_1,\ldots,F_m)'$ may be approximated by an m-dimensional nonsingular normal distribution. Precisely, a random vector variable $\mathbf{X} = (X_1,X_2,\ldots,X_m)$ is said to have a multidimensional nonsingular normal distribution, $N(\boldsymbol{\mu};\Sigma)$, with mean vector $\boldsymbol{\mu} = (\mu_1,\mu_2,\ldots,\mu_m)'$ and $m \times m$ variance-covariance matrix $\Sigma = [\sigma_{i,j}]$ if

$$\mu_i = E\{X_i\}, \qquad \text{for } i = 1,2,\ldots,m, \quad (4.5.3\text{a})$$

and

$$\sigma_{i,j} = E\{(X_i - \mu_i)(X_j - \mu_j)\} \qquad \text{for } i = 1,2,\ldots,m \quad j = 1,2,\ldots,m, \quad (4.5.3\text{b})$$

and the probability density of $X_1,\ldots,X_m$ is

$$p(x_1,x_2,\ldots,x_n) = \frac{1}{(2\pi)^{m/2}\sqrt{\det\Sigma}} \exp\left(-\tfrac{1}{2}(\mathbf{x} - \boldsymbol{\mu})'\,\Sigma^{-1}(\mathbf{x} - \boldsymbol{\mu})\right) \quad (4.5.4)$$

where $\det\Sigma$ represents the determinant of Σ and $\det\Sigma \neq 0$.

It is possible to show that any matrix Σ which satisfies (4.5.3b) is positive semidefinite. Moreover, it is positive definite unless with probability one at least one of the X_i can be written as a linear combination of the others. Therefore, assuming **F** has a *nonsingular* normal distribution is the same as assuming that no branch flow random variable can be written as a linear combination of the other branch flows. This assumption is easily eliminated. In fact, we recall that for the general analysis procedure discussed in Section 4, the *cut-set* variables usually have a singular distribution. We will see that a similar situation is encountered here.

The probability distribution of F is

$$p(\mathbf{f}') = p(f_1,\ldots,f_m) = \frac{1}{(2\pi)^{m/2}\sqrt{\det \Sigma}} \exp\left(-\tfrac{1}{2}(\mathbf{f}-\boldsymbol{\mu})'\,\Sigma^{-1}(\mathbf{f}-\boldsymbol{\mu})\right) \quad (4.5.5)$$

and since $\mathbf{C} = \mathbf{c} - \mathbf{F}$, the vector $\mathbf{C}$ has the distribution $N(\mathbf{c}-\boldsymbol{\mu},\Sigma)$. As in Section 4, we want to find Prob $\{\tau_{s,t} \geq r_{s,t}\}$. From the Max-Flow Min-Cut Theorem, $r_{s,t}$ is attainable if and only if

$$\text{Min}\left[c(A^1_{s,t}),\ldots,c(A^q_{s,t})\right] = \tau_{s,t} \geq r_{s,t}. \quad (4.5.6)$$

Clearly,

$$\text{Prob}\{\tau_{s,t} \geq r_{s,t}\} = \text{Prob}\{c(A^1_{s,t}) \geq r_{s,t}, c(A^2_{s,t}) \geq r_{s,t},\ldots,c(A^q_{s,t}) \geq r_{s,t}\}. \quad (4.5.7)$$

Thus, we must again compute the joint density of $(c(A^1_{s,t}),\ldots,c(A^q_{s,t}))' = |\mathbf{A}|$.

The random vectors $|\mathbf{A}|$ and $\mathbf{C}$ are related by the *s-t* cut-set matrix $\mathscr{A}_{s,t} = [a_{i,j}]$ through the relation

$$|\mathbf{A}| = \mathscr{A}_{s,t}\mathbf{C}. \quad (4.5.8)$$

Thus, $|\mathbf{A}|$ is obtained from $\mathbf{C}$ by means of a *linear* transformation. Linear transformations of normal random variables are themselves normal random variables.[4] Furthermore, since $\mathbf{C}$ has the distribution $N(\mathbf{c}-\boldsymbol{\mu},\Sigma)$, $|\mathbf{A}|$ has the distribution $N(\mathscr{A}_{s,t}(\mathbf{c}-\boldsymbol{\mu}), \mathscr{A}_{s,t}\Sigma\mathscr{A}'_{s,t})$. We can see this if we examine the characteristic function $H(\alpha_1,\ldots,\alpha_m)$ of $\mathbf{C}$:

$$\begin{aligned} H(\boldsymbol{\alpha}') = H(\alpha_1,\ldots,\alpha_m) &= E\left\{\exp\left(\sqrt{-1}\sum_{j=1}^{m}\alpha_j C_j\right)\right\} \\ &= \exp\left(\sqrt{-1}\,(\mathbf{c}-\boldsymbol{\mu})'\boldsymbol{\alpha} - \tfrac{1}{2}\boldsymbol{\alpha}'\Sigma\boldsymbol{\alpha}\right). \end{aligned} \quad (4.5.9)$$

Let $\psi(\beta_1,\ldots,\beta_q)$ be the characteristic function of $|\mathbf{A}|$.

$$\psi(\boldsymbol{\beta}') = \psi(\beta_1,\ldots,\beta_q) = E\left\{\exp\left(\sqrt{-1}\sum_{j=1}^{q}\beta_j c(A^j_{s,t})\right)\right\}. \quad (4.5.10)$$

---

[4] This gives us a further basis for our initial normality assumption. Even if the branch flows are not normal variables, the cut-set variables will, under mild conditions, tend to normal variables. Furthermore, the approximation will improve as the complexity of the network increases.

Since

$$\sum_{j=1}^{q} \beta_j c(A^j_{s,t}) = \sum_{j=1}^{q} \beta_j \sum_{k=1}^{m} a_{j,k} C_k, \tag{4.5.11}$$

it follows that

$$\psi(\beta_1, \ldots, \beta_q) = \exp\left(\sqrt{-1}\,(\mathscr{A}_{s,t}(\mathbf{c} - \boldsymbol{\mu})\boldsymbol{\beta} - \tfrac{1}{2}\boldsymbol{\beta}'\mathscr{A}_{s,t}\Sigma\mathscr{A}'_{s,t}\boldsymbol{\beta}\right). \tag{4.5.12}$$

This is the characteristic function of an $N(\mathscr{A}_{s,t}(\mathbf{c} - \boldsymbol{\mu}), \mathscr{A}_{s,t}\Sigma\mathscr{A}'_{s,t})$ random variable.

If the rank of $\mathscr{A}_{s,t}$ over the real field is q', and $q' < q$, the variance-covariance matrix $\mathscr{A}_{s,t}\Sigma\mathscr{A}'_{s,t}$ will be singular. This means that some of the cut-sets may be expressed as linear combinations of others. If $q' < q$, then $[\mathscr{A}_{s,t}\Sigma\mathscr{A}'_{s,t}]^{-1}$ does not exist. In this case, we can reorder the $A^k_{s,t}$'s such that $c(A^1_{s,t}), \ldots, c(A^{q'}_{s,t})$ are not linear combinations of each other. There exist constants $\xi_{k,q'+h}$ for $k = 1, \ldots, q'$ and $h = 1, \ldots, q - q'$ such that [GN1]

$$c(A^h_{s,t}) = \sum_{k=1}^{q'} \xi_{k,q'+h} c(A^k_{s,t}). \tag{4.5.13}$$

Let $\hat{\mathscr{A}}_{s,t}$ be a $q' \times m$ matrix with respect to the reordered $A^k_{s,t}$'s whose rows correspond to $A^1_{s,t}, A^2_{s,t}, \ldots, A^{q'}_{s,t}$. Then

$$\begin{bmatrix} c(A^1_{s,t}) \\ \cdot \\ \cdot \\ \cdot \\ c(A^{q'}_{s,t}) \end{bmatrix} = \hat{\mathscr{A}}_{s,t}\mathbf{C} \tag{4.5.14}$$

and we can write the probability that the maximum flow is at least $r_{s,t}$ as

$$\text{Prob}\,\{\tau_{s,t} \geq r_{s,t}\} = \int \cdots \int_{\Omega} \frac{1}{(2\pi)^{q'/2}\sqrt{\det(\hat{\mathscr{A}}_{s,t}\Sigma\hat{\mathscr{A}}'_{s,t})}} \exp\left[-\tfrac{1}{2}(\mathbf{x} - \hat{\mathscr{A}}_{s,t}(\mathbf{c} - \boldsymbol{\mu}))'[\hat{\mathscr{A}}_{s,t}\Sigma\hat{\mathscr{A}}'_{s,t}]^{-1}(x - \hat{\mathscr{A}}_{s,t}(\mathbf{c} - \boldsymbol{\mu}))\right] \mathbf{dx}. \tag{4.5.15}$$

where Ω is the convex region defined in (4.4.26).

To compute Prob $\{\tau_{s,t} \geq r_{s,t}\}$ we must evaluate a probability integral of a multidimensional normal distribution. An excellent review of this problem is given by S. Gupta [GU1], and is briefly summarized below.

If $\mathbf{C}$ is $N(\mathbf{c} - \boldsymbol{\mu}, \Sigma)$, the *correlation* matrix of $\mathbf{C}$ is an $m \times m$ matrix $\Xi(\mathbf{C})$ defined by

$$\Xi(\mathbf{C}) = [\zeta_{i,j}(\mathbf{C})], \tag{4.5.16}$$

where

$$\zeta_{i,j}(\mathbf{C}) = \frac{\sigma_{i,j}}{\sqrt{\sigma_{i,i}}\sqrt{\sigma_{j,j}}} \quad (\Sigma = [\sigma_{i,j}]).$$

Clearly, $\zeta_{i,i}(\mathbf{C}) = 1$ for all i and given Σ we can easily find $\Xi(\mathbf{C})$. The variance-covariance matrix of $|A|$ is $\mathscr{A}_{s,t}\Sigma\mathscr{A}'_{s,t}$ and the correlation matrix $\Xi(|\mathbf{A}|)$ of $|\mathbf{A}|$ is also readily obtained. For example, if $C_1, C_2, \ldots, C_m$ are independent with common unity variance, the i-jth entry of $\Xi(|\mathbf{A}|)$ is

$$\zeta_{i,j}(|\mathbf{A}|) = \frac{n_{i,j}}{\sqrt{n_{i,i}}\sqrt{n_{j,j}}}, \tag{4.5.17}$$

where $n_{i,j}$ is the number of branches in $A^i_{s,t} \cap A^j_{s,t}$.

For simplicity, standardize $c(A^i_{s,t})$ and consider the random variable $c(\hat{A}^i_{s,t})$ defined by

$$\frac{c(A^i_{s,t}) - E\{c(A^i_{s,t})\}}{(\operatorname{Var}\{c(A^i_{s,t})\})^{1/2}} = c(\hat{A}^i_{s,t}) \qquad \text{for } i = 1, \ldots, q. \tag{4.5.18}$$

Then, if $\zeta_{\min} = \operatorname*{Min}_{i \neq j}[\zeta_{i,j}]$, and $\zeta_{\max} = \operatorname*{Max}_{i \neq j}[\zeta_{i,j}]$, the following upper and lower bounds can be shown[5] to hold for Prob $\{c(\hat{A}_{s,t}) \geq r_1, \ldots, c(\hat{A}^q_{s,t}) \geq r_q\}$, where

$$r_i = \frac{r_{s,t} - \sum_j a_{i,j}(c_j - u_j)}{[\operatorname{Var} c(A^i_{s,t})]^{1/2}} \qquad \text{for } i = 1, \ldots, q$$

$$\int_{-\infty}^{\infty} \prod_{i=1}^{q} \Phi\left(\frac{\sqrt{\zeta_{\min}}\, y - r_i}{\sqrt{1 - \zeta_{\min}}}\right) \phi(y)\, dy \leq$$

$$\text{Prob}\{c(\hat{A}^1_{s,t}) \geq r_1, \ldots, c(\hat{A}^q_{s,t}) \geq r_q\}$$

$$\leq \int_{-\infty}^{\infty} \prod_{i=1}^{q} \Phi\left(\frac{\sqrt{\zeta_{\max}}\, y - r_i}{\sqrt{1 - \zeta_{\max}}}\right) \phi(y)\, dy \tag{4.5.19}$$

where $\Phi(\cdot)$ and $\phi(\cdot)$ are the distribution and density functions of the one-dimensional normal random variable with zero mean and unity variance, respectively (i.e., the *Standard* Normal Random Variable). Suppose that the correlation constant $\zeta_{i,j}$ can be written as

$$\zeta_{i,j} = \alpha_i \alpha_j \qquad \text{for all } i,j \text{ such that } i \neq j. \tag{4.5.20}$$

This condition holds if, for example, the number of branches in $A^i_{s,t} \cap A^j_{s,t}$ $(i \neq j)$ is constant. Then, an *exact* expression for Prob $\{\tau_{s,t} \geq r_{s,t}\}$ is

$$\text{Prob}\{c(A^1_{s,t}) \geq r_1, \ldots, c(A^q_{s,t}) \geq r_q\} = \int_{-\infty}^{\infty} \prod_{i=1}^{q} \Phi\left(\frac{\alpha_i y - r_i}{\sqrt{1 - \alpha_i^2}}\right) \phi(y)\, dy. \tag{4.5.21}$$

---

[5] Usually the condition that $\zeta_{\min} \geq 0$ is required, but since all of the $c(A^i_{s,t})$ are nonnegatively correlated, this condition is always satisfied.

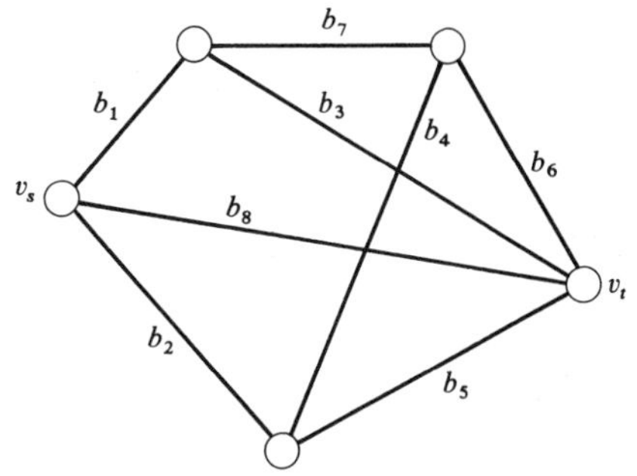

Fig. 4.5.1 Graph for Example 4.5.1.

If the number of branches $n_{i,j}$ in $A^i_{s,t} \cap A^j_{s,t}$ are not equal for all i,j, but are small compared to the $n_{i,i}$ the preceding equation gives an excellent approximation to the actual probability.

Example 4.5.1 To illustrate these ideas, consider the graph shown in Fig. 4.5.1. The s-t cut-set matrix is

$$\mathscr{A}_{s,t} = \begin{array}{c} \begin{matrix} b_1 & b_2 & b_3 & b_4 & b_5 & b_6 & b_7 & b_8 \end{matrix} \\ \begin{bmatrix} 1 & 1 & 0 & 0 & 0 & 0 & 0 & 1 \\ 1 & 0 & 0 & 1 & 1 & 0 & 0 & 1 \\ 0 & 1 & 1 & 0 & 0 & 0 & 1 & 1 \\ 1 & 0 & 0 & 0 & 1 & 1 & 1 & 1 \\ 0 & 1 & 1 & 1 & 0 & 1 & 0 & 1 \\ 0 & 0 & 1 & 1 & 1 & 0 & 1 & 1 \\ 0 & 0 & 1 & 0 & 1 & 1 & 0 & 1 \end{bmatrix} \end{array} \tag{4.5.22}$$

and the number of branches in $A^i_{s,t} \cap A^j_{s,t}$ for $i \leq j$

$$n_{1,1} = 3 \tag{4.5.23a}$$

$$n_{1,2} = 2,\ n_{2,2} = 4 \tag{4.5.23b}$$

$$n_{1,3} = 2,\ n_{2,3} = 1,\ n_{3,3} = 4 \tag{4.5.23c}$$

$$n_{1,4} = 2,\ n_{2,4} = 3,\ n_{3,4} = 2,\ n_{4,4} = 5 \tag{4.5.23d}$$

$$n_{1,5} = 2,\ n_{2,5} = 2,\ n_{3,5} = 3,\ n_{4,5} = 2,\ n_{5,5} = 5 \tag{4.5.23e}$$

$$n_{1,6} = 1,\ n_{2,6} = 3,\ n_{3,6} = 3,\ n_{4,6} = 3,\ n_{5,6} = 3,\ n_{6,6} = 5 \tag{4.5.23f}$$

$$n_{1,7} = 1,\ n_{2,7} = 2,\ n_{3,7} = 2,\ n_{4,7} = 3,\ n_{5,7} = 3,\ n_{6,7} = 3,\ n_{7,7} = 4. \tag{4.5.23g}$$

Consequently, if the variance-covariance matrix of the branch flows is the identity matrix, the correlation matrix of the random cut-set vector $|\mathbf{A}|$ is

$$\Xi(|\mathbf{A}|) = \begin{bmatrix} 1 & \frac{1}{\sqrt{3}} & \frac{1}{\sqrt{3}} & \frac{2}{\sqrt{15}} & \frac{2}{\sqrt{15}} & \frac{1}{\sqrt{15}} & \frac{1}{2\sqrt{3}} \\ \frac{1}{\sqrt{3}} & 1 & \frac{1}{4} & \frac{3}{2\sqrt{5}} & \frac{1}{\sqrt{5}} & \frac{3}{2\sqrt{5}} & \frac{1}{2} \\ \frac{1}{\sqrt{3}} & \frac{1}{4} & 1 & \frac{1}{\sqrt{5}} & \frac{3}{2\sqrt{5}} & \frac{3}{2\sqrt{5}} & \frac{1}{2} \\ \frac{2}{\sqrt{15}} & \frac{3}{2\sqrt{5}} & \frac{1}{\sqrt{5}} & 1 & \frac{2}{5} & \frac{3}{5} & \frac{3}{2\sqrt{5}} \\ \frac{2}{\sqrt{15}} & \frac{1}{\sqrt{5}} & \frac{3}{2\sqrt{5}} & \frac{1}{5} & 1 & \frac{3}{5} & \frac{3}{2\sqrt{5}} \\ \frac{1}{\sqrt{15}} & \frac{3}{2\sqrt{5}} & \frac{3}{2\sqrt{5}} & \frac{3}{5} & \frac{3}{5} & 1 & \frac{3}{2\sqrt{5}} \\ \frac{1}{2\sqrt{3}} & \frac{1}{2} & \frac{1}{2} & \frac{3}{2\sqrt{5}} & \frac{3}{2\sqrt{5}} & \frac{3}{2\sqrt{5}} & 1 \end{bmatrix}. \quad (4.5.24)$$

Hence, $\zeta_{\min} = \frac{1}{4}$, $\zeta_{\max} = 3/(2\sqrt{5})$. If

$$r_i = \left[r_{s,t} - \sum_{j=1}^{m} \alpha_{i,j}(c_j - \mu_j) \right] \Big/ \sqrt{n_{i,i}},$$

we can bound Prob $\{\tau_{s,t} \geq r_{s,t}\}$ by

$$\int_{-\infty}^{\infty} \prod_{i=1}^{7} \Phi\left(\frac{1}{\sqrt{3}} y - \frac{2}{\sqrt{3}} r_i \right) \varphi(y)\, dy$$

$$\leq \text{Prob}\,\{\tau_{s,t} \geq r_{s,t}\} \leq \int_{-\infty}^{\infty} \prod_{i=1}^{7} \Phi\left(\frac{3}{\sqrt{20 - 6\sqrt{5}}} y - \frac{r_i}{\sqrt{1 - 3/2\sqrt{5}}} \right) \phi(y)\, dy. \quad (4.5.25)$$

4.6 APPROXIMATIONS

The computation of the maximum flow probability, even in the case of normally distributed branch flows, is a formidable task. The nature of the exact results indicates that we should search for *qualitative* estimates of the terminal capacity behavior. Such qualitative estimates can be generated by Monte Carlo techniques [HA1].

Monte Carlo methods are concerned with experiments on random numbers. The simplest Monte Carlo method consists of observing random numbers which are selected to simulate directly the physical process under study. The desired solution is then inferred from the behavior of these random numbers. The key to the Monte Carlo approach is the generation of these random numbers, since they must have the same statistical properties as the process which is being studied. Tables are available but it is preferable to use a digital computer to generate random numbers. This is most conveniently done by using fixed sets of rules. Numbers generated by such rules are called *pseudo-random* and have the property that standard statistical tests do not detect any significant departure from randomness. Pseudo-random number generation has the advantage that sequences of numbers can be exactly reproduced and so can be used to repeat experiments.

Monte Carlo methods can be applied to the maximum flow problem in the following manner. We simulate the behavior of the system a large number of times by randomly generating a set of branch flow vectors $\{\mathbf{F}(k);\ k = 1,\ldots,K\}$. We then apply the Max-Flow Min-Cut Theorem via the relation

$$\tau_{s,t}(k) = \operatorname{Min}\ [\mathscr{A}_{s,t}(\mathbf{c} - \mathbf{F}(k))] \tag{4.6.1}$$

to find the maximum flow through the graph for the kth flow vector. We repeat this process K times and then form the function $S_K(z)$ defined by

$$S_K(z) = \frac{1}{K}\sum_{k=1}^{K} h_z(\tau_{s,t}(k)), \tag{4.6.2}$$

where

$$h_z(U) = \begin{cases} 0 & \text{for } U \geq z \\ 1 & \text{for } U < z. \end{cases}$$

Thus, $KS_K(z)$ is the number of $\tau_{s,t}(k)$ that are smaller than z where z is an arbitrary real number.

The function $S_K(z)$ may assume values between zero and one and is a non-decreasing function of z. It is easy to verify that $S_K(z)$ is continuous from the left and satisfies all other properties of a probability distribution function. $S_K(z)$ is a random variable for every z. It is called the *empirical distribution function* of $\tau_{s,t}$ and is very closely related to the true distribution function Prob $\{\tau_{s,t} < z\}$ of $\tau_{s,t}$. The following theorem (Theorem 10.10.1 of [FI1]) exhibits this relationship explicitly.

Theorem 4.6.1 (*Glivenko's Theorem*) *Let $S_K(x)$ be the empirical distribution function of a sample of K elements in which the characteristic X has the theoretical distribution function $P(x)$. The probability that the sequence $S_K(x)$ converges to $P(x)$, as K tends to infinity uniformly in x for $-\infty < x < \infty$, equals one. That is, let Δ_K be the random variable*

$$\Delta_K = \operatorname*{Sup}_x |S_K(x) - P(x)|. \tag{4.6.3a}$$

Then,

$$\text{Prob}\{\lim_{K\to\infty} \Delta_K = 0\} = 1. \tag{4.6.3b}$$

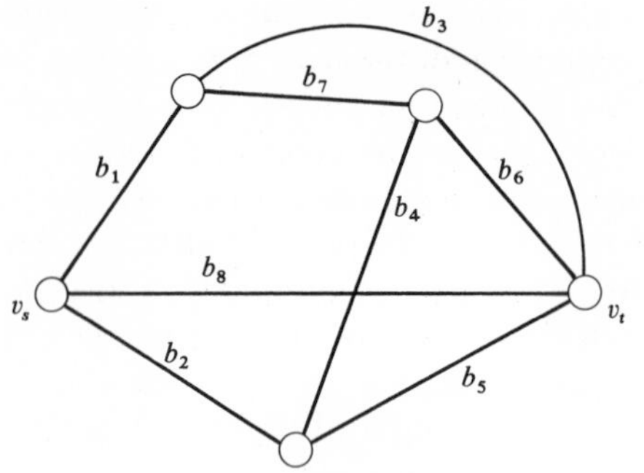

Fig. 4.6.1 Graph G_1.

According to this theorem, if we select a large enough K, the empirical distribution function of $\tau_{s,t}$ will have a high probability of being close to the true probability distribution function of $\tau_{s,t}$.

Example 4.6.1 Consider the graphs G_1 and G_2 shown in Figs. 4.6.1 and 4.6.2. A computer program was written to perform the above simulation [FR3]. For simplicity, branch flows were assumed to be identically and independently[6] distributed normal variables. The effect of truncation made necessary by the finite branch capacities and nonnegative flow was included after branch flows were generated. Capacity was treated as a variable parameter and, for a fixed capacity vector $\mathbf{c} = (c_1, \ldots, c_m)'$, the values of $1 - S_K(z)$ were generated for 100 samples of maximum flow. In some cases, the capacity of each branch was taken to be a constant c; then c was varied over a wide range of values. In other cases, the capacities of a subset of branches of G were varied while the other capacities were held fixed. Two of the graphs analyzed in this manner were G_1 and G_2. In each case, the means of the branch flows were taken to be 3 and the variances were assumed to be unity. Branch capacity

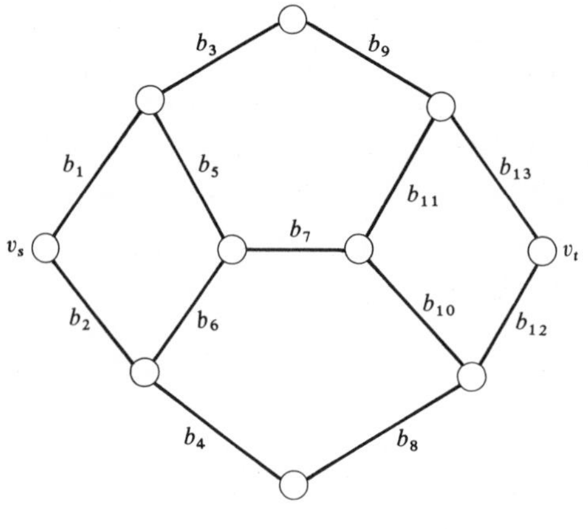

Fig. 4.6.2 Graph G_2.

[6] Since $c(A^1_{s,t}), \ldots, c(A^q_{s,t})$ are *not* independent even if $\mathbf{F}_1, \ldots, \mathbf{F}_m$ are independent, there is no great loss of generality.

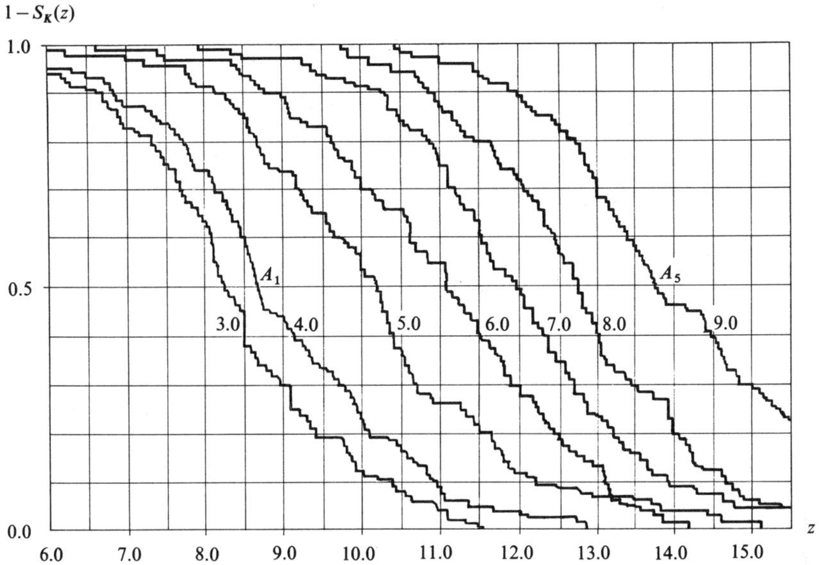

Fig. 4.6.3 Branch b_8 of G_1 is varied from $c_8 = 3.0$ to $c_8 = 9.0$. All other branch capacities equal 7.

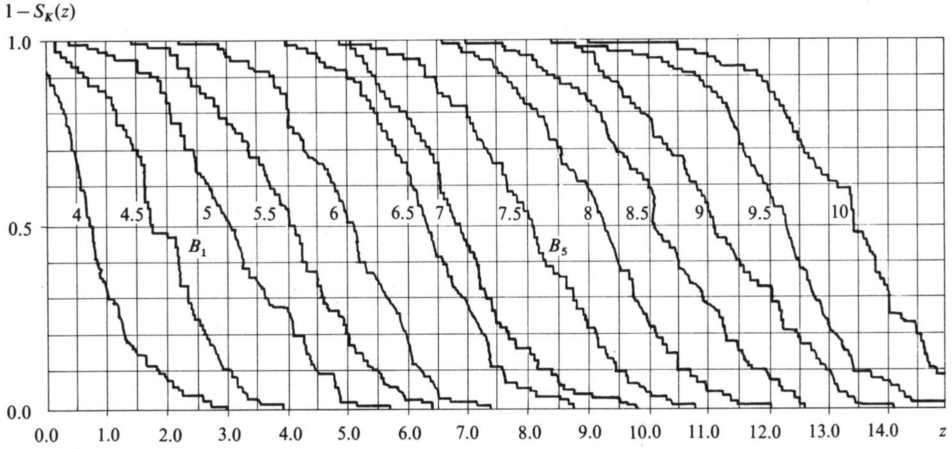

Fig. 4.6.4 All branches of G_2 are varied from $c = 4$ to $c = 8.5$.

was then varied between $c = 4$ and $c = 10$. Figs. 4.6.3 to 4.6.5 show some of the results of the simulation. One significant result that emerged from the simulation was that in every case, $S_K(z)$ could be accurately approximated by a cumulative normal distribution function. This observation is illustrated in Figs. 4.6.6 and 4.6.7, where normal approximations to typical empirical curves are given.

El-Ghoroury [EL1] has investigated the applicability of the normality approximation for large classes of graphs. In his investigation, branch flows were selected to

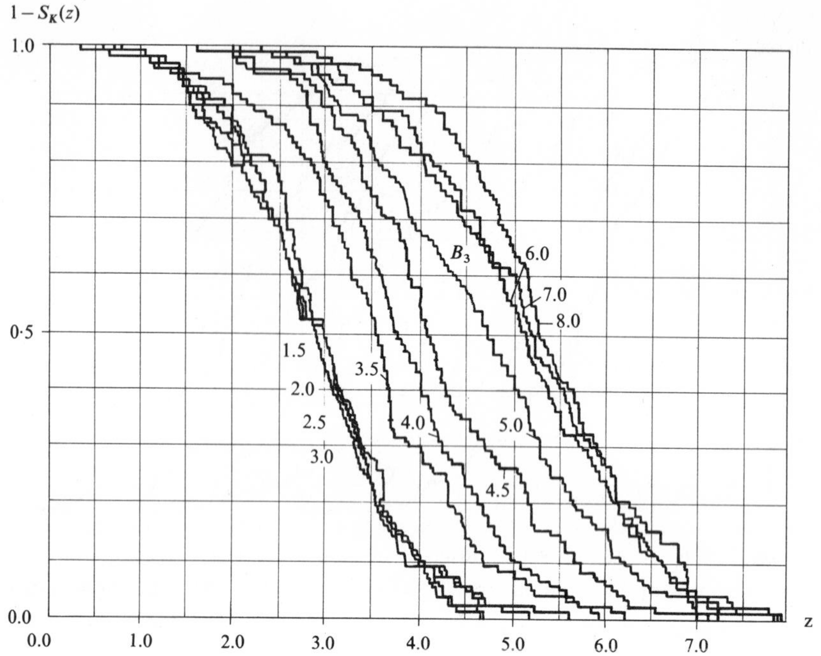

Fig. 4.6.5 Branches b_3, b_4, b_8, b_9 of G_2 are varied from $c = 1.5$ to $c = 8.0$. All other branch capacities equal 6.

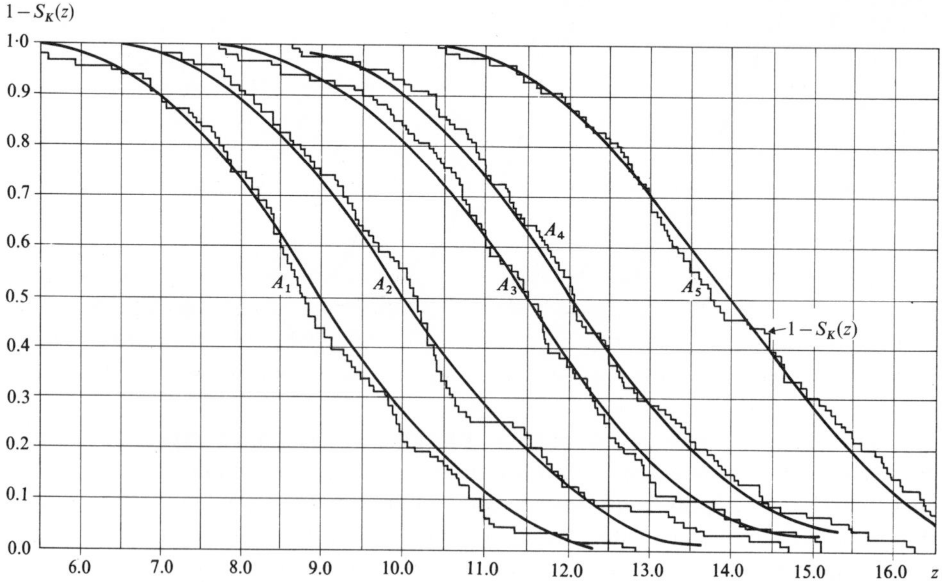

Fig. 4.6.6 Empirical curves generated by Monte Carlo simulation for G_1 are approximated with normal distribution functions with variance ≈ 2.7.

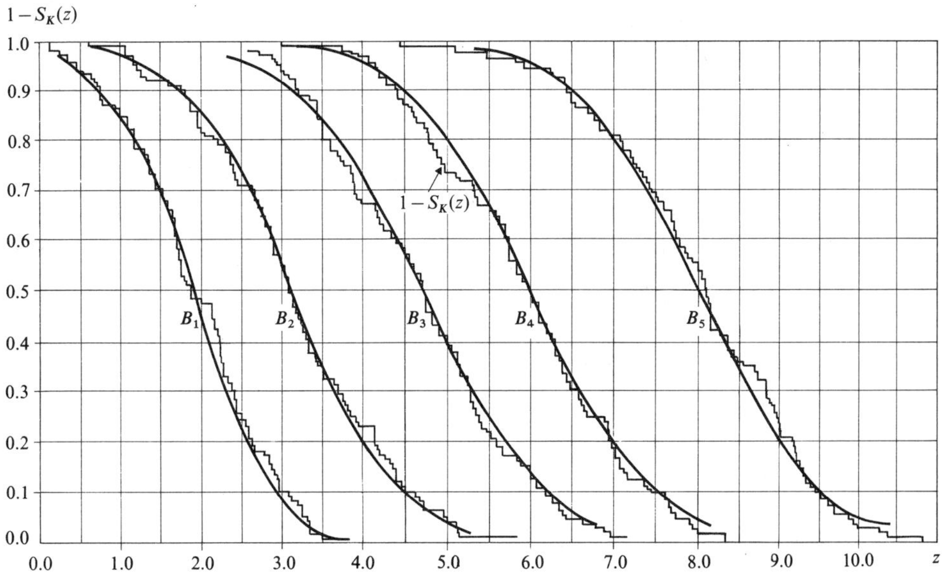

Fig. 4.6.7 Empirical curves generated by Monte Carlo simulation for G_2 are approximated with normal distribution functions. The variances of the approximating curves are: $B_1 = 0.64$; $B_2 = 1.22$; $B_3 = 1.4$; $B_4 = 1.4$; $B_5 = 1.4$.

be either uniformly distributed, normally distributed, or exponentially distributed. To investigate the normality conjecture, the χ^2 goodness-of-fit test for normal distributions [FI1] was applied. These results indicated that the normality assumption improves as the number of branches and vertices in the graph increases, and is even reasonably accurate for graphs with few elements. However, in this case, the exact techniques developed in the preceding sections may be easily applied and there is no need to resort to approximations. The main point to be considered here is that without regard to the limit distribution of $S_K(z)$, reasonable estimates of probabilities can be obtained by normal approximations.

Even if we assume that the maximum flow probability distribution is adequately approximated by a normal distribution, the problem of finding Prob $\{\tau_{s,t} \geq r_{s,t}\}$ is still far from solved. We must now find $E\{\tau_{s,t}\}$ and Var $\{\tau_{s,t}\}$. From the discussion in Section 4 and the integral form of Prob $\{\tau_{s,t} \geq r_{s,t}\}$ given in Section 5, we can see that this is a difficult problem. However, if we use a method due to Clark [CL1], the problem becomes manageable for small graphs.[7]

Let X_1 and X_2 be normal random variables. We can give exact expressions for the moments of the random variable Min $[X_1,X_2]$. Then, if we assume that Min $[X_1,X_2]$ is normal, and if X_3 is normal, we can find the moments of Min $[\text{Min}[X_1,X_2],X_3]$. We continue in this manner; at each step we find the moments of Min $[X_1,\ldots,X_i]$ and then, if we assume that this variable is normally distributed, we find the moments

[7] The remainder of this section is not essential for an understanding of subsequent material.

of Min $[\text{Min}[X_1,\ldots,X_i],X_{i+1}] = \text{Min}\,[X_1,\ldots,X_i,X_{i+1}]$. Thus by induction, we can compute the moments of Min $[X_1,\ldots,X_q]$. Naturally, each such calculation will be inaccurate since Min $[X_1,\ldots,X_i]$ is not normally distributed. However, as will be seen, empirical evidence indicates errors introduced by this procedure do not seem to be significant.

Let X_1, X_2, and X_3 be normally distributed with mean values μ_1, μ_2, and $E\{X_3\}$ and variances σ_1^2, σ_2^2, and Var $\{X_3\}$, respectively. $E\{X_3\}$ and Var $\{X_3\}$ do not enter specifically into the computations and hence are not specified. Let Cor $(\cdot,\cdot)$ denote the coefficient of linear correlation between the variables within the parentheses and define Cor $(X_1,X_2) = \zeta$, Cor $(X_1,X_3) = \zeta_1$, and Cor $(X_2,X_3) = \zeta_2$.

If $\sigma_1 = \sigma_2$ and $\zeta = 1$, then X_1 and X_2 differ by a constant. If so, Min $[X_1,X_2]$ is easily found and the analysis below is not applicable. Let $v_i(2)$ be the ith moment about zero of the random variable Min $[X_1,X_2]$. Furthermore, let

$$a^2 = \sigma_1^2 + \sigma_2^2 - 2\sigma_1\sigma_2\zeta \tag{4.6.4a}$$

and

$$\alpha = (\mu_2 - \mu_1)/a. \tag{4.6.4b}$$

Then, it can be shown that

$$v_1(2) = \mu_1\Phi(\alpha) + \mu_2\Phi(-\alpha) - \alpha\phi(\alpha) \tag{4.6.5}$$

$$v_2(2) = (\mu_1^2 + \sigma_1^2)\Phi(\alpha) + (\mu_2^2 + \sigma_2^2)\Phi(-\alpha) - (\mu_1 + \mu_2)a\phi(\alpha) \tag{4.6.6}$$

and

$$\text{Cor}\,(\text{Min}[X_1,X_2],X_3) = [\sigma_1\zeta_1\Phi(\alpha) + \sigma_2\zeta_2\Phi(-\alpha)]/(v_2 - v_1^2)^{1/2}. \tag{4.6.7}$$

Suppose now, we let $v_i(k)$ be the ith moment about zero of the random variable Min $[X_1,\ldots,X_k]$. Let $k = 3$. To calculate the moments of Min $[X_1,X_2,X_3]$, we find the moments of Min $[\text{Min}[X_1,X_2],X_3]$, using (4.6.5) to (4.6.7). First we find $v_1(2)$ and $v_2(2)$ and consequently, Var $\{\text{Min}[X_1,X_2]\}$. Then, we replace μ_1 by $v_1(2)$, μ_2 by $E\{X_3\}$, and ζ by Cor $(\text{Min}[X_1,X_2],X_3)$ in (4.6.3) to (4.6.7). We can then find

$$v_1(3) = E\{\text{Min}[X_1,X_2,X_3]\}, \tag{4.6.8a}$$

$$v_2(3) = E\{(\text{Min}[X_1,X_2,X_3])^2\}, \tag{4.6.8b}$$

and

$$\text{Cor}\,(\text{Min}[X_1,X_2,X_3],X_k) \qquad \text{for } k = 4,5,\ldots,q. \tag{4.6.8c}$$

At the jth step in the procedure we compute

$$v_1(j) = E\{\text{Min}[X_1,X_2,\ldots,X_j]\}, \tag{4.6.9a}$$

$$v_2(j) = E\{(\text{Min}[X_1,X_2,\ldots,X_j])^2\}, \tag{4.6.9b}$$

and

$$\text{Cor}\,(\text{Min}[X_1,\ldots,X_j],X_{j+k}) \qquad \text{for } k = 1,2,\ldots,q - j. \tag{4.6.9c}$$

To find Cor $(\text{Min}[X_1,\ldots,X_j],X_{j+k})$, we rewrite this correlation coefficient as

$$\text{Cor}\,(\text{Min}[X_1,\ldots,X_j],X_{j+k}) = \text{Cor}\,(\text{Min}[X_1,\ldots,X_{j-1}],X_j],X_{j+k}). \quad (4.6.10)$$

To find this number, we must know Cor $(\text{Min}[X_1,\ldots,X_{j-1}],X_{j+k})$, which will replace ζ_1 in (4.6.7), and Cor (X_j,X_{j+k}), which will replace ζ_2. The second number, $\zeta_{j,j+k}$, is already known, and the first number has been calculated at the $j - 1$ stage of the process. Therefore, we can continue the procedure until we find the moments of Min $[X_1,\ldots,X_q]$.

We can apply the above procedure to the random cut-set variables $c(A^1_{s,t}),\ldots,$ $c(A^q_{s,t})$ to find the mean and variance of the maximum flow $\tau_{s,t}$. Then, using the assumption that Min $[c(A^1_{s,t}),\ldots,c(A^q_{s,t})]$ is normally distributed, we have

$$\text{Prob}\,\{\tau_{s,t} \geq r_{s,t}\} = 1 - \Phi\left(\frac{r_{s,t} - v_1(q)}{\sqrt{v_2(q) - v_1^2(q)}}\right). \quad (4.6.11)$$

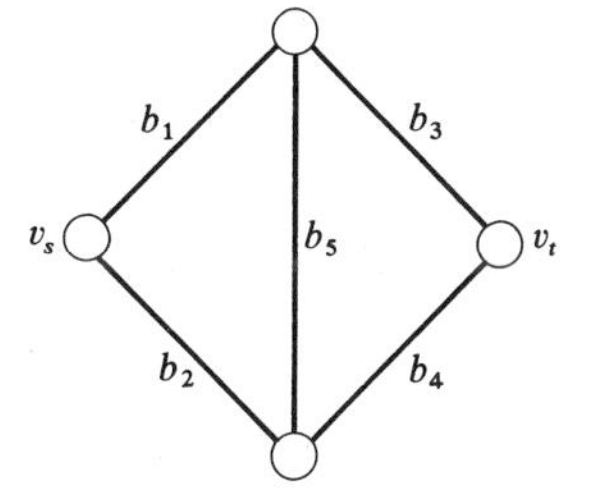

Fig. 4.6.8 Graph for Example 4.6.2.

Example 4.6.2 Let G be the graph shown in Fig. 4.6.8. Suppose that the branch flows are identically and independently distributed random variables with mean vector $\boldsymbol{\mu}' = (3,3,3,3,3)$ and an identity variance-covariance matrix. Suppose that each branch has capacity $c_i = 5$. Then each random branch capacity variable is an independent normal variable with mean 2 and unity variance. The cut-set matrix of G is

$$\mathscr{A}_{s,t} = \begin{bmatrix} 1 & 1 & 0 & 0 & 0 \\ 0 & 0 & 1 & 1 & 0 \\ 1 & 0 & 0 & 1 & 1 \\ 0 & 1 & 1 & 0 & 1 \end{bmatrix}. \quad (4.6.12)$$

We can easily find the correlation matrix of the cut-set by means of (4.5.17). Thus

$$\Xi(|\mathbf{A}|) = \begin{bmatrix} 1 & 0 & 1/\sqrt{6} & 1/\sqrt{6} \\ 0 & 1 & 1/\sqrt{6} & 1/\sqrt{6} \\ 1/\sqrt{6} & 1/\sqrt{6} & 1 & 1/3 \\ 1/\sqrt{6} & 1/\sqrt{6} & 1/3 & 1 \end{bmatrix}. \quad (4.6.13)$$

From (4.6.4),

$$a^2 = 2 + 2 - 2(\sqrt{2})(\sqrt{2})(0) = 4, \tag{4.6.14a}$$

$$\alpha = \frac{4 - 4}{2} = 0. \tag{4.6.14b}$$

Therefore, from (4.6.5) and (4.6.6),

$$v_1(2) = 4\Phi(0) + 4\Phi(0) - 2\phi(0) \doteq 3.2, \tag{4.6.15a}$$

$$v_2(2) = (16 + 2)\Phi(0) + (16 + 2)\Phi(0) - 2(4 + 4)\phi(0) \doteq 11.6, \tag{4.6.15b}$$

and $v_2(2) - v_1^2(2) = 1.27$. Hence

$$\text{Cor}\,(\text{Min}[c(A_{s,t}^1),c(A_{s,t}^2)],c(A_{s,t}^3)) = \left[\sqrt{2}\frac{1}{\sqrt{6}}\Phi(0) + \sqrt{2}\frac{1}{\sqrt{6}}\Phi(0)\right]\Big/\sqrt{1.27} \doteq 0.51. \tag{4.6.16}$$

Also, since $\zeta_{1,3}(|\mathbf{A}|) = \zeta_{1,4}(|\mathbf{A}|)$, we have

$$\text{Cor}\,(\text{Min}[c(A_{s,t}^1),c(A_{s,t}^2)],c(A_{s,t}^4)) \doteq 0.51. \tag{4.6.17}$$

To find the moments of $\text{Min}[c(A_{s,t}^1),c(A_{s,t}^2),c(A_{s,t}^3)]$, we can substitute the numbers just calculated along with $E\{c(A_{s,t}^3)\}$ and $\text{Var}\,\{c(A_{s,t}^3)\}$ into (4.6.4) to (4.6.6). Then,

$$a^2 = 1.27 + 3 - 2\sqrt{1.27}\,\sqrt{3}(0.51) = 2.33, \tag{4.6.18a}$$

$$\alpha = (6 - 3.2)/\sqrt{2.33} \doteq 1.86, \tag{4.6.18b}$$

and

$$v_1(3) = 3.2\Phi(1.86) + 6\Phi(-1.86) - 1.5\phi(1.86) \doteq 3.1, \tag{4.6.19a}$$

$$v_2(3) = (3.2^2 + 1.27)\Phi(1.86) + (6^2 + 3)\Phi(-1.86) - (3.2 + 6)(1.5)\phi(1.86) \doteq 11.4. \tag{4.6.19b}$$

Consequently, $v_2(3) - v_1^2(3) \doteq 1.8$. The correlation between $\text{Min}\,[c(A_{s,t}^1),c(A_{s,t}^2),$ $c(A_{s,t}^3)]$, and $c(A_{s,t}^4)$ is found as

$$\text{Cor}\,(\text{Min}[c(A_{s,t}^1),c(A_{s,t}^2),c(A_{s,t}^3)],c(A_{s,t}^4)) = \text{Cor}\,(\text{Min}[\text{Min}\{c(A_{s,t}^1),c(A_{s,t}^2)\},c(A_{s,t}^3)],c(A_{s,t}^4)). \tag{4.6.20}$$

Since $\text{Cor}\,(\text{Min}[c(A_{s,t}^1),c(A_{s,t}^2)],c(A_{s,t}^4)) = \zeta_1 \doteq 0.51$ we have from (4.6.7) that

$$\text{Cor}\,(\text{Min}[c(A_{s,t}^1),c(A_{s,t}^2),c(A_{s,t}^3)],c(A_{s,t}^4)) = \left[\sqrt{1.27}(0.51)\Phi(1.86) + \sqrt{3}\left(\frac{1}{3}\right)\Phi(-1.86)\right]\Big/\sqrt{1.8} \doteq 0.51. \tag{4.6.21}$$

We can now find the moments of $\tau_{s,t}$. Let

$$a^2 = 1.8 + 3 - 2\sqrt{1.8}\,\sqrt{3}(0.51) \doteq 2.3, \tag{4.6.22a}$$

$$\alpha = (6 - 3.1)/\sqrt{2.3} \doteq 1.9. \tag{4.6.22b}$$

Then

$$E\{\tau_{s,t}\} = 3.1\Phi(1.9) + 6\Phi(-1.9) - (1.53)\phi(1.9) \doteq 3; \tag{4.6.23}$$

$$E\{\tau_{s,t}^2\} = (3.1^2 + 1.8)\Phi(1.9) + (6^2 + 3)\Phi(-1.9) - (3.1 + 6)\sqrt{2.3}\,\phi(1.9) \doteq 11.3; \tag{4.6.24}$$

$$\text{Var}\,\{\tau_{s,t}\} = 11.3 - 3^2 \doteq 2.3. \tag{4.6.25}$$

Consequently,

$$\text{Prob}\,\{\tau_{s,t} \geq r_{s,t}\} \doteq 1 - \Phi\left(\frac{r_{s,t} - 3}{\sqrt{2.3}}\right). \tag{4.6.26}$$

A discussion of the accuracy of the technique is given by Clark. Clark considers the *maximum* of a set of random variables, but since

$$\text{Min}\,[X_1, \ldots, X_q] = -\text{Max}\,[-X_1, \ldots, -X_q],$$

his conclusions are valid for our case. An argument concerning the adequacy of the approximation is given in Table 4.6.1 (adapted from Clark's paper) of the mean of the minimum of q independent standard normal variables. Furthermore, the normal approximations to the empirical distributions shown in Figs. 4.6.6 and 4.6.7 have means and variances found by the above method.

Table 4.6.1

q	Exact $E\{\text{Min}[X_1, \ldots, X_q]\}$	Approximation
2	−0.5642	−0.5642
3	−0.8463	−0.8476
4	−1.0294	−1.0310
5	−1.1630	−1.1643
6	−1.2672	−1.2679
7	−1.3522	−1.3522
8	−1.4236	−1.4230
9	−1.4850	−1.4837
10	−1.5833	−1.5367

* 4.7 PARAMETRIC TERMINAL CAPACITY ANALYSIS WITH NORMALLY DISTRIBUTED MAXIMUM FLOW[8]

We now pursue further the observation that $\tau_{s,t}$ may be approximated by a normal random variable. In Section 3, the case in which the branch flows have unknown probability distributions but are observable was investigated. In this section we investigate the effect of knowledge of the probability distribution of $\tau_{s,t}$ on the testing problem. Thus, we assume that $\mathbf{F}(k) = (f_1(k), \ldots, f_m(k))'$ is a measurement of the

[8] This section is a prerequisite only to Section 12 of Chapter 6.

flows $F_1, \ldots, F_m$ at time k for $k = 1,2,\ldots,K$. As before, $\mathbf{F}(1)$, $\mathbf{F}(2), \ldots, \mathbf{F}(K)$ are assumed to be identically and independently distributed random variables.

We test the hypothesis

$$H_0 : p = \text{Prob}\,\{\tau_{s,t} \geq r_{s,t}\} \geq p_0 \tag{4.7.1a}$$

against the alternative

$$H_1 : p < p_0 \tag{4.7.1b}$$

at level α, where α is the probability of Type I error. By assumption, $\tau_{s,t}$ is a normal random variable with mean $v_1(q)$ and variance of $\sigma^2 = v_2(q) - v_1^2(q)$. For each value of $\mathbf{F}(k)$, we compute

$$\tau_{s,t}(k) = \underset{1 \leq i \leq q}{\text{Min}} \left[\sum_{j=1}^{m} a_{i,j}(c_j - f_j(k)) \right] \qquad \text{for } k = 1,2,\ldots,K. \tag{4.7.2}$$

The random variables $\tau_{s,t}(1), \ldots, \tau_{s,t}(K)$ are identically and independently distributed normal variables with mean $v_1(q)$ and variance σ^2. The probability that $\tau_{s,t}$ is at least $r_{s,t}$ can be written as

$$p = \text{Prob}\,\{\tau_{s,t} \geq r_{s,t}\} = 1 - \Phi\left(\frac{r_{s,t} - v_1(q)}{\sigma}\right). \tag{4.7.3}$$

Consequently,

$$\Phi\left(\frac{r_{s,t} - v_1(q)}{\sigma}\right) = 1 - p \tag{4.7.4a}$$

and

$$\frac{r_{s,t} - v_1(q)}{\sigma} = \Phi^{-1}(1 - p). \tag{4.7.4b}$$

Since Φ and Φ^{-1} are strictly increasing functions, hypothesis H_0 becomes

$$\frac{r_{s,t} - v_1(q)}{\sigma} \leq \Phi^{-1}(1 - p_0) \tag{4.7.5}$$

or, if we define a new random variable $\hat{\tau}_{s,t}(i) = \tau_{s,t}(i) - r_{s,t}$ for $i = 1, \ldots, K$, this becomes

$$-\xi/\sigma \leq \Phi^{-1}(1 - p_0) \tag{4.7.6}$$

where $\xi = v_1(q) - r_{s,t}$. In other words, we test the hypothesis

$$\hat{H}_0 : \frac{\xi}{\sigma} \geq \theta_0 = -\Phi^{-1}(1 - p_0) \tag{4.7.7a}$$

against the alternative

$$\hat{H}_1 : \frac{\xi}{\sigma} < \theta_0 \tag{4.7.7b}$$

where ξ and σ^2 are the mean and variance of the identically and independently distributed normal variables $\hat{\tau}_{s,t}(1), \ldots, \hat{\tau}_{s,t}(K)$.

Before considering the testing problem defined by (4.7.7a) and (4.7.7b), we must introduce several statistical concepts. Suppose that we have a class of probability distributions $\{P_\theta : \theta \in \Omega\}$ such that P_θ is the distribution of the random variable X for some value of θ in Ω. A statistical test must distinguish between true and false values of θ on the basis of a random sample $(x_1, x_2, \ldots, x_K)$ of X. Often, part of the information contained in the sample is irrelevant to the testing problem. For example, if θ is the mean of X, the order of the x_i contains no useful information and may be disregarded. That is, another sample $(x_1^*, x_2^*, \ldots, x_K^*)$ such that there is a one-to-one correspondence between each element of $(x_1, \ldots, x_K)$ and the elements of $(x_1^*, \ldots, x_K^*)$ can be considered to be an "equivalent" sample.

The notion of "equivalent" sample is made precise by the concept of a "sufficient statistic." Basically, a random variable is sufficient for the parameter θ if it contains all of the information about θ that the random sample $(x_1, \ldots, x_K)$ initially contained. For example, if $x_1, \ldots, x_K$ are identically and independently distributed *uniform* random variables on the interval $[0,\theta]$, the joint density of $(X_1, \ldots, X_K)$ at a point $(x_1, \ldots, x_K)$ is

$$p_\theta(\mathbf{x}) = \theta^{-K} u[\max_i x_i, \theta], \tag{4.7.8}$$

where $u[a,b] = 1$ if $a \leq b$ and $u[a,b] = 0$ otherwise. Then, a sufficient statistic for θ is $\operatorname{Max}_i [X_i]$, since the largest value of $X_1, \ldots, X_K$ is the only important factor for estimating θ.

In general, the statistic T is *sufficient* for the parameter θ if the density $p_\theta(\mathbf{x})$ can be factored as

$$p_\theta(\mathbf{x}) = g_\theta(T(\mathbf{x}))h(\mathbf{x}), \tag{4.7.9}$$

where $g_\theta(T(\mathbf{x}))$ is a function of θ and $\mathbf{x}$ but depends on $\mathbf{x}$ only through the statistic $T(\mathbf{x})$ and $h(\mathbf{x})$ is a function of $\mathbf{x}$ independent of θ. An important property of such statistics is that if T is a sufficient statistic for θ, then the conditional distribution of any other statistic, given that $T = t$, is independent of θ and hence does not give any information about θ [LE1].

Example 4.7.1 Let $X_1, X_2, \ldots, X_n$ be identically and independently distributed normal random variables with mean θ_1 and variance θ_2. The probability density of $(X_1, \ldots, X_n)$ evaluated at a point $(x_1, \ldots, x_n)$ is

$$p_{\theta_1,\theta_2}(\mathbf{x}) = \prod_{i=1}^{n} \frac{1}{\sqrt{2\pi\theta_2}} \exp\left\{-\frac{1}{2}\left(\frac{(x_i - \theta_1)^2}{\theta_2}\right)\right\}$$

$$= \frac{1}{(2\pi\theta_2)^{n/2}} \exp\left\{-\frac{1}{2\theta_2}\left[\sum_{i=1}^{n} x_i^2 - 2\theta_1 \sum_{i=1}^{n} x_i + n\theta_1^2\right]\right\}. \tag{4.7.10}$$

Here, $\theta = (\theta_1,\theta_2)$ is a two-dimensional variable, and a sufficient statistic for (θ_1,θ_2) is (T_1,T_2) where

$$T_1(\mathbf{x}) = \sum_{i=1}^{n} x_i \tag{4.7.11a}$$

and

$$T_2(\mathbf{x}) = \sum_{i=1}^{n} x_i^2. \tag{4.7.11b}$$

To see this, we can write $p_{\theta_1,\theta_2}(\mathbf{x})$ as

$$p_{\theta_1,\theta_2}(\mathbf{x}) = \frac{1}{(2\pi\theta_2)^{n/2}} \exp\left\{\frac{1}{2\theta_2}\left[t_2 - 2\theta_1 t_1 + n\theta_1^2\right]\right\} \tag{4.7.12}$$

where $t_1 = T_1(\mathbf{x})$ and $t_2 = T_2(\mathbf{x})$ are the values of T_1 and T_2 at the point $\mathbf{x}$. Thus, we see that $p_\theta(\mathbf{x})$ depends only on $\mathbf{x}$ through $(T_1(\mathbf{x}),T_2(\mathbf{x}))$ and so $h(\mathbf{x}) = 1$, $g_\theta(\mathbf{x}) = p_\theta(\mathbf{x})$ and (T_1,T_2) is sufficient for (θ_1,θ_2).

The utility of a sufficient statistic is that the testing problem may often be simplified if a sufficient statistic is available to reduce the data to a more compact form. We noted above that, in testing the mean of a random variable, the order of the observations was of no importance. Thus, we may consider the testing problem to be *symmetric* with respect to permutations of the samples. Many statistical testing problems exhibit symmetries, which provide natural restrictions on the statistical procedures that are to be employed. For example, if we are testing the hypothesis that sixth-grade boys are taller than sixth-grade girls, a good test should not depend upon the scale of measure that we adopt. That is, if we accept the hypothesis when using a measure based on inches, we should also accept the hypothesis if we convert our measurement to centimeters. The only "good" tests for these problems should be ones which remain "invariant" under suitable changes of scale.

The mathematical expression of symmetry is invariance under some group of transformations [LE1]. Let $\mathscr{X}$ be the sample space of the experiment and let g be a transformation of $\mathscr{X}$. Suppose that g is a one-to-one mapping of $\mathscr{X}$ onto itself. If $\mathbf{X}$ is a random variable, let $g\mathbf{X}$ be the random variable which has value $g(\mathbf{x})$ when $\mathbf{X} = \mathbf{x}$. Suppose that when the distribution of $\mathbf{X}$ is P_θ for $\theta \in \Omega$, the distribution of $g\mathbf{X}$ is $P_{\theta'}$ with $\theta' \in \Omega$. We can then consider that g operating on $\mathscr{X}$ induces a transformation $\bar{g}$ operating on Ω. The parameter set Ω is said to *remain invariant* under g if $\bar{g}\theta \in \Omega$ when $\theta \in \Omega$ and if, in addition, for any $\theta' \in \Omega$ there exists $\theta \in \Omega$ such that $\bar{g}\theta = \theta'$. To denote this condition we write $\bar{g}\Omega = \Omega$.

We shall say that the problem of testing $H_0 : \theta \in \Omega_0$ against $H_1 : \theta \in \Omega_1$ *remains invariant* under a transformation g if

$$\bar{g}\Omega = \Omega \tag{4.7.13a}$$

and

$$\bar{g}\Omega_0 = \Omega_0. \tag{4.7.13b}$$

In other words, both the entire parameter space and the spaces of the hypotheses are unchanged under the transformation g on the sample space. We can define a *group* of transformations $\mathscr{G}$ such that if g and $g' \in \mathscr{G}$, then $g'g\mathbf{x} = g'(g(\mathbf{x}))$ and $g(g^{-1}\mathbf{x}) = \mathbf{x}$ for all $\mathbf{x} \in \mathscr{X}$. It is not difficult to show that $\mathscr{G}$ induces a group of transformations $\bar{\mathscr{G}}$ on the parameter space Ω. Suppose that the problem of testing H_0 against H_1 remains invariant under any g in $\mathscr{G}$. Then we will say that *the problem remains invariant under* $\mathscr{G}$.

If we are given a group of transformations $\mathscr{G}$, we can consider a number of points in the sample space $\mathscr{X}$ to be *equivalent*. For example, in the problem discussed in Example 4.7.1 all points such that

$$\sum_{i=1}^{n} x_i \qquad \text{and} \qquad \sum_{i=1}^{n} x_i^2$$

are constant may be considered equivalent for the testing problem. To be precise, two points $\mathbf{x}_1$ and $\mathbf{x}_2$ are said to be *equivalent under* $\mathscr{G}$ if there exists a $g \in \mathscr{G}$ such that $\mathbf{x}_1 = g\mathbf{x}_2$. Equivalence can be expressed symbolically as $\mathbf{x}_1 \sim \mathbf{x}_2 \pmod{\mathscr{G}}$. For Example 4.7.1, the sufficient statistic (T_1,T_2) will assume the same value for any point $\mathbf{x}_2$ which is obtained from some point $\mathbf{x}_1$ by permuting the components of $\mathbf{x}_1$. Thus, (T_1,T_2) remains invariant under permutations of the x_i. In general, a function T is said to be *invariant under* $\mathscr{G}$ if, for all $\mathbf{x} \in \mathscr{G}$,

$$T(\mathbf{x}) = T(\hat{\mathbf{x}}) \qquad \text{for all} \qquad \hat{\mathbf{x}} \sim \mathbf{x} \pmod{\mathscr{G}}. \tag{4.7.14}$$

In other words, a function is invariant if it assigns the same value to all equivalent points. Clearly, such a function is important if a realistic testing problem is desired. Another desirable feature is that the function T assigns *different* values to points which are not equivalent. In other words; the function assigns a *unique* value to the set of equivalent points. The function T is said to be a *maximal invariant* if T is invariant and if

$$T(\mathbf{x}) = T(\hat{\mathbf{x}}) \qquad \text{implies that } \hat{\mathbf{x}} = g\mathbf{x} \text{ for some } g \in \mathscr{G}. \tag{4.7.15}$$

The importance of maximal invariants is summarized by the following theorem which is proved on page 216 of [LE1]:

Theorem 4.7.1 *Let $T(\mathbf{x})$ be a maximal invariant with respect to the group of transformations $\mathscr{G}$. Then, a necessary and sufficient condition for a test to be invariant under $\mathscr{G}$ is that it depend on $\mathbf{x}$ only through $T(\mathbf{x})$.*

To illustrate these ideas consider the following examples.

Example 4.7.2 Let $\mathbf{x} = (x_1,\ldots,x_n)$ and let $\mathscr{G}$ be the *group of translations* $g(\mathbf{x}) = (x_1 + c, x_2 + c,\ldots,x_n + c)$ with $-\infty < c < \infty$. Let $T(\mathbf{x}) = (x_1 - x_n, x_2 - x_n,\ldots,x_{n-1} - x_n)$. Clearly, T is invariant under $\mathscr{G}$ since

$$\begin{aligned} T(g(\mathbf{x})) &= ((x_1 + c) - (x_n + c),(x_2 + c) - (x_n + c),\ldots,(x_{n-1} + c) - (x_n + c)) \\ &= (x_1 - x_n, x_2 - x_n,\ldots,x_{n-1} - x_n). \end{aligned} \tag{4.7.16}$$

Suppose that $x_i' - x_n' = x_i - x_n$ for $i = 1,2,\ldots,n-1$. Then, if $x_n' - x_n = c$, we have that $x_n' = x_n + c$ and so $x_i' = x_i + c$ for $i = 1,2,\ldots,n$. Hence $T(\mathbf{x})$ is also a maximal invariant.

Example 4.7.3 Let $\mathbf{X} = (X_1,X_2,\ldots,X_n)$ and suppose that we are testing the hypotheses H_0: the density of $\mathbf{X}$ is $f_0(x_1 - \theta,\ldots,x_n - \theta)$ against H_1: the density of $\mathbf{X}$ is $f_1(x_1 - \theta,\ldots,x_n - \theta)$ for finite θ. The problem of testing H_0 against H_1 is invariant under the group G of transformations

$$g(\mathbf{x}) = (x_1 + c,\ldots,x_n + c) \qquad \text{for all finite } c. \tag{4.7.17}$$

In the parameter space, $g(\mathbf{x})$ induces the transformation

$$\bar{g}\theta = \theta + c. \tag{4.7.18}$$

From Example 4.7.2, a maximal invariant under $\mathcal{G}$ is $T = (X_1 - X_n,\ldots,X_{n-1} - X_n)$. The distribution of T is independent of θ, and under H_i has the density

$$\int_{-\infty}^{\infty} f_i(t_1 + z,\ldots,t_{n-1} + z,z)\,dz \qquad \text{for} \qquad i = 0,1. \tag{4.7.19}$$

If we test H_0 against H_1 on the basis of $T(\mathbf{x})$, we are testing a simple hypothesis against a simple alternative and according to the Neyman-Pearson Lemma the Uniformly Most Powerful test has the rejection region given by

$$\frac{\displaystyle\int_{-\infty}^{\infty} f_1(t_1 + z,\ldots,t_{n-1} + z,z)\,dz}{\displaystyle\int_{-\infty}^{\infty} f_0(t_1 + z,\ldots,t_{n-1} + z,z)\,dz} > \hat{K}. \tag{4.7.20}$$

We now return to our original problem, which is to test the hypothesis $\hat{H}_0: \xi/\sigma \geq \theta_0$ against the alternative $\hat{H}_1: \xi/\sigma < \theta_0$, where ξ and σ are the mean and variance of the identically and independently distributed random variables $\hat{\tau}_{s,t}(i) = \tau_{s,t}(i) - r_{s,t}$ for $i = 1,2,\ldots,K$. The problem of testing $\hat{H}_0$ against $\hat{H}_1$ is discussed in Section 6.4 of [LE1]. We will briefly review the solution of the problem. To test $\hat{H}_0$ on the basis of the observations $\hat{\tau}_{s,t}(i) = \tau_{s,t}(i) - r_{s,t}$, we need only be concerned with the pair of variables

$$\bar{M}(K) = \frac{1}{K}\sum_{i=1}^{K} \hat{\tau}_{s,t}(i) \tag{4.7.21a}$$

and

$$S = \left[\sum_{i=1}^{K} (\hat{\tau}_{s,t}(i) - \bar{M})^2\right]^{1/2}. \tag{4.7.21b}$$

These variables are independent and are sufficient statistics for (ξ,σ). Furthermore, it can be shown that the distribution of $\bar{M}$ is $N(\xi,\sigma^2/K)$ and the density $h(u)$ of $v = S/\sigma$ is

$$h(u) = \begin{cases} \dfrac{K^{(K-1)/2}u^{K-2}\exp\{-nu^2/2\}}{2^{(K-3)/2}\Gamma((K-1)/2)} & \text{for } u > 0. \\ 0 & \text{for } u \le 0. \end{cases} \tag{4.7.22}$$

We are examining ξ/σ; a sufficient statistic which depends only on the parameter $\theta = \xi/\sigma$ is $\bar{M}/S$. In fact, the statistic $\bar{M}/S$, or equivalently

$$T = \frac{\sqrt{K}\,\bar{M}(K)}{S\sqrt{K-1}}, \tag{4.7.23}$$

is a maximal invariant under the group $\mathscr{G}$ of transformations, multiplication of $\bar{M}$ and S by a positive constant β. In other words, if $\hat{\tau}_{s,t} = (\hat{\tau}_{s,t}(1),\ldots,\hat{\tau}_{s,t}(K))$ and $g \in \mathscr{G}$,

$$g(\hat{\tau}_{s,t}) = (\beta\hat{\tau}_{s,t}(1),\ldots,\beta\hat{\tau}_{s,t}(K)) \qquad \text{for } \beta > 0, \tag{4.7.24}$$

$$\bar{M}(g(\hat{\tau}_{s,t})) = \beta\frac{1}{K}\sum_{i=1}^{K}\hat{\tau}_{s,t}(i) = \beta\bar{M}(\hat{\tau}_{s,t}), \tag{4.7.25a}$$

$$S(g(\hat{\tau}_{s,t})) = \left[\sum_{i=1}^{K}(\beta\hat{\tau}_{s,t} - \beta\bar{M}(\hat{\tau}_{s,t}))^2\right]^{1/2} = \beta S(\hat{\tau}_{s,t}), \tag{4.7.25b}$$

and

$$T(g(\hat{\tau}_{s,t})) = \frac{\sqrt{K}\,\bar{M}(g(\hat{\tau}_{s,t}))}{S(g(\hat{\tau}_{s,t}))/\sqrt{K-1}} = \frac{\beta\sqrt{K}\,\bar{M}(\hat{\tau}_{s,t})}{\beta S(\hat{\tau}_{s,t})/\sqrt{K-1}} = T(\hat{\tau}_{s,t}). \tag{4.7.26}$$

$\mathscr{G}$ induces the group of transformations $\bar{\mathscr{G}}$ on the parameter space of (ξ,σ) such that for $\bar{g} \varepsilon \bar{\mathscr{G}}$

$$\bar{g}(\xi,\sigma) = (\beta\xi,\beta\sigma). \tag{4.7.27}$$

However, with respect to the space of the parameter $\theta = \xi/\sigma$, $\bar{\mathscr{G}}$ leaves the problem invariant. The original hypotheses are also left invariant since

$$\Phi\left(\frac{r_{s,t} - v_1(q)}{\sigma}\right) = \Phi\left(-\frac{\xi}{\sigma}\right) = \Phi\left(-\frac{\beta\xi}{\beta\sigma}\right). \tag{4.7.28}$$

Consequently, it is reasonable to restrict ourselves to test problems which remain invariant under $\mathscr{G}$ or, in other words, problems which depend on $\hat{\tau}_{s,t}(1), \hat{\tau}_{s,t}(2), \ldots, \hat{\tau}_{s,t}(K)$ only through $T(\hat{\tau}_{s,t})$.

Let $\delta = \sqrt{K}\theta$. The probability density $p_\delta(t)$ of the random variable T can be shown to be

$$p_\delta(t) = \left[2^{K/2}\,\Gamma\left(\frac{K-1}{2}\right)\sqrt{\pi(K-1)}\right]^{-1}$$

$$\cdot \int_0^\infty \exp\left[-\frac{1}{2}\left(t\sqrt{\frac{\omega}{K-1}} - \delta\right)^2\right] \omega^{(K-2)/2} \exp\left(-\frac{\omega}{2}\right) d\omega. \tag{4.7.29}$$

The distribution of T is known as the noncentral Student's distribution, with $K - 1$ degrees of freedom.

Consider the ratio

$$\Lambda(t) = \frac{\int_0^\infty \exp\left[-\frac{1}{2}\left(t\sqrt{\frac{\omega}{K-1}} - \delta_1\right)^2\right] \omega^{(K-2)/2} \exp\left(-\frac{\omega}{2}\right) d\omega}{\int_0^\infty \exp\left[-\frac{1}{2}\left(t\sqrt{\frac{\omega}{K-1}} - \delta_0\right)^2\right] \omega^{(K-2)/2} \exp\left(-\frac{\omega}{2}\right) d\omega}. \tag{4.7.30}$$

It can be shown that $\Lambda(t)$ is an increasing function of t for $\delta_0 < \delta_1$. This means that the family of probability densities $\{p_\delta(t); -\infty < \delta < \infty\}$ constitutes what is known as a *monotone likelihood ratio family*. The ratio $\Lambda(t)$ is the ratio of joint density of $T(\hat{\tau}_{s,t})$, given that $\delta = \delta_1$, to the joint density of $T(\hat{\tau}_{s,t})$, given that $\delta = \delta_0$. By the Neyman-Pearson Lemma, the Most Powerful Test for testing $H_0^*: \delta = \delta_0$ against $H_1^*: \delta = \delta_1 < \delta_0$ rejects H_0^* if $\Lambda(t) \leq \hat{K}$ and accepts H_0^* if $\Lambda(t) > \hat{K}$. However, because the family $\{p_\delta(t), -\infty < \delta < \infty\}$ is a monotone likelihood ratio family, it can be shown ([LE1], Section 3.3) that this test is Uniformly Most Powerful for testing $\hat{H}_0$ against $\hat{H}_1$, among all tests depending on $\hat{\tau}_{s,t}$ only through $T(\hat{\tau}_{s,t})$. In terms of the original variables, consider the Uniformly Most Powerful Invariant Level α Test for testing

$$H_0: \text{Prob}\,\{\tau_{s,t} \geq r_{s,t}\} \geq p_0 \tag{4.7.31a}$$

against

$$H_1: \text{Prob}\,\{\tau_{s,t} \geq r_{s,t}\} < p_0. \tag{4.7.31b}$$

The test rejects hypothesis H_0 if

$$\frac{\sqrt{K}(\bar{M}(K) - r_{s,t})}{\left[\sum_{i=1}^{K} (\tau_{s,t}(i) - \bar{M}(K))^2/(K-1)\right]^{1/2}} \leq \hat{K} \tag{4.7.32a}$$

and accepts H_0 if

$$\frac{\sqrt{K}(\bar{M}(K) - r_{s,t})}{\left[\sum_{i=1}^{K} (\tau_{s,t}(i) - \bar{M}(K))^2/(K-1)\right]^{1/2}} > \hat{K} \tag{4.7.32b}$$

where $\bar{M}(K) = 1/K \sum_{i=1}^{K} \tau_{s,t}(i)$. The constant $\hat{K}$ is determined by

$$\int_{-\infty}^{\hat{K}} p_{\delta_0}(t)\, dt = \alpha. \tag{4.7.32c}$$

Substituting the proper expression for $p_{\delta_0}(t)$ into (4.7.32) gives the constraint

$$\int_{-\infty}^{\hat{K}} \left[\int_0^{\infty} \omega^{(K-2)/2} \exp\left(-\frac{\omega}{2}\right) \exp\left[-\frac{1}{2}\left(t\sqrt{\frac{\omega}{K-1}} - \delta_0\right)^2\right] d\omega \right] dt$$
$$= \alpha 2^{K/2} \Gamma\left(\frac{K-1}{2}\right) \sqrt{\pi(K-1)}. \quad (4.7.32d)$$

Thus, to find $\hat{K}$ we must evaluate the integral on the left-hand side of the preceding equation. As we have already noted, $p_\delta(t)$ is the probability density of Student's noncentral t distribution with $K - 1$ degrees of freedom. A good approximation is

$$\int_{-\infty}^{t} p_\delta(t)\, dt \approx \text{Prob}\left\{\chi < \frac{t\left(1 - \dfrac{1}{4(K-1)}\right) - \delta}{\left(1 + \dfrac{t^2}{2(K-1)}\right)^{1/2}}\right\} \quad (4.7.33)$$

where χ is a Standard Normal Variable [AB1]. Applying this approximation to (4.7.32d) gives

$$\alpha = \text{Prob}\left\{\chi < \frac{\hat{K}\left(1 - \dfrac{1}{4(K-1)}\right) - \delta}{\left(1 + \dfrac{\hat{K}^2}{2(K-1)}\right)^{1/2}}\right\}. \quad (4.7.34)$$

Hence, $\hat{K}$ can be approximately found by solving the quadratic equation

$$\hat{K}^2\left[\left(1 - \frac{1}{4(K-1)}\right)^2 - \frac{[\Phi^{-1}(\alpha)]^2}{2(K-1)}\right]$$
$$- 2\delta\left(1 - \frac{1}{4(K-1)}\right)\hat{K} + \delta^2 - [\Phi^{-1}(\alpha)]^2 = 0, \quad (4.7.35)$$

where $\delta = -\sqrt{K}\Phi^{-1}(1 - p_0)$. Care must be taken to choose the correct root of (4.7.35) as the value of $\hat{K}$. The other root corresponds to the error probability $\alpha' = 1 - \alpha$. (The smaller root corresponds to the smaller of α and $\alpha' = 1 - \alpha$.)

An iterative procedure based on the approximation given above has been developed by Johnson and Welsh [JO1]. Because certain tabulated results (see Table 4.7.1 for an example of these results) are needed, this procedure is directly applicable only for $\alpha = 0.005, 0.01, 0.025, 0.05, 0.1, 0.2, 0.3, 0.4, 0.5$ unless additional tables are compiled. If $\alpha < 0.5$, the method can be described briefly as follows.

For a first approximation, let

$$-\hat{K}_1 = \frac{-\delta + \Phi^{-1}(1-\alpha)\left[1 + \dfrac{\delta^2 - [\Phi^{-1}(1-\alpha)]^2}{2(K-1)}\right]^{1/2}}{1 - \dfrac{[\Phi^{-1}(1-\alpha)]^2}{2(K-1)}}. \quad (4.7.36)$$

Here we have replaced the term $1 - 1/4(K - 1)$ in (4.7.35) by unity and selected the proper root. If

$$\left| \frac{\hat{K}_1}{\sqrt{2(K-1)}} \right| < 0.75,$$

calculate

$$y^* = \frac{\hat{K}_1}{\sqrt{2(K-1)}} \left(1 + \frac{\hat{K}_1^2}{2(K-1)}\right)^{-1/2}. \tag{4.7.37a}$$

Otherwise, calculate

$$y = \left(1 + \frac{\hat{K}_1^2}{2(K-1)}\right)^{-1/2}. \tag{4.7.37b}$$

Then, using the table given in [JO1] with the appropriate value of α, find $\lambda_1(K - 1, -\hat{K}_1, \alpha)$ corresponding to $-\hat{K}_1$ and $K - 1$. The second approximation is

$$-\hat{K}_2 = \frac{-\delta + \lambda_1 \left[1 + \dfrac{\delta^2 - \lambda_1^2}{2(K-1)}\right]^{1/2}}{1 - \dfrac{\lambda_1^2}{2(K-1)}}. \tag{4.7.38}$$

In other words, we have replaced $\Phi^{-1}(1 - \alpha)$ by λ_1. We can continue this procedure until two consecutive $\hat{K}_1$ are within a specified tolerance. If we terminate when two consecutive $\hat{K}_1$ agree to three places after the decimal point, the process almost always terminates after two or three iterations.

Example 4.7.4 We want to test the hypothesis H_0: Prob $\{\tau_{s,t} \geq r_{s,t}\} \geq 0.9$ at confidence level $\alpha = 0.1$. Suppose we must make a decision based on ten values of $\tau_{s,t}$, say $\tau_{s,t}(1), \ldots, \tau_{s,t}(10)$. Then the Uniformly Most Powerful Invariant Test for this problem is: Reject H_0 if

$$\sqrt{10}\left[\frac{1}{10}\sum_{k=1}^{10} (\tau_{s,t}(k)) - r_{s,t}\right] \leq \hat{K}\frac{1}{\sqrt{9}}\left[\sum_{k=1}^{10} (\tau_{s,t}(k) - \frac{1}{10}\sum_{k=1}^{10} (\tau_{s,t}(k))^2\right]^{1/2} \tag{4.7.39}$$

and accept H_0 otherwise.

According to (4.7.35), $\hat{K}$ is approximately equal to a root of the quadratic equation

$$\hat{K}^2\left[\left(1 - \frac{1}{36}\right)^2 - \frac{[\Phi^{-1}(0.1)]^2}{18}\right] - 2(-\sqrt{10}\Phi^{-1}(1 - 0.9))\left(1 - \frac{1}{36}\right)\hat{K}$$
$$+ 10[\Phi^{-1}(1 - 0.9)]^2 - [\Phi^{-1}(0.1)]^2 = 0. \tag{4.7.40}$$

Table 4.7.1 Values of $\lambda(K,\hat{K},\alpha)$ for $\alpha = 0.10$.

$\hat{K}$	y^*	y	$K = 4$	5	6	7	8	9	16	36	144	∞
			$12/\sqrt{K} =$					4	3	2	1	0
$-\infty$	-1.0	0.0	1.116	1.136	1.150	1.161	1.169	1.1765	1.2049	1.2319	1.2575	1.2816
		0.1	1.116	1.136	1.150	1.161	1.170	1.1768	1.2051	1.2321	1.2576	1.2816
		0.2	1.118	1.137	1.151	1.162	1.171	1.1777	1.2057	1.2325	1.2578	1.2816
		0.3	1.121	1.140	1.154	1.164	1.172	1.1793	1.2069	1.2332	1.2581	1.2816
		0.4	1.125	1.143	1.157	1.167	1.175	1.1819	1.2087	1.2343	1.2586	1.2816
		0.5	1.131	1.149	1.162	1.171	1.179	1.1856	1.2114	1.2360	1.2594	1.2816
		0.6	1.140	1.157	1.169	1.178	1.185	1.1912	1.2153	1.2385	1.2606	1.2816
		0.7	1.153	1.168	1.179	1.187	1.194	1.1992	1.2210	1.2421	1.2623	1.2816
Negative	-0.6	0.8	1.173	1.185	1.194	1.201	1.206	1.2110	1.2295	1.2475	1.2649	1.2816
	-0.5		1.191	1.201	1.208	1.214	1.218	1.2222	1.2376	1.2527	1.2673	1.2816
	-0.4		1.209	1.217	1.223	1.227	1.231	1.2338	1.2461	1.2582	1.2700	1.2816
	-0.3		1.228	1.233	1.238	1.241	1.244	1.2458	1.2548	1.2639	1.2728	1.2816
	-0.2		1.246	1.250	1.253	1.255	1.256	1.2578	1.2638	1.2697	1.2757	1.2816
	-0.1		1.264	1.266	1.267	1.268	1.269	1.2698	1.2727	1.2756	1.2786	1.2816
0	0.0	1.0	1.282	1.282	1.282	1.282	1.282	1.2816	1.2816	1.2816	1.2816	1.2816
	0.1		1.298	1.297	1.295	1.294	1.294	1.2929	1.2902	1.2874	1.2845	1.2816
	0.2		1.313	1.310	1.308	1.306	1.305	1.3038	1.2985	1.2930	1.2873	1.2816
	0.3		1.327	1.323	1.320	1.318	1.316	1.3140	1.3064	1.2984	1.2901	1.2816
	0.4		1.340	1.335	1.331	1.328	1.326	1.3233	1.3137	1.3035	1.2927	1.2816
	0.5		1.350	1.345	1.341	1.337	1.334	1.3316	1.3204	1.3082	1.2952	1.2816
	0.6	0.8	1.358	1.353	1.349	1.345	1.342	1.3387	1.3263	1.3124	1.2974	1.2816
Positive		0.7	1.365	1.360	1.355	1.352	1.348	1.3453	1.3320	1.3166	1.2997	1.2816
		0.6	1.367	1.363	1.359	1.355	1.352	1.3492	1.3354	1.3193	1.3012	1.2816
		0.5	1.368	1.365	1.361	1.358	1.354	1.3516	1.3377	1.3210	1.3021	1.2816
		0.4	1.369	1.366	1.362	1.359	1.356	1.3531	1.3392	1.3222	1.3028	1.2816
		0.3	1.369	1.367	1.363	1.360	1.357	1.3541	1.3401	1.3229	1.3032	1.2816
		0.2	1.370	1.367	1.364	1.361	1.358	1.3548	1.3408	1.3234	1.3035	1.2816
		0.1	1.370	1.367	1.364	1.361	1.358	1.3552	1.3411	1.3237	1.3037	1.2816
∞	1.0	0.0	1.370	1.368	1.364	1.361	1.358	1.3554	1.3413	1.3238	1.3038	1.2816

Since $1 - 1/36 \approx 1$, we use the approximation given by (4.7.36) to obtain

$$-\hat{K}_1 = \frac{-1.282\sqrt{10} + 1.282\left(1 + \dfrac{1.282^2}{2}\right)^{1/2}}{1 - \dfrac{1.282^2}{18}}. \tag{4.7.41a}$$

Solving this equation gives

$$\hat{K}_1 = 2.557. \tag{4.7.41b}$$

Now, since $|-2.557/\sqrt{18}| < 0.75$, we calculate

$$y^* = \frac{-2.557}{\sqrt{18}}\left(1 + \frac{2.557^2}{18}\right)^{1/2} = -0.5162. \tag{4.7.42}$$

From Table 4.7.1 for $\alpha = 0.1$

$$\lambda_1(9,-2.557,0.1) = 1.2204, \tag{4.7.43}$$

and therefore

$$-\hat{K}_2 = \frac{1.2204(-1.8414)}{0.91726} = -2.4499 \tag{4.7.44}$$

and

$$y^* = \frac{-2.4499}{18}\left(1 + \frac{2.4499^2}{18}\right)^{-1/2} = -0.500. \tag{4.7.45}$$

Then

$$\lambda_2(9,-2.4499,0.1) = 1.2222 \tag{4.7.46}$$

and, since λ_2 differs from λ_1 in the third decimal place,

$$\hat{K} \doteq \hat{K}_2 = 2.45. \tag{4.7.47}$$

Note that the first and second approximations, $\hat{K}_1$ and $\hat{K}_2$, are also close so we are not far from the initial approximation.

Example 4.7.5 Consider a graph G with fixed branch capacities but random branch flows. We must determine whether a flow of 1 unit can be attained between the pair of vertices v_s and v_t at least 90 percent of the time. Suppose we want the probability of Type I error to be no greater than 0.1. Our statistical problem is to test

$$H_0 : \text{Prob}\,\{\tau_{s,t} \geq 1\} \geq 0.9 \tag{4.7.48a}$$

against

$$H_1 : \text{Prob}\,\{\tau_{s,t} \geq 1\} < 0.9 \tag{4.7.48b}$$

at confidence level 0.1.

Assume there are ten samples of the branch flow vector **F**. We can therefore compute ten values of maximal flow $\tau_{s,t}(1), \ldots, \tau_{s,t}(10)$. Suppose these numbers are: 2.1, 0.3, 3.1, 1.4, 0.6, 4.2, 1.7, 2.8, 3.5, 0.1. Then

$$\bar{M}(K) = \tfrac{1}{10}(21.8) = 2.18 \tag{4.7.49a}$$

$$\left[\sum_{k=1}^{10} (\tau_{s,t}(k) - \bar{M}(K))^2\right]^{1/2} = 3.76 \tag{4.7.49b}$$

and we reject H_0 if

$$\frac{3\sqrt{10}(2.18 - 1)}{3.76} = 2.476 \leq \hat{K}. \tag{4.7.50}$$

From Example 4.7.4, $\hat{K} = 2.45$ and so we accept H_0.

Thus, we have given a procedure for testing, on the basis of a set of observations, whether Prob $\{\tau_{s,t} \geq r_{s,t}\}$ is at least p_0. As in the statistical procedure of Section 3, we note that although the U.M.P. Invariant Test minimizes Type II error for a given Type I error, the error involved may still be large. Hence, if we want stronger control over the probability of deciding that Prob $\{\tau_{s,t} \geq r_{s,t}\} \geq p_0$ when it is actually less than p_0, we should test H'_0 : Prob $\{\tau_{s,t} \geq r_{s,t}\} \leq p_0$ against H'_1 : Prob $\{\tau_{s,t} \geq r_{s,t}\} > p_0$. Since Prob $\{\tau_{s,t} \geq r_{s,t}\} = 1 -$ Prob $\{\tau_{s,t} < r_{s,t}\}$ an equivalent test is H'_0: Prob $\{\tau_{s,t} < r_{s,t}\} \geq 1 - p_0 = p'_0$ against H'_1 : Prob $\{\tau_{s,t} < r_{s,t}\} < p'_0$. This test is similar to the one discussed here and is examined in detail by Lehmann.

The statistical test discussed above seems to be the best testing procedure available. It is "optimal" if the terminal capacity is a normal random variable. However, even if flow is not normally distributed the test gives a reasonable method of determining whether the flow probability is large or small. From (4.7.32b), we accept the hypothesis if the mean of the sample is "high" while the variance of the sample is "low." The optimality of the test stems from the precision with which the concepts of "high" and "low" are defined.

4.8 FURTHER REMARKS

In this chapter, we have given various methods for the probabilistic analysis of maximum flow in networks. These methods are contained primarily in three papers by Frank and Hakimi [FR1–3]. A number of extensions of their results and modifications of their assumptions are possible. For example, in all cases the random variables of interest were assumed to be stationary. In reality, traffic flow is usually time-dependent and this assumption may not be valid. Other assumptions also deserve further investigation, especially those assumptions on which the statistical testing procedures are based. In general, the statistical procedures of Sections 3 and 7 appear to be more promising than the probabilistic ones of Sections 4 and 5. The main difference between the two types of approach is that the probabilistic techniques require that we find all the cut-sets of the graph while the statistical procedures require only that we find the maximum flows through the graph at various instances of time. Since we can find the maximum flow *without* enumerating cut-sets, the statistical procedures are computationally far more efficient. On the other hand, the statistical procedures are subject to error.

Numerous other methods for statistical testing and estimation of flow rate probabilities can be used for the analysis problem. For example, Bayes decision theory [WA1] can easily be applied, as well as a variety of nonparametric [FR4] and parametric [FI1] estimation procedures. Methods of sequential testing [FI1] could also be applied.

PROBLEMS

1. Let $\tau_{s,t}(1), \tau_{s,t}(2), \ldots, \tau_{s,t}(K)$ be identically and independently distributed observations of the maximum flow $\tau_{s,t}$ from v_s to v_t. Develop a statistical test to test the hypothesis

$$H_0 : p = \text{Prob}\,\{\tau_{s,t} \geq r_{s,t}\} \leq p_0$$

against

$$H_1 : p > p_0$$

at confidence level α, when the probability distribution of $\tau_{s,t}$ is unknown. What are the constants associated with the test when $\alpha = 0.1$ and $p_0 = 0.8$?

2. Let $C_1, \ldots, C_m$ be random branch capacities of a graph G. Suppose that each C_i is a discrete random variable which assumes two values, $C_i = c$ and $C_i = 0$ with probabilities p and $1 - p$, respectively. Find a general expression for Prob $\{\tau_{s,t} \geq r_{s,t}\}$. How many operations must be performed to evaluate this expression?

3. a) Investigate the Sequential Probability Ratio Test (SPRT) for testing hypotheses concerning binomial random variables.
   b) Apply the results of (a) to determine a test for

$$H_0 : \text{Prob}\,\{\tau_{s,t} \geq r_{s,t}\} \geq p_0$$

   against

$$H_1 : \text{Prob}\,\{\tau_{s,t} \geq r_{s,t}\} \leq p_1 \qquad (p_1 < p_0)$$

   when the probabilities of Type I and Type II errors are α_1 and α_2, respectively [FI1].
   c) Apply the test developed in (b) to test

$$H_0 : \text{Prob}\,\{\tau_{s,t} \geq 1\} \geq 0.8$$

   against

$$H_1 : \text{Prob}\,\{\tau_{s,t} \geq 1\} \leq 0.5$$

   when $\alpha_1 = \alpha_2 = 0.1$ and the maximum flows observed in order are

   2,0,3,1,0,4,1,2,3,0,1,0,2,2,0,1,3,2,0,2,1,2,4,1.

4. Suppose the Uniformly Most Powerful Invariant Test developed in Section 4.7 is used to test

$$\text{H}_0 : \text{Prob}\,\{\tau_{s,t} \geq r_{s,t}\} \geq 0.8$$

against

$$H_1 : \text{Prob}\,\{\tau_{s,t} \geq r_{s,t}\} < 0.8$$

at level α, when 25 samples of maximum flow are taken. Let $\alpha = 0.1$.
   a) The test rejects H_0 if the statistic given by the left-hand side of (4.7.32a) is no larger than $\hat{K}$. Determine $\hat{K}$.
   b) Apply the U.M.P. Invariant Test defined above to the data given in Problem 3(c). For these data, compare the outcomes of
      i) the fixed-sample size pseudo-parametric test;
      ii) the Sequential Probability Ratio Test of Problem 3;
      iii) the U.M.P. Invariant Test.

5. Let G be the graph shown in Fig. P4.1. Suppose that the capacity C_i of branch b_i is an independent Poisson random variable with probability function

$$\text{Prob}\,\{C_i = k\} = \frac{\lambda^k}{k!} e^{-\lambda} \qquad \text{for } k = 0,1,2,\ldots \text{ and } i = 1,2,3.$$

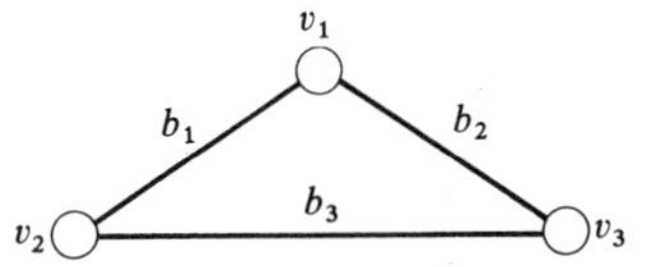

Figure P4.1

a) What is the correlation matrix of the cut-sets?
b) What is Prob $\{\tau_{2,3} \geq k\}$?
c) What is Prob $\{\underset{i,j}{\text{Min}}[\tau_{i,j}] \geq k\}$?

6. Let G be the graph shown in Fig. P4.2. Suppose that each branch b_i contains a random flow F_i with unity variance such that the joint flow distribution is a 3-dimensional normal distribution with mean vector $\boldsymbol{\mu} = (3,3,3)'$ and correlation matrix

$$\begin{bmatrix} 1 & \frac{1}{3} & \frac{1}{3} \\ \frac{1}{3} & 1 & \frac{1}{3} \\ \frac{1}{3} & \frac{1}{3} & 1 \end{bmatrix}.$$

Let $C_i = c_i - F_i$ for $i = 1,2,3$.

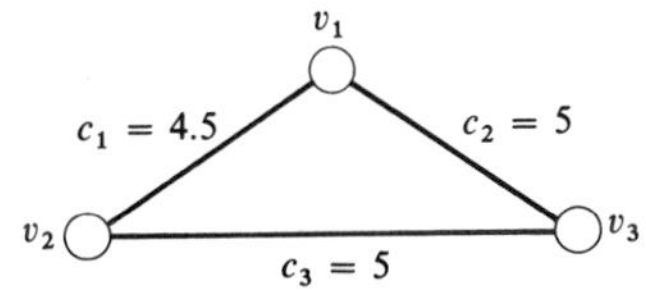

Figure P4.2

a) Find the probability distribution of the cut-set vector when the branch capacities of G are taken to be c_1, c_2, c_3.
b) Express Prob $\{\tau_{2,3} \geq 3\}$ in terms of an integral relationship.
c) Evaluate the integral obtained in (b) by means of tables of the bivariate normal distribution.
d) Use Clark's procedure to find the approximate mean and variance of $\tau_{2,3}$ and Min $[\tau_{1,2}, \tau_{1,3}, \tau_{2,3}]$. Using these numbers, compute Prob $\{\tau_{2,3} \geq 3\}$ and Prob $\{\underset{i,j}{\text{Min}}[\tau_{i,j}] \geq 3\}$.

7. Evaluate the upper and lower bounds given by (4.5.25) in Example 4.5.1.
8. Show that to apply Clark's procedure, we must not only compute

$$\text{Cor}\,((X_1, \ldots, X_j), X_{j+1})$$

but also

$$\text{Cor}\,((X_1, \ldots, X_j), X_{j+k})$$

for $k = 2,3, \ldots, q - j$. What is the total number of computations in Clark's procedure as a function of q?

9. Consider the Sequential Probability Ratio Test discussed in Problem 3. Let the probabilities of Type I and Type II error be $\alpha_1 = \alpha_2 = 0.1$. Find the expected length of the sequential test $H_0 : p \geq 0.8$ against $H_1 : p \leq 0.5$ given that

a) the true value of p is $p = 0.8$;
b) the true value of p is $p = 0.5$.

10. Suppose that the maximal flows $\tau_{s,t}(1), \tau_{s,t}(2), \ldots, \tau_{s,t}(K)$ are random variables generated by K identically and independently distributed measurements of branch flow.

Suppose that the $\tau_{s,t}(i)$ are relabeled as $\hat{\tau}_{s,t}(1), \ldots, \hat{\tau}_{s,t}(K)$ so that

$$\hat{\tau}_{s,t}(1) \leq \hat{\tau}_{s,t}(2) \leq \cdots \leq \hat{\tau}_{s,t}(K).$$

Show that

$$\text{Prob}\,\{\text{Prob}\{\tau_{s,t} < \hat{\tau}_{s,t}(i)\} \leq \hat{p}\} = \int_0^{\hat{p}} h(x_i)\,dx_i$$

where

$$h(x_i) = K!\int_0^{x_1} dx_1 \int_{x_1}^{x_2} dx_2 \cdots \int_{x_{i-2}}^{x_{i-1}} dx_{i-1} \int_{x_{i+1}}^{x_{i+2}} dx_{i+1} \cdots \int_{x_k}^{1} dx_K.$$

11. Let G be the graph shown in Fig. 3.5.1(a). Suppose the branches contain normally distributed random flows. Let the variance of f_i be $\frac{1}{5}c_i$ and the mean of f_i be $\frac{2}{3}c_i$ for $i = 1, \ldots, m$.
    a) Simulate the random flows in G to obtain empirical distribution functions for $\tau_{s,t}$.
    b) Approximate the empirical distribution function by a normal distribution function and apply a "goodness-of-fit" test to the approximation.

12. Let G have branches $b_1, \ldots, b_m$ with random capacities $C_1, \ldots, C_m$. Let Prob $\{C_i = c_{i_0}\} = p_{i_0}$ and Prob $\{C_i = 0\} = 1 - p_{i_0}$ for $i = 1, \ldots, m$.
    a) Derive a general expression for $E\{\tau_{s,t}\}$.
    b) Let b_k be any branch. Then

$$E\{\tau_{s,t}\} = p_{k_0}E\{\tau_{s,t}|C_k = c_{k_0}\} + (1 - p_{k_0})E\{\tau_{s,t}|C_k = 0\}.$$

Show that $\dfrac{\partial E\{\tau_{s,t}\}}{\partial C_{k_0}} > 0$ and hence

$$E\{\tau_{s,t}|C_k = 0\} \leq E\{\tau_{s,t}\} \leq E\{\tau_{s,t}|b_k \text{ is shorted}\}$$

and

$$E\{\tau_{s,t}|p_{k_0} = 0\} \leq E\{\tau_{s,t}\} \leq E\{\tau_{s,t}|p_{k_0} = 1\}.$$

    c) Suppose that p_{k_0} is reduced to $p_{k_0} - \Delta p_k$. Find Δc_k such that, if c_{k_0} is increased to $c_{k_0} + \Delta c_k$, $E\{\tau_{s,t}\}$ is unchanged. Show that there is an upper limit beyond which increasing c_{k_0} will not balance a decrease in p_{k_0}. What conditions determine this limit?

13. Let $C_1, \ldots, C_m$ be independent random branch capacities of $b_1, \ldots, b_m$, respectively. Suppose that C_i takes on discrete values $C_{i_1}, C_{i_2}, \ldots, C_{i_k}$ with probability $p_{i_1}, p_{i_2}, \ldots, p_{i_{k_i}}$ for $i = 1, \ldots, m$. Let $\tau_{s,t}(C_1, \ldots, C_m) = \tau_{s,t}(\mathbf{C})$ be the maximum s-t flow as a function of the random capacity vector $\mathbf{C}$.
    a) Show that $\tau_{s,t}(\mathbf{C})$ is a concave function of each variable C_i [ON1].
    b) Show that $\tau_{s,t}(\mathbf{C})$ is a concave function of the capacity vector $\mathbf{C}$ [ON1].
    c) If $f(x)$ is a concave function of a single variable x, it is well known that

$$E\{f(X)\} \leq f(E\{X\}).$$

A generalization of this statement is: if $(X_1, \ldots, X_m) = f(\mathbf{X})$ is a concave function of $\mathbf{X}$, then

$$E\{f(X_1, X_2, \ldots, X_m)\} \leq E\{f(E\{X_1\}, X_2, \ldots, X_m)\} \leq E\{f(E\{X_1\}, E\{X_2\}, X_3, \ldots, X_m)\}$$
$$\leq \cdots \leq f(E\{X_1\}, E\{X_2\}, E\{X_3\}, \ldots, E\{X_m\}).$$

Use these inequalities to derive upper bounds on $E\{\tau_{s,t}(\mathbf{C})\}$.

d) Compute $E\{\tau_{s,t}\}$ for the graph shown in Fig. P4.3 where

$$\begin{aligned} C_1 &= 0, 2; & \text{Prob}\{C_1 = 2\} &= 0.95. \\ C_2 &= 0, 2; & \text{Prob}\{C_2 = 2\} &= 0.90. \\ C_3 &= 0, 3; & \text{Prob}\{C_3 = 3\} &= 0.90. \\ C_4 &= 0, 2; & \text{Prob}\{C_4 = 2\} &= 0.99. \\ C_5 &= 0, 1; & \text{Prob}\{C_5 = 1\} &= 0.98. \end{aligned}$$

Compare the upper bounds obtained in (c) with $E\{\tau_{s,t}\}$.

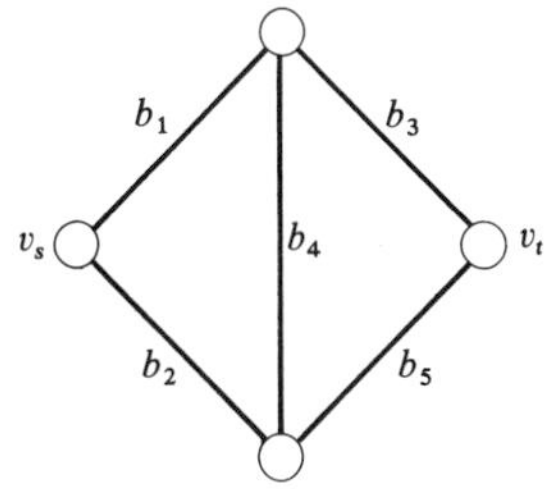

Figure P4.3

14. a) Generalize the analyses given in this chapter to the two-commodity network flow problem.
    b) Suppose both the branches and vertices of G are capacity-limited. Generalize the analysis procedures to graphs of this type.

REFERENCES

AB1 M. Abramowitz and L. A. Stegun (eds.), *Handbook of Mathematical Functions*, National Bureau of Standards, and American Mathematical Society, Washington, D.C., 1964, p. 933.

CL1 C. E. Clark, "The Greatest of a Finite Set of Random Variables," *Operations Res.* **9,** 145–162 (1961).

CR1 H. Cramer, *Mathematical Methods of Statistics*, Princeton University Press, Princeton, 1963.

EL1 H. N. El-Ghoroury, "Maximum Flow in Probabilistic Networks," M.S. Research Project, Department of Electrical Engineering, University of California, 1968.

FE1 W. Feller, *An Introduction to Probability Theory and Its Application*, Vol. 1, Wiley, New York, 1956.

FI1 M. Fisz, *Probability Theory and Mathematical Statistics*, Wiley, New York, 1963.

FR1 H. Frank and S. L. Hakimi, "Probabilistic Flows through a Communication Network," *IEEE Trans. Circuit Theory* **CT-12,** 413–414 (1965).

FR2 H. Frank and S. L. Hakimi, "On the Optimum Synthesis of Statistical Communication Networks—Pseudo Parametric Techniques," *J. Franklin Inst.* **284,** 407–416 (1967).

FR3 H. Frank and S. L. Hakimi, "Parametric Analysis of Statistical Communication Networks," *Quart. Appl. Math.* **26,** 249–263 (1968).

FR4 D. A. S. Fraser, *Nonparametric Methods of Statistics*, Wiley, New York, 1957.

GA1 F. R. Gantmacher, *The Theory of Matrices*, Vol. 1, Chelsea, New York, 1960.

GN1 B. Gnedenko, *Theory of Probability*, Chelsea, New York, 1962.

GU1 S. Gupta, "Probability Integrals of Multivariate Normal and Multivariate *t*," *Ann. Math. Statistics* **34,** 828 (1963).

HA1 J. M. Hammersley and D. C. Handscomb, *Monte Carlo Methods*, Methuen, London, 1965.

JO1 N. L. Johnson and B. L. Welsh, "Applications of the Non-Central *t*-Distribution," *Biometrika* **40,** 362–389 (1939).

KL1 L. Kleinrock, *Communication Networks: Stochastic Message Flow and Delay*, McGraw-Hill, New York, 1964.

LE1 E. L. Lehmann, *Testing Statistical Hypotheses*, Wiley, New York, 1959.

ON1 K. Onaga, "Bounds on the Average Terminal Capacity of Probabilistic Nets," *IEEE Trans. Inform. Theory* **IT-14,** 766–768 (1968).

WA1 A. Wald, *Statistical Decision Functions*, Wiley, New York, 1964.

CHAPTER 5

MULTITERMINAL ANALYSIS AND SYNTHESIS

5.1 INTRODUCTION

In the chapter on maximum deterministic flow, we considered the problem of satisfying several flow requirements simultaneously. We now consider the analysis problem of finding the maximum flows for various source-terminal pairs when each of these flows is to be sent through the graph separately and the corresponding synthesis problem of constructing a graph to satisfy several requirements individually. More formally, we define a requirement matrix $\mathscr{R} = [r_{i,j}]$ with one row and one column corresponding to each vertex in a graph. For $i \neq j$, $r_{i,j} \geq 0$ is the flow requirement from the vertex v_i, to the vertex v_j. We arbitrarily define $r_{i,i}$ to be infinite. We also define a terminal capacity matrix T with one row and column corresponding to each vertex of a graph. For $i \neq j$ the entry $\tau_{i,j}$ is the value of the minimum i-j cut. We define $\tau_{i,i}$ to be infinite. The synthesis problem consists of finding a graph G with terminal capacity matrix T equal to a given requirement matrix $\mathscr{R}$. (G is said to be a realization of $\mathscr{R}$.) The analysis problem consists of finding a terminal capacity matrix T for a given graph G.

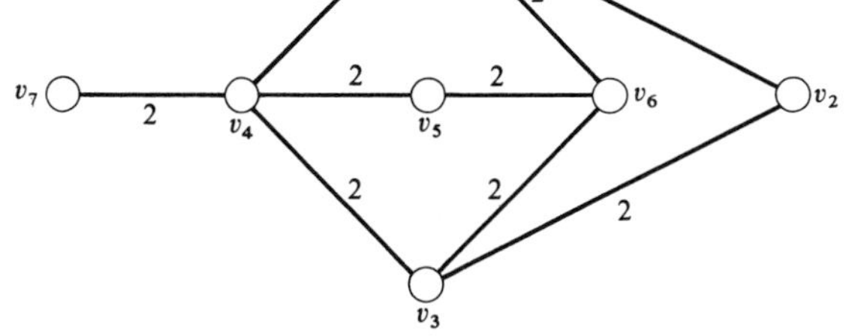

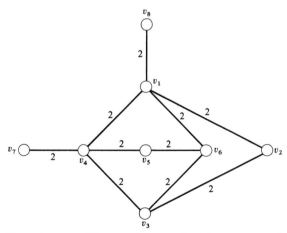

Fig. 5.1.1 Graph illustrating the difference between the Multicommodity and Multiterminal Problems.

To see the difference between the above multiterminal problem and the multicommodity flow problem considered in Chapter 3, examine the graph in Fig. 5.1.1. The terminal capacity matrix for the graph is

$$T = \begin{array}{c} \\ 1 \\ 2 \\ 3 \\ 4 \\ 5 \\ 6 \\ 7 \\ 8 \end{array} \begin{array}{c} \begin{array}{cccccccc} 1 & 2 & 3 & 4 & 5 & 6 & 7 & 8 \end{array} \\ \begin{bmatrix} \infty & 4 & 6 & 6 & 4 & 6 & 2 & 2 \\ 4 & \infty & 4 & 4 & 4 & 4 & 2 & 2 \\ 6 & 4 & \infty & 6 & 4 & 6 & 2 & 2 \\ 6 & 4 & 6 & \infty & 4 & 6 & 2 & 2 \\ 4 & 4 & 4 & 4 & \infty & 4 & 2 & 2 \\ 6 & 4 & 6 & 6 & 4 & \infty & 2 & 2 \\ 2 & 2 & 2 & 2 & 2 & 2 & \infty & 2 \\ 2 & 2 & 2 & 2 & 2 & 2 & 2 & \infty \end{bmatrix} \end{array}. \tag{5.1.1}$$

Hence a flow of value $r_{i,j} = \tau_{i,j}$ can be sent from v_i to v_j as specified by T in (5.1.1). However, we have shown in Chapter 3 that $f^1_{2,5} = 4$, $f^2_{7,6} = 2$, and $f^3_{8,3} = 1$ cannot be satisfied simultaneously.

The concept of terminal capacities arises in a number of physical problems. For example, from the point of view of vulnerability or reliability, the terminal capacity $\tau_{i,j}$ represents the "weakest" section between v_i and v_j and can therefore be used as a reliability measure. Indeed, if all branch weights are unity, the *i-j* entry in T represents the minimum number of branches that must be removed to break all *i-j* paths. As another example, there are systems in which only a single pair of stations can communicate with each other at a time. In this case the entry in the *i-j* position of T represents the actual capacity available for communication from v_i to v_j. One such system is a teletype system in which there is only one sender at any given time.

In Section 2 we present the basic theorems concerning terminal capacity matrices for branch weighted graphs. These are applied to analysis in Section 3 and synthesis in Sections 4 through 8. In Section 9, the results are generalized to include graphs with weighted vertices.

5.2 PROPERTIES OF TERMINAL CAPACITY MATRICES

As a prerequisite to studying both the analysis and synthesis problems several properties of terminal capacity matrices must be derived. Basic to all these properties will be Theorems 5.2.1, 5.2.2, and 5.2.3.

***Definition* 5.2.1** *A matrix M is said to be semiprincipally partitionable or belong to the set of semiprincipally partionable matrices ($M \in SPP$) if after possibly permuting rows and the corresponding columns, the resulting matrix satisfies (5.2.1) below:*

M is square. (5.2.1a)

M has only real nonnegative entries. (5.2.1b)

The entries on the diagonal are infinite. (5.2.1c)

M can be partitioned as in (5.2.1d) where $M_{1,1}$ is a square matrix and all entries in $M_{1,2}$ are equal to the smallest entry in M.

$$M = \left[\begin{array}{c|c} M_{1,1} & M_{1,2} \\ \hline M_{2,1} & M_{2,2} \end{array}\right] \tag{5.2.1d}$$

Every submatrix on the main diagonal resulting from a partitioning as in (5.2.1d) can again be so partitioned until all resulting submatrices on the main diagonal are of order one. (5.2.1e)

As an example, the matrix $\hat{M}$ in (5.2.2a) is semiprincipally partionable, since after interchanging columns 2 and 6 and rows 2 and 6 the resulting matrix M in (5.2.2b) satisfies (5.2.1). The partitioning corresponding to (5.2.1d) and (5.2.1e) is indicated in (5.2.2b).

$$\hat{M} = \begin{bmatrix} \infty & 2 & 3 & 3 & 2 & 5 \\ 15 & \infty & 17 & 18 & 8 & 16 \\ 4 & 2 & \infty & 3 & 2 & 3 \\ 10 & 2 & 4 & \infty & 2 & 12 \\ 5 & 5 & 4 & 2 & \infty & 4 \\ 6 & 2 & 3 & 3 & 2 & \infty \end{bmatrix} \tag{5.2.2a}$$

$$M = \left[\begin{array}{cc|cc|cc} \infty & 5 & 3 & 3 & 2 & 2 \\ 6 & \infty & 3 & 3 & 2 & 2 \\ \hline 4 & 3 & \infty & 3 & 2 & 2 \\ 10 & 12 & 4 & \infty & 2 & 2 \\ \hline 5 & 4 & 4 & 2 & \infty & 5 \\ 15 & 16 & 17 & 18 & 8 & \infty \end{array}\right] \tag{5.2.2b}$$

We now derive the first of two important necessary conditions on T.

Theorem 5.2.1 *If T is the terminal capacity matrix of a graph, then $T \in SPP$.*

Proof. By definition of a terminal capacity matrix (5.2.1a, b, c) are necessary. We need show only that T can be partitioned as in (5.2.1d, e). Let G be a graph with terminal capacity matrix T and let

$$c(X_1,\bar{X}_1) = \min_{X_l \subseteq V} [c(X_l,\bar{X}_l)].$$

That is, $(X_1,\bar{X}_1)$ is a cut of smallest value in G. Then, for all $v_i \in X_1$ and $v_j \in \bar{X}_1$, $\tau_{i,j} = c(X_1,\bar{X}_1)$. Hence, by permuting the rows and corresponding columns of T so the first $|X_1|$ rows and columns correspond to the vertices in X_1, the matrix is partitioned as in (5.2.1d) with all entries in $T_{1,2}$ equal to $c(X_1,\bar{X}_1)$, the smallest entry in T. Next examine a submatrix, say $T_{1,1}$ on the main diagonal. The entries in $T_{1,1}$ represent terminal capacities between pairs of vertices in X_1. In G, let $c(X_2,\bar{X}_2)$ be the value of the smallest cut breaking all directed paths between two vertices in X_1. Then, by further permuting the rows and corresponding columns of T so the first $|X_2 \cap X_1|$ rows and columns correspond to the vertices in X_2, $T_{1,1}$ is again partitioned as in (5.2.1d). Note that this new permutation does not affect the original partitioning. In exactly the same way all resulting submatrices on the main diagonal can be partitioned until only submatrices of order one remain.

To place an $n \times n$ matrix in semiprincipally partitioned form we must partition the matrix exactly $(n - 1)$ times. Every element above the diagonal of T is an element of the submatrix in the off diagonal position of a partitioning. Moreover the above diagonal position matrices have identical entries. There are at most $(n - 1)$ numerically distinct entries above the main diagonal, since each off diagonal submatrix contains at least one element of the first super-diagonal of T. We therefore have:

Corollary 5.2.1 *If T is the terminal capacity matrix of a graph, the maximum number of numerically distinct entries in T is $(n + 2)(n - 1)/2$ (excluding the entries on the diagonal), where n is the number of vertices in the graph.*

Proof. Assume T is in semiprincipally partitioned form. Above the main diagonal the maximum number of distinct entries is $(n - 1)$. The number of distinct entries below the diagonal is $n(n - 1)/2$. The maximum number of distinct entries is therefore $(n + 1) + n(n - 1)/2 = (n + 2)(n - 1)/2$.

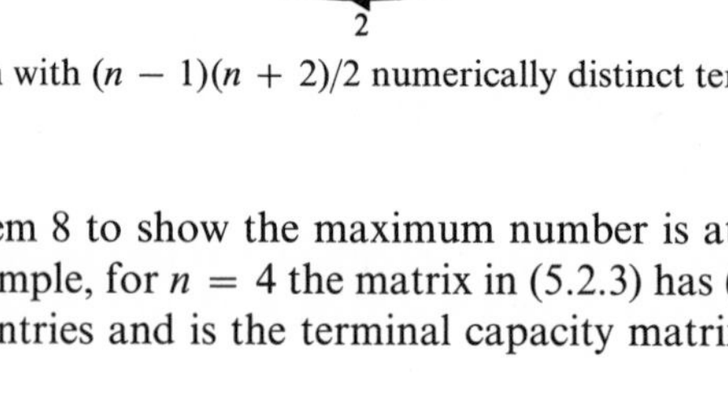

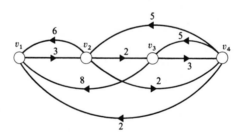

Fig. 5.2.1 Graph with $(n - 1)(n + 2)/2$ numerically distinct terminal capacities.

It is left as Problem 8 to show the maximum number is attainable *for arbitrary values of n*. As an example, for $n = 4$ the matrix in (5.2.3) has $(n + 2)(n - 1)/2 = 9$ numerically distinct entries and is the terminal capacity matrix of the graph in Fig. 5.2.1.

$$T = \begin{bmatrix} \infty & 3 & 3 & 3 \\ 10 & \infty & 4 & 4 \\ 11 & 6 & \infty & 5 \\ 12 & 8 & 7 & \infty \end{bmatrix}. \tag{5.2.3}$$

***Corollary* 5.2.2** *If T is in semiprincipally partitioned form, then for* $i + 1 < j$

$$\tau_{i,j} = \operatorname*{Min}_{k=i,\ldots,(j-1)} [\tau_{k,k+1}]. \tag{5.2.4}$$

Proof. In the semiprinciple partitioning process, every element $\tau_{i,j}$ with $i + 1 < j$ eventually appears in a submatrix in the off diagonal position of a main diagonal submatrix, say, T_1. $\tau_{i,j}$ is then the smallest element in T_1 and is equal to an element of T_1 of the form $\tau_{k,k+1}$ for some k in the range $k = i, \ldots, (j = 1)$.

It is easy to visualize (5.2.4) since the elements $\tau_{k,k+1}$ for $k = i, \ldots, (j - 1)$ are simply the elements on the first superdiagonal of T as shown below.

$$T = \begin{bmatrix} \tau_{i,i+1} & \cdots & & \tau_{i,j} \\ & \tau_{i+1,i+2} & & \vdots \\ & & \ddots \; \tau_{j-2,j-1} & \vdots \\ & & & \tau_{j-1,j} \end{bmatrix}. \tag{5.2.5}$$

Note that the matrix, T, must already be in semiprincipally partitioned form for the conclusion of the theorem to hold. Thus, it does not hold for $\hat{M}$ in (5.2.2a), but is valid for M in (5.2.2b)..

We now obtain another necessary condition on T in Theorem 5.2.2.

***Theorem* 5.2.2** *If T is the terminal capacity matrix of a graph then for all i, j, and k*

$$\tau_{i,j} \geq \text{Min}\,[\tau_{i,k}, \tau_{k,j}]. \tag{5.2.6}$$

Proof. Let $(X,\bar{X})$ be an *i-j* cut with value $\tau_{i,j}$. For any $k \neq i,j$, $v_k \in X$ or $v_k \in \bar{X}$. If $v_k \in \bar{X}$, $(X,\bar{X})$ is also an *i-k* cut, and $\tau_{i,k} \leq \tau_{i,j}$. If $v_k \in X$, $(X,\bar{X})$ is also a *k-j* cut, and $\tau_{k,j} \leq \tau_{i,j}$. In either case, $\tau_{i,j} \geq \text{Min}\,[\tau_{i,k}, \tau_{k,j}]$.

Equation (5.2.6) is called the *triangle inequality*. From Theorem 5.2.2 we immediately have:

***Corollary* 5.2.3** *If T is the terminal capacity matrix of a graph, then for all* v_i, v_j, $v_1, \ldots, v_q$

$$\tau_{i,j} \geq \text{Min}\,[\tau_{i,a_1}, \tau_{a_1,a_2}, \tau_{a_2,a_3}, \ldots, \tau_{a_{q-2},a_{q-1}}, \tau_{a_{q-1},a_q}, \tau_{a_q,j}]. \tag{5.2.7}$$

Proof. The statement is true for $q = 1$ since it reduces to Theorem 5.2.2. Suppose the statement is true for $q = k-1$. Then

$$\tau_{i,j} \geq \text{Min}\,[\tau_{i,a_1}, \tau_{a_1,a_2}, \tau_{a_2,a_3}, \ldots, \tau_{a_{k-2},a_{k-1}}, \tau_{a_{k-1},j}]. \tag{5.2.8}$$

But from (5.2.6) we also have

$$\tau_{a_{k-1},j} \geq \text{Min}\,[\tau_{a_{k-1},a_k}, \tau_{a_k,j}]. \tag{5.2.9}$$

Combining (5.2.8) and (5.2.9) we see that (5.2.7) is true for $q = k$. Hence the proof is complete by induction. (5.2.7) will sometimes be referred to as the *extended triangle inequality*.

In order to establish a third basic property of terminal capacity matrices we need several definitions in terms of an arbitrary square matrix M whose rows and columns correspond to the vertices of a graph.

***Definition* 5.2.2** *For an arbitrary subset of vertices $X_i \subseteq V$, the cut matrix, M_i, is the submatrix of M formed by removing the set of columns corresponding to X_i, and by removing the complementary set of rows. If $m_{k,j}$ is an element of M_i then M_i is said to be a cut matrix of $m_{k,j}$ and may be written explicitly $M_i(k,j)$.*

***Definition* 5.2.3** *If $m_{k,j}$ is a largest entry in M_i then M_i is said to be a min-cut matrix of $m_{k,j}$ and is written as either $\mu_i(k,j)$ or μ_i.*

Example 5.2.1 Consider the matrix

$$M = \begin{bmatrix} \infty & 5 & 5 & 4 & 4 \\ 6 & \infty & 30 & 4 & 4 \\ 6 & 50 & \infty & 4 & 4 \\ 8 & 7 & 7 & \infty & 6 \\ 5 & 10 & 11 & 12 & \infty \end{bmatrix}. \tag{5.2.10}$$

With $X_i = \{v_1,v_4\}$, M_i is a min-cut matrix of $m_{4,3}$. M_i is a cut matrix of $m_{1,2}$ but is not a min-cut matrix of $m_{1,2}$ since $m_{4,3} > m_{1,2}$.

***Theorem* 5.2.3** *If T is the terminal capacity matrix of a graph, then every element of T has at least one min-cut matrix.*

Proof. Consider an arbitrary entry $\tau_{k,j}$. There is a k-j cut, say $(X_1,\bar{X}_1)$ of value $\tau_{k,j}$. Therefore $\tau_{k,j}$ is in the cut matrix $M_1(k,j)$. Furthermore for any entry $\tau_{a,b}$ in $M_1(k,j)$ we have $\tau_{a,b} \leq c(X_1,\bar{X}_1) = \tau_{k,j}$. Hence $M_1(k,j)$ is a min-cut matrix of $\tau_{k,j}$ and the proof is completed.

The necessary conditions on T, expressed in Theorems 5.2.1, 5.2.2, and 5.2.3, will be used throughout this chapter, each being useful in different situations. Indeed the conditions expressed in Theorems 5.2.2 and 5.2.3 are equivalent and either of them imply the conditions in Theorem 5.2.1. We now proceed to establish these facts. We first show that the conditions of Theorems 5.2.2 and 5.2.3 are equivalent.

***Lemma* 5.2.1** *Let M be a real square matrix with entries of infinite value on the diagonal. M satisfies the triangle inequality if and only if every element of M has a min-cut matrix.*

Proof. Suppose every element of M has a min-cut matrix. For an arbitrary element $m_{i,j}$, let M_l be its min-cut matrix, corresponding to a set of vertices X_l. For any vertex v_k, either $v_k \in X_l$ or $v_k \in \bar{X}_l$. Hence, for $k \neq j$ either $m_{k,j}$ or $m_{i,k}$ is in M. Thus either

$$m_{i,j} \geq m_{i,k}, \tag{5.2.11}$$

or

$$m_{i,j} \geq m_{k,j}, \tag{5.2.12}$$

so $m_{i,j} \geq \text{Min}\,[m_{i,k},m_{k,j}]$.

Assume M satisfies the triangle inequality and consider an arbitrary element $m_{i,j}$. We will show that $m_{i,j}$ has a min-cut matrix. Let V_a be the set of vertices such that $v_{a_\alpha} \in V_a$ implies

$$m_{i,j} < m_{i,a_\alpha} \tag{5.2.13}$$

and let V_b be the set of vertices such that $v_{b_\beta} \in V_b$ implies

$$m_{i,j} < m_{b_\beta,j}. \tag{5.2.14}$$

If there is a vertex v_p in both V_a and V_b, then $m_{i,j} < \text{Min}\,[m_{i,p}, m_{p,j}]$, which contradicts the triangle inequality. Therefore, $V_a \cap V_b = \phi$.

Let

$$V_c = V - (V_a \cup V_b). \tag{5.2.15}$$

Let V_d be the subset of V_c consisting of all elements v_{d_δ} such that, for at least one element v_{a_α} of V_a,

$$m_{a_\alpha,d_\delta} > m_{i,j}. \tag{5.2.16}$$

Finally let

$$\bar{V}_d = V_c - V_d. \tag{5.2.17}$$

We will prove that M_k is a min-cut matrix of $m_{i,j}$ with

$$X_k = \{v_i\} \cup V_a \cup V_d. \tag{5.2.18}$$

Since $V_a \cap V_b = \phi$, $\bar{X}_k$ is given by

$$\bar{X}_k = \{v_j\} \cup V_b \cup \bar{V}_d. \tag{5.2.19}$$

Suppose to the contrary that

$$m_{q,r} > m_{i,j} \tag{5.2.20}$$

for some $v_q \in X_k$ and $v_r \in \bar{X}_k$. We show that this leads to a contradiction.

CASE 1. Suppose $v_q \in \{v_i\} \cup V_a$ and $v_r \in \{v_j\} \cup V_b$. From (5.2.13), (5.2.14), and (5.2.20),

$$\begin{aligned} m_{i,j} &< \text{Min}\,[m_{i,q}, m_{q,r}, m_{r,j}] \\ &= \text{Min}\,[m_{i,q}, \text{Min}\,[m_{q,r}, m_{r,j}]] \\ &\leq \text{Min}\,[m_{i,q}, m_{q,j}]. \end{aligned} \tag{5.2.21}$$

This contradicts the triangle inequality.

CASE 2. Suppose $v_q \in \{v_i\} \cup V_a$ and $v_r \in \bar{V}_d$. This contradicts the definition of V_d.

CASE 3. Suppose $v_q \in V_d$ and $v_r \in \{v_j\} \cup V_b \cup \bar{V}_d$. From the triangle inequality, (5.2.16), and (5.2.20),

$$m_{a_\alpha,r} \geq \text{Min}\,[m_{a_\alpha,q}, m_{q,r}] > m_{i,j} \tag{5.2.22}$$

for at least one $v_{a_\alpha} \in V_a$. This led to contradictions in Cases 1 and 2. Hence the proof is completed.

We now relate Theorem 5.2.3 and Theorem 5.2.1.

***Lemma* 5.2.2** *Let M be a real, square matrix with infinite entries on the diagonal. If every off diagonal element of M has a min-cut matrix, then M is semiprincipally partitionable.*

Proof. Let m_1 be an entry of smallest value in M. Let M_i be a min-cut matrix, corresponding to $(X_i, \bar{X}_i)$. Since m_1 is a largest element in M_i and a smallest element in M, all elements in M_i are equal to the smallest element in M. Permute the rows and columns of M so the first $|X_i|$ rows correspond to the vertices in X_i. The resulting matrix $\hat{M}$ can then be partitioned as

$$\hat{M} = \left[\begin{array}{c|c} \hat{M}_{1,1} & \hat{M}_{1,2} \\ \hline \hat{M}_{2,1} & \hat{M}_{2,2} \end{array}\right] \tag{5.2.23}$$

where all elements in $\hat{M}_{1,2}$ are equal to m_1.

Let m_2 be an entry of smallest value in $\hat{M}_{1,1}$ and let $\hat{M}_j$ be a min-cut matrix of m_2, corresponding to $(X_j, \bar{X}_j)$. All elements in $\hat{M}_j$ have value m_2 or m_1. Permute the rows and corresponding columns of M so the first $|X_i \cap X_j|$ rows of $\hat{M}$ correspond to $X_i \cap X_j$. Clearly the partitioning in (5.2.23) is unaffected. The resulting matrix $\hat{M}$ can be partitioned as

$$\hat{\hat{M}} = \left[\begin{array}{cc|c} \begin{array}{c|c} \hat{\hat{M}}_{1,1} & \hat{\hat{M}}_{1,2} \\ \hline \hat{\hat{M}}_{2,1} & \hat{\hat{M}}_{2,2} \end{array} & & \hat{M}_{1,2} \\ \hline \hat{M}_{2,1} & & \hat{M}_{2,2} \end{array}\right] \tag{5.2.24}$$

where all entries in $\hat{\hat{M}}_{1,2}$ are equal to m_2.

This process can be continued until each resulting submatrix on the diagonal is of order one. The resulting matrix is semiprincipally partitioned and the proof is completed.

From Lemma 5.2.1 we see that the following statements are equivalent.

a) Every element in M has a min-cut matrix.
b) M satisfies the triangle inequality.

From Lemma 5.2.2, either of these statements implies that M is semiprincipally partitionable. The converse is not true. Thus the matrix in (5.2.25) is in semiprincipally partitioned form but the element of value 3 does not have a min-cut matrix and the matrix does not satisfy the triangle inequality.

$$M = \left[\begin{array}{cc|c} \begin{array}{c|c} \infty & 2 \\ \hline 5 & \infty \end{array} & & \begin{array}{c} 1 \\ 1 \end{array} \\ \hline 3 \quad 4 & & \infty \end{array}\right] \tag{5.2.25}$$

We now further refine the description of the structure of a matrix in semiprincipally partitioned form. Let $M^{(0)}$ denote a matrix in semiprincipally partitioned form with all entries above the first superdiagonal set equal to zero; that is, $m_{i,j} = 0$ for $i + 1 < j$. For $M^{(0)}$, consider all one-to-one mappings from the set containing $m_{k,j}$ for all k and j with $k \neq j$, to the set containing the elements m_a for $a = 1, \ldots, (n-1)(n+2)/2$ with the requirement

$$a < b \text{ implies } m_a \leq m_b. \tag{5.2.26}$$

For a given mapping we will indicate that $m_{k,j}$ is mapped into m_a by writing

$$m_{k,j} \to m_a \tag{5.2.27a}$$

or

$$m_a \to m_{k,j}. \tag{5.2.27b}$$

This mapping is not unique since we have not specified into which element two equal terminal capacities are mapped. As an example of the mapping in (5.2.26) and (5.2.27), consider the matrix

$$M^{(0)} = \begin{bmatrix} \infty & 3 & 0 & 0 \\ 14 & \infty & 6 & 0 \\ 17 & 15 & \infty & 8 \\ 10 & 10 & 10 & \infty \end{bmatrix}. \tag{5.2.28}$$

Two possible mappings are given in (5.2.29) where m_a is in the $k - j$ position of $M^{(0)}$ if $m_{k,j} \to m_a$. The terms in (5.2.29b) which are different from the terms in (5.2.29a) are enclosed in boxes.

$$M^{(0)} = \begin{bmatrix} \infty & m_1 & 0 & 0 \\ m_7 & \infty & m_2 & 0 \\ m_9 & m_8 & \infty & m_3 \\ m_4 & m_6 & m_5 & \infty \end{bmatrix} \tag{5.2.29a}$$

$$M^{(0)} = \begin{bmatrix} \infty & m_1 & 0 & 0 \\ m_7 & \infty & m_2 & 0 \\ m_9 & m_8 & \infty & m_3 \\ \boxed{m_6} & \boxed{m_5} & \boxed{m_4} & \infty \end{bmatrix} \tag{5.2.29b}$$

***Definition* 5.2.5** *If $m_b \geq m_a$ for all m_a in μ_i then μ_i is a semidistinct min-cut matrix of m_b denoted by $S(b)$. If m_b is the unique largest element in μ_i, then it is also a distinct min-cut matrix of m_b.*

Example 5.2.2 For the matrix in (5.2.10), for $X_i = \{v_1, v_4\}$, M_i is not a distinct min-cut matrix of $m_{4,3}$ since $m_{4,3} = m_{4,2}$. M is a semidistinct min-cut matrix of $m_{4,3}$ or $m_{4,2}$. For $X_i = \{v_4, v_5\}$, M_i is a distinct min-cut matrix of $m_{5,3}$.

Finally, we examine the relationships of particular entries in a matrix with regard to membership in various submatrices. $m_{a,b}$ and $m_{c,d}$ are said to be *coupled* if they are both equal to the largest element in the same min-cut matrix. Otherwise they are *uncoupled.* If $m_{a,b}$ and $m_{c,d}$ are coupled, or they have the same min-cut matrix $M_i(a,b) = M_i(c,d)$, it may be possible to realize $m_{a,b}$ and $m_{c,d}$ by a single cut $(X_i,\bar{X}_i)$; i.e., $\tau_{a,b} = \tau_{c,d} = c(X_i,\bar{X}_i)$, with $v_a,v_c \in X_i$ and $v_b,v_d \in \bar{X}_i$. Next, $m_{a,b}$ and $m_{c,d}$ are *completely coupled* if every min-cut matrix of $m_{a,b}$ is a min-cut matrix of $m_{c,d}$ and every min-cut matrix of $m_{c,d}$ is a min-cut matrix of $m_{a,b}$. Finally, $m_{a,b}$ and $m_{c,d}$ are *min-coupled* if the set of minimum *a-b* cuts is identical to the set of minimum *c-d* cuts. Thus, referring to M given by (5.2.10), we find that $m_{4,2}$ and $m_{4,3}$ are completely coupled since they are both contained in the only min-cut matrix containing either of them, $M_i(4,2)$ with $X_i = \{v_1,v_4\}$. $m_{1,2}$ and $m_{5,1}$ are uncoupled since there are no min-cut matrices containing both of them. In (5.2.28), $m_{4,1}$ and $m_{4,2}$ are min-coupled, since $m_{4,1}$ and $m_{4,2}$ are in the min-cut matrix $\mu_i(4,1)$ with $X_i = \{v_4\}$. They are not completely coupled since for $X_i = \{v_1,v_4\}$, $m_{4,2}$ is in $\mu_i(4,2)$ but $m_{4,1}$ is not.

Finally we reformulate the synthesis problem in terms of the concepts introduced in this section.

***Definition* 5.2.6** *For an n-vertex graph let* $W = [w_{i,j}]$ *be an* $n \times n$ *matrix such that* $w_{i,j} = c(i,j)$ *for* $i \neq j$ *and* $w_{i,i} = \infty$ *for all i.*

***Definition* 5.2.7** *For an arbitrary matrix M, let* $\|M\|$ *denote the sum of the elements in M. We then have:*

***Theorem* 5.2.4** *Let M be a real square symmetric matrix with infinite entries on the diagonal. Then the following two statements are equivalent:*

a) M is the terminal capacity matrix of a graph.
b) There exists a graph G such that for all $\mu_i(k,j)$

$$m_{k,j} \leq \|W_i(k,j)\| \tag{5.2.30a}$$

and for all $m_{k,j}$ *there is at least one* $\mu_i(k,j)$ *such that*

$$m_{k,j} = \|W_i(k,j)\|. \tag{5.2.30b}$$

Proof. We first show that (a) implies (b). Let G be a graph for which $T = M$. By definition $\|W_i(k,j)\|$ is the sum of the capacities of all branches (a,b) such that $v_a \in X_i$ and $b \in \bar{X}_i$, that is,

$$c(X_i,\bar{X}_i) = \|W_i(k,j)\|, \tag{5.2.31}$$

and (b) therefore follows from (a).

We next show that (b) implies (a). Let G be a graph for which W and M satisfy (5.2.30). We show that for $G, T = M$. From (5.2.30a) we have $c(X_i,\bar{X}_i) \geq m_{k,j}$ for any *k-j* cut $(X_i,\bar{X}_i)$ in G. If $(X_i,\bar{X}_i)$ is also an *a-b* cut then $m_{a,b} \leq m_{k,j}$ from Definition 5.2.3. Hence from (5.2.30a), the minimum *a-b* cut is at least as large as $m_{a,b}$. From (5.2.30b) there must be at least one *a-b* cut of value $m_{a,b}$. Hence $m_{a,b} = \tau_{a,b}$ for all $a \neq b$.

Theorem 5.2.4 does not provide an algorithm for deciding realizability. However, it will be useful in deriving such an algorithm.

5.3 ANALYSIS OF PSEUDOSYMMETRIC GRAPHS

For arbitrary directed graphs, there is no general procedure at present to reduce systematically the number of entries in T which must be found to determine all its entries uniquely. We now develop such a procedure for pseudosymmetric graphs, i.e., graphs for which

$$c(i,V) = c(V,i) \quad \text{for all } v_i \in V. \tag{5.3.1}$$

In particular, we develop an algorithm for finding all entries in the terminal capacity matrix of a pseudosymmetric graph by solving only $n - 1$ flow problems. Furthermore, each flow calculation is performed on a graph which is usually simpler, and never more complicated, than the original graph.

For a pseudosymmetric graph

$$\sum_{v_i \in X} c(i,V) = \sum_{v_i \in X} c(V,i), \tag{5.3.2a}$$

or

$$c(X,V) = c(V,X) \quad \text{for any } X \subseteq V. \tag{5.3.2b}$$

Since $V = X \cup \bar{X}$ and $X \cap \bar{X} = \phi$, (5.3.2b) can be written as

$$c(X,X \cup \bar{X}) = c(X,X) + c(X,\bar{X}) = c(X \cup \bar{X},X) = c(X,X) + c(\bar{X},X). \tag{5.3.3}$$

Hence G is pseudosymmetric if and only if

$$c(X,\bar{X}) = c(\bar{X},X) \quad \text{for all } X. \tag{5.3.4}$$

Thus, we have proved:

Lemma 5.3.1 *T is symmetric if G is pseudosymmetric.*

The analysis scheme consists of the following steps.

Multiterminal Analysis Algorithm

Step 1. For any v_i and v_j in a graph G find the smallest i-j cut $(X,\bar{X})$. Represent this cut by two generalized vertices X and $\bar{X}$ with a branch of capacity $\tau_{i,j}$ joining them as in Fig. 5.3.1.

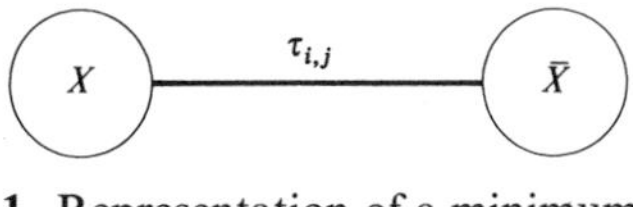

Fig. 5.3.1 Representation of a minimum i-j cut.

Step 2. Choose two vertices v_k and v_l in X (or two in $\bar{X}$) and find the smallest k-l cut in a graph G_a in which $\bar{X}$ (or X) is a single vertex. The resulting cut with value $\tau_{k,l}$ is represented by a branch connecting X_1 and $\bar{X}_1$, the two parts into which X (or $\bar{X}$) is divided. $\bar{X}$ is adjacent to $\bar{X}_1$ or X_1 depending upon which part of the graph G_a it is allocated to by the cut. The result is shown in Fig. 5.3.2, in the case in which v_k and v_l are in X and $\bar{X}_1$ is adjacent to $\bar{X}$.

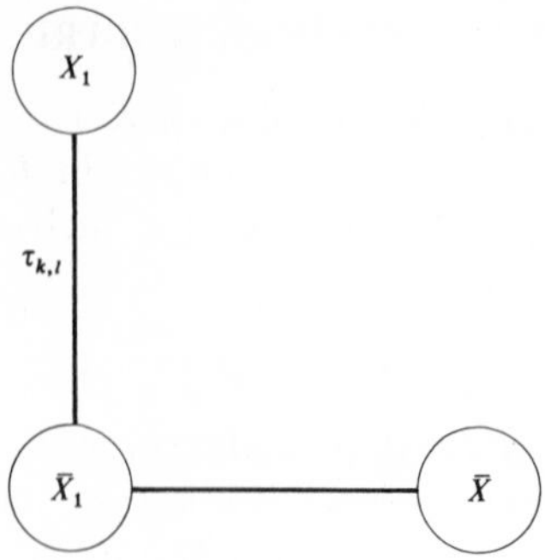

Fig. 5.3.2 Representation of a minimum *i-j* cut and a minimum *k-l* cut.

Step 3. The procedure is then repeated. At each step G_a consists of certain generalized vertices, which may represent many vertices of the original graph, and certain branches connecting them. To proceed with the computation select a generalized vertex X_i and any two original vertices v_s and v_t in X_i. Upon removing all branches which connect to X_i from the remaining part of G_a, the graph of generalized vertices is split into a number of disconnected components. Condense each component in G, except X_i, into a single vertex. This is equivalent to letting $c(i,j) = \infty$ for all branches in the condensed component. Then, solve the flow problem in the graph consisting of these condensed vertices, the original vertices within X_i and with v_s and v_t as source and terminal. The minimum cut obtained by this flow calculation splits X_i into two parts X_{i_1} and X_{i_2}. This is represented in G_a by replacing X_i by two generalized vertices X_{i_1} and X_{i_2} connected by a branch bearing the cut value. All other branches and generalized vertices in G_a are unchanged except those branches which were formerly connected to X_i. Such a branch is now connected to X_{i_1} if its component was on the same side of the cut as the vertices in X_{i_1}, and attached to X_{i_2} if its component is on the other side.

This process is repeated until the generalized vertices of G_a consist of exactly one vertex each. This point is reached after exactly $n - 1$ flow calculations since G_a is a tree at all times. Hence, when the procedure terminates, G_a is an n-vertex tree. Each of the $n - 1$ branches in the tree has been created by solving a flow problem in a graph no larger than the original one. For any v_s and v_t, the terminal capacity $\tau_{s,t}$ in the original graph G is given by the smallest branch capacity in the unique s-t-path in the resulting tree.

The fact that the resulting tree and the original graph have the same terminal capacity matrix is proved in Lemma 5.3.2 and Theorem 5.3.1 below. Before giving the proofs we illustrate the use of the algorithm in Example 5.3.1.

Example 5.3.1 Consider the pseudosymmetric graph in Fig. 5.3.3(a). Arbitrarily select a pair of vertices v_4 and v_2. A minimum 4-2 cut is $(\{v_1,v_4,v_5\},\{v_2,v_3\})$ with value 7. The graph is therefore represented by the tree with two generalized vertices in Fig. 5.3.3(b). Next select vertices v_1 and v_4 and find the smallest 1-4 cut in the graph in Fig. 5.3.3(c) formed by condensing the vertices v_2 and v_3. A minimum 1-4 cut is $(\{v_1\},\{v_2,v_3,v_4,v_5\})$ with value 6. The resulting tree representing the graph is then given in Fig. 5.3.3(d). Now find the minimum 4-5 cut in the graph in Fig. 5.3.3(c) again. The minimum 4-5 cut is $(\{v_1,v_2,v_3,v_4\},\{v_5\})$, with value 5. The resulting tree is given in Fig. 5.3.3(e). Finally, find the minimum 2-3 cut in the graph in

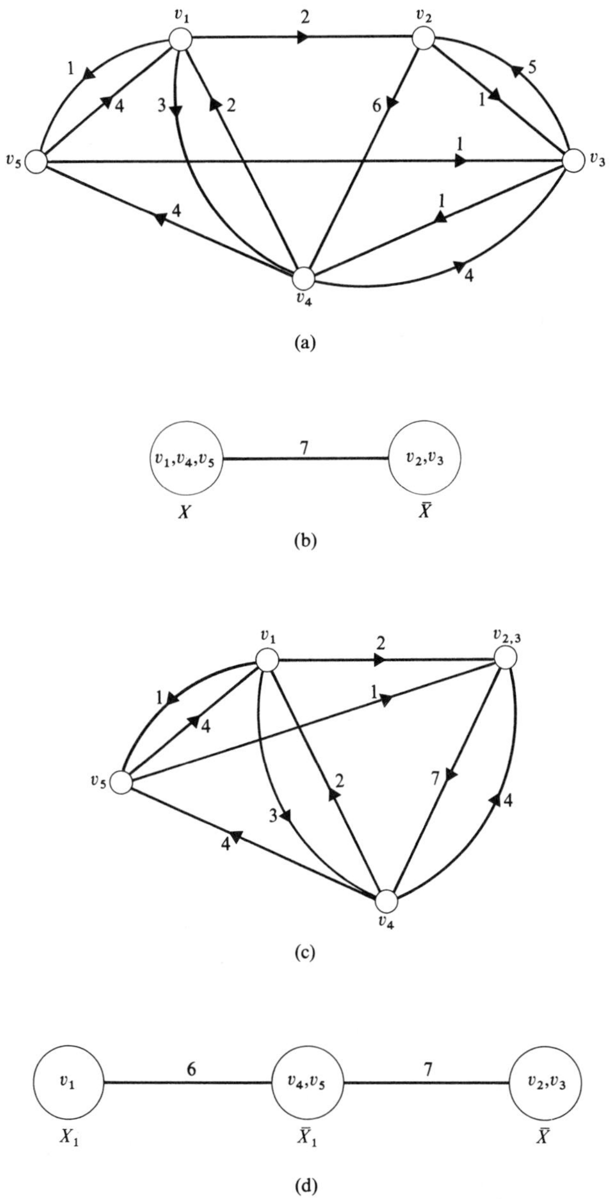

Fig. 5.3.3 Graphs illustrating the Multiterminal Analysis Algorithm.

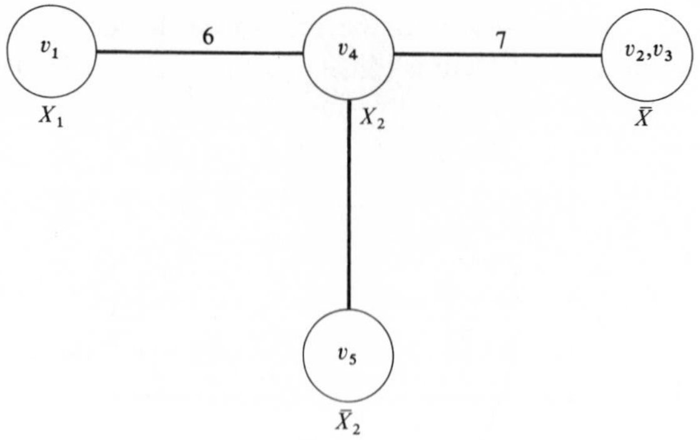

(e)

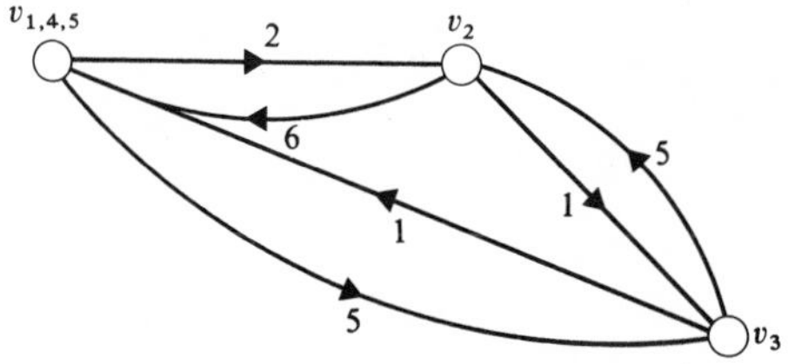

(f)

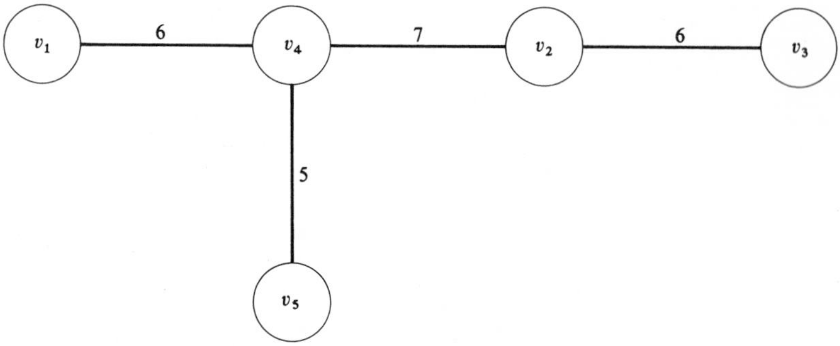

(g)

Figure 5.3.3 (continued)

Fig. 5.3.3(f) obtained by condensing the vertices v_1, v_4, and v_5 in G. The minimum 2-3 cut is $(\{v_1,v_2,v_4,v_5\},\{v_3\})$ and has value 6. The final tree is therefore given in Fig. 5.3.3(g). From this tree we see that the terminal capacity matrix of the original graph is:

$$T = \begin{bmatrix} \infty & 6 & 6 & 6 & 5 \\ 6 & \infty & 6 & 7 & 5 \\ 6 & 6 & \infty & 6 & 5 \\ 6 & 7 & 6 & \infty & 5 \\ 5 & 5 & 5 & 5 & \infty \end{bmatrix}. \tag{5.3.5}$$

Fig. 5.3.4 Graph used to prove Lemma 5.3.2.

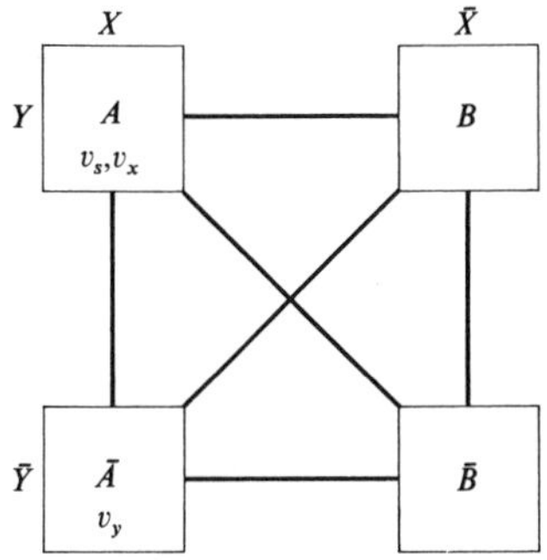

In Lemma 5.3.2 we show that the operation of condensing components of G does not change the values of terminal capacities.

Lemma 5.3.2 *Let G be a pseudosymmetric graph and let $(X,\bar{X})$ be a minimum s-t cut. Let G′ be the graph obtained from G by condensing the vertices in the set $\bar{X}$ into a single vertex and let v_x and v_y be vertices in X. Then the value of the minimum x-y cut, $\tau_{x,y}$, in G is equal to the value of the minimum x-y cut, $\tau'_{x,y}$, in G′.*

Proof. Let $(Y,\bar{Y})$ be a minimum x-y cut in G and define the sets

$$A = X \cap Y,\ \bar{A} = X \cap \bar{Y},\ B = \bar{X} \cap Y \text{ and } \bar{B} = \bar{X} \cap \bar{Y}. \tag{5.3.6}$$

We may assume $v_x \in A$, $v_y \in \bar{A}$ and $v_s \in A$. A pictorial representation of G with vertices partitioned as in (5.3.6) is given in Fig. 5.3.4.

CASE 1. $v_t \in B$. Then,

$$c(X,\bar{X}) = c(A,B) + c(A,\bar{B}) + c(\bar{A},B) + c(\bar{A},\bar{B}), \tag{5.3.7}$$

$$c(Y,\bar{Y}) = c(A,\bar{A}) + c(A,\bar{B}) + c(B,\bar{A}) + c(B,\bar{B}). \tag{5.3.8}$$

Since $(Y,\bar{Y})$ is a minimum x-y cut, and since $(A \cup B \cup \bar{B},\bar{A})$ is an x-y cut,

$$c(Y,\bar{Y}) - c(A \cup B \cup \bar{B},\bar{A}) = c(A,\bar{B}) + c(B,\bar{B}) - c(\bar{B},\bar{A}) \le 0. \tag{5.3.9}$$

Since $(X,\bar{X})$ is a minimum s-t cut, and since $(A \cup \bar{A} \cup \bar{B},B)$ is an s-t cut,

$$c(X,\bar{X}) - c(A \cup \bar{A} \cup \bar{B},B) = c(A,\bar{B}) + c(\bar{A},\bar{B}) - c(\bar{B},B) \le 0. \tag{5.3.10}$$

Adding (5.3.9) and (5.3.10) and using the definition of pseudosymmetry gives

$$\begin{aligned}
&[c(Y,\bar{Y}) - c(A \cup B \cup \bar{B},\bar{A})] + [c(X,\bar{X}) - c(A \cup \bar{A} \cup \bar{B},B)] \\
&= [c(A,\bar{B}) + c(\bar{A},\bar{B}) + c(B,\bar{B})] - [c(\bar{B},A) + c(\bar{B},\bar{A}) \\
&\quad + c(\bar{B},B)] + c(A,\bar{B}) + c(\bar{B},A) \\
&= c(V,\bar{B}) - c(\bar{B},V) + c(A,\bar{B}) + c(\bar{B},A) \\
&= c(A,\bar{B}) + c(\bar{B},A) \le 0.
\end{aligned} \tag{5.3.11}$$

Since $c(A,\bar{B}) \ge 0$ and $c(\bar{B},A) \ge 0$,

$$[c(Y,\bar{Y}) - c(A \cup B \cup \bar{B},\bar{A})] + [c(X,\bar{X}) - c(A \cup \bar{A} \cup \bar{B},B)] = 0. \tag{5.3.12}$$

The sum of the two nonpositive numbers can be zero only if both are zero. Hence, $c(A \cup B \cup \bar{B},\bar{A}) = c(Y,\bar{Y})$ and $(A \cup B \cup \bar{B},\bar{B})$ is a minimum x-y cut.

CASE 2. $v_t \in \bar{B}$. A similar proof will show that $(A,\bar{A} \cup B \cup \bar{B})$ is a minimum x-y cut with value $\tau_{x,y}$ in this case.

Hence, there is always a minimum x-y cut such that the set of vertices $\bar{X}$ is on one side of this cut. Consequently, condensing $\bar{X}$ to a single vertex does not affect the value of the maximum flow from v_x to v_y.

In Theorem 5.3.1 we show that for all i and j, $\tau_{i,j}$ in G is equal to the value of the minimum i-j cut in the tree, G_a, obtained from the algorithm. In G_a, let $c_1,c_2,\ldots,c_\alpha$ be the capacities of the branches $(s,1),(1,2),\ldots,(\alpha\text{-}2,\alpha\text{-}1),(\alpha\text{-}1,t)$ in the unique s-t path. Let $\tau_{s,t}$ be the value of the minimum s-t cut in G. We then have:

Theorem 5.3.1 $\tau_{s,t} = \text{Min}\,[c_1,\ldots,c_\alpha]$.

Proof. Application of the Multiterminal Analysis Algorithm generates a tree whose vertices represent subsets of vertices of G and branch capacities represent capacities of cuts in G. We first show by induction on the number of branches in the tree that if a branch with capacity c_k connects X_i and X_j, then there are vertices $v_i \in X_i$ and $v_j \in X_j$ such that $\tau_{i,j} = c(i,j)$. The statement is certainly true after the first minimum cut. Now assume it is true after finding β minimum cuts. We now show that it is true after the next minimum cut is found.

Consider an X_i about to be cut into X_{i_a} and X_{i_b} thus creating a branch of capacity $\tau_{a,b}$ where $v_a,v_b \in X_i$. Suppose there is an undirected branch of capacity c_k joining X_i to X_j. The sets under consideration are shown before and after the cut in Figs. 5.3.5 and 5.3.6 respectively. Without loss of generality, assume X_j is adjacent to X_{i_a}. Clearly the capacity of the new branch is the terminal capacity from v_a to v_b with $v_a \in X_{i_a}$ and $v_b \in X_{i_b}$. It is also clear by the inductive hypothesis that in Fig. 5.3.5, c_k is equal to the terminal capacity between two vertices $v_i \in X_i$ and $v_j \in X_j$. We must now show that the branch b_k of capacity c_k still has this property in Fig. 5.3.6.

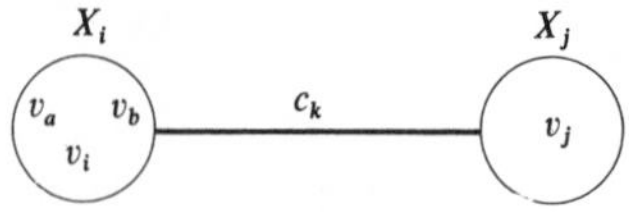

Fig. 5.3.5 Graph illustrating the first step in the Multiterminal Analysis Algorithm.

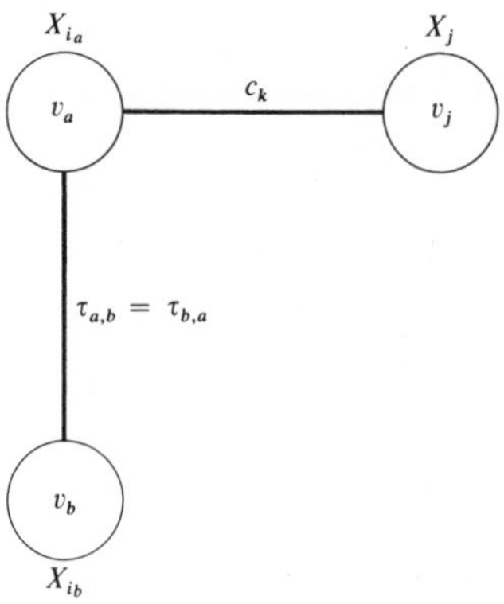

Fig. 5.3.6 Graph illustrating the second step in the Multiterminal Analysis Algorithm.

There are now two cases to consider. If $v_i \in X_{i_a}$, then c_k is still the terminal capacity between $v_i \in X_{i_a}$ and $v_j \in X_j$. If $v_i \in X_{i_b}$, we show that c_k is the terminal capacity between $v_a \in X_i$ and $v_j \in X_j$. From the triangle inequality,

$$\tau_{j,a} \geq \text{Min}\,[\tau_{j,i},\tau_{i,b},\tau_{b,a}]. \tag{5.3.13}$$

From Fig. 5.3.6 and Lemma 5.3.2, $\tau_{j,a}$ is unaffected if X_{i_b} is replaced by a single vertex, that is, if $\tau_{i,b}$ is made infinite. Hence (5.3.13) reduces to

$$\tau_{j,a} \geq \text{Min}\,[\tau_{j,i},\tau_{b,a}]. \tag{5.3.14}$$

But from Fig. 5.3.6, $\tau_{j,i} \leq c_k$. Thus

$$\tau_{j,a} \geq \text{Min}\,[c_k,\tau_{b,a}]. \tag{5.3.15}$$

However, c_k and $\tau_{b,a}$ both are values of *i-j* cuts in G and $c_k = \tau_{j,i}$. Therefore $c_k \leq \tau_{b,a}$ and

$$\tau_{j,a} \geq c_k. \tag{5.3.16}$$

However, from Lemma 5.3.2

$$\tau_{j,a} \leq c_k. \tag{5.3.17}$$

Combining (5.3.16) and (5.3.17) gives $\tau_{j,a} = c_k$ with $v_j \in X_j$ and $v_a \in X_{i_a}$. Therefore, in the final tree G_a, $c(i,j) = \tau_{i,j}$.

From the triangle inequality,

$$\tau_{i,j} \geq \text{Min}\,[\tau_{s,1},\tau_{1,2},\ldots,\tau_{\alpha-1,t}] = \underset{l=1,\ldots,\alpha}{\text{Min}}\,[c_l]. \tag{5.3.18}$$

From Lemma 5.3.2,

$$\tau_{i,j} \leq \underset{l=1,\ldots,\alpha}{\text{Min}}\,[c_l]. \tag{5.3.19}$$

Combining (5.3.18) and (5.3.19),

$$\tau_{i,j} = \underset{l=1,\ldots,\alpha}{\text{Min}}\,[c_l], \tag{5.3.20}$$

and the proof is completed.

5.4 SEMIGRAPHS AND SHIFTING

We now consider further properties that will be of interest in the synthesis problem. We will find it useful to be able to transform a given graph into another graph with a special structure but with the same terminal capacity matrix. Once we fully characterize the transformation used to obtain the special graph, we limit our discussion to only these graphs. The transformation is called "shifting" and the graph with the special structure called a "semigraph."

Definition 5.4.1 *A forward circuit $L_f(i,j)$ is a directed circuit, with more than two branches, containing (i,j).*

A backward circuit $L_b(i,j)$ is a directed circuit, with more than two branches, containing (j,i).

A double circuit L_d is a backward circuit and a forward circuit such that $(k,l) \in L_f(i,j)$ if and only if $(l,k) \in L_b(i,j)$.

For a given (i,j) and a given double circuit, shifting is defined by Steps 1 and 2 below.

Shifting

Step 1. For all $(l,k) \in L_f(i,j)$ increase $c(l,k)$ by a real number $s(i,j)$.

Step 2. For all $(l,k) \in L_b(i,j)$ decrease $c(l,k)$ by $s(i,j)$.

Example 5.4.1 We begin with the graph in Fig. 5.4.1(a) and choose $L_f(3,2) = \{(3,2),(2,5),(5,4),(4,3)\}$. To obtain a double circuit, $L_b(3,2)$ is then uniquely determined by $L_f(3,2)$ as $L_b(3,2) = \{(3,4),(4,5),(5,2),(2,3)\}$. If we choose $s(i,j) = 3$, the graph in Fig. 5.4.1(b) results from shifting. Although branch (4,3) was not present in the graph in Fig. 5.4.1(a) we inserted a branch (4,3) with zero capacity.

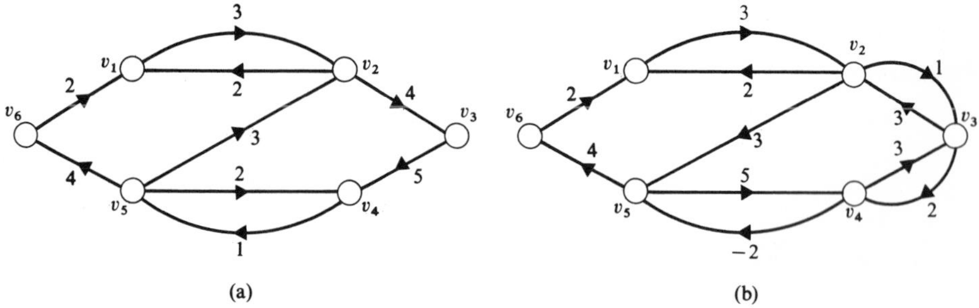

Fig. 5.4.1 Equivalent graphs. In (b) $\tau_{4,5} \neq \text{Max}\,[f_{4,5}]$.

At this point a complication is apparent. The shifting procedure can lead to a negative branch capacity. Thus, in Fig. 5.4.1(b), $c(4,5) < 0$. We will allow such branches in intermediate stages.

In a graph with negative branch capacities, $\tau_{s,t}$ still denotes the value of the smallest s-t cut. However, we choose not to allow negative flows. Hence, $\tau_{s,t}$ is no longer necessarily equal to the maximum s-t flow. As an example, in Fig. 5.4.1(b), $\tau_{4,5} = c\{(4,5),(4,3)\} = 1$. However, a flow of 3 can be sent from 4 to 5 along the path (4,3),(3,2),(2,5). Furthermore, the value of the smallest s-t cut is no longer equal to the value of the smallest s-t cut-set. In Fig. 5.4.2, the smallest 1-2 cut consists of branches (1,2), (4,2), and (4,3) and has value 6. The smallest 1-2 cut-set consists of branch (1,2) and has value 7.

In spite of the fact that negative branch capacities destroy most of the intuitive interpretations of terminal capacity, many of the previous relationships between different terminal capacities are still preserved. Most important, both the triangle inequality and the extended triangle inequality remain valid. This is easily checked

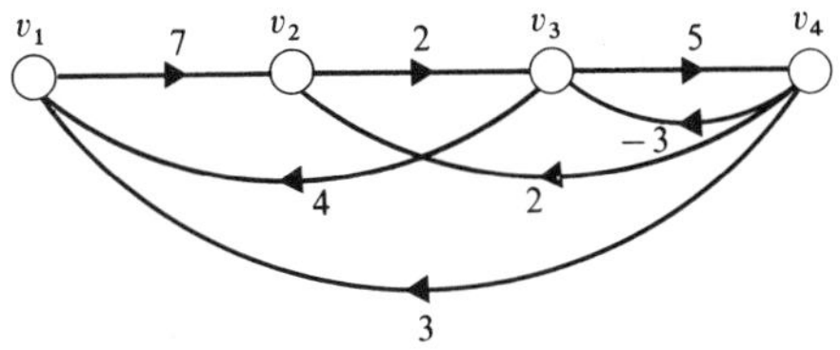

Fig. 5.4.2 Graph in which the value of the minimum 1-2 cut is not equal to the value of the minimum 1-2 cut-set.

since the proofs of Theorem 5.2.2 and Corollary 5.2.3 do not depend upon the signs of branch capacities. We now show that under shifting the values of minimum cuts remain the same.

***Theorem* 5.4.1** *The terminal capacity matrix of a graph is invariant under shifting.*

Proof. We prove the theorem by showing that shifting leaves the value of every cut unchanged. Therefore, consider an arbitrary cut $(X,\bar{X})$. As shown in Chapter 2, $(X,\bar{X})$ and $(\bar{X},X)$ contain an equal number of branches in common with any directed circuit. Therefore, if the capacity of β branches in $(X,\bar{X})$ is increased by $s(i,j)$ in the shifting procedure, the capacity of β branches in $(X,\bar{X})$ is decreased by $s(i,j)$ in the shifting procedure.

***Definition* 5.4.2** *A semigraph is a graph with n vertices,* $v_1,v_2,\ldots,v_n$, *such that:*

$$c(i,j) = 0 \text{ for } j > i + 1, \tag{5.4.1a}$$

$$c(i + 1,i) \text{ is unrestricted in sign for } i = 1,\ldots,(n - 1), \tag{5.4.1b}$$

$$c(i,j) \geq 0 \text{ for all other } i \text{ and } j. \tag{5.4.1c}$$

As an example, the graph in Fig. 5.4.2 is a semigraph.

***Lemma* 5.4.1** *A graph with nonnegative branch capacities can be converted to a semigraph by shifting.*

Proof. For a graph G we apply shifting using all forward circuits of three branches such that

$$L_f(j,i) = \{(j,i),(i,i + 1),(i + 1,j)\} \tag{5.4.2a}$$

and

$$s(j,i) = c(i,j). \tag{5.4.2b}$$

Consider the circuits formed by letting i range from 1 to $n - 2$ in increasing order and, for each i, letting j range from $i + 2$ to n in increasing order. After $L_f(j,i)$ is used, $c(i,j)$ becomes zero for each i and j such that $i + 1 < j$. It is permissible for $c(i + 1,i)$ to become negative, because of (5.4.1b). Finally in using $L_f(j,i)$, the capacity of $c(j,i + 1)$ may become negative for $i + 1 < j$. In this case the new value is $c(j,i + 1) - \alpha$, where α is a real positive constant. In the same shifting step $(i + 1,j)$ receives a capacity $c(i + 1,j) + \alpha$. From the order in which the forward circuits are considered, $L_f(j,i + 1)$ will be chosen for shifting in a subsequent step.

But $(j,i+1)$ then receives capacity $c(j,i+1) - \alpha + c(i+1,j) + \alpha \geq 0$. Hence after all circuits have been considered, $c(j,i+1) \geq 0$ and $c(i+1,j) \geq 0$ for $i+1 \leq j$.

Example 5.4.3 We convert the graph in Fig. 5.4.3(a) to a semigraph. First select $c(1,3) = 11$, $c(3,1) = 9$, $s(3,1) = 11$, and $L_f(3,1) = \{(3,1),(1,2),(2,3)\}$. The resulting graph, after shifting, is given in Fig. 5.4.3(b). Next select $c(1,4) = 12$, $c(4,1) = 8$, $s(4,1) = 12$, and $L_f(4,1) = \{(4,1),(1,2),(2,4)\}$. The resulting graph is given in Fig. 5.3.4(c). Note that $c(4,2)$ has become negative. However, this will be remedied in the next step in which $c(2,4) = 24$, $c(4,2) = -11$, $s(4,2) = 24$, and $L_f(4,2) = \{(4,2),(2,3),(3,4)\}$. The final semigraph is shown in Fig. 5.4.3(d).

From Lemma 5.4.1 we see that any terminal capacity matrix realizable as a graph with nonnegative branch capacities is realizable as a semigraph. We could therefore

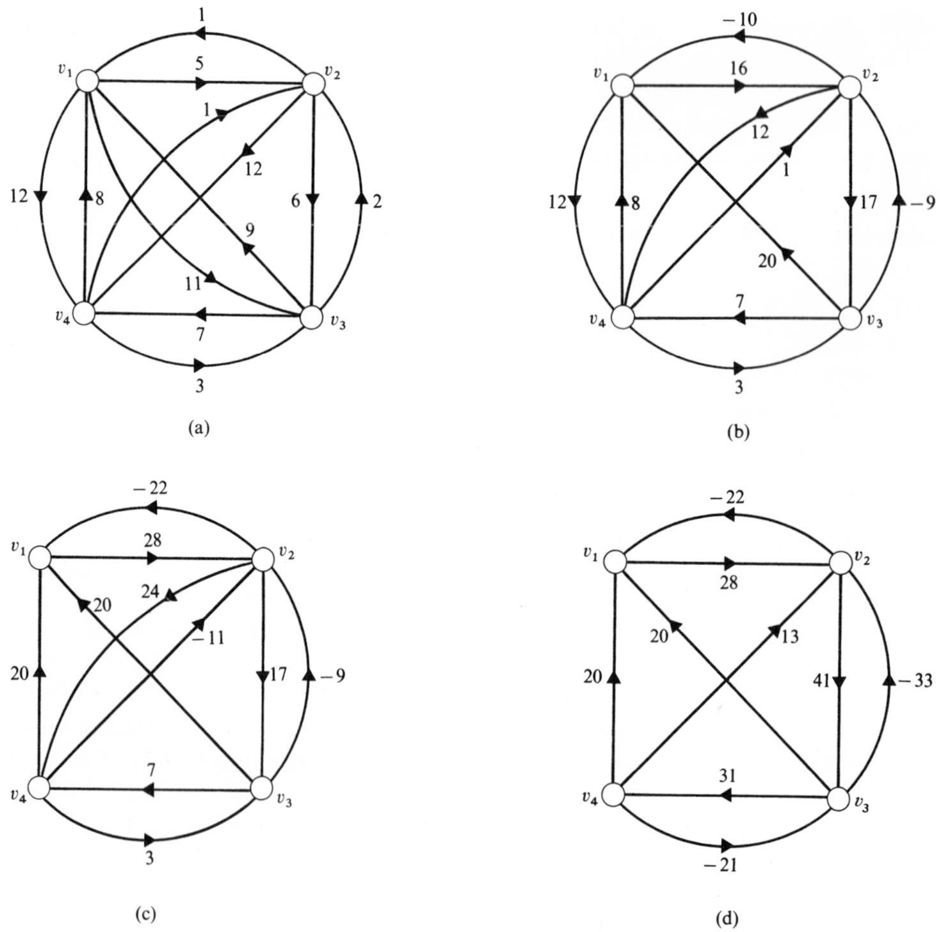

Fig. 5.4.3 Conversion of a graph to a semigraph by shifting.

limit our synthesis problem to the realization of T as a semigraph if we could find a method for converting a semigraph to a graph with nonnegative branch capacities. Once we have converted to a graph with nonnegative branch capacities, we would again have $\tau_{s,t} = \underset{\mathscr{F}}{\text{Max}} \{f_{s,t}\}$.

The required information is provided in Theorem 5.4.2. This theorem shows that the Semigraph Shifting Routine defined below will convert a semigraph to a graph whenever $c(l,k) + c(k,l) \geq 0$ for all l and k such that $l \neq k$.

Semigraph Shifting Routine

For each branch (i,j) with negative capacity $c(i,j)$, apply shifting to all possible double circuits containing (i,j) in the forward circuit. For the double circuit $L_d(i,j)$, let

$$s(i,j) = \text{Max}\,[0, m(i,j)] \tag{5.4.3a}$$

where

$$m(i,j) = \underset{(l,k) \in L_b(i,j)}{\text{Min}} [c(l,k), -c(i,j)]. \tag{5.4.3b}$$

Equation (5.4.3a) makes $s(i,j)$ nonnegative. From (5.4.3b), if $c(i,j) = 0$ then $m(i,j)$ and $s(i,j) = 0$ for every double circuit containing (i,j) in a forward circuit.

Theorem 5.4.2 *A semigraph can be converted into a graph with nonnegative branch capacities by shifting if and only if in the semigraph $c(l,k) + c(k,l) \geq 0$ for all l and k such that $l \neq k$, and all entries in the terminal capacity matrix of the semigraph are nonnegative.*

Proof. We first show that if $c(l,k) \geq 0$ and $c(k,l) \geq 0$ for all (l,k), the necessity part of the theorem is vacuous. If $c(l,k) < 0$ and $c(k,l) \geq 0$ for some (l,k) then, in converting $c(l,k)$ into a nonnegative number by shifting, the value of $c(k,l)$ must be reduced by at least the original value of $c(l,k)$. For $c(k,l)$ to be nonnegative after shifting, we must originally have had $c(k,l) \geq -c(l,k)$ or $c(k,l) + c(l,k) \geq 0$. If $c(l,k) \geq 0$ and $c(k,l) < 0$ the argument is similar. By identical reasoning, the conversion by shifting is impossible if $c(l,k) < 0$ and $c(k,l) < 0$. Clearly if $\tau_{i,j} < 0$ then $\tau_{i,j}$ cannot be obtained with only positive branch capacities.

We now show that if $c(l,k) + c(k,l) \geq 0$ for all (l,k), then, after the Semigraph Shifting Routine, $c(i,j) \geq 0$ for all (i,j). Suppose to the contrary that the final value of $c(i,j)$ is negative. Since $c(i,j)$ has attained its final value there are no more double circuits with (i,j) in the forward circuit and nonzero $s(i,j)$. Hence, from (5.4.3), every directed i-j path (other than the path consisting of only (i,j)), contains a branch of nonpositive capacity. But $\tau_{i,j}$ is invariant under shifting and therefore is negative, contradicting the hypothesis. Finally, we show that during the shifting procedure no nonnegative branch capacity is made negative. Suppose the capacity of (l,k) is increased by the maximum amount $|c(k,l)|$. Then the capacity of (k,l) is decreased by $|c(k,l)|$. But since $c(k,l) + c(l,k) \geq 0$ the capacity of (k,l) remains nonnegative.

Corollary 5.4.1 *A requirement matrix $\mathscr{R}$ is realizable as the terminal capacity matrix of a graph with a nonnegative branch capacity matrix if and only if it is realizable as a semigraph in which $c(l,k) + c(k,l) \geq 0$ for all (l,k).*

Proof. From Lemma 5.4.1, any graph can be converted to a semigraph by shifting. By repeating the shifting steps with the negative of the previous (i,j) the semigraph can be converted to the original graph by shifting. Hence if $\mathscr{R}$ is realizable by a graph it is realizable by a semigraph which can be converted to a graph by shifting. The corollary therefore follows from Theorem 5.4.2.

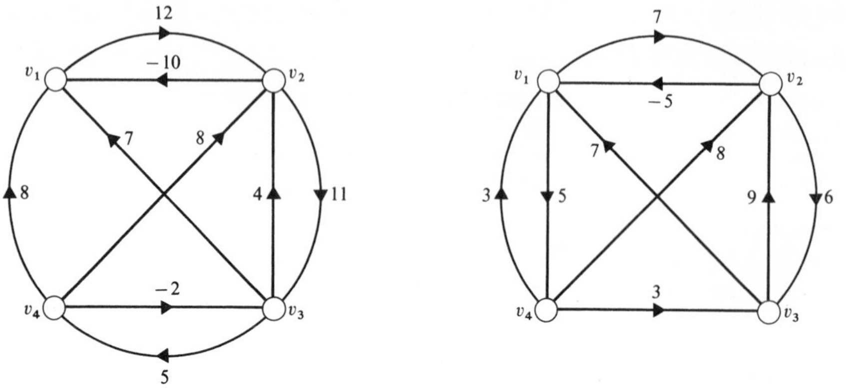

Fig. 5.4.4 Intermediate step in converting a semigraph to a graph by the Semigraph Shifting Routine.

Example 5.4.4 For the semigraph in Fig. 5.4.4(a), $c(j,k) + c(k,j) \geq 0$ for all j and k, and all terminal capacities are nonnegative. Hence, the semigraph can be converted to a graph with nonnegative branch capacities by shifting. The forward circuits containing (2,1) are:

$$L_{f_1}(2,1) = \{(2,1),(1,3),(3,2)\}; \tag{5.4.4a}$$
$$L_{f_2}(2,1) = \{(2,1),(1,4),(4,2)\}; \tag{5.4.4b}$$
$$L_{f_3}(2,1) = \{(2,1),(1,4),(4,3),(3,2)\}; \tag{5.4.4c}$$
$$L_{f_4}(2,1) = \{(2,1),(1,3),(3,4),(4,2)\}. \tag{5.4.4d}$$

We now use the Semigraph Shifting Routine. Using $L_{f_4}(2,1)$ we have $m(2,1) = \text{Min}\,[12,7,-2,0,10]$ and $s(i,j) = \text{Max}\,[0,-2] = 0$. Hence the graph is unchanged by shifting. Using $L_{f_2}(2,1)$, we have $m(2,1) = \text{Min}\,[12,8,5,11,10] = 5$ and $s(i,j) = \text{Max}\,[0,5] = 5$. After shifting, the graph shown in Fig. 5.4.4(b) is obtained. Using $L_{f_1}(2,1)$, we have $m(2,1) = \text{Min}\,[7,7,6,5] = 5$ and $s(i,j) = \text{Max}\,[0,5] = 5$. After shifting, the graph shown in Fig. 5.4.5 is obtained. Since $c(4,3)$ was increased from -2 to $+3$ when $c(2,1)$ was increased, the procedure is completed. Note that in the Semigraph Shifting Routine each forward circuit containing (i,j) is used at most once.

We now prove several facts which will enable us to incorporate the above developments into a synthesis procedure.

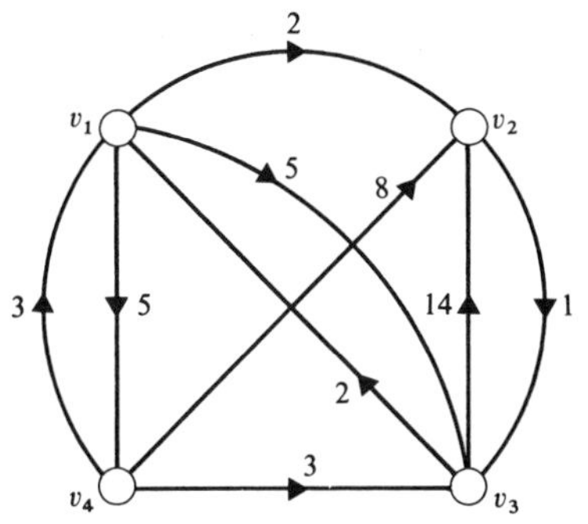

Fig. 5.4.5 Graph obtained by shifting.

Theorem 5.4.3 *Given a matrix $\mathscr{R}$ in semiprincipally partitioned form and a graph such that $\tau_{k,k+1} = r_{k,k+1}$ for $k = 1, \ldots, n-1$ then $\tau_{i,j} = r_{i,j}$ for $i + 1 < j$ if and only if there is an i-j cut with value $\underset{k=1,\ldots,j-1}{\text{Min}} [r_{k,k+1}]$.*

Proof. As previously mentioned, both Corollary 5.2.2 and the triangle inequality are valid for graphs with negative branch capacities. Hence, from Corollary 5.2.2 and the fact that $\tau_{k,k+1} = r_{k,k+1}$ for $k = 1, \ldots, n-1$,

$$r_{i,j} = \underset{k=1,\ldots,j-1}{\text{Min}} [r_{k,k+1}] \tag{5.4.5}$$

for $i + 1 < j$, and necessity is proved.

If there is an i-j cut of value $\underset{k=i,\ldots,j-1}{\text{Min}} [r_{k,k+1}]$,

$$\tau_{i,j} \leq \underset{k=i,\ldots,j-1}{\text{Min}} [r_{k,k+1}]. \tag{5.4.6}$$

However, if $\tau_{i,j} < \underset{k=i,\ldots,j-1}{\text{Min}} [r_{k,k+1}]$, there is a k-$(k+1)$ cut of value less than $\underset{k=i,\ldots,j-1}{\text{Min}} [r_{k,k+1}]$ for some k in the range $i \leq k \leq j-1$. This contradicts the fact that $\tau_{k,k+1} = r_{k,k+1}$ for $k = 1, \ldots, n-1$. Hence

$$\tau_{i,j} = \underset{k=i,\ldots,j-1}{\text{Min}} [r_{k,k+1}] = r_{i,j} \tag{5.4.7}$$

and the proof is completed.

It is important to note that the theorem is not vacuous since it is possible to have $\tau_{i,j} > \underset{k=i,\ldots,j-1}{\text{Min}} [\tau_{k,k+1}]$ for some $i + 1 < j$. Such a case is shown in Example 5.4.2.

Example 5.4.2 In (5.4.8) $\mathscr{R}$ is in semiprincipally partitioned form:

$$\mathscr{R} = \begin{bmatrix} \infty & 4 & 4 \\ 4 & \infty & 4 \\ 5 & 4 & \infty \end{bmatrix}. \tag{5.4.8}$$

The terminal capacity matrix of the semigraph in Fig. 5.4.6 is

$$T = \begin{bmatrix} \infty & 4 & 6 \\ 4 & \infty & 4 \\ 5 & 4 & \infty \end{bmatrix}. \tag{5.4.9}$$

Note that $\tau_{1,2} = r_{1,2}$ and $\tau_{2,3} = r_{2,3}$. However, $\tau_{1,3} > \text{Min}\,[\tau_{1,2},\tau_{2,3}]$ and $\tau_{1,3} \neq r_{1,3}$.

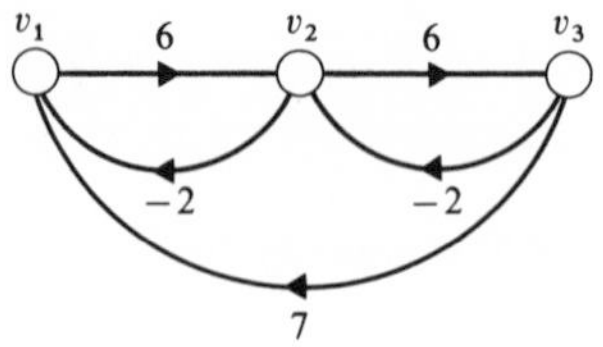

Fig. 5.4.6 Graph in which $\tau_{1,3} > \text{Min}\,[\tau_{1,2},\tau_{2,3}]$.

Since there are at most $n - 1$ numerically distinct entries $r_{i,j}$ in $\mathscr{R}$ for $i + 1 \leq j$, we should be able to realize these elements with only $n - 1$ calculations. Therefore, in our realization algorithms we will consider only the elements in $\mathscr{R}^{(0)}$. To guarantee the realization of $r_{i,j}$ for $i + 1 < j$, we prove:

Corollary 5.4.1 *Given $\mathscr{R}$ in semiprincipal partitioned form and a semigraph such that $\tau_{k,k+1} = r_{k,k+1}$ for $k = 1,\ldots,n-1$ then $\tau_{i,j} = r_{i,j}$ for $i + 1 < j$ if (5.4.10a), (5.4.10b), (5.4.10c), or (5.4.10d) is satisfied.*

Each entry $r_{i,j}$ for $i + 1 < j$ is coupled to some element $r_{a,b} = r_{i,j}$ by a matrix $\mathscr{R}_q^{(0)}$ such that in the semigraph $c(X_q,\bar{X}_q) = r_{a,b}$ where $(X_q,\bar{X}_q)$ is an i-j cut. (5.4.10a)

$r_{k,k+1} = c(k,k+1)$ *for* $k = 1,\ldots,n-1.$ (5.4.10b)

The largest entry in $\mathscr{R}^{(0)}$ above the main diagonal is smaller than all entries below the diagonal. (5.4.10c)

All branch weights in the semigraph are nonnegative. (5.4.10d)

Proof. 1) The sufficiency of (5.4.10a) follows immediately from Theorem 5.4.3.

2) If condition (5.4.10b) holds, the element $r_{i,j}$ for $i + 1 < j$ is coupled to the element $r_{p,p+1} = r_{i,j} = \underset{h=i,\ldots,j-1}{\text{Min}}\,[r_{h,h+1}]$ by the matrix $\mathscr{R}_q^{(0)}$ with $X_q = \{v_1,v_2,v_3,\ldots,v_l\}$ and with $c(X_q,\bar{X}_q) = r_{p,p+1}$. Hence the sufficiency follows from (5.4.10a).

3) If condition (5.4.10c) holds, then $r_{i,j}$ for $i + 1 < j$ is coupled to an element of $\mathscr{R}^{(0)}$ of the form $r_{p,p+1}$ for some $1 < p < n - 1$. $r_{i,j}$ is therefore coupled to $r_{p,p+1} = r_{i,j}$ by the same $\mathscr{R}_q^{(0)}$ as used in part (b) and the sufficiency is proved.

4) If all branch weights in the semigraph are nonnegative, then for $i + 1 < j, \tau_{i,j} = \underset{k=i,\ldots,j-1}{\text{Min}}\,[\tau_{k,k+1}]$.

5.5 UNDIRECTED GRAPHS

In this section, we show that realizability conditions and synthesis procedures become extremely simple if $\mathscr{R}$ is symmetric. We first examine the class of "completely partitioned" matrices.

Definition 5.5.1 *$\mathscr{R}$ is said to be completely partitioned if:*
1) $\mathscr{R}$ is semiprincipally partitioned, and
2) when in semiprincipally partitioned form $\mathscr{R}$ can be further partitioned as

$$\mathscr{R} = \left[\begin{array}{c|c} \mathscr{R}_{1,1} & \mathscr{R}_{1,2} \\ \hline \mathscr{R}_{2,1} & \mathscr{R}_{2,2} \end{array}\right] \tag{5.5.1}$$

such that
a) all elements in $\mathscr{R}_{2,1}$ are identical and equal to the smallest elements in $\mathscr{R}$ below the diagonal, and
b) all resulting submatrices on the main diagonal can be partitioned in the same way until the resulting submatrices on the diagonal are of order one.

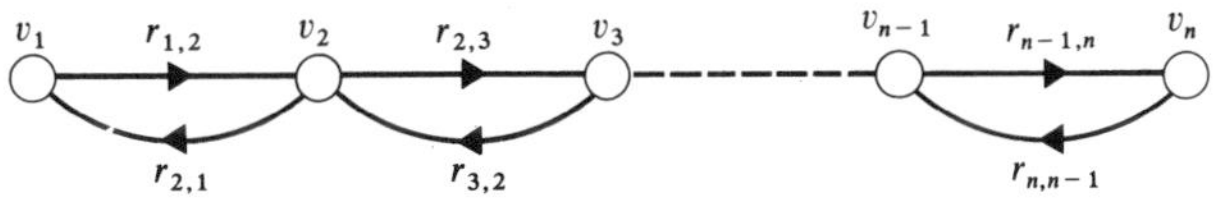

Fig. 5.5.1 Graph realizing a completely partitioned requirement matrix.

If $\mathscr{R}$ is in completely partitioned form, the graph in Fig. 5.5.1 realizes $\mathscr{R}$. In this graph $c(i,j) = r_{i,j}$ for $i = j + 1$ or $i = j - 1$, and is zero otherwise. That this graph realizes $\mathscr{R}$ follows immediately from Corollary 5.4.1. Since all branch capacities in the graph are nonnegative, part (d) of the Corollary shows that $\tau_{i,j} = r_{i,j}$ for $i + 1 < j$. In Fig. 5.5.1, the values of $\tau_{i,j}$ for $i + 1 \leq j$ clearly do not affect the values of $\tau_{i,j}$ for $i + 1 > j$. We therefore increase $r_{i,j}$ for $i + 1 \leq j$ until the transpose of $\mathscr{R}$ is semiprincipally partitioned. By the same reasoning we see that the graph in Fig. 5.5.1 realizes $r_{i,j}$ for $i + 1 > j$. Hence the proof is completed.

Example 5.5.1 Consider the semiprincipally partitioned matrix

$$\mathscr{R} = \begin{bmatrix} \infty & 4 & 3 & 1 & 1 \\ 4 & \infty & 3 & 1 & 1 \\ 4 & 8 & \infty & 1 & 1 \\ 4 & 6 & 6 & \infty & 2 \\ 4 & 6 & 6 & 7 & \infty \end{bmatrix}. \tag{5.5.2}$$

With the same ordering of rows and columns, $\mathscr{R}$ can also be partitioned as

$$\mathscr{R} = \left[\begin{array}{c|cccc} \infty & 4 & 3 & 1 & 1 \\ \hline 4 & \infty & 3 & 1 & 1 \\ 4 & 8 & \infty & 1 & 1 \\ 4 & 6 & 6 & \infty & 2 \\ 4 & 6 & 6 & 7 & \infty \end{array}\right]. \tag{5.5.3}$$

Therefore, $\mathscr{R}$ is completely partitioned and its realization is given in Fig. 5.5.2.

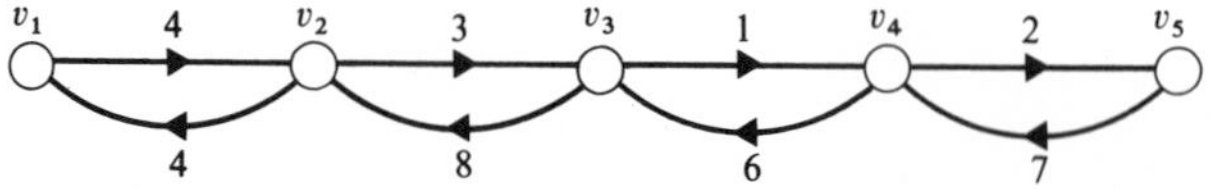

Fig. 5.5.2 Graph for Example 5.5.1.

***Definition* 5.5.2** *If $\mathscr{R}$ is semiprincipally partitioned and is symmetric, it is said to be principally partitioned.*

***Theorem* 5.5.1** *A real symmetric matrix $\mathscr{R}$ is realizable as a terminal capacity matrix of a graph if and only if after possibly permuting rows and corresponding columns, it can be principally partitioned.*

Proof. If $\mathscr{R}$ is principally partitioned, then it is completely partitioned such that the partitionings above and below the main diagonal are identical. $\mathscr{R}$ can then be realized as shown in Fig. 5.5.1 with $r_{i,i+1} = r_{i+1,i}$ for $i = 1, \ldots, n-1$. Hence $\mathscr{R}$ is realizable as the undirected graph in Fig. 5.5.3.

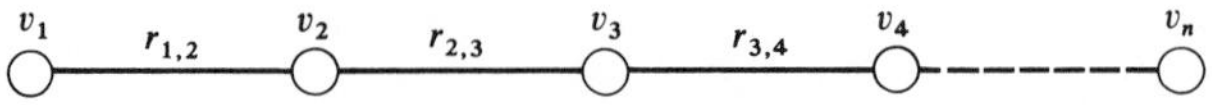

Fig. 5.5.3 Realization of a symmetric matrix.

If a requirement matrix is symmetric and cannot be principally partitioned, it cannot be semiprincipally partitioned and therefore is not realizable.

Example 5.5.2 The matrix $\mathscr{R}$ in (5.5.4) is realized in Fig. 5.5.4.

$$\mathscr{R} = \left[\begin{array}{ccc|cc} \infty & 5 & 3 & 1 & 1 \\ 5 & \infty & 3 & 1 & 1 \\ 3 & 3 & \infty & 1 & 1 \\ \hline 1 & 1 & 1 & \infty & 2 \\ 1 & 1 & 1 & 2 & \infty \end{array}\right] \tag{5.5.4}$$

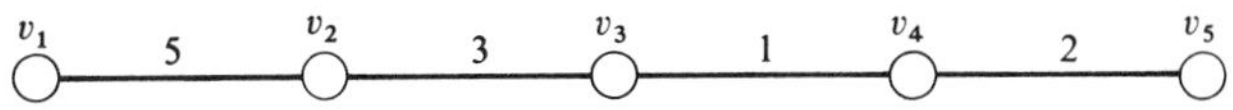

Fig. 5.5.4 Graph realizing $\mathscr{R}$ in (5.5.4).

5.6 MAXIMALLY DISTINCT REQUIREMENT MATRICES

Definition 5.6.1 *A matrix $M \in SPP$ is said to be maximally distinct if all the off diagonal entries in $M^{(0)}$ are numerically distinct.*

If $\mathscr{R}$ is maximally distinct, we can again give a simple and direct realization algorithm. In order to motivate the algorithm, we attempt to realize the maximally distinct matrix $\mathscr{R}$ given in SPP form in (5.6.1) using the theorems and lemmas proved thus far.

$$\mathscr{R} = \begin{bmatrix} \infty & 3 & 3 & 3 \\ 10 & \infty & 4 & 4 \\ 11 & 6 & \infty & 5 \\ 12 & 8 & 7 & \infty \end{bmatrix} \tag{5.6.1}$$

We first form

$$\mathscr{R}^{(0)} = \begin{bmatrix} \infty & 3 & 0 & 0 \\ 10 & \infty & 4 & 0 \\ 11 & 6 & \infty & 5 \\ 12 & 8 & 7 & \infty \end{bmatrix}. \tag{5.6.2}$$

We now attempt to find the branch capacities in a semigraph to realize $\mathscr{R}$.

Step 1. $r_{1,2} = 3$ is the largest element in only one cut matrix, that formed by using $\bar{X}_i = \{v_2,v_3,v_4\}$. Hence, $c(\{v_1\},\{v_2,v_3,v_4\}) = 3$. For the semigraph $c(1,3) = c(1,4) = 0$, and therefore $c(1,2) = 3$. This is the first branch capacity in our realization. Note that we have simultaneously assured that there are also min-cut matrices for $r_{1,3}$ and $r_{1,4}$ of value 3.

Step 2. Using exactly the same reasoning, we consider the entry $r_{2,3}$ to obtain $c(2,3) = 4$ and then consider the entry $r_{3,5}$ to obtain $c(3,5) = 5$.

Step 3. Consider the entry $r_{3,2} = 6$. Only one min-cut matrix contains $r_{3,2}$, that defined by $\bar{X}_i = \{v_2,v_4\}$. Hence, we must have $c(\{v_1,v_3\},\{v_2,v_4\}) = 6$. However, we already know that $c(1,2) = 3$, $c(1,4) = 0$ and $c(3,4) = 5$. Therefore $c(3,2) = -2$.

Note that in each step there was only one branch of unknown capacity in the cut and hence we were able to solve for its value. This will always be the case since we are considering the elements of $\mathscr{R}^{(0)}$ in increasing numerical order. When we consider the kth largest element, its min-cut matrix will contain only smaller entries than the kth element and the branch capacities in the corresponding cut will all have been found except for one.

Step 4. We repeat the process for the entry $r_{4,3}$ and obtain $c(4,3) = 3$.

Step 5. Now consider the entry $r_{4,2} = 8$. For the first time, we encounter the complication that the element has more than one min-cut matrix. One corresponds to $\bar{X}_i = \{v_2\}$ and the other, to $\bar{X}_i = \{v_2,v_3\}$. The sum of the known branch capacities in $(\{v_1,v_3,v_4\},\{v_2\})$ is 1 and the sum of the known branch capacities in $(\{v_1,v_4\},\{v_2,v_3\})$ is 6. If we use $\bar{X}_i = \{v_2\}$ in defining the min-cut matrix to calculate $c(4,2)$, we obtain $c(4,2) = 7$. In the second case, we obtain $c(4,2) = 2$. The question now is which min-cut matrix to use. If we let $c(4,2) = 7$, $c(\{v_1,v_3,v_4\},\{v_2\}) = 8$ and $c(\{v_1,v_4\},\{v_2,v_3\}) = 13$. If we let $c(4,2) = 2$, $c(\{v_1,v_4\},\{v_2,v_3\}) = 8$ and $c(\{v_1,v_3,v_4\},\{v_2\}) = 3$. This value is below the minimum value allowable for a 4-2 cut. We shall see that in general we should always select the min-cut matrix which gives the largest value for the branch capacity.

Step 6. Continuing this development we can calculate the remaining branch capacities to obtain the semigraph in Fig. 5.6.1(a). This semigraph is then shifted to obtain the graph in Fig. 5.6.1(b).

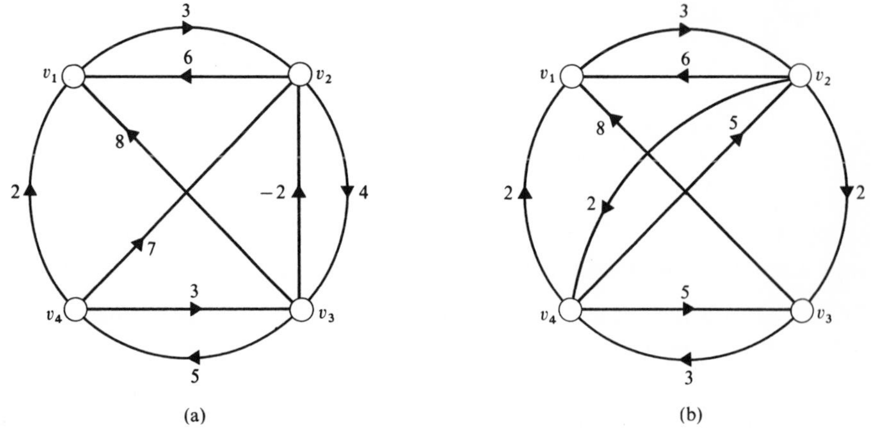

Fig. 5.6.1 Semigraph and graph for $\mathscr{R}$ in (5.6.1).

The steps used to realize $\mathscr{R}$ in (5.6.1) are now formalized as the Substitution Algorithm below. In the algorithm, $r_l^{(k)}$ is the entry in $\mathscr{R}^{(k)}$ in the location of the lth smallest entry in $\mathscr{R}^{(0)}$. $\|\mathscr{R}_i^{(d-1)}\|$ is the sum of the entries in $\mathscr{R}_i^{(d-1)}$.

Substitution Algorithm

Step 1. Find a semiprincipal partitioning of $\mathscr{R}$ and find $\mathscr{R}^{(0)}$.

Step 2. For $d = 1,2,\ldots,(n-1)(n+2)/2$ form $\mathscr{R}^{(d)}$ from $\mathscr{R}^{(d-1)}$ by letting $r_d^{(d)} = 2r_d^{(d-1)} - \operatorname*{Min}_i \|\mu_i^{(d-1)}\|$. The minimization is over the submatrices defined by $\bar{X}_i$ such that $\mathscr{R}_i^{(0)}$ is a min-cut matrix of r_d.

Step 3. Let $W = \mathscr{R}$. (Recall that W is the matrix whose entries are the branch capacities.)

In the Substitution Algorithm, we are performing the steps used in the example with the addition that once we consider $r_{a,b}$ we replace it by $c(a,b)$ to form the next matrix. Hence, in the last step we are left with a matrix in which the entries are the branch capacities. This also enables us to find the values of the cuts, directly from the matrix.

We now prove necessary and sufficient conditions for realizability of a maximally distinct requirement matrix.

Theorem 5.6.1 *A real, maximally distinct square matrix $\mathscr{R}$ with infinite entries on the main diagonal is realizable as the terminal capacity matrix of a graph if and only if every element has at least one min-cut matrix and the Substitution Algorithm yields a semigraph for which*

$$c(i,j) + c(j,i) \geq 0 \qquad \textit{for all } i \textit{ and } j \tag{5.6.3a}$$

and

$$c(k,k+1) = r_{k,k+1} \qquad \textit{for } k = 1,\ldots,n-1. \tag{5.6.3b}$$

Proof. From Lemma 5.2.2 we know that $\mathscr{R}$ is semiprincipally partitionable and $\mathscr{R}^{(0)}$ can be found. We first show that if the Substitution Algorithm yields a semigraph for which $c(i,j) + c(j,i) \geq 0$ for all i and j, then:

for every cut matrix $M_i(k,j)$ with $v_j \in \bar{X}_i$, $v_k \in X_i$ and $k \geq j+1$, $c(X_i,\bar{X}_i) \geq r_{k,j}$, (5.6.4a)

every $r_{k,j}$ with $k \geq j+1$ has at least one min-cut matrix $\mu_i(k,j)$ such that $c(X_i,\bar{X}_i) = r_{k,j}$ with $v_j \in \bar{X}_i$ and $v_k \in X_i$, (5.6.4b)

$\tau_{k,j} = r_{k,j}$ for $k+1 < j$. (5.6.4c)

In view of Theorem 5.2.3, (5.6.4) guarantees that we can find a semigraph which realizes all elements of $\mathscr{R}$. Since we require $c(k,j) + c(j,k) \geq 0$, the semigraph can then be converted to a graph by shifting. Assume $r_{a,b} \to r_d$. From Step 2 of the Algorithm,

$$r_d^{(d)} = 2r_d^{(d-1)} \operatorname*{Min}_i \left\|\mu_i^{(d-1)}(d)\right\|. \tag{5.6.5}$$

Using the facts that

$$\left\|\mu_i^{(d-1)}(d)\right\| = \left\|\mu_i(d)\right\| - c(a,b) + r_d, r_d^{(d-1)} = r_d \text{ and } c(a,b) = r_d^{(d)},$$

(5.6.5) becomes

$$c(a,b) = 2r_d - \operatorname*{Min}_i\left[\sum_{l,m:r_{a,b}\in\mu_i(d)} c(l,m) - c(a,b) + r_d\right]. \tag{5.6.6}$$

Hence

$$\operatorname*{Min}_i\left[\sum_{l,m:r_{a,b}\in\mu_i(d)} c(l,m)\right] = r_d. \tag{5.6.7}$$

Equations (5.6.4a,b) therefore follow from (5.6.7).

If as a result of Steps 1 to 3, $c(k,k+1) = r_{k,k+1}$ for $k = 1, \ldots, n-1$, then from (5.4.10b) of Corollary 5.4.1, (5.6.4c) is satisfied.

Next assume that $\mathscr{R}$ is realizable. We then prove the Substitution Algorithm yields an $\mathscr{R}^{((n-1)(n+2)/2)}$ such that $c(i,j) + c(j,i) \geq 0$ for all i and j. Since $\mathscr{R}$ is realizable, we can find an $\mathscr{R}^{(0)}$. From Theorem 5.2.3 every element must have a min-cut matrix. From (5.2.11a) and (5.2.11b), for any realization of $\mathscr{R}$

$$r_d = \operatorname*{Min}_i \left[\sum_{l,m : r_{a,b} \in \mu_i(d)} c(l,m) \right] \qquad \text{for all } d. \tag{5.6.8}$$

Therefore

$$r_d^{(d-1)} = \operatorname*{Min}_i \left[\| \mu_i^{(d)}(d) \| \right] = \operatorname{Min} \left[\| \mu^{(d-1)}(d) \| \right] - r_d^{(d-1)} + r_d^{(d)} \tag{5.6.9a}$$

or

$$r_d^{(d)} = 2r_d^{(d-1)} - \operatorname*{Min}_i \left[\| \mu_i^{(d-1)}(d) \| \right]. \tag{5.6.9b}$$

Since every element has a min-cut matrix, all branch capacities will exist. Equation (5.6.9b) shows that for the given $\mathscr{R}^{(0)}$, the values of the branch capacities in the semigraph are unique. Hence since $\mathscr{R}^{(0)}$ is realizable as the terminal capacity matrix of a graph we can convert the semigraph to a graph by shifting. This implies $c(i,j) + c(j,i) \geq 0$ for all i,j. To realize $r_{i,j}$ for $i + 1 < j$, $r_{i,j}$ must be coupled to an element $r_{l,l+1}$ with the same value. Hence the min-cut matrix used to determine the value of $r_{l,l+1}$ must be $\mathscr{R}_q^{(0)}$ with $X_q = \{v_1, v_2, \ldots, v_l\}$. Therefore we obtain $r_{k,k+1} = c(k,k+1)$ for $k = 1, \ldots, n-1$.

Example 5.6.1 Applying the Substitution Algorithm to $\mathscr{R}^{(0)}$ in (5.6.2) we have the matrices in (5.6.10). $\mathscr{R}^{(1)}$, $\mathscr{R}^{(2)}$, and $\mathscr{R}^{(3)}$ are the same as $\mathscr{R}^{(0)}$. The resulting semigraph is shown in Fig. 5.6.1(a). Since $c(i,j) + c(j,i) \geq 0$ for all i and j, we can obtain the semigraph in Fig. 5.6.1(b) by shifting.

$$\mathscr{R}^{(4)} = \begin{bmatrix} \infty & 3 & 0 & 0 \\ 10 & \infty & 4 & 0 \\ 11 & -2 & \infty & 5 \\ 12 & 8 & 7 & \infty \end{bmatrix} \tag{5.6.10a}$$

$$\mathscr{R}^{(7)} = \begin{bmatrix} \infty & 3 & 0 & 0 \\ 6 & \infty & 4 & 0 \\ 11 & -2 & \infty & 5 \\ 12 & 7 & 3 & \infty \end{bmatrix} \tag{5.6.10d}$$

$$\mathscr{R}^{(5)} = \begin{bmatrix} \infty & 3 & 0 & 0 \\ 10 & \infty & 4 & 0 \\ 11 & -2 & \infty & 5 \\ 12 & 8 & 3 & \infty \end{bmatrix} \tag{5.6.10b}$$

$$\mathscr{R}^{(8)} = \begin{bmatrix} \infty & 3 & 0 & 0 \\ 6 & \infty & 4 & 0 \\ 8 & -2 & \infty & 5 \\ 12 & 7 & 3 & \infty \end{bmatrix} \tag{5.6.10e}$$

$$\mathscr{R}^{(6)} = \begin{bmatrix} \infty & 3 & 0 & 0 \\ 10 & \infty & 4 & 0 \\ 11 & -2 & \infty & 5 \\ 12 & 7 & 3 & \infty \end{bmatrix} \tag{5.6.10c}$$

$$\mathscr{R}^{(9)} = \begin{bmatrix} \infty & 3 & 0 & 0 \\ 6 & \infty & 4 & 0 \\ 8 & -2 & \infty & 5 \\ 2 & 7 & 3 & \infty \end{bmatrix} \tag{5.6.10f}$$

Table 5.6.1 Substitution Algorithm applied to $\mathscr{R}^{(0)}$ in (5.6.2).

d	$\bar{X}_i$	$r_{h,j}$	$r_d^{(d)}$	$\bar{X}_l$
4	$\{v_2,v_4\}$	$r_{3,2}$	-2	
5	$\{v_3\}$	$r_{4,3}$	3	
6	$\{v_2\}$	$r_{4,2}$	7	$\{v_2,v_3\}$
7	$\{v_1,v_3,v_4\}$	$r_{2,1}$	6	
8	$\{v_1,v_2,v_4\}$	$r_{3,1}$	8	$\{v_1,v_4\}$
9	$\{v_1,v_2,v_3\}$	$r_{4,1}$	2	$\left\{v_1 \bigcup_k \binom{v_2,v_3}{k}\right\} - \{v_1,v_2,v_3\}$

All the information in (5.6.10) is presented in more compact and explicit form in Table 5.6.1. Henceforth, instead of writing out the matrices from succeeding algorithms, we present only tables of a similar form. Column 1 contains the value of d for the step corresponding to the row in the table. Column 2 contains $\bar{X}_i$ for the min-cut matrix used to determine the branch capacities $r_{h,j}$ such that $r_{h,j} \to r_d$. Column 3 contains $r_{h,j}$. Column 4 contains $r_d^{(d)} = c(h,j)$. Column 5 contains $\bar{X}_l$ for all the other min-cut matrices of r_d other than $\bar{X}_i$. In the last row and column,

$$\left\{v_1 \bigcup_k \binom{v_2,v_3}{k}\right\} - \{v_1,v_2,v_3\}$$

indicates the sets formed by taking the union of the sets containing v_1 with all possible sets formed by taking combinations of the elements v_2 and $v_3 k$ at a time in all ways for $k = 1,2$ except $\{v_1,v_2,v_3\}$. Table 5.6.1 does not replace the calculation in terms of the matrix. However, it is useful for bookkeeping.

As a byproduct of Theorem 5.6.1 we can show that the triangle inequality is not sufficient for realizability. For small ε the matrix

$$M = \begin{bmatrix} \infty & 1+2\varepsilon & 1+\varepsilon & 1 \\ 10+\varepsilon & \infty & 1+\varepsilon & 1 \\ 10+2\varepsilon & 1+3\varepsilon & \infty & 1 \\ 10+5\varepsilon & 10+4\varepsilon & 10+3\varepsilon & \infty \end{bmatrix} \tag{5.6.11}$$

is maximally distinct and satisfies the triangle inequality. However, the matrix is not realizable since the Substitution Algorithm yields a graph in which $c(4,1)$ is negative.

*5.7 PERTURBATIONS TO OBTAIN MAXIMALLY DISTINCT MATRICES

We now consider the synthesis of a nonmaximally distinct matrix $\mathscr{R}$. We attempt to convert the matrix to one that is maximally distinct, and again try the Substitution Algorithm. First, we simply perturb the elements which are equal in value, apply the Substitution Algorithm and then let the perturbations go to zero. In other words, we propose the following algorithm.

Perturbation Algorithm

Step 1. From $\mathscr{R}^{(0)}$ form $\mathscr{R}^{(0)}(\boldsymbol{\varepsilon})$ by replacing r_d by $r_d + (l - 1)\varepsilon_d$ for $l = 1, \ldots, n_d$ and $d = 1, \ldots, \zeta$, where $\boldsymbol{\varepsilon}$ is a real n_d-dimensional vector whose dth entry is ε_d, all entries in $\boldsymbol{\varepsilon}$ are arbitrarily small compared to $r_d^{(d)}$ for all d, n_d is the number of entries in $\mathscr{R}^{(0)}$ with the dth smallest value in $\mathscr{R}^{(0)}$, and ζ is the number of numerically distinct values in $\mathscr{R}^{(0)}$.

Step 2. Apply the Substitution Algorithm to realize $\mathscr{R}(\boldsymbol{\varepsilon})$.

Step 3. Let $\boldsymbol{\varepsilon} = \mathbf{0}$.

With the Perturbation Algorithm, we can realize some matrices that cannot be realized with the Substitution Algorithm, as illustrated in Example 5.7.1.

Example 5.7.1 Let

$$\mathscr{R} = \begin{bmatrix} \infty & 3 & 3 & 3 \\ 14 & \infty & 6 & 6 \\ 17 & 15 & \infty & 8 \\ 10 & 10 & 10 & \infty \end{bmatrix}. \tag{5.7.1}$$

A possible $\mathscr{R}(\boldsymbol{\varepsilon})$ is given in (5.7.2) with $r_{4,3} \to r_4$, $r_{4,2} \to r_5$, and $r_{4,1} \to r_6$.

$$\mathscr{R} = \begin{bmatrix} \infty & 3 & 3 & 3 \\ 14 & \infty & 6 & 6 \\ 17 & 15 & \infty & 8 \\ 10 + 2\varepsilon_4 & 10 + \varepsilon_4 & 10 & \infty \end{bmatrix} \tag{5.7.2}$$

Applying the Substitution Algorithm after using the steps in Table 5.7.1 gives $W(\boldsymbol{\varepsilon})$ and the graph in Fig. 5.7.1(a). Letting $\boldsymbol{\varepsilon} = \mathbf{0}$ gives the graph shown in Fig. 5.7.1(b).

$$W(\boldsymbol{\varepsilon}) = \begin{bmatrix} \infty & 3 & 0 & 0 \\ 8 & \infty & 6 & 0 \\ (6 - \varepsilon_4) & (9 - \varepsilon_4) & \infty & 8 \\ (3 + \varepsilon_4) & (3 + \varepsilon_4) & 4 & \infty \end{bmatrix} \tag{5.7.3}$$

Table 5.7.1 Perturbation Algorithm applied to $\mathscr{R}$ in (5.7.1).

d	$\bar{X}_i$	$r_{h,j}$	$r_d^{(d)}$	$\bar{X}_l$
4	$\{v_3\}$	$r_{4,3}$	4	
5	$\{v_2,v_3\}$	$r_{4,2}$	$3 + \varepsilon_1$	
6	$\{v_1,v_2,v_3\}$	$r_{4,1}$	$3 + \varepsilon_1$	
7	$\{v_1,v_3,v_4\}$	$r_{2,1}$	8	$\{v_1,v_3\}$
8	$\{v_2\}$	$r_{3,2}$	$9 - \varepsilon_4$	$\{v_2,v_4\}$
9	$\{v_1\}$	$r_{3,1}$	$6 - \varepsilon_4$	$\left\{v_1 \bigcup_k \binom{v_2,v_4}{k}\right\} - \{v_1\}$

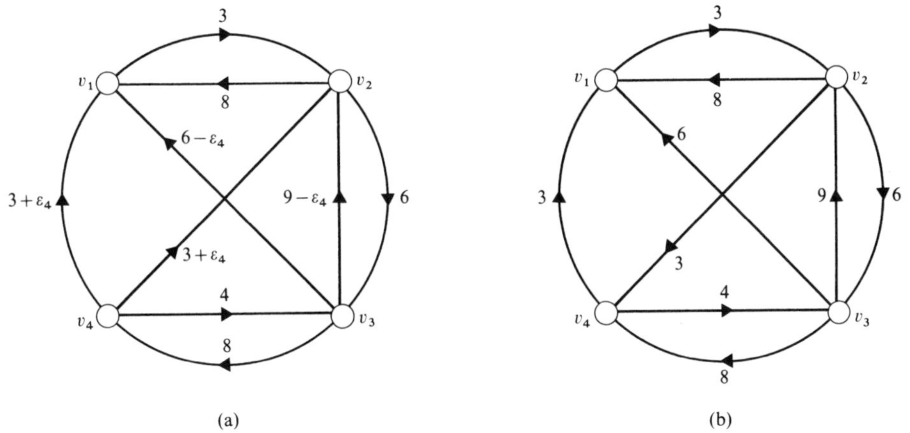

Fig. 5.7.1 Semigraphs for Example 5.7.1.

There are realizable matrices which are not realizable by the Perturbation Algorithm and shifting operations, as illustrated in Examples 5.7.2 and 5.7.3.

Example 5.7.2 The matrix $\mathscr{R}$ in (5.7.4) cannot be realized by the Perturbation Algorithm. No matter how we perturb the elements either $r_{2,1}$ or $r_{3,1}$ will not have a distinct min-cut matrix.

$$\mathscr{R} = \begin{bmatrix} \infty & 5 & 5 & 4 \\ 6 & \infty & 30 & 4 \\ 6 & 50 & \infty & 4 \\ 8 & 7 & 7 & \infty \end{bmatrix} \tag{5.7.4}$$

However, $\mathscr{R}$ is realized by the semigraph in Fig. 5.7.2 and for all k and $j, c(k,j) + c(j,k) \geq 0$.

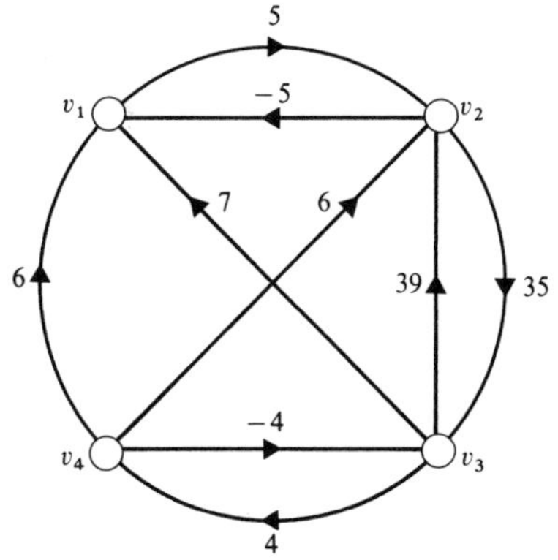

Fig. 5.7.2 Semigraph for Example 5.7.2.

Example 5.7.3 Let

$$\mathscr{R} = \begin{bmatrix} \infty & 6 & 3 & 3 \\ 7 & \infty & 3 & 3 \\ 10 & 10 & \infty & 8 \\ 10 & 10 & 12 & \infty \end{bmatrix}. \tag{5.7.5}$$

The only possible perturbations such that every element in $\mathscr{R}(\varepsilon)$ has at least one distinct min-cut matrix are given in (5.7.6) and (5.7.7).

$$\mathscr{R}(\varepsilon) = \begin{bmatrix} \infty & 6 & 3 & 3 \\ 7 & \infty & 3 & 3 \\ 10 & (10 + \varepsilon_5) & \infty & 8 \\ (10 + 2\varepsilon_5) & (10 + 3\varepsilon_5) & 12 & \infty \end{bmatrix} \tag{5.7.6a}$$

$$\mathscr{R}(\varepsilon) = \begin{bmatrix} \infty & 6 & 3 & 3 \\ 7 & \infty & 3 & 3 \\ (10 + \varepsilon_5) & 10 & \infty & 8 \\ (10 + 2\varepsilon_5) & (10 + 3\varepsilon_5) & 12 & \infty \end{bmatrix} \tag{5.7.6b}$$

$$\mathscr{R}(\varepsilon) = \begin{bmatrix} \infty & 6 & 3 & 3 \\ 7 & \infty & 3 & 3 \\ 10 & (10 + \varepsilon_5) & \infty & 8 \\ (10 + 3\varepsilon_5) & (10 + 2\varepsilon_5) & 12 & \infty \end{bmatrix} \tag{5.7.7a}$$

$$\mathscr{R}(\varepsilon) = \begin{bmatrix} \infty & 6 & 3 & 3 \\ 7 & \infty & 3 & 3 \\ (10 + \varepsilon_5) & 10 & \infty & 8 \\ (10 + 3\varepsilon_5) & (10 + 2\varepsilon_5) & 12 & \infty \end{bmatrix} \tag{5.7.7b}$$

If we apply the Substitution Algorithm to the $\mathscr{R}(\varepsilon)$ in (5.7.6) or (5.7.7), we obtain $c(3,1) = -2$ in Table 5.7.2 and $c(3,2) = -4$ in Table 5.7.3, respectively.

Table 5.7.2 Perturbation Algorithm applied to $\mathscr{R}$ in (5.7.6).

d	$\bar{X}_i$	$r_{h,j}$	$r_d^{(d)}$	$\bar{X}_l$
3	$\{v_1,v_3,v_4\}$	$r_{2,1}$	4	
5	$\{v_1,v_4\}$	$r_{3,1}$	-2	

Table 5.7.3 Perturbation Algorithm applied to $\mathscr{R}$ in (5.7.7).

d	$\bar{X}_i$	$r_{h,j}$	$r_d^{(d)}$	$\bar{X}_l$
3	$\{v_1,v_3,v_4\}$	$r_{2,1}$	4	
5	$\{v_2,v_4\}$	$r_{3,2}$	-4	

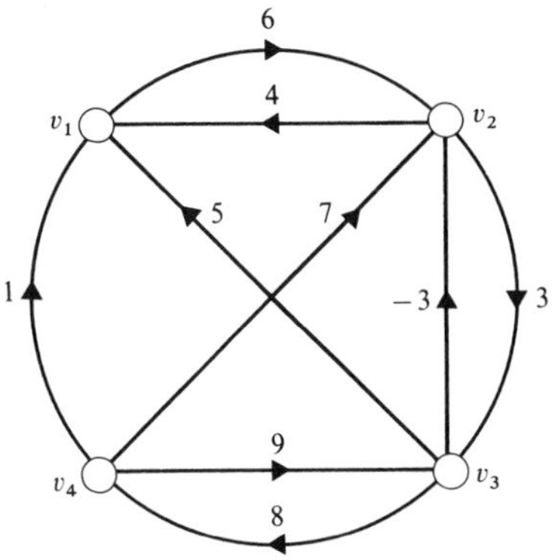

Fig. 5.7.3 Semigraph for Example 5.7.3.

Similarly it can be shown that no other perturbation leads to a realization. However, $\mathcal{R}$ is realized by the semigraph in Fig. 5.7.3 and $c(j,k) + c(k,j) \geq 0$ for all k and j.

Thus any matrix realizable by the Substitution Algorithm is realizable by the Perturbation Algorithm. However, the converse is untrue and there are realizable matrices which cannot be realized by either algorithm. We now derive a "Replacement Algorithm" which includes both the Substitution Algorithm and Perturbation Algorithm. Furthermore, we will see why the Perturbation Algorithm succeeds in some cases but not in others. In the following algorithm, $\operatorname*{Min}_i [\|S_i^{(d-1)}(d)\|] = \infty$ if r_d has no semidistinct min-cut matrices, $S_i^{(d-1)}(d)$.

Replacement Algorithm

Step 1. Given $\mathcal{R}$ find any $\mathcal{R}^{(0)}$.

Step 2. For $d = 1,2,\ldots,(n-1)(n+2)/2$, form $\mathcal{R}^{(d)}$ from $\mathcal{R}^{(d-1)}$ by letting

$$r_d^{(d)} = \begin{cases} \operatorname{Max}[-r_g^{(g)}, 2r_d^{(d-1)}] - \operatorname*{Min}_i [\|S_i^{(d-1)}(d)\|] \text{ if } r_d \to r_{j+1,j} \text{ and } r_g \to r_{j,j+1}. \\ \\ \operatorname{Max}[0, 2r_d^{(d-1)}] - \operatorname*{Min}_i [\|S_i^{(d-1)}(d)\|] \text{ otherwise.} \end{cases} \tag{5.7.8}$$

Step 3. Set $W = \mathcal{R}^{(n-1)(n+2)/2}$.

Note that the only difference between the Substitution and Replacement Algorithm is that in the Replacement Algorithm an element of W can be increased above its value in the Substitution Algorithm.

Theorem 5.7.1 *A real, square matrix Q with elements of infinite value on the diagonal is realizable if every element has at least one min-cut matrix for some ordering of the elements of $\mathcal{R}^{(0)}$, the Replacement Algorithm yields $\mathcal{R}^{(n-1)(n+2)/2}$ which satisfies (5.4.10), and if*

$$r_d^{(d)} > 2r_d^{(d-1)} - \operatorname*{Min}_i [\|S_i^{(d-1)}(d)\|], \tag{5.7.9a}$$

then r_d is coupled to at least one r_q such that

$$r_q^{(q)} = 2r_q^{(q-1)} - \operatorname*{Min}_i [\|S_i^{(q-1)}(q)\|] \tag{5.7.9b}$$

and $r_d^{(d-1)}$ is in $S_i^{(q-1)}(q)$.

Proof. From Lemma 5.2.2, $\mathscr{R}^{(0)}$ exists. We show that if the Replacement Algorithm yields a semigraph and (5.4.10) is satisfied then (5.7.10) below is satisfied.

For every min-cut matrix $\mu_i(k,j)$ with $v_j \in \bar{X}_i$ and $v_k \in X_i$, $c(X_i,\bar{X}_i) \geq r_{k,j}$. (5.7.10a)

Every $r_{k,j}$ with $k + 1 \geq j$ has at least one min-cut matrix $\mu_i(k,j)$ such that $c(X_i,\bar{X}_i) = r_{k,j}$ with $v_j \in \bar{X}_i$, $v_k \in X_i$. (5.7.10b)

$\tau_{k,j} = r_{k,j}$ for $k + 1 < j$. (5.7.10c)

$c(i,j) + c(j,i) \geq 0$ for all i and j. (5.7.10d)

Equations (5.4.10a, b, and c) guarantee that all elements in $\mathscr{R}$ are realized by the semigraph; (5.4.10d) guarantees that the semigraph can be converted to a graph by shifting.

Suppose $r_{a,b} \to r_d$. From (5.7.8)

$$r_d^{(d)} \geq 2r_d^{(d-1)} - \operatorname*{Min}_i [\|S_i^{(d-1)}(d)\|]. \tag{5.7.11}$$

Since $\|S_i^{(d-1)}(d)\| = \|S_i(d)\| - c(a,b) + r_d$, $r_d^{(d-1)} = r_d$ and $c(a,b) = r_d^{(d)}$, Eq. (5.7.11) becomes

$$c(a,b) \geq 2r_d - \operatorname*{Min}_i \left[\sum_{l,m:r_{a,b} \in S_i(d)} c(l,m) - c(a,b) + r_d \right]. \tag{5.7.12}$$

Transposition gives

$$\operatorname*{Min}_i \left[\sum_{l,m:r_{a,b} \in S_i(d)} c(l,m) \right] \geq r_d. \tag{5.7.13}$$

Thus

$$\operatorname*{Min}_i \left[\sum_{l,m:r_{a,b} \in \mu_i(d)} c(l,m) \right] \geq r_d \tag{5.7.14a}$$

or

$$c(X_i,\bar{X}_i) \geq r_d \tag{5.7.14b}$$

for all X_i such that there is a min-cut matrix $\mu_i(k,j)$. Thus (5.7.10a) is proved.

If in (5.7.8)

$$r_d^{(d)} = 2r_d^{(d-1)} - \operatorname*{Min}_i [\|S_i^{(d-1)}(d)\|], \tag{5.7.15}$$

by using similar steps we can show that $c(X_l,\bar{X}_l) = r_d$ with $v_a \in X_l$ and $v_b \in \bar{X}_l$ for some l. If in (5.7.8)

$$r_d^{(d)} > 2r_d^{(d-1)} - \operatorname*{Min}_i [\|S_i^{(d-1)}(d)\|] \tag{5.7.16}$$

then (5.7.9) requires that r_d be coupled to r_q for which

$$r_q^{(q)} = 2r_q^{(q-1)} - \operatorname*{Min}_j [\|S_j^{(q-1)}(q)\|] \tag{5.7.17}$$

and $r_d^{(d-1)}$ is in $S_q^{(q-1)}$. But then we can again show that $c(X_l,\bar{X}_l) = r_d$ with $v_a \in X_l$ and $v_b \in \bar{X}_l$ for some l. Therefore (5.7.10b) is proved. From (5.4.10) it is clear that (5.7.10a) is satisfied. From (5.7.8)

$$r_d^{(d)} \geq -r_g^{(g)} \tag{5.7.18a}$$

or equivalently

$$c(j+1,j) \geq -c(j,j+1) \qquad \text{for } j = 1,2,\ldots,n-1. \tag{5.7.18b}$$

From (5.7.8)

$$r^{(d)}(d) = c(a,b) \geq 0 \qquad \text{for } a > b. \tag{5.7.19}$$

Since $c(a,b) = 0$ for $a + 1 \leq b$ it follows from (5.7.18) and (5.7.19) that $c(a,b) + c(b,a) \geq 0$ for all v_a and v_b. Hence (5.7.10d) is proved.

We first illustrate the use of the Replacement Algorithm and Theorem 5.7.1 for a matrix used a number of times in the remainder of this chapter.

Example 5.7.4 Consider the matrix $\mathscr{R}$ in (5.7.20) in which $r_{4,2}$, $r_{5,2}$, and $r_{5,4}$ are equal in value.

$$\mathscr{R} = \begin{bmatrix} \infty & 3 & 3 & 3 & 3 & 1 \\ 15 & \infty & 6 & 3 & 3 & 1 \\ 20 & 9 & \infty & 3 & 3 & 1 \\ 25 & 10 & 12 & \infty & 6 & 1 \\ 30 & 10 & 11 & 10 & \infty & 1 \\ 60 & 50 & 45 & 40 & 35 & \infty \end{bmatrix} \tag{5.7.20}$$

Table 5.7.4 represents the application of the Replacement Algorithm to $\mathscr{R}$; $r_{j,j+1} = r_d^{(d)}$ for $j = 1,2,3,4,5$ and hence these calculations are omitted from the table.

Table 5.7.4 Replacement Algorithm applied to $\mathscr{R}$ in (5.7.20).

d	$\bar{X}_i$	$r_{h,j}$	$r_d^{(d)}$	$\bar{X}_l$
6	$\{v_2,v_4,v_5,v_6\}$	$r_{3,2}$	3	
7	$\{v_4,v_6\}$	$r_{5,4}$	6	
8	$\{v_2,v_5,v_6\}$	$r_{4,2}$	0	
9	$\{v_2,v_6\}$	$r_{5,2}$	3	$\{v_2,v_4,v_6\}$
10	$\{v_3,v_4,v_6\}$	$r_{5,3}$	-2	$\{v_3,v_6\},\{v_2,v_3,v_6\},\{v_2,v_3,v_4,v_6\}$

The matrix $\mathscr{R}$ is not realizable by the Replacement Algorithm with this ordering of the elements since $c(5,3) < 0$ and $r_{5,3}$ is a distinct entry. On the other hand, if we apply the Replacement Algorithm with the ordering in Table 5.7.5, the matrix is realized as a semigraph since the conditions of Theorem 5.7.1 are satisfied.

Table 5.7.5 Replacement Algorithm applied to $\mathscr{R}$ in (5.7.20).

d	$\bar{X}_i$	$r_{h,j}$	$r_d^{(d)}$	$\bar{X}_l$
6	$\{v_2,v_4,v_5,v_6\}$	$r_{3,2}$	3	
7	$\{v_4,v_6\}$	$r_{5,4}$	6	
8	$\{v_2,v_4,v_6\}$	$r_{5,2}$	0	
9	$\{v_2,v_6\}$	$r_{4,2}$	3	$\{v_2,v_5,v_6\}$
10	$\{v_2,v_3,v_4,v_6\}$	$r_{5,3}$	1	$\{v_3,v_4,v_6\}$
11	$\{v_3,v_6\}$	$r_{4,3}$	4	$\{v_2,v_3,v_6\},\{v_3,v_5,v_6\},\{v_2,v_3,v_5,v_6\}$
12	$\{v_1,v_3,v_4,v_5,v_6\}$	$r_{2,1}$	9	
13	$\{v_1,v_2,v_4,v_5,v_6\}$	$r_{3,1}$	14	$\{v_1,v_4,v_5,v_6\}$
14	$\{v_1,v_2,v_3,v_5,v_6\}$	$r_{4,1}$	12	$\{v_1,v_2,v_5,v_6\},\{v_1,v_3,v_5,v_6\},\{v_1,v_5,v_6\}$
15	$\{v_1,v_2,v_3,v_4,v_6\}$	$r_{5,1}$	22	$\left\{v_1,v_6 \bigcup_k \binom{v_2,v_3,v_4}{k}\right\} - \{v_1,v_2,v_3,v_4,v_6\}$
16	$\{v_5\}$	$r_{6,5}$	29	
17	$\{v_4\}$	$r_{6,4}$	33	$\{v_4,v_5\}$
18	$\{v_3\}$	$r_{6,3}$	24	$\left\{v_3 \bigcup_k \binom{v_4,v_5}{k}\right\} - \{v_3\}$
19	$\{v_2\}$	$r_{6,2}$	41	$\left\{v_1 \bigcup_k \binom{v_2,v_3,v_4,v_5}{k}\right\} - \{v_2\}$
20	$\{v_1\}$	$r_{6,1}$	3	$\left\{v_1 \bigcup_k \binom{v_2,v_3,v_4,v_5}{k}\right\} - \{v_1\}$

This example also points out that the Replacement Algorithm may yield a realization for one ordering of elements but not for another.

Example 5.7.5 Consider the matrices in Examples 5.7.2 and 5.7.3 for which the Perturbation Algorithm failed. Applying the Replacement Algorithm to $\mathscr{R}$ in (5.7.4), we have the steps in Table 5.7.6 and the resulting semigraph in Fig. 5.7.2 which satisfies the conditions of Theorem 5.7.1; $r_1^{(1)}$ and $r_2^{(2)}$ which are $r_{1,2}$ and $r_{3,4}$ respectively, are not included in the table.

Applying the algorithm to $\mathscr{R}$ in (5.7.5) yields the steps in Table 5.7.7 and the resulting semigraph in Fig. 5.7.3 which satisfies the conditions of Theorem 5.7.1; $r_1^{(1)}$ $r_2^{(2)}$, and $r_4^{(4)}$ are equal to $r_{j,j+1}$ for $j = 1$, 2, and 3, respectively, and are not included in the table.

Table 5.7.6 Replacement Algorithm applied to $\mathscr{R}$ in (5.7.4).

d	$\bar{X}_i$	$r_{h,j}$	$r_d^{(d)}$	$\bar{X}_l$
3	—	$r_{2,1}$	-5	
4	$\{v_1,v_4\}$	$r_{3,1}$	7	
5	—	$r_{4,3}$	-4	
6	$\{v_2,v_3\}$	$r_{4,2}$	6	
7	$\{v_1\}$	$r_{4,1}$	6	$\{v_1,v_2,v_3\}$
8	$\{v_1,v_3,v_4\}$	$r_{2,3}$	35	$\{v_3,v_1\},\{v_3,v_4\},\{v_1,v_3,v_4\}$
9	$\{v_2,v_4\}$	$r_{3,2}$	41	$\{v_2,v_1\},\{v_2\},\{v_1,v_2,v_4\}$

Table 5.7.7 Replacement Algorithm applied to $\mathscr{R}$ in (5.7.5).

d	$\bar{X}_i$	$r_{h,j}$	$r_d^{(d)}$	$\bar{X}_l$
3	$\{v_1,v_3,v_4\}$	$r_{2,1}$	4	
5	$\{v_2,v_4\}$	$r_{3,2}$	-3	
6	$\{v_1,v_2,v_4\}$	$r_{3,1}$	5	$\{v_1,v_4\}$
7	$\{v_1\}$	$r_{4,1}$	1	
8	$\{v_1,v_2\}$	$r_{4,2}$	7	$\{v_2\}$
9	$\{v_3\}$	$r_{4,3}$	9	$\{v_1,v_3\},\{v_2,v_3\},\{v_1,v_2,v_3\}$

We can now relate the Replacement Algorithm to the Perturbation Algorithm. It is clear that in the Perturbation Algorithm, the only use of the ε_i's is to order the elements of equal value and to make all min-cut matrices distinct. Thus, any matrix realizable by the Perturbation Algorithm can be realized by the same graph by the Replacement Algorithm. On the other hand, consider the possible ways in which the Perturbation Algorithm can fail to realize a realizable matrix. First, in the perturbed matrix an element may fail to have a semidistinct min-cut matrix as in Example 5.7.2; i.e., two elements may be completed coupled. Such a matrix might still be realizable by the Replacement Algorithm as shown for the same matrix in Example 5.7.4. Second, the Perturbation Algorithm might yield a semigraph with $c(j,k) + c(k,j) < 0$ as in Example 5.7.3. Such a matrix might still be realizable by the Replacement Algorithm as shown for the same matrix in Example 5.7.5. Finally, it may be necessary to consider matrices other than semidistinct min-cut matrices. In this case both the Perturbation and Replacement Algorithms will fail even though the matrix is realizable. From the above discussion we have the following theorem.

***Theorem* 5.7.2** *$\mathscr{R} \in SPP$ is realizable by the Perturbation Algorithm only if it is realizable by the Replacement Algorithm. The converse is not true.*

As illustrated by the next example, there are realizable matrices which cannot be realized by the Replacement Algorithm:

Example 5.7.6 Consider $\mathscr{R}$ in (5.7.21) obtained from $\mathscr{R}$ in (5.7.20) by increasing $r_{4,3}$ from 12 to 30.

$$\mathscr{R} = \begin{bmatrix} \infty & 3 & 3 & 3 & 3 & 1 \\ 15 & \infty & 6 & 3 & 3 & 1 \\ 20 & 9 & \infty & 3 & 3 & 1 \\ 25 & 10 & 30 & \infty & 6 & 1 \\ 30 & 10 & 11 & 10 & \infty & 1 \\ 60 & 50 & 45 & 40 & 35 & \infty \end{bmatrix} \tag{5.7.21}$$

In (5.7.20) the change of $r_{4,3}$ does not affect the realization of $r_{5,4}$, $r_{4,2}$, and $r_{5,2}$. In addition, $r_{4,3} > 10$. Thus from Example 5.7.4 the order of realization in Table 5.7.4 is impossible. Continuing the realization in Table 5.7.5 we have the steps in Table 5.7.8.

Table 5.7.8 Replacement Algorithm applied to $\mathscr{R}$ in (5.7.17).

d	$\bar{X}_i$	$r_{h,j}$	$r_d^{(d)}$	$\bar{X}_l$
7	$\{v_1,v_3,v_4,v_5,v_6\}$	$r_{2,1}$	9	
8	$\{v_1,v_2,v_4,v_5,v_6\}$	$r_{3,1}$	14	$\{v_1,v_4,v_5,v_6\}$
9	$\{v_1,v_2,v_5,v_6\}$	$r_{4,1}$	-1	$\{v_1,v_5,v_6\}$

Since $c(4,1) = r_9^{(9)}$ is negative, the matrix cannot be realized with the ordering in Tables 5.7.5 and 5.7.8. The remaining orderings for realizing $r_{4,2}$, $r_{5,2}$, and $r_{5,4}$ are all similar to the previous orderings; i.e., either $c(4,2) = 0$, $c(5,2) = 3$, and $c(5,4) = -1$ or $c(4,2) = 3$, $c(5,4) = 0$, and $c(4,2) = -1$. Similarly, no realizations can be obtained for other semiprincipal partitionings and hence the matrix is not realizable by the Replacement Algorithm. However, the graph in Fig. 5.7.4 realizes $\mathscr{R}$.

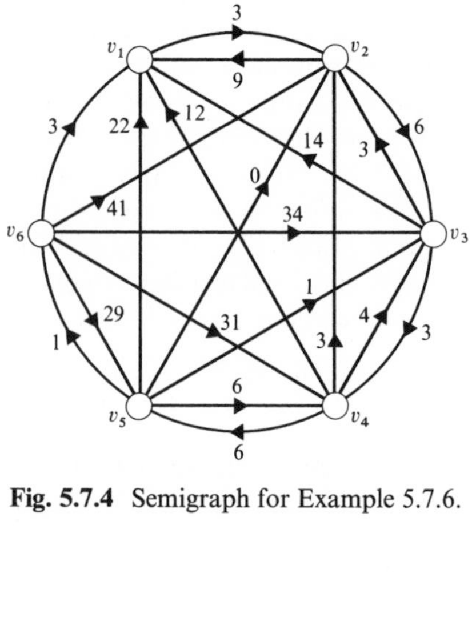

Fig. 5.7.4 Semigraph for Example 5.7.6.

*5.8 A GENERAL REALIZATION ALGORITHM

We can now extend the results of the previous section to give a general realization algorithm and, in terms of this algorithm, necessary and sufficient conditions for realizability of $\mathscr{R}$. In the algorithm α_d is a nonnegative constant for each d.

Terminal Capacity Realization Algorithm

Step 1. Given $\mathscr{R}$, find any $\mathscr{R}^{(0)}$.

Step 2. For $d = 1,2,\ldots,(n-1)(n+2)/2$, form $\mathscr{R}^{(d)}$ from $\mathscr{R}^{(d-1)}$ by letting

$$r_d^{(d)} = \begin{cases} \operatorname{Max}\left[-r_g^{(g)}, 2r_d^{(d-1)} - \operatorname*{Min}_i\left[\|S_i^{(d-1)}(d)\|\right]\right] + \alpha_d, \text{ if } r_d \rightarrow r_{j+1,j} \\ \quad\text{and } r_g \rightarrow r_{j,j+1} \\ \\ \operatorname{Max}\left[0, 2r_d^{(d-1)} - \operatorname*{Min}_i\left[\|S_i^{(d-1)}(d)\|\right]\right] + \alpha_d, \text{ otherwise.} \end{cases} \tag{5.8.1}$$

Step 3. Set $W = \mathscr{R}^{((n-1)(n+2)/2)}$.

The only difference between the Replacement Algorithm and the Terminal Capacity Realization Algorithm is the introduction of α_d, which allows branch capacities to be increased.

Theorem 5.8.1 *A real square requirement matrix $\mathscr{R}$ with infinite entries on the main diagonal is realizable if and only if every element has at least one min-cut matrix and for some ordering of the elements there exists a set of α_d such that the Terminal Capacity Realization Algorithm yields an $\mathscr{R}^{((n-1)(n-2)/2)}$ satisfying (5.4.10a).*

Proof. The sufficiency part of the proof is exactly the same as the proof of Theorem 5.7.1 with all reference to Eq. (5.7.8) being replaced by references to (5.8.1). All that remains to be proved is the necessity of the theorem. Assume that $\mathscr{R}$ is realizable. From Theorem 5.2.2 every element must have a min-cut matrix. We now show that there exists a set of α_d such that there is an $\mathscr{R}^{((n-1)(n+2)/2)}$. Since $\mathscr{R} \in$ SPP we can find an $\mathscr{R}^{(0)}$. For any realization of $\mathscr{R}$, satisfying (5.4.10) and (5.7.9),

$$r_d \leq \operatorname*{Min}_i\left[\sum_{a,b:r_{a,b} \in \mu_i(d)} c_{a,b}\right] \quad \text{for all } d \tag{5.8.2}$$

and hence

$$r_d \leq \operatorname*{Min}_i\left[\sum_{a,b:r_{a,b} \in S_i(d)} c_{a,b}\right]. \tag{5.8.3}$$

From (5.8.3)

$$r_d^{(d-1)} \leq \operatorname*{Min}_i\left[\|S_i^d(d)\|\right] = \operatorname*{Min}_i\left[\|S_i^{(d-1)}(d)\|\right] - r_d^{(d-1)} + r_d^{(d)} \tag{5.8.4a}$$

or

$$r_d^{(d)} \geq 2r_d^{(d-1)} - \operatorname*{Min}_i\left[\|S_i^{(d-1)}(d)\|\right]. \tag{5.8.4b}$$

We now use (5.8.4) to find the α_d in (5.8.1). We consider two cases depending upon whether or not $r_d \to r_{j+1,j}$ for some j.

CASE 1. Assume $r_d \to r_{j+1,j}$ for some j.

Since $\mathcal{R}$ is realizable as a semigraph which is to be converted to a graph by shifting, we must have

$$c(j + 1,j) + c(j,j + 1) \geq 0. \tag{5.8.5}$$

If $r_{j,j+1} \to r_g$ then (5.8.5) becomes

$$r_d^{(d)} \geq -r_g^{(g)}. \tag{5.8.6}$$

Combining (5.8.4) and (5.8.6) gives

$$r_d^{(d)} \geq \text{Max}\,[-r_g^{(g)}, 2r_d^{(d-1)} - \underset{i}{\text{Min}}\,[\|S_i^{(d-1)}(d)\|]]. \tag{5.8.7}$$

Therefore, if we define α_d as in (5.8.8), (5.8.7) yields (5.8.1).

$$\alpha_d = r^{(d)} - \text{Max}\,[-r_g^{(g)}, 2r_d^{(d-1)} - \underset{i}{\text{Min}}\,[\|S_i^{(d-1)}(d)\|]] \geq 0. \tag{5.8.8}$$

CASE 2. Assume $r_d \to r_{a,b}$ with $a \neq b + 1$.

a) If $r_d = r_d^{(d)} \to r_{k,k+1}$ for some k, then $r_d \geq 0$ from (5.8.1).

b) If r_d does not map into $r_{k,k+1}$ for any k,

$$c(a,b) + c(b,a) \geq 0. \tag{5.8.9}$$

But in a semigraph $c(b,a) = 0$. Equation (5.8.9) therefore shows

$$c(a,b) = r_d^{(d)} \geq 0. \tag{5.8.10}$$

In either case,

$$r_d^{(d)} \geq 0. \tag{5.8.11}$$

Combining (5.8.4) and (5.8.11), we get

$$r_d^{(d)} \geq \text{Max}\,[0, 2r_d^{(d-1)} - \underset{i}{\text{Min}}\,[\|S_i^{(d-1)}(d)\|]]. \tag{5.8.12}$$

Therefore if we define α_d as in (5.8.13), (5.8.12) yields (5.8.1).

$$\alpha_d = r_d^{(d)} - \text{Max}\,[0, 2r_d^{(d-1)} - \underset{i}{\text{Min}}\,[\|S_i^{(d-1)}(d)\|]] > 0. \tag{5.8.13}$$

The necessity of condition (5.4.10) follows immediately from Corollary 5.4.1, (5.2.11b), and the fact that we must realize $r_{a,b}$ for $a + 1 < b$. The proof is therefore completed.

In applying the Terminal Capacity Realization Algorithm to a given matrix, inequalities on the α_d result from certain steps. In particular, if an element r_d has several semidistinct min-cut matrices $S_i^{(d-1)}(d)$ whose elements contain α_q's, the decision as to which is the min-cut matrix of smallest value may be an arbitrary one which restricts the value of some of the α_q's. Moreover, the conditions $r_d^{(d)} \geq -r_g^{(g)}$ and $r_d^{(d)} \geq 0$ in (5.8.1) also place constraints on the α_q's. Finally, (5.4.10a) forces some of the α_q's to equal zero. Hence, at the completion of the algorithm we arrive at a set of α_q's which are subject to inequality constraints derived from Theorem 5.8.1. If we can find a set of α_q's satisfying these constraints, then $\mathscr{R}$ is realizable. If there are no α_q's satisfying any set of inequalities resulting from the algorithm, the matrix is not realizable.

These remarks and the use of the algorithm and theorem are illustrated in the next example.

Example 5.8.1 Consider the matrix

$$\mathscr{R} = \begin{bmatrix} \infty & 3 & 3 & 3 & 3 & 1 \\ 16 & \infty & 6 & 3 & 3 & 1 \\ 20 & 9 & \infty & 3 & 3 & 1 \\ 25 & 10 & 30 & \infty & 6 & 1 \\ 30 & 10 & 11 & 10 & \infty & 1 \\ 60 & 50 & 45 & 40 & 35 & \infty \end{bmatrix}. \tag{5.8.14}$$

$\mathscr{R}^{(0)}$ is formed by letting $r_{k,j} = 0$ for $k + 1 < j$. The elements of $\mathscr{R}^{(0)}$ are ordered as

$$\mathscr{R}^{(0)} = \begin{bmatrix} \infty & r_2 & 0 & 0 & 0 & 0 \\ r_{11} & \infty & r_4 & 0 & 0 & 0 \\ r_{12} & r_6 & \infty & r_3 & 0 & 0 \\ r_{13} & r_9 & r_{15} & \infty & r_5 & 0 \\ r_{14} & r_8 & r_{10} & r_7 & \infty & r_1 \\ r_{20} & r_{19} & r_{18} & r_{17} & r_{16} & \infty \end{bmatrix}. \tag{5.8.15}$$

The results of applying the Terminal Capacity Realization Algorithm to $\mathscr{R}^{(0)}$ in (5.8.15) are summarized in Table 5.8.1. The table differs from the previous tables used only in the addition of columns 6 and 7. Column 6 contains the inequalities on α_d resulting from the fact that $r_d^{(d)} \geq -r_g^{(g)}$ and $r_d^{(d)} \geq 0$ in (5.8.1). Column 7 contains the inequalities on α_d arising from the arbitrary selection of a semidistinct min-cut matrix $S_i^{(d-1)}$ as the smallest of all the semidistinct min-cut matrices of $r_d^{(d-1)}$.

To illustrate the use of the table, the row with $d = 10$ indicates that r_{10} has two semidistinct min-cut matrices $S_i^{(d-1)}(d)$ with $\bar{X}_i = \{v_2,v_3,v_4,v_6\}$ and $S_l^{(d-1)}(d)$ with $X_l = \{v_3,v_4,v_6\}$. The decision is arbitrarily made that $\|S_i^{(d-1)}(d)\| \leq \|S_l^{(d-1)}(d)\|$ resulting in the inequality in column 7, $13 + \alpha_7 \geq 10 + \alpha_7 + \alpha_8$. In the row with $d = 9$, the requirement that $r_{4,2} \geq 0$ leads to the restriction $3 - \alpha_8 \geq 0$ in column 6.

In performing the steps in the algorithm α_d is set to zero in the table unless r_d is min-coupled to some r_q with $q > d$. Otherwise (5.7.9) forces α_d to equal zero. Using the inequalities from column 6, we obtain from the row with $d = 10$

$$1 \geq \alpha_7 + \alpha_8. \tag{5.8.16a}$$

From the row with $d = 13$

$$\alpha_8 \geq 1. \tag{5.8.16b}$$

From the row with $d = 20$

$$15 \geq \alpha_8 + \alpha_{14}. \tag{5.8.16c}$$

Table 5.8.1

1	2	3	4	5	6	7
d	$\bar{X}_j$	$r_{h,j}$	$r_d^{(d)} = c(h,j)$	$\bar{X}_l$	$c(h,j) \geq -c(j,h)$ For $h \geq j+1$	$\lVert S_i^{(d-1)}(d)\rVert \leq \lVert S_l^{(d-1)}(d)\rVert$
6	$\{v_2,v_4,v_5,v_6\}$	$r_{3,2}$	3			
7	$\{v_4,v_6\}$	$r_{5,4}$	$6+\alpha_7$			
8	$\{v_2,v_4,v_6\}$	$r_{5,2}$	α_8			
9	$\{v_2,v_6\}$	$r_{4,2}$	$3-\alpha_8$	$\{v_2,v_5,v_6\}$	$3-\alpha_8 \geq 0$	$12 \geq 7+\alpha_8$
10	$\{v_2,v_3,v_4,v_6\}$	$r_{5,3}$	$1-\alpha_8-\alpha_7$	$\{v_3,v_4,v_6\}$	$1-\alpha_8-\alpha_7 \geq 0$	$13+\alpha_7 \geq 10+\alpha_7+\alpha_8$
11	$\{v_1,v_3,v_4,v_5,v_6\}$	$r_{2,1}$	9			
12	$\{v_1,v_2,v_4,v_5,v_6\}$	$r_{3,1}$	14	$\{v_1,v_4,v_5,v_6\}$		
13	$\{v_1,v_2,v_5,v_6\}$	$r_{4,1}$	$-1+\alpha_8$	$\{v_1,v_5,v_6\}$	$-1+\alpha_8 \geq 0$	$29 \geq 26-\alpha_8$
14	$\{v_1,v_2,v_3,v_4,v_6\}$	$r_{5,1}$	$22+\alpha_{14}$	$\left\{v_1,v_6 \bigcup_k \binom{v_2,v_3,v_4}{k}\right\} - \{v_1,v_2,v_3,v_4,v_6\}$		$16-\alpha_8-\alpha_7 \geq 8$ $23-\alpha_8 \geq 8$
15	$\{v_3,v_6\}$	$r_{4,3}$	$22+\alpha_8+\alpha_7$	$\left\{v_3,v_6 \bigcup_k \binom{v_1,v_2,v_5}{k}\right\} - \{v_3,v_5\}$		
16	$\{v_5\}$	$r_{6,5}$	29			
17	$\{v_4\}$	$r_{6,4}$	$31-\alpha_7$	$\{v_4,v_5\}$	$31-\alpha_7 \geq 0$	
18	$\{v_3\}$	$r_{6,3}$	$16+\alpha_7$	$\left\{v_3 \bigcup_k \binom{v_4,v_5}{k}\right\} - \{v_2\}$		
19	$\{v_2\}$	$r_{6,2}$	41	$\left\{v_2 \bigcup_k \binom{v_3,v_4,v_5}{k}\right\} - \{v_2\}$		$44-\alpha_8 \geq 9$ $85+\alpha_8-\alpha_7 \geq 9$
20	$\{v_1\}$	$r_{6,1}$	$15-\alpha_8-\alpha_{14}$	$\left\{v_1 \bigcup_k \binom{v_2,v_3,v_4,v_5}{k}\right\} - \{v_1\}$	$15-\alpha_8-\alpha_{14} \geq 0$	$\left.\begin{matrix}73\\ 57+\alpha_8\\ 84+\alpha_{14}+\alpha_7-\alpha_8\end{matrix}\right\} \geq 44+\alpha_8+\alpha_{14}$

Hence,

$$\alpha_8 = 1, \tag{5.8.17a}$$

$$\alpha_7 = 0, \tag{5.8.17b}$$

$$14 \geq \alpha_{14}. \tag{5.8.17c}$$

Using the inequalities in column 7 we obtain no new restrictions. Finally applying (5.7.9), $\alpha_{14} = 0$ since $\alpha_8 \neq 0$. Hence $\alpha_7 = 0$, $\alpha_{14} = 0$, $\alpha_8 = 1$ and the resulting graph is given in Fig. 5.8.1.

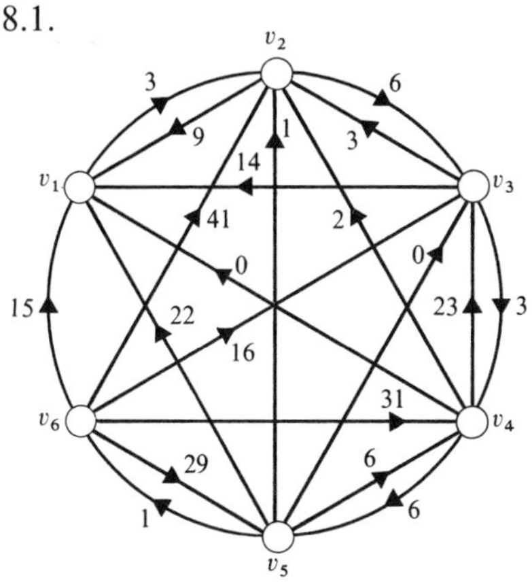

Fig. 5.8.1 Semigraph for $\mathscr{R}$ in (5.8.14).

It should be emphasized that Table 5.8.1 contains a great deal of very specific information, in readily accessible form. For example, for all s and t the locations and values of the minimum s-t cuts in the realization are easily checked. For example, consider $r_{4,3}$. From row 15, the minimum 4-3 cut is given by $(X_i, \bar{X}_i) = (\{v_1, v_2, v_4, v_5\}, \{v_3, v_6\})$. Examining the graph, we see that $c(X_i, \bar{X}_i) = 30$ as required.

The efficiency of the algorithm is greatly enhanced as the number of numerically distinct entries in $\mathscr{R}^{(0)}$ becomes large since the number of variables α_q becomes correspondingly small. The number of variables becomes zero when $\mathscr{R}^{(0)}$ is maximally distinct, since in this case the algorithm reduces to the Substitution Algorithm. If the number of numerically distinct entries is small, it may be advantageous to attempt the Replacement Algorithm first, since it is extremely fast (it is simply the Terminal Capacity Realization Algorithm with $\alpha_q = 0$ for all q) and it provides a quick sufficiency test for realizability.

The realization as a semigraph obtained from the Terminal Capacity Realization Algorithm is not unique since the ordering of the elements $r_{i,j} \to r_d$ is not unique, and in general the values of α_d are not unique. As an example the semigraphs in Fig. 5.8.2 both have the terminal capacity matrix given in (5.8.18). The semigraphs are different and furthermore cannot be obtained from one another by shifting.

$$T = \begin{bmatrix} \infty & 3 & 5 \\ 5 & \infty & 5 \\ 1 & 1 & \infty \end{bmatrix} \tag{5.8.18}$$

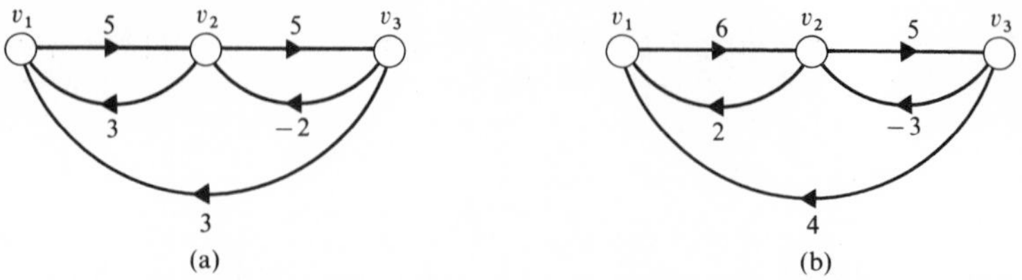

Fig. 5.8.2 Semigraphs which cannot be obtained from one another by shifting.

5.9 VERTEX-WEIGHTED GRAPHS

In a network connecting manufacturers and consumers, the limitation on the amount of goods transferred may be due to the supply of the manufacturer or the demand of the consumer rather than to the capacity of the distribution system. In a radio communication system the limitations are often due to the maximum power of the transmitter. In such cases, the graph which models the system must be vertex-weighted rather than branch-weighted. The branches are assumed to have infinite capacity and the vertices have finite capacity. For vertex-weighted graphs, the multiterminal problem can be formulated in a fashion similar to that used for the branch-weighted case.

***Definition* 5.9.1** *Let G be a vertex-weighted graph. The terminal capacity $\tau_{i,j}$ between vertices v_i and v_j for $i \neq j$ is given by*

$$\tau_{i,j} = \text{Min}\,[c(v_i), c(v_j), \underset{k}{\text{Min}}\,[c(A^k_{i,j})]], \tag{5.9.1}$$

where $c(v_i)$ is the capacity of v_i and $c(A^k_{i,j})$ is the capacity of the kth i-j vertex cut-set $A^k_{i,j}$.

***Definition* 5.9.2** *Let G be a vertex-weighted graph with n vertices. The terminal capacity matrix, T, of G is an $n \times n$ matrix such that for $i \neq j$ the i-j entry is $\tau_{i,j}$. $\tau_{i,i}$ is arbitrarily taken to be infinite for $i = 1, \ldots, n$.*

The simplest analysis problem for a vertex-weighted graph consists of finding T. The corresponding synthesis problem consists of obtaining a vertex-weighted graph whose terminal capacity matrix is equal to a given requirement matrix $\mathcal{R}$. The formulation of an efficient analysis method for vertex-weighted graphs is difficult. In order to apply the labeling algorithm to find $\tau_{i,j}$, the vertex-weighted graph must be converted to a branch-weighted graph by the methods given in Section 6 of Chapter 3. However, the resulting graph is directed, even if the original graph is undirected. Hence, the results of Section 3 are not directly applicable. Suppose that in an undirected vertex-weighted graph G, $\tau_{i,j} = c(A^k(i,j))$ and the removal of $A^k(i,j)$ leaves v_i in a component G_i and v_j in a component G_j. Ali [AL2] has shown that if v_a and v_b are in G_j then $\tau_{a,b}$ can be calculated in the graph formed by condensing all vertices in G_i into a single vertex. However, the result cannot, in general, be represented by a tree structure as in Section 3. More will be said in Chapter 7 regarding the analysis of vertex-weighted graphs in which all weights are unity. In this case, further simplifications are possible.

Little is known regarding the synthesis of directed vertex-weighted graphs. The remainder of this section is devoted to the synthesis of undirected vertex-weighted graphs, for which a complete solution is known. We first derive two necessary conditions on a terminal capacity matrix T.

***Lemma* 5.9.1** *Let T be the terminal capacity matrix of an undirected vertex-weighted graph G. Then for all x,y,w,*

$$\tau_{x,y} \geqq \text{Min}\,[\tau_{x,w},\tau_{w,y}]. \tag{5.9.2}$$

Proof

CASE 1. If $\tau_{x,y} < \text{Min}\,[c(v_x),c(v_y)]$, there is a minimum x-y cut-set $A_{x,y}$ with value $\tau_{x,y}$. Suppose the removal of $A_{x,y}$ breaks the graph into several components one of which is G_x, containing v_x, and another of which is G_y, containing v_y. If v_w is not in G_x, then $\tau_{x,y} \geqq \tau_{x,w}$, and if v_w is not in G_y, then $\tau_{x,y} \geqq \tau_{w,y}$. In either case (5.9.2) is true.

CASE 2. For $\tau_{x,y} = \text{Min}\,[c(v_x),c(v_y)]$, $c(v_x) \geq \tau_{x,w}$ and $c(v_y) \geq \tau_{w,y}$. Hence we again have $\tau_{x,y} \geq \text{Min}\,[\tau_{x,w},\tau_{w,y}]$.

From Lemmas 5.2.1 and 5.2.2, T is semiprincipally partitionable. Hence T is principally partitionable, since G is undirected.

***Lemma* 5.9.2** *If T is the terminal capacity matrix of an undirected vertex-weighted graph G, then for $i \neq j$ any $\tau_{i,j}$ which is less than the maximum number in both row i and column j must be the sum of some of the entries $\tau_{i,k}$ in row i, each of which is the maximum number in an off diagonal position in its column k.*

Proof. Suppose $\tau_{i,j}$ is less than the maximum number in both its row and its column in T. Then,

$$\tau_{i,j} < \text{Min}\,[c(v_i),c(v_j)]. \tag{5.9.3}$$

Hence $\tau_{i,j}$ is the value of a minimum i-j cut-set, say $A^k_{i,j}$. Suppose $A^k_{i,j}$ consists of vertices $v_{l_1},v_{l_2},\ldots,v_{l_a}$. Then $\tau_{i,l_1},\tau_{i,l_2},\ldots,\tau_{i,l_a}$ are equal to $c(v_{l_1}),c(v_{l_2}),\ldots,c(v_{l_a})$, respectively. Hence $\tau_{i,l_1},\tau_{i,l_2},\ldots,\tau_{i,l_a}$ are all the largest entries in their corresponding columns.

Example 5.9.1 Consider the graph G in Fig. 5.9.1. The vertex weights are the capacities of the vertices. The terminal capacity matrix for G is

$$T = \begin{bmatrix} \infty & 1 & 1 & 1 & 1 & 1 \\ 1 & \infty & 1 & 1 & 1 & 1 \\ 1 & 1 & \infty & 3 & 2 & 2 \\ 1 & 1 & 3 & \infty & 2 & 2 \\ 1 & 1 & 2 & 2 & \infty & 3 \\ 1 & 1 & 2 & 2 & 3 & \infty \end{bmatrix}. \tag{5.9.4}$$

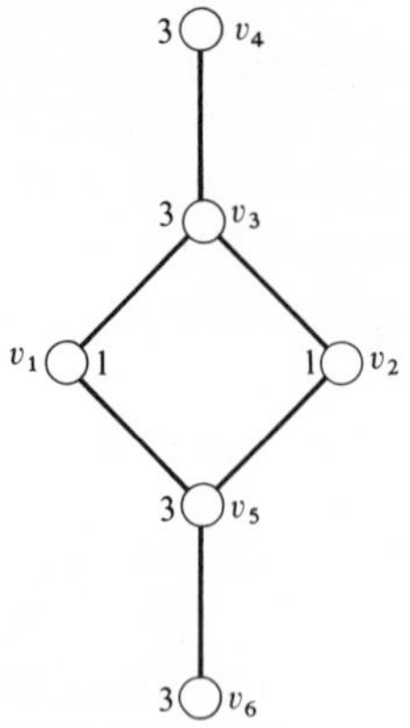

Fig. 5.9.1 Graph for Example 5.9.1.

The element $\tau_{3,5}$ is less than the maximum value in row 3 and in column 5. But $\tau_{3,5} = \tau_{3,1} + \tau_{3,2}$ where $\tau_{3,1}$ is a largest element in column 1 and $\tau_{3,2}$ is a largest element in column 2.

We now consider possible approaches to synthesize a given requirement matrix $\mathscr{R}$.

Lemma 5.9.3 *If a symmetric $n \times n$ matrix $\mathscr{R}$ is realizable by a graph G with n vertices in which $c(v_i) > \operatorname{Max}_j [r_{i,j}]$, then the graph will realize $\mathscr{R}$ if $c(v_i)$ is reduced to $\operatorname{Max}_j [r_{i,j}]$.*

Proof. Suppose, contrary to the lemma, that it is necessary to have $c(v_i) > \alpha = \operatorname{Max}_j [r_{i,j}]$. Then v_i must be in a minimum cut-set $A_{x,y}$ for some $y \neq i$ such that to send a maximum flow from v_x to v_y of value $\tau_{x,y} > \alpha$, $\beta > \alpha$ units of flow must pass through vertex v_i. This implies that β units can be sent from v_i to v_y. Hence $\tau_{i,y} \geqq \beta > \alpha = \operatorname{Max}_j [\tau_{i,j}]$, and we have a contradiction.

As a consequence of Lemma 5.9.3, we see that if we are to realize a matrix $\mathscr{R}$ then we can always set $c(v_i) = \operatorname{Max}_j [r_{i,j}]$ for all i.

Using this fact we now show, by means of a counter-example, that the conditions of Lemmas 5.9.1 and 5.9.2 are not sufficient for realizability.

Example 5.9.2 The matrix $\mathscr{R}$ in (5.9.5), shown in a principally partitioned form,

$$\mathscr{R} = \begin{bmatrix} \infty & 1 & 1 & 1 & 1 & 1 & 1 & 1 & 1 & 1 \\ 1 & \infty & 1 & 1 & 1 & 1 & 1 & 1 & 1 & 1 \\ 1 & 1 & \infty & 3 & 2 & 2 & 1 & 1 & 1 & 1 \\ 1 & 1 & 3 & \infty & 2 & 2 & 1 & 1 & 1 & 1 \\ 1 & 1 & 2 & 2 & \infty & 3 & 1 & 1 & 1 & 1 \\ 1 & 1 & 2 & 2 & 3 & \infty & 1 & 1 & 1 & 1 \\ 1 & 1 & 1 & 1 & 1 & 1 & \infty & 3 & 2 & 2 \\ 1 & 1 & 1 & 1 & 1 & 1 & 3 & \infty & 2 & 2 \\ 1 & 1 & 1 & 1 & 1 & 1 & 2 & 2 & \infty & 3 \\ 1 & 1 & 1 & 1 & 1 & 1 & 2 & 2 & 3 & \infty \end{bmatrix} \tag{5.9.5}$$

satisfies the triangle inequality and the conditions of Lemma 5.9.2. We now show that all attempts to realize $\mathcal{R}$ by a graph with ten vertices must fail.

Using Lemma 5.9.3, we first assign to each vertex v_i the weight $c(v_i) = \underset{j}{\text{Max}}\,[r_{i,j}]$. We will require a terminal capacity of two units between a vertex in $\{v_3,v_4\}$ and a vertex in $\{v_5,v_6\}$, or between a vertex in $\{v_7,v_8\}$ and a vertex in $\{v_9,v_{10}\}$, and a terminal capacity of one unit between a vertex in $\{v_3,v_4,v_5,v_6\}$ and a vertex in $\{v_7,v_8,v_9,v_{10}\}$. Therefore, the vertices must be grouped as shown in Fig. 5.9.2(a). The only way in which we can obtain $\tau_{3,4} = \tau_{5,6} = \tau_{7,8} = \tau_{9,10} = 3$ is to add the branches in Fig. 5.9.2(b). In order to form the cut-set of value two between $\{v_3,v_4\}$ and $\{v_5,v_6\}$ we must add at least the branches b_1,b_2,b_3,b_4 as shown in Fig. 5.9.2(c). Finally, to form the cut-set of value two between $\{v_7,v_8\}$ and $\{v_9,v_{10}\}$ we must add at least the branches b_5,b_6,b_7,b_8. However, once these branches are added the minimum cut-set separating, say, v_4 and v_7 has value two rather than one as required. Hence it is impossible to realize $\mathcal{R}$ even though it satisfies the conditions of Lemmas 5.9.1 and 5.9.2.

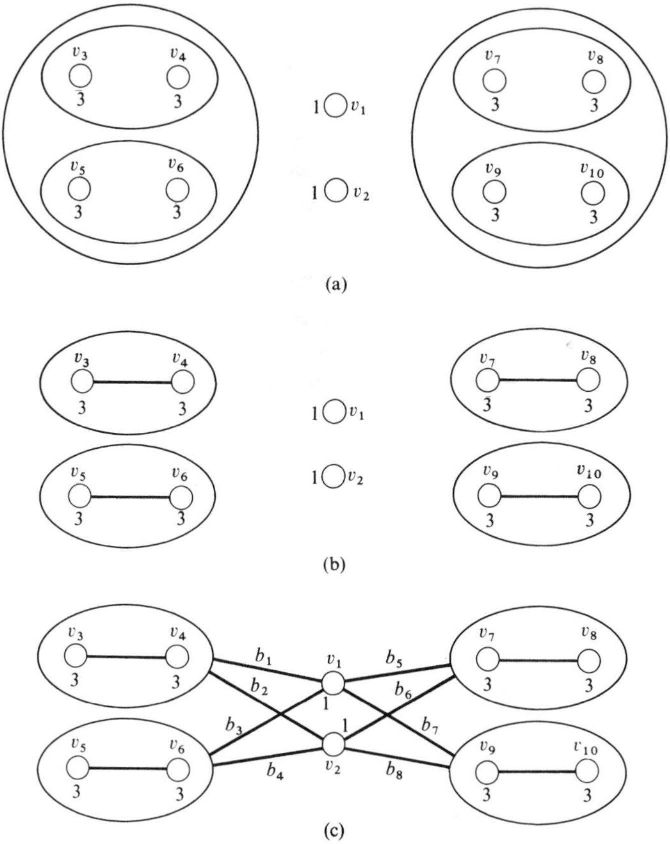

Fig. 5.9.2 Graphs for Example 5.9.2.

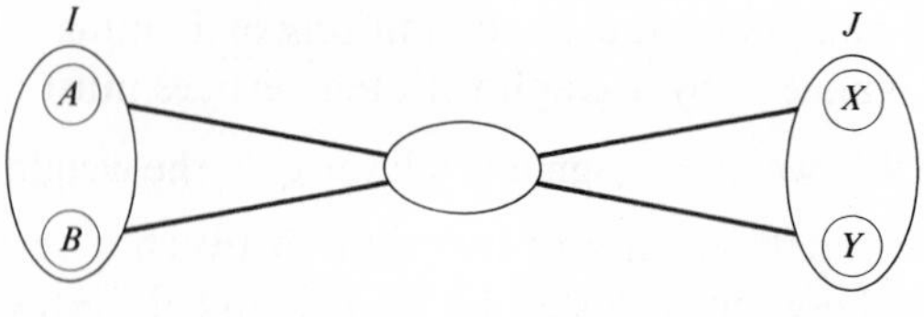

Fig. 5.9.3 Generalization of counter-example in Example 5.9.2.

The general condition highlighted by Example 5.9.2 is illustrated as in Fig. 5.9.3. I and J are subsets of vertices with

$$A \subseteq I, \tag{5.9.6a}$$

$$B \subseteq I, \tag{5.9.6b}$$

$$X \subseteq J, \tag{5.9.6c}$$

$$Y \subseteq J. \tag{5.9.6d}$$

In Example 5.9.2, $I = \{v_3,v_4,v_5,v_6\}$, $J = \{v_7,v_8,v_9,v_{10}\}$, $A = \{v_3,v_4\}$, $B = \{v_5,v_6\}$, $X = \{v_7,v_8\}$, and $Y = \{v_9,v_{10}\}$. The point of the example is that if an arbitrary vertex v_i is in a minimum A-B cut-set and a minimum X-Y cut-set, then it must be in a minimum I-J cut-set. In the counterexample we required either v_1 or v_2 to be in a minimum A-B cut-set and a minimum X-Y cut-set but not in a minimum I-J cut-set. To state necessary and sufficient conditions for realizability of a requirement matrix, we must be able to state this new condition in terms of the elements of a given matrix $\mathscr{R}$. For this purpose we must consider in more detail the partitioning of a terminal capacity matrix.

***Definition* 5.9.3** *An ordered principal partitioning of a matrix M is a principal partitioning such that if, in M or in any submatrix on the diagonal resulting from the partitioning, there are rows in which all entries are equal, then these rows are the first rows in the matrix or submatrix and the corresponding columns are the first columns.*

The rows and corresponding columns of a principally partitionable matrix can always be permuted to give an ordered principal partitioning. For example, the matrix M in (5.9.8) has an ordered principal partitioning, obtained by permuting the rows and columns of the principally partitioned matrix M^* in (5.9.7). For the moment the shading in the matrix can be ignored.

***Definition* 5.9.4** *A cut array of a matrix in ordered principally partitioned form consists of all equal elements above the diagonal from consecutive rows and columns, each of which is less than the maximum value of all other elements in the same row and column.*

$M^* =$

	1*	9*	10*	5*	6*	4*	7*	8*	3*	11*	12*	13*	14*	2*
1*	∞	1	1	1	1	1	1	1	1	1	1	1	1	1
9*	1	∞	3	1	1	1	1	1	1	1	1	1	1	1
10*	1	3	∞	1	1	1	1	1	1	1	1	1	1	1
5*	1	1	1	∞	3	2	2	2	1	1	1	1	1	1
6*	1	1	1	3	∞	2	2	2	1	1	1	1	1	1
4*	1	1	1	2	2	∞	2	2	1	1	1	1	1	1
7*	1	1	1	2	2	2	∞	3	1	1	1	1	1	1
8*	1	1	1	2	2	2	3	∞	1	1	1	1	1	1
3*	1	1	1	1	1	1	1	1	∞	1	1	1	1	1
11*	1	1	1	1	1	1	1	1	1	∞	3	2	2	1
12*	1	1	1	1	1	1	1	1	1	3	∞	2	2	1
13*	1	1	1	1	1	1	1	1	1	2	2	∞	3	1
14*	1	1	1	1	1	1	1	1	1	2	2	3	∞	1
2*	1	1	1	1	1	1	1	1	1	1	1	1	1	∞

(5.9.7)

$M =$

	1	2	3	4	5	6	7	8	9	10	11	12	13	14
1	∞	1	1	1	1	1	1	1	1	1	1	1	1	1
2	1	∞	1	1	1	1	1	1	1	1	1	1	1	1
3	1	1	∞	1	1	1	1	1	1	1	1	1	1	1
4	1	1	1	∞	2	2	2	2	1	1	1	1	1	1
5	1	1	1	2	∞	3	2	2	1	1	1	1	1	1
6	1	1	1	2	3	∞	2	2	1	1	1	1	1	1
7	1	1	1	2	2	2	∞	3	1	1	1	1	1	1
8	1	1	1	2	2	2	3	∞	1	1	1	1	1	1
9	1	1	1	1	1	1	1	1	∞	3	1	1	1	1
10	1	1	1	1	1	1	1	1	3	∞	1	1	1	1
11	1	1	1	1	1	1	1	1	1	1	∞	3	2	2
12	1	1	1	1	1	1	1	1	1	1	3	∞	2	2
13	1	1	1	1	1	1	1	1	1	1	2	2	∞	3
14	1	1	1	1	1	1	1	1	1	1	2	2	3	∞

(5.9.8)

The matrix M in (5.9.8) has a cut array consisting of ones, indicated by the region shaded with lines of positive slope and two cut arrays composed of twos, indicated by the regions shaded with lines of negative slope.

Definition 5.9.5 *The lead element of a cut array (or of a submatrix on the diagonal) resulting from an ordered principal partitioning of M is the element* $m_{i,j}$, $i < j$, *in the array (or submatrix) with the smallest value of i and largest value of j. In other words, it is the element in the upper right-hand corner of the array (or submatrix).*

Definition 5.9.6 *Let* $m_{i,j}$ *be the lead element of a cut array of M and consider the submatrix M on the diagonal composed of the elements common to the rows and columns i through j. If an element* $m_{a,b}$ *is in the cut array, then set* $m_{a,b}$ *and* $m_{b,a}$ *equal to zero;* $\hat{M}$ *then becomes a direct sum of submatrices on the diagonal, the lead elements of which are said to be subtended by* $m_{i,j}$. *Also we say that the cut array subtends each of the submatrices in the direct sum.*

In the matrix M in (5.9.8), the lead element $m_{4,14}$ of the cut array of ones subtends the lead elements $m_{4,8}$, $m_{9,10}$, and $m_{11,14}$, and the cut array subtends the submatrices on the diagonal corresponding to the $\{v_4,v_5,v_6,v_7,v_8\}$, $\{v_9,v_{10}\}$, and $\{v_{11},v_{12},v_{13},v_{14}\}$.

Definition 5.9.7 *A cut array is an A-B cut array if it contains all elements* $m_{i,j}$, $i < j$, *such that all diagonal elements corresponding to* $v_i \in A$ *are in one of the submatrices subtended by the cut array and all diagonal elements corresponding to* $v_j \in B$, *in another.*

The shaded region of ones in (5.9.8), for example, is an A_1-B_1 cut array for $A_1 = \{v_4,v_5,v_6,v_7,v_8\}$, $B = \{v_{11},v_{12},v_{13},v_{14}\}$ and an A_2-B_2 cut array for $A_2 = \{v_5,v_6\}$, $B_2 = \{v_{11},v_{12}\}$.

Finally divide all off-diagonal elements of a matrix M into the sets Q and S where $Q = \{q_{i,j}\}$ consists of those entries $q_{i,j} = m_{i,j}$ which are less than the largest elements in the same row and column. The remaining elements are in $S = \{s_{i,j}\}$ with $s_{i,j} = m_{i,j}$.

Theorem 5.9.1 *An* $n \times n$ *matrix M is the terminal capacity matrix of a vertex-weighted graph G with n vertices if and only if*

M is principally partitionable. (5.9.9a)

With M in ordered principally partitioned form, any lead element $q_{i,j}$ *in a cut array can be written as*

$$q_{i,j} = \sum_{k:s_{i,j} \in S_{i,j}} s_{i,k} \qquad (5.9.9b)$$

for some subset $S_{i,j}$ *of S.*

Let the lead element $q_{a,b}$ *be in an A-B cut array, the lead element* $q_{x,y}$ *be in an X-Y cut array, and the lead element* $q_{i,j}$ *be in an I-J cut array such that* $A \cup B \subseteq I$, $X \cup Y \subseteq J$ *and* $q_{i,j} < \text{Min}\,[q_{a,b},q_{x,y}]$. *Then, for all* β,a,b,x,y,i,j, *such that* $s_{a,\beta} \in S_{a,b}, s_{x,\beta} \in S_{x,y}$, *we must have* $s_{i,\beta} \in S_{i,j}$. (5.9.9c)

The necessity of (5.9.9a) follows from Lemma 5.9.1 and the necessity of (5.9.9b) follows from Lemma 5.9.2. The remainder of the proof of the theorem requires a lengthy development and is omitted. The interested reader is referred to the paper by Frisch and Shein [FR2]. In the proof, (5.9.9c) is required to prevent the situation in Fig. 5.9.3 where there is a vertex in the minimum A-B cut-set, the minimum X-Y cut-set and not the minimum *I-J* cut-set. Sufficiency is proved by showing that the algorithm below yields a realization whenever (5.9.9) is satisfied.

Suppose that $\mathscr{R}$ has δ cut arrays with lead elements $q_{i_1,j_1} \geqq q_{i_2,j_2} \geqq \cdots \geqq q_{i_\lambda,j_\lambda} \geqq \cdots \geqq q_{i_\delta,j_\delta}$. Denote by $\mathscr{R}_p(i_\lambda,j_\lambda)$ for $p = 1,2,\ldots,p_\lambda$, the submatrices subtended by the cut array with lead element q_{i_λ,j_λ}. The vertices corresponding to the rows of these submatrices are denoted by $V[\mathscr{R}_p(i_\lambda,j_\lambda)]$. The procedure for obtaining a graph realizing M is given below.

Vertex-Weighted Terminal Capacity Algorithm

Step 1. Let $c(v_i) = \operatorname{Max}_j [r_{i,j}]$ for $i = 1,2,\ldots,n$.

Step 2. Realize each submatrix on the diagonal having no cut arrays, resulting from the ordered principal partitioning of $\mathscr{R}$, by a linear tree with vertex weights arranged in nondecreasing order.

Step 3. Set $\lambda = 1$.

Step 4. For $p = 1,2,\ldots,p_\lambda$, add a branch from any isolated vertex $v_f \in V[\mathscr{R}_p(i_\lambda,j_\lambda)]$ to v_g, where $r_{f,f+1},\ldots,r_{f,g}$ are all largest entries in row f of $\mathscr{R}_p(i_\lambda,j_\lambda)$. If $\lambda > \delta$, terminate; otherwise go to Step 5.

Step 5. For $p = 1,2,\ldots,p_\lambda$, connect each vertex v_k such that $s_{i_\lambda,k} \in S_{i_\lambda,j_\lambda}$ to v_l, whenever the lead element $r_{l,z}$ of $\mathscr{R}_p(i_\lambda,j_\lambda)$ is subtended by q_{i_λ,j_λ}, except when v_k is already connected by a branch to a vertex in $V[\mathscr{R}_p(i_\lambda,j_\lambda)]$. Increase λ by one and return to Step 4.

Example 5.9.3 We illustrate the realization Steps 1 to 5 for the matrix M in (5.9.8), which has three cut arrays. The lead elements of these cut arrays are $q_{5,8}$, $q_{11,14}$, and $q_{4,14}$. Note that $q_{5,8}$ subtends $m_{5,6}$ and $m_{7,8}$; $q_{11,14}$ subtends $m_{11,12}$ and $m_{13,14}$; and $q_{4,14}$ subtends $m_{4,8}$, $m_{9,10}$, and $m_{11,14}$. The lead elements can be written to satisfy condition (5.9.9c) as:

$$q_{5,8} = 2 = s_{5,2} + s_{5,3} = 1 + 1; \tag{5.9.10a}$$

$$q_{11,14} = 2 = s_{11,1} + s_{11,2} = 1 + 1; \tag{5.9.10b}$$

$$q_{4,14} = 1 = s_{4,2} = 1. \tag{5.9.10c}$$

Steps 1 and 2 give the graph shown in Fig. 5.9.4(a). The first application of Steps 3 to 5 results in the graph of Fig. 5.9.4(b). Branches [1,11], [1,13], [2,11], and [2,13] are added in the second iteration. In the final iteration, Step 4 produces [4,8] and Step 5 adds branch [2,9], resulting in the graph containing 15 branches shown in Fig. 5.9.4(c).

An alternate realization requiring only 14 branches is given in Fig. 5.9.4(d) and corresponds to

$$q_{5,8} = 2 = s_{5,4} = 2; \tag{5.9.11a}$$

$$q_{11,14} = 2 = s_{11,1} + s_{11,2} = 1 + 1; \tag{5.9.11b}$$

$$q_{4,14} = 1 = s_{4,2} = 1. \tag{5.9.11c}$$

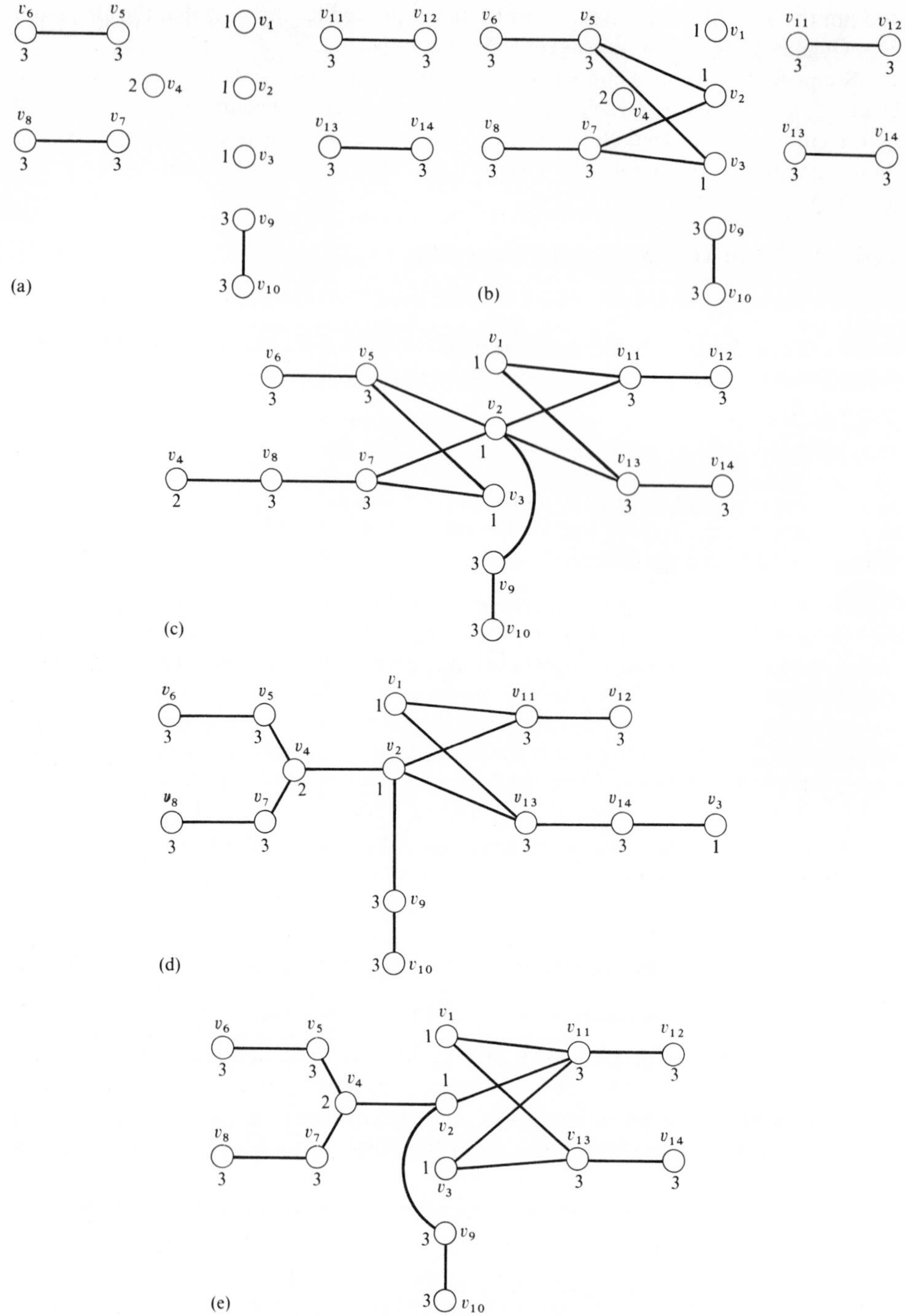

Fig. 5.9.4 Graphs for Example 5.9.3.

Here, [4,5] and [4,7] are added in the first iteration. In the second iteration, [1,11], [1,13], [2,11], and [2,13] are inserted. The third iteration produces the branches [2,4] and [2,9] in Step 5, and the branch [2,14] is added in the final iteration of Step 4. Another 14-branch realization, shown in Fig. 5.9.4(e), is obtained from

$$q_{5,8} = 2 = s_{5,4} = 2; \tag{5.9.12a}$$

$$q_{11,14} = 2 = s_{11,1} + s_{11,3} = 1 + 1; \tag{5.9.12b}$$

$$q_{4,14} = 1 = s_{4,2} = 1. \tag{5.9.12c}$$

Here the branches [1,11], [1,13], [3,11], and [3,13] are added in the second iteration and the branches [2,4], [2,9], and [2,11] are inserted in the final iteration.

5.10 FURTHER REMARKS

The multiterminal model and the terminal capacity matrix were first introduced by Mayeda [MA1], who proved the necessity of the principal partitioning condition for the undirected case. Tang and Chien subsequently proved the necessity of the semi-principal partitioning condition for directed graphs [TA1]. Gomory and Hu [GO1] proved the triangle inequality and introduced the analysis scheme for undirected graphs which was extended to pseudosymmetric graphs by Gupta [GU1]. Lemmas 5.2.1 and 5.2.2 are due to Ali [AL2]. The Multiterminal Analysis Algorithm in Section 3 is taken from references [GO1] and [GU1].

The idea of shifting was introduced independently by Sen and Frisch [SE1, SE2] and by Resh [RE2] (under the name *k*-equivalence). The semigraph was introduced by Sen and Frisch [SE2] in conjunction with the theorems concerning its properties and methods of obtaining semigraphs by shifting. The min-cut matrix and related concepts were first used by Mayeda [MA2] and were employed in references [RE2] and [SE2] in realizing terminal capacity matrices.

The problem of the synthesis of a terminal capacity matrix was first published by Mayeda [MA1], who gave a realization method for an undirected terminal capacity matrix. Gomory and Hu [GO1] gave a minimum cost realization, to be presented in the next chapter. Other minimum cost realizations for the undirected case were provided by Chien [CH1] and Wing and Chien [WI1]. The results of Gomory and Hu also appear in the book by Ford and Fulkerson [FO1], and those of Mayeda, Chien and Wing, and Chien also appear in the book by Kim and Chien [KI1].

Tang and Chien [TA1] were the first to publish results concerning the realization of unsymmetric requirement matrices. They gave conditions under which two graphs can be superimposed so that their terminal capacity matrices can be added. Mayeda [MA2] considered the special case of completely partitioned matrices and Barnard [BA1] considered the minimum cost synthesis realization of these matrices. Mayeda and Jelinek [JE1] showed that for any n, a graph could be constructed with $(n - 1)(n + 2)/2$ numerically distinct entries. Sen [SE1], Sen and Frisch [SE2], and Resh [RE1, RE2] then considered maximally distinct terminal capacity matrices and independently derived the shifting technique and a synthesis algorithm for maximally distinct terminal capacity matrices. Resh considered the general problem of synthesis with negative branch capacities. Finally Frisch and Sen [FR1] gave the general algorithm for synthesis of terminal capacity matrices.

A number of variations of the main problem were also considered by various authors. Gomory and Hu have considered minimal cost synthesis schemes in terms of linear programs [GO2]. Ali [AL1] has given necessary and sufficient conditions for the realization of special classes of graphs such as trees and circuits.

Yau [YA1, YA2, YA3] discussed the problem of synthesizing undirected graphs with vertex capacities and no branch capacities and gave necessary conditions for realizability of a terminal capacity matrix. Tapia and Meyers [ME1] derived realization schemes for trees. Both Tapia and Meyers [TA2] and Ali [AL2] gave several results for graphs with vertex and branch weights. The necessary and sufficient conditions for realizability of a terminal capacity matrix by an undirected vertex-weighted graph were given by Frisch and Shein [FR2].

PROBLEMS

1. Do the following matrices satisfy the triangle inequality? Are they semiprincipally partitionable?

$$M_1 = \begin{bmatrix} \infty & 3 & 5 & 3 & 3 \\ 4 & \infty & 8 & 4 & 4 \\ 1 & 1 & \infty & 1 & 1 \\ 5 & 3 & 4 & \infty & 6 \\ 0 & 0 & 0 & 0 & \infty \end{bmatrix}$$

$$M_2 = \begin{bmatrix} \infty & 1 & 7 & 6 & 1 \\ 1 & \infty & 1 & 1 & 1 \\ 7 & 1 & \infty & 6 & 2 \\ 6 & 1 & 6 & \infty & 2 \\ 2 & 1 & 2 & 2 & \infty \end{bmatrix}$$

2. Find all cut matrices, min-cut matrices, and semidistinct min-cut matrices for the element $m_{1,2}$ in the matrix below.

$$M = \begin{bmatrix} \infty & 1 & 1 & 1 & 1 \\ 6 & \infty & 2 & 2 & 2 \\ 6 & 5 & \infty & 3 & 3 \\ 5 & 7 & 9 & \infty & 4 \\ 4 & 11 & 10 & & \infty \end{bmatrix}$$

3. Find the terminal capacity matrices for the graphs in Figs. P5.1 and P5.2.

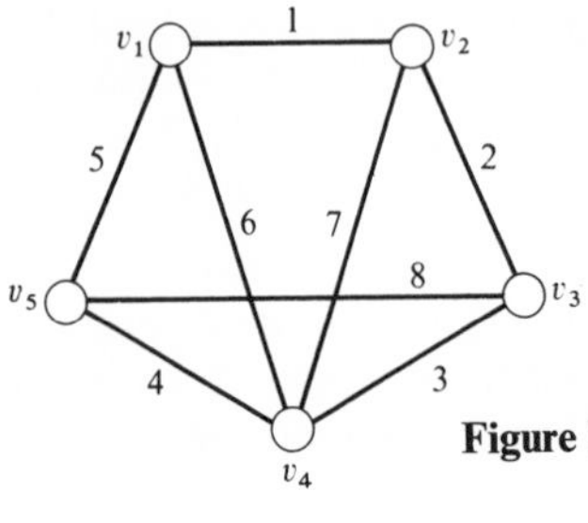

Figure P5.1

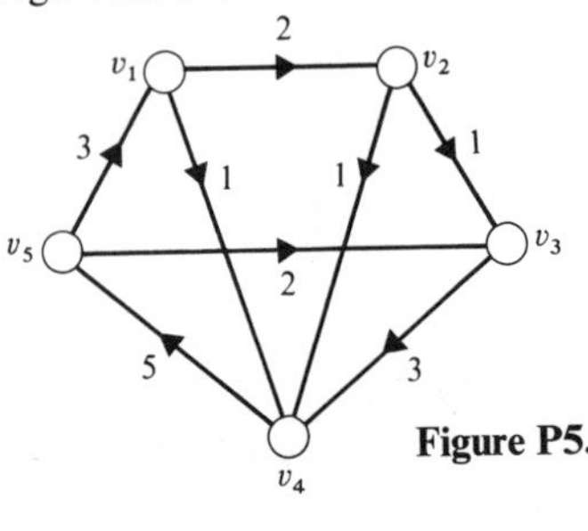

Figure P5.2

4. Convert the semigraph in Fig. P5.3 to a graph by shifting.

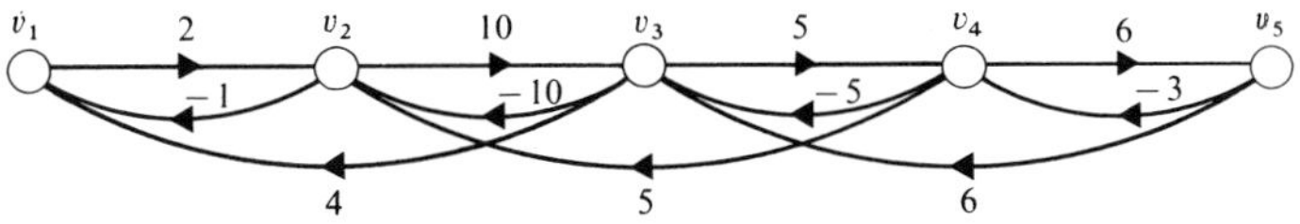

Figure P5.3

5. Convert the graph in Fig. P5.4 to a semigraph by shifting.

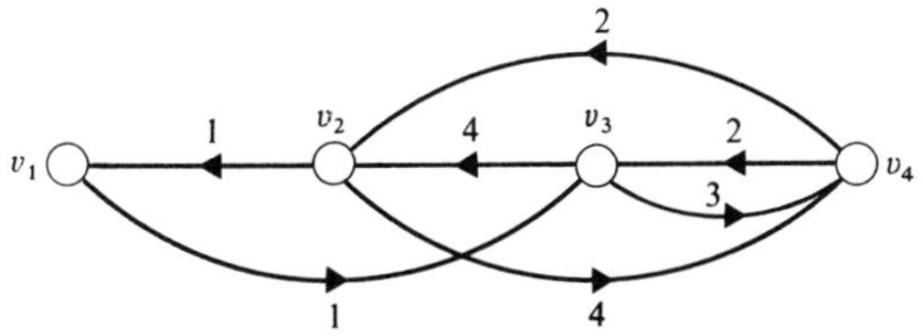

Figure P5.4

6. Which of the requirement matrices below are realizable as terminal capacity matrices of branch-weighted graphs? Realize the realizable requirement matrices.

$$\mathscr{R}_1 = \begin{bmatrix} \infty & 4 & 3 & 3 \\ 4 & \infty & 3 & 3 \\ 10 & 9 & \infty & 5 \\ 13 & 10 & 5 & \infty \end{bmatrix}$$

$$\mathscr{R}_2 = \begin{bmatrix} \infty & 2 & 2 & 2 & 2 \\ 2 & \infty & 3 & 6 & 3 \\ 2 & 3 & \infty & 3 & 5 \\ 2 & 6 & 3 & \infty & 3 \\ 2 & 3 & 5 & 3 & \infty \end{bmatrix}$$

$$\mathscr{R}_3 = \begin{bmatrix} \infty & 5 & 4 & 3 & 3 & 3 \\ 5 & \infty & 4 & 3 & 3 & 3 \\ 4 & 4 & \infty & 3 & 3 & 3 \\ 3 & 3 & 3 & \infty & 2 & 2 \\ 3 & 3 & 3 & 2 & \infty & 2 \\ 3 & 3 & 3 & 2 & 2 & \infty \end{bmatrix}$$

$$\mathscr{R}_4 = \begin{bmatrix} \infty & 4 & 3 & 3 \\ 6 & \infty & 3 & 3 \\ 3 & 3 & \infty & 5 \\ 2 & 2 & 2 & \infty \end{bmatrix}$$

$$\mathscr{R}_5 = \begin{bmatrix} \infty & 1 & 1 & 1 \\ 2 & \infty & 3 & 5 \\ 2 & 6 & \infty & 7 \\ 2 & 4 & 3 & \infty \end{bmatrix}$$

7. Prove that Steps 1 to 3 below realize a real, symmetric matrix M, with nonnegative entries as the terminal capacity matrix of an undirected branch-weighted graph if and only if all diagonal entries are infinite and, for all i,j and k,

$$m_{i,j} \geq \text{Min}\,[m_{i,k}, m_{k,j}].$$

Step 1. Let $m_{a,b}$ be an entry of largest value in M. Connect a branch $[a,b]$ of capacity $c[a,b] = m_{a,b}$ between v_a and v_b.

Step 2. Let V_1 be the vertices of the graph included thus far and let $\bar{V}_1$ be the vertices not yet included. Let v_β be a vertex in $\bar{V}_1$ and let v_α be a vertex in V_1 such that $m_{\alpha,\beta} \geq m_{\delta,\lambda}$ for all $v_\delta \in V_1$ and all $v_\lambda \in \bar{V}_1$. Add branch $[\alpha,\beta]$ with $c[\alpha,\beta] = m_{\alpha,\beta}$.

Step 3. Repeat Step 2 until all vertices are included in V_1 [GO1].

8. Consider a directed branch-weighted graph with n vertices labeled from v_1 to v_n. For all $i \neq j$ let

$$c(i,j) = \begin{cases} (n-i) & \text{for } j = i+1, \\ 0 & \text{for } j > i+1, \\ n^{n-i+2} & \text{for } j < i. \end{cases}$$

Show that for any given n, the terminal capacity matrix of such a graph is maximally distinct [JE1].

9. Show that the two graphs in Fig. P5.5 have the same terminal capacities with respect to vertices v_1, v_2, and v_3 [AK1].

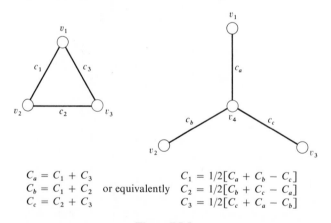

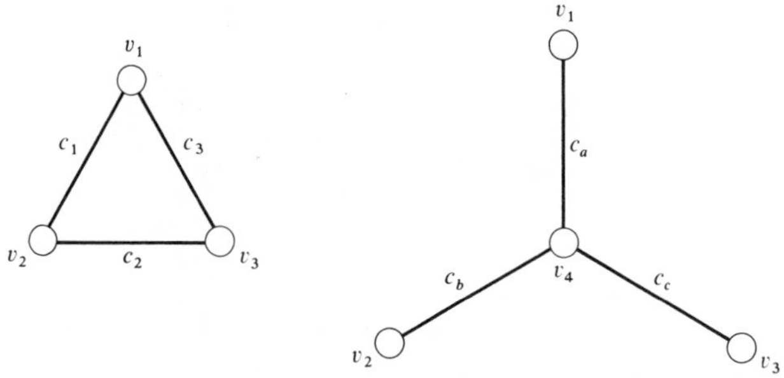

$$C_a = C_1 + C_3 \qquad\qquad C_1 = 1/2[C_a + C_b - C_c]$$
$$C_b = C_1 + C_2 \quad \text{or equivalently} \quad C_2 = 1/2[C_b + C_c - C_a]$$
$$C_c = C_2 + C_3 \qquad\qquad C_3 = 1/2[C_c + C_a - C_b]$$

Figure P5.5

10. Let a "double tree" be a branch-weighted connected directed graph containing no directed circuit of more than two branches. Prove that a matrix M is realizable as the terminal capacity matrix of a "double tree" if and only if M is semiprincipally partitionable and in every matrix in the 1-2 position resulting from the partitioning there is at least one element $m_{i,j}$ such that for every other element in the submatrix, say $m_{l,k}$,

$$m_{l,k} = \text{Min}\,[m_{l,i}, m_{i,j}, m_{j,k}].$$

[AL1]

11. Prove the necessity of conditions (5.9.9) in Theorem 5.9.1.

12. Is the matrix M below in ordered principally partitioned form? If not then find an ordered principal partitioning.

$$M = \begin{bmatrix} \infty & 1 & 1 & 1 & 1 & 1 & 1 & 1 & 1 & 1 & 1 \\ 1 & \infty & 1 & 1 & 1 & 1 & 1 & 1 & 1 & 1 & 1 \\ 1 & 1 & \infty & 1 & 1 & 1 & 1 & 1 & 1 & 1 & 1 \\ 1 & 1 & 1 & \infty & 3 & 2 & 2 & 1 & 1 & 1 & 1 \\ 1 & 1 & 1 & 3 & \infty & 2 & 2 & 1 & 1 & 1 & 1 \\ 1 & 1 & 1 & 2 & 2 & \infty & 3 & 1 & 1 & 1 & 1 \\ 1 & 1 & 1 & 2 & 2 & 3 & \infty & 1 & 1 & 1 & 1 \\ 1 & 1 & 1 & 1 & 1 & 1 & 1 & \infty & 3 & 2 & 2 \\ 1 & 1 & 1 & 1 & 1 & 1 & 1 & 3 & \infty & 2 & 2 \\ 1 & 1 & 1 & 1 & 1 & 1 & 1 & 2 & 2 & \infty & 3 \\ 1 & 1 & 1 & 1 & 1 & 1 & 1 & 2 & 2 & 3 & \infty \end{bmatrix}$$

13. Is the matrix M in the last problem realizable as the terminal capacity matrix of an undirected vertex-weighted graph with eleven vertices? If it is, realize it.

14. Prove the following theorem [SH1]:
    An $n \times n$ matrix $\mathscr{R}$ is the terminal capacity matrix of an undirected vertex-weighted tree with n vertices, if and only if, after possibly permuting rows and corresponding columns,

    1) $\mathscr{R}$ is principally partitionable;
    2) Each resulting submatrix in a 2-1 position contains an element $r_{i,j}$ of value

    $$r_{i,j} = \text{Min}\,[r_i, r_j],$$

    where

    $$r_i = \text{Max}\,[\underset{j:j \neq i}{\text{Max}}[r_{i,j}], \underset{j:j \neq i}{\text{Max}}[r_{j,i}]].$$

15. Consider a terminal capacity matrix T of an undirected vertex-weighted graph G. Suppose T is in ordered principally partitioned form. With the rows and columns of the connection matrix K arranged in the same order as in T, prove that if $\tau_{i,j}$ is in a cut array in T then $k_{i,j} = 0$ [YA3].

16. Prove that for a branch- and vertex-weighted tree in which all branches are directed, the rows and corresponding columns of the terminal capacity matrix T can be permuted so that T is lower triangular (i.e., all entries above the diagonal are zero [SH1]).

17. In an undirected vertex-weighted tree let v_m be the vertex of largest weight. The tree is said to be "concave" if, in any i-m path, the vertex weights are monotonically decreasing from v_i to v_m. Prove that a matrix is realizable as a concave tree if and only if for all (i,j)

    $$m_{i,j} = \underset{k,k \neq i}{\text{Max}}\,[m_{i,k}] = \underset{k,k \neq j}{\text{Max}}\,[m_{k,j}].$$

    [TA2]

REFERENCES

AK1 S. B. Akers, Jr., "Use of the Wye-Delta Transformation in Network Simplification," *Operations Res.* **8,** 311–323 (1960).

AL1 A. Ali, "Realizability Conditions of Special Types of Oriented Communication Nets," *IEEE Trans. Circuit Theory* **CT-12,** 417 (1965).

AL2 A. Ali, "Synthesis of Communication Nets," Ph.D. Thesis, University of London, 1965.

BA1 H. M. Barnard, "Note on Completely Partitionable Terminal Capacity Matrices," *IEEE Trans. Circuit Theory* **CT-12,** 122 (1965).

CH1 R. T. Chien, "Synthesis of a Communication Net," *IBM J. Res. Develop.* **4,** 311–320 (1960).

FO1 L. R. Ford, Jr., and D. R. Fulkerson, *Flows in Networks*, Princeton University Press, Princeton, 1962.

FR1 I. T. Frisch and D. K. Sen, "Algorithms for Synthesis of Oriented Communication Nets," *IEEE Trans. Circuit Theory* **CT-14,** 370–379 (1967).

FR2 I. T. Frisch and N. P. Shein, "Necessary and Sufficient Conditions for Realizability of Vertex Weighted Nets," *IEEE Trans. Circuit Theory* **CT-16,** 496–502 (1969).

GO1 R. E. Gomory and T. G. Hu, "Multiterminal Network Flows," *J. Soc. Ind. Appl. Math.* **9,** 551–570 (1961).

GO2 R. E. Gomory and T. G. Hu, "Synthesis of a Communication Network," *J. Soc. Ind. Appl. Math.* **12,** 348–369 (1964).

GU1 R. P. Gupta, "On Flows in Pseudosymmetric Networks," *J. Soc. Ind. Appl. Math.* **14,** 215–225 (1966).

JE1 F. Jelinek and W. Mayeda, "On the Maximum Number of Different Entries in the Terminal Capacity Matrix of an Oriented Communication Net," *IEEE Trans. Circuit Theory* **CT-10,** 308–309 (1963).

KI1 W. H. Kim and R. T. Chien, *Topological Analysis and Synthesis of Communication Networks*, Columbia University Press, New York, 1962.

MA1 W. Mayeda, "Terminal and Branch Capacity Matrices of a Communication Net," *IRE Trans. Circuit Theory* **CT-8,** 260–269 (1960).

MA2 W. Mayeda, "On Oriented Communication Nets," *IRE Trans. Circuit Theory* **CT-9,** 261–267 (1962).

ME1 B. R. Meyers and M. A. Tapia, "Analysis and Synthesis of Node Weighted Networks," in Proceedings of the Conference on Electrical Network Theory, University of Newcastle-upon-Tyne, England, 1966.

RE1 J. A. Resh, "On the Synthesis of Oriented Communication Nets," *IEEE Trans. Circuit Theory* **CT-12,** 540–546 (1965).

RE2 J. A. Resh, "Semi-Cuts and Communication Nets," Ph.D. Thesis, University of Illinois, July 1963.

SE1 D. K. Sen, "Synthesis of Oriented Communication Nets," Ph.D. Thesis, University of California, Berkeley, June 1966.

SE2 D. K. Sen and I. T. Frisch, "Synthesis of Oriented Communication Nets," in Proceedings of the IEEE Symposium on Signal Transmission and Processing, Columbia University, New York, May 1965, pp. 90–101.

SH1 N. P. Shein and I. T. Frisch, "Some Sufficient Conditions for Realizability of Vertex Weighted Directed Graphs," in Proceedings of the Third Annual Princeton Conference on Information Sciences and Systems, 1969, pp. 16–20.

TA1 D. T. Tang and R. T. Chien, "Analysis and Synthesis Techniques of Oriented Communication Nets," *IRE Trans. Circuit Theory* **CT-8,** 39–44 (1961).

TA2 M. A. Tapia and B. R. Myers, "Generation of Concave Node Weighted Trees," *IEEE Trans. Circuit Theory* **CT-2,** 229–230 (1967).

WI1 O. Wing and R. T. Chien, "Optimal Synthesis of a Communication Net," *IRE Trans. Circuit Theory* **CT-8,** 44–49 (1961).

YA1 S. S. Yau, "On the Structure of a Communication Net," *IRE Trans. Circuit Theory* **CT-8,** 365 (1961).

YA2 S. S. Yau, "A Generalization of the Cut-Set," *J. Franklin Inst.* **274,** 31–48 (1962).

YA3 S. S. Yau, "Synthesis of Radio-Communication Nets," *IRE Trans. Circuit Theory* **CT-9,** 62–68 (1962).

CHAPTER 6

MINIMUM COST PROBLEMS

6.1 INTRODUCTION

In the preceding chapters, we considered the problems of maximum flow and multiterminal analysis and synthesis without regard to a major factor, cost. In this chapter, we consider a number of fundamental minimum cost design problems. The first to be studied is finding shortest paths in a graph. This problem plays a central role in most minimum cost syntheses. In Section 2, the problems of finding the shortest and Nth shortest paths are considered. Finding the most reliable path is shown to be equivalent to finding the shortest path. An algorithm is then given for finding maximum "capacity" paths with specified reliabilities. In Section 3 an algorithm is developed to find a "shortest" tree in a graph. This problem is also central to many minimum cost design problems. The Shortest Path Algorithm derived in Section 2 and the results of Chapter 3 are used in Section 4 to develop a method of maximizing flow at minimum cost. In Sections 5 and 6 the concept of the shortest path is used to define the distances on a graph. Then minimum cost realizations of distance matrices are considered. The optimum location of points on a network to minimize various cost functions is developed in Sections 7 and 8. Deterministic problems are considered in Section 7 and their analogous probabilistic versions in Section 8.

In Section 9, the Shortest Tree Algorithm is used to extend the results in Chapter 5 to minimum cost realizations of symmetric terminal capacity matrices. The problem of improving an existing deterministic network, at minimum cost, is also considered. In Sections 10 and 11 the same problem is considered for graphs with probabilistic requirements. In Section 12 the results on shortest paths are applied to graphs with gains. A method is presented for obtaining maximum output for minimum input.

6.2 SHORTEST PATHS, Nth SHORTEST PATHS, AND GENERALIZATIONS

A number of practical problems are mathematically equivalent to finding the "shortest paths in a graph"; for example, finding the shortest route between two cities, finding a communication path with the fewest relay stations or channels, finding the quickest way out of a labyrinth, and so on. In these same situations the knowledge of an Nth best path is also helpful. Thus for example if the $(N - 1)$ shortest communication paths are occupied it might be desirable to utilize the Nth best path. We shall also see that the problem of sending minimum cost flows in a graph can also be formulated as a shortest path problem in which the lengths of branches correspond, in some sense, to costs of branches.

We therefore introduce the concept of a length of a branch (i,j) which we designate as $l(i,j)$. This need not be a Euclidean distance; we make no restriction on the sign of $l(i,j)$ and we allow $l(i,j) \neq l(j,i)$. For a directed s-t path π_k we let $l(\pi_k)$ denote

$$\sum_{(i,j)\in\pi_k} l(i,j).$$

A shortest directed s-t path π_1 satisfies $l(\pi_1) = \underset{k}{\text{Min}}\,[l(\pi_k)]$ where k ranges over all s-t paths. Although we allow negative lengths we do make the restriction that for any directed circuit L_1,

$$\sum_{(i,j)\in L_1} l(i,j) \geq 0.$$

We shall see that this assumption appears crucially in the proof of finiteness of the proposed algorithms. We also restrict $l(i,j)$ to be finite for all (i,j).

The algorithm for finding a shortest directed s-a path for all $a \neq s$ is as follows.

Shortest Path Algorithm

Step 1. Assign all vertices v_i labels of the form $(\cdot,\delta(i))$ where $\delta(s) = 0$ and $\delta(a) = \infty$ for $a \neq s$.

Step 2. Find a branch (i,j) such that $\delta(i) + l(i,j) < \delta(j)$. If such a branch is found, change the label on vertex v_j to $(v_i,\delta(i) + l(i,j))$. Repeat this operation until no such branch can be found.

Step 3. To identify the vertices in a shortest directed s-a path for $a \neq s$:

a) Let $i = a$.
b) Identify v_k from the label $(v_k,\delta(i))$ on vertex v_i. If v_k does not exist then there is no s-a path in the graph.
c) Let $i = k$. If $i = s$ terminate. Otherwise return to (b).

We will show in Theorems 6.2.1 and 6.2.2 that the Shortest Path Algorithm terminates in a finite number of steps and finds the shortest directed s-a path. Lemma 6.2.1 below, is used in establishing Theorem 6.2.1.

Lemma 6.2.1 *Let $v_2,\ldots,v_x$ be a sequence of vertices labeled in Step 2 as $(v_1,\delta(1) + l(1,2)), (v_2,\delta(2) + l(2,3)),\ldots,(v_{x-1},\delta(x-1) + l(x-1,x))$, respectively. Then v_x is not identical with v_1.*

Proof. Suppose that v_x is identical to v_1. Suppose the vertices received labels in the order $v_2,v_3,\ldots,v_x$. Just before v_x was labeled as $(v_{x-1},\delta(x-1) + l(x-1,x))$, we had labels indicated by primed quantities as follows

$$\delta'(x) > \delta'(x-1) + l(x-1,x) \tag{6.2.1}$$

and

$$\delta'(2) \geq \delta'(1) + l(1,2). \tag{6.2.2}$$

After the label is changed, since v_x and v_1 are identical,

$$\delta(2) > \delta(1) + l(1,2). \tag{6.2.3}$$

For all $i \neq 1$ the second entry in the label of any vertex v_i on the circuit may already be $\delta(i - 1) + l(i - 1,i)$ or may become so in a subsequent step. In either case

$$\delta(i + 1) \geq \delta(i) + l(i,i + 1). \tag{6.2.4}$$

Adding (6.2.3) and (6.2.4) for all $i \neq 1$ and v_i on the circuit, after simplification, gives

$$\sum_{i=1}^{x-1} l(i,i + 1) < 0. \tag{6.2.5}$$

But this contradicts the assumption that there are no circuits of negative length and the proof is completed. //

Using Lemma 6.2.1, we now show:

DID IN CLASS

Theorem 6.2.1 *Step 2 of the Shortest Path Algorithm terminates after a finite number of labelings.*

Proof. It is clear that for any vertex $v_y, \delta(y)$ is either decreased by an integer or is unchanged with each application of Step 2 of the Shortest Path Algorithm and that with each application of Step 2 the weight on at least one vertex is reduced. We complete the proof by showing that $\delta(y)$ is bounded from below.

If $\delta(y)$ remains infinite then clearly it is bounded below by any finite positive integer. If $\delta(y) < \infty$, then there is a directed s-y path which, after possibly renumbering vertices, will contain $v_1, v_2, \ldots, v_\alpha$, with $v_s = v_1$ and $v_\alpha = v_y$. For these vertices

$$\delta(x) + l(x,x + 1) \leq \delta(x + 1) \qquad \text{for } x = 1, \ldots, \alpha - 1. \tag{6.2.6}$$

Summing these inequalities yields

$$\delta(s) + \sum_{r=1}^{\alpha-1} l(r,r + 1) \leq \delta(y). \tag{6.2.7}$$

But $\delta(s) = 0$ after Step 1 and is never reduced in Step 2, since otherwise the vertices in a circuit would be identified in Step 3 contrary to Lemma 6.2.1. Hence

$$\delta(y) \geq \sum_{r=1}^{\alpha-1} l(r,r + 1). \tag{6.2.8}$$

Therefore $\delta(y)$ is bounded below if $l(i,j) < \infty$ for all i and j and the theorem is proved.

Lemma 6.2.2 *If there is an s-a path of finite length, then v_k can be found in each iteration of Step 3(b).*

Proof. At the termination of Step 2, $\delta(a) < \infty$. For arbitrary $x \neq s$ suppose $\delta(x) < \infty$. But from Step 2 of the algorithm it is clear that there are one or more vertices, say $v_{i_1}, v_{i_2}, \ldots, v_{i_\alpha}$, such that $\delta(x) = l(i_r, x) + \delta(i_r)$, for $r = 1, \ldots, \alpha$. Hence the label on v_x is $(v_{i_\beta}, \delta(x))$ for some β with $1 \leq \beta \leq \alpha$. Furthermore $\delta(i_\beta) = \delta(x) - l(i_1, x) < \infty$. Hence $v_k = v_{i_\beta}$ and the proof is completed by induction.

From Lemmas 6.2.1 and 6.2.2 we have:

Lemma 6.2.3 *The Shortest Path Algorithm terminates in a finite number of steps.*

Finally we can show:

Theorem 6.2.2 *At the termination of the Shortest Path Algorithm, the s-a path found in Step 3 is a shortest s-a path.*

Proof. From Theorem 6.2.1, the algorithm terminates and from Lemma 6.2.2, a path π_1 consisting of vertices $v_1, v_2, \ldots, v_\alpha$ (with $v_1 = v_s$ and $v_\alpha = v_a$) is found in Step 3. Suppose contrary to the theorem that there is a shorter *s-a* path than π_1, namely π_2, consisting of vertices $v_{1'}, v_{2'}, \ldots, v_{\alpha'}$ (with $v_{1'} = v_s, v_{\alpha'} = v_a$). Then

$$\delta(1') + \sum_{x=1}^{\alpha-1} l(x', x+1') < \delta(\alpha'). \tag{6.2.9}$$

But since the algorithm has terminated,

$$\delta(x') + l(x', x+1') \geq \delta(x+1') \qquad \text{for } x = 1, \ldots, \alpha - 1, \tag{6.2.10}$$

or

$$\delta(1') + \sum_{x=1}^{\alpha-1} l(x', x+1') \geq \delta(\alpha'). \tag{6.2.11}$$

Hence there is a contradiction and the theorem is proved.

Once labeling of vertices is achieved by means of the Shortest Path Algorithm it is clear that, for any vertex v_j, a unique *s-j* path can be identified in Step 3. Hence we have found a tree in which the path from v_s to any vertex v_j is the shortest *s-j* path and the weight $\delta(j)$ is the length of that path.

Example 6.2.1 In Fig. 6.2.1, consider the branches in the following order: (1,2), (2,4), (1,3), (3,4), (3,2), (2,4). The resulting labels are shown in Fig. 6.2.1. The final labels are checked and the final tree indicating shortest paths from v_1 to any other

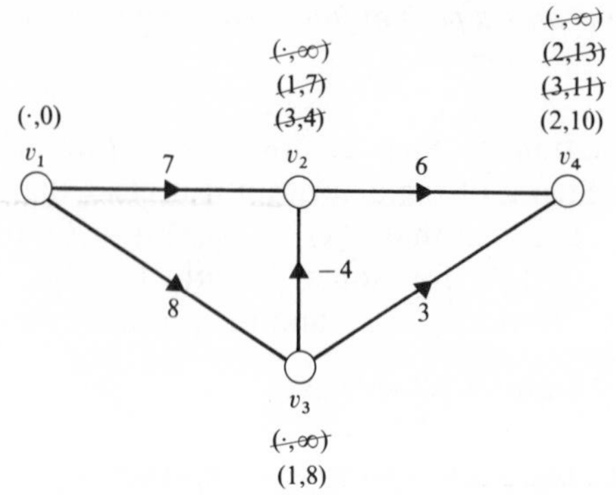

Fig. 6.2.1 Graph for Example 6.2.1.

vertex consists of branches (1,3), (3,2), and (2,4). The length of the shortest path from v_1 to v_3 is 8, from v_1 to v_2 is 4, and from v_1 to v_4 is 10.

Example 6.2.2 For the graph in Fig. 6.2.2 the Shortest Path Algorithm can be used to identify the tree consisting of branches (s,3), (s,4), (3,6), (3,8), (4,5), (5,8), (5,10), (6,7), and (8,t). The labels assigned to the vertices will be discussed shortly in Example 6.2.3.

Finally we note that in the resulting tree we need not retain the first labels on each vertex. We can identify a shortest s-a path by simply inspecting $\delta(i)$ for each i. In particular, if $v_i, v_{i+1}, \ldots, v_a$ is a sequence of consecutive vertices in a shortest s-a path then (k,i) is in a shortest s-a path if $\delta(i) - \delta(k) = l(k,i)$ and v_k is not in

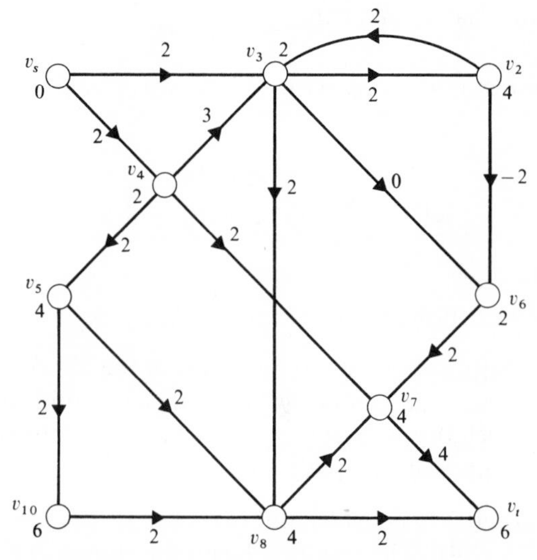

Fig. 6.2.2 Graph for Examples 6.2.2 and 6.2.3; branch weights are costs.

$\{v_i,v_{i+1},\ldots,v_a\}$. Furthermore, if the shortest s-a path is not unique, all shortest s-a paths can be found in this way.

Example 6.2.3 In Fig. 6.2.2 the weights assigned to the vertices are the $\delta(i)$ from the Shortest Path Algorithm. Selecting those branches (k,i) for which $\delta(i) - \delta(k) = l(k,i)$ we can immediately identify a tree. However, since $\delta(6) - \delta(3) = l(6,3)$ and $\delta(6) - \delta(4) = l(4,6)$ either (2,6) or (3,6) can be in a shortest s-6 path.

We now wish to find the Nth shortest s-t path. In the procedure, we will identify sequences of vertices which form circuits as well as s-t paths. We therefore introduce the concept of a chain.

Definition 6.2.1 *A directed s-t chain, or simply a chain, is an ordered set of vertices and branches* $v_{i_1},b_1,v_{i_2},b_2,\ldots,v_{i_{k-1}},b_{k-1},v_k$ *such that for* $j = 1,\ldots,k-1$, $b_j = (i_j,i_{j+1})$.

Branches and vertices may be repeated in a chain. We will denote a chain by the symbol Λ.

Let ϕ_N be the set of all Nth shortest chains. That is, ϕ_1 is the set of s-t paths with smallest length, ϕ_2 is the set of chains with the next smallest length, and so on. We let $l(\phi_N)$ be the common length of all s-t chains in ϕ_N. As an example, if in a graph there are ten s-t chains Λ_i, $i = 1,\ldots,10$ with lengths less than or equal to eight as specified in Table 6.2.1, then ϕ_N is given in Table 6.2.2 for all N.

Definition 6.2.2 *A deviation from an s-t chain* Λ_1 *at vertex* v_{k+1} *to vertex* v_k *is another s-t chain* Λ_2 *containing* $v_s,v_1,v_2,\ldots,v_k,v_{k+1},\ldots,v_t$ *such that*

a) $v_s,\ldots,v_k$ *is a shortest s-k chain. As a special case we allow* $k = s$.

b) $(k,k+1)$ *is not in* Λ_1 *and* v_{k+1} *is in* Λ_1.

c) $v_{k+1},\ldots,v_t$ *is in* Λ_1.

The branch $(k,k+1)$ above is called a *deviation branch.* Note that in Definition 6.2.2, $v_s,\ldots,v_k$ may or may not be in Λ_1. If Λ_2 is a deviation from Λ_1 at vertex v_{k+1} to vertex v_k we write $\Lambda_2 = \text{dev}(\Lambda_1)$ or $\Lambda_2 = \text{dev}(\Lambda_1,k+1,k)$.

Table 6.2.1

i	$l(\pi_i)$
1	1
2	2
3	3
4	4
5	5
6	5
7	6
8	6
9	7
10	8

Table 6.2.2

N	ϕ_N
1	$\{\pi_1\}$
2	$\{\pi_2\}$
3	$\{\pi_3\}$
4	$\{\pi_4\}$
5	$\{\pi_5, \pi_6\}$
6	$\{\pi_7, \pi_8\}$
7	$\{\pi_9\}$
8	$\{\pi_{10}\}$

Lemma 6.2.4 *If Λ_2 is not a shortest s-t chain then there is an s-t chain Λ_1 such that* $\Lambda_2 = \text{dev}(\Lambda_1)$ *where* $l(\Lambda_1) < l(\Lambda_2)$.

Proof. Let Λ_2 be an s-t chain with $l(\Lambda_2)$ greater than the length of the shortest s-t chain. Then there must be a vertex v_{k+1} in Λ_2 such that

a) $v_s, \ldots, v_k, v_{k+1}$ is contained in Λ_2;
b) $v_s, \ldots, v_k$ is a shortest s-k chain;
c) $v_s, \ldots, v_k, v_{k+1}$ is not a shortest s-$k + 1$ chain.

Let Λ_3 be a shortest s-$k + 1$ chain and let Λ_4 consist of $v_s, \ldots, v_k, v_{k+1}$ and let Λ_5 consist of $v_{k+1}, \ldots, v_t$ such that $\Lambda_2 = \Lambda_4 \cup \Lambda_5$. Then

$$\begin{aligned} l(\Lambda_2) &= l(\Lambda_4) + l(\Lambda_5) \\ &> l(\Lambda_3) + l(\Lambda_5) \\ &= l(\Lambda_1), \end{aligned} \tag{6.2.12}$$

where Λ_1 is the s-t chain with the vertices in Λ_5 and in Λ_3. Since $\Lambda_2 = \text{dev}(\Lambda_1)$, the lemma is proved.

From Lemma 6.2.4 we immediately have the Nth Shortest Chain Algorithm below for finding ϕ_N.

Nth Shortest Chain Algorithm

Step 1. Use the Shortest Path Algorithm to find all shortest s-j paths for all j. Find the paths in ϕ_1. Let $N = 2$. Superimpose on the graph the tree resulting from the application of the Shortest Path Algorithm to find the shortest paths originating at v_s. Each vertex v_i is therefore labeled by a single label $\delta(i)$ so that all shortest paths from vertex v_s can be found.

Step 2. Find all chains $\{\Lambda_{N_j}\}$ in ϕ_N by repeating the following step:

$$\Lambda_{N_j} \in \phi_N \text{ if } l(\Lambda_{N_j})$$
$$= \text{Min}\left[l[\text{dev}(\Lambda_r, k + 1, k)] \text{ such that } l[\text{dev}(\Lambda_r, k + 1, k)] > l[\phi_{N-1}]\right], \tag{6.2.13}$$

with the minimum taken over all $\Lambda_r \in \phi_h$, $h < N$ and all possible deviation branches $(k + 1, k)$ for Λ_r.

Step 3. If the deviations of all chains previously generated have been found then terminate. Otherwise replace N by $N + 1$ and return to Step 2.

Note that if there is a circuit of zero length, there may be an infinite number of chains of a given length.

Example 6.2.4 Consider the graph in Fig. 6.2.2 with the vertices already weighted to indicate the shortest s-j paths for all j. The shortest s-t path Λ_1 is identified in (6.2.14). Since the shortest s-t path is unique, ϕ_1 in (6.2.15) consists of a single path Λ_1. The deviations from Λ_1 are next given in (6.2.16) to (6.2.22). Note that in (6.2.22) a chain, and not a path, results from finding a deviation from Λ_1. The shortest of these deviations are contained in ϕ_2 which is identified in (6.2.23). There are only four new deviations of chains in ϕ_2 and these are given in (6.2.24) to (6.2.27). Then ϕ_3 consists of Λ_7, as in (6.2.28), and Λ_7 has one new deviation in (6.2.29). Hence ϕ_4 is given in (6.2.30). The set of Nth best chains, including the set of Nth best paths, are given for $N = 1,2,3,4$ in (6.2.15), (6.2.23), (6.2.28), and (6.2.30). The process can be continued to obtain ϕ_N for $N > 4$, and is left as Problem 2.

$$\pi_1 = v_s,v_3,v_8,v_t \qquad l(\pi_1) = 6 \tag{6.2.14}$$

$$\phi_1 = \{\pi_1\} \tag{6.2.15}$$

$$\pi_2 = \text{dev}\,(\pi_1) = v_s,v_3,v_2,v_6,v_7,v_t \qquad l(\pi_2) = 8 \tag{6.2.16}$$

$$\pi_3 = \text{dev}\,(\pi_1) = v_s,v_3,v_6,v_7,v_t \qquad l(\pi_3) = 8 \tag{6.2.17}$$

$$\pi_4 = \text{dev}\,(\pi_1) = v_s,v_4,v_7,v_t \qquad l(\pi_4) = 8 \tag{6.2.18}$$

$$\pi_5 = \text{dev}\,(\pi_1) = v_s,v_4,v_5,v_8,v_t \qquad l(\pi_5) = 8 \tag{6.2.19}$$

$$\pi_6 = \text{dev}\,(\pi_1) = v_s,v_4,v_5,v_{10},v_8,v_t \qquad l(\pi_6) = 10 \tag{6.2.20}$$

$$\pi_7 = \text{dev}\,(\pi_1) = v_s,v_4,v_3,v_8,v_t \qquad l(\pi_7) = 9 \tag{6.2.21}$$

$$\Lambda_8 = \text{dev}\,(\pi_1) = v_s,v_3,v_2,v_3,v_8,v_t \qquad l(\Lambda_8) = 10 \tag{6.2.22}$$

$$\phi_2 = \{\pi_2,\pi_3,\pi_4,\pi_5\} \tag{6.2.23}$$

$$\pi_9 = \text{dev}\,(\pi_2) = v_s,v_4,v_3,v_2,v_6,v_7,v_t \qquad l(\pi_9) = 11 \tag{6.2.24}$$

$$\Lambda_{10} = \text{dev}\,(\pi_2) = v_s,v_3,v_2,v_3,v_2,v_6,v_7,v_t \qquad l(\Lambda_{10}) = 12 \tag{6.2.25}$$

$$\pi_{11} = \text{dev}\,(\pi_3) = v_s,v_4,v_3,v_6,v_7,v_8,v_t \qquad l(\pi_{11}) = 11 \tag{6.2.26}$$

$$\Lambda_{12} = \text{dev}\,(\pi_3) = v_s,v_3,v_2,v_3,v_6,v_7,v_t \qquad l(\Lambda_{12}) = 12 \tag{6.2.27}$$

$$\phi_3 = \{\pi_7\} \tag{6.2.28}$$

$$\Lambda_{13} = \text{dev}\,(\pi_7\} = v_s,v_4,v_3,v_2,v_3,v_8,v_t \qquad l(\Lambda_{13}) = 13 \tag{6.2.29}$$

$$\phi_4 = \{\pi_6,\Lambda_8\} \tag{6.2.30}$$

The results on shortest paths are easily extended to find most reliable paths. Consider a graph G with n vertices and m branches. For each branch b_k a reliability p_k may be specified. This reliability may represent the probability that the channel b_k is operative (i.e., not destroyed or jammed), it may be the probability that the channel is unoccupied, or it may be the probability that a character (or symbol) is transmitted correctly through the channel. Suppose our objective is to route flow between vertices as reliably as possible. The simplest problem we can consider is to find the most reliable s-t path in G.

Let $\{\pi_1,\pi_2,\ldots,\pi_l\}$ be the set of directed paths from v_s to v_t. A reliability may be assigned to each path π_k equal to the product of the reliabilities of the branches in the path. In other words, the reliability P_k of π_k is defined as

$$P_k = \prod_{b_i \in \pi_k} p_i. \tag{6.2.31}$$

Our problem is to find a set $\boldsymbol{\pi} = \{\pi_k\}$ of s-t paths such that for $\pi_k \in \boldsymbol{\pi}$,

$$P_k = \operatorname*{Max}_i\,[P_i]. \tag{6.2.32}$$

Consider the same graph G and instead of associating reliability p_k with branch b_k, associate length q_k with b_k for $k = 1,2,\ldots,m$, where

$$q_k = -\ln p_k. \tag{6.2.33}$$

The *length* of any path π_k is

$$l(\pi_k) = \sum_{b_i \in \pi_k} q_i. \tag{6.2.34}$$

We now show that the most reliable path in G when the branches of G are weighted with the probabilities $p_1,p_2,\ldots,p_m$ is the shortest path in G when the branches of G are weighted by the lengths $q_1,q_2,\ldots,q_m$ where $q_i = -\ln p_i$ for all i. This fact follows from (6.2.31).

$$\ln P_k = \ln\left(\prod_{b_i \in \pi_k} p_i\right) = \sum_{b_i \in \pi_k} \ln p_i. \tag{6.2.35}$$

Hence,

$$-\ln P_k = l(\pi_k). \tag{6.2.36}$$

Therefore, if

$$\ln P_j = \operatorname*{Max}_i\,[\ln P_i], \tag{6.2.37}$$

then

$$l(\pi_j) = -\ln P_j = \operatorname*{Min}_i\,[-\ln P_i] = \operatorname*{Min}_i\left[\sum_{b_k \in \pi_i} q_k\right], \tag{6.2.38}$$

which proves our contention.

Finding the most reliable path may not be sufficient to solve even the simplest routing problem since this path may not be able to accommodate a specified amount of flow. Thus we must assume that each branch b_k of G is weighted with the pair (c_k,p_k), where c_k is the capacity of b_k and p_k is its reliability. The capacity of a path is then the smallest branch capacity of any branch in the path. For an undirected graph we will set a lower bound on the acceptable reliability of a path and then look for a maximum capacity path whose reliability is not less than this bound. Note that in general, if the lower bound is too high, no such path may exist. One way of solving this problem is to generate all paths in the graph G and then find the best one; however, we will give a method which requires finding at most k paths where k is the number of different branch capacities in the graph. This method depends on finding the "principal subgraph" of G. Let π_α be an s-t path of maximum capacity.

The *principal subgraph* of an undirected graph, with respect to vertices v_s and v_t, is the subgraph consisting of all branches of G with capacity not less than the capacity $c(\pi_\alpha)$ of π_α. The s-t path of maximum reliability in the principal subgraph can easily be found. If the reliability of this path is not less than the specified lower bound, the procedure terminates. Otherwise, the subgraph is found which contains all paths with the next largest capacity and the path with maximum reliability is again found for this graph. The procedure continues until a path with reliability not less than the given lower bound is found or it is determined that no such path exists. In the latter case, the lower bound must be reduced. An efficient algorithm for finding the principal subgraph is available. This algorithm is based on Lemmas 6.2.5 and 6.2.6 below.

Lemma 6.2.5 *An s-t path π_k, with capacity $c(\pi_k) \geq \beta$, exists in the graph G if and only if every s-t cut-set contains at least one branch b_i with capacity $c_i \geq \beta$.*

Proof. 1) Suppose there is a path π_k of capacity β. Since each *s-t* cut-set must contain at least one branch from every *s-t* path, every *s-t* cut-set must contain at least one branch with capacity β or more.

2) Suppose all *s-t* paths have capacity less than β. This means that every *s-t* path contains a branch with capacity less than β. Then the graph contains an *s-t* cut-set in which every branch has capacity less than β. This cut-set is contained in the set obtained by taking one branch with capacity less than β from each path.

We say that vertices v_p and v_q are *shorted* if v_p and v_q are made into a single vertex. The branch between v_p and v_q is eliminated and all resulting branches in parallel are replaced by a single branch whose capacity is the sum of the capacities of the parallel branches.

Lemma 6.2.6 *If $b_k = [p,q]$ and if v_p and v_q are shorted,*
a) the resulting graph G′ contains all cut-sets which in G do not contain b_k,
b) G′ contains no cut-sets with respect to v_s and v_t that were not contained in G.

Fig. 6.2.3 Graph in which b_k is not in $A_{s,t}$.

Proof. 1) Consider a cut-set $A_{s,t}$ in G which does not contain b_k as in Fig. 6.2.3; b_k must be in the subgraph G_1 or G_2. Therefore, the shorting operation will maintain $A_{s,t}$ as an *s-t* cut-set in G'.

2) Suppose, contrary to part (b) that a new cut-set is created in G' from G by shorting v_s and v_t. Consider the possible relationships of b_k and $A_{s,t}$ in G. The two possible cases are shown in Figs. 6.2.4 and 6.2.5. In Case 1, if b_k is removed in shorting v_s and v_t, $A_{s,t}$ cannot be a cut-set in G'. In Case 2, $A_{s,t}$ will be a cut-set in G', but it is also a cut-set in G. Hence the lemma is proved.

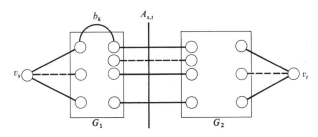

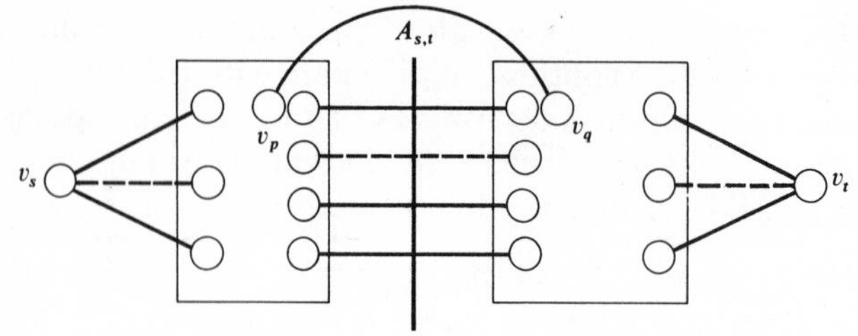

Fig. 6.2.4 Graph in which branches in $A_{s,t}$ no longer correspond to a cut-set.

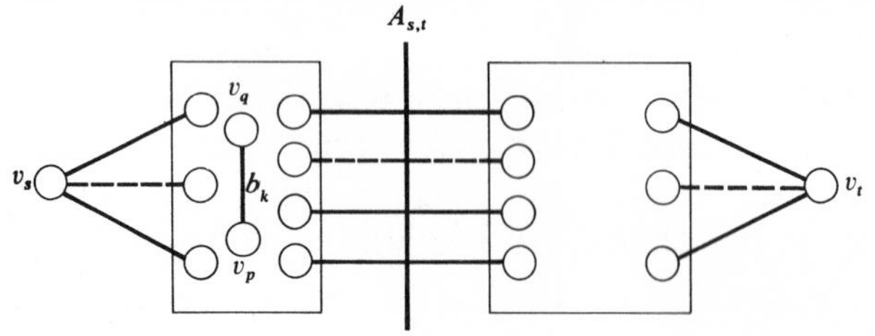

Fig. 6.2.5 Graph in which branches in $A_{s,t}$ correspond to a cut-set.

We can now give Algorithm 6.2.1 to construct a principal subgraph.

Algorithm 6.2.1

Step 1. Select any *s-t* cut-set $A_{s,t}$ in G,

Step 2. If in the last graph constructed (G in the first iteration, otherwise the graph constructed in Step 2 of the previous iteration), there is a branch $[p,q]$ with capacity greater than or equal to the largest branch capacity in the last cut-set selected (in Step 1 in the first iteration, otherwise in Step 4), short vertices v_p and v_q.

Step 3. Suppose in the graph resulting from Step 2 in the last iteration, vertices v_s and v_t are shorted. Then the subgraph, consisting of the branches that have been removed from G by repeated iterations of Step 2, and the vertices of G comprise the principal subgraph.

Step 4. If v_s and v_t are not shorted in Step 2 in the last iteration, select any cut-set $A_{s,t}$ in the last graph constructed and return to Step 2.

The use of Algorithm 6.2.1 is illustrated in the following example.

Example 6.2.5 The problem is to find the principal subgraph with respect to vertices v_s and v_t in the graph in Fig. 6.2.6. Algorithm 6.2.1 is applied as follows. Arbitrarily select the cut-set $A_{s,t}^1$ shown in Fig. 6.2.6. The largest branch capacity in the cut-set is 15. Hence all branches with capacity larger than or equal to 15 are removed after shorting their end vertices. The resulting reduced graph G' is shown in Fig. 6.2.7. Vertices v_s and v_t are not yet shorted. Next cut-set $A_{s,t}^2$ in the reduced graph G' is

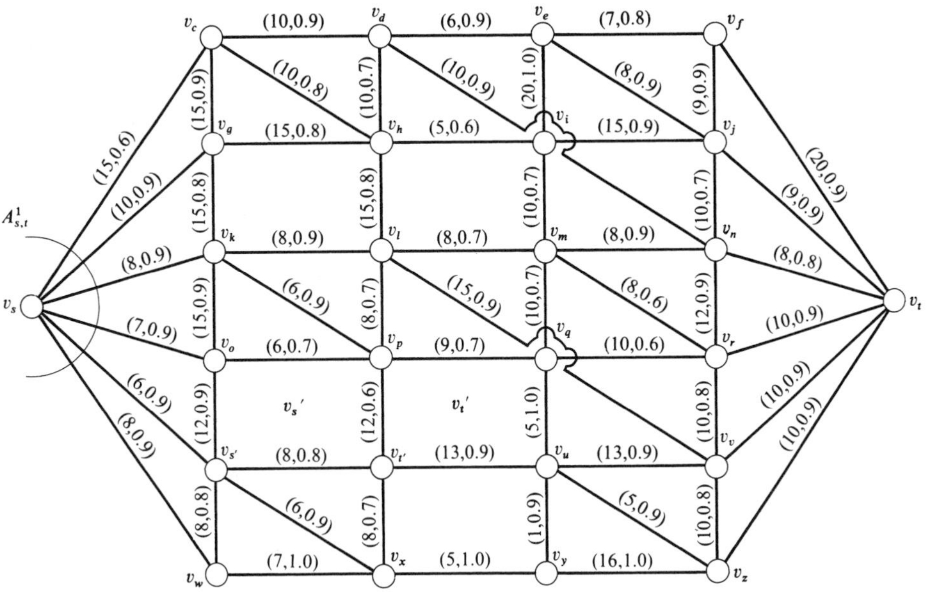

Fig. 6.2.6 Graph for Example 6.2.5.

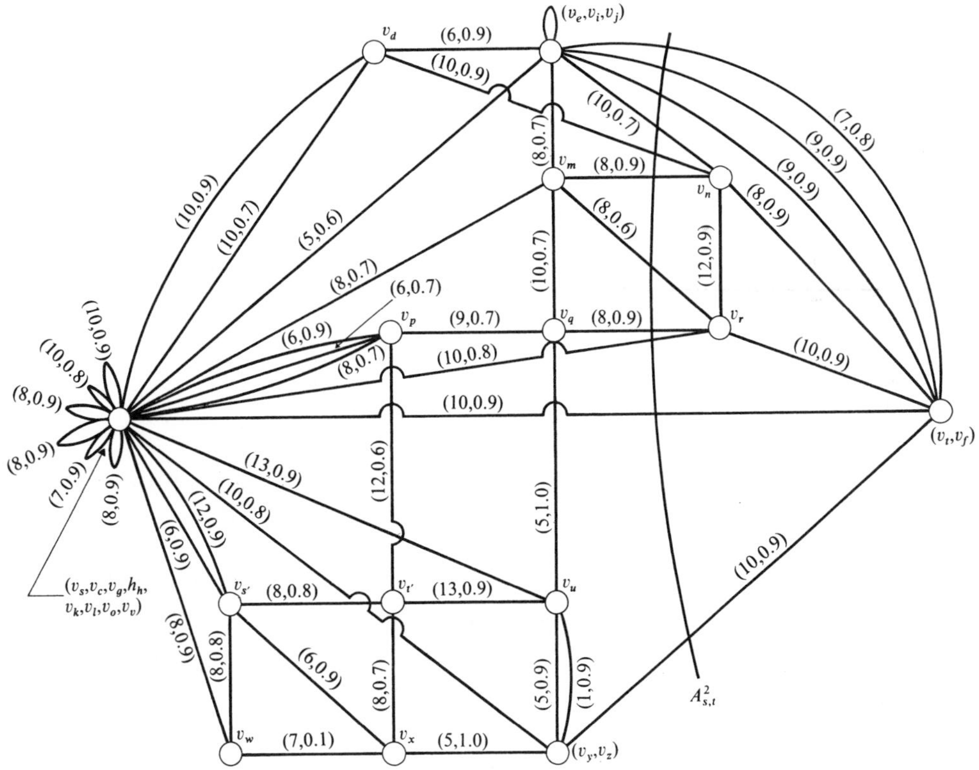

Fig. 6.2.7 Branches with capacity greater than or equal to 15 are shorted.

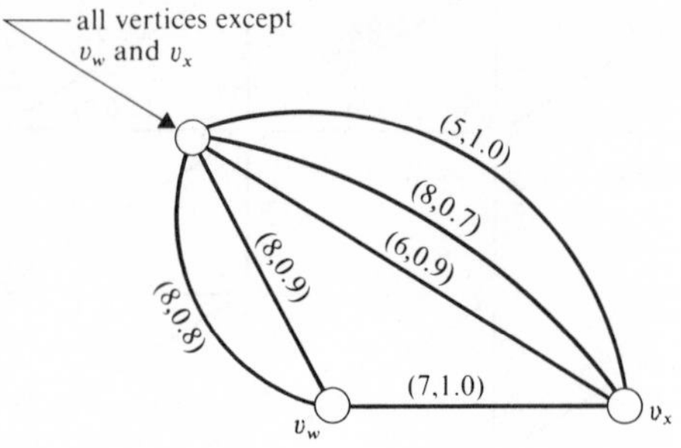

Fig. 6.2.8 Branches with capacity greater than or equal to 10 are shorted.

selected. The largest branch capacity in $A_{s,t}^2$ is 10. Hence all branches with capacity larger than or equal to 10 are removed after shorting their end vertices. The new reduced graph G'' is shown in Fig. 6.2.8. Vertices v_s and v_t are shorted. Hence the principal subgraph containing all paths from v_s to v_t with capacity 10 is formed from all the branches and the vertices incident with them that have been removed from G. The principal subgraph is shown in Fig. 6.2.9.

Note that with a more careful selection of cut-sets the principal subgraph could have been obtained with one iteration rather than two. In general, the graph in which the cut-set is to be selected becomes simpler after each iteration. Thus, in

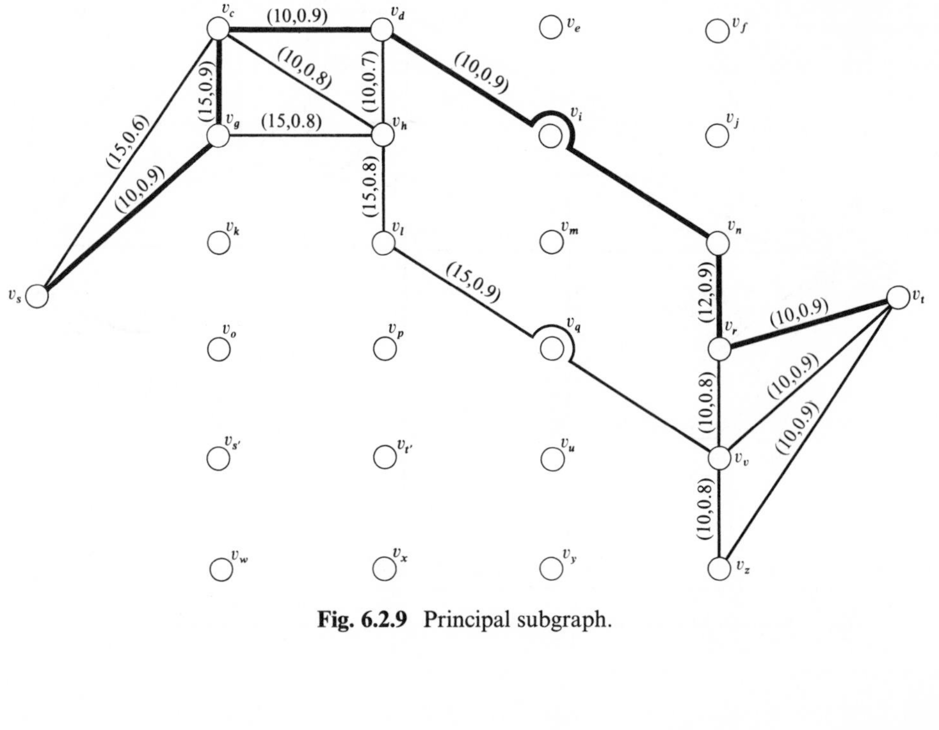

Fig. 6.2.9 Principal subgraph.

Example 6.2.5, after selecting the first cut-set $A^1_{s,t}$ and applying Step 2, the number of vertices is reduced from 26 to 15 and the number of branches from 58 to 42.

With the above algorithm as a basis, an algorithm can now be given for finding the optimum path between a given pair of vertices v_s and v_t.

Algorithm 6.2.2

Step 1. Find the s-t path with maximum reliability in the principal subgraph. If the reliability of the path exceeds or equals the threshold reliability the procedure terminates, since the last path that has been found is the optimum path.

Step 2. If the reliability of the resulting path from Step 1 is less than the threshold reliability, add to the last subgraph used (the principal subgraph after Step 1, otherwise the subgraph from the previous iteration of Step 2) all branches in G with the largest branch capacity that does not exceed the smallest branch capacity in the last subgraph.

Step 3. Find the s-t path with maximum reliability in the subgraph resulting from the last iteration of Step 2. If this reliability exceeds the threshold reliability, the path found is the optimum path; otherwise return to Step 2.

Example 6.2.6 The problem is to find the optimum path from v_s to v_t in the graph G in Fig. 6.2.6, first with threshold reliability 0.5 and then with threshold 0.6. The solution using Algorithm 6.3.2 is as follows. In the principal subgraph found in

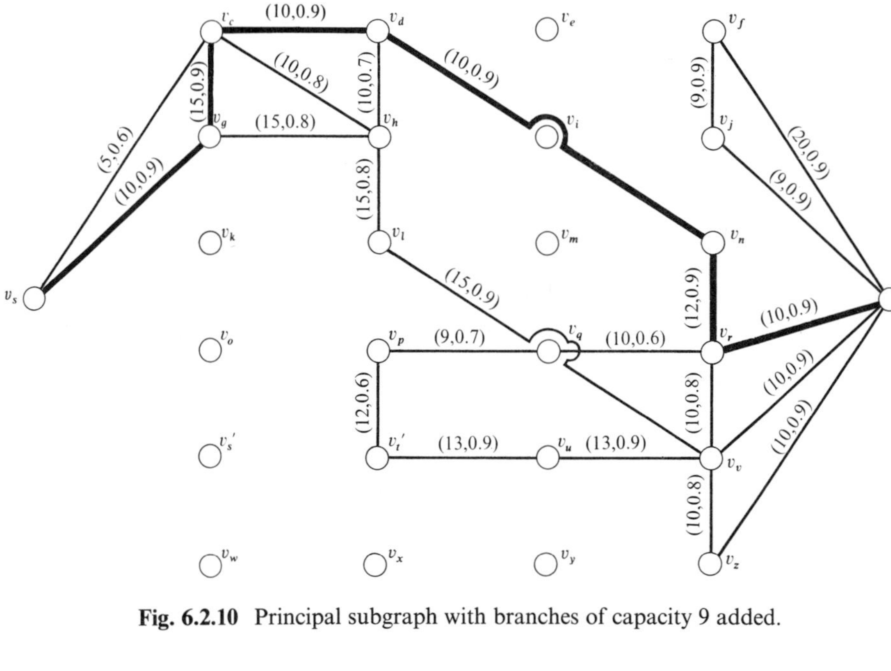

Fig. 6.2.10 Principal subgraph with branches of capacity 9 added.

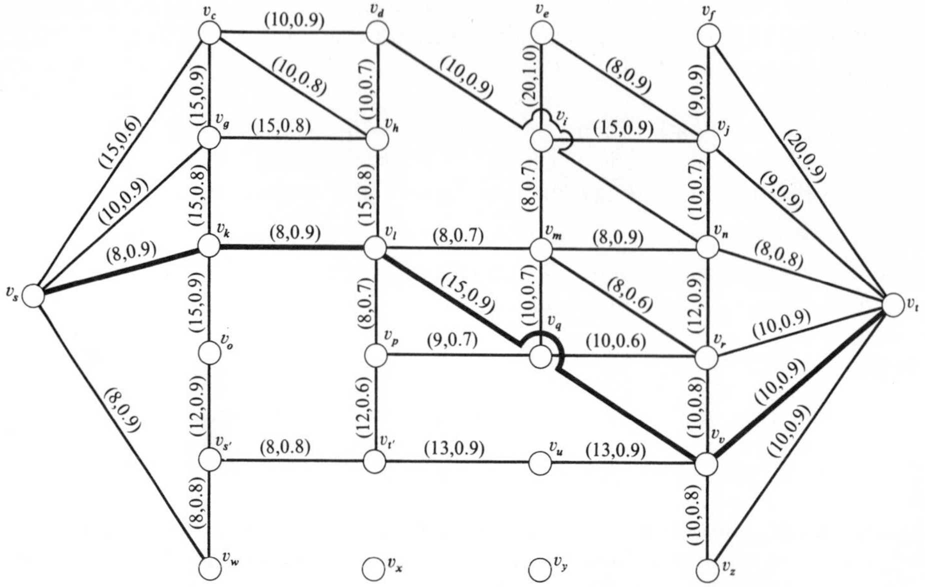

Fig. 6.2.11 Principal subgraph with branches of capacity 9 and 8 added.

Example 6.2.5 and shown in Fig. 6.2.9 the path with maximum reliability is shown in heavy lines. The reliability of this path is 0.54. Hence it is the optimum path for threshold reliability 0.5 and the capacity of the optimum path is 10. However, for a threshold reliability of 0.6 this path will not suffice and we must proceed to Step 2 of Algorithm 6.2.2. By adding all branches with capacity 9 to the principal subgraph, the graph in Fig. 6.2.10 is obtained. The path with maximum reliability in this subgraph is 0.54, which is less than 0.6. All branches with capacity 8 are next added and the resulting graph is shown in Fig. 6.2.11. The path with maximum reliability in the new graph is shown in heavy lines. Since the reliability of this path is 0.66, which exceeds 0.6, the path is the optimum path and its capacity is 8.

6.3 SHORTEST TREES

A number of physical situations can be envisioned in which a minimum length tree is desired to connect a set of vertices. For example, it might be necessary to build a system of roads to interconnect cities or a communication system to interconnect stations. Usually, of course, many other constraints must be met. Even in its most primitive form, however, the problem is of interest and, as we shall see in the next section, a number of seemingly unrelated combinatoric problems can be reduced to the problem of finding the shortest connecting network.

In this section we consider an undirected graph with branch lengths $l[i,j]$ which may be negative. We define the length of a graph G to be

$$l(G) = \sum_{[i,j] \in G} l[i,j].$$

We then seek a tree G_1 containing all vertices such that $l(G_1) = \text{Min}\,[l(G_i)]$ where the minimization is over all trees. If a branch is not in G we can add the branch and assign it infinite length.

(a)   (b)   (c)

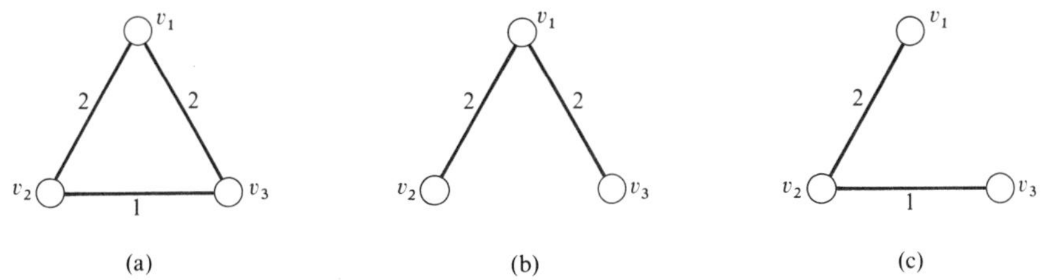

Fig. 6.3.1 Graphs illustrating the difference between shortest paths and shortest trees.

It is important to realize that the shortest tree we shall find is not the same as the ⊛
tree found in Section 2 by the Shortest Path Algorithm. As an example, for the graph in Fig. 6.3.1(a), the tree found by the Shortest Path Algorithm for paths from v_1 is given in Fig. 6.3.1(b). The shortest tree is shown in Fig. 6.3.1(c). Moreover, since we allow negative lengths, when we can solve the shortest tree problem we can also solve the longest tree problem. All we need do is multiply each branch length by -1 and find the shortest tree.   → there may be several nearest neighbors

Finally, we remark that if all branch capacities are positive then the shortest tree will be the shortest subgraph of G containing all vertices. In this case the problem of finding the shortest tree is the same as finding the shortest connecting graph.

We now give an efficient algorithm to find the shortest tree in an undirected graph. An *isolated vertex* is a vertex which, at a given stage of the construction, has zero degree. A *nearest neighbor* to a vertex v_i is a vertex v_j such that $l[i,j] = \underset{k}{\text{Min}}\,[l[i,k]]$.

Let $G_1 = (V_1, \Gamma_1)$ be a component of G. The length from a vertex v_i in $\bar{V}_1$ to G_1 is given by $\underset{v_j \in V_1, v_i \in V_i}{\text{Min}}[l[i,j]]$. The *nearest neighbor to a component* is a vertex not in the component with shortest length to the component. Then the repeated application of   ⊛
any combination of Operations 6.3.1 and 6.3.2 as often as possible in any order will yield a shortest tree.

Operation 6.3.1. Connect any isolated vertex to a nearest neighbor.

Operation 6.3.2. Connect any component to a nearest neighbor by a shortest possible branch.

The validity of Operations 6.3.1 and 6.3.2 is established by proving Lemmas 6.3.1, 6.3.2, and 6.3.3 below.

Lemma 6.3.1 *Every vertex in a shortest tree is adjacent to at least one nearest neighbor.*

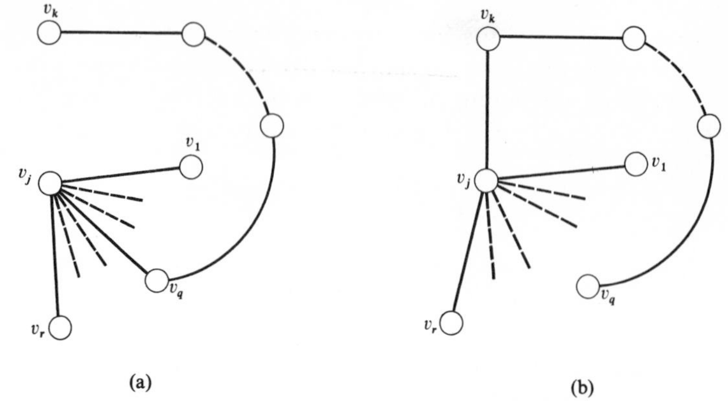

Fig. 6.3.2 Graphs used in the proof of Lemma 6.3.1.

Proof. Suppose there is a shortest tree for which the lemma is untrue. Then there is a vertex v_j which is not adjacent to any nearest neighbor, one of which is, say, v_k. Since the tree is a connected graph, there is a j-k path in the tree. Let (j,q) be a branch in the path. If we remove (j,q) from the tree and add (j,k) we have another tree which is shorter than the original tree since $l(j,q) > l(j,k)$. Hence we have a contradiction and the proof is completed.

Lemma 6.3.2 *Every proper subgraph of a shortest tree is connected to at least one nearest neighbor by a shortest possible branch.*

Proof. The proof of Lemma 6.3.1 carries over exactly if v_j is thought of as a proper subgraph and $[j,k]$ is the shortest branch from v_k to any vertex in the subgraph.

We have shown in Lemmas 6.3.1 and 6.3.2 that Operations 6.3.1 and 6.3.2 yield trees which have some properties required for a shortest tree. We now show that any tree having these properties has the same length. Hence, using Operations 6.3.1 and 6.3.2, we construct a shortest tree.

Lemma 6.3.3 *The length of the tree resulting from Operation 6.3.1 and 6.3.2 is independent of the nearest neighbor chosen for a vertex.*

Proof. At each stage of the construction using Operations 6.3.1 or 6.3.2, at which a choice is to be made among two or more nearest neighbors $v_1, \ldots, v_r$ of an isolated vertex (or component) v_k, subtract a small positive quantity, ε from any of the distances $l[k,1], l[k,2], \ldots, l[k,r]$. The construction will be uniquely determined. The total length L of the resulting shortest tree will depend on ε as well as on the lengths in the original graph.

The dependence on ε is continuous because the minimum of a finite set of continuous functions of ε (the set of lengths of all trees of the modified problem) is itself a continuous function of ε. Hence, as ε is made vanishingly small, L approaches the minimum length regardless of which "nearest neighbor" branches were chosen.

Example 6.3.1 Consider the graph in Fig. 6.3.3 with lengths as indicated. Suppose that the shortest tree has been partially constructed as in Fig. 6.3.4. Then a possible sequence of steps in constructing the shortest tree consists of adding [5,6], [7,9], [4,5], and [6,7] in any order. The final shortest tree is shown in Fig. 6.3.5.

Several variations of Operations 6.3.1 and 6.3.2 are possible for constructing a shortest tree. In particular Constructions 6.3.1, 6.3.2, and 6.3.3 below will all yield shortest trees for a graph $G = (V,\Gamma)$.

Construction 6.3.1. Perform the following step as many times as possible. Among the branches not yet chosen select the shortest branch which does not form a circuit with those branches already chosen.

Construction 6.3.2. Let $G_1 = (V_1,\Gamma_1)$ be a specified shortest tree on the vertices $V_1 \subseteq V$ with V_1 not empty. Perform the following step as many times as possible.

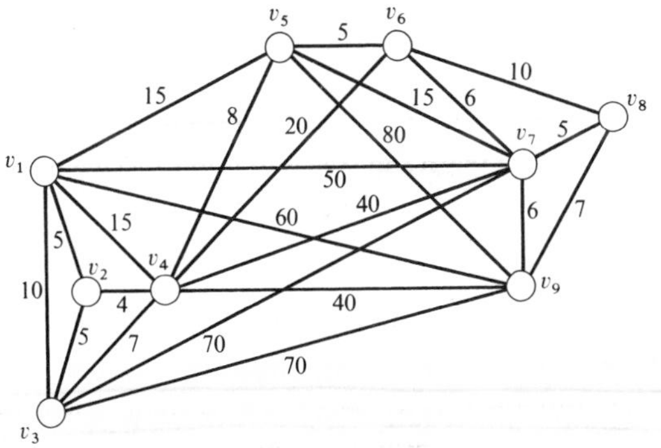

Fig. 6.3.3 Graph for Example 6.3.1.

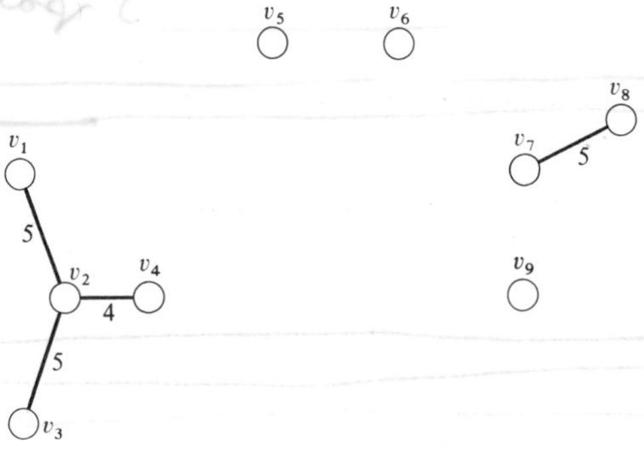

Fig. 6.3.4 Partial construction of shortest tree for G in Fig. 6.3.3.

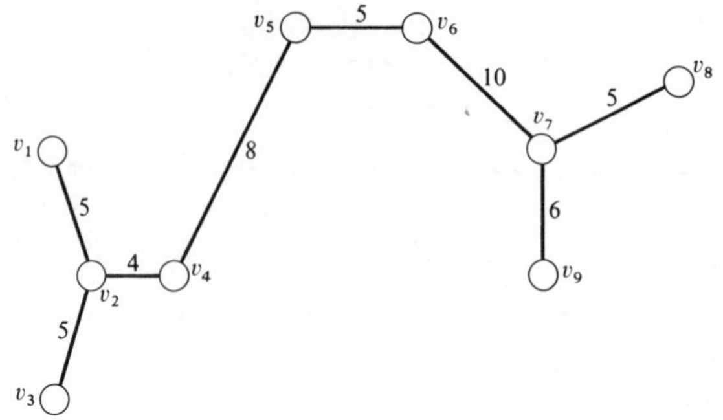

Fig. 6.3.5 Shortest tree for G in Fig. 6.3.3.

Let $\hat{\Gamma}$ be the set of branches of G which are not yet chosen, but which are connected either to a vertex of V_1 or to a branch already chosen. Choose the shortest branch from $\hat{\Gamma}$ which does not form any circuits with the branches already chosen.

Construction 6.3.3. Starting with a complete graph, perform the following step as many times as possible. Among the branches not yet removed, remove the longest branch whose removal will not disconnect the graph.

Construction 6.3.1 is a variation of Operations 6.3.1 and 6.3.2, and Construction 6.3.2 is a special case of Operation 6.3.2. The validity of Construction 6.3.3 follows as a corollary to Lemmas 6.3.1 and 6.3.2 and is left as an exercise.

6.4 MINIMUM COST FLOWS

In our treatment of maximizing flow in Chapter 3, we did not consider cost as a parameter. The previous techniques are now modified to yield flow patterns which maximize $f_{s,t}$ at minimum cost subject to integer capacity constraints and conservation of flow. Specifically we associate with each branch (i,j) a nonnegative cost per unit of flow $h(i,j)$ and define the cost function

$$\sum_{(i,j)\in\Gamma} h(i,j)f(i,j),$$

denoted by $h(G;\mathscr{F})$ or simply $h(\mathscr{F})$. We then seek a feasible integer flow pattern $\mathscr{F}$ to maximize $f_{s,t}$ with a minimum value for the cost $h(\mathscr{F})$. The resulting algorithm can also be used with minor modification to obtain any integer value $f_{s,t} < \tau_{s,t}$ at minimum cost. We shall also see that with minor changes the same flow algorithm can be used if $h(f(i,j))$ is a nonnegative *convex* function of $f(i,j)$ with $h(f(i,j)) = 0$ for $f(i,j) = 0$.

We will show that to maximize the flow at minimum cost we can increase flow along the least cost augmentation path, where the cost of a branch is defined in an appropriate fashion relative to the flow in the graph at a given stage of the calculation. We can first make a simplification in the nature of the allowable flows. In the previous flow problems there was no objection to having $f(i,j)$ and $f(j,i)$ both nonzero

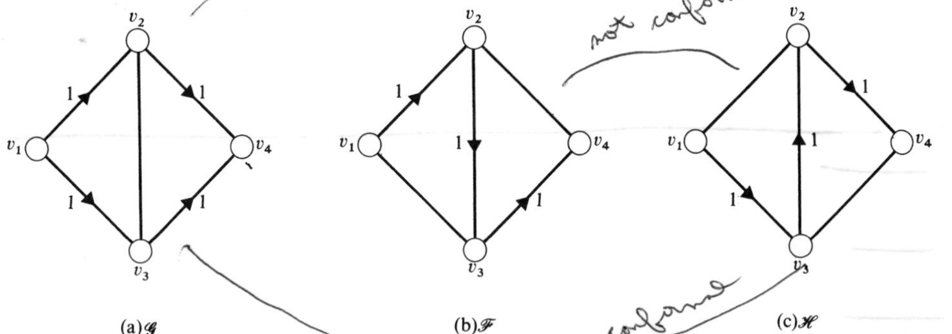

Fig. 6.4.1 $\mathscr{F}$ and $\mathscr{H}$ are not conformal.

for a given flow pattern. However, for the minimum cost flow problem it is clear that if $f(i,j) - f(j,i) > 0$ then the two flows should be replaced by a single flow $g(i,j) = f(i,j) - f(j,i)$ with $g(j,i) = 0$. In this way conservation of flow and the capacity constraints are still satisfied and the cost of the flow is lowered. We therefore assume that $f(i,j) \cdot f(j,i) = 0$ for all (i,j). Thus, either $f(i,j) \neq 0$ or $f(j,i) \neq 0$, but not both. With this assumption we can define two flow patterns $\mathscr{F}$ and $\mathscr{G}$ as being *conformal* if for any (i,j), $f(i,j) > 0$ then $g(j,i) = 0$. It is important to realize that if $\mathscr{F}$ and $\mathscr{G}$ are conformal and $\mathscr{G}$ and $\mathscr{H}$ are conformal then $\mathscr{F}$ and $\mathscr{H}$ are not necessarily conformal. As an example, $\mathscr{F}$ and $\mathscr{G}$ and $\mathscr{H}$ and $\mathscr{G}$ of Fig. 6.4.1 are conformal, but $\mathscr{F}$ and $\mathscr{H}$ are not conformal. When we say that $\mathscr{F}_1, \mathscr{F}_2, \ldots, \mathscr{F}_k$ are conformal we mean that any pair of the flow patterns is conformal. This concept of conformal flow patterns will be crucial in the proof of the optimality of the minimum cost algorithm to be presented.

We now transform the problem into a shortest path problem. Let the *up-cost* $u(i,j)$ of a branch (i,j) be the cost of sending an additional unit of flow through the branch and let the *down-cost* $\Delta(i,j)$ be the cost (a negative number) of reducing the flow in (i,j) by one unit. These costs are

$$u(i,j) = \begin{cases} h(i,j) & \text{if } 0 \leq f(i,j) < c(i,j) \text{ and } f(j,i) = 0, \\ \infty & \text{otherwise.} \end{cases} \tag{6.4.1a}$$

$$\Delta(i,j) = \begin{cases} -h(i,j) & \text{if } f(j,i) > 0, \\ \infty & \text{otherwise.} \end{cases} \tag{6.4.1b}$$

Definition 6.4.1 *Let $l(i,j)$ be the effective length of (i,j) relative to a flow pattern $\mathscr{F}$ and s-t path π with value*

$$l(i,j) = u(i,j) \qquad \textit{if } f(i,j) \textit{ is a forward flow in } \pi, \tag{6.4.2a}$$

$$l(i,j) = \Delta(i,j) \qquad \textit{if } f(i,j) \textit{ is a backward flow in } \pi. \tag{6.4.2b}$$

The above definition of length is unusual in several respects. The lengths depend upon the flows and hence must be recalculated at each stage of the algorithm. To make this fact explicit, we sometimes write $l(i,j)$ as $l((i,j);\mathscr{F})$. Similarly, the length of a path π may be written as $l(\pi;\mathscr{F})$. Moreover, the length of a branch depends upon whether it is a backward or forward branch in a path. Finally some branches have infinite lengths. $l(i,j) = \infty$ will mean that $f(i,j)$ is not to be changed at a certain stage of the algorithm.

We can now state an algorithm for obtaining the minimum cost maximum flow pattern.

Unit Path Minimum Cost Flow Algorithm

Step 1. Assign any feasible minimum cost s-t flow pattern $\mathscr{F}_o$. The flow pattern in which all flows are zero is permissible.

Step 2. Using the Shortest Path Algorithm find an s-t path π_1 with the shortest length $l(\pi_i;\mathscr{F}_i)$. (Since $\mathscr{F}_i$ is minimum cost there are no circuits of negative length. Hence the Shortest Path Algorithm can be used.) In the algorithm the length of a branch is its up-cost if it is used as a forward branch and its down-cost if it is used as a backward branch. If no s-t path of finite length exists, terminate. Otherwise proceed to Step 3.

Step 3. Construct a flow pattern $\mathscr{F}_{i+1} = \mathscr{F}_i \oplus 1(\pi_i)$. Replace i by $i + 1$ and return to Step 2.

We now illustrate the use of the algorithm by calculating a minimum cost maximum flow pattern. The validity of the algorithm will then be proved.

Example 6.4.1 Consider the graph in Fig. 6.4.2(a) with flow pattern $\mathscr{F}_i$. Associated with each branch (k,j) are the following five numbers in order: $c(k,j)$, $f(k,j)$, $h(k,j)$, $u(k,j)$, and $\Delta(k,j)$. Using Definition 6.4.1 of length, the Shortest Path Algorithm finds the s-t path v_s,v_2,v_1,v_t, as the "shortest" augmentation path. The flow pattern $\mathscr{F}_{i+1}$, derived by Step 3 of the Unit Path Minimum Cost Flow Algorithm, is given in

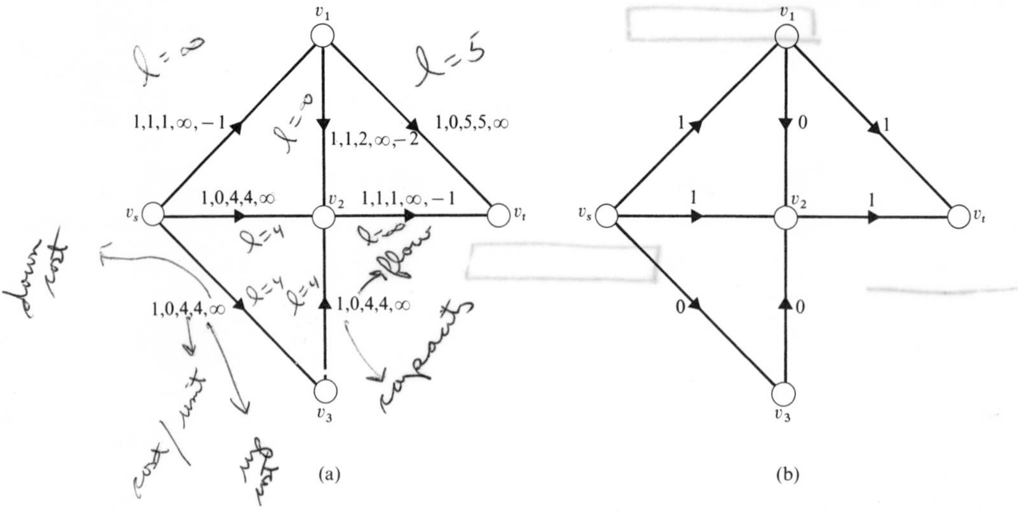

Fig. 6.4.2 Graphs for Example 6.4.1.

Fig. 6.4.2(b). The Labeling Algorithm alternatively might have detected the path v_s, v_3, v_2, v_t and resulted in a flow pattern different from that shown in Fig. 6.4.2(b), but with the same cost.

The validity of the procedure is established by Theorems 6.4.1 and 6.4.2. As a preliminary to Theorem 6.4.2 we first prove Lemmas 6.4.1, 6.4.2, and 6.4.3, which are of interest themselves. Theorem 6.4.1 deals with feasibility and Theorem 6.4.2 deals with cost.

Theorem 6.4.1 *A feasible flow pattern $\mathscr{F}$ is maximum if and only if $l(\pi_{s,t};\mathscr{F}) = \infty$ for every s-t path $\pi_{s,t}$.*

Proof. If $l(\pi;\mathscr{F})$ is finite no forward branch of π is saturated and no backward branch is empty. Therefore, the flow in each branch of π can be changed by at least one unit. Therefore, $\mathscr{F} \oplus 1(\pi)$ is feasible if $l(\pi;\mathscr{F})$ is finite. Conversely if $\mathscr{F} \oplus 1(\pi)$ is feasible no forward branch of π is saturated, no backward branch is empty, and $l(\pi;\mathscr{F})$ is finite.

We then immediately have:

Lemma 6.4.1 *If $\mathscr{F}$ is a feasible s-t flow pattern and $\mathscr{G}_1, \mathscr{G}_2, \ldots, \mathscr{G}_k$ are conformal s-t flow patterns such that $\mathscr{G}_1 \oplus \mathscr{G}_2 \oplus \cdots \oplus \mathscr{G}_k \oplus \mathscr{F}$ is feasible then $\mathscr{F} + \mathscr{S}$ is also feasible if $\mathscr{S} = \mathscr{G}_{i_1} \oplus \mathscr{G}_{i_2} \oplus \cdots \oplus \mathscr{G}_{i_j}$ where the $\mathscr{G}_{i_q}$'s are any subset of the $\mathscr{G}_i$'s.*

Proof. For a given branch (i,j), let $\hat{f} = f(i,j) - f(j,i)$, $\hat{s} = s(i,j) - s(j,i)$, and $\hat{g}_k = g_k(i,j) - g_k(j,i)$. We must have either

$$\hat{f} \leq \hat{f} + \hat{s} \leq \hat{f} + \hat{g}_1 + \hat{g}_2 + \cdots + \hat{g}_k \tag{6.4.3}$$

or

$$\hat{f} + \hat{g}_1 + \hat{g}_2 + \cdots + \hat{g}_k \leq \hat{f} + \hat{s} \leq \hat{f}. \tag{6.4.4}$$

The feasibility assumption also implies that

$$-c(j,i) \leq \hat{f} \leq c(i,j) \tag{6.4.5}$$

and

$$-c(j,i) \leq \hat{f} + \hat{g}_1 + \hat{g}_2 + \cdots + \hat{g}_k \leq c(i,j). \tag{6.4.6}$$

Hence

$$-c(j,i) \leq \hat{f} + \hat{s} \leq c(i,j). \tag{6.4.7}$$

But since (i,j) is an arbitrary branch this shows that $\mathscr{F} \oplus \mathscr{S}$ is feasible.

Lemma 6.4.2 *For a graph with nonnegative costs per unit flow, if $\mathscr{F}$, $\mathscr{G}$, and $\mathscr{K}$ are s-t flow patterns and $\mathscr{G}$ and $\mathscr{K}$ are conformal, then*

$$h(\mathscr{F} \oplus \mathscr{G} \oplus \mathscr{K}) - h(\mathscr{F} \oplus \mathscr{G}) \geq h(\mathscr{F} \oplus \mathscr{K}) - h(\mathscr{F}). \tag{6.4.8}$$

Proof. Let

$$Q = h(\mathscr{F} \oplus \mathscr{G} \oplus \mathscr{K}) - h(\mathscr{F} \oplus \mathscr{G}) - h(\mathscr{F} \oplus \mathscr{K}) \oplus h(\mathscr{F}). \tag{6.4.9}$$

We wish to prove that $Q \geq 0$ under the assumption of the lemma. For a branch (i,j), let $\hat{f} = f(i,j) - f(j,i)$, $\hat{g} = g(i,j) - g(j,i)$, and $\hat{k} = k(i,j) - k(j,i)$. Define $Q(i,j)$ as

$$Q(i,j) = |\hat{f} + \hat{g} + \hat{k}|\bar{h}(i,j) - |\hat{f} + \hat{g}|\bar{h}(i,j) - |\hat{f} + \hat{k}|\bar{h}(i,j) + |\hat{f}|\bar{h}(i,j) \quad (6.4.10)$$

where $\bar{h}(i,j) = h(i,j)$ if $f(i,j) \geq 0$ and $\bar{h}(i,j) = h(j,i)$ if $f(j,i) > 0$.

$Q(i,j)$ represents the contribution of (i,j) and (j,i) to Q. Hence it is sufficient to show that $Q(i,j) \geq 0$. Assume, without loss of generality, that $\hat{f}(i,j) > 0$. Either (6.4.11a) or (6.4.11b) is true since $\mathscr{G}$ and $\mathscr{K}$ are conformal:

$$f(i,j) \geq 0,\ g(i,j) \geq 0,\ k(i,j) \geq 0; \quad (6.4.11a)$$

$$f(i,j) \geq 0,\ g(j,i) \geq 0;\ k(j,i) \geq 0. \quad (6.4.11b)$$

If (6.4.11a) holds, then

$$\begin{aligned} Q(i,j) = {} & (f(i,j) + g(i,j) + k(i,j))h(i,j) - (f(i,j) + g(i,j))h(i,j) \\ & -(f(i,j) + k(i,j))h(i,j) + f(i,j)h(i,j) = 0, \end{aligned} \quad (6.4.12)$$

and the lemma is proved in this case.

If (6.4.11b) holds and

$$f(i,j) - g(j,i) - k(j,i) \geq 0, \quad (6.4.13)$$

then

$$f(i,j) - g(j,i) \geq 0 \quad (6.4.14a)$$

and

$$f(i,j) - k(j,i) \geq 0. \quad (6.4.14b)$$

Hence $Q(i,j)$ is again zero as in (6.4.12).

The only remaining possibility is that (6.4.11b) holds and $f(i,j) - g(j,i) - k(j,i) \leq 0$. In this case, $Q(i,j)$ reduces to one of the following expressions:

$$-[\hat{f} + \hat{g} + \hat{k}]h(j,i) - [\hat{f} + \hat{g}]h(i,j) - [\hat{f} + \hat{k}]h(i,j) + [\hat{f}]h(i,j); \quad (6.4.15)$$

$$-[\hat{f} + \hat{g} + \hat{k}]h(j,i) + [\hat{f} + \hat{g}]h(j,i) - [\hat{f} + \hat{k}]h(i,j) + [\hat{f}]h(i,j); \quad (6.4.16)$$

$$-[\hat{f} + \hat{g} + \hat{k}]h(j,i) - [\hat{f} + \hat{g}]h(i,j) + [\hat{f} + \hat{k}]h(j,i) + [\hat{f}]h(i,j); \quad (6.4.17)$$

$$-[\hat{f} + \hat{g} + \hat{k}]h(j,i) + [\hat{f} + \hat{g}]h(j,i) + [\hat{f} + \hat{k}]h(j,i) + [\hat{f}]h(i,j). \quad (6.4.18)$$

When simplified, (6.4.15) to (6.4.18) reduce to (6.4.19) to (6.4.22), respectively.

$$-[\hat{f} + \hat{g} + \hat{k}][h(i,j) + h(j,i)]; \quad (6.4.19)$$

$$-[\hat{k}][h(i,j) + h(j,i)]; \quad (6.4.20)$$

$$-[\hat{g}][h(i,j) + h(j,i)]; \quad (6.4.21)$$

$$+[\hat{f}][h(i,j) + h(j,i)]. \quad (6.4.22)$$

From (6.4.11) and the facts that

$$[\hat{f} + \hat{g} + \hat{h}] < 0 \tag{6.4.23a}$$

and

$$[h(i,j) + h(j,i)] \geq 0 \tag{6.4.23b}$$

we see that all expressions in (6.4.19) to (6.4.22) are nonnegative. Hence the proof is completed.

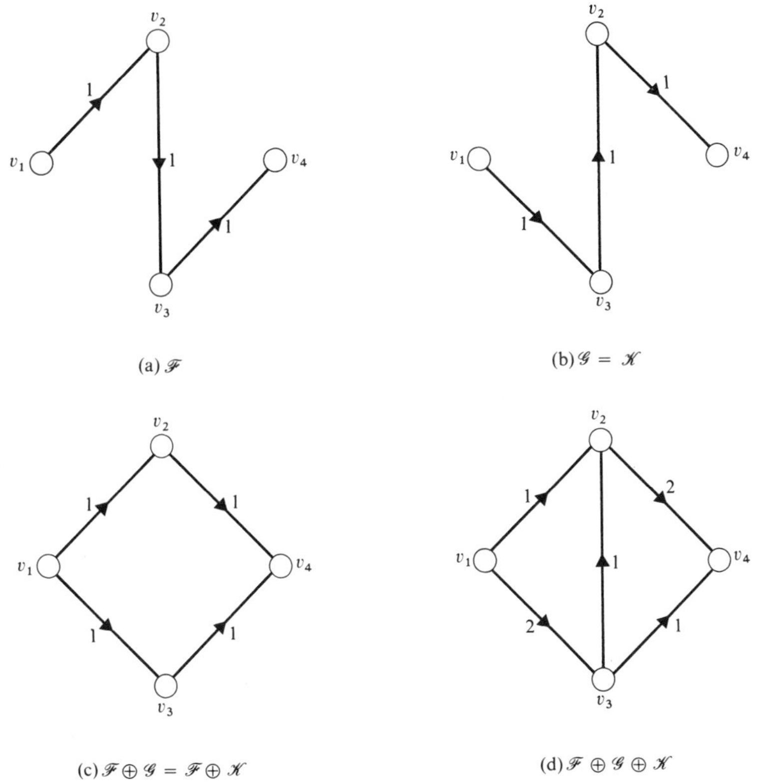

Fig. 6.4.3 Graphs for Example 6.4.2.

Example 6.4.2 As an illustration of Lemma 6.4.2, consider the flow patterns $\mathscr{F}$ in Fig. 6.4.3(a) and $\mathscr{G} = \mathscr{K}$ in Fig. 6.4.3(b). Assume that $h(i,j) = 1$ for all (i,j). Then $\mathscr{F} \oplus \mathscr{G} = \mathscr{F} \oplus \mathscr{K}$ are shown in Fig. 6.4.3(c) and $\mathscr{F} \oplus \mathscr{G} \oplus \mathscr{K}$ is in Fig. 6.4.3(d). We have

$$h(\mathscr{F} \oplus \mathscr{G} \oplus \mathscr{K}) - h(\mathscr{F} \oplus \mathscr{G}) = 7 - 4 = 3 \tag{6.4.24a}$$

and

$$h(\mathscr{F} \oplus \mathscr{K}) - h(\mathscr{F}) = 4 - 3 = 1. \tag{6.4.24b}$$

Lemma 6.4.3 *If $\mathscr{F}$ is a feasible flow pattern and π is an s-t path such that $l(\pi;\mathscr{F})$ is finite, then*

$$h(\mathscr{F} \oplus 1(\pi)) - h(\mathscr{F}) = l(\pi;\mathscr{F}). \tag{6.4.25}$$

Proof. Consider any forward branch (i,j) in π. If $f(i,j) \geq 0$, the flow in (i,j) corresponding to $\mathscr{F} \oplus 1(\pi)$ is $f(i,j) + 1$ and $h(\mathscr{F})$ is increased in the branch by the amount $h(i,j)$. If $f(i,j) > 0$ in a backward branch, the flow in (i,j) corresponding to $\mathscr{F} \oplus 1(\pi)$ is $f(i,j) - 1$. Hence $h(\mathscr{F})$ is decreased in this branch by $h(i,j)$. In either case the increase in $h(\mathscr{F})$ with respect to (i,j) is $l((i,j);\mathscr{F})$ as defined and the proof is completed.

From Lemmas 6.4.1, 6.4.2 and 6.4.3 we can prove Theorem 6.4.2.

Theorem 6.4.2 *If $\mathscr{F}$ is a minimum cost s-t flow pattern with value $f_{s,t} < \tau_{s,t}$ and π is a shortest s-t path relative to $\mathscr{F}$ then $\mathscr{F} \oplus 1(\pi)$ is a minimum cost s-t flow pattern with value $f_{s,t} + 1$.*

Proof. Let $\mathscr{G}$ be a minimum cost s-t flow pattern with value $f_{s,t} + 1$. By the Semipath Decomposition Algorithm $\mathscr{G} \ominus \mathscr{F}$ can be written as $1(\pi_1) \oplus \mathscr{Z}_1$ where π_1 is an s-t path, $\mathscr{Z}_1$ is a zero flow pattern, and $1(\pi_1)$ and $\mathscr{Z}_1$ are conformal.[1] Since $\mathscr{F} \oplus 1(\pi_1) \oplus \mathscr{Z}_1$ is feasible, Lemma 6.4.1 asserts that $\mathscr{F} \oplus 1(\pi_1)$ and $\mathscr{F} \oplus \mathscr{Z}_1$ are feasible. Therefore from Lemma 6.4.2

$$h(\mathscr{G}) - h(\mathscr{F} \oplus 1(\pi_1)) \geq h(\mathscr{F} \oplus \mathscr{Z}_1) - h(\mathscr{F}). \tag{6.4.26}$$

Since $\mathscr{F} \oplus \mathscr{Z}_1$ is feasible and by assumption $\mathscr{F}$ is feasible and minimum cost,

$$h(\mathscr{F} \oplus \mathscr{Z}_1) - h(\mathscr{F}) \geq 0. \tag{6.4.27}$$

Hence, from (6.4.26) and (6.4.27),

$$h(\mathscr{G}) \geq h(\mathscr{F} \oplus 1(\pi_1)). \tag{6.4.28}$$

But by Lemma 6.4.2 $\mathscr{F} \oplus 1(\pi_1)$ is also feasible with value $f_{s,t} + 1$. It follows that $\mathscr{F} \oplus 1(\pi_1)$ is also a minimum cost feasible s-t flow pattern. Therefore π_1 must minimize $l(\pi_1;\mathscr{F})$. Otherwise there exists an s-t path π_2 such that $l(\pi_2;\mathscr{F}) < l(\pi_1;\mathscr{F})$ and Lemma 6.4.3 would imply that $l(\mathscr{F} \oplus 1(\pi_2)) < l(\mathscr{F} \oplus 1(\pi_1))$. Also from Lemma 6.4.3, if π_i is any other path with $l(\pi_i;\mathscr{F}) = l(\pi_1;\mathscr{F})$ then $\mathscr{F} \oplus 1(\pi_i)$ is a minimum cost feasible s-t flow pattern whose value is $f_{s,t} + 1$. Hence the proof is completed.

Theorems 6.4.1 and 6.4.2 comprise a justification of the validity of the Unit Path Minimum Cost Flow Algorithm. We now show that we need not seek unit paths for augmentation at each stage of the calculation; in other words we can use the Minimum Cost Flow Algorithm below to find the minimum cost maximum s-t flow. The Minimum Cost Flow Algorithm is obtained from the Unit Path Minimum Cost Flow Algorithm by replacing Step 3 by the following:

Step 3′. Construct a flow pattern $\mathscr{F}_{i+a} = \mathscr{F}_i \oplus a(\pi_i)$ where $a \leq c(\pi_i)$. Replace i by $i + a$ and return to Step 2.

---

[1] It should be noted that $\mathscr{F} \ominus \mathscr{G}$ is not necessarily a feasible flow.

The justification of the algorithm is in Corollary 6.4.1.

Corollary 6.4.1 *If $\mathscr{F}$ is a minimum cost s-t flow pattern of value $k < \tau_{s,t}$ and π is any s-t path of shortest length relative to $\mathscr{F}$, then $\mathscr{F} \oplus a(\pi)$ is a minimum cost flow pattern of value $k + a$, where $a = 1,2,\ldots,c(\pi)$.*

Proof. By sending $a = 1,2,\ldots,c(\pi)$ units of flow along π, the resulting flow pattern $\mathscr{F} \oplus a(\pi)$ is clearly feasible and has value $k + a$. The only question is whether it is minimum cost. Assume there is an $a < c(\pi)$ such that there exists a path π' for which

$$h[\mathscr{F} \oplus a(\pi)] > h[\mathscr{F} \oplus (a - 1)(\pi) \oplus 1(\pi')]. \tag{6.4.29}$$

Then, for $f_{s,t} = k + a - 1$, π' is shorter than π. But for all $a \le c(\pi)$, the cost of adding a unit flow to π is the same and hence at flow k, $h(\pi') < h(\pi)$, which is a contradiction.

We now observe that by simply extending the definition of up-costs and down-costs we can apply the Unit Path Minimum Cost Flow Algorithm to find the integer minimum cost maximum flow for a graph in which the branch costs $h(f(i,j))$ are nonnegative convex functions of $f(i,j)$ such that $f(i,j) = 0$ implies $h(f(i,j)) = 0$.

Consider a particular branch (i,j); $h(f)$ will be an abbreviation for $h(f(i,j))$. Define the *up-cost* $u(i,j)$ of branch (i,j) as the cost of sending an additional unit of flow in the branch (i,j) and the *down-cost* $\Delta(i,j)$ as the cost of removing a unit of flow in (i,j).

$$u(i,j) = \begin{cases} h(f+1) - h(f) & \text{if } 0 \le f(i,j) < c(i,j), \\ \infty & \text{otherwise.} \end{cases} \tag{6.4.30}$$

$$\Delta(i,j) = \begin{cases} -h(f) + h(f-1) & \text{for } f(i,j) > 0, \\ \infty & \text{otherwise.} \end{cases} \tag{6.4.31}$$

The intuitive reason the algorithm works with these definitions is that the up-cost is always nonnegative and the down-cost is always nonpositive. To see this, consider the meaning of convexity: $h(f)$ is convex if for all $\lambda_1,\lambda_2 \ge 0$ such that $\lambda_1 + \lambda_2 = 1$,

$$\lambda_1 h(x_1) + \lambda_2 h(x_2) \ge h(\lambda_1 x_1 + \lambda_2 x_2). \tag{6.4.32}$$

Let

$$x_2 = f + 1,\ x_1 = 0,\ \lambda_2 = \frac{f}{f+1},\ \text{and } \lambda_1 = \frac{1}{f+1}. \tag{6.4.33}$$

Substituting these values in (6.4.32) gives

$$\frac{f}{f+1} h(f+1) \ge h(f). \tag{6.4.34}$$

Hence $h(f + 1) \ge h(f)$ and the up-cost is nonnegative. A similar argument can be used to show the down-cost is nonpositive.

Definition 6.4.1 of length remains unchanged. However $u(i,j)$ and $\Delta(i,j)$ are now given in (6.4.30) and (6.4.31) rather than (6.4.1) and (6.4.2). An illustration of the application of the Unit Path Minimum Cost Flow Algorithm is given in the next example.

Example 6.4.3 Consider the graph in Fig. 6.4.4(a) in which the weight assigned to each branch (i,j) is $\alpha(i,j)$ and the cost function for each branch (i,j) is $\alpha(i,j)f(i,j)^2$. Assume that all branch capacities are unity and that we are given the initial minimum cost flow pattern in Fig. 6.4.4(b). The up-costs and down-costs of every branch are given in Fig. 6.4.4(c) with the up-cost as the first number in each case.

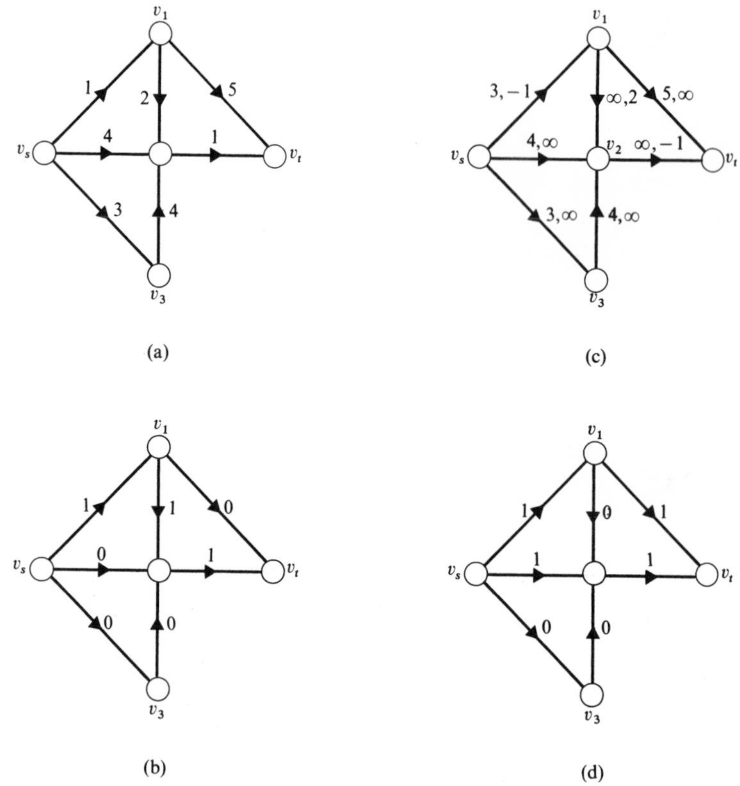

Fig. 6.4.4 Graphs for Example 6.4.3.

For example, the up-cost of $(1,t)$ is $5(1)^2 - 5(0)^2 = 5$. We therefore send one additional unit of flow along the branches $(s,2)$, $(2,1)$, and $(1,t)$. The effective length of this path is $4 + (-2) + 5 = 7$. This yields the final flow pattern shown in Fig. 6.4.4(d).

In order to prove that the Unit Path Minimum Cost Flow Algorithm yields a minimum cost integer flow pattern we recognize that the only proof which must be changed is that of Lemma 6.4.2.

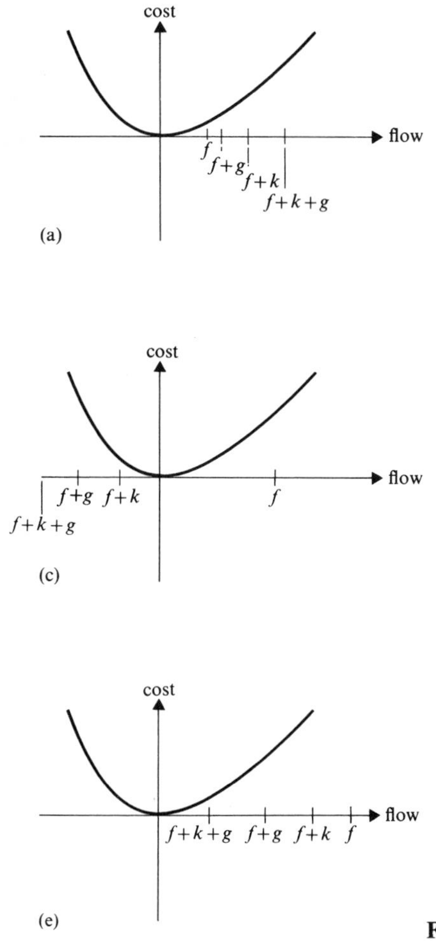

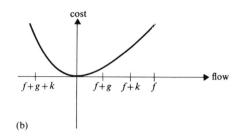

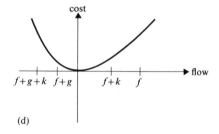

Fig. 6.4.5 Graphs for proof of Lemma 6.4.5.

Lemma 6.4.5 *For all* $(i,j) \in \Gamma$, *let the branch cost of* (i,j) *be a convex function* $h(f(i,j))$ *with* $h(0) = 0$. *If* $\mathscr{F}$, $\mathscr{G}$, *and* $\mathscr{K}$ *are s-t flow patterns and* $\mathscr{G}$ *and* $\mathscr{K}$ *are conformal, then*

$$h(\mathscr{F} \oplus \mathscr{G} \oplus \mathscr{K}) - h(\mathscr{F} \oplus \mathscr{G}) \geq h(\mathscr{F} \oplus \mathscr{K}) - h(\mathscr{F}). \tag{6.4.35}$$

Proof. Depending upon the directions and relative values of j, k, and g (defined as in Lemma 6.4.1) the flows and cost functions can have the possible appearances in cases (a), (b), (c), (d), and (e) in Fig. 6.4.5. In Fig. 6.4.5 the cost functions are convex and g and k can be interchanged.

In case (a) first apply (6.4.31) with

$$x_1 = f, x_2 = f + g + k, \lambda_1 = \frac{g}{g + k}, \lambda_2 = \frac{k}{g + k} \tag{6.4.36}$$

and then with

$$x_1 = f, x_2 = f + g + k, \lambda_1 = \frac{k}{g + k}, \lambda_2 = \frac{g}{g + k}. \tag{6.4.37}$$

The results are

$$\frac{g}{g + k} h(f + g + k) + \frac{k}{g + k} h(f) \geq h(f + g) \tag{6.4.38a}$$

and

$$\frac{k}{g + k} h(f + g + k) + \frac{g}{g + k} h(f) \geq h(f + k). \tag{6.4.38b}$$

Adding (6.4.38a) and (6.4.38b) yields

$$h(f + g + k) + h(f) \geq h(f + g) + h(f + k). \tag{6.4.39}$$

In cases (b) through (e) first apply (6.4.32) with

$$x_1 = f + g + k, \quad x_2 = f, \quad \lambda_1 = \frac{g}{g + k}, \quad \text{and} \quad \lambda_2 = \frac{k}{g + k} \tag{6.4.40a}$$

and then with

$$x_1 = f + g + k, \quad x_2 = f, \quad \lambda_1 = \frac{k}{g + k}, \quad \text{and} \quad \lambda_2 = \frac{g}{g + k}. \tag{6.4.40b}$$

We again obtain (6.4.39) and the proof is completed.

Finally, it should be noted that with the up-cost and down-cost as defined in (6.4.30) and (6.4.31) we have essentially approximated the cost function by a staircase function in which the distance between steps is one unit of flow.

6.5 DISTANCES ON A GRAPH

Consider a graph G which is to be the model of a communications system or highway or traffic network. That is, the branches of G correspond to wires or highways or roads and the vertices correspond to switching centers, intersections, communities, or cities. For such systems it is natural to associate real, nonnegative numbers with the branches of G such that the weight l_i of branch b_i for $i = 1, \ldots, m$ is the length of that branch. Let π_i consisting of $b_{i_1}, \ldots, b_{i_r}$ be a path between an arbitrary pair of vertices v_j and v_k. As in Section 2 we will say that the length of path π_i, written $l(\pi_i)$, is the sum of the weights of the branches in the path. In other words,

$$l(\pi_i) = \sum_{j=1}^{r_i} l_{i_j}. \tag{6.5.1}$$

The idea of the distance between two points may now be examined. In a physical situation, we tend to think of the distance between two points in terms of the length of the shortest path between these points. This idea generalizes naturally to the idea of distance on a graph. Here, we will say that the distance between any two vertices v_j and v_k is the length of the shortest path connecting these vertices.

Definition 6.5.1 *Let $\{\pi_1, \pi_2, \ldots, \pi_{q_{j,k}}\}$ be the set of paths connecting vertex v_j to vertex v_k. Then, the distance $d(v_j, v_k)$ from v_j to v_k is*

$$d(v_j, v_k) = \min_{1 \le i \le q_{j,k}} [l(\pi_i)]. \tag{6.5.2}$$

Note that, if the graph G is undirected, any path from v_j to v_k is also a path from v_k to v_j and hence $d(v_j, v_k) = d(v_k, v_j)$. For directed graphs this is not necessarily true. Also, we adopt the convention that $d(v_j, v_j) = 0$ for all j. Finally, if there is no j-k path in G, we will say that $d(v_j, v_k) = \infty$.

Suppose that shortest s-t paths of G are generated to compute the distance between each pair of vertices. This information may be tabulated in a $n \times n$ matrix array, $\mathscr{D}$, which we will call a *distance matrix.* $\mathscr{D} = [d_{i,j}]$ has the properties that its main diagonal elements are identically zero and its i-jth element $d_{i,j}$ is the distance $d(v_i, v_j)$ between vertices v_i and v_j. Furthermore, if G is undirected then $\mathscr{D}$ is symmetric. As an example, consider the undirected graph G shown in Fig. 6.5.1. The distance matrix $\mathscr{D}$ of G is found by the techniques of Section 2 to be

$$\mathscr{D} = \begin{bmatrix} 0 & 1 & 2 & 3 & 2 \\ 1 & 0 & 3 & 2 & 1 \\ 2 & 3 & 0 & 2 & 3 \\ 3 & 2 & 2 & 0 & 1 \\ 2 & 1 & 3 & 1 & 0 \end{bmatrix}. \tag{6.5.3}$$

Given a graph G, one may find its distance matrix $\mathscr{D}$. Alternatively, given an $n \times n$ matrix $\mathscr{D}$, one may ask, under what conditions does there exist a graph G with distance matrix $\mathscr{D}$? Hence, we pose the question: Given a matrix $\mathscr{D}$ with real, nonnegative entries, when is $\mathscr{D}$ a distance matrix? This question is answered by the following theorem.

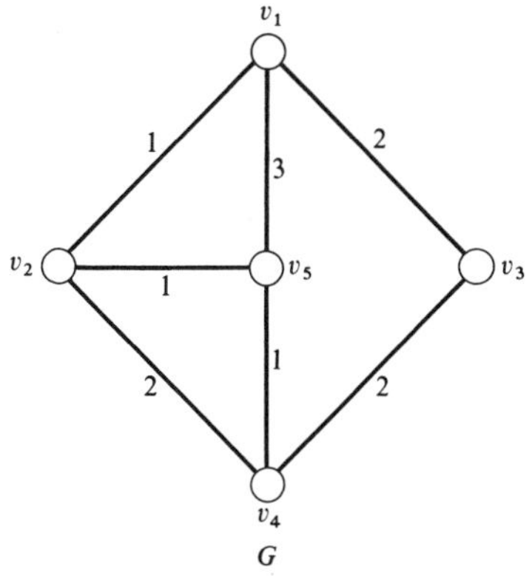

Fig. 6.5.1 Realization of a distance matrix.

Theorem 6.5.1 *The necessary and sufficient conditions for an $n \times n$ matrix $\mathscr{D} = [d_{i,j}]$ with real, nonnegative entries to be a distance matrix of a graph are:*

$$a)\ d_{i,i} = 0 \qquad \text{for } i = 1,2,\ldots,n, \tag{6.5.4a}$$

$$b)\ d_{i,j} + d_{j,k} \geq d_{i,k} \qquad \text{for } i,j, \text{ and } k = 1,\ldots,n. \tag{6.5.4b}$$

Proof. The necessity of the above condition (6.5.4) simply states that distance must satisfy the usual axioms of a metric, including a triangle inequality for distances. The necessity of (6.5.4a) is a consequence of the convention $d(v_i,v_i) = 0$ for all i. To see the necessity of (6.5.4b), suppose G is a graph with distance matrix $\mathscr{D}$ and for some i, j, and k, $d_{i,j} + d_{j,k} < d_{i,k}$. This means that in G the minimum length path from v_i to v_k is longer than the sum of the minimum length paths from v_i to v_j and from v_j to v_k. This is impossible since these two paths themselves form a path from v_i to v_k (via v_j) with length less than $d_{i,k}$. To prove that conditions (a) and (b) imply the existence of a graph with distance matrix $\mathscr{D}$, we construct an n-vertex graph G' as follows. Pick n points and label them $v_1,v_2,\ldots,v_n$. For each entry $d_{i,j}$ of $\mathscr{D}$, connect a branch (i,j) of weight $d_{i,j}$ from v_i to v_j. Let G' have a distance matrix $\mathscr{D}' = [d'_{i,j}]$. Clearly, $d'_{i,j} \leq d_{i,j}$, and if G' is not a realization of $\mathscr{D}$, then for some i and j, $d'_{i,j} < d_{i,j}$. This implies there exists a path $\pi' = \pi'_{i,j}$ in G' such that $l(\pi') < d_{i,j}$.

Suppose $\pi'_{i,j}$ is the shortest i-j path in G' for all i,j. We will show $l(\pi'_{i,j}) = d_{i,j}$ by induction on the number of branches in π'. Any path $\pi'_{i,j}$ with only one branch consists of (i,j), and $l(\pi'_{i,j}) = d'_{i,j} = d_{i,j}$. Suppose that $l(\pi'_{i,j}) = d_{i,j}$ for all shortest paths which contain $m - 1$ branches for $m \leq n$, and suppose π' contains m branches. Then, $\pi'_{i,j} = \pi'_{i,k} \cup \{(k,j)\}$. By the inductive hypothesis $l(\pi'_{i,k}) = d_{i,k}$ and since the weight of (k,j) is $d_{k,j}$ we have that $l(\pi'_{i,j}) = d_{i,k} + d_{k,j}$. Since $d_{i,k} + d_{k,j} \geq d_{i,j}$ by hypothesis, it follows that $d'_{i,j} \geq d_{i,j}$. Hence, $d'_{i,j} = d_{i,j}$ and $\mathscr{D}' = \mathscr{D}$. //

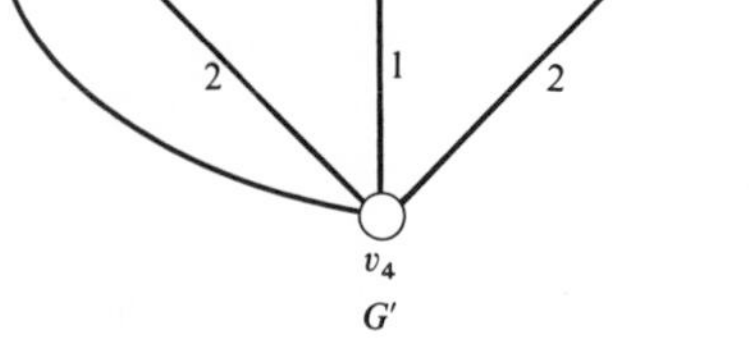

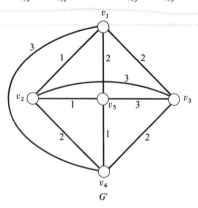

Fig. 6.5.2 Alternate realization of a distance matrix.

Figure 6.5.2 shows the graph G', obtained from the distance matrix $\mathscr{D}$ of G in Fig. 6.5.1, using the technique given in the proof of sufficiency of Theorem 6.5.1. It is clear that since G and G' are not identical, a distance matrix (i.e., any matrix satisfying the hypotheses of Theorem 6.5.1) *does not* have a unique realization. This

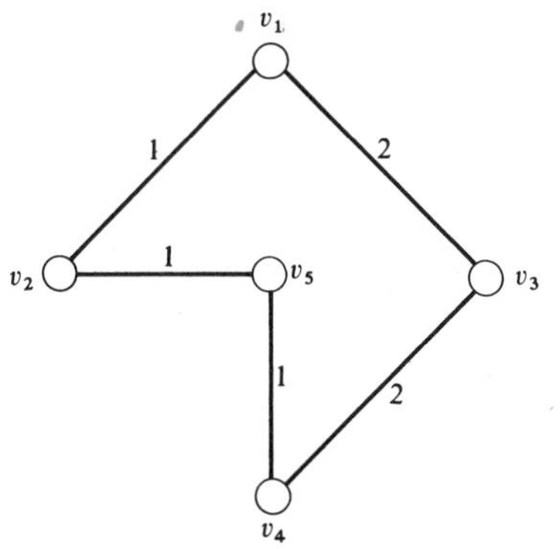

Fig. 6.5.3 Realization of a distance matrix with fewer branches than in Fig. 6.5.1.

result is not surprising. In fact, given a graph G with distance matrix $\mathscr{D}$ such that $(i,j) \notin \Gamma$, we can always construct a graph G' with the same distance matrix by adding branch (i,j), if the weight of (i,j) is at least $d_{i,j}$. All distances in G' will be computed without regard to (i,j) since the shortest i-j path still has length $d_{i,j}$ and all other shortest paths are unaffected.

In view of the above facts, we will say that a branch (i,j), with weight $l(i,j)$, is *redundant* if there exists a vertex v_k for $k \neq i,j$, such that $l(i,j) \geq d_{i,k} + d_{k,j}$. It should be clear that if (i,j) is redundant, it can be removed from G without affecting the distance matrix of G. For example, elements [1,5] and [2,4] in G of Fig. 6.5.1 may be removed from G to obtain the graph G' shown in Fig. 6.5.3 with the same distance matrix. It is now logical to pose the question: Is a realization of $\mathscr{D}$ without redundant elements unique? The answer is in the affirmative and is given in the next theorem.

Theorem 6.5.2 *If G is an n-vertex realization of the $n \times n$ distance matrix D, without redundant elements, then G is unique.*

Proof. Suppose G is not unique. Then, there exists an n-vertex graph G' without redundant branches such that G' and G are not identical. This means that either (a) at least one corresponding pair of branches $b(i,j)$ in G and $b'(i,j)$ in G' have different nonzero weights or (b) there is at least one branch $b(i,j)$ in G [or $b'(i,j)$ in G'] with no corresponding branch in G' (or G).

a) Since neither G nor G' contains redundant branches then $l'(i,j) = d_{i,j}$ and $l(i,j) = d_{i,j}$.
b) If $b(i,j)$ is not present in G, and $b'(i,j)$ *is* present in G', then there exists a vertex v_k such that $d_{i,k} + d_{k,j} = d_{i,j}$. This is true for both G and G' since they both have the same distance matrix. Then $l'[i,j] = d_{i,k} + d_{k,j}$ and $b'[i,j]$ is redundant.

DO

6.6 MINIMUM COST REALIZATIONS OF DISTANCE MATRICES OF UNDIRECTED GRAPHS

Theorem 6.5.2 states that an $n \times n$ distance matrix $\mathscr{D}$ has a unique n-vertex realization without redundant elements. This realization is, in a sense, the "best" possible one since all others must use more branches with a greater total length (more concrete

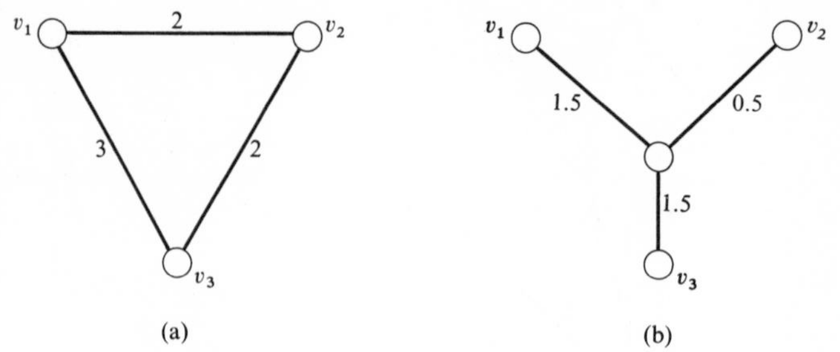

Fig. 6.6.1 Realization of $\mathscr{D}$ in (6.6.1).

must be used to construct a highway system with redundant branches). However, suppose the number of vertices of G is not restricted to be equal to the number of rows of the distance matrix. In other words, suppose we are given the distance matrix for only a set of "external" vertices of interest and are allowed to add additional "internal" vertices (intersections) at our discretion. For example, the graphs shown in Fig. 6.6.1 are both realizations of

$$\mathscr{D} = \begin{bmatrix} 0 & 2 & 3 \\ 2 & 0 & 2 \\ 3 & 2 & 0 \end{bmatrix}. \tag{6.6.1}$$

Let $G(\mathscr{D})$ be the family of graphs with distance matrix $\mathscr{D}$. Then, since in general, $G(\mathscr{D})$ has more than one element, it is meaningful to speak of an optimum realization of $\mathscr{D}$. Let $l(G)$ be the sum of the weights of the branches of G. We will say that G^* is an *optimum realization* of $\mathscr{D}$ if

$$l(G^*) = \min_{G \in G(\mathscr{D})} [l(G)]. \tag{6.6.2}$$

Finding an optimum realization of $\mathscr{D}$ is an unsolved problem. We will discuss some partial results for undirected graphs. We first note that given a realization of $\mathscr{D}$, we can construct an infinite number of new realizations by breaking branches and adding internal vertices of degree two. Similarly, internal vertices of degree one are not of interest. Therefore, assume that in any realization of $\mathscr{D}$, each internal vertex is adjacent to at least three external vertices. In this case, the following theorem is true.

Theorem 6.6.1 *If $\mathscr{D}$ is realizable as an undirected tree T, then T is the only tree realization of $\mathscr{D}$.*

Proof. We proceed by induction on the order of $\mathscr{D}$. If $\mathscr{D}$ is a 2×2 matrix the theorem is trivially true. Assume the theorem is true for $\mathscr{D}$ an $n - 1 \times n - 1$ matrix. Let $\mathscr{D}$ be an $n \times n$ matrix, supposing the vertices have been numbered so that v_n has degree one. Let $\mathscr{D}^*$ be the $n - 1 \times n - 1$ leading principal submatrix of $\mathscr{D}$. Suppose $\mathscr{D}$ is realizable as a tree T. Clearly this implies that $\mathscr{D}^*$ is realizable as a tree T' and by the inductive hypothesis T' is unique.

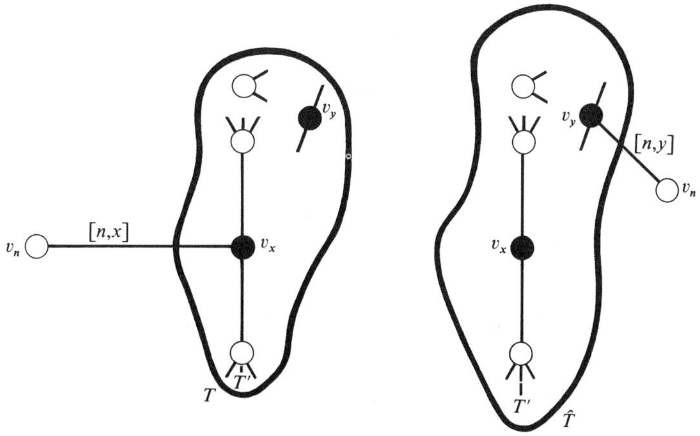

Fig. 6.6.2 Graphs used in proof of Theorem 6.6.1.

We obtain T from T' by adding the vertex v_n to T' and connecting v_n to T' at a point v_x by the branch $[n,x]$. This procedure is illustrated in Fig. 6.6.2. T is unique if the point v_x is unique. Let v_n be connected to a point v_y on T' by the branch $[n,y]$ such that the resulting tree $\hat{T}$ is a realization of $\mathscr{D}$. Suppose that the weights of $[n,x]$ and $[n,y]$ are $l[n,x]$ and $l[n,y]$, respectively.

It should be clear that on T', there is a vertex v_i such that on T

$$d_{n,i} = l[n,x] + d(v_x,v_i) \tag{6.6.3}$$

and on $\hat{T}$

$$d_{n,i} = l[n,y] + d(v_y,v_x) + d(v_x,v_i). \tag{6.6.4}$$

Combining these two equalities, we obtain

$$l[n,y] + d(v_y,v_x) = l[n,x]. \tag{6.6.5}$$

Similarly, we can select a vertex v_j on T' such that

$$d_{n,j} = l[n,y] + d(v_y,v_j) \tag{6.6.6a}$$

and on $\hat{T}$

$$d_{n,j} = l[n,x] + d(v_x,v_y) + d(v_y,v_j). \tag{6.6.6b}$$

Hence,

$$l[n,x] + d(v_y,v_x) = l[n,y]. \tag{6.6.7}$$

Since $d(v_x,v_y) = d(v_y,v_x)$, we find from (6.6.5) and (6.6.7) that

$$d(v_x,v_y) = 0 \text{ and } l[n,x] = l[n,y]. \tag{6.6.8}$$

Thus, T is unique.

We now give a procedure called the *elementary reduction cycle* which may be repeatedly applied to reduce the total weight of a realization. However, there is no guarantee that this procedure will result in an optimum realization. On the other hand, a modification of this procedure can be used to obtain the unique tree realization (if it exists), and, as will be seen, this realization is optimum.

Suppose that in the realization G of $\mathscr{D}$ there are three branches $[i,j]$, $[j,k]$, and $[k,i]$ such that $l[i,j] + l[j,k] > l[i,k]$, $l[i,j] + l[i,k] > l[j,k]$, and $l[i,k] + l[j,k] > l[i,j]$. Then, there exists a graph G' such that $l(G) > l(G')$. We obtain G' from G by first deleting the aforementioned branches and adding the vertex v_{n+1}. We then connect v_{n+1} to v_i, v_j, and v_k with branches $[n+1,i]$, $[n+1,j]$, and $[n+1,k]$ with weights $\frac{1}{2}\{l[i,j] + l[i,k] - l[j,k]\}$, $\frac{1}{2}\{l[i,j] + l[j,k] - l[i,k]\}$, and $\frac{1}{2}\{l[i,k] + l[j,k] - l[i,j]\}$, respectively. Clearly G' realizes $\mathscr{D}$ and $l(G') = l(G) - \frac{1}{2}\{l[i,j] + l[j,k] + l[i,k]\}$.

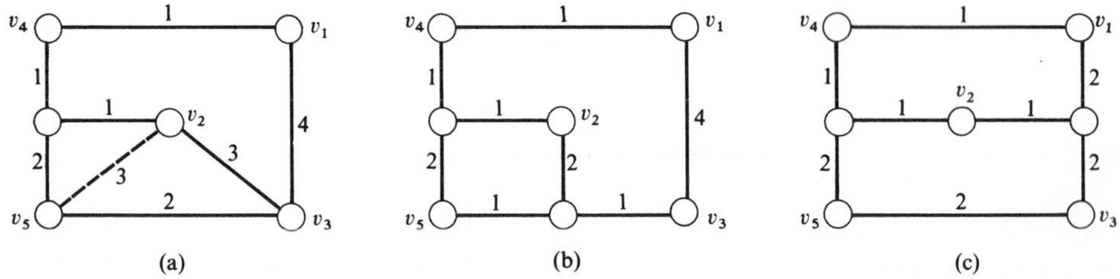

Fig. 6.6.3 Elementary reduction cycles.

Even if three branches $[i,j]$, $[j,k]$, and $[i,k]$ with the above properties are not present in the graph, it is occasionally possible still to apply the idea of the reduction cycle by adding a fictitious branch to the graph to obtain the proper relations.

For example, [2,5] in Fig. 6.6.3(a) is fictitious. Application of the reduction cycle results in the graph shown in Fig. 6.6.3(b). At this point, the reduction cycle can no longer be applied but the graph G' is not optimum, as evidenced by the graph G^* of Fig. 6.6.3(c), which realizes the same distance matrix with $l(G^*) = 12 < 13 = l(G')$. For these graphs,

$$\mathscr{D} = \begin{bmatrix} 0 & 3 & 4 & 1 & 4 \\ 3 & 0 & 3 & 2 & 3 \\ 4 & 3 & 0 & 5 & 2 \\ 1 & 2 & 5 & 0 & 3 \\ 4 & 3 & 2 & 3 & 0 \end{bmatrix}. \tag{6.6.9}$$

If the distance matrix $\mathscr{D}$ is realizable as a tree we may use Theorem 6.6.1 to find the unique realization. The procedure for realizing $\mathscr{D}$ is identical to the steps used in the proof of the theorem. Let $\mathscr{D}^{(i)}$ be the $i \times i$ leading principal submatrix of $\mathscr{D}$. Then if $\mathscr{D}$ is realizable as a tree, $\mathscr{D}^{(i)}$ is also realizable as a tree. Starting with $\mathscr{D}^{(2)}$, realize the tree $T^{(2)}$ as two vertices v_1 and v_2 connected by a branch of weight $d_{1,2}$. Then add the vertex v_3 and try to find a point v_x on [1,2] such that $\mathscr{D}^{(3)}$ is realized. Continue until at some point $\mathscr{D}^{(i)}$ is not realizable as a tree or $\mathscr{D}$ is realized.

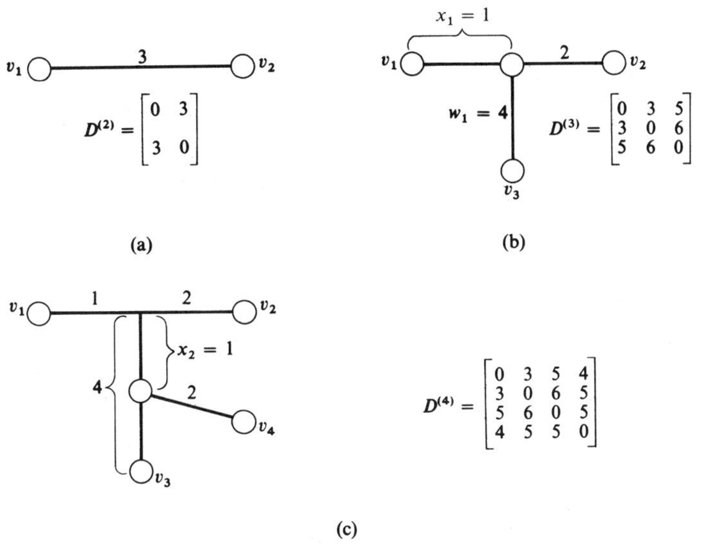

Fig. 6.6.4 Graphs for Example 6.6.1.

Example 6.6.1 Let $\mathscr{D}$ be given by (6.6.10) below. We first realize the 2×2 leading principal matrix of $\mathscr{D}$. The result is shown in Fig. 6.6.4(a). We then add vertex v_3 and realize the 3×3 leading principal matrix by the graph shown in Fig. 6.6.4(b). We must now add vertex v_4 to the realization and so on.

$$\mathscr{D} = \begin{bmatrix} 0 & 3 & 5 & 4 & 5 \\ 3 & 0 & 6 & 5 & 6 \\ 5 & 6 & 0 & 5 & 4 \\ 4 & 5 & 5 & 0 & 3 \\ 5 & 6 & 4 & 3 & 0 \end{bmatrix}. \tag{6.6.10}$$

We note that first two steps are trivial, but to add v_4 we must first locate the branch on which the internal vertex is to be added. This is done by assuming x_2 is on the path between v_1 and v_2 and then solving the equations

$$x_2 + w_2 = 4(= d_{1,4}), \tag{6.6.11a}$$

$$3 - x_2 + w_2 = 5(= d_{2,4}), \tag{6.6.11b}$$

which give $w_2 = 3$ and $x_2 = 1$. This implies that if x_2 is on the 1-2 path it is located 2 units from v_2. We now examine $d(v_3,v_4)$ given by $d(v_3,v_4) = 4 + w_2 = 7 \neq d_{3,4}$. This solution does not satisfy the remaining distance specification, and we must solve another pair of equations which are written with the assumption that an internal vertex will be created on the 1-3 path. These equations are

$$x_2 + w_2 = 4(= d_{1,4}), \tag{6.6.12a}$$

$$5 - x_2 + w_2 = 5(= d_{3,4}). \tag{6.6.12b}$$

These give $w_2 = 2$ and $x_2 = 2$; moreover, with this location, $d(v_2,v_4) = 2 + 1 + w_2 = d_{2,4}$ and so the third distance requirement is satisfied. The four-vertex tree realization which has just been found is shown in Fig. 6.6.4(c).

In order to add v_5, we solve the equations

$$x_3 + w_3 = 5(= d_{1,5}) \tag{6.6.13a}$$

$$3 - x_3 + w_3 = 6(= d_{2,5}), \tag{6.6.13b}$$

or

$$x_3 + w_3 = 5(= d_{1,5}) \tag{6.6.14a}$$

$$5 - x_3 + w_3 = 4(= d_{3,5}), \tag{6.6.14b}$$

or

$$x_3 + w_3 = 5(= d_{1,5}) \tag{6.6.15a}$$

$$4 - x_3 + w_3 = 3(= d_{4,5}). \tag{6.6.15b}$$

The solution of (6.6.13) is $w_3 = 4$ and $x_3 = 1$. But then $d(v_4,v_5) = 4 + w_3 = 8 \neq d_{4,5}$ so that we cannot create an internal vertex on the 1-2 path. The solution of (6.6.14) is $w_3 = 2$ and $x_3 = 3$. This means $d(v_2,v_5) = 2 + 2 + w_3 = 6 = d_{2,5}$ but $d(v_4,v_5) = 2 + 1 + w_3 = 5 \neq d_{4,5}$. Hence, we cannot create an internal vertex on the 1-3 path. Finally, the solution of (6.6.15) is $w_3 = 2$ and $x_3 = 3$. Then $d(v_2,v_5) = w_3 + 1 + 1 + 2 = 6 = d_{2,5}$ but $d(v_3,v_5) = 2 + 1 + w_3 = 5 \neq d_{3,5}$. Hence, although $\mathscr{D}^{(4)}$ is realizable as a tree, the distance matrix $\mathscr{D}$ is not.

Less intuitive, but more efficient realization techniques are possible for trees: one technique [HA5] utilizes the elementary reduction cycle; another is based on algebraic operations on $\mathscr{D}$ and seems to be the best available [BO1]. We now show that a tree realization of a given distance matrix (if it exists) is the optimum realization. Hence, we have a procedure for arriving at optimum realization for a class of distance matrices.

Given a realization G of the distance matrix $\mathscr{D}$, it may be that an elementary reduction cycle can be used to reduce the total weight of the graph. However, suppose that the reduction cycle cannot be applied. It is important to know whether or not the given realization is optimal. This question is in general unanswered. However, a strong necessary condition for optimality is available.

A vertex v_p in G is said to be a *compact* vertex of G, if there exists an entry $d_{i,j}$ in $\mathscr{D}$ such that

$$d_{i,j} = d(v_i,v_p) + d(v_p,v_j) \text{ for } i,j \neq p. \tag{6.6.16}$$

Theorem 6.6.2 *If G is an optimum realization of the distance matrix $\mathscr{D}$, then every external vertex of degree greater than one is compact.*

Proof. 1) Assume that v_p is at least of degree 2 but is not compact. Let v_q and v_r be adjacent to v_p. First, we show that if the branch $[q,r]$ exists in G, then G is not optimum. Since v_p is not compact, $d(v_q,v_r) < d(v_q,v_p) + d(v_p,v_r)$. Then, either $d(v_q,v_p) < d(v_p,v_r) + d(v_r,v_q)$ and $d(v_p,v_r) < d(v_q,v_p) + d(v_r,v_q)$ or there is a redundant element. In the latter case, G is clearly not optimum. In the former case, the conditions which allow an elementary reduction cycle are present and, consequently, G cannot be optimum.

2) Since $[q,r]$ is not in G, we can form a new graph G', by adding $[q,r]$ with weight w_0 to G where

$$w_0 = \operatorname*{Max}_{i,j} [d(v_i,v_j) - d(v_i,v_q) - d(v_j,v_r), 0]. \tag{6.6.17}$$

Let G' have distance matrix $\mathscr{D}' = [d'_{i,j}]$. Clearly $d'_{i,j} \leq d_{i,j}$. Moreover, $d'_{i,j} = d_{i,j}$ for all i,j. If not, there is at least one $d'_{i,j}$, say $d'_{s,t} < d_{s,t}$. In order that the addition of $[q,r]$ reduces $d_{s,t}$ to $d'_{s,t}$, we must have

$$d'_{s,t} = d(v_s,v_q) + w_0 + d(v_r,v_t). \tag{6.6.18}$$

By definition,

$$w_0 \geq d(v_t,v_s) - d(v_s,v_q) - d(v_t,v_r). \tag{6.6.19}$$

Combining (6.6.18) and (6.6.19), we obtain

$$d'_{s,t} \geq d(v_s,v_t) = d_{s,t} \tag{6.6.20}$$

and, hence, a contradiction. Consequently, $\mathscr{D}' = \mathscr{D}$.

3) We will now show that $w_0 < d(v_q,v_p) + d(v_p,v_r)$. Since v_p is not compact,

$$d(v_s,v_t) < d(v_s,v_p) + d(v_p,v_t) \tag{6.6.21}$$

and, by Theorem 6.5.1,

$$d(v_s,v_p) \leq d(v_s,v_q) + d(v_q,v_p) \tag{6.6.22a}$$

and

$$d(v_p,v_t) \leq d(v_p,v_r) + d(v_r,v_t). \tag{6.6.22b}$$

Hence,

$$d(v_s,v_t) < d(v_s,v_q) + d(v_q,v_p) + d(v_p,v_r) + d(v_r,v_t) \tag{6.6.23a}$$

or

$$d(v_s,v_t) - d(v_s,v_q) - d(v_t,v_r) < d(v_q,v_p) + d(v_p,v_r). \tag{6.6.23b}$$

By the definition of w_0, v_s and v_t may be picked such that the left-hand side of (6.6.23b) is at least w_0.

4) Now, consider G' with distance matrix $\mathscr{D}$. Since $w_0 < d(v_q,v_p) + d(v_p,v_r)$, by the same argument as in (1), either $[q,p]$ or $[r,p]$ is redundant or an elementary reduction cycle can be applied. In the first case, remove the redundant branch from G' and obtain a graph with smaller total weight than G. In the second case, the weight of G' can be reduced by $\frac{1}{2}(w_0 + l[q,p] + l[r,p])$ and since $w_0 < l[q,p] + l[r,p]$, the reduced graph has total weight less than $l(G)$. Hence, G is not optimum and the theorem is proved.

The method of proof of Theorem 6.6.2 may be used to reduce the weight of a graph. Suppose that G has a noncompact vertex v_p. Then (ii) suggests that we can add a new element of weight w_0 without disturbing the distance matrix of the graph. The total weight of the graph may now be reduced by removing a redundant branch or applying an elementary reduction cycle.

Thus, we have demonstrated a necessary condition for a given realization to be optimum. We will now give a sufficient condition for an optimum realization. This condition essentially entails generating a new graph and distance matrix with lower weight from the graph G. Then, if this new graph is optimum, the original graph is also optimum.

Let $\mathscr{D}_i(a)$ be a matrix obtained from $\mathscr{D}$ by subtracting a nonnegative number "a" from all entries of the ith row and column of $\mathscr{D}$ with the exception of the diagonal element. In other words, if $\mathscr{D}_i(a) = [d^i_{p,q}]$ then

$$d^i_{p,q} = \begin{cases} d_{p,q} & \text{if } p \text{ and } q \neq i, \\ d_{p,q} - a & \text{if } p \text{ or } q = i \text{ and } p \neq q, \\ 0 & \text{if } p = q. \end{cases} \tag{6.6.24}$$

Then the first problem which arises is to determine the range of values of "a" such that $\mathscr{D}_i(a)$ is a distance matrix.

Lemma 6.6.1 *$\mathscr{D}_i(a)$ is a distance matrix if and only if $a \leq \frac{1}{2}(d_{p,i} + d_{i,r} - d_{p,r})$ for $p,r = 1,2,\ldots,n$ and $p,r \neq i$.*

Proof. The necessity of the lemma is obvious. If the inequality is violated $2a > d_{p,i} + d_{i,r} - d_{p,r}$ or $d_{p,r} > (d_{p,i} - a) + (d_{i,r} - a)$. This implies that in $\mathscr{D}_i(a)$, $d^i_{p,r} > d^i_{p,i} + d^i_{i,r}$, which violates the conditions of Theorem 6.6.1.

To prove the sufficiency, we show that for every q

$$d^i_{p,q} + d^i_{q,r} \geq d^i_{p,r} \qquad \text{for } p,r = 1,\ldots,n. \tag{6.6.25}$$

CASE 1. $q \neq i$. If $p \neq i, r \neq i$, (6.6.25) is automatically satisfied since $\mathscr{D}$ is a distance matrix. For $p = i$ or $q = i$, (6.6.25) is again satisfied regardless of the value of a. Finally, if $p = r = i$, then (6.6.25) may be rewritten as

$$(d_{i,q} - a) + (d_{q,i} - a) \geq 0 \tag{6.6.26a}$$

or

$$d_{i,q} \geq a \qquad \text{for all } q \neq i. \tag{6.6.26b}$$

Clearly, $d_{i,q} \geq a$ is implied by the hypothesis by letting $p = r$ (i.e., $d_{p,r} = 0$) in the statement of the lemma.

CASE 2. $q = i$. For any $p \neq i$, $r \neq i$, (6.6.25) may be written as $(d_{p,i} - a) + (d_{i,r} - a) \geq d_{p,r}$, which is identical to the inequality of the hypothesis. Finally, for $p = i$ or $r = i$, (6.6.25) is again satisfied for all values of a, and if $p = r = i$, (6.6.25) is trivially satisfied.

The above lemma shows that if we define a number a_0 as

$$a_0 = \min_{\substack{p,r=1,\ldots,n \\ p,r \neq i}} [(d_{p,i} + d_{i,r} - d_{p,r})/2], \tag{6.6.27}$$

then $\mathscr{D}_i(a)$ is a distance matrix if and only if $0 \leq a \leq a_0$.

Suppose we are given $G_i(a)$, a realization of the matrix $\mathscr{D}_i(a)$. We can easily construct a graph $\hat{G}$ which is a realization of $\mathscr{D}$. To do this, add a vertex $v_{i'}$ to $G_i(a)$ and connect $v_{i'}$ to v_i by branch $[i',i]$ of weight a. Then, if we treat v_i as an internal vertex and $v_{i'}$ as the ith external vertex, we have realized G.

Theorem 6.6.3 *For $0 \leq a \leq a_0$ let $G_i(a)$ be an optimum realization of $\mathscr{D}_i(a)$. Let $\hat{G}$ be obtained from $G_i(a)$ by the above process. Then $\hat{G}$ is an optimum realization of $\mathscr{D}$.*

Proof. If $a = 0$ the theorem is trivial. Suppose that $a > 0$. First we show for any optimum realization G^* of $\mathscr{D}$, the ith vertex v_i must be of degree one. Suppose v_i has degree two or more. By the definition of a_0,

$$a_0 \leq (d_{p,i} + d_{i,r} - d_{p,r})/2 \qquad \text{for } p,r = 1,\ldots,i-1,i+1,\ldots,n \tag{6.6.28}$$

and since $0 < a \leq a_0$

$$d_{p,i} + d_{i,r} > d_{p,r}. \tag{6.6.29}$$

Then v_i is *not* compact and thus G^* cannot be optimum (by Theorem 6.6.2); but since G^* is optimum by assumption, $v_{i'}$ must be of degree one.

Now, assume that the graph $\hat{G}$ is not optimum. Then, if G^* is an optimum realization of $\mathscr{D}$

$$l(G^*) < l(\hat{G}) = l(G_i(a)) + a. \tag{6.6.30}$$

In G^*, let the vertex adjacent to $v_{i'}$ be v_{x^*}. Then, $d(v_{i'},v_{x^*}) = a_1 < a$ since, if not, we can pick a point v_{i^*} on $[i',x^*]$ such that $d(v_{i'},v_{i^*}) = a$. Now, removing $[i',i^*]$ from G^*, let G_1^* be the resulting graph. Clearly, G_1^* is a realization of $\mathscr{D}_i(a)$ and since

$$l(G_1^*) = l(G^*) - a < l(\hat{G}) - a = l(G_i(a)) \tag{6.6.31}$$

we have contradicted our original hypothesis that $G_i(a)$ is optimum. Hence $d(v_{i'}, v_{x^*}) = a_1 < a$.

Since $d(v_{i'},v_{x^*}) < a$, v_{x^*} must be an internal vertex. Thus, it must be of degree at least three. Consider the graph obtained by removing $[i',x^*]$ from G^*. If this graph is denoted by $G_i^*(a_1)$ and v_{x^*} of G^* is the ith vertex of $G_i^*(a_1)$, $G_i^*(a_1)$ realizes $\mathscr{D}_i(a_1)$. Now, since $0 < a_1 < a$, we again find that

$$d_{p,i} + d_{i,r} > d_{p,r} \qquad \text{for } p,r = 1,\ldots,n \text{ and } p,r \neq i. \tag{6.6.32}$$

Hence v_{x^*} in $G_i^*(a_1)$ is not compact; therefore, $G_i^*(a_1)$ is not an optimum realization of $\mathscr{D}_i(a_1)$. Hence G^* cannot be an optimum realization of $\mathscr{D}$. Since this is a contradiction, G is an optimum realization of $\mathscr{D}$.

Theorem 6.6.4 *If $\mathscr{D}$ has a tree realization T, then T is an optimum realization of $\mathscr{D}$.*

Proof. The theorem is true if $\mathscr{D}$ is of order 2. Suppose it is also true if $\mathscr{D}$ is of order $n - 1$. Let $\mathscr{D}^{(n)}$ be an $n \times n$ distance matrix with the tree realization T_n where the vertices are numbered so that v_n has degree one. Then, the $n - 1 \times n - 1$ leading principal submatrix of $\mathscr{D}^{(n)}$, denoted by $D^{(n-1)}$, has a tree realization which we will call T_{n-1}.

Tree T_n is constructed from T_{n-1} by connecting a branch $[n,x]$ of weight w_0 to a point x on T_{n-1}. Let x on T_{n-1} be labeled $v_{n'}$ and the new tree called T'_{n-1}. Let us find the $n \times n$ distance matrix $\mathscr{D}^{(n)'}$ of T'_n. The matrix $\mathscr{D}^{(n)'}$ is actually $\mathscr{D}^{(n)}(w_0)$. By the induction hypothesis T_{n-1} is an optimum realization of $\mathscr{D}^{(n-1)}$, and hence T'_{n-1} is an optimum realization of $\mathscr{D}^{(n)}(w_0)$. Then, by Theorem 6.6.3, T_n is an optimum realization of $\mathscr{D}^{(n)}$.

Thus, if a distance matrix has a tree realization, this realization is unique and an optimum realization. Furthermore, we have given a constructive procedure to find a tree realization, if one exists. By combining the ideas of the proofs of Theorems 6.6.3 and 6.6.4, it is also possible to show that a tree realization of a distance matrix is the *unique* optimum realization; this is left as Problem 13. Therefore, synthesis techniques for this class of distance matrix are complete.

6.7 CENTERS AND MEDIANS OF A GRAPH

In our present model the vertices of G represent switching centers, intersections, or communities. The branch weights have been considered to be lengths, but they could just as well have been branch transit times or traffic cost factors. Suppose that graph G represents an existing highway system and the weight $w(i,j)$ of each branch (i,j) is the average time required to traverse that branch. Consider the following problem. We would like to build a hospital in a community of G. It has been found that a major factor affecting the mortality rate of emergency cases is the time required to reach the hospital from the scene of the emergency. The hospital board of directors considers this the prime concern in locating the hospital. They would like to minimize the travel time in either direction between the hospital and any community of G. Clearly, such a criterion is impossible to satisfy, unless G has only one vertex. Hence, the directors will accept a solution which minimizes the worst possible case; in other words a solution which minimizes distance to the furthest point.

Any solution of the above problem is known as a "center" of G. This term corresponds to physical intuition since we may consider the center of G as the most "centrally" located vertex of G.

Definition 6.7.1 *Let $G = (V,\Gamma)$ be an n-vertex graph; $v_c \in V$ is a center of G if*

$$\max_{1\le j\le n} [d(v_c,v_j)] \le \max_{1\le j\le n} [d(v_k,v_j)] \quad \textit{for any } v_k \in V. \tag{6.7.1}$$

If the graph G is undirected and connected it is easily seen that G must contain at least one center. This is not necessarily true for directed graphs since it may be that no vertex has directed paths to all other vertices. We will therefore consider undirected graphs in this and the next section. Finding a center of G is routine once the meaning of the definition is understood. Suppose a number d_i^m is associated with each vertex v_i, such that d_i^m is the maximum distance from v_i to any other vertex of G. Let r_0 be the minimum of the d_i^m. Then, any vertex v_i with $d_i^m = r_0$ is a center of G. If

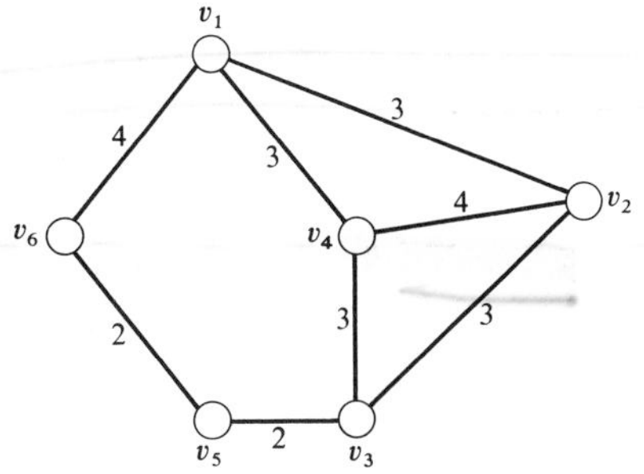

Fig. 6.7.1 Graph for Example 6.7.1.

$\mathscr{D}$ is the distance matrix of G, d_i^m corresponds to the maximum entry in the ith column, and if $r_0 = d_j^m$, then v_j is a center of G. The number r_0 is known as the *radius* of G.

Example 6.7.1 To clarify the above ideas, we find the radius and centers of the graph shown in Fig. 6.7.1. For this graph,

$$\mathscr{D} = \begin{bmatrix} 0 & 3 & 6 & 3 & 6 & 4 \\ 3 & 0 & 3 & 4 & 5 & 7 \\ 6 & 3 & 0 & 3 & 2 & 4 \\ 3 & 4 & 3 & 0 & 5 & 7 \\ 6 & 5 & 2 & 5 & 0 & 2 \\ 4 & 7 & 4 & 7 & 2 & 0 \end{bmatrix} \tag{6.7.2}$$

and $d_1^m = 6$, $d_2^m = 7$, $d_3^m = 6$, $d_4^m = 7$, $d_5^m = 6$, and $d_6^m = 7$. Hence, the radius r_0 of G is 6 and vertices v_1, v_3, and v_5 are the centers of G.

Now consider the following problem. The graph G represents a communication system and the branches of G cables interconnecting various terminals. Each cable will be reserved for communication between a specified pair of points. The weight of each branch corresponds to its length or cost. Suppose in this communication system all messages must be processed by a central station, represented by v_m, before they can be sent to their proper destinations. In locating the switching center, we should like to minimize the cost of the wires needed to connect each station to the switching center. Any vertex of G which minimizes the total length of wire in the system is called a "median" of the graph.

***Definition* 6.7.2** *v_m of G is a median of G if for any v_k*

$$\sum_{j=1}^{n} d(v_m,v_j) \leq \sum_{j=1}^{n} d(v_k,v_j). \tag{6.7.3}$$

If v_m is a median of G, then we say that

$$R_0 = \sum_{j=1}^{n} d(v_m,v_j)$$

is the *median length* of G. Again, it is obvious that finding a median and median

How!?

length of G is routine. If $\mathscr{D}$ is the distance matrix of G, let d_i be the sum of the entries of the ith column of $\mathscr{D}$. Then, if d_j is the minimum of these sums, v_j is a median of G and $R_0 = d_j$. To illustrate this, again consider the graph in Fig. 6.7.1. From its distance matrix, $d_1 = 22$, $d_2 = 22$, $d_3 = 18$, $d_4 = 22$, $d_5 = 20$, and $d_6 = 24$. Hence, the median length of G is 18, and v_3 is the median of G.

add priorities

At this point, we have not assigned any "priorities" to the vertices of G. For example, in the hospital problem all vertices were considered to be of equal importance in terms of travel times. In reality, this may not be the case. For example, it may be that some vertices of G represent areas with higher frequencies of accidents. In the second model, one station may have more average traffic than another. Hence, it would be necessary to connect more wires from the switching center to this vertex than to the other.

The preceding development is easily generalized to include these considerations. Let $h(1), \ldots, h(n)$ be fixed, nonnegative numbers assigned to vertices $v_1, \ldots, v_n$ of G, respectively. The number $h(i)$ represents the average amount of traffic originating at v_i, the number of wires between v_i and the switching center, the average time necessary to reach v_i from a center of G, or some other measure of importance. We can now redefine the concepts of center and median.

Definition 6.7.1′ *v_c is a center of G, if for any $v_k \in V$*

$$\max_{1 \le i \le n} [h(i)d(v_c,v_i)] \le \max_{1 \le i \le n} [h(i)d(v_k,v_i)]. \tag{6.7.4a}$$

Definition 6.7.2′ *v_m is a median of G, if for any k*

$$\sum_{i=1}^{n} h(i)d(v_m,v_i) \le \sum_{i=1}^{n} h(i)d(v_k,v_i). \tag{6.7.4b}$$

We may also define radius and median length to be

$$r_0 = \min_{1 \le i \le n} \left[\max_{1 \le j \le n} [h(i)d(v_i,v_j)] \right] \tag{6.7.5a}$$

and

$$R_0 = \min_{1 \le j \le n} \left[\sum_{i=1}^{n} h(i)d(v_j,v_i) \right] \tag{6.7.5b}$$

respectively. The original Definitions (6.7.1) and (6.7.2) of center and median correspond to the new Definitions (6.7.1′) and (6.7.2′) when $h(1) = h(2) = \cdots = h(n) = 1$.

The center and median of G are the best *vertex* locations for the hospital and switching center problems. However, it may be that there are better locations for the center and median of G than the ones described above; namely, locations other than at vertices. If X_0 is a point on a branch of G, we will say that X_0 is on G.

Let us reexamine Example 6.7.1. The graph G shown in Fig. 6.7.2 has three *centers* located at vertices v_1, v_3, and v_5, and the radius of G is 6. However, selecting

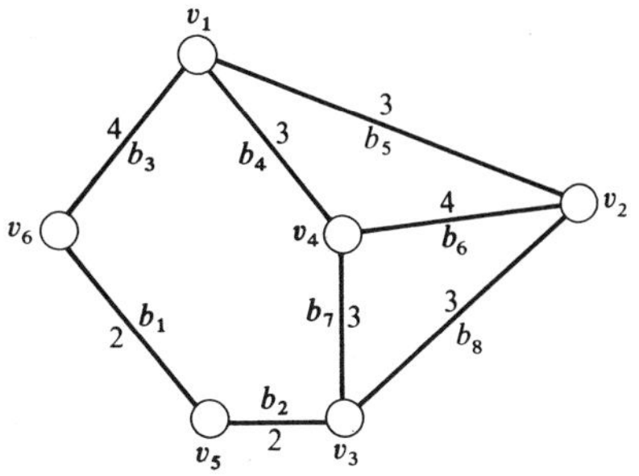

Fig. 6.7.2 Illustration of absolute center.

X_0 on b_7, one unit from v_3, we find that

$$\operatorname*{Max}_{1 \le i \le n} [d(X_0,v_i)] = 5 < \operatorname*{Max}_{1 \le i \le n} [d(v_k,v_i)] \qquad \text{for all } k. \tag{6.7.6}$$

Hence a better location for the hospital is at X_0. We formalize these concepts by means of the following two definitions:

Definition 6.7.3 *A point X_0 on a branch of G is an absolute center of G, if for every point X on G,*

$$\operatorname*{Max}_{1 \le i \le n} [h(i)d(X_0,v_i)] \le \operatorname*{Max}_{1 \le i \le n} [h(i)d(X,v_i)]. \tag{6.7.7}$$

Definition 6.7.4 *A point Y_0 on a branch of G is an absolute median of G, if for every point Y on G,*

$$\sum_{i=1}^{n} h(i)d(Y_0,v_i) \le \sum_{i=1}^{n} h(i)d(Y,v_i). \tag{6.7.8}$$

As before, we may also define the concepts of "radius" and "median length": r_0 is the *absolute radius* of G if, for X_0 an absolute center of G,

$$r_0 = \operatorname*{Max}_{1 \le i \le n} [h(i)d(X_0,v_i)]; \tag{6.7.9}$$

R_0 is the *absolute median length* of G if, for Y_0 an absolute median of G,

$$R_0 = \sum_{i=1}^{n} h(i)d(Y_0,v_i). \tag{6.7.10}$$

Equivalent expressions for these quantities are

$$r_0 = \operatorname*{Min}_{X \text{ on } G} \left[\operatorname*{Max}_{1 \le i \le n} [h(i)d(X,v_i)] \right], \tag{6.7.11a}$$

$$R_0 = \operatorname*{Min}_{Y \text{ on } G} \left[\sum_{i=1}^{n} h(i)d(Y,v_i) \right]. \tag{6.7.11b}$$

Before demonstrating a procedure to find the absolute centers and medians of G, we introduce one more elementary concept. A point X_{ok} on branch b_k of G is a *local center* of G, if, for all points X on b_k, Max $[h(i)d(X_{ok},v_i)] \leq$ Max $[h(i)d(X,v_i)]$. Thus, the local center on b_k corresponds to the best point on that branch to locate the hospital. Naturally, once the set of local centers of G is known, the absolute center may easily be found by selecting the best points from this set. In symbols, if $X_{o1},\ldots,X_{om}$ are the local centers of G, X_o is an absolute center of G if

$$\max_{1\leq i\leq n} [h(i)d(X_0,v_i)] = \min_{1\leq k\leq m} \Big[\max_{1\leq i\leq n} [h(i)d(X_{ok},v_i)]\Big]. \tag{6.7.12}$$

The absolute radius of G is

$$r_0 = \min_{1\leq k\leq m} \Big[\max_{1\leq i\leq n} [h(i)d(X_{ok},v_i)]\Big]. \tag{6.7.13}$$

We can define a local median in an analogous fashion. Consequently, the problem of finding absolute centers or medians of G has been reduced to m possibly simpler problems.

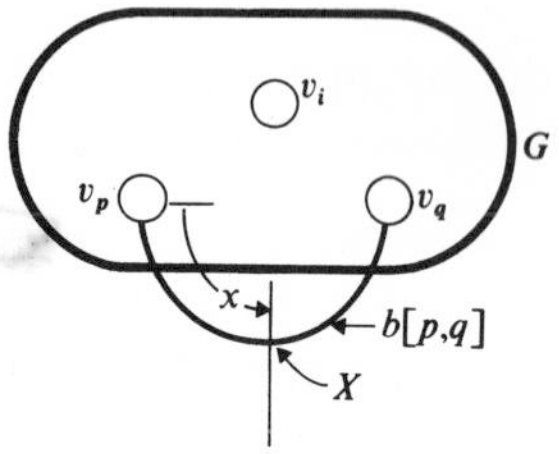

Fig. 6.7.3 Finding all local centers.

We now outline an algorithm for finding the absolute centers of G. We first generate all of the local centers of G. Once these are available, finding the absolute centers is simple. Consider the graph G of Fig. 6.7.3. Let $b_k = [p,q]$ be an arbitrary branch of G and let v_i be an arbitrary vertex. Let X be an arbitrary point on b_k such that X is x units from vertex v_p. The distance from X to v_i is measured relative to either a path passing through v_p or a path passing through v_q. In fact,

$$d(X,v_i) = \text{Min}\,[d(X,v_p) + d(v_p,v_i),d(X,v_q) + d(v_q,v_i)] \tag{6.7.14a}$$

and, if $w_k = w[p,q]$ is the weight of b_k,

$$d(X,v_i) = \text{Min}\,[x + d(v_p,v_i),w_k - x + d(v_q,v_i)]. \tag{6.7.14b}$$

Hence, $d(X,v_i)$ may be computed as a function of x for all X on b_k for $0 \leq x \leq w_k$. A plot of $d(v_i,X)$ versus x is shown in Fig. 6.7.4.

From Fig. 6.7.4, we see that for $0 \leq x \leq a_{k,i}$, $d(v_i,X)$ is measured relative to a path passing through v_p and x satisfies the linear constraint $d(X,v_i) = x + d(v_p,v_i)$. For $a_{k,i} < x \leq w_k$, $d(X,v_i)$ is measured relative to a path passing through v_q and $d(X,v_i) = w_k - x + d(v_q,v_i)$. Clearly, $h(i)d(X,v_i)$ is also a piecewise linear function, which is obtained from $d(X,v_i)$ by a scale change. Thus,

$$h(i)d(X,v_i) = \begin{cases} h(i)(x + d(v_p,v_i)) & \text{for } 0 \leq x \leq a_{k,i} \\ h(i)(w_k - x + d(v_q,v_i)) & \text{for } a_{k,i} < x \leq w_k \end{cases} \tag{6.7.15}$$

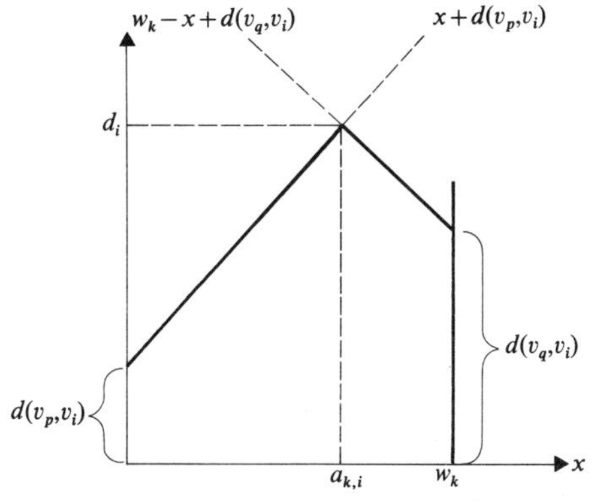

Fig. 6.7.4 $d(v_i,x)$ versus x for G in Fig. 6.7.3.

where

$$a_{k,i} = \begin{cases} 0 & \text{if } w_k + d(v_q,v_i) - d(v_p,v_i) < 0 \\ \frac{1}{2}[w_k + d(v_q,v_i) - d(v_p,v_i)] & \text{if } 0 \le w_k + d(v_q,v_i) - d(v_p,v_i) \le 2w_k \\ w_k & \text{if } d(v_q,v_i) - d(v_p,v_i) > w_k. \end{cases} \tag{6.7.16}$$

Let $M_k(X) = \underset{1 \le i \le n}{\text{Max}} [h(i)d(X,v_i)]$. To find a local center, we must find a point X on b_k which minimizes $M_k(X)$. To do this, first plot the functions $h(1)d(X,v_1)$, $h(2)d(X,v_2), \ldots, h(n)d(X,v_n)$ as a function of x as shown in Fig. 6.7.5. The maximum of these functions, $M_k(X)$, is then easily found and plotted. Any point where $M_k(X)$ attains a minimum is a local center on b_k of G. Hence, repeating this procedure for each branch of G, we may find the set of local centers of G. The absolute center is then known.

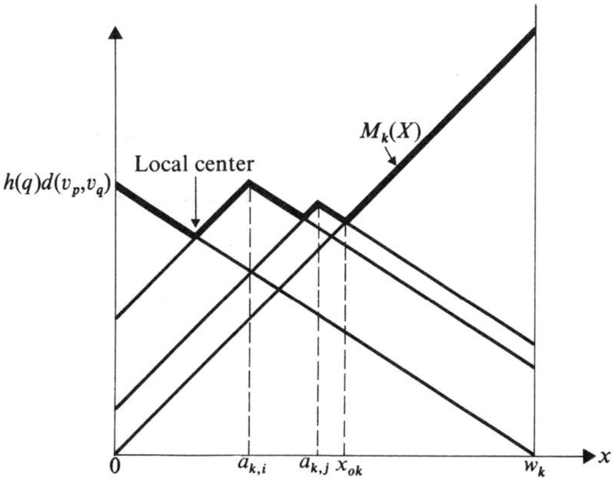

Fig. 6.7.5 $h(i)d(x,v_i)$ versus x.

Example 6.7.2 To illustrate the procedure, consider the graph of Fig. 6.7.1. Assume for simplicity that $h(1) = h(2) = \cdots = h(n) = 1$. We must thus find $M_k(X) = \underset{1 \le i \le n}{\text{Max}} [d(X,v_i)]$ for X on b_k and $k = 1,2,\ldots,8$.

For this example,

$$\underset{X \text{ on } b_1}{\text{Min}} [M_1(X)] = 5.5, \tag{6.7.17a}$$

$$\underset{X \text{ on } b_3}{\text{Min}} [M_3(X)] = 5.5,$$

$$\underset{X \text{ on } b_7}{\text{Min}} [M_7(X)] = 5 \quad \text{when } d(X,v_3) = 1,$$

STOP

and

$$\underset{X \text{ on } b_8}{\text{Min}} [M_8(X)] = 5 \quad \text{when } d(X,v_3) = 1. \tag{6.7.17b}$$

By inspection $M_2(X) \ge 6$, $M_4(X) \ge 6$, $M_5(X) \ge 6$, $M_6(X) \ge 7$ and therefore we do not compute their minima. Hence, the graph has two absolute centers, one on b_7 and one on b_8, and the absolute radius of G is $r_0 = 5$. These results are clearly shown by the curves in Fig. 6.7.6.

We have seen that by allowing centers at points other than vertices, the minimum maximum distance from the center to the vertices of G could be reduced. It seems reasonable to expect the same result for median locations. However, it will be seen that the median of G, located at a vertex of G, is actually as good as the absolute median of G. In other words, the median length and the absolute median length are identical. This result is given by:

Theorem 6.7.1 *There exists at least one vertex of G which is an absolute median.*

Proof. Let Y be an arbitrary point on G. We will show that there always exists a vertex v_m in G such that

$$\sum_{i=1}^{n} h(i)d(Y,v_i) \ge \sum_{i=1}^{n} h(i)d(v_m,v_i). \tag{6.7.18}$$

Suppose that Y is on branch $[p,q]$ of G and suppose the vertices of G are renumbered such that for $i = 1,2,\ldots,r$, $d(Y,v_i)$ is measured relative to a path through vertex v_p and for $i = r+1,\ldots,n$, $d(Y,v_i)$ is measured relative to a path through v_q. Thus,

$$d(Y,v_i) = \begin{cases} d(Y,v_p) + d(v_p,v_i) & \text{for} \quad 1 \le i \le r, \\ d(Y,v_q) + d(v_q,v_i) & \text{for} \quad r+1 \le i \le n. \end{cases} \tag{6.7.19}$$

Then,

$$\sum_{i=1}^{n} h(i)d(Y,v_i) = \sum_{i=1}^{r} h(i)[d(Y,v_p) + d(v_p,v_i)] + \sum_{i=r+1}^{n} h(i)[d(Y,v_q) + d(v_q,v_i)]. \tag{6.7.20}$$

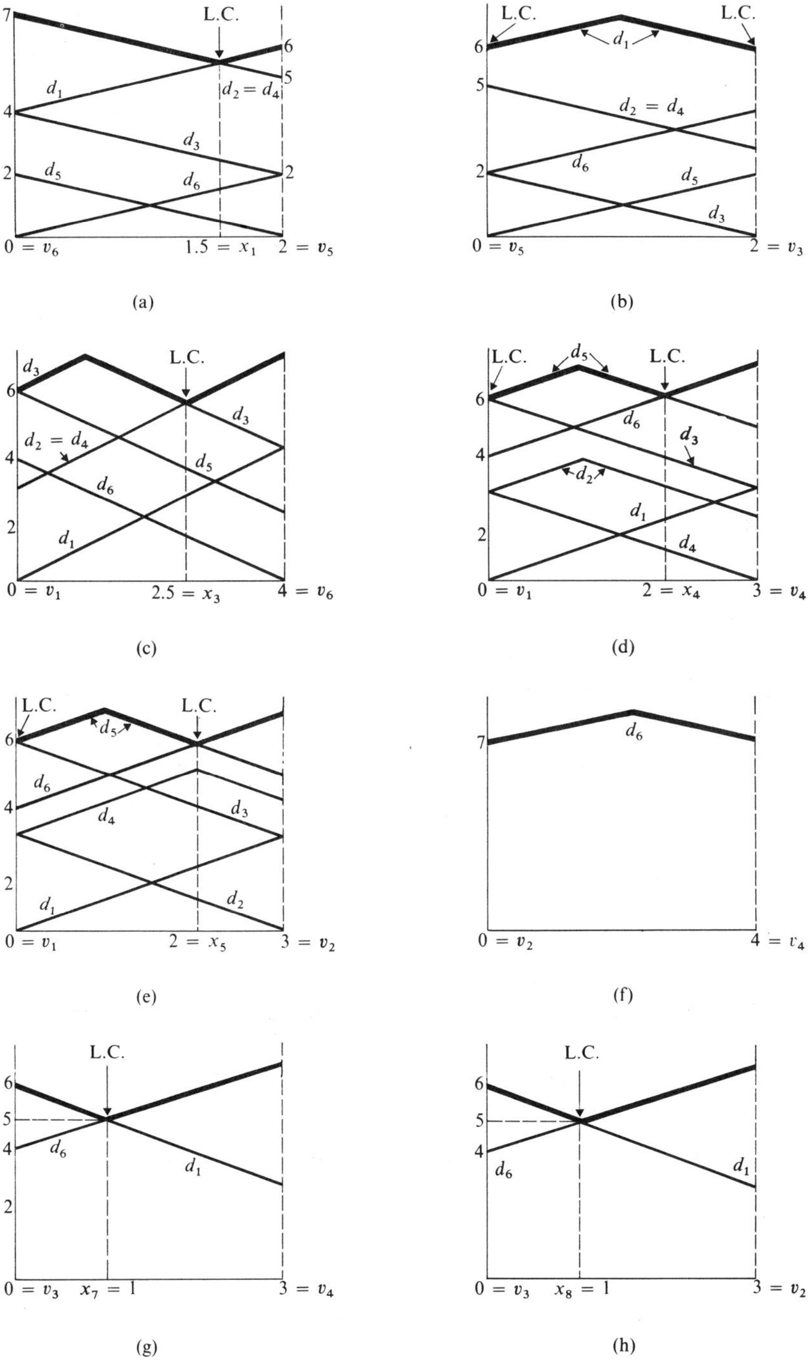

Fig. 6.7.6 Plots for Example 6.7.2

Without loss of generality, assume

$$\sum_{i=1}^{r} h(i) \geq \sum_{i=r+1}^{n} h(i). \tag{6.7.21}$$

Since $d(Y,v_q) = d(v_p,v_q) - d(Y,v_p)$, by substituting this relation into (6.7.20), we obtain

$$\sum_{i=1}^{n} h(i)d(Y,v_i) = \sum_{i=1}^{r} h(i)[d(Y,v_p) + d(v_p,v_i)]$$
$$+ \sum_{i=r+1}^{n} h(i)[d(v_p,v_q) - d(Y,v_p) + d(v_q,v_i)]. \tag{6.7.22}$$

By the triangle inequality for distances (Theorem 6.5.1), $d(v_p,v_q) + d(v_q,v_i) \geq d(v_p,v_i)$. Using this relationship together with (6.7.21), we obtain the inequality

$$\sum_{i=1}^{n} h(i)d(Y,v_i) \geq \sum_{i=1}^{r} h(i)[d(Y,v_p) + d(v_p,v_i)] + \sum_{i=r+1}^{n} h(i)[d(v_p,v_i) - d(Y,v_p)]. \tag{6.7.23}$$

Rearranging the right-hand side of (6.7.22) yields

$$\sum_{i=1}^{n} h(i)d(Y,v_i) \geq \sum_{i=1}^{n} h(i)d(v_p,v_i) + \left[\sum_{i=1}^{r} h(i) - \sum_{i=r+1}^{n} h(i)\right] d(Y,v_p). \tag{6.7.24}$$

Using (6.7.21) we get

$$\sum_{i=1}^{n} h(i)d(Y,v_i) \geq \sum_{i=1}^{n} h(i)d(v_p,v_i), \tag{6.7.25}$$

which is the desired result. Similarly, if the inequality in (6.7.21) were reversed, we could show that vertex v_q is at least as "good" as point Y.

From Theorem 6.7.1, there is always an optimum location of a switching center at a vertex. A natural generalization of this concept is next considered. We allow a number of switching centers, say $Y_{m_1}, Y_{m_2}, \ldots, Y_{m_t}$, on G. Each switching center Y_{m_i} may be located on any branch of G. All traffic in the network must arrive at *one* of the switching centers before being sent to its proper destination. We assume the cost of interconnecting switching centers is negligible. The problem is then to select switching centers so that the total length of wires from stations to switching centers is minimum.

Let $\mathbf{Y}_t = \{Y_{m_1}, Y_{m_2}, \ldots, Y_{m_t}\}$. By the distance $d(v_i,\mathbf{Y}_t)$ from a point v_i to $\mathbf{Y}_t$, we mean the distance from v_i to the *closest* element of $\mathbf{Y}_t$. Thus,

$$d(v_i,\mathbf{Y}_t) = \text{Min}\,[d(v_i,Y_{m_1}), d(v_i,Y_{m_2}), \ldots, d(v_i,Y_{m_t})]. \tag{6.7.26}$$

The problem stated in the last paragraph may be formulated as: Given a number t, find a set of t points $\mathbf{Y}_t^* = \{Y_1^*, \ldots, Y_t^*\}$ such that

$$\sum_{i=1}^{n} h(i)d(v_i, \mathbf{Y}_t^*) \leq \sum_{i=1}^{n} h(i)d(v_i, \mathbf{Y}_t), \tag{6.7.27}$$

where $\mathbf{Y}_t = \{Y_1, \ldots, Y_t\}$ is any set of t points on G. Any set $\mathbf{Y}_t^*$ which satisfies (6.7.27) will be called a *t-median of G*.

No t-median is better than an optimum set of t vertices of G. Thus, to find a t-median, one can in principal examine all subsets of vertices containing t elements. The proof of this statement is left as an exercise.

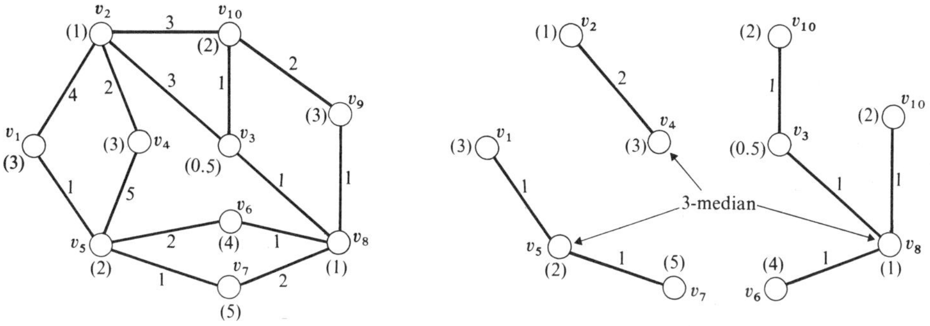

(a) (b)

Fig. 6.7.7 Graph for Example 6.7.3.

Example 6.7.3 Consider the graph shown in Fig. 6.7.7. It can be shown that the vertices v_4, v_5, and v_8 represent a 3-median such that

$$\sum_{i=1}^{10} h(i)d(v_i, V_3^*) = 21.5. \tag{6.7.28}$$

6.8 OPTIMUM LOCATIONS ON GRAPHS WITH PROBABILISTIC VERTEX WEIGHTS

In the preceding development, we associated fixed nonnegative numbers $h(1), \ldots, h(n)$ with the vertices $v_1, \ldots, v_n$. These numbers may represent the average number of messages or accidents originating at these vertices. In reality, the traffic occurring at a vertex is not a fixed number, but rather a random number with a possibly known probability distribution. Consequently, an optimum point found by deterministic methods will vary with different realizations of the random events. For this case, the concepts of absolute centers and medians must be generalized.

Suppose the absolute centers and absolute median are found according to the techniques already discussed. A natural question is: Are these points "optimum" when the random weights of G are considered? The interpretation of "optimum" for deterministic graphs is clear. The absolute center will be no farther than r_0 from any vertex of G; the median length of G will be no greater than R_0. If $h(i)$ is the deterministic weight of v_i, interpretations of these relations are

$$\text{Prob}\{\max_{1\leq i\leq n}[h(i)d(v_i,X_{0e})] > r\} = 0 \qquad \text{for } r \geq r_0\,, \tag{6.8.1}$$

$$\text{Prob}\left\{\sum_{i=1}^{n} h(i)d(v_i,Y_{0e}) > R\right\} = 0 \qquad \text{for } R \geq R_0, \tag{6.8.2}$$

where X_{0e} and Y_{0e} are respectively the absolute center and median of G. If $h(i)$ is a random number, there may be no values of r_0 and R_0 for which (6.8.1) and (6.8.2) are true.

Let $H(i)$ be a nonnegative random variable corresponding to the weight of v_i for $i = 1,2,\ldots,n$. Given a number r, a goal is to find an X_0 on G, such that $\text{Prob}\{\text{Max}[H(i)d(v_i,X_0)] \geq r\}$ is minimized. In other words, we want to find a point X_0 such that the greatest weighted distance stays within an allowable limit with maximum probability. These ideas lead to the following definitions.

Definition 6.8.1 *X_0 on a branch of G is a maximum probability absolute r center (MPArC) of G if for every point X on G*

$$\text{Prob}\{\max_{1\leq i\leq n}[H(i)d(v_i,X_0)] \geq r\} \leq \text{Prob}\{\max_{1\leq i\leq n}[H(i)d(v_i,X)] \geq r\}. \tag{6.8.3}$$

Definition 6.8.2 *Y_0 on a branch of G is a maximum probability absolute R median (MPARM) of G, if for every point Y on G*

$$\text{Prob}\left\{\sum_{i=1}^{n} H(i)d(v_i,Y_0) \geq R\right\} \leq \text{Prob}\left\{\sum_{i=1}^{n} H(i)d(v_i,Y) \geq R\right\}. \tag{6.8.4}$$

The above definitions can be written in expanded form by again using the concept of a *local* optimum point. For example, let X_{0j} be a point on branch b_j such that

$$\text{Prob}\{\max_{1\leq i\leq n}[H(i)d(v_i,X_{0j})] \geq r\} = \min_{X \text{ on } b_j}[\text{Prob}\{\max_{1\leq i\leq n}[H(i)d(v_i,X)] \geq r\}]. \tag{6.8.5}$$

Thus

$$\text{Prob}\{\max_{1\leq i\leq n}[H(i)d(v_i,X_0)] \geq r\} = \min_{1\leq j\leq m}[\text{Prob}\{\max_{1\leq i\leq n}[H(i)d(v_i,X_{0j})] \geq r\}], \tag{6.8.6}$$

and X_{0j} is a local MPArC with X_0 being the value of X_{0j} which minimizes the left-hand side of (6.8.5). Consequently, if this set of local optimum points of G is available, the optimum point is easily obtained.

Consider again the graph of Fig. 6.7.3. If X is an arbitrary point on branch b_k,

$$d(v_i,X) = \text{Min}[x + d(v_p,v_i),\, w_k - x + d(v_q,v_i)]. \tag{6.8.7}$$

The statement $H(i)d(v_i,X) < r$ is equivalent to $H(i) < r/d(v_i,X)$, hence, it is easily

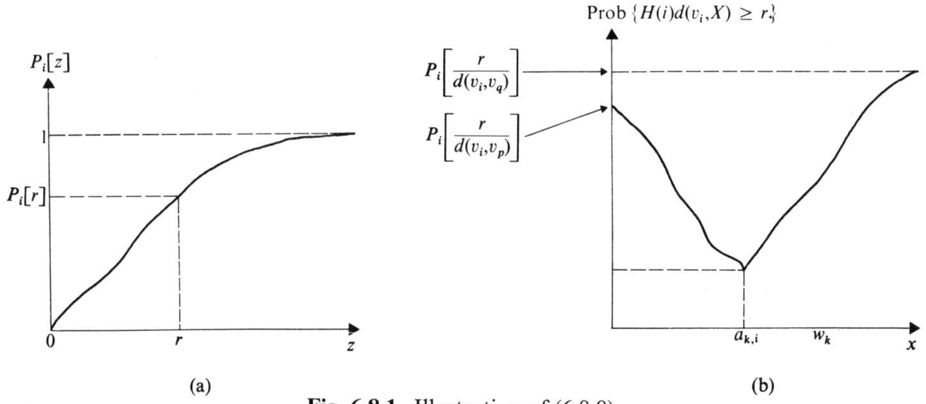

Fig. 6.8.1 Illustration of (6.8.8).

seen that if $\tilde{P}_i(x) = \text{Prob}\,\{H(i)d(v_i,X) < r\}$, then

$$\tilde{P}_i(x) = \begin{cases} P_i\left[\dfrac{r}{x + d(v_p,v_i)}\right] & \text{for } 0 \leq x \leq a_{k,i} \\ P_i\left[\dfrac{r}{w_k - x + d(v_q,v_i)}\right] & \text{for } a_{k,i} < x \leq w_k \end{cases} \tag{6.8.8}$$

where $P_i[z]$ is the cumulative probability distribution function of $H(i)$ for $i = 1,\ldots,n$ and $a_{k,i} = \frac{1}{2}[w_k + d(v_q,v_i) - d(v_p,v_i)]$. The above relationship is illustrated in Fig. 6.8.1.

The problem of finding a local MPArC is equivalent to finding an x which maximizes

$$\prod_i \tilde{P}_i(x).$$

The difficulty of performing this maximization will depend on the functional forms of the probability distributions in question. However, if the random variables $H(1),\ldots,H(n)$ are discrete, the following interesting result is obtained.

Theorem 6.8.1 *Let $H(1),\ldots,H(n)$ be discrete independent random variables. Then, there exists a (possibly degenerate) interval (X^*,X^{**}) on each branch of G, such that any $X \in (X^*,X^{**})$ is a local maximum probability absolute r center of G.*

Proof. Let $P_i[z]$ for $i = 1,\ldots,n$ be the cumulative probability distribution function of $H(i)$, such that

$$P_i[z] = k_{i,j} \text{ for } j - 1 \leq z < j \text{ and } j = 1,2,\ldots,k_{i_0}. \tag{6.8.9}$$

The probability distribution of $d(v_i,X)H(i)$ is

$$P_i\left[\frac{z}{d(v_i,X)}\right].$$

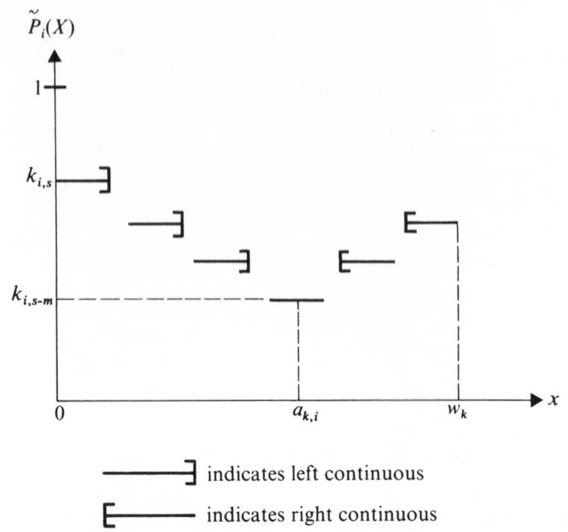

Fig. 6.8.2 Illustration of $\tilde{P}_i(x)$.

If $z = r$, the distribution is a function $\tilde{P}_i(x)$ shown in Fig. 6.8.2 and given by (6.8.10a) for $0 \le x \le a_{k,i}$ and by (6.8.10b) for $a_{k,i} < x \le w_k$.

$$\tilde{P}_i(x) = \begin{cases} k_{i,s} = P\left[\dfrac{r}{d(v_i,v_p)}\right] & \text{for } 0 \le x \le \dfrac{r}{s-1} - d(v_i,v_p) \\ k_{i,(s-j)} & \text{for } \dfrac{r}{s-j} - d(v_i,v_p) < x \le \dfrac{r}{s-j-1} - d(v_i,v_p) \\ 0 & \text{for } j \ge s \end{cases} \tag{6.8.10a}$$

$$\tilde{P}_i(x) = \begin{cases} k_{i,(s-m)} = P_i(a_{k,i}) & \text{for } a_{k,i} < x \le w_k + d(v_i,v_q) - \dfrac{r}{s-m+1} \\ k_{i,(s-m+j)} & \text{for } w_k + d(v_i,v_q) - \dfrac{r}{s-m+j} \le x \\ & < w_k + d(v_i,v_q) - \dfrac{r}{s-m+j+1} \end{cases} \tag{6.8.10b}$$

Let $\eta_i = \{x_{i_0} = 0, x_{i1}, \ldots, x_{iv_i}\}$ for $i = 1, \ldots, n$ be the set of jump points of $\tilde{P}_i(x)$. Let $\eta = \{\omega_0 = 0, \omega_1, \ldots, \omega_v\}$ for $v \le \Sigma v_i$, be the union of the η_i such that $\omega_{j-1} < \omega_j$ for $j = 1, \ldots, v$. Clearly, $\tilde{P}_i(x)$ is constant for all i on the open interval (ω_{j-1}, ω_j) for $j = 1, \ldots, v$. Therefore, Prob $\{\text{Max}[H(i)d(v_i,X)] < r\}$ is also constant on each such interval. Then either there exists at least one interval, say $(\omega_{\mu-1}, \omega_\mu)$, where this probability is maximum or the maximum occurs only at an ω_j of η.

Note that a maximum probability point must occur at an $\omega_j \in \eta$, regardless of the existence of a maximum probability interval. Hence, local MPArC's could readily be found for a graph G by computing the jump points of the $\tilde{P}_i(x)$.

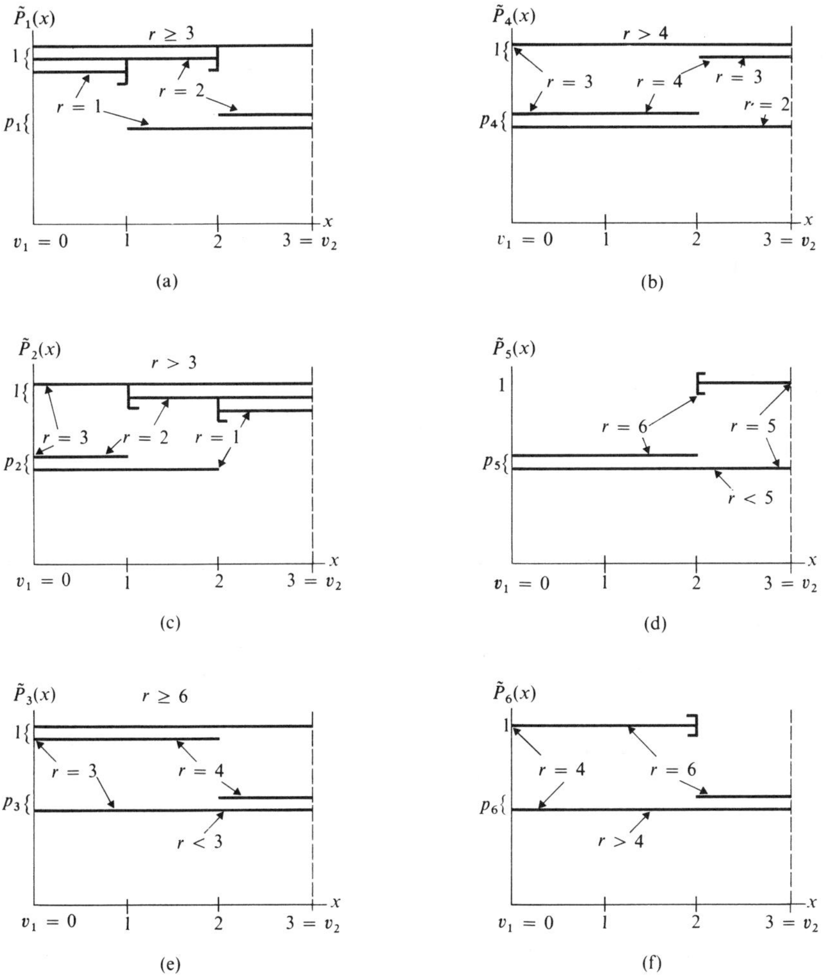

Fig. 6.8.3 Plots for Example 6.8.1.

Example 6.8.1 To illustrate the application of Theorem 6.8.1, consider the graph of Fig. 6.7.1 and let the $H(i)$ be zero-one random variables such that Prob $\{H(i) = 0\} = p_i$. We will find a local maximum probability r center on branch b_5 for several values of r. Figure 6.8.3 shows plots of $\tilde{P}_i(x)$ versus x.

From Fig. 6.8.3, all jump points of $\tilde{P}_i(x)$ are at $x = 0,1,2,3$ for any value of r. We can now easily compute $p = \text{Prob}\{\max_{1 \leq i \leq 6} [H(i)d(v_i,X)] < r\}$ for the values of r used in computing the graphs of Fig. 6.8.3. For example, Table 6.8.1 gives values of p for $r = 1,3,6$.

We now examine the maximum probability absolute R medians of G. Consider the quantity $\sum_{i=1}^{n} H(i)d(v_i,Y)$. Suppose n is large and the $H(i)$ are independent. For

Table 6.8.1 Values of Prob $\{\underset{1 \le i \le 6}{\text{Max}}[H(i)d(v_i X) < r\}$ as a function of X and r.

	$x = 0$	$x = 1$	$x = 2$	$x = 3$
$r = 1$	$p_2p_3p_4p_5p_6$	$p_2p_3p_4p_5p_6$	$p_1p_3p_4p_5p_6$	$p_1p_3p_4p_5p_6$
$r = 3$	$p_2p_5p_6$	$p_3p_4p_5p_6$	$p_3p_4p_5p_6$	$p_3p_4p_5p_6$
$r = 6$	p_5	p_5	1	p_6

each fixed Y, we can approximate the sum by a normal random variable Z with mean $\mu(Y)$ and variance $\sigma^2(Y)$ where

$$\mu(Y) = \sum_{i=1}^{n} E\{H(i)\}d(v_i, Y) \tag{6.8.11a}$$

and

$$\sigma^2(Y) = \sum_{i=1}^{n} \text{Var}\,\{H(i)\}d^2(v_i, Y). \tag{6.8.11b}$$

Therefore,

$$\text{Prob}\,\{Z < R\} = \Phi\left(\frac{R - \mu(Y)}{\sigma(Y)}\right) \tag{6.8.12}$$

where Φ is the standard normal distribution function. Since Φ is a strictly increasing function, $Y = Y_o$ maximizes Prob $\{Z < R\}$ if and only if $Y = Y_o$ maximizes

$$\frac{R - \mu(Y)}{\sigma(Y)} = \frac{R - \sum_{i=1}^{n} E\{H(i)\}d(v_i, Y)}{\left(\sum_{i=1}^{n} \text{Var}\,\{H(i)\}d^2(v_i, Y)\right)^{1/2}}. \tag{6.8.13}$$

Thus, if the normal approximation is used, MPARM's of G can be located without great theoretical difficulty. If exact expressions are required, n-fold convolutions are encountered and the solution is considerably more complicated.

In finding an optimum location for a median of G, the designer may be satisfied in knowing that the probabilistic demands for the system stay close to a nominal value. Hence, it is desirable to find a point Y_{ov} on G such that the variance of

$$\sum_{i=1}^{n} H(i)d(v_i, Y_{ov})$$

is minimum.

Let $H(1), \ldots, H(n)$ be independent random variables with variances $\sigma^2(1), \ldots,$ $\sigma^2(n)$, respectively. For Y on branch $b_k = [p,q]$

$$\operatorname{Var}\left\{\sum_{i=1}^{n} H(i)d(v_i,Y)\right\} = \sum_{i=1}^{n} \sigma^2(i)\,\{\operatorname{Min}[y + d(v_p,v_i), w_k - y + d(v_q,v_i)]\}^2. \quad (6.8.14)$$

The minimization of the left-hand side of (6.8.14) is considerably simplified if b_k is a cut-set.

Theorem 6.8.2 *Let b_k be a cut-set of G, which if deleted divides G into subgraphs G_1 and G_2. Then, there exists a point Y_{ov} on $b_k = [p,q]$ such that*

$$\operatorname{Var}\left\{\sum_{i=1}^{n} H(i)d(v_i,Y_{ov})\right\} = \operatorname*{Min}_{Y \text{ on } b_k}\left[\operatorname{Var}\left\{\sum_{i=1}^{n} H(i)d(v_i,Y)\right\}\right] \quad (6.8.15a)$$

and

$$d(v_p,Y_{ov}) = \begin{cases} 0 & \text{for } Q \le 0 \\ Q & \text{for } 0 < Q < w_k \\ w_k & \text{for } w_k \le Q \end{cases} \quad (6.8.15b)$$

where

$$Q = \frac{\sum_{v_i \in G_2} \sigma^2(i)d(v_p,v_i) - \sum_{v_i \in G_1} \sigma^2(i)d(v_p,v_i)}{\sum_{i=1}^{n} \sigma^2(i)}. \quad (6.8.15c)$$

Proof. Since b_k is a cut-set of G, there are no paths from any vertex of G_1 to any vertex of G_2, except through b_k. Let $v_p \in G_1$ and $v_q \in G_2$. Then, for $v_i \in G_1$, $v_j \in G_2$, and Y on b_k,

$$d(v_i,Y) = d(v_i,v_p) + d(v_p,Y) \quad (6.8.16a)$$

and

$$d(v_j,Y) = d(v_j,v_q) + d(v_q,Y). \quad (6.8.16b)$$

Referring to Fig. 6.7.3, we find

$$d(v_i,Y) = d(v_i,v_p) + y \quad (6.8.17a)$$

$$\begin{aligned} d(v_j,Y) &= d(v_j,v_q) + w_k - y \\ &= d(v_j,v_p) - y. \end{aligned} \quad (6.8.17b)$$

Therefore, from (6.8.17),

$$\operatorname{Var}\left\{\sum_{i=1}^{n} H(i)d(v_i,Y)\right\} = \sum_{v_i\in G_1} \sigma^2(i)(d(v_i,v_p) + y)^2 + \sum_{v_j\in G_2} \sigma^2(j)(d(v_j,v_p) - y)^2$$

$$= y^2 \sum_{i=1}^{n} \sigma^2(i) + 2y\left[\sum_{v_i\in G_1} \sigma^2(i)d(v_i,v_p) - \sum_{v_j\in G_2} \sigma^2(j)d(v_j,v_p)\right]$$

$$+ \sum_{i=1}^{n} \sigma^2(i)d^2(v_i,v_p). \tag{6.8.18}$$

Taking the derivative with respect to y of both sides gives

$$\frac{d}{dy}\operatorname{Var}\left\{\sum_{i=1}^{n} H(i)d(v_i,Y)\right\} = 2y\sum_{i=1}^{n} \sigma^2(i) + 2\sum_{v_i\in G_1} \sigma^2(i)d(v_i,v_p) - 2\sum_{v_j\in G_2} \sigma^2(j)d(v_j,v_p). \tag{6.8.19}$$

Consequently,

$$\operatorname{Var}\left\{\sum_{i=1}^{n} H(i)d(v_i,Y)\right\}$$

attains an *extremum* at $y = Q$. By taking second derivatives we see that a *minimum* occurs at this point. If $0 \le Q \le w_k$, then Y_{ov}, located on b_k by $y = Q$, is a local minimum variance median. If $Q < 0$, then Y_{ov} is at v_p and if $Q > w_k$ then Y_{ov} is at v_q. If G is a tree, every branch of G is a cut-set of G. Hence we have:

Corollary 6.8.1 *If G is a tree, the minimum variance absolute median is located on a branch $b_k = [p,q]$ with $d(v_p,Y_{ov})$ given by (6.8.15b).*

The proof of Theorem 6.8.2 uses the fact that if Y is on a cut-set of G, then $d(v_i,Y) = d(v_i,v_s) + d(v_s,Y)$ where $s = p$ or $s = q$. Let G be a connected graph and b_k any branch of G. From (6.8.7), $d(v_i,Y) = d(v_i,v_p) + y$ for $0 \le y \le a_{k,i}$ and $d(v_i,Y) = d(v_i,v_q) + w_k - y$ for $a_{k,i} < y < w_k$. Introduce a new vertex at $y = a_{k,i}$ and label the subdivided branch as $b_{k,1}$ and $b_{k,2}$. For the purposes of computing $d(v_i,Y)$, (6.8.16a) and (6.8.16b) hold on each branch segment. Generate all $a_{k,i}$ for $i = 1,\ldots,n$ and reorder them such that $a_{k,i} \le a_{k,(i+1)}$. The $a_{k,i}$ divide b_k into $w \le n + 1$ segments, and on each such segment (6.8.16a) and (6.8.16b) hold. Consequently, the appropriate modification of Theorem 6.8.2 can be used to find the minimum variance point on each segment. The local minimum variance absolute median is easily found among the set of these points. The same technique can be applied to find MPARM's if the normal approximation discussed above is used.

Example 6.8.2 As an example of Theorem 6.8.2, and its generalization, consider again the graph in Fig. 6.7.1. For simplicity, let $\sigma^2(1) = \sigma^2(2) = \cdots = \sigma^2(6) = 1$. We will find a local minimum variance median on branch b_5 of G. We first find the

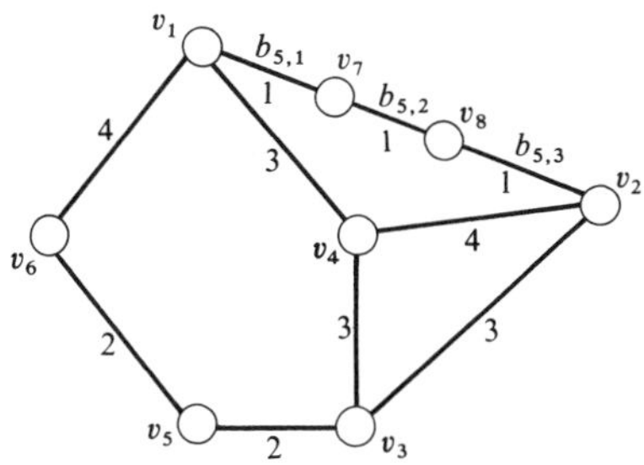

Fig. 6.8.4 Graph for Example 6.8.4.

$a_{5,i}$ to be

$$a_{5,3} = 0, a_{5,4} = 2, a_{5,5} = 1, a_{5,6} = 3. \tag{6.8.20}$$

Hence branch b_5 should be divided into three segments $b_{5,1}$, $b_{5,2}$, and $b_{5,3}$, as shown in Fig. 6.8.4, to find a local minimum variance point.

On $b_{5,1}$, v_4, v_5, v_6 are in the subgraph in G_1 of the theorem; i.e., the distance from any point on $b_{5,1}$ to v_4, v_5, or v_6 is measured with respect to a path passing through v_1. Vertices v_2 and v_3 are in G_2. Let

$$Q_1 = \frac{d(v_1,v_2) + d(v_1,v_3) - d(v_1,v_4) - d(v_1,v_5) - d(v_1,v_6)}{6}. \tag{6.8.21}$$

Then $Q_1 < 0$ and the local minimum variance median on $b_{5,1}$ is at v_1.

On $b_{5,2}$, G_1 contains the vertices v_1, v_4, and v_6 and G_2 the vertices v_2, v_3, and v_5. Let

$$Q_2 = \frac{d(v_7,v_2) + d(v_7,v_3) + d(v_7,v_5) - d(v_7,v_6) - d(v_7,v_1) - d(v_7,v_4)}{6}. \tag{6.8.22}$$

Then $Q_2 = \frac{2}{3}$ and consequently the local minimum variance median on $b_{5,2}$ is two thirds of a unit from v_7.

On $b_{5,3}$, G_1 contains vertices v_1 and v_6 while G_2 contains v_2, v_3, v_4, and v_5. Let

$$Q_3 = \frac{d(v_8,v_2) + d(v_8,v_3) + d(v_8,v_4) + d(v_8,v_5) - d(v_8,v_1) - d(v_8,v_6)}{6}. \tag{6.8.23}$$

Then, $Q_3 = \frac{4}{3}$ and consequently the minimum variance point on $b_{5,3}$ is located at v_2.

Now, to determine the minimum variance median on b_5, we simply compare the variance at the three points identified above. At v_1, $\sigma^2 = 106$; at v_2, $\sigma^2 = 108$; and $\frac{5}{3}$ units from v_1, $\sigma^2 = 117\frac{1}{3}$. Hence, the local minimum variance median on b_5 is at v_1.

A problem analogous to that considered above is to find a point X on G which minimizes Var $\{\text{Max}[H(i)d(v_i,X)]\}$. This problem is far more complicated because of the difficulty of obtaining analytic expressions for the moments of a finite set of nonidentical random variables. Even if normality assumptions are made, the approximate expressions encountered are complicated and the problem is exceedingly cumbersome (see Chapter 4).

Often, it is unreasonable to assume that the probability distributions of the $H(i)$ are known. A more reasonable assumption is that a set of observations

$h_1(i), \ldots, h_{m_i}(i)$ of $H(i)$ is available. Let the random variable $H_j(i)$ for $i = 1, \ldots, n$, and $j = 1, \ldots, m_i$, correspond to the jth observation of $H(i)$ and suppose that $H_1(i), \ldots, H_{m_i}(i)$ are identically and independently distributed. Furthermore, assume that $P_i[z]$, the unknown probability distribution of $H(i)$, is strictly increasing; that is, $H(i)$ is a continuous random variable.

On the basis of the observed values of $H(i)$, we would like to locate the optimum points of G. Let $S_{m_i}(z)$ be the empirical distribution function of $H(i)$ [FI1]. In other words,

$$S_{m_i}(z) = \frac{1}{m_i} \sum_{j=1}^{m_i} f_z(H_j(i)) \tag{6.8.24}$$

where

$$f_z(U) = \begin{cases} 0 & \text{for } U \geq z, \\ 1 & \text{for } U < z. \end{cases}$$

Thus, $m_i S_{m_i}(z)$ is the number of samples of $H(i)$ which are smaller than z. The following theorem shows that if the m_i are large, local maximum probability absolute centers found by using S_{m_i} as the true distributions of the $H(i)$ will be "close" to the actual centers of G. Since the proof of this theorem is lengthy the reader may wish to omit it on first reading.

Theorem 6.8.3 *Let $S_{m_i}(z)$ be the empirical distribution function of the continuous independent random variable $H(i)$ whose true distribution is $P_i[z]$ for $i = 1, \ldots, n$. Let $X_0(m_1, \ldots, m_n)$ be a local MPArC on branch b_k, found with respect to the S_{m_i}. If $\delta > 0$, there exists an integer M, and a point X_0 on b_k, such that if $m_i \geq M$ for $i = 1, \ldots, n$, then*

$$\text{Prob}\,\{d(X_0, X_0(m_1, \ldots, m_n)) < \delta\} \to 1 \text{ as } M \to \infty \tag{6.8.25}$$

where X_0 is a local maximum probability absolute r center of G.

Proof. 1) We first prove that for any $\varepsilon > 0$, there exists an M such that for $m_i > M$ with $i = 1, \ldots, n$,

$$\text{Prob}\left\{\prod_{i=1}^{n} S_{m_i}(z) - \prod_{i=1}^{n} P_i[z] < \varepsilon\right\} \to 1 \text{ as } M \to \infty. \tag{6.8.26}$$

By Glivenko's Theorem [FI1], for $\beta > 0$,

$$\text{Prob}\,\{\sup_z |S_{m_i}(z) - P_i[z]| < \beta\} \to 1 \text{ as } m_i \to \infty. \tag{6.8.27}$$

This implies

$$\text{Prob}\,\{P_i[z] - \beta < S_{m_i}(z) < P_i[z] + \beta\} \to 1 \text{ as } m_i \to \infty. \tag{6.8.28}$$

For M an integer and $m_i \geq M$ with $i = 1, \ldots, n$

$$\text{Prob}\left\{\prod_{i=1}^{n} (P_i[z] - \beta) < \prod_{i=1}^{n} S_{m_i}(z) < \prod_{i=1}^{n} (P_i[z] + \beta)\right\} \to 1 \text{ as } M \to \infty. \tag{6.8.29}$$

From the inequalities

$$\prod_{i=1}^{n} (P_i[z] + \beta) \leq \prod_{i=1}^{n} P_i[z] - 1 + (1 + \beta)^n = \prod_{i=1}^{n} P_i[z] + \sum_{k=1}^{n} \beta^k \binom{n}{k} \quad (6.8.30)$$

and

$$\prod_{i=1}^{n} (P_i[z] - \beta) \geq \prod_{i=1}^{n} P_i[z] + (-\beta)^n \geq \prod_{i=1}^{n} P_i[z] - \sum_{k=1}^{n} \beta^k \binom{n}{k}, \quad (6.8.31)$$

it follows that

$$\text{Prob}\left\{-\sum_{k=1}^{n} \binom{n}{k} \beta^k < \prod_{i=1}^{n} S_{m_i}(z) - \prod_{i=1}^{n} P_i[z] < \sum_{k=1}^{n} \binom{n}{k} \beta^k\right\} \to 1 \text{ as } M \to \infty. \quad (6.8.32)$$

If $\beta = (\varepsilon + 1)^{1/n} - 1$, (6.8.26) follows.

2) Let $X = X_o$ maximize

$$P(X) = \prod_{i=1}^{n} P_i\left[\frac{r}{d(v_i,X)}\right] \quad (6.8.33)$$

and let $X = X_{om} = X_o(m_1, \ldots, m_n)$ maximize the random variable

$$S(X) = \prod_{i=1}^{n} S_{m_i}\left(\frac{r}{d(v_i,X)}\right). \quad (6.8.34)$$

Then, we will prove that for $m_i > M$

$$\text{Prob}\{P(X_{om}) > P(X_o) - 2\varepsilon\} \to 1 \text{ as } M \to \infty. \quad (6.8.35)$$

From (6.8.27)

$$\text{Prob}\{P(X_o) - \varepsilon < S(X_o) < P(X_o) + \varepsilon\} \to 1. \quad (6.8.36)$$

Let $P(X_{om}) = P(X_o) - \delta$, where δ is a nonnegative random variable. Then, from (6.8.36)

$$\text{Prob}\{P(X_{om}) + \delta - \varepsilon < S(X_o)\} \to 1. \quad (6.8.37)$$

By the assumption of (6.8.34), $\text{Prob}\{S(X_o) \leq S(X_{om})\} = 1$, and from (6.8.27) we also have

$$\text{Prob}\{P(X_{om}) - \varepsilon < S(X_{om}) < P(X_{om}) + \varepsilon\} \to 1. \quad (6.8.38)$$

Hence,

$$\text{Prob}\{P(X_{om}) + \delta - \varepsilon < S(X_o) \leq S(X_{om}) < P(X_{om}) + \varepsilon\} \to 1 \quad (6.8.39)$$

and therefore

$$\text{Prob}\,\{(X_{om}) + \delta - \varepsilon < P(X_{om}) + \varepsilon\} \to 1. \tag{6.8.40}$$

Consequently,

$$\text{Prob}\,\{\delta < 2\varepsilon\} \to 1 \text{ as } M \to \infty, \tag{6.8.41}$$

and (6.8.35) follows.

3) Let x_o and x_{om} be the locations of X_o and X_{om} on b_k, respectively, and assume

$$\frac{d}{dx} P(x)\bigg|_{x=x_0} \neq 0.$$

If this is not the case, the arguments to follow are easily modified by considering the endpoints of the flat interval. Then, since $P(x_o)$ is maximum, within a sufficiently small ε region of x_o, $P(X)$ must be strictly increasing for $x < x_o$ and strictly decreasing for $x > x_o$. Consequently, within this region, the inverse function of $P(x)$ exists and is continuous. Thus, for any $\gamma > 0$, we can find an $\alpha > 0$ such that

$$|P(x_o) - P(x)| < \alpha \tag{6.8.42a}$$

implies

$$|x_o - x| < \gamma. \tag{6.8.42b}$$

Note that $|x_o - x_{om}| = d(X_o,X_{om})$ for sufficiently small γ. From (6.8.35), if $\alpha = 2\varepsilon$,

$$\text{Prob}\,\{|P(X_o) - P(X_{om})| < \alpha\} \to 1 \tag{6.8.43}$$

and therefore

$$\text{Prob}\,\{d(X_o,X_{om}) < \gamma\} \to 1 \text{ as } M \to \infty. \tag{6.8.44}$$

This completes the proof of the theorem.

The problem of finding the maximum probability absolute medians of G on the basis of the population samples requires, in general, an n-fold convolution of the empirical distributions. However, the problem may be considerably simplified if we make the assumption that

$$\sum_{i=1}^{n} H(i)d(v_i,Y)$$

is a normal random variable. In this case, we must maximize

$$\frac{R - \mu(Y)}{\sigma(Y)} = \frac{R - \displaystyle\sum_{i=1}^{n} E\{H(i)\}d(v_i,Y)}{\left(\displaystyle\sum_{i=1}^{n} \text{Var}\,\{H(i)\}d^2(v_i,Y)\right)^{1/2}}. \tag{6.8.45}$$

Let $\bar{H}_{m_i}$ and $V_{m_i}^2$ be random variables defined by

$$\bar{H}_{m_i} = \frac{1}{m_i} \sum_{j=1}^{m_i} H_j(i) \qquad \text{for } i = 1,2,\ldots,n \tag{6.8.46a}$$

and

$$V_{m_i}^2 = \frac{1}{m_i} \sum_{j=1}^{m_i} (H_j(i) - \bar{H}_{m_i})^2 \qquad \text{for } i = 1,2,\ldots,n. \tag{6.8.46b}$$

The sample mean and variance, $\bar{H}_{m_i}$ and $V_{m_i}^2$, are well-behaved estimates of the population mean $E\{H(i)\}$ and variance $\sigma^2(i)$. For example, it can be shown [FI1] that for $\varepsilon > 0$,

$$\lim_{m_i \to \infty} \text{Prob}\,\{|\bar{H}_{m_i} - E\{H(i)\}| > \varepsilon\} = 0, \tag{6.8.47a}$$

and

$$\lim_{m_i \to \infty} \text{Prob}\,\{|V_{m_i}^2 - \sigma_{v_i}^2| > \varepsilon\} = 0. \tag{6.8.47b}$$

In other words, the sample mean and variance converge *stochastically* to the population mean and variance.

We base our optimization procedure on the H_{m_i} and the $V_{m_i}^2$. Thus, we find a point Y on G which maximizes

$$\frac{R - \hat{\mu}(Y)}{\hat{\sigma}(Y)} = \frac{R - \sum_{i=1}^{n} \bar{H}_{m_i} d(v_i, Y)}{\left(\sum_{i=1}^{n} V_{m_i}^2 d^2(v_i, Y)\right)^{1/2}}. \tag{6.8.48}$$

To see that this is a reasonable procedure, consider the quantity

$$S = \frac{R - \sum_{i=1}^{n} \bar{H}_{m_i} d(v_i, Y)}{\left(\sum_{i=1}^{n} V_{m_i}^2 d^2(v_i, Y)\right)^{1/2}} - \frac{R - \sum_{i=1}^{n} E\{H(i)\} d(v_i, Y)}{\left(\sum_{i=1}^{n} \sigma^2(i) d^2(v_i, Y)\right)^{1/2}}. \tag{6.8.49}$$

We will show that for $m_i > M$ and $i = 1,2,\ldots,n$

$$\lim_{M \to \infty} \text{Prob}\left\{\left|\frac{R - \hat{\mu}(Y)}{\hat{\sigma}(Y)} - \frac{R - \mu(Y)}{\sigma(Y)}\right| > \varepsilon\right\} = 0. \tag{6.8.50}$$

Let $\bar{H}_{m_i} = E\{H(i)\} + \delta_{1i}$ and $V_{m_i}^2 = \sigma^2(i) + \delta_{2i}$, where δ_{1i} and δ_{2i} are random variables. To simplify matters, consider the quantity

$$S' = \frac{[R - \hat{\mu}(Y)]^2}{\hat{\sigma}^2(Y)} - \frac{[R - \mu(Y)]^2}{\sigma^2(Y)}. \tag{6.8.51}$$

Then, substituting the values of $\bar{H}_{m_i}$ and $V_{m_i}^2$

$$S' = \frac{R^2 + \left[\sum_{i=1}^{n} (E\{H(i)\} + \delta_{1i})d(v_i,Y)\right]^2 - 2R\left[\sum_{i=1}^{n} (E\{H(i)\} + \delta_{1i})d(v_i,Y)\right]}{\sum_{i=1}^{n} (\sigma^2(i) + \delta_{2i})d^2(v_i,Y)}$$

$$- \frac{R^2 + \left[\sum_{i=1}^{n} E\{H(i)\}d(v_i,Y)\right]^2 - 2R\left[\sum_{i=1}^{n} E\{H(i)\}d(v_i,Y)\right]}{\sum_{i=1}^{n} \sigma^2(i)d^2(v_i,Y)}. \tag{6.8.52}$$

If we combine the quantity on the right-hand side of (6.8.52) into a single fraction and simplify the expression, we find that the *numerator* N of S' is

$$N = \sigma^2(Y)\mu(Y)\left[\sum_{i=1}^{n} \delta_{1i}d(v_i,Y)\right] + \left[\sum_{i=1}^{n} \delta_{1i}d(v_i,Y)\right]^2 \sigma^2(Y)$$

$$+ 2R\mu(Y)\left[\sum_{i=1}^{n} \delta_{2i}d^2(v_i,Y)\right] - 2R\sigma^2(Y)\sum_{i=1}^{n} \delta_{1i}d(v_i,Y)$$

$$- R^2\sum_{i=1}^{n} \delta_{2i}d^2(v_i,Y) - [\mu(Y)]^2\sum_{i=1}^{n} \delta_{2i}d^2(v_i,Y). \tag{6.8.53}$$

By the stochastic convergence of $\bar{H}_{m_i}$ and $V_{m_i}^2$,

$$\lim_{M\to\infty} \text{Prob}\,\{|\delta_{1i}| > \varepsilon\} = \lim_{M\to\infty} \text{Prob}\,\{|\delta_{2i}| > \varepsilon\} = 0 \qquad \text{for } i = 1,2,\ldots,n. \tag{6.8.54}$$

Hence, if we define the real, positive number β as

$$\beta = \varepsilon\sigma^2(Y)\mu(Y)\sum_{i=1}^{n} d(v_i,Y) + \varepsilon^2\sigma^2(Y)\left[\sum_{i=1}^{n} d(v_i,Y)\right]^2$$

$$+ \varepsilon 2R\mu(Y)\sum_{i=1}^{n} d^2(v_i,Y) + \varepsilon 2R\sigma^2(Y)\sum_{i=1}^{n} d(v_i,Y)$$

$$+ \varepsilon R^2\sum_{i=1}^{n} d^2(v_i,Y) + \varepsilon[\mu(Y)]^2\left[\sum_{i=1}^{n} d^2(v_i,Y)\right], \tag{6.8.55}$$

we see that

$$\lim_{M\to\infty} \text{Prob}\left\{\frac{-\beta}{[\sigma^2(Y)]^2 + \varepsilon\sigma^2(Y)} \le S' \le \frac{\beta}{[\sigma^2(Y)]^2 + \varepsilon\sigma^2(Y)}\right\} = 1. \quad (6.8.56)$$

Since β can be made arbitrarily small by making ε arbitrarily small, the procedure outlined above is reasonable.

6.9 MINIMUM COST REALIZATION OF TERMINAL CAPACITIES

In Chapter 5 we gave methods of realizing terminal capacity matrices without reference to cost. In general, the realization of a terminal capacity matrix at minimum cost with linear cost functions can be done only by resorting to linear programming formulations. However, for the case of symmetric matrices a simple solution is available when the cost function is the sum of the branch capacities in the resulting graph. We begin the procedure by realizing a terminal capacity matrix by a tree as in Section 5 of Chapter 5 and then proceed systematically to modify the graph so as to reduce the cost while leaving the terminal capacity matrix unchanged. In this procedure, we require the operations of Uniform Tree Formation and Circuit Formation, introduced below. A *uniform tree* is a tree in which all branches have the same capacity. A *linear tree* is a tree which is a path.

Uniform Tree Formation Given a linear tree G, construct a linear tree G_1 and a graph G_2 as follows: let G_1 have the same number of vertices and branches as G; let G_2 have the same number of vertices as G; let the vertices of G_1 and G_2 be numbered as in G.

a) For all $[i,j]$ in G_1, $c[i,j]$ is equal to the smallest nonzero capacity of any branch in G.
b) For all $[i,j]$ in G_2, $c[i,j]$ is equal to the capacity of $[i,j]$ in G minus the capacity of $[i,j]$ in G_1.

Circuit Formation Given a uniform linear tree, with common branch capacity c,

a) Reduce the capacity of each branch to $c/2$.
b) Add a branch with capacity $c/2$ between the two vertices of degree one.

The procedure is then as follows:

Minimum Cost Terminal Capacity Algorithm

Step 1. Repeat the operation of Uniform Tree Formation on the given linear tree and on all resulting graphs until every graph is a uniform linear tree on $k \le n$ vertices.

Step 2. Apply Circuit Formation to every uniform linear tree of more than one branch resulting from Step 1.

Step 3. Form an s-graph consisting of the union of all vertices and branches generated in Step 2. Form the final graph by combining parallel branches into a single branch whose capacity is the sum of the capacities of the parallel branches.

The application of the algorithm is illustrated in the next example.

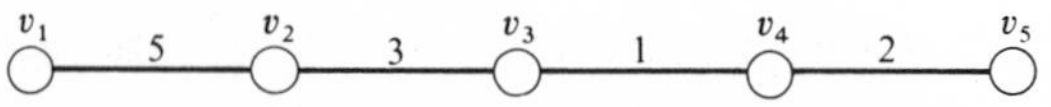

Fig. 6.9.1 Linear tree realization of a symmetric terminal capacity matrix.

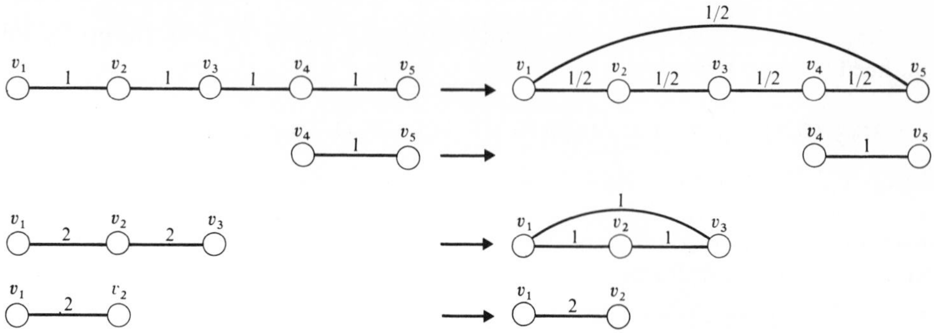

Fig. 6.9.2 Graphs illustrating the Minimum Terminal Capacity Algorithm.

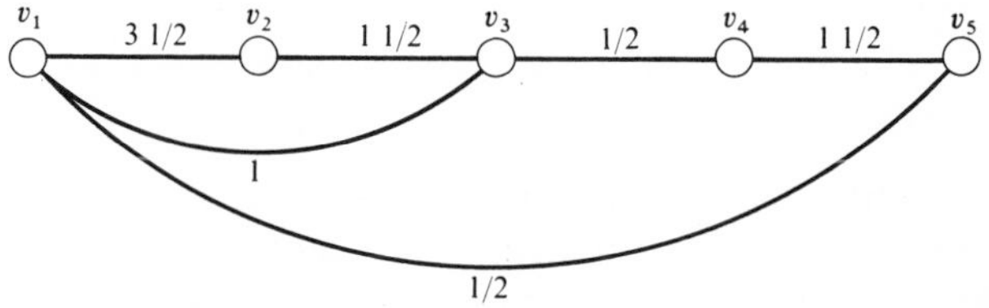

Fig. 6.9.3 Minimum cost realization.

Example 6.9.1 For the uniform linear tree in Fig. 6.9.1, Step 1 of the Minimum Cost Terminal Capacity Algorithm yields the uniform linear trees in Fig. 6.9.2 which are transformed by Step 2 into the circuits shown. The final graph resulting from Step 3 is shown in Fig. 6.9.3. It has a total capacity of 8.5 as compared to a total capacity of 11 for the graph in Fig. 6.9.1.

We now show that if a linear tree realizes a terminal capacity matrix, the graph resulting from the Minimum Cost Terminal Capacity Algorithm also realizes the matrix. We require Lemma 6.9.1 as an intermediate step. The lemma is proved for directed graphs although we will use it only for the undirected case.

Let $T^{(i)}$ be the terminal capacity matrix of graph $G^{(i)}$ for $i = 1,2,3$ where $G^{(1)}$, $G^{(2)}$, and $G^{(3)}$ each contain n vertices. Furthermore, let $c^{(3)}(k,j) = c^{(1)}(k,j) + c^{(2)}(k,j)$ where $c^{(i)}(k,j)$ is the capacity of branch (k,j) in $G^{(i)}$.

Lemma 6.9.1

$$T^{(3)} = T^{(2)} + T^{(1)}, \tag{6.9.1}$$

if and only if for each distinct v_k and v_j, there is a k-j cut $(X,\bar{X})$ which is a minimum k-j cut in $G^{(1)}$, $G^{(2)}$, and $G^{(3)}$.

Proof. Suppose $(X,\bar{X})$ is a minimum k-j cut in $G^{(1)}$, $G^{(2)}$, and $G^{(3)}$. Then

$$\tau_{k,j}^{(1)} = c^{(1)}(X,\bar{X}), \tag{6.9.2a}$$

$$\tau_{k,j}^{(2)} = c^{(2)}(X,\bar{X}), \tag{6.9.2b}$$

$$\tau_{k,j}^{(3)} = c^{(3)}(X,\bar{X}). \tag{6.9.2c}$$

But by definition

$$c^{(3)}(X,\bar{X}) = c^{(1)}(X,\bar{X}) + c^{(2)}(X,\bar{X}). \tag{6.9.3}$$

From (6.9.2) and (6.9.3)

$$\tau_{k,j}^{(3)} = \tau_{k,j}^{(2)} + \tau_{k,j}^{(1)}. \tag{6.9.4}$$

Suppose next

$$\tau_{k,j}^{(3)} = \tau_{k,j}^{(2)} + \tau_{k,j}^{(1)} \tag{6.9.5}$$

and there is no k-j cut which is a minimum k-j cut in $G^{(1)}$ and $G^{(2)}$ and $G^{(3)}$. Let $(X,\bar{X})$ be a minimum k-j cut in $G^{(3)}$. Then either

$$c^{(2)}(X,\bar{X}) > \tau_{k,j}^{(2)} \tag{6.9.6a}$$

or

$$c^{(1)}(X,\bar{X}) > \tau_{k,j}^{(1)}. \tag{6.9.6b}$$

Hence

$$\tau_{i,j}^{(3)} = c^{(1)}(X,\bar{X}) + c^{(2)}(X,\bar{X}) > \tau_{k,j}^{(1)} + \tau_{k,j}^{(2)} \tag{6.9.7}$$

Equation (6.9.7) contradicts (6.9.5) and the proof is completed.

We can now show:

Theorem 6.9.1 *If the Minimum Cost Terminal Capacity Algorithm is applied to a linear tree with terminal capacity matrix T, the resulting graph G also has a terminal capacity matrix T.*

Proof. The operation of Uniform Tree Formation clearly does not affect T. If a circuit is formed from a linear tree by Circuit Formation, clearly the maximum i-j flow is the same in the tree and the circuit for all i,j. Moreover, from Lemma 6.9.1 the terminal capacity matrix of the graph resulting from Step 3 is the sum of the terminal capacity matrices of the circuits generated in Step 3. The proof is therefore completed.

It is obvious that, if a requirement matrix $\mathscr{R}$ is to be realized by a graph,

$$\sum_{v_j \in V} c[i,j] \geq \operatorname*{Max}_{j} [r_{i,j}] \text{ for all } i. \tag{6.9.8}$$

We complete the justification of the algorithm by showing that it yields a minimum cost realization.

Theorem 6.9.2 *The Minimum Cost Terminal Capacity Algorithm yields a graph with*

$$\sum_{v_j \in V} c[i,j] \leq \operatorname*{Max}_j [r_{i,j}]. \tag{6.9.9}$$

Proof. The algorithm begins with a linear tree realization of $\mathscr{R}$. In this realization, if the vertices are numbered in the order they appear in the tree, then

$$\sum_{v_j \in V} c[i,j] = r_{i-1,i} + r_{i,i+1} \tag{6.9.10}$$

for $i = 2, \ldots, n-1$. In the next steps branches incident at v_i, of total capacity $2 \operatorname{Min} [r_{i-1,i}, r_{i,i+1}]$, appear in some uniform tree of more than two vertices. Hence, when the trees are converted to circuits, this sum becomes $\operatorname{Min} [r_{i-1,i}, r_{i,i+1}] = \alpha_i$; α_i is then added to the remaining unchanged portion, Δ_i, of the capacity of all branches at v_i where

$$\Delta_i = \operatorname{Max} [r_{i-1,i}, r_{i,i+1}] - \operatorname{Min} [r_{i-1,i}, r_{i,i+1}]. \tag{6.9.11}$$

The final total is therefore

$$\begin{aligned} \sum_{v_j \in V} c[i,j] &= \operatorname{Min} [r_{i-1,i}, r_{i,i+1}] + \operatorname{Max} [r_{i-1,i}, r_{i,i+1}] - \operatorname{Min} [r_{i-1,i}, r_{i,i+1}] \\ &= \operatorname{Max} [r_{i-1,i}, r_{i,i+1}] \leq \operatorname*{Max}_{v_j \in V} [r_{i,j}] \end{aligned} \tag{6.9.12}$$

and the theorem is proved for $i = 2, \ldots, n-1$.

For $i = 1$ or $i = n$ the value of $\sum_{v_j \in V} c[i,j]$ is unchanged throughout the algorithm and has a value of $r_{1,2}$ and $r_{n-1,n}$ respectively. Thus we again have

$$\sum_{v_j \in V} c[i,j] \leq \operatorname*{Max}_{v_j \in V} [r_{i,j}]. \tag{6.9.13}$$

The theorem is therefore proved.

We can now obtain a realization with minimum total capacity, for any realizable symmetric matrix $\mathscr{R}$. Suppose next we are given a symmetric requirement matrix $\mathscr{R}$ which is not realizable. In this case we seek a minimum cost graph G with terminal capacity matrix T such that

$$T \geq \mathscr{R}. \tag{6.9.14}$$

In other words,

$$\tau_{i,j} \geq r_{i,j} \qquad \text{for all } i,j \text{ with } i \neq j. \tag{6.9.15}$$

The following algorithm will yield such a graph.

Minimum Excess Terminal Capacity Algorithm

Step 1. Construct a graph $\hat{G}$ with n vertices such that the length $l[i,j] = r[i,j]$ for all $i \neq j$.

Step 2. Let $\tilde{G}$ be a longest tree in G. In $\tilde{G}$ let $c[i,j] = l[i,j]$.

Step 3. Find a linear tree G with the same terminal capacity matrix as $\tilde{G}$.

Step 4. Apply the Minimum Cost Terminal Capacity Algorithm to G.

To prove that the resulting graph has minimum total capacity we first show:

Theorem 6.9.3 *For a tree $\tilde{G}$, $T \geq \mathscr{R}$ if and only if for any branch $[i,j]$ in $\tilde{G}$, $c[i,j] \geq l[i,j]$.*

Proof. In order to have $\tau_{i,j} \geq r_{i,j}$ in a tree containing $[i,j]$ we must have $c[i,j] \geq r[i,j] = l[i,j]$. If $c[i,j] \geq l[i,j] \geq r[i,j]$ for all branches in $\tilde{G}$,

$$\tau_{x,z} \geq \text{Min}\,[\tau_{x,y},\tau_{y,u},\ldots,\tau_{w,z}] \tag{6.9.16}$$

where $v_x,v_y,v_u,\ldots,v_w,v_z$ represents an x-z path in the tree. Hence,

$$\tau_{x,z} \geq \text{Min}\,[r_{x,y},r_{y,u},\ldots,r_{w,z}]. \tag{6.9.17}$$

But

$$\text{Min}\,[r_{x,y},r_{y,u},\ldots,r_{w,z}] \geq r_{x,z} \tag{6.9.18}$$

since $\tilde{G}$ is a longest tree. Therefore

$$\tau_{x,z} \geq r_{x,z} \tag{6.9.19}$$

and the proof is completed.

The fact that the final graph has minimum total capacity follows from (6.9.8).

Example 6.9.2 To illustrate the use of the Minimum Excess Terminal Capacity Algorithm consider the unrealizable requirement matrix

$$\mathscr{R} = \begin{bmatrix} \infty & 1 & 2 & 3 \\ 1 & \infty & 4 & 5 \\ 2 & 4 & \infty & 6 \\ 3 & 5 & 6 & \infty \end{bmatrix}. \tag{6.9.20}$$

$\hat{G}$ is given in Fig. 6.9.4(a). $\tilde{G}$ is given in Fig. 6.9.4(b). The linear tree of Step 3 is in Fig. 6.9.4(c) and the final graph resulting from Step 4 is in Fig. 6.9.4(d).

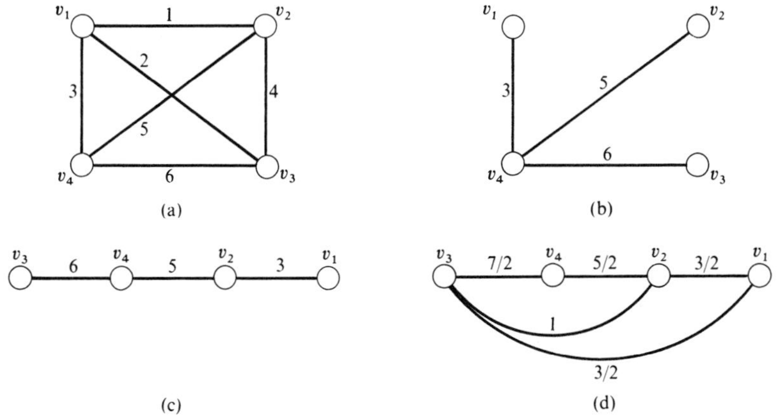

Fig. 6.9.4 Graphs for Example 6.9.2.

In the methods presented thus far we had complete freedom in selecting a graph and branch capacities. We now consider the minimum cost modification of an existing graph to satisfy a given requirement. Suppose we are given a graph G with fixed branch capacities $c_{01}, c_{02}, \ldots, c_{0m}$. Suppose that $\tau_{s,t} < r_{s,t}$ where $r_{s,t}$ is a given flow requirement. One way of obtaining $\tau_{s,t} \geq r_{s,t}$ is to increase the capacities of the branches of G. Naturally, we would like to accomplish this with minimum cost.

Let the cost of increasing the capacity of branch b_i by Δc_i be $h_i \Delta c_i$ where h_i is a nonnegative number. Then the problem may be stated as: find $\Delta c_1, \ldots, c_m$ such that

$$\tau_{s,t} \geq r_{s,t} \tag{6.9.21}$$

and

$$\sum_{i=1}^{m} h_i \Delta c_i \text{ is minimized.} \tag{6.9.22}$$

This problem can be easily solved by the minimum cost maximum flow method discussed in Section 4. The cost function for b_k is shown in Fig. 6.9.5(a). After applying the algorithm to the graph to obtain a minimum cost flow of value $r_{s,t}$, Δc_k is given by

$$\Delta c_k = \text{Max}\,[0,(X_k - c_{0k})]. \tag{6.9.23}$$

In addition, the same algorithm can be used to solve the problem of optimally adding branch capacity when the cost function $h_k(f_k)$ for b_k is convex, as shown in Fig. 6.9.5(b).

In the remainder of this section we will formulate the problem defined by (6.9.21) and (6.9.22) as a linear program. The advantage of this alternative formulation is that it can be extended to the more difficult problem discussed in the following two sections.

To satisfy (6.9.21),

$$\sum_{k=1}^{m} a_{i,k}(c_{0k} + \Delta c_k) \geq r_{s,t} \qquad \text{for } i = 1,2,\ldots,q, \tag{6.9.24}$$

where $\mathscr{A}_{s,t} = [a_{i,j}]$ is the s-t cut-set matrix of G and q is the number of s-t cut-sets.

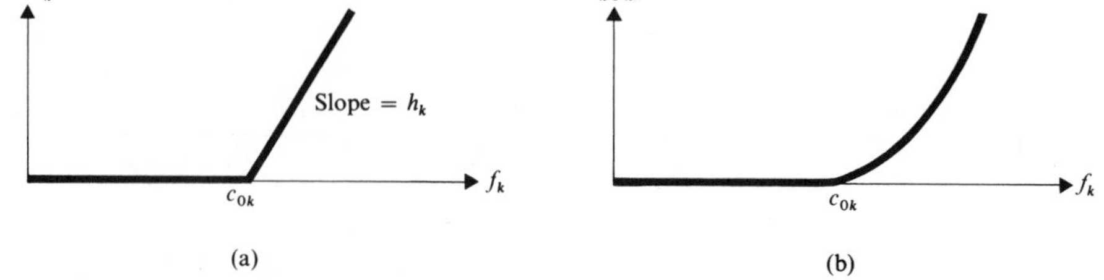

Fig. 6.9.5 Cost functions for flow in branch b_k.

The problem can be formulated as the linear program: Find a vector $\Delta\mathbf{c} \geq \mathbf{0}$ to minimize $\mathbf{h}'\Delta\mathbf{c}$ subject to

$$\mathscr{A}_{s,t}(\mathbf{c}_0 + \Delta\mathbf{c}) \geq \mathbf{r}_{s,t} \qquad (6.9.25)$$

where $\mathbf{h}' = (h_1,h_2,\ldots,h_m)$ and $\mathbf{r}_{s,t}$ is a $q \times 1$ matrix whose entries are all equal to $r_{s,t}$.

Rather than solve this problem, we can give a better formulation. Assign arbitrary directions to the branches of G and add a directed branch b^*_{m+1} directed from v_t to v_s. Call the resulting graph G^*. Let the capacity of branch b^*_k in G^* be $c_{0k} = c_{0k}$ if $k = 1,2,\ldots,m$ and infinity if $k = m + 1$. Denote the branch flow vector of G^* as

$$\mathbf{X} = (X_1,\ldots,X_{m+1})'.$$

Then, in G, $\tau_{s,t} \geq r_{s,t}$ if and only if we can find a value of $\mathbf{X}$ in G^* such that $X_{m+1} \geq r_{s,t}$.

Let $\mathscr{U}^*$ be the incidence matrix of G^*. Any flow in G^*, with source v_s and terminal v_t, must satisfy

$$\mathscr{U}^*\mathbf{X} = \mathbf{0} \qquad (6.9.26a)$$

and

$$-c^*_k \leq X_k \leq c^*_k \text{ for } k = 1,\ldots,m. \qquad (6.9.26b)$$

Let

$$\mathbf{X}_1 = (X_1,X_2,\ldots,X_m)'.$$

Then, we can achieve $\tau_{s,t} \geq r_{s,t}$ with minimum cost by using the following linear program: Find $\mathbf{X}$ and $\Delta\mathbf{c} \geq \mathbf{0}$ to minimize $\mathbf{h}'\Delta\mathbf{c}$ subject to

$$\mathscr{U}^*\mathbf{X} = \mathbf{0}, \qquad (6.9.27a)$$

$$X_{m+1} \geq r_{s,t}, \qquad (6.9.27b)$$

$$-\mathbf{c}^*_0 - \Delta\mathbf{c} \leq \mathbf{X}_1 \leq \mathbf{c}^*_0 + \Delta\mathbf{c}. \qquad (6.9.27c)$$

We can introduce a change of variables and write the above problem in terms of nonnegative variables only. Let

$$\mathbf{X}_1 = \mathbf{X}_{1,1} - \mathbf{X}_{1,2}.$$

The new program is then: find $\mathbf{X}_{1,1} \geq 0$, $\mathbf{X}_{1,2} \geq 0$, and $\Delta\mathbf{c} \geq \mathbf{0}$ to minimize $\mathbf{h}'\Delta\mathbf{c}$ subject to

$$\mathscr{U}^*\begin{bmatrix}\mathbf{X}_{1,1} - \mathbf{X}_{1,2}\\ X_{m+1}\end{bmatrix} = \mathbf{0} \qquad (6.9.28a)$$

$$X_{m+1} \geq r_{s,t} \qquad (6.9.28b)$$

$$\mathbf{X}_{1,1} + \mathbf{X}_{1,2} \leq \mathbf{c}^*_0 + \Delta\mathbf{c}. \qquad (6.9.28c)$$

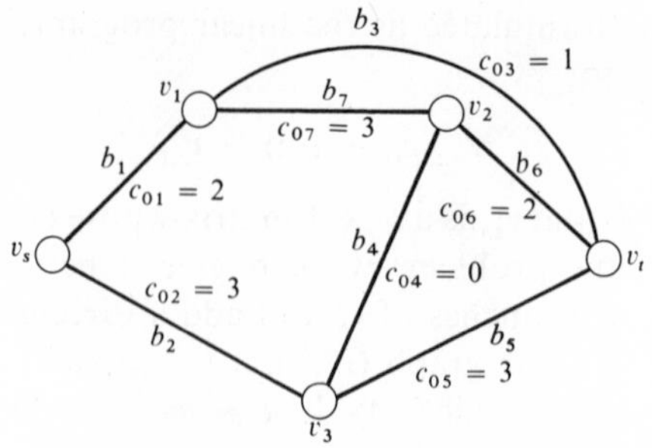

Fig. 6.9.6 Original graph G for Example 6.9.3.

Example 6.9.4 Consider the graph G shown in Fig. 6.9.6. Again, let the capacity of each branch in G be c_{0i} with

$$\mathbf{c}_0 = \begin{bmatrix} 2 \\ 3 \\ 1 \\ 0 \\ 3 \\ 2 \\ 3 \end{bmatrix}. \tag{6.9.29}$$

Let G^* be the graph shown in Fig. 6.9.7 obtained from G by assigning arbitrary directions to the branches of G and then adding the branch b_8. The capacity c^*_{0k} of each branch b^*_k in G^* is c_{0k} if $k = 1, \ldots, 7$ and infinity if $k = 8$.

The incidence matrix $\mathscr{U}^*$ of G^* is

$$\mathscr{U}^* = \begin{matrix} & \begin{matrix} b_1 & b_2 & b_3 & b_4 & b_5 & b_6 & b_7 & b_8 \end{matrix} & \\ & \begin{bmatrix} -1 & 0 & 1 & 0 & 0 & 0 & 1 & 0 \\ 0 & 0 & 0 & 1 & 0 & 1 & -1 & 0 \\ 0 & -1 & 0 & -1 & 1 & 0 & 0 & 0 \\ 1 & 1 & 0 & 0 & 0 & 0 & 0 & -1 \\ 0 & 0 & -1 & 0 & -1 & -1 & 0 & 1 \end{bmatrix} & \begin{matrix} v_1 \\ v_2 \\ v_3 \\ v_s \\ v_t \end{matrix} \end{matrix}. \tag{6.9.30}$$

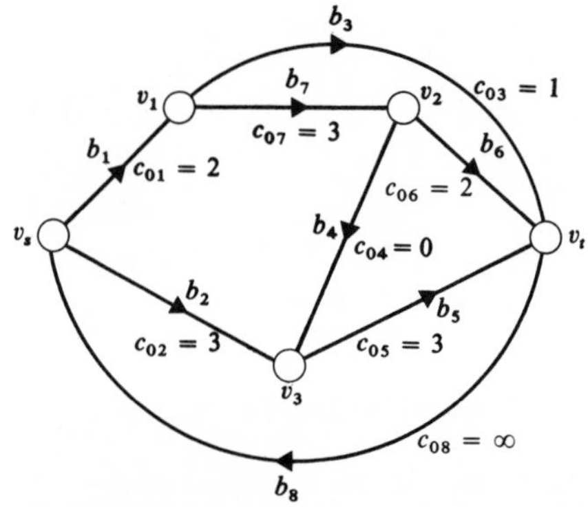

Fig. 6.9.7 Graph G^* obtained from G in synthesis procedure.

Suppose $h_i = 1$ for $i = 1, \ldots, 7$. If X_i is the flow in b_i^* of G^*, the linear program which guarantees $r_{s,t} = 7$ with minimum cost is: Minimize $\Delta c_1 + \Delta c_2 + \Delta c_3 + \Delta c_4 + \Delta c_5 + \Delta c_6 + \Delta c_7$ subject to

$$\begin{bmatrix} -1 & 0 & 1 & 0 & 0 & 0 & 1 & 0 \\ 0 & 0 & 0 & 1 & 0 & 1 & -1 & 0 \\ 0 & -1 & 0 & -1 & 1 & 0 & 0 & 0 \\ 1 & 1 & 0 & 0 & 0 & 0 & 0 & -1 \end{bmatrix} \begin{bmatrix} X_1 \\ X_2 \\ X_3 \\ X_4 \\ X_5 \\ X_6 \\ X_7 \\ X_8 \end{bmatrix} = 0; \qquad (6.9.31a)$$

$$X_8 \geq 7; \qquad (6.9.31b)$$

$$\begin{bmatrix} -2 - \Delta c_1 \\ -3 - \Delta c_2 \\ -1 - \Delta c_3 \\ 0 - \Delta c_4 \\ -3 - \Delta c_5 \\ -2 - \Delta c_6 \\ -3 - \Delta c_7 \end{bmatrix} \leq \begin{bmatrix} X_1 \\ X_2 \\ X_3 \\ X_4 \\ X_5 \\ X_6 \\ X_7 \end{bmatrix} \leq \begin{bmatrix} 2 + \Delta c_1 \\ 3 + \Delta c_2 \\ 1 + \Delta c_3 \\ 0 + \Delta c_4 \\ 3 + \Delta c_5 \\ 2 + \Delta c_6 \\ 3 + \Delta c_7 \end{bmatrix}. \qquad (6.9.31c)$$

Note that in (6.9.31a), we have deleted the last row of $\mathscr{U}^*$ since $\mathscr{U}^*$ has rank $n - 1 = 4$.

The system of equations given above can be solved using routine linear programming techniques. An optimal solution is $\Delta c_1 = 2$, $\Delta c_6 = 1$, $\Delta c_2 = \Delta c_3 = \Delta c_4 = \Delta c_5 = \Delta c_7 = 0$, with $X_1 = 4$, $X_2 = 3$, $X_3 = 1$, $X_4 = 0$, $X_5 = 3$, $X_6 = 4$, $X_7 = 3$, and $X_8 = 7$. The resulting flows and graph are shown in Fig. 6.9.8.

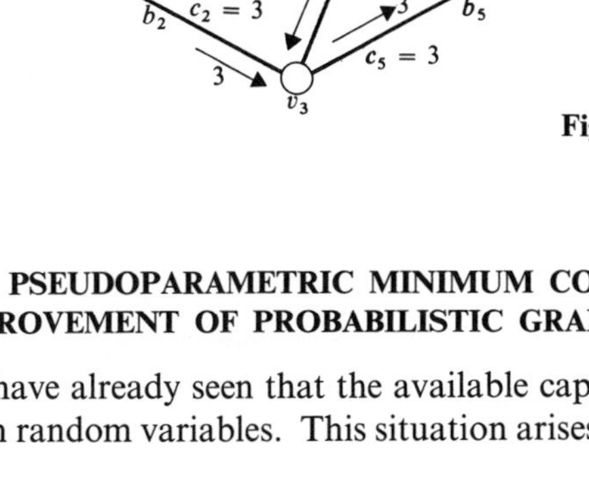

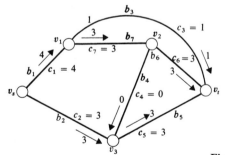

Fig. 6.9.8 Optimum graph and optimum flows.

6.10 PSEUDOPARAMETRIC MINIMUM COST IMPROVEMENT OF PROBABILISTIC GRAPHS

We have already seen that the available capacities of the branches in a network are often random variables. This situation arises if the branches are unreliable or if there are existing flows in the network which are themselves random variables. A natural goal in the analysis of such networks is to find the probability distribution of the maximum flow between various pairs of vertices. This problem was discussed in Chapter 4.

Suppose that an *existing* network is analyzed and it is found that the probability of establishing a flow of $r_{s,t}$ units between a given pair of vertices is "too small" to meet the demands or requirements of the system. The designer must then determine how the network may be modified to increase the flow rate probability to an acceptable level. Specifically, we will examine the following problem. We are given a network with random branch flows which are either observable or have known probability distributions. If, on the basis of our analysis procedure, we decide that the flow rate probability is too small, we must find a *minimum cost increase* of the branch capacities so that a specified flow rate probability is satisfied.

The synthesis problem we have just posed has the following interpretation. If we are given a network for which the probability of attaining a specified flow rate is small, the transmission scheme of waiting until enough free paths are available between source and sink is inherently poor. It may be better to transmit flow partially through the network to an intermediate vertex where it can be stored until enough capacity is available to complete transmission. On the other hand, the cost of storing messages and the cost of delaying transmission may be prohibitive. In this case, we must guarantee that enough channels are available at any one time to transmit at the specified flow rate. One way to guarantee this event is to increase the capacities of the existing branches or to add new branches to the network.

Let G be an n-vertex, m-branch graph for the network in question and let c_{0i} be the existing capacity of branch b_i for $i = 1, \ldots, m$. Suppose the cost of increasing the branch capacity of b_i to $c_i = c_{0i} + \Delta c_i$ is a linear function $h_i \Delta c_i$ of Δc_i, where h_i is a nonnegative constant for $i = 1, \ldots, m$. Then the optimum synthesis problem can be stated formally as: Given constants $r_{s,t} > 0$ and p_0 where $0 < p_0 < 1$, find $\Delta \mathbf{c} = (\Delta c_1, \ldots, \Delta c_m)'$ such that if $\mathbf{c}_0 = (c_{01}, \ldots, c_{0m})'$ and

$$\mathbf{c} = \mathbf{c}_0 + \Delta \mathbf{c} \tag{6.10.1a}$$

and

$$\Delta \mathbf{c} \geq \mathbf{0}, \tag{6.10.1b}$$

then

$$\text{Prob}\,\{\tau_{s,t} \geq r_{s,t}\} \geq p_0 \tag{6.10.2}$$

and

$$\sum_{i=1}^{m} h_i \Delta c_i \text{ is minimum.} \tag{6.10.3}$$

An alternative problem might be to guarantee that the terminal capacity $\tau_{s,t}$ between *any* pair of vertices v_s and v_t for $s,t = 1, \ldots, n$ is at least some positive number

r with at least probability p_0. In other words, we might want to find $\Delta\mathbf{c}$ such that

$$\sum_{i=1}^{m} h_i \Delta c_i$$

is minimized and

$$\text{Prob}\,\{\underset{s,t}{\text{Min}}[\tau_{s,t}] \geq r\} \geq p_0. \tag{6.10.4}$$

We can attack this problem with the same methods as the previously stated problem and so, for simplicity, we will discuss only the problem defined by (6.10.1) to (6.10.3).

We found in Chapter 4 that the methods of analysis and the forms of the solutions of the analysis problems depend very strongly on the nature of the statistical information available about the random network flows. Naturally, the same conclusion is true for the synthesis problem. In this section, we investigate the above synthesis problem for the case where the random branch flows have unknown probability distributions, but are observable. We will make the same assumptions as in Section 4.3. That is, $\mathbf{F}(k) = (f_1(k), \ldots, f_m(k))'$ is a measurement of the branch flows at time k and $\mathbf{F}(1), \ldots, \mathbf{F}(K)$ are identically and independently distributed random variables. Recall that the Uniformly Most Powerful level α test for testing $H_0 : p = \text{Prob}\,\{\tau_{s,t} \geq r_{s,t}\} \geq p_0$ against $H_1 : p < p_0$ is: Reject H_0 if $N > \hat{K}$, where N is the number of $\tau_{s,t}(k)$ which are less than $r_{s,t}$ and $\hat{K}$ is a constant given by (4.3.12).

On the basis of $\mathbf{F}(1), \ldots, \mathbf{F}(K)$, we might conclude that hypothesis H_0 is true. If this is the case, our decision will be the best one possible without additional information about the system. Consequently, there is no need to modify the system. On the other hand, if we reject H_0, the level of the test implies that

$$\text{Prob}\,\{\text{Prob}\{\tau_{s,t} \geq r_{s,t}\} \geq p_0\} \leq \alpha \tag{6.10.5}$$

and, if α is small, there is a high probability that the flow specification is not satisfied.

If we reject H_0, we must modify the network. The only data available for guidance in modifying the network are the observed flows $\mathbf{F}(k)$ and the corresponding maximum flows $\tau_{s,t}(k)$. In the following approach, we determine the least cost modification that can be made in order to make it appear that H_0 is accepted if the $\mathbf{F}(k)$ were used as observed flow vectors for the modified network. This approach has the advantage that it leads to tractable computational procedures. It will also be seen that this approach can be used for other improvement criteria.

We will reject H_0 if there are N observations of the maximum flow, say $\tau_{s,t}(i_1), \ldots, \tau_{s,t}(i_N)$ such that

$$\tau_{s,t}(i_j) < r_{s,t} \qquad \text{for } j = 1,2,\ldots,\ N \tag{6.10.6}$$

where $N \geq \hat{K}$. Let $N = \hat{K} + \hat{k} - 1$ for $\hat{k} \geq 1$. If $\hat{k}$ or more of the $\tau_{s,t}(i_j)$ had been at least $r_{s,t}$, the test would have accepted H_0. Hence, if we could modify $\tau_{s,t}(i_j)$ until $\hat{k}$ or more are at least $r_{s,t}$, the probability constraint $\text{Prob}\,\{\tau_{s,t} \geq r_{s,t}\} \geq p_0$ will be accepted on the basis of all the available data (here, for simplicity, we have not used a randomized test; that is, $\gamma = 1$).

Example 6.10.1 Consider a simple graph consisting of a single branch between vertices v_s and v_t. If the capacity of this branch is c_0 and the branch flows $f_{s,t}(1)$, $f_{s,t}(2), \ldots, f_{s,t}(K)$ have been observed, the maximum flow rates at times $1,2,\ldots,K$ are

$$\begin{aligned} \tau_{s,t}(1) &= c_0 - f_{s,t}(1); \\ \tau_{s,t}(2) &= c_0 - f_{s,t}(2); \\ &\;\vdots \\ \tau_{s,t}(K) &= c_0 - f_{s,t}(K). \end{aligned} \tag{6.10.7}$$

Hypothesis H_0 will be rejected if there are at least $\hat{K}$ values of $\tau_{s,t}(i)$ such that $\tau_{s,t}(i_j) < r_{s,t}$ or, equivalently, if there are at least $\hat{K}$ values of the $f_{s,t}(i)$ such that

$$f_{s,t}(i_j) > c_0 - r_{s,t} \qquad \text{for } j = 1, \ldots, N. \tag{6.10.8}$$

Reorder these $f(i_j)$ such that

$$f_{s,t}(i_1) \le f_{s,t}(i_2) \le \cdots \le f_{s,t}(i_N) \tag{6.10.9}$$

and let $N = \hat{K} + \hat{k} - 1$. Now, if we increase c_0 to $c = c_0 + \Delta c$ such that

$$f_{s,t}(i_{\hat{k}}) = c_0 + \Delta c - r_{s,t}, \tag{6.10.10}$$

it follows that

$$f_{s,t}(i_j) \le c_0 + \Delta c - r_{s,t} \qquad \text{for } j = 1,2,\ldots,\hat{k} - 1 \tag{6.10.11a}$$

and

$$f_{s,t}(i_j) \ge c_0 + \Delta c - r_{s,t} \qquad \text{for } j = \hat{k} + 1, \ldots, N. \tag{6.10.11b}$$

Clearly, at least $\hat{k}$ of the $\tau_{s,t}(i_j)$ which were initially less than $r_{s,t}$ are now at least $r_{s,t}$. In addition, Δc is the smallest possible positive number for which this statement holds. Consequently, the capacity of the single branch was increased by the smallest amount possible in order to accept hypothesis H_0.

Now, consider a more complicated graph where $\tau_{s,t}(1), \ldots, \tau_{s,t}(K)$ are the maximum flows at times $1, \ldots, K$ and $\tau_{s,t}(i_j) < r_{s,t}$ for $j = 1, \ldots, N$, with $N = \hat{K} + \hat{k} - 1$ for $\hat{k} > 1$. In this case H_0 is rejected. The condition $\tau_{s,t}(i_j) < r_{s,t}$ is equivalent to

$$\operatorname*{Min}_{1 \le l \le q} \left[\sum_{k=1}^{m} a_{l,k}(c_{0k} - f_k(i_j)) \right] < r_{s,t}. \tag{6.10.12}$$

By a procedure similar to that used in Example 6.10.1, we can increase the capacities of the branches of G until at least an additional $\hat{k}$ of the maximal flows are at least $r_{s,t}$. This is known as the problem of $\hat{k}$-fold alternatives [DA2]. We have a set of N constraints of the form

$$\mathscr{A}_{s,t}[\mathbf{c}_0 + \Delta\mathbf{c} - \mathbf{F}(i_j)] \ge \mathbf{r}_{s,t} \qquad \text{for } j = 1, \ldots, N \tag{6.10.13}$$

of which at least $\hat{k}$ must be *simultaneously* satisfied. This problem can be solved by linear programming if we add a set of auxiliary variables $\delta_1, \delta_2, \ldots, \delta_N$ which may assume only *integer* values. The program we need has the formulation:

Find a vector $\Delta \mathbf{c} \geq \mathbf{0}$ and integers $\delta_1, \delta_2, \ldots, \delta_N$ such that

$$\mathbf{h}'\Delta\mathbf{c} \text{ is minimized} \tag{6.10.14}$$

and

$$\mathscr{A}_{s,t}(\mathbf{c}_0 + \Delta\mathbf{c}) \geq \mathbf{r}_{s,t} + \delta_1 \mathscr{A}_{s,t}\mathbf{F}(i_1); \tag{6.10.15a}$$

$$\vdots$$

$$\mathscr{A}_{s,t}(\mathbf{c}_0 + \Delta\mathbf{c}) \geq \mathbf{r}_{s,t} + \delta_N \mathscr{A}_{s,t}\mathbf{F}(i_N); \tag{6.10.15n}$$

$$0 \leq \delta_i \leq 1 \qquad \text{for } i = 1, \ldots, N; \tag{6.10.16}$$

$$\sum_{i=1}^{N} \delta_i \geq \hat{k}. \tag{6.10.17}$$

By introducing the integer variables $\delta_1, \ldots, \delta_N$ which may assume only the values 0 and 1, we force the constraint (6.10.15j) to become either

$$\mathscr{A}_{s,t}(\mathbf{c}_0 + \Delta\mathbf{c}) \geq \mathbf{r}_{s,t} \tag{6.10.18a}$$

or

$$\mathscr{A}_{s,t}(\mathbf{c}_0 + \Delta\mathbf{c}) \geq \mathbf{r}_{s,t} + \mathscr{A}_{s,t}\mathbf{F}(i_j). \tag{6.10.18b}$$

However, since $\mathscr{A}_{s,t}$ and $\mathbf{F}(i_j)$ have only nonnegative entries, if any constraint, say (6.10.15k), is satisfied with $\delta_k = 1$, all constraints with $\delta_i = 0$ must automatically be satisfied.

The constraint given by (6.10.17) forces *at least* $\hat{k}$ of the δ_i to be nonzero (i.e., equal to one). Thus, we are requiring that at least $\hat{k}$ of the constraints given by (6.10.13) must simultaneously be satisfied. Therefore, we are changing at least $\hat{k}$ of the maximum flows which were originally less than $r_{s,t}$ to numbers which are now at least $r_{s,t}$. Since we are minimizing the cost function $\mathbf{h}'\Delta\mathbf{c}$, we are actually finding the $\hat{k}$ "cheapest" maximal flows to change. (We are able to do this by inspection in Example 6.10.1 by simply arranging the $\tau_{s,t}(k)$'s in nondecreasing order.) Moreover, instead of improving $\hat{k}$ of the maximum flows, the above formulation can be used to improve any number of these maximum flows. If we want to increase all the maximum flows to at least $r_{s,t}$ the problem becomes much simpler. In this case, all of the δ_i equal unity and we have a standard linear programming problem.

Naturally, the same simplifications that are possible in the deterministic problem discussed in the last section are possible here. An equivalent but far more efficient linear program is to find vectors $\mathbf{X}_{1,1} \geq \mathbf{0}$, $\mathbf{X}_{1,2} \geq \mathbf{0}$, and integers $\delta_1, \delta_2, \ldots, \delta_N$ such that

$$\mathbf{h}'\Delta\mathbf{c} \text{ is minimized} \tag{6.10.19}$$

and

$$\mathscr{U}^* \begin{bmatrix} \mathbf{X}_{1,1} - \mathbf{X}_{1,2} \\ X_{m+1} \end{bmatrix} = \mathbf{0}; \tag{6.10.20}$$

$$X_{m+1} \geq r_{s,t}; \tag{6.10.21}$$

$$\mathbf{X}_{1,1} + \mathbf{X}_{1,2} \leq \mathbf{c}_0 + \Delta\mathbf{c} - \delta_1 \mathbf{F}(i_1); \tag{6.10.22a}$$

$$\vdots$$

$$\mathbf{X}_{1,1} + \mathbf{X}_{1,2} \leq \mathbf{c}_0 + \Delta\mathbf{c} - \delta_N \mathbf{F}(i_N); \tag{6.10.22n}$$

$$0 \leq \delta_i \leq 1 \qquad \text{for } i = 1,2,\ldots,N; \tag{6.10.23}$$

$$\sum_{i=1}^{N} \delta_i \geq \hat{k}. \tag{6.10.24}$$

Example 6.10.2 Let G be the graph shown in Fig. 6.10.1. Suppose that the branch flows F_1,F_2,F_3,F_4,F_5 are observable random variables with unknown probability distributions. Let $\mathbf{F}(1),\ldots,\mathbf{F}(10)$ be past values of the flow vector $\mathbf{F} = (F_1,\ldots,F_5)$. Consider the following situation. We are testing hypothesis H_0 : Prob $\{\tau_{s,t} \geq 3\} \geq 0.9$ against alternative H_1 : Prob $\{\tau_{s,t} \geq 3\} < 0.9$ at level $\alpha = 0.1$. Suppose that three values of maximum flow, corresponding to the observations $\mathbf{F}(1),\mathbf{F}(2),\mathbf{F}(3)$, are less than 3. Then, if the cost of increasing any branch b_i is $h_i = 1$, find $\Delta c_1,\ldots,\Delta c_5$ such that $\Delta c_1 + \Delta c_2 + \Delta c_3 + \Delta c_4 + \Delta c_5$ is minimized and H_0 is accepted. The flows $\mathbf{F}(1),\mathbf{F}(2),\mathbf{F}(3)$ are

$$\mathbf{F}(1) = \begin{bmatrix} 2 \\ 2 \\ 1 \\ 1 \\ 2 \end{bmatrix}, \qquad \mathbf{F}(2) = \begin{bmatrix} 1 \\ 2 \\ 3 \\ 0 \\ 2 \end{bmatrix}, \qquad \mathbf{F}(3) = \begin{bmatrix} 3 \\ 1 \\ 1 \\ 2 \\ 2 \end{bmatrix}. \tag{6.10.25}$$

To test H_0: Prob $\{\tau_{s,t} \geq 3\} \geq 0.9$ against H_1 : Prob $\{\tau_{s,t} \geq 3\} < 0.9$ at confidence level $\alpha = 0.1$, we determine N, the number of $\tau_{s,t}(k)$ which are less than 3. If $N > \hat{K}$ we reject the hypothesis H_0. The constant $\hat{K}$ is found from the equation

$$\text{Prob}\,\{\text{reject } H_0 | H_0 \text{ is true}\} = 0.1. \tag{6.10.26}$$

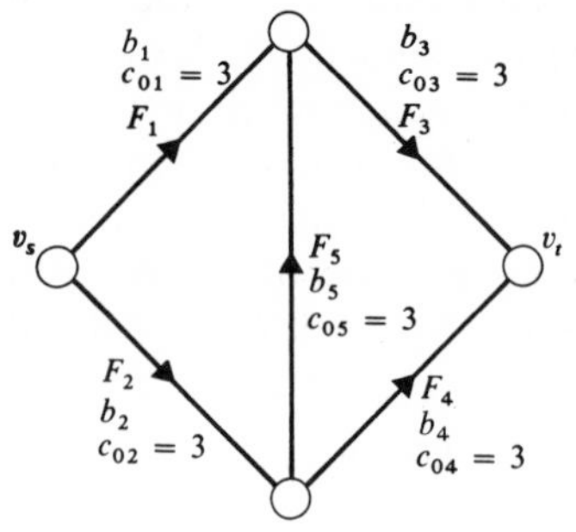

Fig. 6.10.1 Graph for Example 6.10.2.

We saw in Chapter 4 that (6.10.26) can be written as

$$\sum_{k=\hat{K}+1}^{K} \frac{e^{-\lambda}\lambda^k}{k!} + \gamma \frac{e^{-\lambda}\lambda^{\hat{K}}}{\hat{K}!} = 0.1 \tag{6.10.27a}$$

or

$$\sum_{k=0}^{\hat{K}-1} \frac{e^{-\lambda}\lambda^k}{k!} + (1-\gamma) \frac{e^{-\lambda}\lambda^{\hat{K}}}{\hat{K}!} = 0.9 \tag{6.10.27b}$$

where $\lambda = K(1 - p_0) = 10(0.1) = 1$ and γ is the probability that we reject H_0 if $N = \hat{K}$. From (6.10.27b), $\hat{K} = 2$ and $\gamma = 0.212$.

The cut-set matrix of G is

$$\mathscr{A}_{s,t} = \begin{bmatrix} 1 & 1 & 0 & 0 & 0 \\ 1 & 0 & 0 & 1 & 1 \\ 0 & 1 & 1 & 0 & 1 \\ 0 & 0 & 1 & 1 & 0 \end{bmatrix} \tag{6.10.28}$$

and from (6.10.15), the constraints are

$$\begin{bmatrix} 6 + \Delta c_1 + \Delta c_2 \\ 9 + \Delta c_1 + \Delta c_4 + \Delta c_5 \\ 9 + \Delta c_2 + \Delta c_3 + \Delta c_5 \\ 6 + \Delta c_3 + \Delta c_4 \end{bmatrix} \geq \begin{bmatrix} 3 \\ 3 \\ 3 \\ 3 \end{bmatrix} + \delta_i \begin{bmatrix} F_1(i) + F_2(i) \\ F_1(i) + F_4(i) + F_5(i) \\ F_2(i) + F_3(i) + F_5(i) \\ F_3(i) + F_4(i) \end{bmatrix} \quad \text{for } i = 1,2,3; \tag{6.10.29a}$$

$$0 \leq \delta_i \leq 1; \tag{6.10.29b}$$

$$\delta_1 + \delta_2 + \delta_3 \geq \hat{k} = 2. \tag{6.10.29c}$$

Equation (6.10.29a) is equivalent to

$$\begin{bmatrix} 3 + \Delta c_1 + \Delta c_2 \\ 6 + \Delta c_1 + \Delta c_4 + \Delta c_5 \\ 6 + \Delta c_2 + \Delta c_3 + \Delta c_5 \\ 3 + \Delta c_3 + \Delta c_4 \end{bmatrix} \geq \delta_i \begin{bmatrix} S_1(i) \\ S_2(i) \\ S_3(i) \\ S_4(i) \end{bmatrix} \tag{6.10.30}$$

where

$$[S_1(1),S_2(1),S_3(1),S_4(1)] = [4,5,5,2],$$
$$[S_1(2),S_2(2),S_3(2),S_4(2)] = [3,3,7,3],$$

and

$$[S_1(3),S_2(3),S_3(3),S_4(3)] = [4,7,5,3].$$

A number of the constraints in (6.10.30) are not active. After removing these, the synthesis problem is: Find numbers $\Delta c_1,\Delta c_2,\Delta c_3,\Delta c_4,\Delta c_5$ and integers $\delta_1,\delta_2,\delta_3$ such that

$$\Delta c_1 + \Delta c_2 + \Delta c_3 + \Delta c_4 + \Delta c_5 \text{ is minimized} \tag{6.10.31a}$$

and

$$\begin{bmatrix} \Delta c_1 + \Delta c_2 \\ \Delta c_2 + \Delta c_3 + \Delta c_5 \\ \Delta c_1 + \Delta c_2 \\ \Delta c_1 + \Delta c_4 + \Delta c_5 \end{bmatrix} \geq \begin{bmatrix} \delta_1 \\ \delta_2 \\ \delta_3 \\ \delta_3 \end{bmatrix} \tag{6.10.31b}$$

$$\Delta c_i \geq 0 \qquad \text{for } i = 1,2,3,4,5; \tag{6.10.31c}$$

$$0 \leq \delta_i \leq 1 \qquad \text{for } i = 1,2,3; \tag{6.10.31d}$$

$$\delta_1 + \delta_2 + \delta_3 \geq 2. \tag{6.10.31e}$$

The solution of this problem is, by inspection, $\delta_1 = \delta_3 = 1$, $\delta_2 = 0$, and $\Delta c_1 = 1$ with $\Delta c_2 = \Delta c_3 = \Delta c_4 = \Delta c_5 = 0$ or $\delta_1 = \delta_2 = 1$, $\delta_3 = 0$, and $\Delta c_2 = 1$ with $\Delta c_1 = \Delta c_3 = \Delta c_4 = \Delta c_5 = 0$. These are the only optimal solutions to the problem.

Given a set of K observations of the branch flows, we may decide that Prob $\{\tau_{s,t} \geq r_{s,t}\} < p_0$. We can then use linear integer programming to modify the network with minimum cost to reverse our conclusion. The formulation given by (6.10.19) to (6.10.24) *does not require enumeration of the cut-sets of G*. This is important since the number of cut-sets in a large system could be enormous. The analysis procedure requires that, for each $\mathbf{F}(k)$, we find the maximum flow rate $\tau_{s,t}(k)$. In our discussion, we solved this problem by finding the minimum cut-set of G when the capacities of the branches of G were taken to $c_{0i} - f_i(k)$. In reality, we could have used more practical methods to find $\tau_{s,t}(k)$. For example, the labeling algorithm (see Chapter 3) gives a very efficient computational procedure to find this flow.

As a final note, we again mention that the hypothesis and alternative could be reversed. In this case, we would test $H'_0 : p \leq p_0$ against $H'_1 : p > p_0$. Now, the synthesis procedure will be applied if H'_0 is accepted. This means that the probability of improving the graph when it does *not* need improvement is minimum, while the probability of not improving the graph when it does need improvement is bounded by α. If it is important that we improve the graph when it needs improvement, we should set α small, and use the test H'_0 against H'_1.

*6.11 PARAMETRIC MINIMUM COST IMPROVEMENT OF PROBABILISTIC GRAPHS

We have shown in Sections 6 and 7 of Chapter 4 that Prob $\{\tau_{s,t} \geq r_{s,t}\}$ can often be expressed with reasonable accuracy as

$$\text{Prob}\,\{\tau_{s,t} \geq r_{s,t}\} = 1 - \Phi\left(\frac{r_{s,t} - v_1(q)}{\sqrt{v_2(q) - v_1^2(q)}}\right) \tag{6.11.1}$$

where $v_1(q)$ and $v_2(q)$ are the unknown first and second absolute moments of $\text{Min}\,[c(A^1_{s,t}),\ldots,c(A^q_{s,t})]$. Thus the constraint, $\text{Prob}\,\{\tau_{s,t} \geq r_{s,t}\} \geq p_0$ is equivalent to

$$1 - \Phi\left(\frac{r_{s,t} - v_1(q)}{\sqrt{v_2(q) - v_1^2(q)}}\right) \geq p_0 \tag{6.11.2}$$

or

$$\frac{v_1(q) - r_{s,t}}{\sqrt{v_2(q) - v_1^2(q)}} \geq \theta_0 = -\Phi^{-1}(1 - p_0). \tag{6.11.3}$$

Assume that a set of branch flow observations $\{\mathbf{F}\} = \{\mathbf{F}(k), k = 1,2,\ldots,K\}$ is available. We have seen that if the maximum flow is normally distributed,[2] a Uniformly Most Powerful Invariant Test for testing

$$H_0 : p = \text{Prob}\,\{\tau_{s,t} \geq r_{s,t}\} \geq p_0 \tag{6.11.4a}$$

against

$$H_1 : p < p_0 \tag{6.11.4b}$$

at confidence level α is

$$\text{reject } H_0 \text{ if } t(\{\mathbf{F}\}) \leq \hat{K} \tag{6.11.5a}$$

$$\text{accept } H_0 \text{ if } t(\{\mathbf{F}\}) > \hat{K}, \tag{6.11.5b}$$

where

$$t(\{\mathbf{F}\}) = \sqrt{K(K-1)}\,\frac{(\bar{M}(K) - r_{s,t})}{\left[\sum_{k=1}^{K} (\tau_{s,t}(k) - \bar{M}(K))^2\right]^{1/2}};$$

$$\tau_{s,t}(k) = \underset{1 \leq i \leq q}{\text{Min}} \left[\sum_{j=1}^{m} a_{i,j}(c_j - f_j(k))\right];$$

$$\bar{M}(K) = \frac{1}{K}\sum_{k=1}^{K} \tau_{s,t}(k),$$

and $\hat{K}$ is a constant determined by

$$\int_{-\infty}^{\hat{K}} \left[\int_0^{\infty} \omega^{(K-2)/2} \exp\left(-\frac{\omega}{2}\right) \exp\left(t\sqrt{\frac{\omega}{K-1}} + \sqrt{K}\Phi^{-1}(1 - p_0)\right)^2 d\omega\right] dt$$
$$= \alpha 2^{K/2}\Gamma\left(\frac{K-1}{2}\right)\sqrt{\pi(K-1)}.$$

Suppose that on the basis of the observed data we reject H_0. Then, as in the preceding section, we can increase the capacities of the branches of G so that H_0 is accepted in the least cost manner.

---

[2] If the maximum flow is not normally distributed, the test is not optimal, but may still be quite good.

Hypothesis H_0 will be accepted if

$$\left[\sum_{k=1}^{K} (\tau_{s,t}(K) - \bar{M}(K))^2\right]^{-1/2} (\bar{M}(K) - r_{s,t}) \geq \frac{1}{\sqrt{K(K-1)}} \hat{K} = \sqrt{K^*} \qquad (6.11.6)$$

or, equivalently,[3] if $\hat{K} > 0$,

$$\bar{M}(K) \geq r_{s,t} \qquad (6.11.7a)$$

and

$$(1 + K^*K)\bar{M}(K)^2 + r_{s,t}^2 - 2\bar{M}(K)r_{s,t} - K^* \sum_{k=1}^{K} \tau_{s,t}^2(k) \geq 0. \qquad (6.11.7b)$$

Since

$$\tau_{s,t}(k) = \min_{1 \leq i \leq q} \left[\sum_{j=1}^{m} a_{i,j}(c_{0j} + \Delta c_j - f_j(k)) \right],$$

we would like to find values of the Δc_j such that (6.11.7) is satisfied. After some manipulation, we find that the constraint given in (6.11.7b) is equivalent to

$$(\tau_{s,t}(1), \ldots, \tau_{s,t}(K))W \begin{bmatrix} \tau_{s,t}(1) \\ \cdot \\ \cdot \\ \cdot \\ \tau_{s,t}(K) \end{bmatrix} + 2\bar{M}(K)r_{s,t} \leq r_{s,t}^2 \qquad \text{if } \hat{K} > 0 \qquad (6.11.8)$$

where W is a $K \times K$ matrix given by

$$W = K^*I - \frac{K^*K + 1}{K^2} \begin{bmatrix} 1 & 1 & \cdot & \cdot & \cdot & 1 \\ 1 & 1 & & & & 1 \\ \cdot & \cdot & & & & \cdot \\ \cdot & \cdot & & & & \cdot \\ 1 & 1 & \cdot & \cdot & \cdot & 1 \end{bmatrix}$$

where I is the $K \times K$ identity matrix.

Now consider the following problem. Let $\mathbf{Y} = (y_1, \ldots, y_k)'$. We want to find $\mathbf{Y} \geq \mathbf{0}$ and $\Delta\mathbf{c} \geq \mathbf{0}$ such that

$$\mathbf{h}'\Delta\mathbf{c} \text{ is minimized} \qquad (6.11.9a)$$

and

$$y_k \leq \sum_{j=1}^{m} a_{i,j}(c_{0j} + \Delta c_j - f_j(k)) \qquad \text{for } i = 1, \ldots, q; \qquad (6.11.9b)$$

$$\sum_{k=1}^{K} y_k \geq Kr_{s,t}; \qquad (6.11.9c)$$

$$\mathbf{Y}'W\mathbf{Y} + \frac{2r_{s,t}}{K} \sum_{k=1}^{K} y_k \leq r_{s,t}^2. \qquad (6.11.9d)$$

[3] If $\hat{K}$ is negative an analogous procedure can be followed.

If we could find a solution to this problem, we would have a solution to our synthesis problem provided $\bar{Y}$ is a feasible flow vector. We can guarantee that $\mathbf{Y}$ is a feasible flow vector and at the same time *eliminate the dependency on the cut-set matrix* $\mathscr{A}_{s,t}$ by using the technique discussed in Section 10. Again define the graph G^* obtained from G by directing the branches of G and adding a branch directed from v_t to v_s. Then, if $\mathscr{U}^*$ is the incidence matrix of G^*, the synthesis problem is: Find $\mathbf{Y} \geq 0$, $\mathbf{X}_{1,1k} \geq 0$, $\mathbf{X}_{1,2k} \geq 0$ for $k = 1, \ldots, K$ and $\Delta\mathbf{c} \geq \mathbf{0}$ such that

$$\mathbf{h}'\Delta\mathbf{c} \text{ is minimized} \tag{6.11.10a}$$

and

$$\mathscr{U}^* \begin{bmatrix} \mathbf{X}_{1,1k} - \mathbf{X}_{1,2k} \\ y_k \end{bmatrix} = \mathbf{0} \quad \text{for } k = 1, \ldots, K; \tag{6.11.10b}$$

$$\mathbf{X}_{1,1k} + \mathbf{X}_{1,2k} \leq \mathbf{c}_0 + \Delta\mathbf{c} - \mathbf{F}(k) \qquad \text{for } k = 1, \ldots, K; \tag{6.11.10c}$$

$$\sum_{k=1}^{K} y_k \geq K r_{s,t}; \tag{6.11.10d}$$

$$\mathbf{Y}'W\mathbf{Y} + \frac{2r_{s,t}}{K} \sum_{k=1}^{K} y_k \leq r_{s,t}^2. \tag{6.11.10e}$$

The synthesis procedure is in the form of a nonlinear program. Note that constraint (6.11.10e) is quadratic, and is, in fact, the only nonlinear constraint in the program. Also, its order is equal to the number of samples. The remaining constraints are linear and there are no dependencies on the cut-set matrix of G. The only matrix related to G is the incidence matrix $\mathscr{U}^*$ of G^*. Since $\mathscr{U}^*$ is an $n \times m$ matrix, reasonably large graphs can be treated. The nonlinear program thus has constraints which are amenable to solution.

Example 6.11.1 Let G be the graph shown in Fig. 6.11.1. We wish to guarantee with probability $p_0 = 0.9$ that a flow rate of at least 3 units can be attained between vertices v_s and v_t. Assume that the maximum flow rate $\tau_{s,t}$ is a normal random variable and that ten values of the branch flow vector $\mathbf{F}$ have been observed. On the basis of these flow vectors we must decide whether or not Prob $\{\tau_{s,t} \geq 3\} \geq 0.9$. If we con-

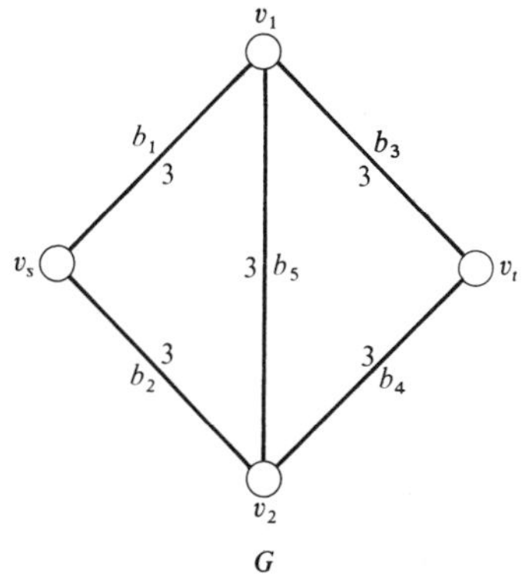

Fig. 6.11.1 Graph G for Example 6.11.1.

clude that Prob $\{\tau_{s,t} \geq 3\} < 0.9$ we must modify the graph to reverse the conclusions of the test at minimum cost. For simplicity, assume that the cost function is $\Delta c_1 + \Delta c_2 + \Delta c_3 + \Delta c_4 + \Delta c_5$. Furthermore, we do not require the probability, α of modifying the graph when it is already acceptable, to be smaller than 0.1.

Suppose that the observed branch flow vectors are

$$\mathbf{F}(1) = \begin{bmatrix} 2 \\ 2 \\ 1 \\ 1 \\ 2 \end{bmatrix}, \mathbf{F}(2) = \begin{bmatrix} 1 \\ 2 \\ 3 \\ 0 \\ 2 \end{bmatrix}, \mathbf{F}(3) = \begin{bmatrix} 3 \\ 1 \\ 1 \\ 2 \\ 2 \end{bmatrix}, \mathbf{F}(4) = \begin{bmatrix} 1 \\ 1 \\ 1 \\ 1 \\ 1 \end{bmatrix}, \mathbf{F}(5) = \begin{bmatrix} 2 \\ 1 \\ 1 \\ 2 \\ 2 \end{bmatrix}$$

$$\mathbf{F}(6) = \begin{bmatrix} 1 \\ 2 \\ 2 \\ 1 \\ 2 \end{bmatrix}, \mathbf{F}(7) = \begin{bmatrix} 1 \\ 1 \\ 2 \\ 0 \\ 2 \end{bmatrix}, \mathbf{F}(8) = \begin{bmatrix} 0 \\ 3 \\ 0 \\ 3 \\ 3 \end{bmatrix}, \mathbf{F}(9) = \begin{bmatrix} 3 \\ 0 \\ 1 \\ 2 \\ 1 \end{bmatrix}, \mathbf{F}(10) = \begin{bmatrix} 1 \\ 1 \\ 2 \\ 1 \\ 1 \end{bmatrix}. \tag{6.11.11}$$

We test the hypothesis $H_0 : p = \text{Prob}\,\{\tau_{s,t} \geq 3\} \geq 0.9$ against the alternative $H_1 : p < 0.9$ at confidence level $\alpha = 0.1$. From Example 4.7.4, a Uniformly Most Powerful Invariant test for this problem is: Reject H_0 if and only if

$$3\sqrt{10}\,\frac{\dfrac{1}{10}\left(\displaystyle\sum_{k=1}^{10} \tau_{s,t}(k) - 3\right)}{\left[\displaystyle\sum_{k=1}^{K} \tau_{s,t}(k)^2 - \frac{1}{10}\left(\sum_{k=1}^{10} \tau_{s,t}(k)\right)^2\right]^{1/2}} \leq \hat{K} = 2.45. \tag{6.11.12}$$

The maximum flows corresponding to $\mathbf{F}(1), \ldots, \mathbf{F}(10)$ can be shown to be 2,2,2,4,3,3,4,3,3,3. Hence, the left-hand side of (6.11.12) is

$$3\sqrt{10}\,\frac{(2.9 - 3)}{\sqrt{4.9}} = 4.29(2.9 - 3) = -0.429. \tag{6.11.13}$$

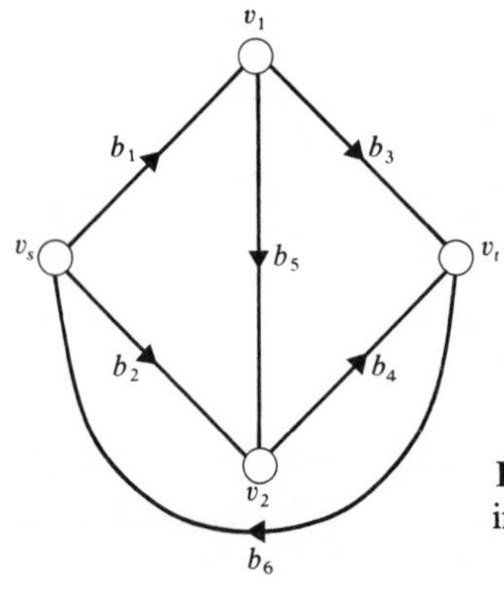

Fig. 6.11.2 Graph G^* obtained from G in synthesis procedure.

Therefore, we reject the hypothesis. Note that, if the sample mean were 3.73 or greater and the sample variance were no larger, we would accept the hypothesis.

In order to meet the flow rate specification, we must increase the capacities of the branches in G. Let G^* be the graph shown in Fig. 6.11.2, obtained from G by arbitrarily directing the branches of G and adding branch $b_6(t,s)$. The incidence matrix $\mathscr{U}^*$ of G^* with one row deleted is

$$\mathscr{U}^* = \begin{bmatrix} +1 & +1 & 0 & 0 & 0 & -1 \\ 0 & 0 & -1 & -1 & 0 & +1 \\ -1 & 0 & +1 & 0 & +1 & 0 \end{bmatrix} \tag{6.11.14}$$

and the constraints given by (6.11.10b) are, for $k = 1,2,\ldots,10$,

$$\begin{aligned} X_{1,1k} - X_{1,2k} + X_{2,1k} - X_{2,2k} - y_k &= 0 \\ -X_{3,1k} + X_{3,2k} - X_{4,1k} + X_{4,2k} + y_k &= 0 \\ -X_{1,1k} + X_{1,2k} + X_{3,1k} - X_{3,2k} + X_{5,1k} - X_{5,2k} &= 0. \end{aligned} \tag{6.11.15}$$

Furthermore, it is clear that $X_{1,2k} = X_{2,2k} = X_{3,2k} = X_{4,2k} = 0$. Consequently, the flow constraints from (6.11.10b) are

$$\begin{aligned} X_{1,1k} + X_{2,1k} &= y_k, \\ X_{3,1k} + X_{4,1k} &= y_k, \\ -X_{1,1k} + X_{3,1k} + X_{5,1k} - X_{5,2k} &= 0. \end{aligned} \tag{6.11.16}$$

The remaining constraints, (6.11.10c) to (6.11.10e), are

$$\mathbf{X}_{1,k} + \begin{bmatrix} 0 \\ 0 \\ 0 \\ 0 \\ X_{5,2k} \end{bmatrix} \le \begin{bmatrix} 3 \\ 3 \\ 3 \\ 3 \\ 3 \end{bmatrix} + \begin{bmatrix} \Delta c_1 \\ \Delta c_2 \\ \Delta c_3 \\ \Delta c_4 \\ \Delta c_5 \end{bmatrix} - \mathbf{F}(k) \qquad \text{for } k = 1,\ldots,10 \tag{6.11.17a}$$

$$\sum_{k=1}^{10} y_k \ge 30 \tag{6.11.17b}$$

$$(y_1,\ldots,y_{10})W \begin{bmatrix} y_1 \\ \cdot \\ \cdot \\ \cdot \\ y_{10} \end{bmatrix} + 0.6 \sum_{k=1}^{10} y_k \le 9 \tag{6.11.18}$$

where W is a 10×10 matrix with main diagonal entries equal to 0.05 with off-diagonal entries equal to -0.017.

From this point on, the problem can be attacked by using conventional nonlinear programming methods [HA2]. One possible method would be to solve a sequence of linear problems generated by approximating the nonlinear constraint by linear ones. Gradient methods can also be used to find reasonable approximate solutions.

A number of extensions of the results of the last two sections are possible. For example, there are other synthesis goals besides minimizing the cost to achieve a given flow with a specified probability. We may be allowed to spend a fixed sum of money to improve an existing network. Naturally, in this case we want to buy the maximum possible improvement and so we might want to maximize the flow rate probability. Consequently, assuming linear cost gives the new synthesis problem: find $\Delta\mathbf{c} \geq \mathbf{0}$ to maximize Prob $\{\tau_{s,t} \geq r_{s,t}\}$ such that $\mathbf{h}'\Delta\mathbf{c}$ is constant.

We can use the techniques developed in the preceding sections to attack this problem. For example, under the normality assumptions of the parametric statistical test, the problem reduces to: find $\Delta\mathbf{c} \geq \mathbf{0}$ to maximize

$$(\bar{M}(K) - r_{s,t})\left[\sum_{k=1}^{K} (\tau_{s,t}(k) - \bar{M}(K))^2\right]^{-1/2}$$

such that $\mathbf{h}'\Delta\mathbf{c}$ is constant.

Often, it is adequate to maximize the expected maximum flow. Then, if the variance is small, we can expect Prob $\{\tau_{s,t} \geq r_{s,t}\}$ to be large. Thus, we could consider the problem: maximize $E\{\tau_{s,t}\}$ with $\mathbf{h}'\Delta\mathbf{c}$ constant. In terms of the statistical test we want to maximize

$$\sum_{k=1}^{K} \tau_{s,t}(k)$$

and thus, we can solve this problem by linear programming.

Another objective could be to minimize variance. We might be willing to accept a lower average maximum flow if we were sure that the actual performance of the system will not deviate far from the average. For the statistical problem, we then minimize the sample variance

$$\sum_{k=1}^{K} \tau_{s,t}(k)^2 - K\bar{M}(K)^2.$$

Thus, we must solve a quadratic programming problem.

The ideas discussed in the last two sections could be used to synthesize a network as well as to improve an existing network. To begin the procedure, it would be necessary to estimate the traffic requirements between pairs of stations. Then, a set of minimum branch capacities could be selected so that the requirements are satisfied with a given probability. Once this is done, we can proceed to consider the terminal capacities between various points in the graph. In this case, instead of considering Prob $\{\tau_{s,t} \geq r_{s,t}\} \geq p_0$, we might wish Prob $\{I \geq t\} \geq p_0$ where

$$I = \sum_{j=i+1}^{n} \sum_{i=1}^{n} \tau_{i,j}.$$

An alternate criterion could be Prob $\{\operatorname{Min}_{s,t}[\tau_{s,t}] \geq r\}$. Then instead of finding the s-t cut-set matrix $\mathscr{A}_{s,t}$ we would find the full cut-set matrix of G and attempt to guarantee that the minimum cut-set of G is at least r. We can attack this problem in the same manner as our simpler problem.

6.12 GRAPHS WITH GAINS

An assumption that has prevailed throughout the book is that flow does not vary along a branch. For a number of problems this assumption must be abandoned. If the branches of the graph represent a road system along which goods are transported, items may be damaged or lost en route, indicating a loss of flow along the branch. Similarly, in a system of pipes carrying water or wires transmitting power, leakage in the system indicates loss of flow along branches. In a communication system, information may be corrupted by noise or enhanced by means of repeaters in a line, indicating loss of flow in the first case and gain in the second case. If branches represent operations in a manufacturing process and flow represents the value of goods, then the value of goods may be increased by a branch, again indicating a gain of flow greater than unity along the branch.

In order to model networks with gains, we must define the *input flow*, $\tilde{f}(i,j)$, and the *output flow*, $\underset{\sim}{f}(i,j)$ for an arbitrary branch (i,j). Then $\tilde{f}(i,j)$ and $\underset{\sim}{f}(i,j)$ are related by the real, positive *gain* $a(i,j)$ of branch (i,j) as

$$\underset{\sim}{f}(i,j) = a(i,j)\tilde{f}(i,j). \tag{6.12.1}$$

A flow pattern $\mathscr{F} = \{\tilde{f}(i,j), \underset{\sim}{f}(i,j)\}$ is then said to be feasible if it satisfies constraints (6.12.2) and (6.12.3) below.

$$\text{For all } v_i, \qquad \sum_{(i,j)\in\Gamma} \tilde{f}(i,j) - \sum_{(j,i)\in\Gamma} \underset{\sim}{f}(j,i) = \begin{cases} \tilde{f}_{s,t} & \text{for } i = s, \\ 0 & \text{for } i \neq s,t, \\ -\underset{\sim}{f}_{s,t} & \text{for } i = t, \end{cases} \tag{6.12.2}$$

where the source v_s and the terminal v_t are specified. We use $\tilde{f}_{s,t}$ to denote a real nonnegative number called the *source flow* and $\underset{\sim}{f}_{s,t}$ to denote real nonnegative number called the *terminal flow*. For all $(i,j) \in \Gamma$,

$$\tilde{f}(i,j) \leq c(i,j), \tag{6.12.3}$$

where $c(i,j)$ is a real nonnegative number called the *input capacity* of branch (i,j). We will simply refer to $c(i,j)$ as the capacity of branch (i,j).

As in the case of graphs in which all gains are unity, a reasonable goal is to maximize the terminal flow $\underset{\sim}{f}_{s,t}$. We therefore introduce:

Definition 6.12.1 *In a graph with gains, a feasible flow pattern $\mathscr{F}$ is said to be maximum if it yields a maximum value of $\underset{\sim}{f}_{s,t}$ over all feasible flow patterns.*

In graphs with gains it is possible to achieve a maximum value of $\underset{\sim}{f}_{s,t}$ with varying "cost," namely with different values $\tilde{f}_{s,t}$.

Definition 6.12.2 *A feasible flow pattern $\mathscr{F}_o$ with source flow $\tilde{f}_{s,t,o}$ and terminal flow $\underset{\sim}{f}_{s,t,o}$, is said to be optimum if for any other feasible flow pattern $\mathscr{F}$ either (6.12.4) or (6.12.5) is true:*

$$\underset{\sim}{f}_{s,t,o} \geq \underset{\sim}{f}_{s,t} \quad \text{if} \quad \tilde{f}_{s,t,o} = \tilde{f}_{s,t}, \tag{6.12.4}$$

$$\tilde{f}_{s,t,o} \leq \tilde{f}_{s,t} \quad \text{if} \quad \underset{\sim}{f}_{s,t,o} = \underset{\sim}{f}_{s,t}. \tag{6.12.5}$$

It should be noted that the intuitive concept of optimality corresponding to (6.12.4) is

$$\underset{\sim}{f}_{s,t,o} \geq \underset{\sim}{f}_{s,t} \qquad \text{if} \qquad \tilde{f}_{s,t,o} \geq \tilde{f}_{s,t}.$$

However, suppose $\underset{\sim}{f}_{s,t,o} < \underset{\sim}{f}_{s,t}$ for $\tilde{f}_{s,t,o} > \tilde{f}_{s,t}$. Then we can surely decrease flows until $\tilde{f}_{s,t,o} = \tilde{f}_{s,t}$ while $\underset{\sim}{f}_{s,t,o}$ decreases, thereby violating (6.12.4). Thus (6.12.4) is all that is required. Similarly, we can argue that (6.12.5) is sufficient as it stands.

Definition 6.12.3 *A feasible flow pattern $\mathscr{F}_o$ is optimum maximum if it is a maximum flow pattern and an optimum flow pattern.*

These concepts are illustrated in the following example.

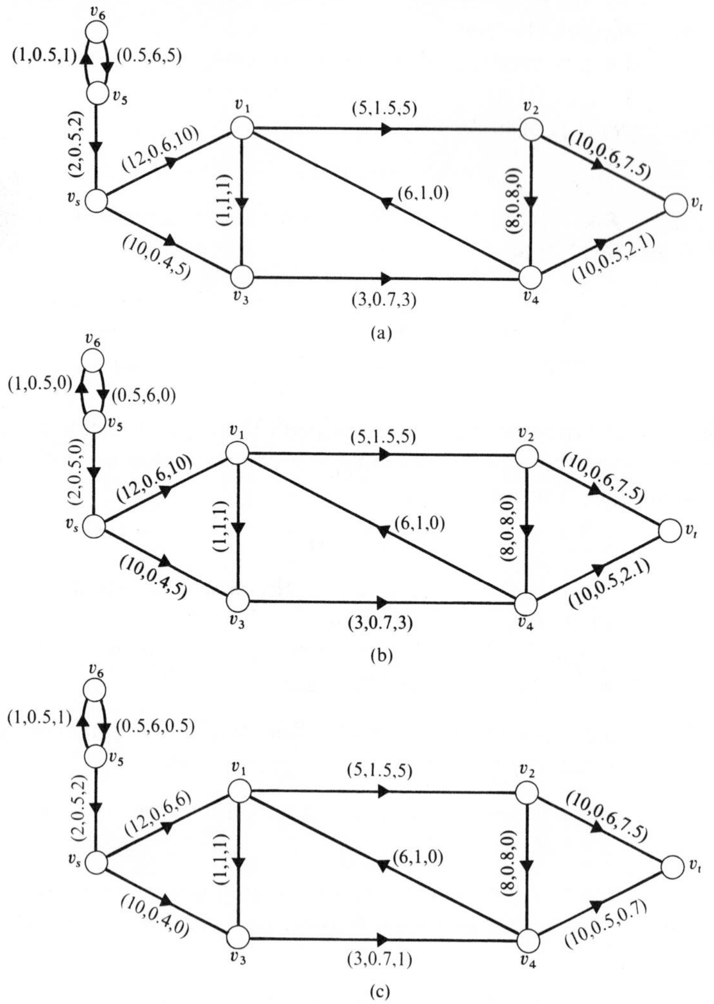

Fig. 6.12.1 (a) Optimum maximum flow pattern; (b) maximum nonoptimum flow pattern; (c) optimum nonmaximum flow pattern.

Example 6.12.1 In the graph in Fig. 6.12.1 and those to follow in this section, the branches are weighted by ordered triples. The first entry for branch (i,j) is $c(i,j)$, the second entry is $a(i,j)$, and the third entry is $\tilde{f}(i,j)$. The flow pattern in Fig. 6.12.1(a) is maximum and will be shown to be optimum.

The flow pattern in Fig. 6.12.1(b) is therefore also maximum, since it yields the same terminal flow as achieved by the flow pattern in Fig. 6.12.1(a). It is not optimum since the value of the source flow is greater than the value of the source flow for the flow pattern in Fig. 6.12.1(a). The flow pattern in Fig. 6.12.1(c) is not maximum since the value of the terminal flow is lower than the value of the terminal flow for the flow patterns in Figs. 6.12.1(a) and (b). However, it will be shown to be optimum.

In order to understand the basic mechanism by which $\underset{\sim}{f}_{s,t}$ and $\tilde{f}_{s,t}$ can be independently varied we must study the concepts relating to graphs with gains which are analogous to augmentation paths and other concepts previously considered.

Consider a chain, i.e., a sequence of vertices and branches $v_{i_1},b_1,v_{i_2},b_2,\ldots,v_{i_{k-1}},b_k,v_{i_k}$ such that for $j = 1,\ldots,k-1$, $b_j = (i_j,i_{j+1})$ or $b_j = (i_{j+1},i_j)$. A direction is assigned to a chain by tracing the vertices and branches in order from v_{i_1} to v_{i_k}. A *forward branch* in a chain is a branch (l,j) such that v_l is the common vertex between (l,j) and the preceding branch in the chain. All other branches are *backward branches*. Let S be a chain, let S_F be the subset of forward branches in S, and let S_B be the subset of backward branches.

Definition 6.12.4 *The gain a_S of a chain S is*

$$a_S = \prod_{b_l \in S_F} a_l \cdot \prod_{b_q \in S_B} 1/a_q \tag{6.12.6}$$

where a_l is the gain of b_l and a_q is the gain of b_q.

To define the "residual capacity" of a chain we require the concept of an "ε_1-increment."

Definition 6.12.5 *Let ε_1 be a real number and let S be a chain. A flow pattern $\mathscr{F}$ is an ε_1-increment for S if it satisfies the recursive relation in (6.12.7) for $i = 2,\ldots,k-1$, in terms of ε_1. Assume the vertices are numbered so that in the sequence defining the chain, b_i is incident with v_i and v_{i+1} for all i.*

For $b_i \in S_F$

$$\tilde{f}(i,i+1) = |\varepsilon_i|; \tag{6.12.7a}$$

$$\underset{\sim}{f}(i,i+1) = a_i|\varepsilon_i|; \tag{6.12.7b}$$

$$\varepsilon_{i+1} = a_i|\varepsilon_i|. \tag{6.12.7c}$$

For $b_i \in S_B$

$$\tilde{f}(i+1,i) = -\frac{|\varepsilon_i|}{a_i}; \tag{6.12.7d}$$

$$\underset{\sim}{f}(i+1,i) = -|\varepsilon_i|; \tag{6.12.7e}$$

$$\varepsilon_{i+1} = -\frac{|\varepsilon_i|}{a_i}. \tag{6.12.7f}$$

An arbitrary branch b_i may appear more than once in S. Therefore it may contain $r_i > 1$ components of the ε_1-increment $\mathscr{F}$, say $\tilde{f}_1(b_i), \tilde{f}_2(b_i), \ldots, \tilde{f}_{r_i}(b_i)$. To simplify notation, for a given ε_1-increment $\mathscr{F}$ we define a new flow pattern $\mathscr{F}_{\varepsilon_1} = \{\tilde{f}_{\varepsilon_1}(b_i), \underset{\sim}{f}_{\varepsilon_1}(b_i)\}$ as

$$\tilde{f}_{\varepsilon_1}(b_i) = \sum_{j=1,\ldots,r_i} \tilde{f}_j(b_i), \tag{6.12.8a}$$

$$\underset{\sim}{f}_{\varepsilon_1}(b_i) = \sum_{j=1,\ldots,r_i} \underset{\sim}{f}_j(b_i). \tag{6.12.8b}$$

If G has a flow pattern $\hat{\mathscr{F}}$ the *residual capacity* of a chain S is the maximum value of ε_1 for which $\mathscr{F}_{\varepsilon_1} \oplus \hat{\mathscr{F}}$ is feasible. If the residual capacity of a chain is zero, the chain is said to be *saturated*; otherwise it is *unsaturated*. These concepts are illustrated in the following example.

Example 6.12.2 For the path in Fig. 6.12.2(a), the addition of ε_1 units of flow to $\tilde{f}_{s,t}$ yields the flow pattern in Fig. 6.12.2(b). If ε_1 is made equal to 2, branch $(s,1)$, and hence the path, becomes saturated. Therefore, the residual capacity of the path is 2. For the circuit in Fig. 6.12.2(c), if $\tilde{f}_{s,t}$ is increased by ε_1, the resulting flow pattern is that shown in Fig. 6.12.2(d). If $\varepsilon_1 = 1/2$, then (1,2) becomes saturated. Hence, the residual capacity of the circuit is 1/2. Next, consider the chain $v_s(s,1)v_1(1,2)v_2(2,1)v_1(1,s)v_s$ in Fig. 6.12.2(e). If we first increase $\tilde{f}_{s,t}$ by ε_1, we obtain the flow pattern in Fig. 6.12.2(f). When flow is returned along $(1,s)$, $f(s,1)$ is first changed to ε_1 and then to $\varepsilon_1 - 0.08\varepsilon_1$. For $\varepsilon_1 = 2.5$, (1,2) and hence the chain becomes saturated. Similarly, for the flow pattern in Fig. 6.12.2(g), a flow can be sent along the chain $v_t(t,1)v_1(1,2)v_2(2,1)v_1(1,t)v_t$ until it is saturated as shown in Fig. 6.12.2(h). Note that in Fig. 6.12.2(f) and (h) flow is sent in both directions through certain branches.

From the definitions and Example 6.12.2, it is clear that a *path* is saturated if and only if every forward branch is saturated and every backward branch is empty. Similarly, we define a cut $(X,\bar{X})$ to be saturated if every branch in $(X,\bar{X})$ is saturated and every branch in $(\bar{X},X)$ is empty. A circuit is said to be saturated if and only if every forward branch is saturated and every backward branch is empty. However, the criterion for saturation of chains cannot be reduced to such simple terms since flow is sent both ways through certain branches. We must therefore study chains more carefully.

Definition 6.12.6 *A return chain is a chain in which the first and last vertices are identical. If the first vertex in a return chain is v_i it is called a v_i return chain. If $v_i = v_s$ it is called a source return chain and if $v_i = v_t$ it is called a terminal return chain. A return chain is said to be simple if the multiplicities of the branches in the chain are at most two and the chain contains exactly one circuit.*

The chains in Fig. 6.12.2(e) and (g) are examples of simple source and terminal return chains, respectively.

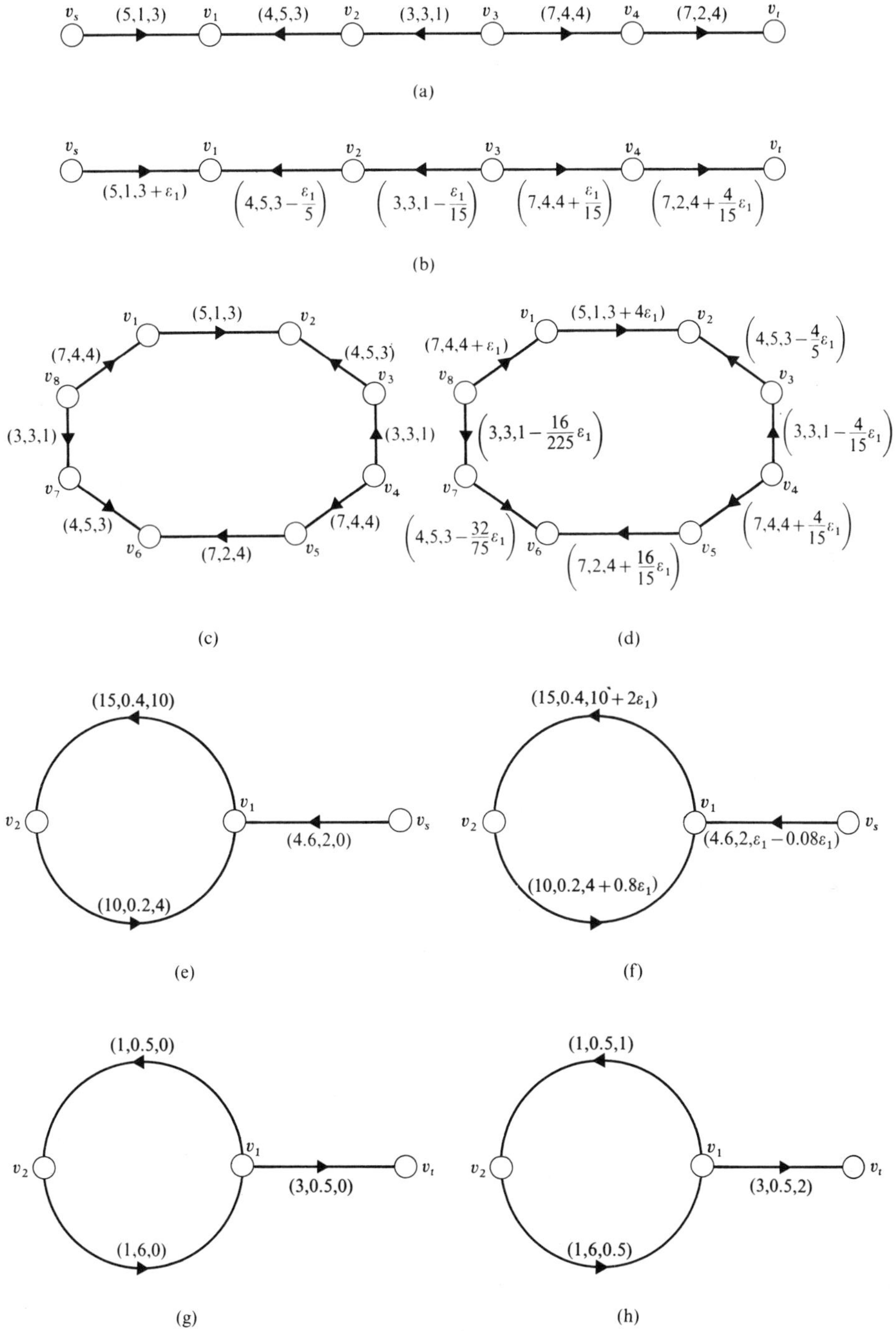

Fig. 6.12.2 Increasing flows along chains.

We finally consider the mechanism by which $f_{s,t}$ and $\underset{\sim}{f}_{s,t}$ can be independently varied.

Definition 6.12.7 *A circuit is said to be active if*

a) it has gain greater than one,

and

b) it is part of an unsaturated simple source or terminal return chain.

If a circuit is not active it is said to be inactive. A circuit is inactive if

a) it has gain less than or equal to one,

or

b) it is isolated, i.e., each path from a vertex in the circuit to v_s or v_t is saturated,

or

c) the circuit is saturated.

Consider the graphs in Fig. 6.12.3: a_i represents the gain of a path, a_{L_i}, the gain of a circuit, and ρ_i some unit of flow. The weights along the arrows represent flows in the directions of the arrows. If a circuit is active, $\tilde{f}_{s,t}$ can be decreased as shown in Fig. 6.12.3(a) and (b) or $\underset{\sim}{f}_{s,t}$ can be increased as shown in Fig. 6.12.3(c) and (d). The circuit in Fig. 6.12.2(g) is an example of an active terminal circuit.

We now show that unsaturated s-t paths of highest gain and active source or terminal chains are the only chains which need to be considered in finding optimum flows.

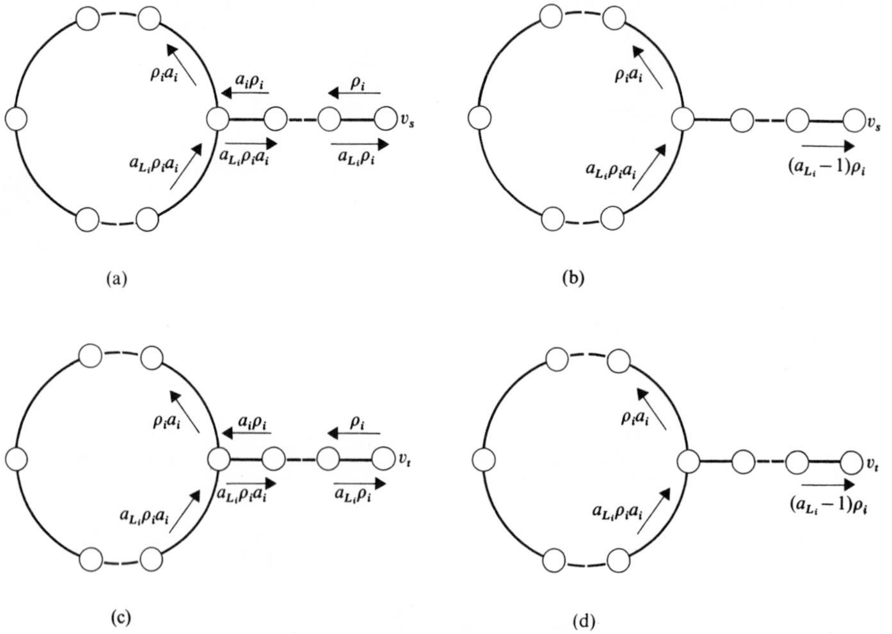

Fig. 6.12.3 Simple return chains.

Lemma 6.12.1 *If $\mathscr{F}_o$ is an optimum flow pattern and $\mathscr{F}$ is another flow pattern such that $\tilde{f}_{s,t,o} < \tilde{f}_{s,t}$ and $f_{s,t,o} = f_{s,t}$, then there is an active source circuit in $\mathscr{F}$.*

Proof. Let

$$\mathscr{G}_1 = \mathscr{F} \ominus \mathscr{F}_o. \tag{6.12.9}$$

Then

$$\tilde{g}_{s,t,1} = \tilde{f}_{s,t} - \tilde{f}_{s,t,o} > 0 \tag{6.12.10}$$

and

$$g_{s,t,1} = 0. \tag{6.12.11}$$

Now consider the original graph G with the flow pattern $\mathscr{G}_1$; $\mathscr{G}_1$ is not necessarily a feasible flow pattern since some branches may have negative flow. Such branches will be called *negative* branches; the remaining branches are called *positive.*

Since $\tilde{g}_{s,t,1} > 0$, there must be a positive branch (s,j) carrying flow from v_s. By conservation, there must be a positive branch (j,k) or a negative branch (k,j). Continuing in this fashion, we can start with v_s and trace a chain of positive forward branches and negative backward branches with each branch having a vertex in common with the previous branch. Since $g_{s,t,1} = 0$, some vertex in the chain will eventually be repeated. At this point, by considering the branches in the chain in reverse order, we will have found a simple return chain S_1 for $\mathscr{G}_1$.

The flow to v_s can now be altered by sending an ε_1-increment of flow along S_1 until S_1 is saturated. The new flow pattern $\mathscr{G}_2$ has

$$\tilde{g}_{s,t,2} = \tilde{g}_{s,t,1} - (a_{S_1} - 1)\delta_1 \tag{6.12.12}$$

and

$$g_{s,t,2} = g_{s,t,1} \tag{6.12.13}$$

where a_{S_1} is the gain of S_1 and δ_1 is nonzero. Next remove any branches which have a zero flow of $\mathscr{G}_2$.

If $\tilde{g}_{s,t,2} > 0$, we can find another return chain S_2 in $\mathscr{G}_2$ by means of which we can decrease $\tilde{g}_{s,t,2}$ by $(a_{S_2} - 1)\delta_2$. The resulting flow pattern is denoted by $\mathscr{G}_3$ with source flow $\tilde{g}_{s,t,3}$ and terminal flow $g_{s,t,3}$. We again remove all empty branches. We can continue in this fashion until we have $\mathscr{G}_l$ such that v_s is isolated or all branches into v_s carry positive flow and all branches out of v_s carry negative flow. Thus

$$\tilde{g}_{s,t,l} = \tilde{f}_{s,t} - \tilde{f}_{s,t,o} - \sum_{i=1}^{l} (a_{S_i} - 1)\delta_i \leq 0. \tag{6.12.14}$$

By assumption, $\tilde{f}_{s,t} - \tilde{f}_{s,t,o} > 0$. Therefore, at least one term $(1 - a_{S_i})\delta_i$ in (6.12.14) is less than 0. In other words, for at least one i, $a_{S_i} > 1$. However, an active source circuit in $\mathscr{G}_l$ is also an active source circuit in $\mathscr{F}$ since a positive branch of $\mathscr{G}_l$ is a nonempty branch of $\mathscr{F}$ and a negative branch of $\mathscr{G}_l$ is an unsaturated branch of $\mathscr{F}$. Hence, there is an active circuit in $\mathscr{F}$ and the proof is completed.

Lemma 6.12.2 *If there is an optimum flow pattern $\mathscr{F}_o$ and a flow pattern $\mathscr{F}$ such that $\tilde{f}_{s,t,o} = \tilde{f}_{s,t}$ and $\underset{\sim}{f}_{s,t,o} > \underset{\sim}{f}_{s,t}$, then there is an active terminal circuit in $\mathscr{F}$.*

Proof. The proof of Lemma 6.12.2 is the same as the proof of Lemma 6.12.1 except that we find an active terminal circuit, rather than an active source circuit, and increase the terminal flow at each iteration.

Theorem 6.12.1 *A flow pattern $\mathscr{F}$ is optimum if and only if it does not contain an active source or terminal circuit.*

Proof. Suppose the optimum flow pattern $\mathscr{F}$ contains an active circuit, L_i. Then $\tilde{f}_{s,t}$ can be decreased by $(a_{L_i} - 1)\delta_i > 0$ without decreasing $\underset{\sim}{f}_{s,t}$ (as in Fig. 6.12.3c) or $\underset{\sim}{f}_{s,t}$ can be increased by $(a_{L_i} - 1)\delta_i$ without increasing $\tilde{f}_{s,t}$ (as in Fig. 6.12.3d). Hence, we contradict the fact that $\mathscr{F}$ is optimum and necessity is proved.

If $\mathscr{F}$ is not optimum there is an optimum flow pattern $\mathscr{F}_o$ such that $\tilde{f}_{s,t,o} = \tilde{f}_{s,t}$ and $\underset{\sim}{f}_{s,t,o} < \underset{\sim}{f}_{s,t}$ or $\underset{\sim}{f}_{s,t,o} = \underset{\sim}{f}_{s,t}$ and $\tilde{f}_{s,t,o} > \tilde{f}_{s,t}$. From Lemma 6.12.1 and 6.12.2, there must be an active circuit in $\mathscr{F}$.

By definition of an active circuit we have:

Corollary 6.12.1 *A flow pattern $\mathscr{F}$ is optimum if and only if every source and terminal circuit has gain less than or equal to one, or is saturated, or is isolated.*

We now derive an important property of optimum flow patterns.

Theorem 6.12.2 *Let $\mathscr{F}_o$ be an optimum flow pattern. If flow is increased along the unsaturated s-t path of highest gain, then the resulting flow pattern $\mathscr{F}_1$ is optimum.*

Proof. Assume, contrary to the statement of the theorem, that $\mathscr{F}_1$ is not optimum. From Corollary 6.12.1, it follows that in $\mathscr{F}_1$ there is an active circuit which was either isolated or saturated in $\mathscr{F}_o$. We consider these possibilities under Case 1 and Case 2 below and show that each leads to a contradiction.

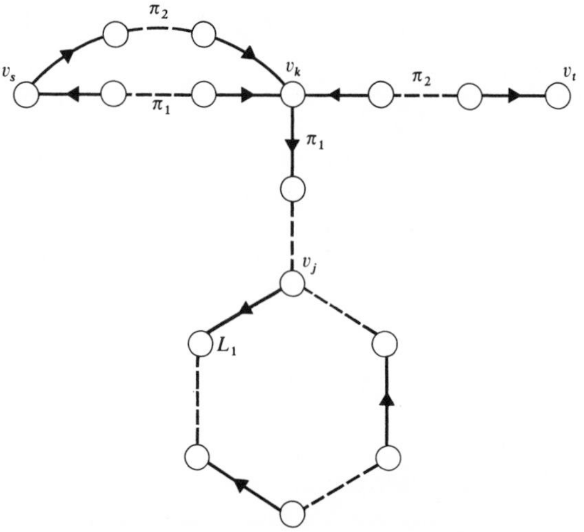

Fig. 6.12.4 Graph used in Case 1 of proof of Theorem 6.12.2.

CASE 1. Suppose that in $\mathscr{F}_o$ there is an isolated circuit L_1 which is active in $\mathscr{F}_1$. This means that for some $v_j \in L_1$ there is a *j-s* or *j-t* path which is saturated by $\mathscr{F}_o$ but not by $\mathscr{F}_1$. Suppose, without loss of generality, that it is a *j-s* path, π_1, as shown in Fig. 6.12.4. But then, there must be an *s-t* path π_2 along which flow is sent in changing $\mathscr{F}_o$ to $\mathscr{F}_1$, which unsaturates π_1. As shown in Fig. 6.12.4, let v_k be the first vertex in π_1 also in π_2. Then, the section of π_1 between v_j and v_k together with the section of π_2 between v_k and v_t comprises an unsaturated *j-t* path for $\mathscr{F}_o$. This is a contradiction.

CASE 2. Suppose in $\mathscr{F}_o$ there is a saturated circuit L_1 which is active with respect to $\mathscr{F}_1$. Suppose further that $\mathscr{F}_1$ was obtained from $\mathscr{F}_o$ by sending flow along an *s-t* path π_1. This flow must then have reduced the flow in some branches in L_1 which were saturated in $\mathscr{F}_o$. This is illustrated in Fig. 6.12.5. The circuit L_1 consists of branches $b_1, b_2, \ldots, b_{l-1}, b_l, b_{l+1}, \ldots, b_{k-1}, b_k$ with the direction of L_1 specified by this ordering of branches; π_a is the path from v_s to the first vertex v_1 in L_1 which is also in π_1; π_b is the path to v_t from the last vertex v_l in L_1 which is also in π_1. The bars on the branches in Fig. 6.12.5 indicate those *saturated* branches which are both backward branches in π_1 and forward branches in L_1.

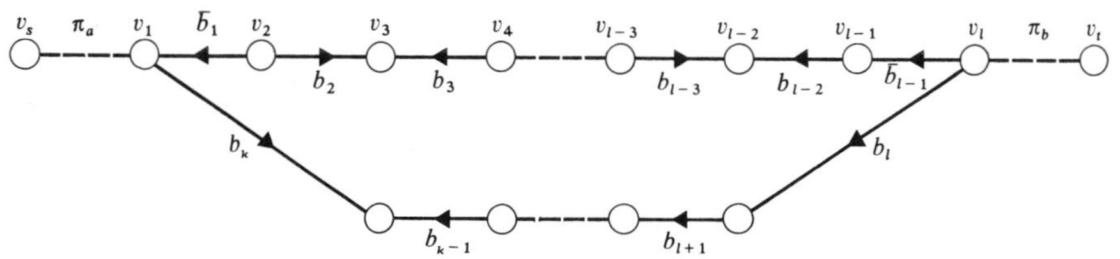

Fig. 6.12.5 Graph used in Case 2 of proof of Theorem 6.12.2.

Let π_{1F} and L_{1F} be the set of forward branches in π_1 and L_1, respectively; and let π_{1B} and L_{1B} be the set of backward branches in π_1 and L_1, respectively. Then the gain of L_1 and π_1 are given in (6.12.15) and (6.12.16) respectively.

$$a_{L_1} = \frac{\prod_{\bar{b}_i \in L_1 \cap \pi_1} a_i \prod_{b_g \in L_{1F} - \pi_{1B}} a_g}{\prod_{b_p \in L_{1B}} a_p} \tag{6.12.15}$$

and

$$a_{\pi_1} = \frac{\prod_{b_j \in \pi_{1F}} a_j}{\prod_{b_k \in \pi_{1B} - L_{1F}} a_k \prod_{\bar{b}_k \in L_1 \cap \pi_1} a_k}. \tag{6.12.16}$$

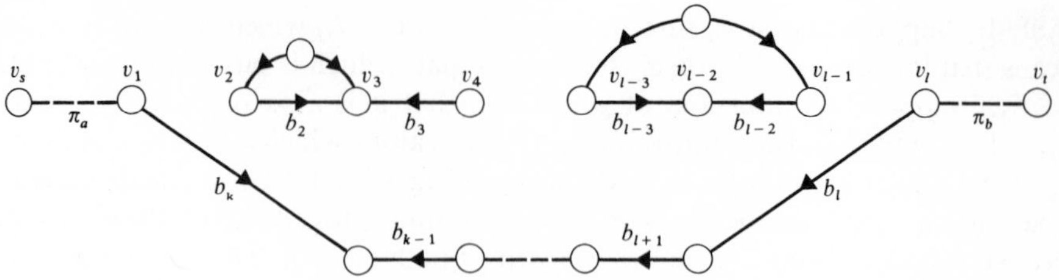

Fig. 6.12.6 Subgraph of $\pi_1 \cup L_1$.

We therefore have

$$a_{L_1} \cdot a_{\pi_1} = \frac{\prod_{b_g \in L_{1F} - \pi_{1B}} a_g \prod_{b_j \in \pi_{1F}} a_j}{\prod_{b_p \in L_{1B}} a_p \prod_{b_k \in \pi_{1B} - L_{1F}} a_k}. \tag{6.12.17}$$

By assumption,

$$a_{L_1} > 1. \tag{6.12.18}$$

Hence

$$a_{L_1} a_{\pi_1} > a_{\pi_1}. \tag{6.12.19}$$

We now construct a subgraph $\hat{G}$ of $\pi_1 \cup L_1$ by eliminating from $\pi_1 \cup L_1$ those branches which were saturated backward branches in π_1 with respect to $\mathscr{F}_o$. Among these are the branches with bars in Fig. 6.12.5. Typically, $\hat{G}$ has the appearance in Fig. 6.12.6 for L_1 in Fig. 6.12.5; $\hat{G}$ is a branch disjoint union of the following subgraphs:

1) The *s-t* path π_2 containing $\pi_a, b_k, b_{k-1}, \ldots, b_{l+1}, b_l, \pi_b$.
2) Circuits such as those containing b_2 and b_{l-2} in Fig. 6.12.6.
3) Branches such as b_3 which are not in π_2 or in circuits. These branches are those unsaturated branches which were both forward branches in L_1 and backward branches in π_1.

If we neglect the isolated branches such as b_3 then $a_{L_1} a_{\pi_1}$ in (6.12.19) is the gain of $\hat{G}$. The gains of branches such as b_3 cancel out of (6.12.19). But the gain of each of the circuits in Fig. 6.12.6 is less than or equal to one since otherwise the circuits would have been active with respect to $\mathscr{F}_o$. Thus

$$a_{\pi_2} \geq a_{L_1} a_{\pi_1} > a_{\pi_1}. \tag{6.12.20}$$

This contradicts the fact that π_1 was the unsaturated path of highest gain with respect to $\mathscr{F}_o$. The proof is therefore completed.

Theorem 6.12.3 *A flow pattern $\mathscr{F}_o$ is maximum if and only if at least one s-t cut $(X, \bar{X})$ is saturated and any circuit with vertices only in $\bar{X}$ is inactive.*

Proof. Suppose $\mathscr{F}_o$ is maximum. Then, in every *s-t* path there must be a backward branch which is empty or a forward branch which is saturated; otherwise $f_{s,t}$ could be increased by sending flow along the path. Selecting one of these branches from each path gives a set of branches containing a saturated s-t cut, say $(X,\bar{X})$. Clearly, since $\mathscr{F}_o$ is maximum, there cannot be an active terminal circuit. But an active circuit with vertices only in $\bar{X}$ cannot be an active source circuit since it would contain a saturated forward branch or an empty backward branch. Hence, there cannot be any active circuit with vertices only in $\bar{X}$ and the proof of necessity is completed.

Since there is a saturated s-t cut, $(X,\bar{X})$, an increase of source flow will leave the terminal flow unchanged. We can therefore assume the source flow $\tilde{f}_{s,t}$ is constant. Lemma 6.12.2 implies that if there is no active terminal circuit then $\tilde{f}_{s,t}$ has been maximized. Since $(X,\bar{X})$ is saturated, any active terminal circuit can contain vertices only in $\bar{X}$ and the proof is completed.

The following algorithm to obtain an optimum maximum flow is suggested by Theorem 6.12.1 to 6.12.3.

Optimum Flow Algorithm

Step 1. Assign any feasible flow pattern $\mathscr{F}_1$ to the graph. (The flow pattern in which all flows are zero is acceptable.)

Step 2. Isolate or saturate all active circuits.

Step 3. Set $i = 1$.

Step 4. For the flow pattern $\mathscr{F}_i$, find the unsaturated s-t path of highest gain and increase the flow along the path until the path is saturated. Let the resulting flow pattern be $\mathscr{F}_{i+1}$.

Step 5. If there is no unsaturated s-t path then terminate. Otherwise replace i by $i + 1$ and return to Step 4.

Example 6.12.3 Consider the graph in Fig. 6.12.7(a) with all initial flows equal to zero. The only active circuit consists of branches (1,2),(2,4),(4,1) with gain 1.2. Saturation of this circuit results in the flow pattern in Fig. 6.12.7(b) with $\tilde{f}_{s,t} = 0$ and $\underset{\sim}{f}_{s,t} = 0.75$. There are now new active circuits and the unsaturated s-t path of highest gain, 0.45, consists of branches $(s,1),(1,4),(4,2),(2,t)$. Saturating this path yields a flow pattern with $\tilde{f}_{s,t} = 8.33$ and $\underset{\sim}{f}_{s,t} = 0.45$, as shown in Fig. 6.12.7(c). The unsaturated path of highest gain is now $(s,1)(1,3),(3,4),(4,t)$. Saturating this path yields the flow pattern in Fig. 6.12.7(d) with $\tilde{f}_{s,t} = 10$ and $\underset{\sim}{f}_{s,t} = 4.85$. The new unsaturated path of highest gain consists of branches $(s,3),(3,4),(4,t)$. Saturation of this path yields the optimum maximum flow pattern in Fig. 6.12.7(c) with $\tilde{f}_{s,t} = 15$ and $\underset{\sim}{f}_{s,t} = 5.55$.

The Optimum Flow Algorithm is an efficient computational algorithm. If the algorithm terminates, then Theorems 6.12.1 to 6.12.3 show that the resulting flow pattern is optimum and maximum. However, since flows may be nonintegral the algorithm may not terminate in a finite number of steps, as shown by Johnson [JO1]. However, Johnson has also shown that if only those flows are allowed which correspond to basic solutions of the associated linear program, then convergence is guaranteed.

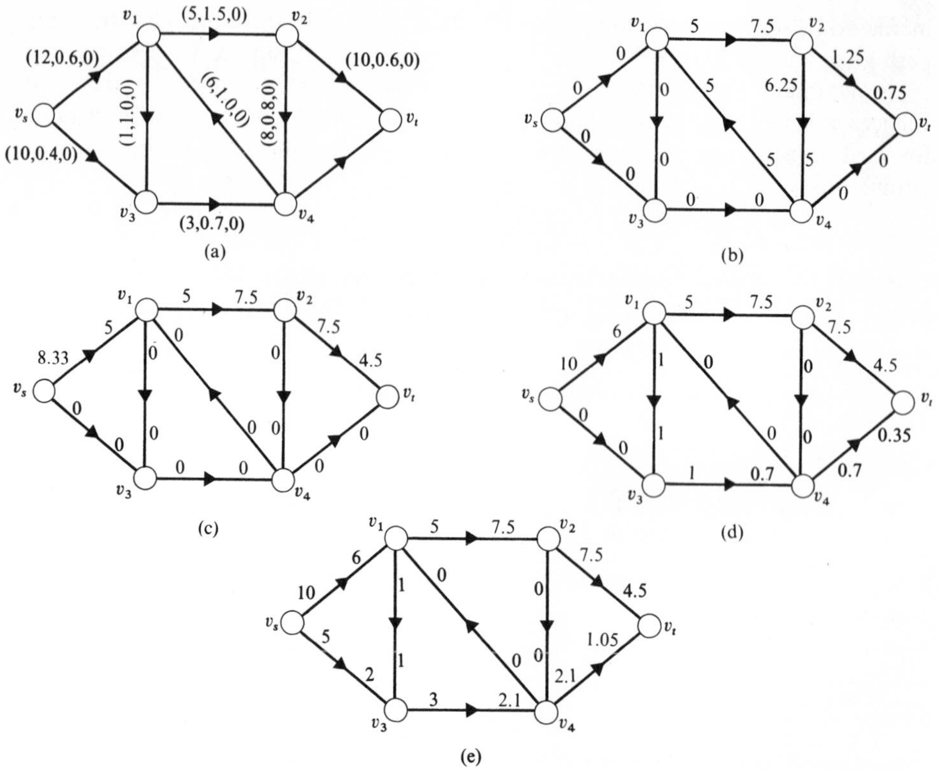

Fig. 6.12.7 Graphs for Example 6.12.3.

A labeling algorithm for finding active circuits, required for step 2 of the optimum flow algorithm, is available in the paper by Jewell [JE2]. (If all the flows in the initial flow pattern are zero then, of course, active circuits need never be found.) Once active circuits have been isolated or saturated a simple method is available for finding the unsaturated s-t path of maximum gain in Step 4. We simply find the shortest s-t path in the graph as in Section 2 where the up-cost $u(i,j)$ and down-cost $\Delta(i,j)$ are defined as

$$u(i,j) = \begin{cases} -\ln a(i,j) & \text{if } \tilde{f}(i,j) < c(i,j), \\ \infty & \text{otherwise;} \end{cases} \tag{6.12.21a}$$

$$\Delta(i,j) = \begin{cases} -\ln \dfrac{1}{a(i,j)} & \text{if } 0 < \tilde{f}(i,j), \\ \infty & \text{otherwise.} \end{cases} \tag{6.12.21b}$$

Logarithms have been taken, as was done in Section 2 for the problem of finding most reliable paths, in order to convert products into sums. The Shortest Path Algorithm can now be used since there are no circuits of negative length. Circuits of negative length correspond to circuits with gain greater than one and these will not be labeled by the Shortest Path Algorithm because they are isolated or saturated.

6.13 FURTHER REMARKS

There is a huge literature on the problems of finding the shortest path and the *N*th shortest paths in a graph. The Shortest Path Algorithm used in this chapter is due to Ford and Fulkerson [FO1]. A number of authors have presented algorithms which are based on the same principle as this one, namely sequentially reducing the labels on vertices by examining branches. Some of these algorithms are applicable only when all lengths are nonnegative [DA1]. Other methods are available using replacement of branches in trees [DA1], dynamic programming [BE2, JE1], matrix multiplication [SH2, HU1], analogs [MI1, RA1] and assignment problems [PE1, PA1, PA2]. The most efficient method for finding shortest paths is a variation of the method in this chapter developed by Floyd [FL1] and Murchland [MU1]. Other of Murchland's results appear in a series of papers, most of which are listed in the work of Yen [YE1]. Decomposition techniques for shortest paths have also been developed by several authors, with the most efficient version due to Hu [HU2]. Finally a general method of incorporating heuristic information into shortest path algorithms is given by Hart *et al.* [HA6]. Extensive bibliographies on the shortest path problem can be found in the references [HU1, YE1, PO1, and DR1]. A number of authors have considered the problem of determining *N*th shortest paths [CL1, PO2, HO1, BE3]. The algorithm in this chapter is due to Hoffman and Pavley [HO1]. The techniques developed in Chapter 4 to study the probabilistic maximum flow problem have been applied by Frank [FR1] to test hypotheses concerning shortest paths in graphs with random branch lengths.

An algorithm for finding the most reliable path in a graph has been presented by Wing [WI1]. The algorithm used in this chapter is due to Parikh and Frisch [PA3]. The problem of finding the most reliable path with a given capacity was proposed as part of a larger problem by Amara *et al.* [AM1]. The solution in this chapter is due to Frisch [FR7].

The concepts of up-cost and down-cost were first used by Beale [BE1]. The Unit Path Minimum Cost Flow Algorithm is due to Busacker and Gowen [BU1] and the extension to graphs with convex cost functions was made by Hu [HU3]. The Minimum Cost Flow Algorithms presented in this chapter are conceptually the simplest. For constant costs, the "out-of-kilter method" [FU2] is also available.

Some fundamental properties of distances on a directed graph are discussed by Berge [BE4] in his Chapters 12 and 13. However, the major results on realizability of a distance matrix are contained in a paper by Hakimi and Yau [HA5]. Other results on distance matrices have been obtained by Boesch [BO1], Goldman [GO1], and Pereira [PE2]. The center of a directed graph is defined in Chapter 12 of Berge and necessary and sufficient conditions for its existence are given; Ore first defined the median of a graph [OR1]. Hakimi [HA3, HA4] extended the concepts of center and median to absolute center, absolute median, and multimedian and Singer [SI1] has given algorithms to find multimedians and multicenters. Goldman [GO2] has significantly generalized the results of Theorem 6.7.1. The extension of absolute centers and medians to graphs with random vertex weights is due to Frank [FR2, FR3].

Kruskal [KR1] presented four different methods for finding a shortest tree, three of which are our Constructions (6.3.1), (6.3.2), and (6.3.3). Prim [PR1] unified these constructions by introducing the operations (6.3.1) and (6.3.2) and presented the most general algorithm for finding shortest trees. Some computational aspects of the problem are presented by Gomory and Hu [GO3].

The Minimum Cost Terminal Capacity Algorithm and the Minimum Excess Terminal Capacity Algorithm were first presented by Gomory and Hu [GO3]. Lemma 6.9.1 is due to Tang and Chien [TA1]. Other minimum cost synthesis algorithms have been developed by Chien [CH1] and Wing and Chien [WI2]. Minimum cost procedures for improving an existing network were given by Gomory and Hu[GO4] and Deo and Hakimi [DE1]. The minimum cost synthesis procedures for networks with random branch flows were developed by Frank and Hakimi [FR5, FR6].

Jewell [JE2] was the first to consider flows in graphs with gains. His model was more general than that considered in this section in that branch costs were also included. The general solution presented by Jewell involves a primal-dual solution of a linear program and included the concepts of active circuits. The most comprehensive graph theoretic treatment is due to Onaga [ON1], who introduced the concept of optimum flow and proved Lemmas 6.12.1 and 6.12.2 and Theorems 6.12.1 and 6.12.2. The Optimum Flow Algorithm is due in part to Fujisawa [FU1]. Mayeda and Van Valkenburg [MA1] proved Theorem 6.12.3 and proved a version of the triangle inequality for graphs with gains.

PROBLEMS

1. In the directed graph shown in Fig. P6.1 the weights on the branches represent lengths. Find the shortest paths from v_1 to all other vertices. Find the second, third, and fourth shortest paths from v_1 to v_9.

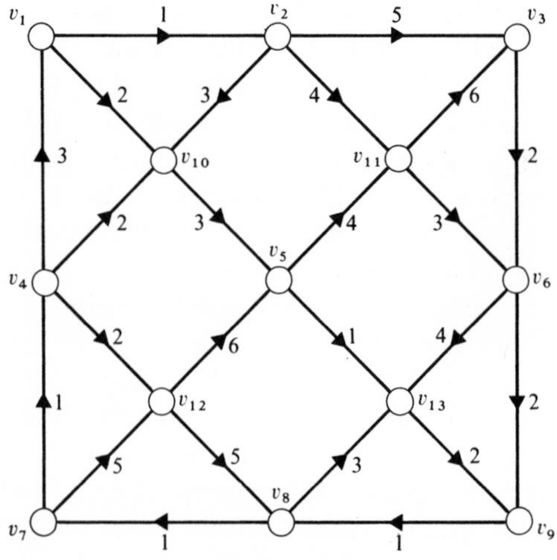

Figure P6.1

2. For the graph used in Example 6.2.4 find ϕ_5 and ϕ_6.
3. Let G be a directed graph with n vertices in which each branch is associated with a length and assume there are no directed circuits of negative length. If $(i,j) \notin \Gamma$, assume $l(i,j) = \infty$.

   The following algorithm has been proposed by Floyd to find the length of the shortest i-j path for all i and j with $i \neq j$.

   Floyd's Algorithm [FL1]

   Step 1. Let $i = 1$.

   Step 2. Let $j = 1$.

   Step 3. Let $\alpha(i,j) = \text{Min}\,[l(i,j), \text{Min}_k[l(i,k) + l(k,j)]]$. Let the new value of $l(i,j)$ be $\alpha(i,j)$.

   Step 4. If $i < n$ increase i and return to Step 3. Otherwise go to Step 5.

   Step 5. If $j < n$ increase j and return to Step 3. Otherwise stop. For all (i,j) with $i \neq j$, $\alpha(i,j)$ is the length of the shortest i-j path.

   Prove that the algorithm yields the length of the shortest i-j path for all i and j with $i \neq j$.
4. In the graph shown in Fig. P6.2 the first weight associated with a branch is its capacity and the second weight is its reliability. Using principal subgraphs find the paths of greatest capacity with reliability at least 0.7.

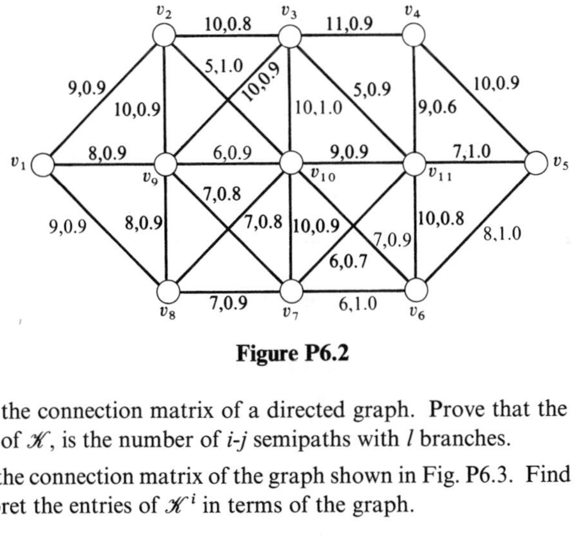

Figure P6.2

5. a) Let $\mathscr{K}$ be the connection matrix of a directed graph. Prove that the i,j entry in $\mathscr{K}^l$, the lth power of $\mathscr{K}$, is the number of i-j semipaths with l branches.

   b) Let $\mathscr{K}$ be the connection matrix of the graph shown in Fig. P6.3. Find $\mathscr{K}^i$ for $i = 1, \ldots, 6$ and interpret the entries of $\mathscr{K}^i$ in terms of the graph.

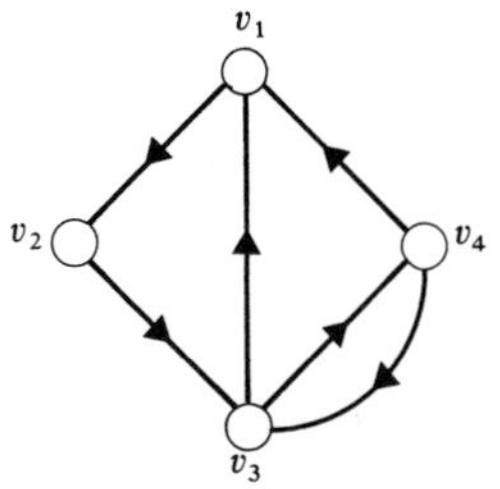

Figure P6.3

6. Consider a graph in which the branch capacities are to be increased so that the maximum s-t flow is also increased. Suppose the cost of increasing $c(i,j)$ by one unit is given by a constant $h(i,j)$ for all i and j with $i \neq j$. We then wish to maximize $f_{s,t}$ for a given allowable increase in cost

$$h = \sum_{\substack{i,j \\ i \neq j}} h(i,j)y(i,j)$$

where $y(i,j)$ is the number of units by which $c(i,j)$ is increased.

Define up-costs and down-costs so that this problem can be formulated as a minimum cost flow problem in a graph with convex branch costs per unit of flow, as a function of flow with characteristics passing through the origin.

7. Consider a directed graph $G = (V,\Gamma)$ in which convex costs per unit of flow are associated with each branch. Prove that an s-t flow pattern is minimum cost for a given value of $f_{s,t}$, if and only if there are no circuits with negative down-cost.

8. Find a minimum cost maximum s-t flow pattern in the graph in Fig. P6.4. The first weight for each branch is its capacity and the second is its cost per unit of flow.

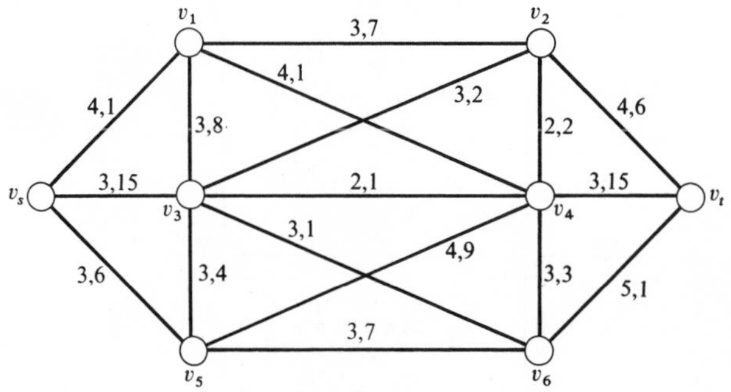

Figure P6.4

9. Prove that a symmetric $n \times n$ matrix $\mathscr{D} = [d_{i,j}]$ with nonnegative integer entries is realizable as the distance matrix of an n-vertex graph for which all branches have unit length if and only if [HA5]

a) $d_{i,i} = 0$ for $i = 1,2,\ldots,n$.

b) For every entry $d_{i,j}$ in $\mathscr{D}$, $d_{i,j} \leq d_{i,p} + d_{p,j}$ for $p = 1,2,\ldots,n$.

c) For every entry $d_{i,j} > 1$, there exists an integer $k < n$ such that $d_{i,k} = 1$ and $d_{i,j} = d_{i,k} + d_{k,j}$.

10. Show that if $\mathscr{D}$ is the distance matrix of a tree, and if

$$d_{l,k} = \underset{i,j}{\text{Max}}\, [d_{i,j}]$$

then v_l and v_k are vertices of degree one.

11. a) Prove that any 3×3 distance matrix has a tree realization.

b) Find necessary and sufficient conditions for a 4×4 distance matrix to have a tree realization [PE2].

[*Note:* Pereira has shown that a necessary and sufficient condition for an $n \times n$ distance matrix to be realizable as a tree is that all its 4×4 principal submatrices be realizable as trees.]

12. Define a double tree of directed graph G to be a connected subgraph of G containing all vertices of G and no circuit or directed circuit of more than two branches.
    a) Show (by example) that a double tree realization of nonsymmetric distance matrix $\mathscr{D}$ is not necessarily unique.
    b) If $\mathscr{D}$ is realizable as a double tree, show that it is realized by a unique graph (i.e., only the weights of the corresponding branches may differ).
    c) Show that every double tree realization of a nonsymmetric distance matrix has the same total weight [SH1].

13. Prove that a tree realization (if it exists) of a symmetric distance matrix $\mathscr{D}$ is the *unique* optimal realization (i.e., the tree has minimum total weight over all graphs which realize $\mathscr{D}$). [*Hint:* Use induction on the order of $\mathscr{D}$.]

14. For the graph G shown in Fig. P6.5, find
    a) The distance matrix of G;
    b) The centers of G;
    c) The medians of G;
    d) The local absolute centers on branch [4,3].

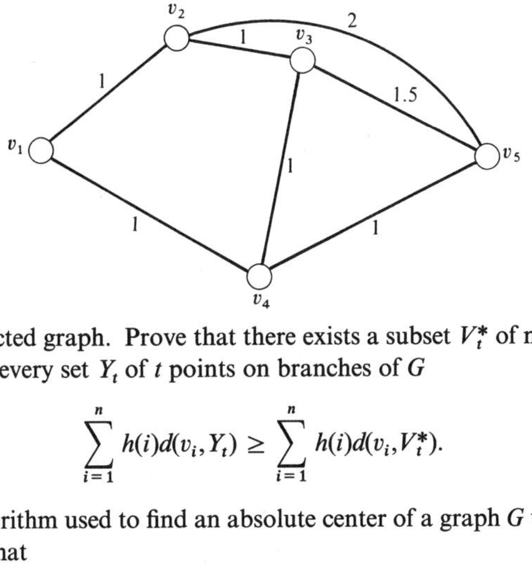

Figure P6.5

15. Let G be an undirected graph. Prove that there exists a subset V_t^* of no more than t vertices of G, such that for every set Y_t of t points on branches of G

$$\sum_{i=1}^{n} h(i)d(v_i, Y_t) \geq \sum_{i=1}^{n} h(i)d(v_i, V_t^*).$$

16. Generalize the algorithm used to find an absolute center of a graph G to find a point Y_0 on a branch of G such that

$$\operatorname*{Max}_{X \text{ on } G} [d(X, Y_0)] \leq \operatorname*{Max}_{X \text{ on } G} [d(X, Y)]$$

for all Y on G [FR4].

17. Let G be an undirected graph and let $H(1), \ldots, H(n)$ be discrete, independent, nonnegative random variables. Suppose there exists a nondegenerate interval (X^*, X^{**}) on a branch of G such that any X on this interval is an MPAr_1C. Show that there exists an $\varepsilon > 0$ such that if $|r_1 - r_2| < \varepsilon$, there exists an $\hat{X} \varepsilon (X^*, X^{**})$ and $\hat{X}$ is a local MPAr_2C of G.

18. a) Let $H(1),\ldots,H(n)$ have a joint nonsingular multidimensional normal distribution. Generalize the discussion of MPARM's and minimum variance medians to this case [FR3].

    b) Let $[p,q]$ be a cut-set of G. Let v_p be in G_1, let v_q be in G_2, and let the vertices of G be relabeled so that $v_1,\ldots,v_\alpha$ are in G_1 and $v_{\alpha+1},\ldots,v_n$ are in G_2. Let $\mu_i = E\{H(i)\}$ for $i = 1,\ldots,\alpha$, and assume the $H(i)$ as in (a). Suppose

$$\sum_{i=1}^{n} \mu_i = \sum_{j=\alpha+1}^{n} \mu_j.$$

    Show that a local minimum variance absolute median is a local maximum probability R median for all positive values of R.

    c) Use the result of (b) to construct a graph for which the MPARM does not occur at a vertex.

19. Prove that Construction (6.9.3) yields a shortest tree.

20. In the graph shown in Fig. P6.6 the branch weights are lengths. Find a shortest tree first using Operations (6.9.1) and then using Operation (6.9.2).

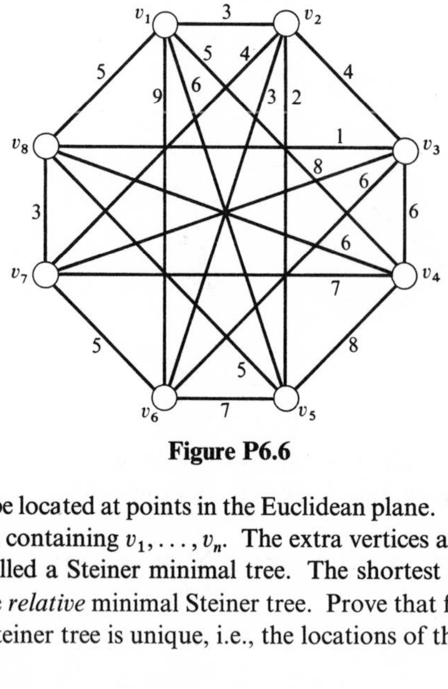

Figure P6.6

21. Let vertices $v_1,\ldots,v_n$ be located at points in the Euclidean plane. We wish to find the shortest tree on $k \geq n$ vertices, containing $v_1,\ldots,v_n$. The extra vertices are called Steiner points and the resulting tree is called a Steiner minimal tree. The shortest tree with a given incidence matrix is said to be the *relative* minimal Steiner tree. Prove that for a given incidence matrix the relative minimal Steiner tree is unique, i.e., the locations of the Steiner points are unique [GI1].

22. Prove that no pair of lines in a minimal Steiner tree meet at an angle less than 120°. (It follows as a corollary that no vertex of a minimal Steiner tree can have a degree greater than three.)

23. Consider the graphs G_a, G_b, and G_c shown in Fig. P7.7. The terminal capacity matrix T_a of G_a is

$$T_a = \begin{bmatrix} \infty & 0 & 0 \\ 2 & \infty & 1 \\ 2 & 3 & \infty \end{bmatrix}.$$

The terminal capacity matrix T_b of G_b is

$$T_b = \begin{bmatrix} \infty & 1 & 1 \\ 1 & \infty & 1 \\ 1 & 1 & \infty \end{bmatrix}.$$

Find the terminal capacity matrix T_c of G_c in terms of T_a and T_b.

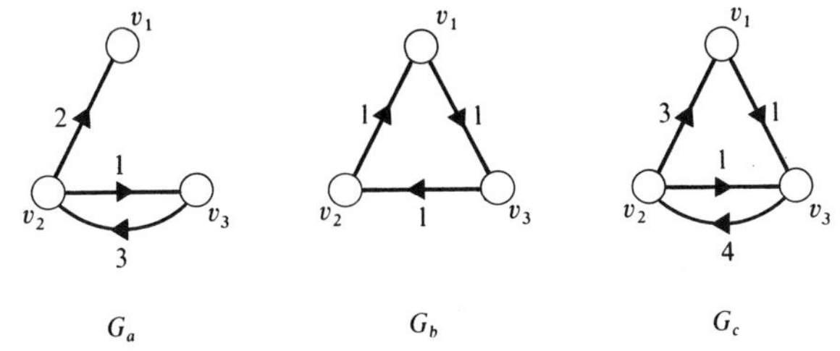

Figure P6.7

24. Realize the terminal capacity matrix below with a graph which has minimum total branch capacity.

$$T = \begin{bmatrix} \infty & 1 & 1 & 1 & 1 \\ 1 & \infty & 3 & 2 & 2 \\ 1 & 3 & \infty & 2 & 2 \\ 1 & 2 & 2 & \infty & 3 \\ 1 & 2 & 2 & 3 & \infty \end{bmatrix}.$$

25. Consider the graph shown in Fig. P6.8. Suppose we are given the sequence of observations of branch flow $\mathbf{F}(1), \mathbf{F}(2), \ldots, \mathbf{F}(5)$ shown below. Formulate and solve the following problem as a linear program. Maximize the average value of $\tau_{4,5}$ subject to the constraint that \$500 is available to improve the network. Let $h = (5,7,10,15,8,6,6)'$ and $\mathbf{c}_0 = (100,48,26,13,42,65,52)'$.

$$\mathbf{F}(1) = (76,20,15,7,28,60,47)'$$
$$\mathbf{F}(2) = (49,31,19,9,31,40,29)'$$
$$\mathbf{F}(3) = (83,37,24,11,40,31,36)'$$
$$\mathbf{F}(4) = (57,31,17,4,25,45,41)'$$
$$\mathbf{F}(5) = (61,25,6,7,17,55,50)'$$

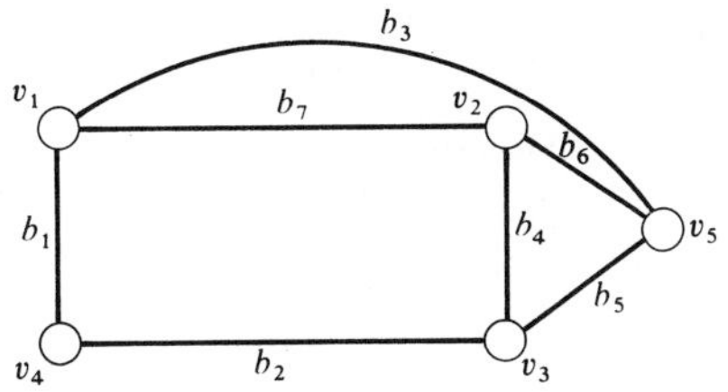

Figure P6.8

26. a) Let G be the graph discussed in Example 6.11.1 and let $h_1 = h_2 = h_3 = h_4 = h_5 = 1$. Suppose we can increase the branch capacity of only a single branch in order to guarantee that hypothesis $H_0 : p = \text{Prob}\{\tau_{s,t} \geq 3\} \geq 0.9$ is accepted at confidence level $\alpha = 0.1$ under the U.M.P. Invariant Test of Section 11. Determine the optimum branch and its final branch capacity.

    b) Reconsider the above problem under the assumption that the branch flows and maximum flow have unknown probability distributions. Use the synthesis procedure discussed in Section 10 and compare the results of this synthesis with the results of part (a).

27. In the graph shown in Fig. P6.9, the first weight of each branch is its capacity and the second is its gain. Find an optimum maximum s-t flow pattern.

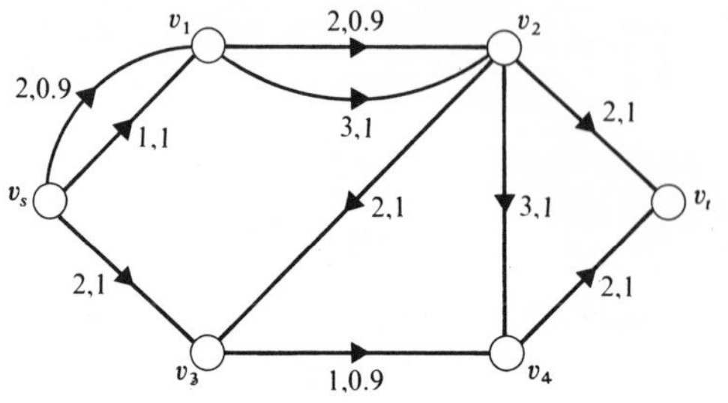

Figure P6.9

28. Repeat the last problem for the graph shown in Fig. P6.10.

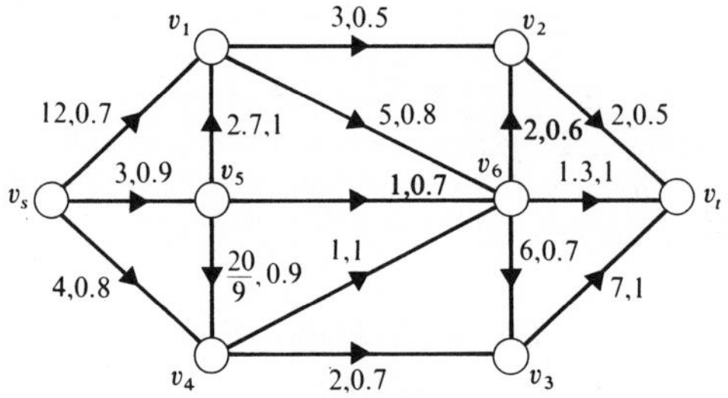

Figure P6.10

REFERENCES

AM1 R. C. Amara, H. Lindgren, and M. Pollack, "Link Error Control and Network Route Selection," *IRE Trans. Commun. Systems* **CS-9,** 328–334 (1961).

BE1 E. M. L. Beale, "An Algorithm for Solving the Transportation Problem When the Shipping Cost over Each Route Is Convex," *Naval Res. Logistics Quart.* **6,** 43–46 (1959).

BE2 R. Bellman, "On a Routing Problem," *Quart. Appl. Math.* **16,** 87–90 (1958).

BE3 R. Bellman and R. Kalaba, "On kth Best Policies," *J. Soc. Ind. Appl. Math.* **8,** 582–588 (1960).

BE4 C. Berge, *The Theory of Graphs*, Methuen, London, 1962.

BO1 F. T. Boesch, "Properties of the Distance Matrix of a Graph," *Quart. Appl. Math.* **26,** 607–610 (1969).

BU1 R. G. Busacker and P. J. Gowen, *A Procedure for Determining a Family of Minimal-Cost Network Flow Patterns*, Operations Research Office, Technical Paper 15 (1961).

CH1 R. T. Chien, "Synthesis of a Communication Net," *IBM J. Res. Develop.* **4,** 311–320 (1960).

CL1 S. Clarke, A. Krikorian, and J. Rausen, "Computing the N Best Loopless Paths in a Network," *J. Soc. Ind. Appl. Math.* **11,** 1096–1102 (1963).

DA1 G. B. Dantzig, "On Shortest Routes through Networks," *Management Sci.* **6,** 187–190 (1960).

DA2 G. B. Dantzig, *Linear Programming and Extensions*, Princeton University Press, Princeton, 1963.

DE1 N. Deo and S. L. Hakimi, "Minimum Cost Increase of the Terminal Capacity of a Communication Network," *IEEE Trans. Commun. Technol.* **COM-14,** 63–64 (1966).

DR1 S. E. Dreyfus, "An Appraisal of Some Shortest Path Algorithms," *Operations Res.* **17,** 395–412 (1969).

FI1 M. Fisz, *Probability Theory and Mathematical Statistics*, Wiley, New York, 1963.

FL1 R. W. Floyd, "Algorithm 97: Shortest Path," *Commun. Assoc. Computing Machinery* **5,** 345 (1962).

FO1 L. R. Ford, Jr., and D. R. Fulkerson, *Flows in Networks*, Princeton University Press, Princeton, 1962.

FR1 H. Frank, "Shortest Paths in Probabilistic Graphs," *Operations Res.* **17,** 583–599 (1969).

FR2 H. Frank, "Optimum Locations on a Graph with Probabilistic Demands," *Operations Res.* **14,** 409–421 (1966).

FR3 H. Frank, "Optimum Locations on a Graph with Correlated Normal Demands," *Operations Res.* **15,** 552–557 (1967).

FR4 H. Frank, "A Note on a Graph Theoretic Game of Hakimi's," *Operations Res.* **15,** 567–570 (1967).

FR5 H. Frank and S. L. Hakimi, "On the Optimum Synthesis of Statistical Communication Networks—Pseudo Parametric Techniques," *J. Franklin Inst.* **284,** 407–416 (1967).

FR6 H. Frank and S. L. Hakimi, "Parametric Synthesis of Statistical Communication Networks," *Quart. Appl. Math.* **27,** 105–120 (1969).

FR7 I. T. Frisch, "Optimum Routes in Communication Systems with Channel Capacities and Channel Reliabilities," *IEEE Trans. Commun. Systems* **CS-11,** 241–244 (1963).

FU1 T. Fujisawa, "Maximum Flows in a Lossy Network," in Proceedings of Allerton Conference on Circuit and System Theory, Urbana, Ill., pp. 385–393 (1963).

FU2 D. R. Fulkerson, "An Out of Kilter Method for Minimal-Cost Flow Problems," *J. Soc. Ind. Appl. Math.* **9,** 18–27 (1961).

GI1 E. N. Gilbert and M. O. Pollack, "Steiner Minimal Trees," *SIAM J. Appl. Math.* **16,** 1–29 (1968).

GO1 A. Goldman, "Realizing Distance Matrices in a Graph," *J. Res. Natl. Bur. Std.* **70B,** 153–154 (1966).

GO2 A. Goldman, "Optimal Locations for Centers in a Network," *Transportation Sci.* **3,** 352–360 (1969).

GO3 R. E. Gomory and T. C. Hu, "Multiterminal Network Flows," *J. Soc. Ind. Appl. Math.* **9,** 551–570 (1961).

GO4 R. E. Gomory and T. C. Hu, "An Application of Generalized Linear Programming to Network Flows," *J. Soc. Ind. Appl. Math.* **10,** 260–283 (1962).

HA1 G. Hadley, *Linear Programming*, Addison-Wesley, Reading, Mass., 1962.

HA2 G. Hadley, *Nonlinear and Dynamic Programming*, Addison-Wesley, Reading, Mass., 1964.

HA3 S. L. Hakimi, "Optimum Locations of Switching Centers and the Absolute Centers and Medians of a Graph," *Operations Res.* **12,** 450–459 (1964).

HA4 S. L. Hakimi, "Optimum Distribution of Switching Centers in a Communications Network and Some Related Problems," *Operations Res.* **13,** 462–475 (1965).

HA5 S. L. Hakimi and S. S. Yau, "Distance Matrix of a Graph and Its Realizability," *Quart. Appl. Math.* **22,** 305–317 (1965).

HA6 P. E. Hart, N. J. Nilson, and B. Raphael, "A Formal Basis for the Heuristic Determination of Minimum Cost Paths," *IEEE Trans. Systems Sci. Cybernetics* **SSCS-4,** 100–107 (1968).

HO1 W. Hoffman and R. Pavley, "A Method for the Solution of the *N*th Best Path Problem," *J. Assoc. Computing Machinery* **6,** 506–514 (1959).

HU1 T. C. Hu, "Revised Algorithms for Shortest Paths," *J. Soc. Ind. Appl. Math.* **15,** 207 (1967).

HU2 T. C. Hu, "A Decomposition Algorithm for Shortest Paths in a Network," *Operations Res.* **16,** 91–102 (1968).

HU3 T. C. Hu, "Minimum Convex Cost Flow in Networks," *Naval Res. Logistics Quart.* **13,** 1–9 (1966).

JE1 R. P. Jefferis, III, and K. A. Fegley, "Application of Dynamic Programming to Routing Problems," *IEEE Trans. Systems Sci. Cybernetics* **SSCS-1,** 21–26 (1965).

JE2 W. S. Jewell, "Optimal Flow through Networks with Gains," *Operations Res.* **10,** 476–499 (1962).

JO1 E. Johnson, "Networks and Basic Solutions," *Operations Res.* **14,** 619–624 (1966).

KR1 J. B. Kruskal, Jr., "On the Shortest Spanning Subtree of a Graph and the Travelling Salesman Problem," *Proc. Am. Math. Soc.* **7,** 48–50 (1956).

MA1 W. Mayeda and M. E. Van Valkenburg, "Properties of Lossy Communication Nets," *IEEE Trans. Circuit Theory* **CT-12,** 334–338 (1965).

MI1 G. J. Minty, "A Comment on the Shortest Route Problem," *Operations Res.* **5,** 724 (1957).

MU1 J. D. Murchland, *Bibliography of the Shortest Route Problem*, London School of Economics, Technical Report TNT-6, August 3, 1967.

ON1 K. Onaga, "Optimum Flows in General Communication Networks," *J. Franklin Inst.* **283,** 308–327 (1967).

OR1 O. Ore, *Theory of Graphs*, Am. Math. Soc. Colloq. Publication, Providence, R.I., 1962.

PA1 S. N. N. Pandit, "The Shortest-Route Problem—An Addendum," *Operations Res.* **9,** 129–132 (1961).

PA2 S. N. N. Pandit, "Some Observations on the Routing Problem," *Operations Res.* **10,** 726–727 (1962).

PA3 S. C. Parikh and I. T. Frisch, "Finding the Most Reliable Routes in a Communication System," *IEEE Trans. Commun. Systems* **CS-11,** 402–407 (1963).

PE1 R. M. Peart, P. H. Randolph, and T. E. Bartlett, "The Shortest Route Problem," *Operations Res.* **8,** 866–868 (1960).

PE2 J. M. S. Simoes Pereira, "On the Tree Realization of a Distance Matrix," unpublished paper.

PO1 M. Pollack and W. Wiebenson, "Solutions of the Shortest-Route Problem—A Review," *Operations Res.* **8,** 224–230 (1960).

PO2 M. Pollack, "Solution of the kth Best Route through a Network. A Review," *J. Math. Analysis Appl.* **3,** 547–559 (1961).

PR1 R. C. Prim, "Shortest Connection Networks and Some Generalizations," *B.S.T.J.*, **36,** 1389–1402 (1957).

RA1 H. Rapaport and P. A. Abramson, "An Analog Computer for Finding an Optimum Route through a Communication Network," *IRE Trans. Commun. Systems* **CS-7,** 37–42 (1959).

SH1 B. P. Shay, "Some Considerations of Distances in a Linear Directed Graph," M.S. Thesis, Department of Electrical Engineering, Northwestern University, Evanston, Ill., June 1965.

SH2 A. Shimbel, "Structure in Communication Networks," in Proceedings of the Symposium on Information Networks, Polytechnic Institute of Brooklyn, pp. 199–203 (1955).

SI1 S. Singer, "Multi-Centers and Multi-Medians of a Graph with an Application to Optimal Warehouse Location," unpublished paper, Dunlap and Associates, Inc., Darien, Conn., 1968.

TA1 D. T. Tang and R. T. Chien, "Analysis and Synthesis Techniques of Oriented Communication Nets," *IRE Trans. Circuit Theory* **CT-8,** 39–44 (1961).

WI1 O. Wing, "Algorithms to Find the Most Reliable Path in a Network," *IRE Trans. Circuit Theory* **CT-8,** 78 (1961).

WI2 O. Wing and R. T. Chien, "Optimal Synthesis of a Communication Net," *IRE Trans. Circuit Theory* **CT-8,** 44–49 (1961).

YE1 J. Y. Yen, *Some Algorithms for Finding the Shortest Routes through a General Network*, Graduate School of Business Administration, University of California, Berkeley, February 1, 1968.

CHAPTER 7

CONNECTIVITY AND VULNERABILITY OF DETERMINISTIC GRAPHS

7.1 INTRODUCTION

In modeling networks by graphs several significant parameters have not yet been considered. One of these, the vulnerability of the network, is discussed in this chapter and in Chapter 8. By vulnerability we mean the susceptibility of the network to attack. For example, if a network is modeled by the graph in Fig. 7.1.1 it is reasonable to conclude that the system is highly vulnerable, since removal of the single station represented by vertex v_1 interrupts all communications. The concept of vulnerability is usually associated with systems subjected to attack by an adversary. However, the problem can be thought of as a reliability problem if damage can be caused by natural disturbances.

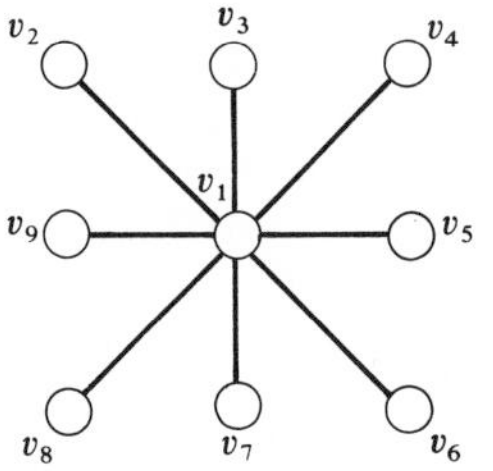

Fig. 7.1.1 Highly vulnerable graph.

The adversary may have a deterministic attack strategy and complete knowledge of the system. In this case we may be able to develop a deterministic network model. On the other hand, the branches and vertices could be destroyed in some random fashion, such that the branches and vertices can be assigned probabilities of destruction. In this case the adversary could have either complete or partial knowledge of the system. In any event, the analysis problem consists of determining the degree of vulnerability of an existing or proposed system and the synthesis problem consists of constructing a system with the least vulnerability. Deterministic problems are discussed in this chapter and a probabilistic approach is taken in the next chapter.

The solution to either the analysis or synthesis problem depends upon the definition of "destruction." For example, a system modeled by a graph may be considered destroyed if, when vertices or branches are removed, the resulting graph G satisfies one or more of the following conditions:

G contains at least two components. (7.1.1)

There are no directed s_i-t_i paths for specified sets of vertices $\{v_{s_i}\}$ and $\{v_{t_i}\}$. (7.1.2)

The number of vertices in the largest component of G is less than some specified number. (7.1.3)

The shortest s_i-t_i path is longer than some specified number. (7.1.4)

We first consider the criterion in (7.1.1) since it is one of the simplest and most fundamental indications of vulnerability. Let v be the minimum number of branches and vertices which must be removed from a graph to break all paths for at least one pair of vertices. The designers' objective is to maximize v. For example, for G_1 (Fig. 7.1.2a) $v = 2$ and for G_2 (Fig. 7.1.2b) $v = 1$, although G_1 and G_2 have the same number of vertices and branches. To see the difficulty of finding v consider the graph G in Fig. 7.1.3(a). There is no obvious way to compute v, although the answer is readily seen once G is redrawn (Fig. 7.1.3b). The removal of either v_3 or v_{10} disconnects G.

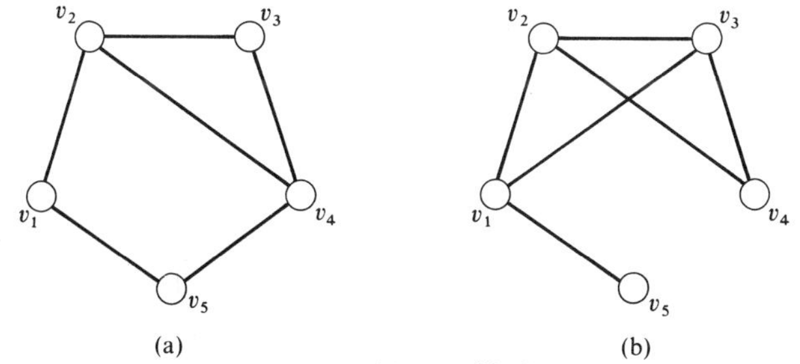

Fig. 7.1.2 (a) G_1; (b) G_2.

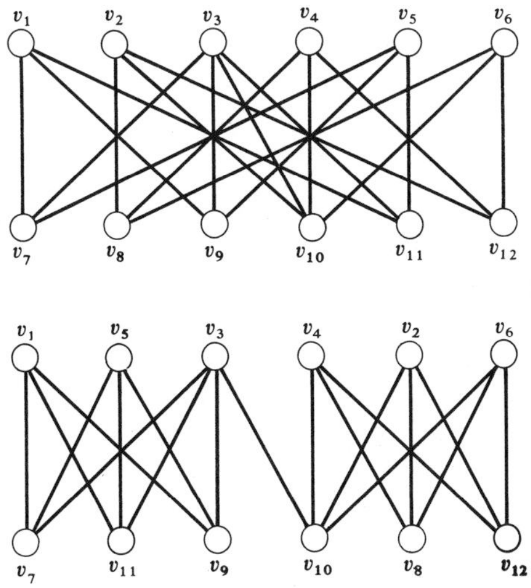

Fig. 7.1.3 Illustration of the difficulty of finding the smallest vertex cut-set in a graph G.

In the next three sections we present the theoretical basis for an analysis method. In Section 2, we show that ν can be determined by finding the minimum number of vertices which must be removed to break all directed paths between specified pairs of vertices. In Section 3 we present an efficient algorithm for finding the number of vertices which must be removed to break all s-t paths for given v_s and v_t. A method is developed in Section 4 to simplify the calculations as v_s and v_t are varied. A procedure is then developed in Section 5 to realize graphs for which a maximum number of branches must be removed to disconnect the graph. A more general version of this problem is discussed in Section 6. In Sections 7 and 8 we give methods to construct graphs with a maximum value of ν. The previous synthesis problems are reconsidered in Section 9 under the additional constraint that cost be minimized. Efficient approximate solutions are given which involve repeated use of the analysis methods developed in Sections 2, 3, and 4. Finally, in Section 10 we give methods of constructing graphs for which a maximum number of elements must be removed to break all paths from a specified number of vertices to the remaining vertices.

7.2 VERTEX AND MIXED CUT-SETS

In order to begin the study of the analysis problem, we must introduce several definitions. The results are given for directed graphs, although they apply to undirected graphs with only slight modification.

***Definition* 7.2.1** *For a directed graph G let $\omega_{s,t}$ be the number of vertices in a smallest s-t vertex cut-set. Let $\theta_{s,t}$ be the number of branches in the smallest s-t branch cut-set.*

Note that if all branch capacities are unity then $\tau_{s,t} = \theta_{s,t}$.

***Definition* 7.2.2** *Let G be a directed graph. An s-t mixed cut-set is a minimal set of branches and vertices, other than v_s and v_t, whose removal from G breaks all directed s-t paths. Let $\sigma_{s,t}$ be the number of elements in the smallest s-t mixed cut-set.*

***Definition* 7.2.3** *For a directed incomplete[1] graph G let*

$$\theta = \operatorname*{Min}_{\substack{v_s, v_t \in V \\ s \neq t}} [\theta_{s,t}] \tag{7.2.1a}$$

and

$$\omega = \operatorname*{Min}_{\substack{v_s, v_t \in V \\ (s,t) \notin \Gamma \\ s \neq t.}} [\omega_{s,t}] \tag{7.2.1b}$$

If G is complete, define ω as $n - 1$, where n is the number of vertices.

In this section we prove that ν is equal to ω and, thus, to find ν we need only consider an s-t vertex cut-set if v_s and v_t are not joined by a branch.

***Lemma* 7.2.1** *For a directed graph G with $(s,t) \notin \Gamma$,*

$$\omega_{s,t} = \operatorname{Min}[\omega_{s,t}, \sigma_{s,t}, \theta_{s,t}]. \tag{7.2.2}$$

---

[1] A graph is *complete* if for all i and j with $i \neq j$ we have $(i,j) \in \Gamma$. Otherwise the graph is *incomplete.*

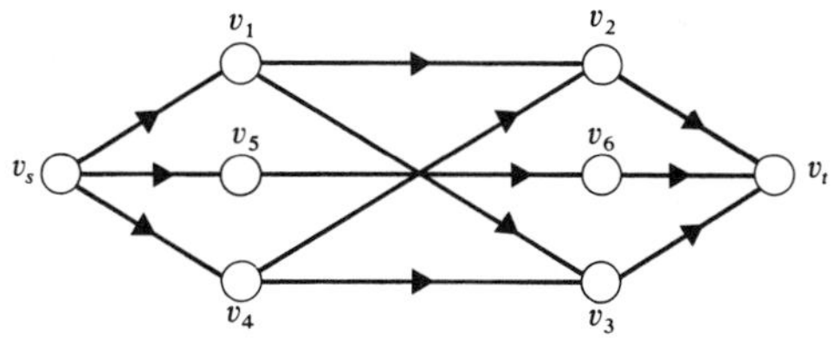

Fig. 7.2.1 Graphs illustrating steps in the proof of Lemma 7.2.1.

Proof. Let $\mathcal{M}$ be a mixed s-t cut-set (a branch cut-set is a special case). If $(i,j) \in \mathcal{M}$, then $i \neq t$ and $j \neq s$. Also, either $i \neq s$ or $j \neq t$, since $(s,t) \notin \Gamma$. If $i \neq s$, form a new s-t cut-set $\hat{\mathcal{M}}$ by adding v_i to $\mathcal{M}$ and removing from $\mathcal{M}$ all branches directed from v_i. If $i = s$, form the new s-t cut-set $\hat{\mathcal{M}}$ by adding v_j to $\mathcal{M}$ and removing from $\mathcal{M}$ all branches directed to v_j. In either case, the resulting cut-set has fewer branches than $\mathcal{M}$ and satisfies $|\hat{\mathcal{M}}| \leq |\mathcal{M}|$. This process can be repeated until an s-t vertex cut-set with no more elements than $|\mathcal{M}|$ is obtained.

As an example of the procedure used in the proof consider the graph in Fig. 7.2.1. Select the s-t mixed cut-set $\{v_5\} \cup \{(1,2),(1,3),(4,2),(4,3)\}$. The resulting vertex cut-set $\hat{\mathcal{M}}$ is $\{v_1,v_4,v_5\}$.

To break all s-t paths if $(s,t) \in \Gamma$, it is insufficient to remove a vertex cut-set. However, after (s,t) is removed, we can apply Lemma 7.2.1 to the resulting graph $\tilde{G}$.

Let $\tilde{X}$ be the s-t vertex cut-set in $\tilde{G}$ with the minimum number of elements. Define $\tilde{\omega}_{s,t}$ by

$$|\tilde{X}| = \tilde{\omega}_{s,t}. \tag{7.2.3}$$

The minimum number of elements which must be removed to break all directed s-t paths if $(s,t) \in \Gamma$ is $\tilde{\omega}_{s,t} + 1$. We now derive the main result of this section.

***Theorem* 7.2.1** *For a directed graph*

$$\nu = \omega. \tag{7.2.4}$$

Proof. Let G be an incomplete graph and assume, contrary to the theorem, that there exist two vertices v_s and v_t such that $(s,t) \in \Gamma$ and there is a mixed s-t cut-set $\mathcal{M}_3 = V_3 \cup \{(s,t)\}$ with $|\mathcal{M}_3| < \omega$. We show this assumption leads to a contradiction. Define the graph $G_3 = (V_3,\Gamma_3)$ shown in Fig. 7.2.2(a) such that Γ_3 is the set of all branches joining two vertices in V_3. Remove V_3 from G and let V_1 be the set of vertices in the remaining subgraph to which there is a directed path from v_s. Let V_2 be the set of vertices in the remaining graph from which there is a directed path to v_t. Form graphs G_1 and G_2 by adding to V_1 and V_2 all branches between two vertices in V_1 and in V_2 respectively. The set of remaining vertices is denoted by V_4. Graph G_4 consists of these vertices and the branches connecting them. In Fig. 7.2.2, lines marked with $\sim$ represent one or more possible branches. In G there are no directed s-t paths containing any vertices in V_4. There are no branches directed from vertices in V_1 to vertices in V_4 or from vertices in V_4 to vertices in V_2, as seen from the definitions of V_1 and V_2. Suppose now that V_4 is not empty. Then the only directed paths from a vertex v_i in V_4 to v_t must pass through G_3 or v_s. Hence the removal of V_3 and v_s breaks all directed i-t paths. Since $|V_3 \cup \{v_s\}| = |\mathscr{M}_3|$ this is a contradiction. Therefore, assume that V_4 is empty and G can be redrawn as in Fig. 7.2.2(b).

Suppose that there is at least one vertex v_i in V_1 other than v_s. Then the removal of V_3 and v_s breaks all directed i-t paths. However, $|V_3 \cup \{v_s\}| = |\mathscr{M}_3|$ and we have a contradiction. Thus V_1 consists of a single vertex v_s. Similarly, we can show that V_2 contains only v_t.

The only remaining possibility is $V_1 = \{v_s\}$ and $V_2 = \{v_t\}$ so there are $|V_3| + 2$ vertices in G. If G is incomplete there must be two vertices separated by a vertex cut-set of less than or equal to $|V_3|$ vertices, which again contradicts the assumption. Hence the proof is completed for this case. If G is complete the theorem is a consequence of the definition of ω.

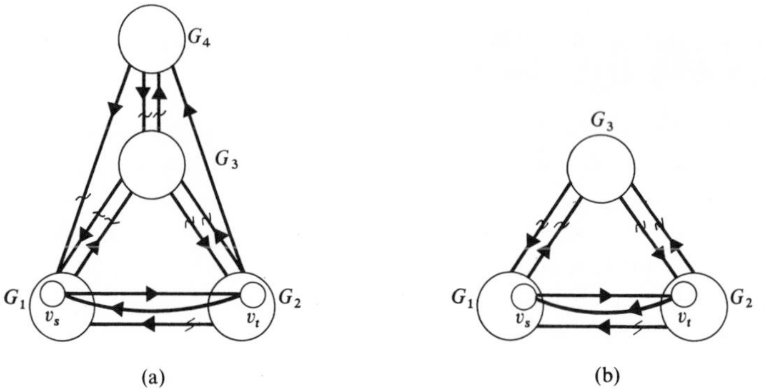

Fig. 7.2.2 Graphs for the proof of Theorem 7.2.1.

7.3 DETERMINATION OF SMALLEST VERTEX CUT-SETS

We show first that the number of vertices in the smallest s-t vertex cut-set is equal to the maximum number of vertex disjoint s-t paths. We then give an algorithm for finding these paths, which is more efficient for this problem than the Labeling Algorithm.

Theorem 7.3.1 *Let G be a directed graph. The number of vertices $\omega_{s,t}$ in the smallest s-t vertex cut-set is equal to the maximum number of vertex disjoint s-t paths.*

Proof. Assign unit capacities to all vertices except v_s and v_t. Assign infinite capacity to v_s and v_t. Clearly the value of the maximum s-t flow is not greater than $\omega_{s,t}$. Assign a maximum s-t flow pattern. Perform a semipath decomposition of this flow pattern and eliminate all isolated circuits of flow. The resulting flow paths identify $\omega_{s,t}$ vertex disjoint s-t paths. Hence the total number of such paths is greater than or equal to $\omega_{s,t}$ and the proof is completed.

To find $\omega_{s,t}$ we can find the maximum value of $f_{s,t}$ subject to the constraints

$$f(i,V) - f(V,i) = \begin{cases} f_{s,t} & \text{for } i = s, \\ 0 & \text{for } i \neq s,t, \\ -f_{s,t} & \text{for } i = t, \end{cases} \tag{7.3.1a}$$

$$f(i,V) \leq 1 \qquad \text{for } i \neq s,t, \tag{7.3.1b}$$

$$f(i,j) \geq 0 \qquad \text{for all } (i,j) \in \Gamma. \tag{7.3.1c}$$

We could convert the original graph G into a new graph G' and then apply the Labeling Algorithm as explained in Section 3.3. However, this method would increase the size of the graph to be analyzed and, as we shall see, leads to inefficiencies. In this section we give an algorithm to eliminate these difficulties. It is called the $\omega_{s,t}$ Labeling Algorithm and consists of a Labeling Routine to find an augmentation path and an $\omega_{s,t}$ Augmentation Routine to increase flow along the augmentation path. The $\omega_{s,t}$ Labeling Algorithm is a modification of the Labeling Algorithm. We motivate this modification by means of an example.

Example 7.3.1 We wish to find $\omega_{s,t}$ for the graph G in Fig. 7.3.1(a). We can therefore let $c(v_i) = 1$ for all $i \neq s,t$ and $c(v_s) = c(v_t) = \infty$. The maximum s-t flow is equal to $\omega_{s,t}$. Assume an initial flow pattern as shown in Fig. 7.3.1(a). Break each vertex into two vertices and add directed branches to obtain G' in Fig. 7.3.1(b). The labels assigned by the Labeling Algorithm are shown in Fig. 7.3.1(b). The labels are only ordered pairs; $\varepsilon(v_i)$ has been omitted because $\varepsilon(v_i)$ equals 0 or 1 for $i \neq s$. We would like to develop an algorithm which assigns labels directly to G without increasing the number of vertices. This could be done by simply assigning the labels on $v_{i'}$ and $v_{i''}$ in G' to v_i in G. However, this is obviously inefficient. For example, the label on $v_{3''}$ does not carry any useful information not already contained in the label on $v_{3'}$. The $\omega_{s,t}$ Labeling Routine will assign the labels shown in Fig. 7.3.1(c). Note that only v_5 is assigned two labels.

The labels to be assigned will be ordered triples. The first entry in the triple will as usual be the vertex scanned in order to assign the label. The second entry will be a + or − to indicate a possible increase or decrease of one unit of flow. The third entry is a marker M of value zero or one whose meaning will be explained shortly. We will call a label *strong* if the second entry is a minus and *weak* if it is a plus. A vertex may be assigned both a weak and a strong label.

$\omega_{s,t}$ Labeling Algorithm

$\omega_{s,t}$ Labeling Routine

Step 1. Assign a weak label $(s,+,0)$ to v_s; v_s is now weakly labeled and unscanned. All other vertices are unlabeled and unscanned.

Step 2. Select any labeled unscanned vertex v_i. If there are none, terminate.

a) Suppose v_i is not strongly labeled.

i) If $f(i,V) = 0$, weakly label by $(i,+,0)$ all unlabeled v_g such that $(i,g) \in \Gamma$; v_g is now weakly labeled and unscanned.

ii) If $f(i,V) \neq 0$, strongly label v_g by $(i,-,0)$, for vertex v_g, for which $f(g,i) = 1$ and v_g is not strongly labeled; v_g is now strongly labeled and unscanned. Encircle the $+$ in the weak label on v_i; v_i is now scanned.

b) Suppose v_i is strongly labeled.
Weakly label by $(i,+,M)$, all unlabeled v_g such that $(i,g) \in \Gamma$. v_g is now weakly labeled and unscanned. Strongly label v_g by $(i,-,M)$ if $f(g,i) = 1$ and v_g is not strongly labeled; v_g is now strongly labeled and unscanned. If v_i is weakly labeled as well as strongly labeled let $M = 1$; otherwise let $M = 0$.
Encircle the $-$ in the strong label on v_i; v_i is now scanned.

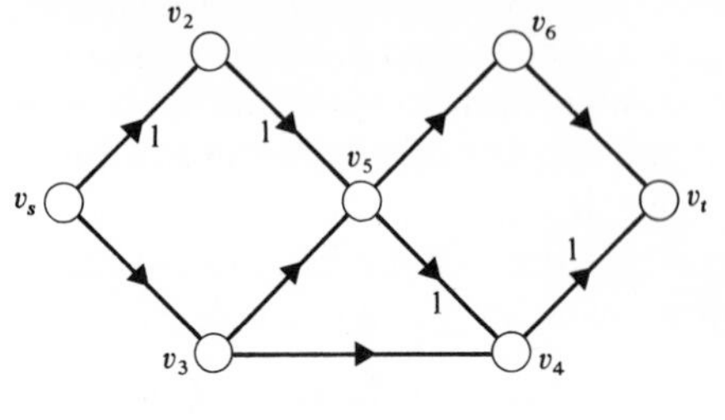

(a)

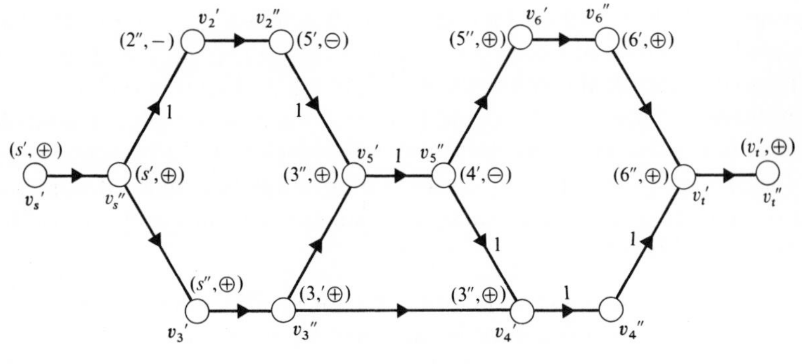

(b)

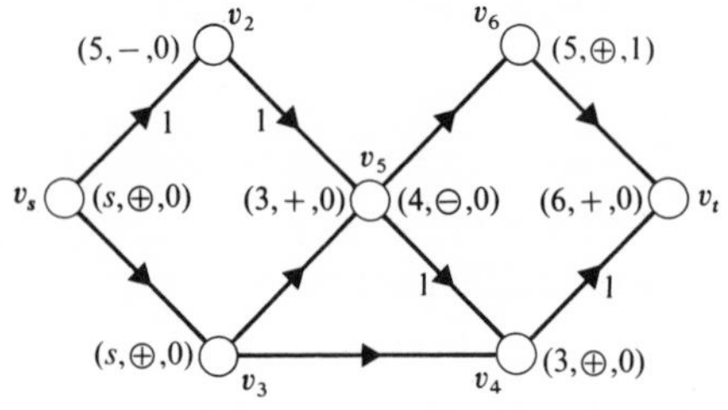

(c)

Fig. 7.3.1 Illustration of development of $\omega_{s,t}$ Labeling Algorithm.

Step 3. If v_t is labeled, proceed to the $\omega_{s,t}$ Augmentation Routine. Otherwise return to Step 2.

$\omega_{s,t}$ *Augmentation Routine*

Step 1. Let $z = t$ and go to Step 2.

Step 2.

a) If v_z is weakly labeled by $(q,+,M)$, increase $f(q,z)$ by one.

b) If v_z is strongly labeled by $(q,-,M)$, decrease $f(z,q)$ by one. Let $N = M$.

Step 3. If $q = s$ return to Step 1 of the $\omega_{s,t}$ Labeling Routine. Otherwise proceed to Step 4.

Step 4. Let $z = q$. If $N = 1$ return to Step 2(b). Otherwise return to Step 2(a).

Example 7.3.2 The $\omega_{s,t}$ Labeling Algorithm yields the labels shown in Figs. 7.3.1(c), 7.3.2, and 7.3.3 for the given flow patterns. In Fig. 7.3.1(c), v_t is labeled and flow can be increased along the path $v_s, v_3, v_4, v_5, v_6, v_t$. In Figs. 7.3.2 and 7.3.3 flow cannot be increased since v_t cannot be labeled.

The $\omega_{s,t}$ Labeling Algorithm must terminate since we are increasing $f_{s,t}$ by one with each iteration and the integer $f_{s,t}$ is bounded above by the integer $\omega_{s,t}$. We next show that the $\omega_{s,t}$ Labeling Algorithm terminates if and only if an s-t vertex cut-set is saturated with all flow directed toward the set of unlabeled vertices. At the termination of the algorithm let U be the set of unlabeled vertices and let L be the set of labeled vertices. L is not empty since v_s is labeled. U is not empty since v_t is unlabeled. Let L_U be the set of labeled vertices connected to an unlabeled vertex by a branch directed from the labeled vertex. That is,

$$L_U = \{v_k : v_y \in V \quad \text{with} \quad (k,y) \in (L,U)\}. \tag{7.3.2}$$

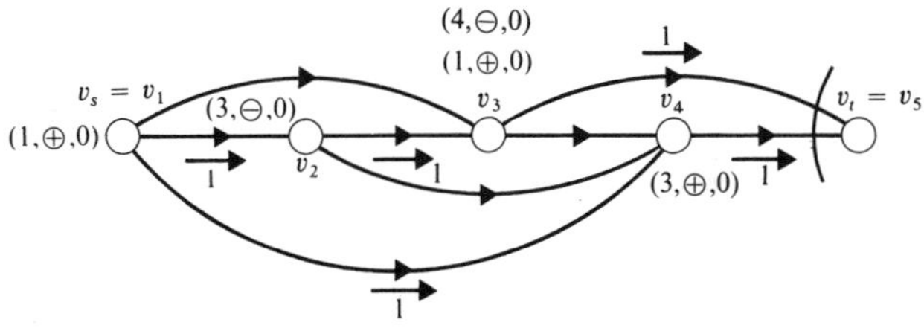

Fig. 7.3.2 Graph with $v_s \notin L_u$.

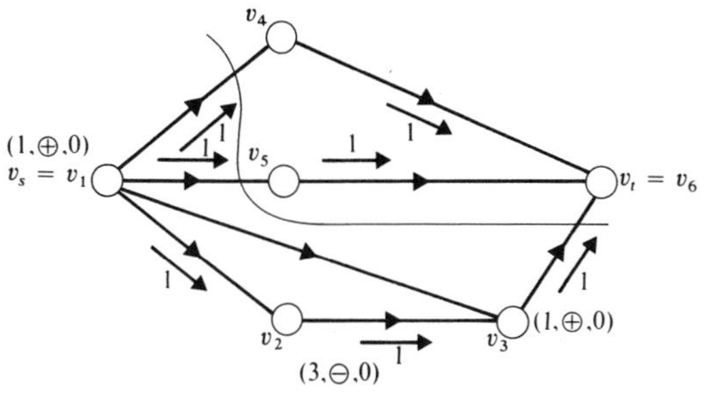

Fig. 7.3.3 Graph with $v_s \in L_u$.

If G is not connected, L_U may be empty.

We will show that a saturated s-t vertex cut-set X_M is

$$X_M = L_U \quad \text{if} \quad v_s \notin L_U \tag{7.3.3}$$

or

$$X_M = (L_U - \{v_s\}) \cup \{v_h:(s,h) \in (L,U)\} \quad \text{if} \quad v_s \in L_U. \tag{7.3.4}$$

Before giving the proof, we illustrate the meanings of (7.3.3) and (7.3.4) in the next example.

Example 7.3.3 In Figs. 7.3.2 and 7.3.3 the lines cutting the graphs separate the vertices into the sets L (to the left) and U (to the right). Thus in Fig. 7.3.2, $L = \{v_1,v_2,v_3,v_4\}$, $U = \{v_5\}$, and $L_U = \{v_3,v_4\}$. Hence for this graph $X_M = L_U = \{v_3,v_4\}$. In Fig. 7.3.3, $L = \{v_1,v_2,v_3\}$, $U = \{v_4,v_5,v_6\}$, and $L_U = \{v_1,v_3\}$. Since $v_s \in L_U$, $X_M = (L_U - \{v_s\}) \cup \{v_h:(s,h) \in (L,U)\} = \{v_3,v_4,v_5\}$.

***Lemma* 7.3.1** *If $v_g \in L_U$ and $g \neq s$, then v_g is not strongly labeled.*

Proof. Suppose v_g is strongly labeled. Since $v_g \in L_U$, there is a branch (g,h) with v_h unlabeled. Then v_h can be labeled unless $f(g,h) = 1$. Since $f(g,V) \leq 1$, (g,U) contains only branch (g,h) and $f(g,h) = +1$. Since v_g is strongly labeled, v_h must already be labeled and is in fact the vertex which was scanned to strongly label v_g. This contradiction proved the lemma.

***Lemma* 7.3.2**

$$\textit{If } v_g \in L_U \textit{ and } g \neq s \textit{ then } f(g,U) = +1 \textit{ and } f(U,g) = 0. \tag{7.3.5a}$$

$$\textit{If } (s,h) \in (L,U) \textit{ then } f(s,h) = +1 \textit{ and } f(h,s) = 0. \tag{7.3.5b}$$

Proof. a) From Lemma 7.3.1, v_g is weakly labeled and not strongly labeled. If $v_l \in U$ such that $f(l,g) = 1$, v_l can be labeled. Hence, $f(U,g) = 0$. Suppose $f(g,U) = 0$. If $f(g,V) + f(V,g) = 0$, any vertex v_h such that $(g,h) \in (g,U)$ can be labeled as shown in Fig. 7.3.4. Therefore assume $v_a \in L$ and $v_b \in L$ with $f(a,g) = 1$ and $f(g,b) = 1$ as shown in Fig. 7.3.4. Then v_g can be labeled $(b,-,M)$ which contradicts Lemma 7.3.1. Hence (7.3.5a) is proved.

b) If $(s,h) \in \Gamma$ and $f(s,h) = 0$, or $(h,s) \in \Gamma$ and $f(h,s) = 1$, v_h can be labeled. But if $(s,h) \in (L,U)$, this is a contradiction. Hence $f(s,h) = +1$ and $f(h,s) = 0$.

From Lemma 7.3.2, $f_{s,t} = |X_M|$. We now show X_M is an s-t vertex cut-set and hence we have proved our original statement.

***Lemma* 7.3.3** *For a connected graph, X_M is an s-t vertex cut-set.*

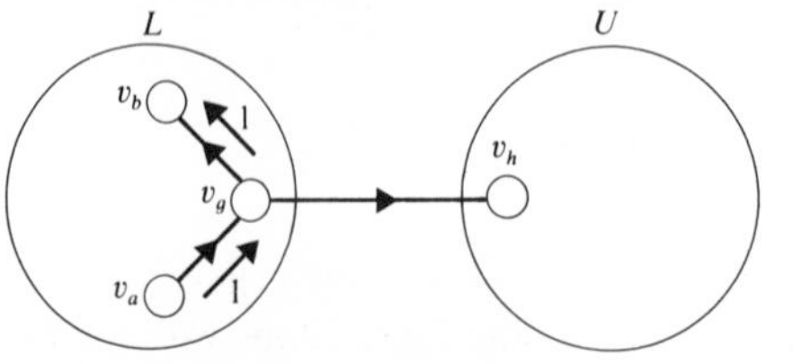

Fig. 7.3.4 Graph used in proof of Lemma 7.3.2.

Proof. Suppose $v_s \notin L_U$. Then $X_M = L_U$ and the removal of L_U from G breaks all directed paths from $(L - L_U)$ to U. But $v_s \in (L - L_U)$ and $v_t \in U$. Hence the lemma is proved for this case.

Suppose $v_s \in L_U$. Then the removal of $X_M = (L_U - \{v_s\})$ from G breaks all directed s-t paths except those which include any vertex v_h such that $(s,h) \in (L,U)$. The remaining paths are then broken by removing $\{v_h:(s,h) \in (L,U)\}$.

From Lemmas 7.3.2 and 7.3.3 we therefore have:

***Theorem* 7.3.2** *The $\omega_{s,t}$ Labeling Algorithm yields an s-t flow pattern of value* $f_{s,t} = \omega_{s,t}$.

7.4 MULTIPLE MINIMUM CUT-SET CALCULATIONS

We showed in Section 2 that v can be found by calculating $\omega_{s,t}$ for all s and t such that $(s,t) \notin \Gamma$. In Section 3 we gave an efficient method to calculate $\omega_{s,t}$. We now wish to reduce the number of pairs v_s,v_t for which $\omega_{s,t}$ must be found and to simplify the calculation of $\omega_{s,t}$ once $\omega_{a,b}$ has been found for $s \neq a$ and $t \neq b$.

We showed in Section 5.3, that for a branch-weighted pseudosymmetric graph, if $\tau_{a,b} = c(X,\bar{X})$ with $v_a \in X$ and $v_b \in \bar{X}$, and $v_s,v_t \in X$, then $\tau_{s,t}$ can be calculated in the graph formed by condensing $\bar{X}$ into a single vertex. Furthermore, only $n - 1$ flow calculations need be done to find $\tau_{s,t}$ for all s,t such that $s \neq t$, where n is the number of vertices in the graph. On the other hand, if the graph is not pseudosymmetric no such results are available. For example, consider the branch-weighted directed graph in Fig. 7.4.1(a). The s'-t cut of minimum value consists of branches (1,2) and (5,4). However, the minimum 2-3 cut cannot be calculated from the subgraph formed by condensing vertices v_1 and v_5 into one vertex. Indeed, the minimum 2-3 cut consists of branches (2,3) and (1,5). For the problem of finding ω, condensation of vertices is not permissible even if the graph is undirected. For the graph in Fig. 7.4.1(b), the minimum s'-t' vertex cut-set consists of v_1, v_2, v_3, and v_4. But the minimum s-t vertex cut-set consists of $v_{s'}$ and $v_{t'}$ and hence condensation cannot be used.

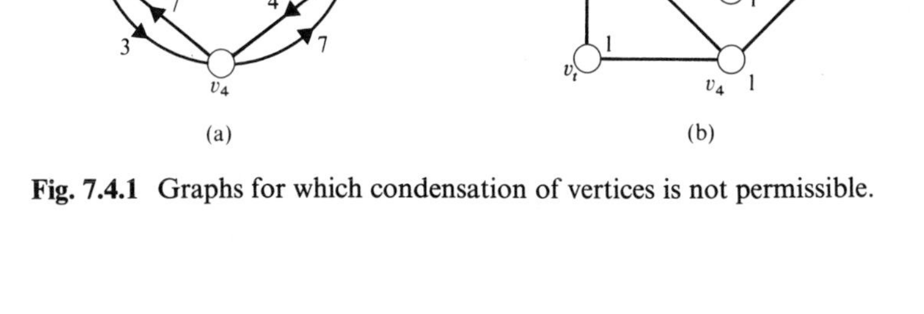

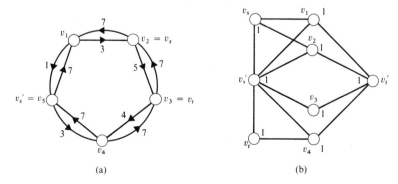

Fig. 7.4.1 Graphs for which condensation of vertices is not permissible.

In this section we prove several theorems which enable us to reduce the number of pairs of vertices for which $\omega_{s,t}$ must be calculated to determine ω. We then introduce a Flow Variation Algorithm to reduce the work in calculating $\omega_{s,t}$ once $\omega_{s',t}$ has been calculated.

***Theorem* 7.4.1** *Let G be an undirected graph and let* $v_a, v_b, v_{i_1}, v_{i_2}, \ldots, v_{i_k}$ *be a set of distinct vertices. If*

$$\sigma_{a,i_j} \geq k \quad \textit{for} \quad j = 1,2,\ldots,k \tag{7.4.1}$$

and

$$\sigma_{b,i_j} \geq k \quad \textit{for} \quad j = 1,2,\ldots,k \tag{7.4.2}$$

then

$$\sigma_{a,b} \geq k. \tag{7.4.3}$$

Proof. Suppose $\sigma_{a,b} < k$ and assume first that v_a and v_b are not adjacent. Then there exists an a-b vertex cut-set $A_{a,b}$ with $h < k$ vertices. Suppose the removal of $A_{a,b}$ leaves v_a in a component G_a and leaves v_b in another component G_b. If v_{i_j} is in G_a then $A_{a,b}$ is an i_j-b cut-set, which contradicts the fact that $\sigma_{b,i_j} \geq k$. Similarly, we cannot have v_{i_j} in G_b. Therefore $\{v_{i_1}, \ldots, v_{i_k}\} \subset A_{a,b}$, contradicting the fact that $A_{a,b}$ contains only h vertices. If v_a and v_b are adjacent, $[a,b]$ can be removed and the above argument can be repeated for the resulting graph.

To determine whether $\sigma \geq k$, and hence $\omega \geq k$, we can use Theorem 7.4.1 to reduce the number of vertex pairs which need to be checked. Instead of performing approximately

$$n(n-1)/2 = \sum_{k=1}^{n-1} (n-k) \tag{7.4.4}$$

calculations, we can find $\sigma_{i,j}$ for $j = i + 1, \ldots, n$ and $i = 1, \ldots, k$. This is a total of

$$(n-1) + (n-2) + \cdots + (n-k) = kn - k(k+1)/2$$

pairs of vertices to be examined.

***Theorem* 7.4.2** *Let G be an undirected graph. To verify the existence of r vertex disjoint paths between each pair of vertices, we need only perform the following operations.*

Step 1. Choose any vertex, say v_1, *and verify the existence of r vertex disjoint paths from* v_1 *to all other vertices. Let* $i = 1$.

Step 2. Remove v_i *to form* G_i. *Choose a vertex* v_{i+1} *and verify the existence of r-i vertex disjoint paths from* v_{i+1} *to all other vertices in* G_i.

Step 3. If $i = r$ *stop. Otherwise increase the value of i by one and return to Step 2.*

Proof. If there are not r vertex disjoint i-j paths, there exists a mixed i-j cut-set of $r - 1$ or fewer elements. If there are r such i-1 paths and r such j-1 paths, v_1 must be in the cut-set. Thus, if we cannot find a mixed cut-set of $r - 2$ or fewer elements which separates v_i and v_j in G_1, we cannot find an $r - 1$ or fewer element cut-set in G and there must exist at least r vertex disjoint i-j paths. Hence verification of the existence of r vertex disjoint paths between v_1 and every other vertex in G means that we need only find $r - 1$ vertex disjoint paths between vertices in G_1 to verify the existence of r such paths in G. Iteration of this argument proves the theorem.

Theorem 7.4.3 *Let G be an undirected graph with at least l vertices, $v_{j_1}, v_{j_2}, \ldots, v_{j_l}$, adjacent to v_j. Then, there are r vertex disjoint paths from v_i to v_j if the following conditions are satisfied:*

a) There are r vertex disjoint paths from v_i to each of $v_{j_1}, v_{j_2}, \ldots, v_{j_l}$.

b) There are $r - l$ vertex disjoint paths from v_i to v_j in the graph formed from G by removing $v_{j_1}, v_{j_2}, \ldots, v_{j_l}$.

Proof. Suppose we have verified the existence of r vertex disjoint paths between v_i and each of $v_{j_1}, \ldots, v_{j_l}$. Then each of $v_{j_1}, \ldots, v_{j_l}$ must be in any i-j cut-set of $r - 1$ or fewer elements. To verify the existence of r i-j paths, we must show that there is no $(r - l - 1)$ element mixed cut-set separating these vertices in the graph $\hat{G}$ formed by removing $v_{j_1}, \ldots, v_{j_l}$. This is done by finding $r - l$ paths in $\hat{G}$.

Theorem 7.4.3 is already stated in a form applicable to graphs for which a general requirement matrix $\mathscr{R} = [r_{i,j}]$ is specified. In Theorem 7.4.2, $r_{i,j} = r$ for all i and j. The more general form is as follows.

Theorem 7.4.4 *Let G be an undirected graph. If there are at least $r_{i,j}$ vertex disjoint i-1 paths and at least $r_{i,j}$ vertex disjoint j-1 paths, then to verify the existence of $r_{i,j}$ vertex disjoint i-j paths we need only show there are $r_{i,j} - 1$ such i-j paths in the graph formed by removing v_1.*

We now attempt to further reduce the work needed to evaluate the different maximum flows. This reduction is accomplished by the introduction of the Flow

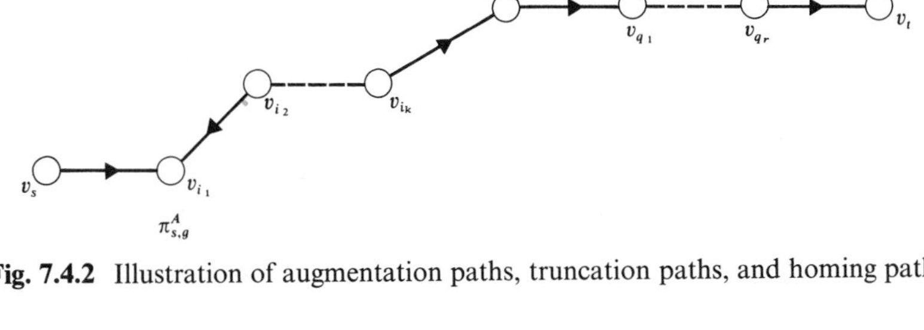

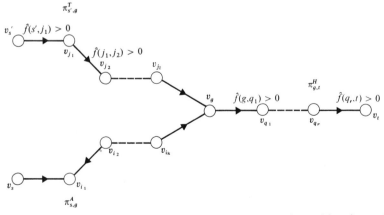

Fig. 7.4.2 Illustration of augmentation paths, truncation paths, and homing paths.

Variation Algorithm [FR2]. The basic ideas underlying the algorithm are given in the remainder of this section.

Rather than apply the Labeling Algorithm to independently maximize $f_{s',t}$ and $f_{s,t}$ we maximize $f_{s',t}$ and then use the resulting flow pattern $\mathscr{F}(s',t)$ to maximize $f_{s,t}$. For convenience, denote the s'-t flow pattern by $\hat{\mathscr{F}}$, with $\hat{f}(i,j) \in \hat{\mathscr{F}}$. Next apply the Labeling Algorithm to maximize $f_{s,t}$ in the usual manner until a vertex v_g is labeled for which $\hat{f}(g,V) > 0$. Then instead of continuing to search for an s-t augmentation path by the Labeling Algorithm, find a directed s'-t path $\pi_{s',t}$ containing vertex v_g such that $\hat{f}(i,j) > 0$ for all $(i,j) \in \pi_{s',t}$. This is accomplished by a "Homing Routine." The section of the path from $v_{s'}$ to v_g is called the "truncation path" and is denoted by $\pi^T_{s',g}$. The remaining section of the path is the "homing path" and is denoted by $\pi^H_{g,t}$. The path from v_s to v_g found by the Labeling Algorithm is called the "augmentation path" and is denoted by $\pi^A_{s,g}$. The various paths are illustrated in Fig. 7.4.2. Let $r(\pi^A_{s,g})$ be the residual capacity of $\pi^A_{s,g}$ and let $\hat{f}(\pi^T_{s',g})$ be the minimum of $\hat{f}(i,j)$ for all (i,j) in $\pi^T_{s',g}$. $\hat{f}(\pi^H_{s,t})$ is defined similarly, and h is defined by

$$h = \text{Min}\,[r(\pi^A_{s,g}), \hat{f}(\pi^T_{s',g}), \hat{f}(\pi^H_{s,t})]. \tag{7.4.5}$$

We first increase $f(\pi^A_{s,g})$ by h; this is called "augmentation." We then decrease $\hat{f}(\pi^T_{s',g})$ by h; this is called "truncation." We next increase $f(\pi^H_{g,t})$ by h; this is called "tracing." Finally, we decrease $\hat{f}(\pi^H_{g,t})$ by h; this is called "erasing." The "trace" and "erase" steps comprise the "homing steps." We then erase all labels and repeat this procedure until $f_{s,t}$ is maximized. In other words instead of continuing to search for the vertices in an s-t augmentation path we merely search for an s-g augmentation path and then "home in" on a g-t path.

Let

$$\hat{\mathscr{F}}_k = \{\hat{f}_k(i,j)\} \text{ and } \mathscr{F}_k = \{f_k(i,j)\}$$

be the flows after the $(k - 1)$st change in the value of $f_{s,t}$. Let the corresponding values of the terminal flows be $\hat{f}_{s',t}(k)$ and $f_{s,t}(k)$. We now illustrate the algorithm.

Example 7.4.1 Consider the graph in Fig. 7.4.3(a). A feasible flow pattern which maximizes $\hat{f}_{s',t}$ is indicated by the solid arrows parallel to the branches in Fig. 7.4.3(b). We now maximize $f_{s,t}$ using the Labeling Algorithm in conjunction with the Homing Routine.

a) The Labeling Algorithm labels vertex v_s. Since $\hat{f}_1(s,V) > 0$ we enter the Homing Routine and identify $\pi^T_{s',g}$ as $v_{s'},v_g$ and $\pi^H_{g,t} = v_g,v_t$ with $s = g$. Reduce $\hat{f}_1(3,1)$ by $h = 1$ and increase $f_1(3,1)$ by 1. Next decrease $\hat{f}_1(s',s)$ by 1. The augmentation step is not used since $g = s$. Thus the feasible flow patterns $\hat{\mathscr{F}}_2$ and $\mathscr{F}_2$ are shown by the solid and dashed arrows respectively in Fig. 7.4.3(c), and $f_{s,t}(2) = 1$.

b) All labels are erased and the Labeling Routine now labels vertex v_s. Since $\hat{f}_2(s,V) > 0$ we again enter the Homing Routine and identify $\pi^T_{s',g}$ as $v_{s'},v_g$ and $\pi^H_{g,t} = v_g,v_2,v_t$ with $g = s$. Decrease $\hat{f}_2(3,2)$ and $\hat{f}_2(2,t)$ by $h = 1$ and increase

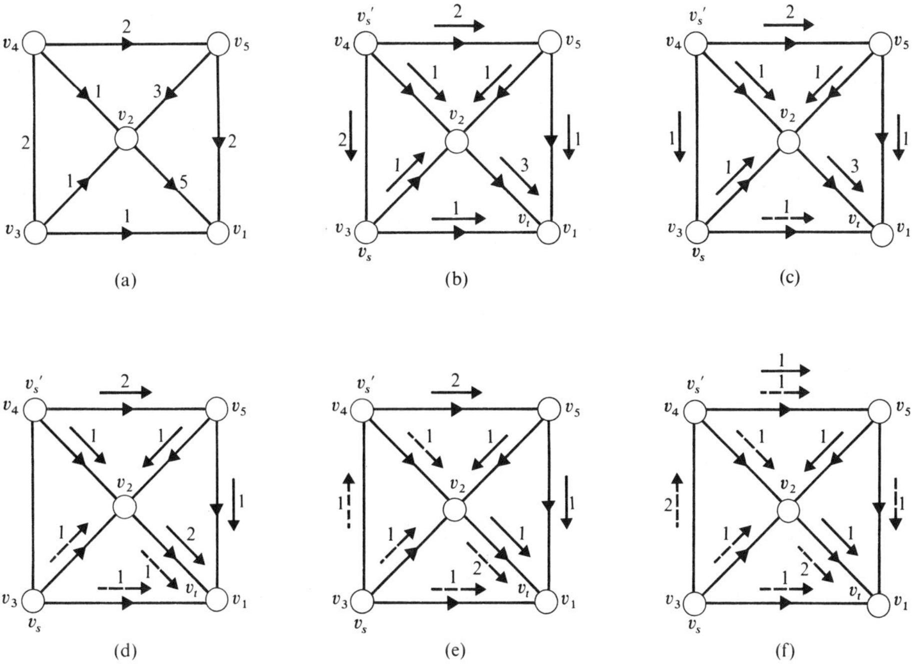

Fig. 7.4.3 Illustration of flow variation.

$f_2(3,2)$ and $f_2(2,t)$ by 1. Finally, decrease $\hat{f}_2(s',s)$ by 1. Augmentation is again not used. The new feasible flow pattern $\hat{\mathscr{F}}_3$ is shown by the dashed lines in Fig. 7.4.3(d) and $\mathscr{F}_3$ by the solid lines; $f_{s,t}(3)$ now is 2. We return to the Labeling Algorithm after erasing.

c) The Labeling Algorithm now labels v_s and then $v_{s'}$ for $\hat{f}(s',V) > 0$, to obtain $\pi^H_{g,t} = v_g,v_2,v_t$ with $g = s'$. There is no truncation path since $g = s'$. Increase $f_3(s',2)$ and $f_3(2,t)$ by $h = 1$ and decrease $\hat{f}_3(s',2)$ and $\hat{f}_3(2,t)$ by 1. The augmentation step increases $f_3(s,s')$ by 1. The new $\hat{\mathscr{F}}_4$ and $\mathscr{F}_4$ are shown respectively as the solid and dashed arrows in Fig. 7.4.3(e); $f_{s,t}(4)$ is now 3. We return to the Labeling Algorithm after erasing all labels.

d) Finally, label vertices v_s and $v_{s'}$ and identify $\pi^H_{g,t} = v_g,v_5,v_t$ with $g = s'$. Again there is no truncation path. Increase $f_4(s',5)$ and $f_4(5,t)$ by $h = 1$ and decrease $\hat{f}_4(s',5)$ and $\hat{f}_4(5,t)$ by 1. Then increase $f(s,s')$ along the augmentation path by 1. The cut $(\{v_3\}, \{v_1,v_2,v_4,v_5\})$ is saturated and $f_{s,t}$ has its maximum value.

A proof of convergence of the algorithm and a computer program for the Flow Variation Algorithm are given in reference FR2.

7.5 MAXIMIZATION OF THE SMALLEST BRANCH CUT-SET

In this section we begin the consideration of the synthesis of invulnerable networks. In particular, we seek graphs with a given number of branches and vertices for which the smallest branch cut-set is as large as possible.

We first derive an elementary upper bound on θ and ω. If $m < n - 1$, clearly $\theta = \omega = 0$. Therefore assume $m \geq n - 1$. From Theorem 7.2.1, $\omega \leq \theta$. Furthermore since every branch is incident with two vertices,

$$2m = \sum_{i=1}^{n} d(i). \tag{7.5.1}$$

Hence,

$$2m \geq n \operatorname*{Min}_{i} [d(i)] \tag{7.5.2}$$

or

$$\operatorname{Min} [d(i)] \leq [2m/n], \tag{7.5.3}$$

where $[x]$ indicates the largest integer not greater than x. Since

$$\theta \leq \operatorname*{Min}_{i} [d(i)], \tag{7.5.4}$$

it follows that

$$\omega \leq \theta \leq [2m/n], \tag{7.5.5}$$

Therefore a graph with $\theta = [2m/n]$ has a maximum value of θ. Furthermore if $\omega = [2m/n]$ then both ω and θ are maximum. We now develop an iterative technique for synthesizing a graph with $\theta = [2m/n]$, given m and n. We begin with any graph and systematically interchange two branches at a time until θ is maximized.

We first investigate the conditions under which a set of integers may correspond to the degrees of the vertices of an undirected graph. Recall that graphs do not contain parallel branches. Let $(k_{\alpha_1}, k_{\alpha_2}, \ldots, k_{\alpha_n})$ be an n-tuple of nonnegative integers. We would like to find the conditions under which there exists an n-vertex graph such that $d(i) = k_{\alpha_i}$ for $i = 1, \ldots, n$.

***Definition* 7.5.1** *A set of nonnegative integers $k_1 \geq k_2 \geq \cdots \geq k_n$, with r of the integers positive, is said to be compressible if $k_1 \leq r - 1$. The set of integers is compressed by performing the operations in (7.5.6a to c) below.*

$$\textit{Subtract one from } k_2, k_3, \ldots, k_{k_1+1}. \tag{7.5.6a}$$

$$\textit{Let } k_1 = 0. \tag{7.5.6b}$$

$$\textit{Relabel the resulting integers as } k_1 \geq k_2 \geq \cdots \geq k_n. \tag{7.5.6c}$$

The resulting set of integers is said to be the compressed set or the set obtained by compression.

We now give necessary and sufficient conditions for the existence of a graph with degrees $k_1, \ldots, k_n$.

***Theorem* 7.5.1** *A set of nonnegative integers* $k_1 \geq k_2 \geq \cdots \geq k_n$ *is realizable as the degrees of an n-vertex undirected graph G, if and only if the given set of integers* $\{k_1, \ldots, k_n\}$ *is compressible and every set of integers resulting from successive compressions is again compressible, until all integers are zero.*

Proof. (*Necessity*) Suppose we have a graph G such that $d(i) = k_i$ for $i = 1, \ldots, n$. The k_1 branches incident with v_1 must each be incident with a set of vertices among $v_2, \ldots, v_r$ where r is the number of positive k_i. Since there are no parallel branches,

$$k_1 \leq r - 1. \tag{7.5.7}$$

Hence the set $\{k_1, \ldots, k_n\}$ is compressible.

If v_1 is adjacent to $v_2, \ldots, v_{k_1+1}$, the resulting compressed integers represent the degrees of vertices in the graph obtained from G by removing v_1 and all branches incident with it. If v_1 is not adjacent to $v_2, \ldots, v_{k_1+1}$, we now show that G can be transformed into a graph $\hat{G}$ with the same set of degrees as G and in $\hat{G}$, v_1 is adjacent to $v_2, \ldots, v_{k_1+1}$. The resulting set of integers would again be compressible. The argument can then be repeated for every successive set of compressed integers.

Suppose there is a vertex v_j in G with $d(j) > d(i)$ such that v_i is adjacent to v_1 and v_j is not. Let V_j be the set of all such vertices. Since $d(j) > d(i)$, there exists a vertex v_k in G adjacent to v_j but not to v_i. Furthermore, $i \neq j$ since v_j and v_1 are not adjacent. Then remove $[i,1]$ and $[k,j]$ and add $[i,k]$ and $[j,1]$. In the resulting graph the degrees of the vertices are the same as in G. Moreover the size of V_j has decreased. By repeating this operation we eventually obtain $\hat{G}$.

(*Sufficiency*) Let $k_1 \geq k_2 \geq \cdots \geq k_n$ be a set of compressible integers such that every successive set of integers resulting from compression is also compressible. Let $(k_{\alpha_1}, k_{\alpha_2}, \ldots, k_{\alpha_n})$ be any permutation of the n-tuple $(k_1, k_2, \ldots, k_n)$ such that β is the map which takes $(1,2,\ldots,n)$ to $(\alpha_1, \alpha_2, \ldots, \alpha_n)$. We give a method for constructing a graph with vertices $v_1, \ldots, v_n$ such that $d(i) = k_{\alpha_i}$ for $i = 1, \ldots, n$. Let G_1 consist of the vertices $v_1, \ldots, v_n$ and branches $[\beta(1),\beta(2)],[\beta(1),\beta(3)],\ldots,[\beta(1),\beta(k_1 + 1)]$; G_1 exists since $k_1 \leq r - 1$. To construct graphs $G_2, G_3, \ldots,$ repeat this operation on each successive set of compressed integers. Each step is possible since each set of integers is compressible. Eventually all integers will be zero since otherwise there would be an incompressible set of integers. By taking the union of all the resulting graphs to obtain the graph G, the proof is completed.

From the proof we immediately have the algorithm below to realize a set of integers as the degrees of an undirected graph.

Degree Algorithm

Step 1. Let $V = \{v_1, \ldots, v_n\}$.

Step 2. Compress the integers in the given n-tuple $(k_{\alpha_1}, \ldots, k_{\alpha_n})$ with $k_1 \geq k_2 \geq \cdots \geq k_n$ where $(\alpha_1, \ldots, \alpha_n)$ is a permutation β of $(1, \ldots, n)$.

Step 3. Add branches $[\beta(1),\beta(2)],[\beta(1),\beta(3)],\ldots,[\beta(1),\beta(k_1 + 1)]$. If all integers in the compressed set are zero, terminate. Otherwise return to Step 2.

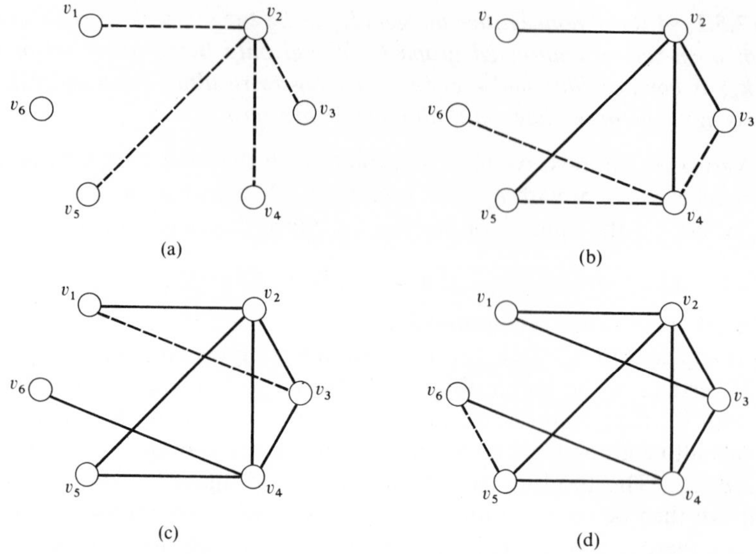

Fig. 7.5.1 Graphs used for Example 7.5.1.

Example 7.5.1 We are given the 6-tuple of integers $(k_5,k_1,k_4,k_2,k_3,k_6) = (2,4,3,4,3,2)$. Applying compression to the 6-tuple and each successive set of compressed integers we have:

$$(k_5,k_1,k_4,k_2,k_3,k_6) = (2,4,3,4,3,2), \tag{7.5.8a}$$

$$(k_5,k_6,k_2,k_1,k_3,k_4) = (1,0,2,3,2,2), \tag{7.5.8b}$$

$$(k_1,k_6,k_2,k_5,k_3,k_4) = (1,0,1,0,1,1), \tag{7.5.8c}$$

$$(k_3,k_5,k_4,k_6,k_1,k_2) = (0,0,0,0,1,1), \tag{7.5.8d}$$

$$(k_1,k_2,k_3,k_4,k_5,k_6) = (0,0,0,0,0,0). \tag{7.5.8e}$$

The realization is obtained from (7.5.8), as shown in Fig. 7.5.1. The branches resulting from the compressions applied to (7.5.8a, b, c, and d) are shown dashed in Fig. 7.5.1(a), (b), (c), and (d), respectively.

We now relate these results to the vulnerability problem.

***Theorem* 7.5.2** *A set of integers $\{k_1,\ldots,k_n\}$ with $n > 1$ is realizable as the set of degrees of vertices of an n-vertex undirected graph G and with $\theta = k$ if and only if*

$$\{k_1,\ldots,k_n\} \text{ is successively compressible,} \tag{7.5.9}$$

and

$$k_i \geq k \quad \text{for} \quad i = 1,\ldots,n, \tag{7.5.10}$$

and, if $k = 1$, then

$$\sum_{i=1}^{n} k_i \geq 2(n-1). \tag{7.5.11}$$

Proof. (*Necessity*) Equation (7.5.9) is required since the vertices in the graph must have degrees $k_1, \ldots, k_n$. Since $|[\{v_i\}, V]| = k_i$ (7.5.10) is necessary for $\theta = k$. If $k = 1$, G must be connected and hence must contain a tree as a subgraph. Therefore $m \geq n - 1$. But

$$\sum_{i=1}^{n} d(i) = 2m. \tag{7.5.12}$$

Thus

$$\sum_{i=1}^{n} d(i) = \sum_{i=1}^{n} k_i = 2m \geq 2(n-1). \tag{7.5.13}$$

(*Sufficiency*) If (7.5.9), (7.5.10), and (7.5.11) are satisfied, we present an algorithm to construct a graph with the given degrees for which $\theta = k$. From (7.5.9), there is a graph G with $d(i) = k_i$ for $i = 1, \ldots, n$. For G, assume $\theta = h < k$. We now show that if (7.5.10) and (7.5.11) are satisfied, we can repeatedly transform G so that the vertex degrees remain the same, but the number of cuts with fewer than k branches is decreased. Since there can be only a finite number of cuts with fewer than k branches, eventually the number of branches in the smallest cut will be at least k.

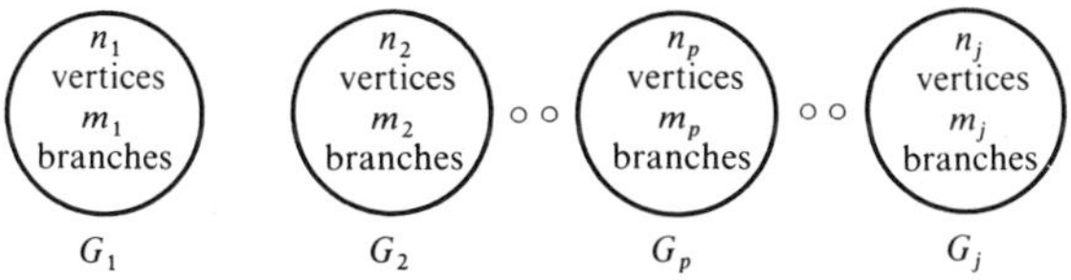

Fig. 7.5.2 Disjoint connected subgraphs used in the proof of Theorem 7.5.2.

CASE 1. Let $k = 1$. Since $h < k$, we have $h = 0$, and G is not connected. Assume that G consists of j components $G_1, \ldots, G_j$ where G_i has n_i vertices and m_i branches (Fig. 7.5.2).

By hypothesis (7.5.11)

$$2m = \sum_{i=1}^{n} k_i \geq 2(n-1) \tag{7.5.14}$$

and

$$m \geq (n-1). \tag{7.5.15}$$

Since $m = \sum_{i=1}^{j} m_i$ and $n = \sum_{i=1}^{j} n_i$,

$$\sum_{i=1}^{j} m_i \geq \left(\sum_{l=1}^{j} n_l\right) - 1 > \sum_{l=1}^{j} (n_l - a_l) \qquad \text{where } a_l = 1 \text{ for } l = 1, \ldots, j \tag{7.5.16}$$

or

$$\sum_{i=1}^{j} [m_i - (n_i - a_i)] > 0. \tag{7.5.17}$$

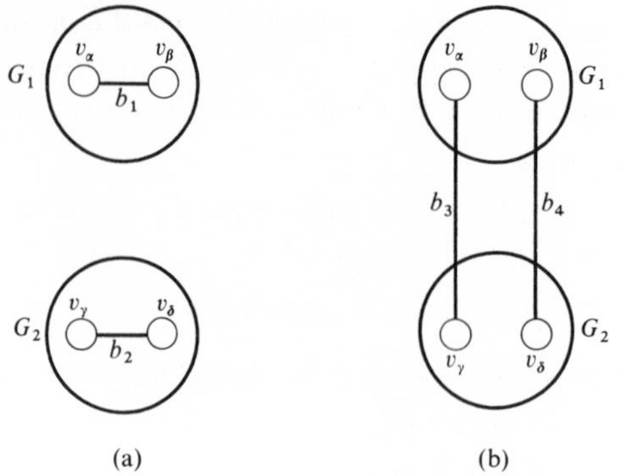

Fig. 7.5.3 Transformation of a graph for Case I in the proof of Theorem 7.5.2: (a) original graph; (b) transformed graph.

Hence, there is at least one subgraph G_1 for which $m_1 > n_1 - 1$. Since by hypothesis G_1 is connected, it contains a tree with $n_1 - 1$ branches and at least one branch $b_1 = [\alpha,\beta]$ not in the tree. Also, any other subgraph G_2 is connected and therefore contains a tree. Pick any branch in this tree, $b_2 = [\gamma,\delta]$, and apply the following transformation: Remove b_1 and b_2 from G and add $b_3 = [\alpha,\gamma]$ and $b_4 = [\beta,\delta]$ as shown in Fig. 7.5.3.

The transformation does not change the degree of any vertex. To complete the proof for Case 1, we show that the transformation reduces the number of cuts with value zero; i.e., the number of components of G is reduced. When b_1 is removed from G_1 the number of components of G is not increased since G_1 still has a tree. When b_2 is removed from G_2, we may increase the number of components in G since G_2 may be broken into two disjoint connected subgraphs, G_{2a} and G_{2b}. However, when b_3 and b_4 are added, G_{2a}, G_{2b}, and G_1 become one connected subgraph of G and hence the overall effect is to decrease the number of components in G by one.

CASE 2. Assume $k > 1$. From (7.5.9), we can construct a graph with $d(i) = k_i$ for $i = 1,\ldots,n$. From (7.5.11),

$$\sum_{i=1}^{n} k_i \geq n - 1.$$

If G is not connected we can use repeatedly the transformation used in the proof for Case 1 to convert G into a connected graph. Therefore, assume G is connected and hence $\theta \geq 1$. We now give a transformation which leaves the degrees of G invariant but decreases the number of cuts with less than k branches. By assumption, there is a cut $[S,\bar{S}]$ of minimum value $h < k$. Find the subset A of S with fewest vertices

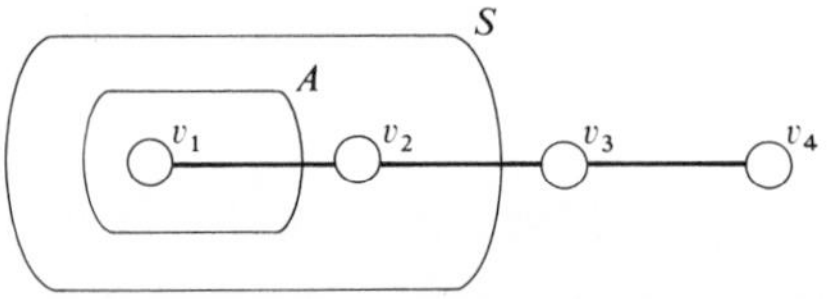

Fig. 7.5.4 Illustration of a nonminimal cut $[S,\bar{S}]$.

such that $|[A,\bar{A}]| = h$. $[A,\bar{A}]$ is a minimal cut such that $|[A,\bar{A}]| = h < k$. For example in Fig. 7.5.4, $|[S,\bar{S}]| = 1$, but $[S,\bar{S}]$ is not minimal since $|[A,\bar{A}]| = 1$ and $A \subset S$.

Select an arbitrary branch $b_0 = [a,\bar{a}]$ in $[A,\bar{A}]$. Since $|[\{v_a\},\bar{A}]| \leq |[A,\bar{A}]|$ we have $1 \leq |[\{v_a\},\bar{A}]| \leq h$. By Hypothesis (7.5.10), $d(a) \geq k$. Hence there are at least k-h branches connecting v_a to vertices in A. Of these vertices at least one, say $v_{a'}$, is not adjacent to a vertex in $\bar{A}$. Otherwise there would be k branches in $[A,\bar{A}]$. Similarly, there is a branch $[\bar{a},\bar{a}']$ in $\bar{A}$ such that $v_{a'}$ is not adjacent to any vertex in A. Designate $[a,a']$ as b_a and $[\bar{a},\bar{a}']$ as $b_{\bar{a}}$. The transformation consists of removing b_a and $b_{\bar{a}}$ from G and adding $b_c = [a',\bar{a}]$ and $b_d = [a,\bar{a}']$ as shown in Fig. 7.5.5.

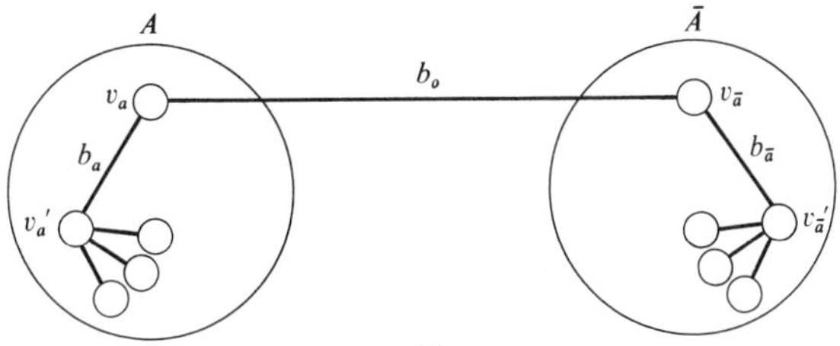

(a)

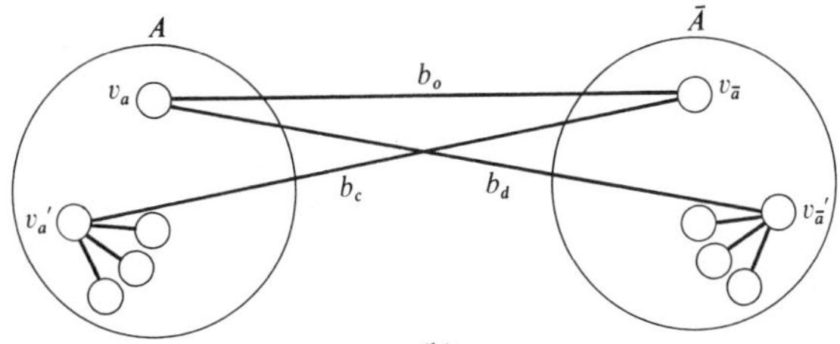

(b)

Fig. 7.5.5 Transformation of a graph used in the proof of Case 2 of Theorem 7.5.2: (a) original graph; (b) transformed graph.

The degrees of the vertices are unaffected by the transformation. To complete the proof we show that the number of cuts with capacity less than k is decreased. Let $\hat{G}$ be the graph resulting from the transformation. Clearly in $\hat{G}$, $|[A,\bar{A}]| = h + 2 > h$.

We now show that no cuts are reduced in value to h or less. Suppose, on the contrary, that for some B the value of $[B,\bar{B}]$ is reduced from a value greater than h in G to h or less in $\hat{G}$. Since only two branches are removed from G the only way in which $|[B,\bar{B}]|$ can decrease is if it does not contain both b_c and b_d in $\hat{G}$. Hence, either

both v_a and $v_{\bar{a}'}$ are in B (or $\bar{B}$)

or

both $v_{\bar{a}}$ and $v_{a'}$ are in $\bar{B}$ (or B)

or both statements are true.

Without loss of generality, assume v_a and $v_{\bar{a}'}$ are in B and $v_{\bar{a}}$ and $v_{a'}$ are in $\bar{B}$. $\hat{G}$ can then be drawn as shown in Fig. 7.5.6, where the only branches indicated are b_o, b_c, and b_d. Now consider the number of elements in several sets of branches and vertices chosen from G.

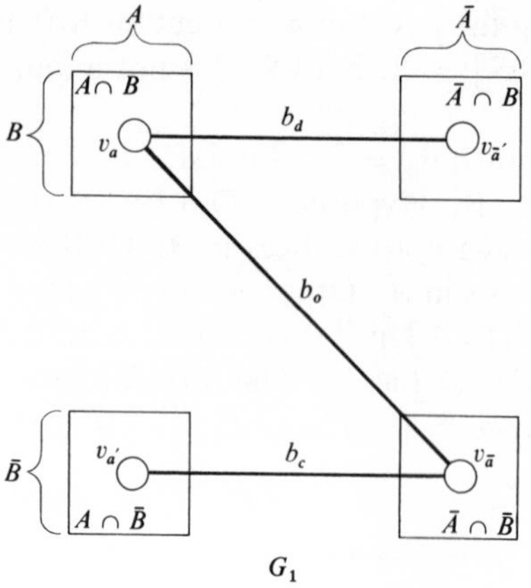

Fig. 7.5.6 Representation of cuts after transformation.

Since $b_0 \in [A \cap B, \bar{A} \cap \bar{B}]$, we have

$$|[A \cap B, \bar{A} \cap \bar{B}]| \geq 1. \tag{7.5.18}$$

But, by assumption,

$$|[B,\bar{B}]| = |[A \cap B, A \cap \bar{B}]| + |[\bar{A} \cap B, \bar{A} \cap \bar{B}]| + |[A \cap B, \bar{A} \cap \bar{B}]| + |[\bar{A} \cap B, A \cap \bar{B}]| \leq h. \tag{7.5.19}$$

Hence from (7.5.18) and (7.5.19),

$$|[A \cap B, A \cap \bar{B}]| + |[\bar{A} \cap B, \bar{A} \cap \bar{B}]| \leq h - 1. \tag{7.5.20}$$

Therefore, either

$$|[A \cap B, A \cap \bar{B}]| \leq \left[\frac{h-1}{2}\right] \tag{7.5.21}$$

or

$$|[\bar{A} \cap B, \bar{A} \cap \bar{B}]| \leq \left[\frac{h-1}{2}\right] \tag{7.5.22}$$

where $[(h - 1)/2]$ denotes the largest integer less than or equal to $(h - 1)/2$. Without loss of generality, suppose (7.5.21) holds. We show this leads to a contradiction. By definition of $[A,\bar{A}]$ and $[B,\bar{B}]$ we have

$$|[A,\bar{A}]| = |[A \cap B, \overline{A \cap B}]| + |[A \cap \bar{B}, \overline{A \cap \bar{B}}]| - 2|[A \cap B, A \cap \bar{B}]| \tag{7.5.23}$$

or

$$|[A \cap B, \overline{A \cap B}]| + |[A \cap \bar{B}, \overline{A \cap \bar{B}}]| = |[A,\bar{A}]| + 2|[A \cap B, A \cap \bar{B}]|. \tag{7.5.24}$$

Substituting (7.5.21) into (7.5.24) and using the fact that $|[A,\bar{A}]| = h$, we have

$$|[A \cap B, \overline{A \cap B}]| + |[A \cap B, \overline{A \cap \bar{B}}]| < h + 2\left(\left[\frac{h-1}{2}\right] + 1\right). \qquad (7.5.25)$$

From (7.5.25), either

$$|[A \cap B, \overline{A \cap B}]| \le \left[\frac{h}{2}\right] + \left[\frac{h-1}{2}\right] + 1 = h \qquad (7.5.26)$$

or

$$|[A \cap B, \overline{A \cap \bar{B}}]| \le \left[\frac{h}{2}\right] + \left[\frac{h-1}{2}\right] + 1 = h. \qquad (7.5.27)$$

This contradicts the fact that $[A,\bar{A}]$ is a minimal cut such that $|[A,\bar{A}]| = h$. Hence the proof is completed.

Note that (7.5.10) and (7.5.11) are independent conditions: (7.5.11) does not imply (7.5.10) since (7.5.11) applies only in the special case in which $k = 1$. Furthermore, (7.5.10) does not imply (7.5.11). For example, the requirements $k_1 = 1$, $k_2 = 1$, $k_3 = 1$, $k_4 = 1$, and $k = 1$ satisfy (7.5.10) but not (7.5.11) since

$$\sum_{i=1}^{n} k_i = 4$$

and $2(n - 1) = 6$.

We now give an algorithm to construct a graph with m branches, n vertices, and $\theta = k$.

Branch Exchange Algorithm

Step 1. Starting with any graph G with m branches and n vertices and all degrees greater than or equal to k, repeat the operation given below until G is connected and then proceed to Step 2.

Find a component of G, say G_1, with n_1 vertices. Select a tree of G_1 and find a branch b_1 of G_1 not in the tree. Select any other branch b_2 not in G_1. If $b_1 = [\alpha,\beta]$ and $b_2 = [\gamma,\delta]$, add branches $b_3 = [\alpha,\gamma]$ and $b_4 = [\beta,\delta]$ to G and remove b_1 and b_2.

Step 2. Repeat the operation given below until $\theta = k$.

Find a minimal cut $[A,\bar{A}]$ such that $c[A,\bar{A}] = h < k$ and select an arbitrary branch $b_0 \in [A,\bar{A}]$. If $b_0 = [a,\bar{a}]$, find $b_a = [a,a']$ and $b_{\bar{a}} = [\bar{a},\bar{a}']$ where $v_{a'} \in A, v_{\bar{a}'} \in \bar{A}, v_{a'}$ is not adjacent to any vertex in $\bar{A}$, and $v_{\bar{a}'}$ is not adjacent to any vertex in A. Add branches $b_c = [a',\bar{a}]$ and $b_d = [a,\bar{a}']$ to G and remove b_a and $b_{a'}$.

The operation of removing b_1 and b_2 or b_a and $b_{\bar{a}}$ and adding b_3 and b_4 or b_c and b_d is called a *branch exchange*.

Clearly, if we start with a structure with no parallel branches, the final structure will not have parallel branches.

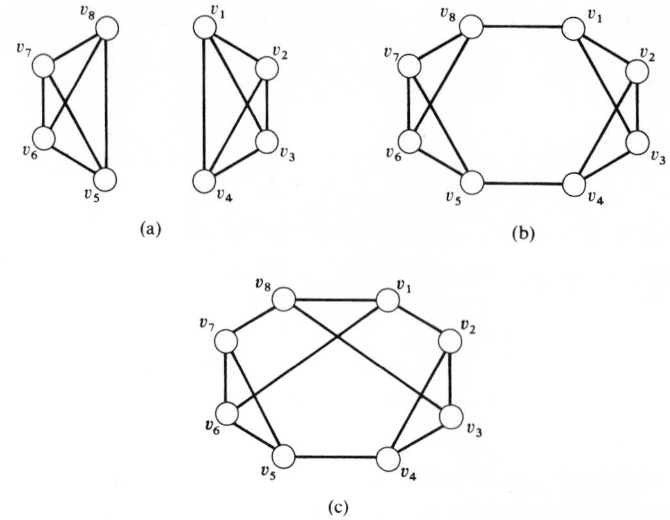

Fig. 7.5.7 Graphs illustrating use of Branch Exchange Algorithm in Example 7.5.2.

Example 7.5.2 We illustrate the Branch Exchange Algorithm by increasing θ to $\theta = 3$ for the graph shown in Fig. 7.5.7(a). Since the graph is initially disconnected, apply Step 1. Take G_1 as the component containing vertices v_5, v_6, v_7, and v_8, with $b_1 = [8,5]$ and $b_2 = [1,4]$. The resulting graph is shown in Fig. 7.5.7(b), where $b_3 = [8,1]$ and $b_4 = [5,4]$. In this graph, $\theta = 2$ since the removal of $[8,1]$ and $[5,4]$ results in a disconnected graph. Therefore, use Step 2. Choose $[A,\bar{A}] = [\{v_5,v_6,v_7,v_8\},\{v_1,v_2,v_3,v_4\}]$, $b_0 = [8,1]$, $b_a = [8,6]$, and $b_{\bar{a}} = [1,3]$. The resulting graph is shown in Fig. 7.5.7(c) with $b_c = [6,1]$ and $b_d = [8,3]$. For the resulting graph $\theta = 3$, and hence the algorithm terminates.

We now show in Theorem 7.5.3 that if the degree of each vertex is at least $[n/2]$, no matter how we arrange the branches $\theta = \operatorname{Min}_i [d(i)]$. However, we cannot guarantee $\theta = [2m/n]$ unless $\operatorname{Min}_i [d(i)] = [2m/n]$. We first introduce Lemma 7.5.1 as an intermediate result.

***Lemma* 7.5.1** *Let G be an undirected graph with n vertices. If $d(i) \geq [n/2]$ for all i, then G is connected.*

Proof. Suppose G is not connected. Each component contains at least $[n/2] + 1$ vertices, namely, an arbitrary vertex v_i and the $[n/2]$ vertices adjacent to v_i. Since there are at least two components, there are at least $2([n/2] + 1)$ vertices in G. However, $2([n/2] + 1) \geq n + 1$. This contradiction proves the lemma.

***Theorem* 7.5.3** *Let G be an undirected graph. If $d(i) \geq [n/2]$ for all i, then $\theta = \operatorname{Min}_i [d(i)]$.*

Proof. Obviously $\theta \leq \operatorname{Min}_i [d(i)]$. We complete the proof by showing that $\theta \geq$

$\operatorname*{Min}_i [d(i)]$. Suppose that $\theta < \operatorname*{Min}_i [d(i)]$. There must be a cut $[X,\bar{X}]$ such that

$$\theta = |[X,\bar{X}]| < \operatorname*{Min}_i [d(i)]. \tag{7.5.28}$$

Suppose the branches in $[X,\bar{X}]$ are incident with q vertices in X and p vertices in $\bar{X}$, where $q \leq \theta$ and $p \leq \theta$. From Lemma 7.5.1, $q > 0$.

Suppose $|X| = q$, that is, all vertices in X are incident with branches in $[X,\bar{X}]$. Then the number of branches incident with two vertices in X is at least

$$\tfrac{1}{2}(q \operatorname*{Min}_i [d(i)] - \theta). \tag{7.5.29}$$

But, by assumption, $\theta < \operatorname*{Min}_i [d(i)]$. Hence

$$\tfrac{1}{2}(q \operatorname*{Min}_i [d(i)] - \theta) > \tfrac{1}{2}(q \operatorname*{Min}_i [d(i)]) - \operatorname*{Min}_i [d(i)]$$

$$= \tfrac{1}{2} \operatorname*{Min}_i [d(i)](q - 1) > \tfrac{1}{2}q(q - 1) = \binom{q}{2}. \tag{7.5.30}$$

Equation (7.5.29) therefore indicates that there are more than $(q/2)$ branches joining q vertices. This is a contradiction and the theorem is proved for this case. We can give a similar proof for the case in which $|\bar{X}| = q$.

The only remaining possibility is that $|X| > q$ and $|\bar{X}| > q$. This means that there are vertices in both X and $\bar{X}$ adjacent only to vertices in X and $\bar{X}$, respectively. Then X and $\bar{X}$ each contain at least $\operatorname*{Min}_i [d(i)] + 1$ vertices, and G contains $2 \operatorname*{Min}_i [d(i)] + 2$ vertices. But $2 \operatorname*{Min}_i [d(i)] + 2 > 2[n/2] + 2 \geq n + 1$. We again have a contradiction and the theorem is proved.

***Corollary* 7.5.1** *If all vertices in G have the same degree $d(i) \geq [n/2]$, then $\theta = d(i)$.*

Finally, note that Theorem 7.5.3 is the strongest possible statement since, if $\operatorname*{Min}_i [d(i)] < [n/2]$, it is possible to construct a graph with $\theta < \operatorname*{Min}_i [d(i)]$. Thus, for the graph in Fig. 7.5.8, the degree of each vertex is $3 < [n/2] = 4$ and $\theta = 2$.

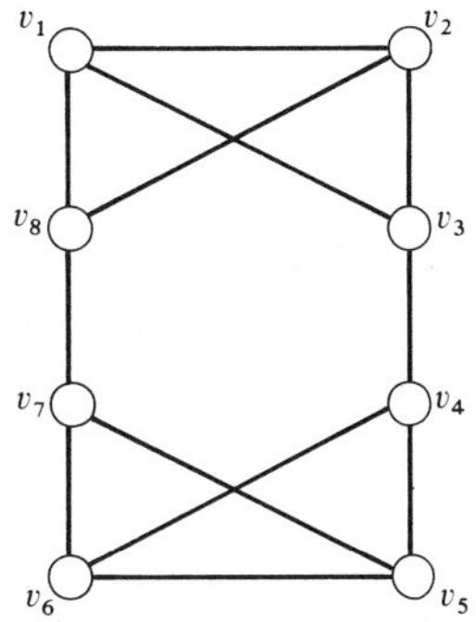

Fig. 7.5.8 Graph in which $\theta < \operatorname{Min} [d(i)] < [n/2]$.

7.6 GRAPHS WITH SPECIFIED BRANCH REDUNDANCIES

In this section we reconsider the problem of constructing graphs which will survive after branch damage. Specifically, suppose an $n' \times n'$ symmetric matrix $\mathscr{R} = [r_{i,j}]$ of nonnegative integers is specified such that $r_{i,i} = 0$ for $i = 1,2,\ldots,n'$. The problem to be considered is: Construct an undirected graph G with a minimum number of branches such that there are at least $r_{i,j}$ branch-disjoint i-j paths for all i,j with $i \neq j$. Assume the rows and columns of $\mathscr{R}$ have been ordered so that

$$\operatorname*{Max}_{1\le i\le n'} [r_{i,1}] \ge \operatorname*{Max}_{1\le i\le n'} [r_{i,2}] \ge \cdots \ge \operatorname*{Max}_{1\le i\le n'} [r_{i,n'}]. \tag{7.6.1}$$

Then $\mathscr{R}$ will be called the *redundancy matrix*. An n-vertex graph G is said to *satisfy* $\mathscr{R}$ if G contains a subset of n' vertices $v_1,\ldots,v_{n'}$ such that there are at least $r_{i,j}$ branch-disjoint paths between v_i and v_j for $i,j = 1,\ldots,n'$.

If G is an n-vertex graph satisfying $\mathscr{R}$ and if $r_{s,t} = \operatorname*{Max}_{i,j} [r_{i,j}]$, then $d(s) \ge r_{s,t}$. Hence v_s (and v_t) must be adjacent to at least $r_{s,t}$ other vertices. This implies there are at least $r_{s,t} + 1$ vertices in G. Therefore

$$n \ge \operatorname{Max}\,[n', \operatorname*{Max}_{i,j} [r_{i,j}] + 1]. \tag{7.6.2}$$

In this section, we give two algorithms to solve special cases of the problem of constructing a graph with a minimum number of branches to satisfy $\mathscr{R}$. An interesting aspect of these algorithms is that they utilize only the maximum entries in the columns of $\mathscr{R}$.

Let $d_i = \operatorname*{Max}_{1\le j\le n'} [r_{i,j}]$. If $n' \le d_1$, no n'-vertex graph satisfies $\mathscr{R}$. To construct a graph that satisfies $\mathscr{R}$, at least $d_1 + 1 - n'$ "redundant" vertices (i.e., vertices with no specified requirements) must be used. The graphs obtained from the application of Algorithm 7.6.1 contain exactly $d_1 + 1 - n'$ redundant vertices. This algorithm is very similar to the Degree Realization Algorithm discussed in Section 7.5.

The following algorithm is applicable when $d_1 \geqq n' - 1$.

Algorithm 7.6.1

Step 1. If $n' = d_1 + 1$, let $n = n' = d_1 + 1$. If $n' < d_1 + 1$, add redundant vertices $v_{n'+1},\ldots,v_n$ with $d_{n'+1} = d_{n'+2} = \cdots = d_n = 0$. Let

$$G_1 = (V,\Gamma_1) \quad \text{with} \quad V = \{v_1,v_2,\ldots,v_n\}, \tag{7.6.3a}$$

$$\Gamma_1 = \{[1,2]\} \cup \{[2,j],[1,j] \quad \text{for} \quad j = 2,3,\ldots,n\}. \tag{7.6.3b}$$

Let $k = 1$ and $v_1 \in V$.

Step 2. Let $d_k(i)$ be the degree of v_i in G_k and define

$$d_i^{k+1} = d_i - d_k(i). \tag{7.6.4}$$

Assume the vertices in V_k are labeled as $v_{g_k(1)},v_{g_k(2)},\ldots,v_{g_k(d_1-k)}$ such that

$$d^k_{g_k(1)} \ge d^k_{g_k(2)} \ge \cdots \ge d^k_{g_k(d_1-k)}. \tag{7.6.5}$$

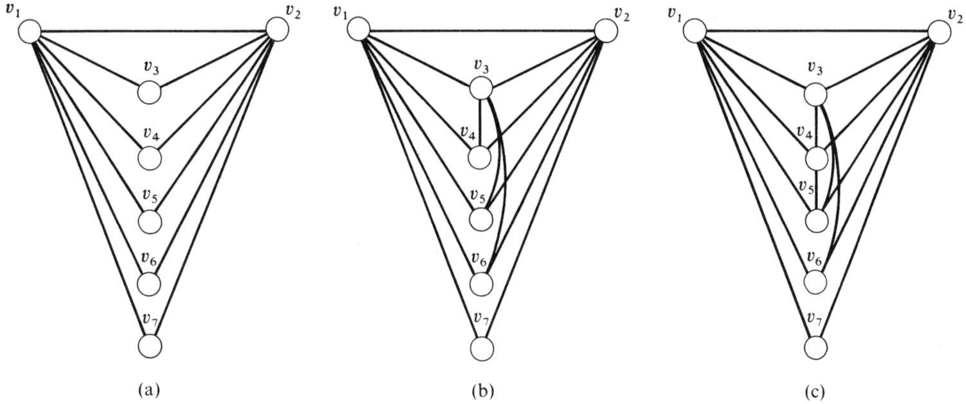

Fig. 7.6.1 Graphs used in Example 7.6.1.

If $d^k_{g_k(1)} > 0$, go to Step 3 where we compress the set of integers $\{d^k_{g_k(i)}\}$. Otherwise let $K = k$, $G = (V,\Gamma_K)$ and stop.

Step 3. Let $G_{k+1} = (V,\Gamma_{k+1})$ where

$$\Gamma_{k+1} = \Gamma_k \cup \{[g_k(1),g_k(i)] \quad \text{for} \quad i = 2,3,\ldots,d_{g_k(1)} + 1\}. \tag{7.6.6}$$

Let $V_{k+1} = V_k - \{v_{g_k(1)}\}$, increase the value of k by 1, and return to Step 2.

Example 7.6.1 Let $\{d_i\} = \{6,6,5,4,3,2\}$. Since $d_1 = 6$ and $n' = 6$, $d_1 + 1 > n'$. Figure 7.6.1(a) shows G_1 constructed by using Step 1. Then

$$d_1^2 = d_1 - d_1(1) = 6 - 6 = 0, \tag{7.6.7a}$$

$$d_2^2 = d_2 - d_1(2) = 6 - 6 = 0, \tag{7.6.7b}$$

$$d_3^2 = d_3 - d_1(3) = 5 - 2 = 3, \tag{7.6.7c}$$

$$d_4^2 = d_4 - d_1(4) = 4 - 2 = 2, \tag{7.6.7d}$$

$$d_5^2 = d_5 - d_1(5) = 3 - 2 = 1, \tag{7.6.7e}$$

$$d_6^2 = d_6 - d_1(6) = 2 - 2 = 0, \tag{7.6.7f}$$

$$d_7^2 = d_7 - d_1(7) = 0 - 2 = -2; \tag{7.6.7g}$$

$$v_{g_1(1)} = v_3, \tag{7.6.8a}$$

$$v_{g_1(2)} = v_4, \tag{7.6.8b}$$

$$v_{g_1(3)} = v_5, \tag{7.6.8c}$$

$$v_{g_1(4)} = v_6, \tag{7.6.8d}$$

$$v_{g_1(5)} = v_7. \tag{7.6.8e}$$

Next, apply Step 3 to obtain $G_{k+1} = G_2$ where

$$\Gamma_2 = \{[3,4],[3,5],[3,6]\} \cup \Gamma_1 \tag{7.6.9}$$

(G_2 is shown in Fig. 7.6.1b). An additional iteration of this procedure yields the graph shown in Fig. 7.6.1(c). Note that the final graph contains 7 vertices and 15 branches even though

$$(1/2)\sum_{i=1}^{6} d_i = 13.$$

The proof of the optimality of Algorithm 7.6.1 consists of two parts: first it is shown that the algorithm terminates with a graph which satisfies $\mathscr{R}$; then it is shown that the graph contains a minimum number of branches.

***Lemma* 7.6.1** *Let G be a graph with at least $r_{i,j}$ branch-disjoint i-j paths and at least $r_{i,k}$ branch-disjoint i-k paths. Then there are at least Min $[r_{i,j}, r_{i,k}]$ branch-disjoint j-k paths.*

Proof. Assign unit capacities to the branches in G. From Theorem 7.3.1 the terminal capacities $\tau_{i,j}$ and $\tau_{i,k}$ satisfy the relationships

$$\tau_{i,j} \geqq r_{i,j} \tag{7.6.10}$$

$$\tau_{i,k} \geqq r_{i,k}. \tag{7.6.11}$$

From the triangle inequality (see Chapter 5)

$$\tau_{j,k} \geqq \text{Min}\,[\tau_{i,j}, \tau_{i,k}]. \tag{7.6.12}$$

Hence

$$\tau_{j,k} \geqq \text{Min}\,[r_{i,j}, r_{i,k}]. \tag{7.6.13}$$

Since $\tau_{j,k}$ is equal to the number of branch-disjoint j-k paths the lemma follows.

Proof of Optimality of Algorithm 7.6.1

1) From Step 1, there are exactly $r_{1,2}$ branch-disjoint 1-2 paths. Let v_i be any vertex in G, $i \neq 1,2$. By Step 1, the branches $[i,1]$ and $[i,2]$ are in G. By Step 3, v_i is connected to $d_i - 2$ other vertices in G. All of these $d_i - 2$ vertices are also adjacent to v_1 (and v_2). Hence there are d_i branch-disjoint paths from any vertex v_i to v_1. A similar statement is true for any other vertex v_j. Therefore, from Lemma 7.6.1, there are at least Min $[d_i, d_j]$ branch-disjoint i-j paths. Moreover, since $d_i \geqq r_{i,k}$ for all k and $d_j \geqq r_{j,l}$ for all l, G satisfies $\mathscr{R}$.

2) Recall that $d_{n'+1} = d_{n'+2} = \cdots = d_n = 0$. For the integers in the set $\{d_1, \ldots, d_n\}$ to be realizable as the degrees of a graph they must be compressible and every set of integers resulting from successive compressions must be compressible. If the original set is not compressible, it is necessary to increase some of the d_i's or increase the number of vertices until a compressible set is obtained. If $d_K(i) = d_i$ for $i = 1, \ldots, n$ at the termination of the algorithm, the number of branches is clearly minimized. If $d_K(i) - d_i > 0$ for some i, then let $k_j^* = d_K(j) - d_j$

for $j = 1,2,\ldots,n$. The number of branches in G_K is

$$\frac{1}{2}\sum_{j=1}^{n} d_K(j) = \frac{1}{2}\left[\sum_{i=1}^{n'} d_i + \sum_{i=1}^{n} k_i^*\right]. \tag{7.6.14}$$

It is shown below that

$$\sum_{i=1}^{n} |(d_i - d_K(i))| = \text{Min}\left[\sum_{i=1}^{n} k_i\right]$$

such that if $k_1,k_2,\ldots,k_n$ are nonnegative integers, the set of integers $\{d_1 + k_1,\ldots, d_n + k_n\}$ is realizable as the degrees of a graph.

To realize $\{d_1,\ldots,d_{n'}\}$ as the degrees of a graph, this set must be compressible. To compress $\{d_1,\ldots,d_{n'}\}$ the set must contain at least $d_1 + 1$ nonzero integers. Let $Z(0)$ be the largest integer such that $d_{Z(0)} > 0$. If $Z(0) < d_1 + 1$, we must connect v_1 to *at least* $d_1 + 1 - Z(0)$ vertices whose degree requirements have already been met[2] (initially, these vertices are among $\{v_{n'+1},\ldots,v_n\}$). This number is independent of which vertices we actually connect to v_1. We then consider the realizability of a new set of integers, those obtained from compressing $\{d_1,\ldots,d_{n'}\}$. If $Z(0) \geq d_1 + 1$, we must consider the realizability of the set $\{d_2 - 1,d_3 - 1,\ldots,d_{d_1+1} - 1,d_{d_1+2},\ldots,d_{n'},0,0,\ldots,0\}$ and if $Z(0) < d_1 + 1$, we must consider the realizability of the set $\{d_2 - 1,d_3 - 1,\ldots,d_{Z(0)} - 1,0,\ldots,0\}$. The set must be compressible, and hence must contain at least one more nonzero entry than the value of the largest element.

At the jth step of the procedure, we are considering the set $\{d_{g_j(1)}^{j+1},d_{g_j(2)}^{j+1},\ldots\}$. If $Z(j)$ is the largest integer such that $d_{g_j(Z(j))}^{j+1} > 0$, then, by an identical argument, if $Z(j) < d_{g_j(1)}^{j+1} + 1$, we must connect at least $d_{g_j(1)}^{j+1} + 1 - Z(j)$ branches from $v_{g_j(1)}$ to vertices whose degree requirements have already been satisfied, regardless of which vertices these are. After connecting $v_{g_j(1)}$ to $d_{g_j(1)}^{j+1}$ vertices, we then must consider the realizability of a new set of integers. If $Z(j) \geq d_{g_j(1)}^{j+1} + 1$, this set is $\{d_{g_j(2)}^{j+1} - 1,\ldots, d_{g_j(d_{g_j(1)}+1)}^{j+1} - 1,d_{g_j(d_{g_j(1)}+2)}^{j+1},\ldots,\}$ and if $Z(j) < d_{g_j(1)}^{j+1} + 1$, this set is $\{d_{g_j(2)}^{j+1} - 1,\ldots, d_{Z(j)}^{j+1} - 1,0,\ldots,0\}$.

As a consequence of the above facts, to guarantee the successive compressibility of $\{d_i + k_i\}$, it is necessary that

$$\sum_{i=1}^{n} k_i \geq \sum_{j=1}^{K} \text{Max}\left[d_{g_j(1)}^{j+1} + 1 - Z(j),0\right] + \text{Max}\left[d_1 + 1 - Z(0),0\right]. \tag{7.6.15}$$

In other words, if G satisfies $\mathscr{R}$,

$$\sum_{i=1}^{n} d(i) = \sum_{i=1}^{n} (d_i + k_i) \geq \sum_{i=1}^{n} d_j + \sum_{j=1}^{K} \text{Max}\left[d_{g_j(1)}^{j+1} + 1 - Z(j),0\right]$$
$$+ \text{Max}\left[d_1 + 1 - Z(0),0\right]. \tag{7.6.16}$$

[2] We follow the degree realizability algorithm of Section 5 which is necessary and sufficient for the existence of a graph with prescribed degrees.

Algorithm 7.6.1 results in a graph which satisfies $\mathscr{R}$ and has exactly the sum on the right-hand side of the above inequality if this sum is even. If this sum is odd, then the sum of the vertex degrees of G_K exceeds this sum by one. However, the sum of the degrees of the vertices of any graph must be even. Therefore, G_K is optimal.

The case in which $d_1 < n' - 1$ is more complicated. If $d_k > 1$ and $d_{k+1} = \cdots = d_{n'} = 1$, we can optimally realize the entire set of integers by optimally realizing the set $\{d_1, \ldots, d_r\}$ for $r = \text{Max}\,[d_{d_1+1}, k]$ and then arbitrarily attaching $v_{r+1}, \ldots, v_{n'}$ as vertices of degree one. Hence, assume $d_i \geq 2$ for all i. For the case where $d_1 < n' - 1$ and $\{d_1, \ldots, d_{d_1+1}\}$ is successively compressible, a conceptually straightforward realization for the entire system is given below. For the case where $\{d_1, \ldots, d_{d_1+1}\}$ is not successively compressible, the realization algorithm and its proof are contained in reference FR1.

The following algorithm is applicable when $d_1 < n' - 1$ and $\{d_1, d_2, \ldots, d_{d_1+1}\}$ is successively compressible.

Algorithm 7.6.2

Step 1. Let $G_0 = (V_0, \Gamma_0)$ where $V_0 = \{v_1, v_2, \ldots, v_{d_1+1}\}$ and Γ_0 is obtained by the application of Algorithm 7.6.1 to the set of integers $\{d_1, d_2, \ldots, d_{d_1+1}\}$.

Step 2. Let $\hat{G}_0 = (V, \Gamma_0)$ where $V = \{v_1, \ldots, v_n\}$ and $n = n'$. We generate a new graph by adding and deleting branches so that the degrees of vertices in V_0 remain invariant and the degree requirements of vertices $v_{d_1+2}, \ldots, v_{n'}$ are eventually met. Let $\{v_{0_1}, \ldots, v_{0_q}\}$ be the set of vertices with odd degree requirements in $\{v_{d_1+2}, \ldots, v_{n'}\}$. Recursively define, for $x_{i,j}$ and $y_{i,j}$ specified by (7.6.18), the graphs

$$\hat{G}_i(j) = (V, \Gamma_i(j)) \tag{7.6.17}$$

in (i) to (iii) below.

i) Form $\Gamma_1(1)$ from Γ_0 by removing $[x_{1,1}, y_{1,1}]$ and adding $[x_{1,1}, d_1 + 1 + 1]$ and $[y_{1,1}, d_1 + 1 + 1]$.

ii) In general, form $\Gamma_i(1)$ from $\Gamma_{i-1}(l)$ by removing $[x_{i,1}, y_{i,1}]$ and adding $[x_{i,1}, d_1 + 1 + i]$ and $[y_{i,1}, d_1 + 1 + i]$ where $l = [(d_{d_1+i})/2]$. If d_{d_1+1+i} is odd, then $v_{d_1+1+i} = v_{0_k}$ for some k. If k is even also add $[0_{k-1}, 0_k]$.

iii) $\Gamma_i(j)$ is formed from $\Gamma_i(j-1)$ by removing $[x_{i,j}, y_{i,j}]$ and adding $[x_{i,j}, d_1 + 1 + i]$ and $[y_{i,j}, d_1 + 1 + i]$.

When forming $\hat{G}_i(j)$, the vertices $v_{x_{i,j}}$ and $v_{y_{i,j}}$ are any vertices in $\{v_1, \ldots, v_{d_1+i}\}$ such that

$$v_{x_{i,j}} \neq v_{y_{i,j}} \tag{7.6.18a}$$

and

$$v_{x_{i,j}}, v_{y_{i,j}} \notin \{v_{x_{i,1}}, \ldots, v_{x_{i,j-1}}, v_{y_{i,j}}, \ldots, v_{y_{i,j-1}}\}. \tag{7.6.18b}$$

If d_{d_1+1+i} is odd, then for some $k, v_{d_1+1+i} = v_{0_k}$, and the additional restrictions on $v_{x_{i,j}}$ and $v_{y_{i,j}}$ must also apply:

$$v_{x_{i,j}} \neq v_{0_{k-1}} \tag{7.6.18c}$$
$$v_{y_{i,j}} \neq v_{0_{k-1}}. \tag{7.6.18d}$$

Step 3. If q is even, the final graph $G = (V,\Gamma)$ is defined by $\Gamma = \Gamma_{n-(d_1+1)}([d_n/2])$. If q is odd, $\Gamma = \Gamma_{n-(d_1+1)}([d_n/2]) \cup \{[0_q,x]\}$ where v_x is any vertex such that $[0_q,x] \notin \Gamma_{n-(d_1+1)}([d_n/2])$.

Example 7.6.2 Let $\{d_i\} = \{6,6,6,4,4,4,3,3\}$. Since $\{d_1,\ldots,d_{d_1+1}\}$ is compressible, we use Algorithm 7.6.2. The graph in Fig. 7.6.2(a) is the result of Step 1. The graphs in Figs. 7.6.2(b) and (c) are the results of applying Step 2 to the graph in Fig. 7.6.2(a). Then v_8 is inserted into the graph at the stage of the algorithm corresponding to Fig. 7.6.2(b) and v_9 and v_{10} are inserted at the stage represented by Fig. 7.6.2(c). The graph shown in Fig. 7.6.2(d) is the final result.

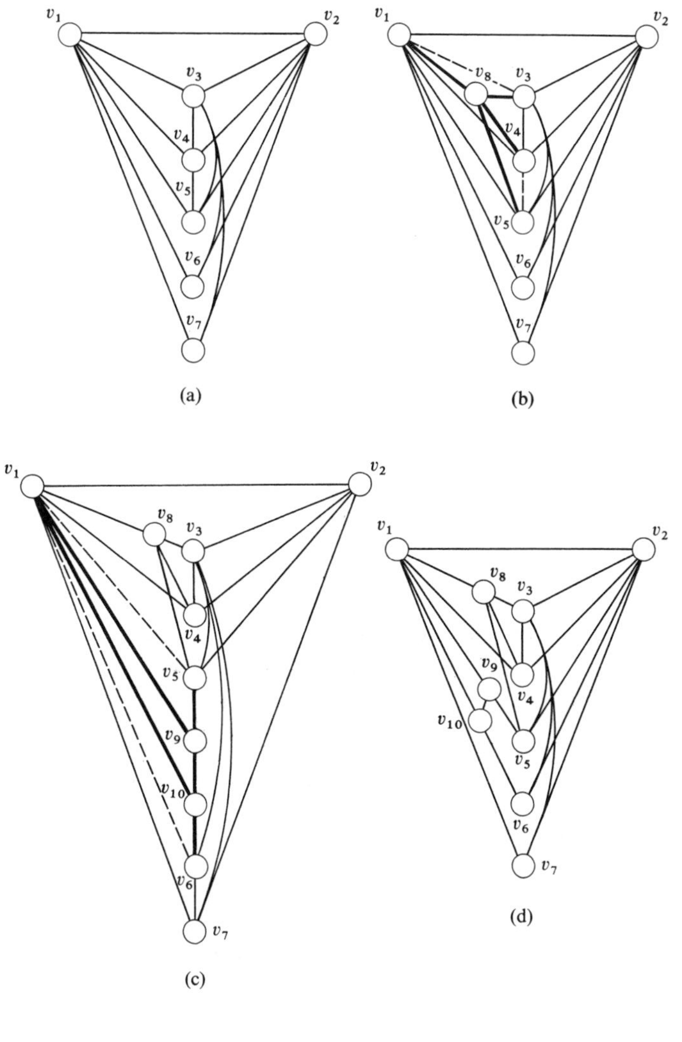

Fig. 7.6.2 Graphs used in Example 7.6.2.

Proof of Optimality of Algorithm 7.6.2

1) Since $\{d_1,d_2,\ldots,d_{d_1+1}\}$ is successively compressible, a graph with these degrees can be realized by applying Algorithm 7.6.1. From Part (a) of the proof of optimality of Algorithm 7.6.1, there are Min $[d_i,d_j]$ branch-disjoint paths from v_i to v_j for $1 \leqq i \leqq d_1 + 1$ and $1 \leqq j \leqq d_1 + 1$. Step 2 of Algorithm 7.6.2 adds branches to the vertices in the set $\{v_{d_1+1+i}\}$. Consider the construction of $\hat{G}_i(j)$ in Step 2. Let v_{x_i} and $v_{y_i} \in \{v_1,\ldots,v_{d_1+1}\}$. At this stage, branch $[x_i,y_i]$ is removed from Γ_0 and replaced by the branches $[x_i,d_1 + 1 + i]$ and $[y_i,d_1 + 1 + i]$. In this operation, the number of branch-disjoint paths between v_{x_i} and v_{y_i} remains the same, as do the degrees of these vertices. If d_{d_1+1+i} is even, at the completion of Step 2, vertex v_{d_1+1+i} is adjacent to d_{d_1+1+i} vertices in $\{v_1,\ldots,v_{d_1+i}\}$. Hence, there are at least d_{d_1+1+i} branch-disjoint path between v_1 and v_{d_1+1+i}. If d_{d_1+1+i} is odd, and $v_{d_1+1+i} = v_{0_k}$ with k even there are $d_{d_1+1+i} - 1$ branch-disjoint paths from v_{d_1+1+i} to v_1 before $[0_{k-1},0_k]$ is added. After $[0_{k-1},0_k]$ is added, an additional branch-disjoint path has been created from $v_{0_{k-1}}$ to v_{0_k} and from v_{0_k} to v_1. For q odd, an additional branch-disjoint path is added between v_{0_q} and v_1 during Step 3.

2) Degree specifications $\{d_1,\ldots,d_n\}$ cannot be realized exactly unless

$$\sum_{i=1}^{n} d_i$$

is even (and $\{d_1,\ldots,d_n\}$ is successively compressible). In this case,

$$(1/2)\sum_{i=1}^{n} d_i$$

branches are required. If

$$\sum_{i=1}^{n} d_i$$

is odd, to realize a graph whose degrees are *at least* $\{d_1,\ldots,d_n\}$ requires at least

$$(1/2)\left[\sum_{i=1}^{n} d_i + 1\right]$$

branches. The graph obtained by application of Algorithm 7.6.2 contains exactly this number of branches.

7.7 MAXIMIZATION OF THE SMALLEST VERTEX CUT-SET

In this section we consider the problem of maximizing the smallest cut-set in a graph with given number of branches and vertices. We find graphs for which $\omega = [2m/n]$ and hence both θ and ω are simultaneously maximized. For $n \leq 2$ there are no vertex cut-sets. Therefore, assume $m \geq n - 1$ and $n > 2$.

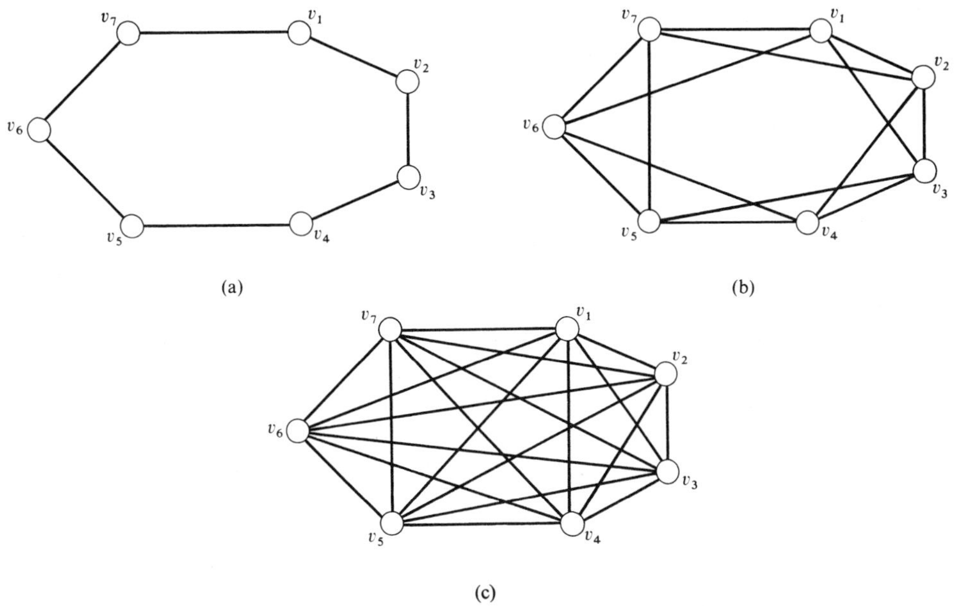

Fig. 7.7.1 Graphs for Example 7.7.1.

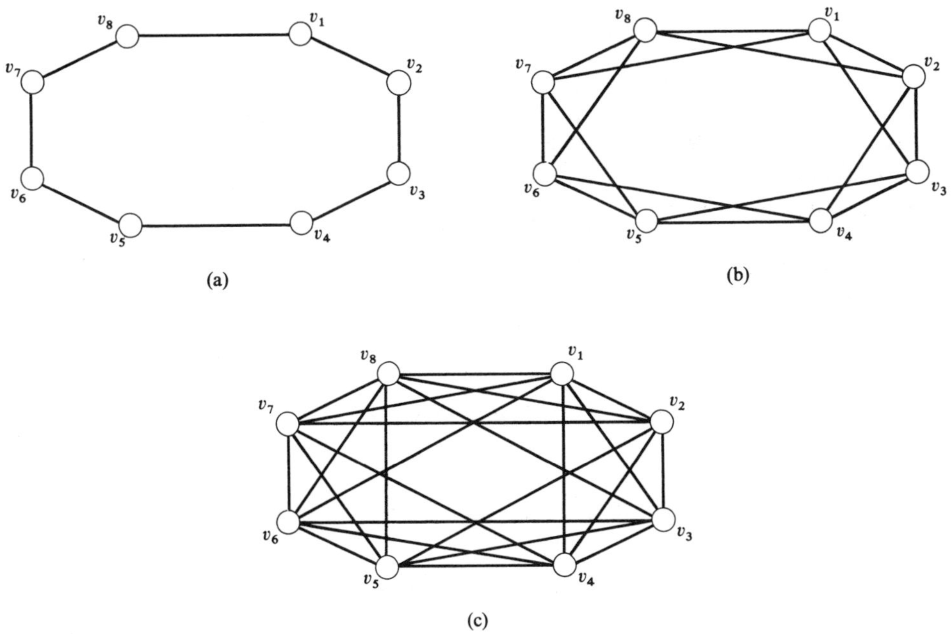

Fig. 7.7.2 Graphs for Example 7.7.2.

Let $\eta = 2m/n$. If η is an integer, let $G_{\eta,n}$ be the graph with vertices $v_1, \ldots, v_n$ such that $[i,j] \in \Gamma$ if and only if for integer values of β in the interval $1 \leq \beta \leq [\eta/2]$

$$i - j \equiv \beta \bmod n. \tag{7.7.1}$$

The notation $i - j \equiv \beta \bmod n$ means that when $i - j$ is divided by n the remainder is β.

Example 7.7.1 $G_{2,7}$,$G_{4,7}$, and $G_{6,7}$ are given in Figs. 7.7.1(a), (b), and (c), respectively. $G_{2,8}$,$G_{4,8}$, and $G_{6,8}$ are given in Figs. 7.7.2(a), (b), and (c), respectively.

We now show that in $G_{\eta,n}$ there are $2k$ vertex-disjoint 1-x paths for any vertex v_x. The integers k, q, r, and r_x are defined by

$$m = kn + r, \quad r \geq 0 \tag{7.7.2}$$

and

$$x = qk + r_x, \quad r_x \geq 0 \tag{7.7.3}$$

where k is the largest integer such that $m = kn + r$ for nonnegative r.

Using this notation we enumerate a set of k paths from v_1 to v_x for $x > k + 1$ as

$$\pi_{1,x}^i = \begin{cases} v_1, v_i, v_{i+k}, \ldots, v_{i+qk}, v_x & \text{for } 2 \leq i < r_x; \quad (7.7.4) \\ v_1, v_i, v_{i+k}, \ldots, v_{i+[(x-i)/k]k}, v_x & \text{for } r_x \leq i \leq k+1. \quad (7.7.5) \end{cases}$$

Theorem 7.7.1 *For $G_{\eta,n}$, $\omega \geq 2k$.*

Proof. The paths in (7.7.4) and (7.7.5) contain no vertices in common other than v_1 and v_x. From (7.7.1), all branches used in (7.7.4) and (7.7.5) are in $G_{\eta,n}$. From (7.7.2) and (7.7.3) the paths in (7.7.4) and (7.7.5) use only vertices v_i with i in the range $1 \leq i \leq x$. Hence for $k < x \leq n$ in $G_{\eta,n}$ there are k vertex-disjoint 1-x paths containing vertex v_i for $1 \leq i \leq x$. Similarly, there are k vertex-disjoint 1-x paths containing v_i with $x \leq i \leq n$. Hence there are $2k$ vertex-disjoint 1-x paths. From Theorem 7.2.1, we need not find 1-x paths for $x < k + 1$ since $[1,x] \in \Gamma$ from (7.7.1). Finally, since any vertex may be numbered v_1, the proof is completed.

Example 7.7.2 For $G_{4,8}$ (as shown in Fig. 7.7.2b) let $x = 4$. Then,

$$k = 2, \tag{7.7.6a}$$

$$r = 0, \tag{7.7.6b}$$

$$q = 2, \tag{7.7.6c}$$

$$r_x = 0. \tag{7.7.6d}$$

From (7.7.4) and (7.7.5)

$$\pi_{1,4}^1 = v_1, v_2, v_4; \tag{7.7.7a}$$

$$\pi_{1,4}^2 = v_1, v_3, v_4. \tag{7.7.7b}$$

For $G_{6,8}$ in Fig. 7.7.2(c) let $x = 5$. Then,

$$k = 3, \tag{7.7.8a}$$

$$r = 0, \tag{7.7.8b}$$

$$q = 1, \tag{7.7.8c}$$

$$r_x = 2. \tag{7.7.8d}$$

From (7.7.4) and (7.7.5),

$$\pi_{1,4}^1 = v_1,v_2,v_5, \tag{7.7.9a}$$

$$\pi_{1,4}^2 = v_1,v_3,v_5, \tag{7.7.9b}$$

$$\pi_{1,4}^3 = v_1,v_4,v_5. \tag{7.7.9c}$$

***Corollary* 7.7.1** *If η is an even integer then for $G_{\eta,n}$, $\omega = 2m/n$.*

Proof. Clearly

$$\omega \leq \frac{2m}{n}. \tag{7.7.10}$$

From Theorem 7.7.1

$$\omega \geq 2k. \tag{7.7.11}$$

Since $\eta = 2m/n$ is an even integer, m/n is an integer equal to k in (7.7.2). Hence the proof is completed.

If η is an odd integer, $\omega < \eta$ since the degrees of the vertices in $G_{\eta,n}$ are all even. Hence $\omega = 2k = \eta - 1$. To construct a graph with $\omega = \eta$ for odd η, let $\hat{G}_{\eta,n}$ be the graph formed from $G_{\eta,n}$ by adding all $[i,j]$ such that

$$i - j = \left[\frac{n+1}{2}\right]. \tag{7.7.12}$$

To show that $\hat{G}_{\eta,n}$ yields $\omega = \eta$, we derive a property of the smallest vertex cut-sets in $G_{\eta,n}$. Let $V_{1,x} = \{v_i\text{:}1 < i \leq x\}$ and let $V_{x,n} = \{v_i\text{:}x < i \leq n\}$.

***Theorem* 7.7.2** *For $x > k$ let $A_{1,x}$ be a smallest 1-x vertex cut-set in $G_{\eta,n}$. Then $A_{1,x}$ contains a set of k consecutively numbered vertices in $V_{1,x}$ and a set of k consecutively numbered vertices in $V_{x,n}$.*

Proof. From (7.7.4) and (7.7.5), $A_{1,x}$ consists of k vertices from $V_{1,x}$ and k vertices from $V_{x,n}$.

Without loss of generality consider the k vertices in $V_{1,x}$ and assume they are not consecutively numbered. Divide these k vertices into two nonempty sequences of vertices $v_{i'}, \ldots, v_{j'}$ and $v_{i''}, \ldots, v_{j''}$, such that the vertices in the sequences are numbered in increasing order and

$$i'' > j' + 1. \tag{7.7.13}$$

Let $V' = \{v_{i'}, \ldots, v_{j'}\}$ and $V'' = \{v_{i''}, \ldots, v_{j''}\}$. Since

$$|V' \cup V''| = k, \tag{7.7.14}$$

clearly

$$|V'| < k \tag{7.7.15}$$

and

$$|V''| < k. \tag{7.7.16}$$

Hence, from (7.7.1), $v_{i'-1}$ and $v_{j'+1}$ are adjacent, as are $v_{i''-1}$ and $v_{j''+1}$. From Theorem 7.7.1, there is a 1-($i' - 1$) path with vertices from the set $\{v_1, \ldots, v_{i'-1}\}$; there is a ($j' + 1$)-($i'' - 1$) path with vertices from the set $\{v_{j'+1}, \ldots, v_{i''-1}\}$, and there is a ($j'' + 1$)-$x$ path with vertices in the set $\{v_{j''+1}, \ldots, v_x\}$. Hence, after the removal of $A_{1,x}$ there is still a 1-x path in $G_{\eta,n}$. This is a contradiction and the proof is completed.

***Corollary* 7.7.2** *For $\hat{G}_{\eta,n}$, $\omega = \eta = 2m/n$.*

Proof. For $G_{\eta,n}$, $\omega = n - 1$. However, a 1-x cut-set of k vertices in $G_{\eta,n}$ cannot be a 1-x cut-set in $\hat{G}_{\eta,n}$ because the cut-set in $\hat{G}_{\eta,n}$ consists of two sets of k consecutive vertices, one in $V_{1,x}$ and one in $V_{x,n}$. These are not cut-sets in $\hat{G}_{\eta,n}$, since $[i,j] \in \Gamma$ if $i - j = [(n + 1)/2]$. Hence $\omega \geq \eta + 1$ and the proof is completed.

Example 7.7.3 $\hat{G}_{3,6}$ is given in Fig. 7.7.3.

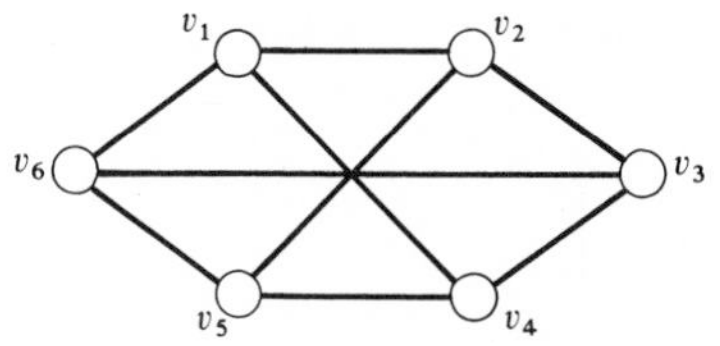

Fig. 7.7.3 $\hat{G}_{3,6}$.

Finally, consider the case in which η is not an integer. If $[\eta]$ and n are both even, the optimal graph is formed by adding to $G_{[\eta],n}$, $m - [\eta]n/2$ arbitrary branches. Thus, the optimal graph with $m = 7$ and $n = 6$ is obtained from $G_{2,6}$ as in Fig. 7.7.4. Moreover ω is unchanged if the last $m - [\eta]n/2$ branches are omitted.

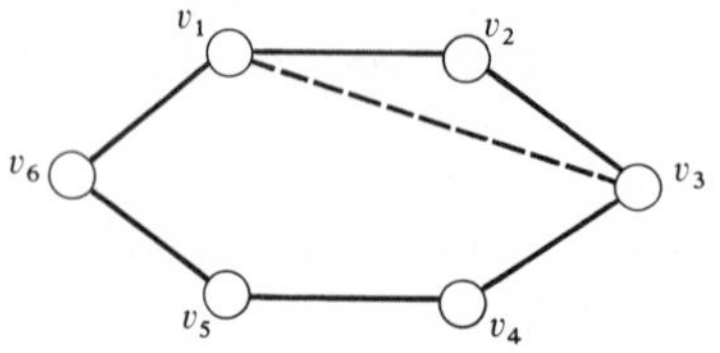

Fig. 7.7.4 Graph with $m = 7$ and $\omega = 2$.

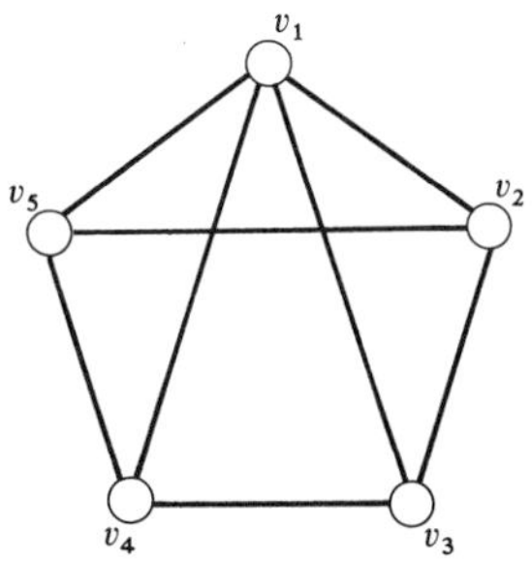

Fig. 7.7.5 $\hat{G}_{3,5}$.

If η is not an integer and either $[\eta]$ or n is odd, then the optimal graph is $\hat{G}_{[\eta],n}$. Thus, for $m = 8$ and $n = 5$ the optimal graph is $\hat{G}_{3,5}$ (Fig. 7.7.5). The proof for this case is the same as the proof of Theorem 7.7.2 and Corollary 7.7.2.

We conclude this section with a property of the graph $G_{\eta,n}$. From (7.7.1), $G_{\eta,n}$ is given by

$$\bigcup_{\beta=1}^{[\eta/2]} G_{\eta,n}^{\beta} \tag{7.7.17}$$

where $G_{\eta,n}^{\beta}$ is a graph with vertices $v_1, \ldots, v_n$ and branches $[i,j]$ such that

$$i - j \equiv \beta \bmod n. \tag{7.7.18}$$

***Theorem** 7.7.3 If n and β are relatively prime,*[3] *$G_{\eta,n}^{\beta}$ is a Hamilton circuit.*

Proof. Let Γ be the set of branches of $G_{\eta,n}^{\beta}$ for which n and β are relatively prime.

Suppose, contrary to the theorem, that $G_{\eta,n}^{\beta}$ contains a circuit with $l < n$ branches. Without loss of generality, assume the circuit contains vertex v_1. Then we have

$$1 + l\beta \equiv 1 \tag{7.7.19}$$

or

$$l\beta \equiv 0. \tag{7.7.20}$$

Hence

$$\frac{l\beta}{n} = q \tag{7.7.21}$$

for some integer q. Thus

$$\frac{\beta}{n} = \frac{q}{l}. \tag{7.7.22}$$

Since β and n are relatively prime and $l < n$ this is a contradiction and the proof is completed.[4]

---

[3] Two integers are relatively prime if their greatest common divisor is unity.
[4] We have essentially proved that the integers $|i - j| = \beta \bmod n$ for $[i,j] \in \Gamma$ form a complete system of residues mod n.

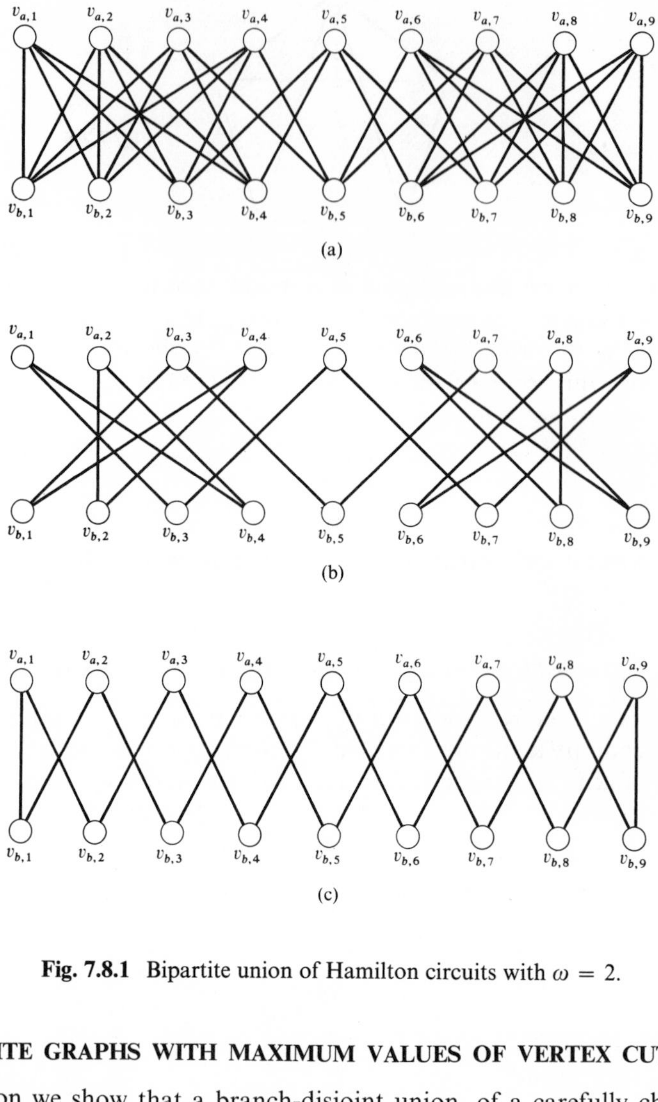

Fig. 7.8.1 Bipartite union of Hamilton circuits with $\omega = 2$.

7.8 BIPARTITE GRAPHS WITH MAXIMUM VALUES OF VERTEX CUT-SETS

In this section we show that a branch-disjoint union, of a carefully chosen set of Hamilton circuits, yields a graph with maximum value for ω.[5] In particular we restrict our attention to a bipartite graph G such that $V = A \cup B$ with $[A,B] = \Gamma$. Figure 7.8.1(b) and (c) shows two branch-disjoint Hamilton circuits whose union is the graph shown in Fig. 7.8.1(a). Clearly the graph is not optimal since

$$\omega = 2 = \left|\{v_{a,5},v_{b,5}\}\right| \quad \text{and} \quad \omega < [2m/n] = 4.$$

[5] For example, at the end of Section 7.7 we saw that for certain values of n and k the resulting graphs were branch-disjoint unions of Hamilton circuits.

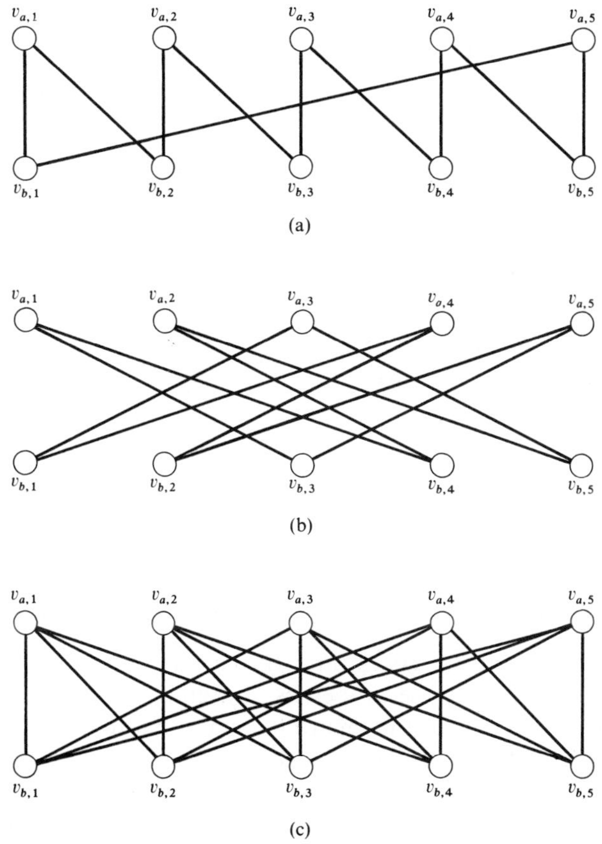

Fig. 7.8.2 Bipartite union of Hamilton circuits with maximum ω.

Hence not every bipartite branch disjoint union of Hamilton circuits gives a graph with maximum ω. However, the bipartite graph in Fig. 7.8.2(c), constructed from the Hamilton circuits in Fig. 7.8.2(a) and (b), is optimal. We now consider the selection of the Hamilton circuits in detail.

Definition 7.8.1 *Let K be the connection matrix of the bipartite graph G with the first $|A|$ rows corresponding to the vertices in A. The bipartite connection matrix $Q = [q_{i,j}]$ is the submatrix of K consisting of the entries in K in the first $|A|$ rows and last $|B|$ columns.*

Denote the vertices in A by $v_{a,1}, \ldots, v_{a,|A|}$ and denote the vertices in B by $v_{b,1}, \ldots, v_{b,|B|}$. Then in Q, $q_{i,j} = 1$ if $[v_{a,i}, v_{b,j}] \in \Gamma$ and $q_{i,j} = 0$ otherwise.

Unless otherwise stated we assume in this section that n is even and

$$|A| = |B| = n/2.$$

***Definition* 7.8.2** *Let $Q_{(1)}$ denote the bipartite connection matrix of a bipartite graph G with $|A| = |B| = n/2$ such that*

$$q_{i,i} = 1 \quad \textit{for} \quad i = 1,2,\ldots,n/2, \tag{7.8.1a}$$

$$q_{i,i+1} = 1 \quad \textit{for} \quad i = 1,2,\ldots,n/2 - 1, \tag{7.8.1b}$$

$$q_{n/2,1} = 1, \tag{7.8.1c}$$

$$q_{i,j} = 0 \quad \textit{otherwise.} \tag{7.8.1d}$$

The nonzero elements are said to constitute a pair of adjacent diagonals in the bipartite connection matrix.

Example 7.8.1 $Q_{(1)}$ is given in (7.8.2) for $n = 10$ and is the bipartite connection matrix of the graph in Fig. 7.8.2(a).

$$Q_{(1)} = \begin{bmatrix} 1 & 1 & 0 & 0 & 0 \\ 0 & 1 & 1 & 0 & 0 \\ 0 & 0 & 1 & 1 & 0 \\ 0 & 0 & 0 & 1 & 1 \\ 1 & 0 & 0 & 0 & 1 \end{bmatrix}. \tag{7.8.2}$$

***Lemma* 7.8.1** *$Q_{(1)}$ is the bipartite connection matrix of a Hamilton circuit.*

Proof. The Hamilton circuit passes through the vertices in the order $v_{b,1}, v_{a,1}, v_{b,2}, v_{a,2}, \ldots, v_{b,n}, v_{a,n}, v_{b,1}$. Clearly the properties of a graph are independent of permutations of rows or columns of the bipartite connection matrix. In particular, any matrix obtained from $Q_{(1)}$ by cyclically permutating all rows represents a Hamilton circuit. Matrix K is not permuted because of the required symmetry of K.

Let $Q_{(j)}$ be the connection matrix with elements[6]

$$q_{i,(i+2j-2 \bmod n/2)} = 1 \tag{7.8.3a}$$

$$\text{for } i = 1,2,\ldots,n/2$$

$$q_{i,(i+2j-1 \bmod n/2)} = 1 \tag{7.8.3b}$$

and all other elements zero.

Example 7.8.2 For $n = 10$

$$Q_{(2)} = \begin{bmatrix} 0 & 0 & 1 & 1 & 0 \\ 0 & 0 & 0 & 1 & 1 \\ 1 & 0 & 0 & 0 & 1 \\ 1 & 1 & 0 & 0 & 0 \\ 0 & 1 & 1 & 0 & 0 \end{bmatrix}. \tag{7.8.4}$$

$Q_{(j)}$ is obtained from Q_1 by cyclic row permutations and thus also represents a Hamilton circuit. Furthermore, for $i \neq j$, the graphs with bipartite connection matrices $Q_{(i)}$ and $Q_{(j)}$ have no branches in common. Using these facts we can give a simple method for constructing a bipartite graph with a maximum value of ω.

[6] $k \bmod l$ denotes the remainder when k is divided by l.

Let G_k for $k = 1,2,\ldots,[n/4]$ be the bipartite graph on n vertices, with connection matrix

$$Q = \sum_{j=1}^{k} Q_{(j)}. \tag{7.8.5}$$

Example 7.8.3 The bipartite connection matrix for G_2 with 10 vertices is given in (7.8.6) and G_2 is given in Fig. 7.8.2(c).

$$Q = Q_{(1)} + Q_{(2)} = \begin{array}{c} \\ v_{a,1} \\ v_{a,2} \\ v_{a,3} \\ v_{a,4} \\ v_{a,5} \end{array} \begin{array}{c} \begin{array}{ccccc} v_{b,1} & v_{b,2} & v_{b,3} & v_{b,4} & v_{b,5} \end{array} \\ \begin{bmatrix} 1 & 1 & 1 & 1 & 0 \\ 0 & 1 & 1 & 1 & 1 \\ 1 & 0 & 1 & 1 & 1 \\ 1 & 1 & 0 & 1 & 1 \\ 1 & 1 & 1 & 0 & 1 \end{bmatrix} \end{array} \tag{7.8.6}$$

We now sketch the proof that G_k is optimal. Since it is difficult to enumerate all vertex-disjoint paths explicitly, we choose an indirect proof, as opposed to the enumeration method used to prove Theorem 7.7.1.

***Theorem* 7.8.1** *For G_k, $\omega = 2k$.*

Proof. Consider G_k with bipartite connection matrix Q where Q is $n/2 \times n/2$ and $k \leq n/4$. Let the rows of Q be numbered in order with odd integers $1,3,\ldots,n/2 - 1$. Let the columns be numbered in order with even integers $2,4,\ldots,n/2$. Then

$$\text{for } i \text{ odd} \qquad [i,l] \in \Gamma \text{ if } l = i, i + 3, \ldots, i + 4k - 1 \bmod n, \tag{7.8.7a}$$

$$\text{for } i \text{ even,} \qquad [i,l] \in \Gamma \text{ if } l = i - 1, i - 3, \ldots, i - (4k - 1) \bmod n. \tag{7.8.7b}$$

Suppose there is a vertex cut-set V_1 with $|V_1| = \alpha < 2k - u$ with $u \geq 1$. We show that this assumption leads to a contradiction. Let the vertices in $V - V_1$ be numbered $i_1, i_2, \ldots, i_\alpha$ where

$$i_1 < i_2 < \cdots < i_\alpha. \tag{7.8.8}$$

Since the removal of V_1 disconnects G_1, there is an integer p such that v_{i_p} and $v_{i_{p+1}}$ represent vertices in different components after V_1 is removed. Suppose p is odd. (The proof follows similarly for even p.)

To simplify the notation, cyclically permute the vertex-subscripts so that i_p is mapped to 1 and i_{p+1} is mapped to some integer j. With the new numbering, v_β is in V_1 if $1 < \beta < j$ since i_p and i_{p+1} are consecutive integers in (7.8.8) and incidence relations are preserved under the cyclic permutation.

Define q by the relation $j = q + 2$ where $q \leq 2k - 1$. Thus

$$j \leq 2k + 1. \tag{7.8.9}$$

If j is even then from (7.8.7), $[1,j] \in \Gamma$ since $2k + 1 < 4k$. This is impossible since V_1 is a 1-j vertex cut-set. Hence j must be odd. Thus we can let

$$q = 2r - 1, \tag{7.8.10}$$

where r is a fixed integer in the range $1 \leq r \leq k$.

Since v_1 and v_j are both numbered with odd integers they must both be adjacent to the vertices numbered with the even integers $i = 2r + 2, \ldots, 4k$. Hence these vertices must be in V_1. Combining these vertices with the q vertices $v_2, v_3, \ldots, v_{j-1}$ gives a total of at least

$$(2k - r) + (2r - 1) = 2k + r - 1 \tag{7.8.11}$$

vertices in V_1. Since $r \geq 1$ this contradicts the fact that $|V_1| < 2k$ and the proof is completed.

For the case in which n is odd we can still construct an optimal bipartite graph although we no longer restrict our attention to unions of Hamilton circuits. A bipartite graph is said to be *complete* if every vertex in A is adjacent to every vertex in B. Let G^* be the complete bipartite graph on n vertices with $|A| = n_a = [n/2]$ and $[n/2](n - [n/2])$ branches. G^* is shown in Fig. 7.8.2(d) for $n = 10$. For $n = 9$, G^* is shown in Fig. 7.8.3.

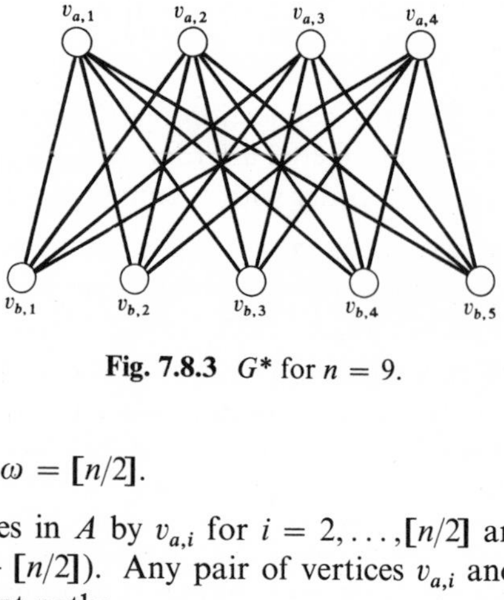

Fig. 7.8.3 G^* for $n = 9$.

***Theorem* 7.8.2** *For* G^*, $\omega = [n/2]$.

Proof. Label the vertices in A by $v_{a,i}$ for $i = 2, \ldots, [n/2]$ and the vertices in B by $v_{b,i}$ for $i = 1, 2, \ldots, (n - [n/2])$. Any pair of vertices $v_{a,i}$ and $v_{a,k}$ are joined by the $(n - [n/2])$ vertex-disjoint paths

$$v_{a,i}, v_{b,j}, v_{a,k} \qquad \text{for } j = 1, 2, \ldots, (n - [n/2]). \tag{7.8.12}$$

Any pair of vertices $v_{b,i}$ and $v_{b,k}$ are joined by the $[n/2]$ vertex-disjoint paths

$$v_{b,i}, v_{a,j}, v_{b,k} \qquad \text{for } j = 1, 2, \ldots, [n/2]. \tag{7.8.13}$$

Any pair of vertices $v_{a,i}$ and $v_{b,k}$ are joined by the $[n/2]$ vertex-disjoint paths

$$v_{a,i}, v_{b,j}, v_{a,j}, v_{b,k} \qquad \text{for } k \neq i, j, \tag{7.8.14}$$

$$v_{a,i}, v_{b,i}, v_{a,k}, v_{b,k}, \tag{7.8.15}$$

and

$$v_{a,i}, v_{b,k}. \tag{7.8.16}$$

For n even, $2m/n = 2(n/2)(n/2)/n = n/2$. Hence G^* is optimal. For n odd and greater than or equal to three,

$$\left\lceil \frac{2m}{n} \right\rceil = \left\lceil \frac{2[n/2](n - [n/2])}{n} \right\rceil = [n/2 - 1/2n] = [n/2] \tag{7.8.17}$$

and G^* is again optimal.

7.9 APPROXIMATIONS TO MINIMUM COST INVULNERABLE GRAPHS

In this section we describe computer techniques for designing invulnerable graphs which are approximately minimum cost. We assume a given set of vertices are distributed on a surface and the cost of a realization is the sum of the lengths of its branches, where the length of a branch is the Euclidean distance between its end vertices. This assumption is adopted for convenience. The techniques to be given can be applied with more realistic cost functions.

We first consider the minimum cost realization of graphs with specified values n and θ, where θ is an even number. Clearly, a Hamilton circuit has a maximum value of θ for a given $n = m$ since $2m/n = \theta = 2$. Furthermore, if we could find a shortest Hamilton circuit we would have a minimum cost graph for $\theta = 2$. However, the problem of finding a shortest Hamilton circuit is extremely difficult; it is known classically as the Traveling Salesman Problem and is unsolved in general. We therefore use a method due to Lin [LI1] to find an *approximation* to a minimum cost Hamilton circuit. The result of an application of Lin's method is shown in Fig. 7.9.1. To generalize this procedure to realize a graph with any even value of θ, simply add to the existing graph the shortest Hamilton circuit not using any branches already in the graph. For the resulting graph $\theta = 2m/n$ since, each time a Hamilton circuit is added, two i-j paths are added for all i and j not containing branches in the previous i-j paths. The graph may not have minimum cost but will be a low-cost realization. The result of adding a second Hamilton circuit to the graph in Fig. 7.9.1 is shown in Fig. 7.9.2.

A union of arbitrary branch disjoint Hamilton circuits does not, in general, yield a maximum value of ω. To maximize ω we can find all vertex cut-sets of value less

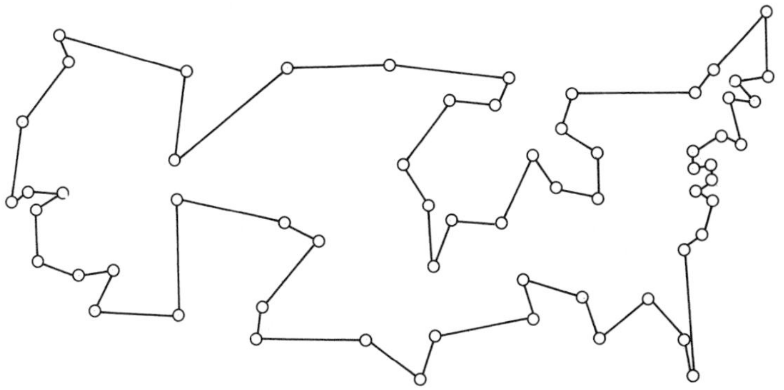

Fig. 7.9.1 Hamilton circuit with approximately minimum length.

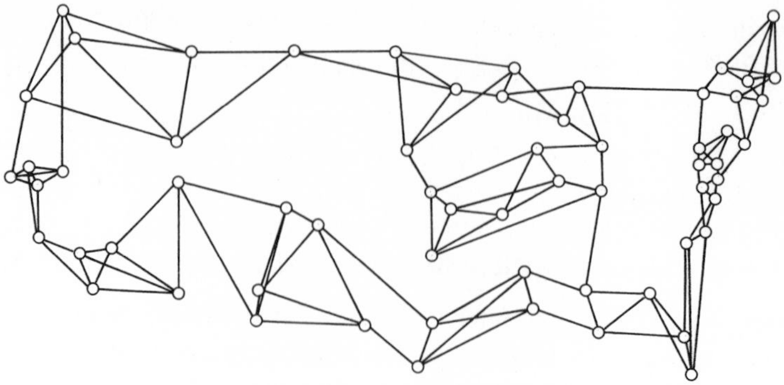

Fig. 7.9.2 Union of two Hamilton circuits.

than $[2m/n]$ and then increase the size of such a cut-set by means of a branch replacement. If G contains $[a,b]$ and $[j,l]$ but does not contain $[a_1,b_1]$ or $[j_1,l_1]$, a branch replacement is generated by removing $[a,b]$ and $[j,l]$ and adding $[a_1,b_1]$ and $[j_1,l_1]$. A given branch replacement is proper only if it increases the value of a vertex cut-set without decreasing the value of any other below $2m/n$. The following theorem shows that, in order to determine whether a branch replacement is proper, the values of the cut-sets between only two pairs of vertices need be evaluated.

***Theorem* 7.9.1** *Suppose G' is obtained from G by a branch replacement which removes $[a_1,b_1]$ and $[j_1,l_1]$ and adds $[a_2,b_2]$ and $[j_2,l_2]$. Furthermore, assume that $\sigma_{x,y}$ is reduced to value $h < k$ by the branch replacement. Then in G' either*

$$\sigma_{a_1,b_1} < k \tag{7.9.1}$$

or

$$\sigma_{j_1,l_1} < k. \tag{7.9.2}$$

Proof. Suppose $[x,y]$ is not in G. Then in G' there is an x-y vertex cut-set $A^q_{x,y}$ of value $h < k$. $A^q_{x,y}$ is either an a_1-b_1 cut-set or a j_1-l_1 cut-set, since otherwise $\sigma_{x,y}$ would not be reduced by the branch replacement. If $A^{kq}_{x,y}$ is an a_1-b_1 cut-set, in G'

$$\sigma_{a_1,b_1} \leq \sigma_{x,y} < k. \tag{7.9.3}$$

$$\sigma_{j_1,l_1} \leq \sigma_{x,y} < k. \tag{7.9.4}$$

If $[x,y]$ is in G it can be removed and the argument repeated for the resulting graph, thus completing the proof.

An algorithm for constructing a low-cost graph with large ω consists of the following steps:

1) repeatedly add a shortest Hamilton circuit not containing any branches already used;

2) find vertex cut-sets of value less than $[2m/n]$;

3) use branch replacement to increase cut-sets with value less than $[2m/n]$.

Example 7.9.1 For the graph G in Fig. 7.9.2, $|\{v_5,v_6\}| = 2 < [2m/n]$. Remove [2,1] and [3,4] and add [2,3] and [1,4]. The number of vertex cut-sets of two vertices has been decreased by one.

In addition to using branch replacements to increase ω, these can also be used to reduce cost while keeping ω or θ unchanged. Thus, we could continue to apply branch replacements so long as ω and θ are unchanged and

$$h[a_2,b_2] + h[j_2,l_2] < h[a_1,b_1] + h[j_1,l_1], \tag{7.9.5}$$

where $h[x,y]$ is the cost of $[x,y]$.

Of course, the results of applying branch replacements to increase ω and decrease costs will lead to realizations which are only *locally* optimum with respect to the transformation of branch replacement. To overcome this handicap it is possible to apply the branch replacements to starting graphs generated at random to provide a wide sample of the local optima in the solution space. A technique used effectively by Steiglitz *et al.* [ST1] is to order the vertices randomly before generating the initial graph. Then the initial graph can be composed of low-cost branch-disjoint unions of Hamilton circuits or graphs generated by any other heuristic methods. One advantage of this approach is that it produces a number of different solutions with low cost.

This same procedure can be applied with minor modifications to the case in which we wish to guarantee $\omega_{i,j} > k_{i,j}$ for specified integers $k_{i,j}$ at minimum cost. The heuristics for selecting the initial graph are different and Theorem 7.9.1 does not allow us to restrict the analysis to finding only σ_{a_1,b_1} and σ_{j_1,l_1}. However, it does mean that after a branch replacement we need test only $\sigma_{a,b}, \sigma_{j,l}$ and those $\sigma_{\alpha,\beta}$ for which

$$k_{\alpha,\beta} \geq \text{Min}\,[k_{a_1,b_1}, k_{j_1,l_1}]. \tag{7.9.6}$$

7.10 MAXIMIZATION OF THE MINIMUM n_1 DEGREE

In this section we assume that stations are hardened to attack so that only channels can be destroyed. Hence we allow only branches to be removed from a graph. We then seek to maximize the minimum number of branches which must be removed from a graph in order to break all paths from a specified number n_1 of vertices to the remaining n-n_1 vertices.

***Definition* 7.10.1** *If V_1 is a subset of n_1 vertices, let $d(V_1)$ be the number of branches incident with exactly one vertex in V_1. The minimum n_1-degree $\delta(n_1)$ is*

$$\delta(n_1) = \underset{V_1:|V_1|=n_1}{\text{Min}} [d(V_1)]. \tag{7.10.1}$$

$\delta(n_1)$ is the minimum number of branches which must be removed from a graph to separate exactly n_1 vertices from the remaining vertices.

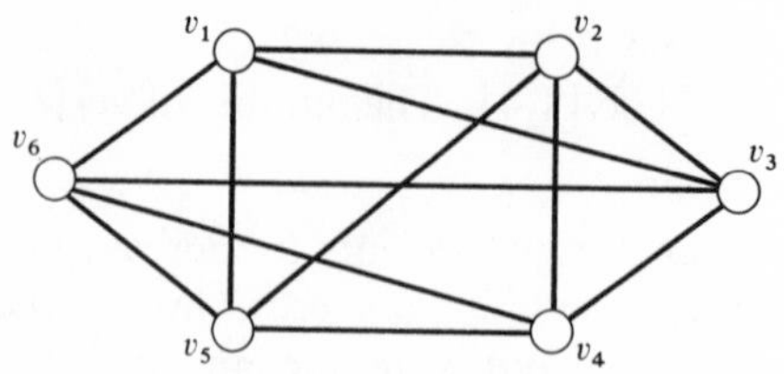

Fig. 7.10.1 Graph for Examples 7.10.1 and 7.10.2.

We now attempt to construct graphs which maximize $\delta(n_1)$ for a given n_1. Clearly $\delta(n_1) = \delta(n - n_1)$. Hence we can assume $n_1 \geq n/2$. We will say that G is *a homogeneous graph of degree* α if all its vertices have degree α.

***Definition* 7.10.2** *Let v_j be a vertex in V_1. The external degree $d_e(V_1(j))$ of v_j is equal to the number of branches incident with v_j and with no other vertex in V_1. The internal degree $d_i(V_1(j))$ of v_j is equal to the number of branches incident with v_j and another vertex in V_1. Thus for $v_j \in V_1$*

$$d(j) = d_i(V_1(j)) + d_e(V_1(j)). \tag{7.10.2}$$

Example 7.10.1 The graph G in Fig. 7.10.1 is a homogeneous graph of degree 4. Let $n_1 = 3$ and $V_1 = \{v_3,v_4,v_5\}$. Then $d(V_1) = d(\bar{V}_1) = 8$. $\delta(3)$ is given by

$$\delta(3) = d(\{v_1,v_2,v_3\}) = 6 = \operatorname*{Min}_{V_1:|V_1|=n_1} [d(V_1)]. \tag{7.10.3}$$

From the definitions of external and internal degrees, for $V_1 = \{v_3,v_4,v_5\}$,

$$d_e(V_1(3)) = 3, \tag{7.10.4a}$$

$$d_e(V_1(4)) = 2, \tag{7.10.4b}$$

$$d_e(V_1(5)) = 3, \tag{7.10.4c}$$

$$d_i(V_1(3)) = 1, \tag{7.10.4d}$$

$$d_i(V_1(4)) = 2, \tag{7.10.4e}$$

$$d_i(V_1(5)) = 1. \tag{7.10.4f}$$

We first derive a simple lower bound on $\delta(n_1)$ for a homogeneous graph.

***Lemma* 7.10.1** *Let G be a homogeneous graph of degree α with $V_1 \subset V, |V_1| = n_1$, and $V_2 = \bar{V}_1, |V_2| = n_2$. Then*

$$\delta(n_1) \geq n_2(\alpha - n_2 + 1). \tag{7.10.5}$$

Proof.

$$d(V_1) = \sum_{v_j \in V_2} d_e(V_2(j)). \tag{7.10.6}$$

Hence

$$d(V_1) = \sum_{v_j \in V_2} [\alpha - d_i(V_2(j))] \tag{7.10.7}$$

or

$$d(V_1) = \alpha n_2 - \sum_{v_j \in V_2} d_i(V_2(j)). \tag{7.10.8}$$

But

$$\sum_{v_j \in V_2} d_i(V_2(j))$$

is equal to twice the number of branches with both end vertices in V_2. Hence

$$\sum_{v_j \in V_2} d_i(V_2(j)) \leq n_2(n_2 - 1). \tag{7.10.9}$$

Combining (7.10.8) and (7.10.9) yields

$$\delta(V_1) \geq \alpha n_2 - n_2(n_2 - 1) = n_2(\alpha - n_2 + 1). \tag{7.10.10}$$

Example 7.10.2 For the graph in Fig. 7.10.1, (7.10.5) yields

$$\delta(1) \geq 5(4 - 5 + 1) = 0, \tag{7.10.11a}$$

$$\delta(2) \geq 4(4 - 4 + 1) = 4, \tag{7.10.11b}$$

$$\delta(3) \geq 3(4 - 3 + 1) = 6, \tag{7.10.11c}$$

$$\delta(4) \geq 2(4 - 2 + 1) = 6, \tag{7.10.11d}$$

$$\delta(5) \geq 1(4 - 1 + 1) = 4, \tag{7.10.11e}$$

$$\delta(6) \geq 0(4 - 0 + 1) = 0. \tag{7.10.11f}$$

In Theorem 7.10.2 we derive an upper bound on $\delta(n_1)$ for a homogeneous graph. As an intermediate step we require the following lemma.

***Lemma* 7.10.2** *Let G be a homogeneous graph of degree α with $V_1 \subset V$ and $d(V_1) = \delta(n_1)$. Then for $v_i \in V_1$ and $v_j \in V_2 = \bar{V}_1$,*

$$\alpha \geq d_e(V_1(i)) + d_e(V_2(j)) \qquad \textit{if } v_i \textit{ and } v_j \textit{ are not adjacent}, \tag{7.10.12}$$

$$2 + \alpha \geq d_e(V_1(i)) + d_e(V_2(j)) \qquad \textit{if } v_i \textit{ and } v_j \textit{ are adjacent}. \tag{7.10.13}$$

Proof. Consider the set $\hat{V}_1$ defined by

$$\hat{V}_1 = \{v_j\} \cup V_1 - \{v_i\}. \tag{7.10.14}$$

If v_i and v_j are not adjacent,

$$d(\hat{V}_1) = \delta(n_1) - d_e(V_1(i)) - d_e(V_2(j)) + d_i(V_1(i)) + d_i(V_2(j)) \tag{7.10.15}$$

or

$$d(\hat{V}_1) - \delta(n_1) = d_i(V_1(i)) + d_i(V_2(j)) - d_e(V_1(i)) - d_e(V_2(j)). \tag{7.10.16}$$

From (7.10.14),

$$|\hat{V}_1| = n_1. \tag{7.10.17}$$

Hence

$$d(\hat{V}_1) \geq \delta(n_1), \tag{7.10.18}$$

and (7.10.16) and (7.10.18) yield

$$d_i(V_1(i)) - d_e(V_1(i)) + d_i(V_2(j)) - d_e(V_2(j)) \geq 0. \tag{7.10.19}$$

From (7.10.2),

$$d_i(V_1(i)) = \alpha - d_e(V_1(i)) \tag{7.10.20}$$

and

$$d_i(V_2(j)) = \alpha - d_e(V_2(j)). \tag{7.10.21}$$

Substituting (7.10.20) and (7.10.21) into (7.10.19) yields

$$\alpha \geq d_e(V_1(i)) + d_e(V_2(j)). \tag{7.10.22}$$

If v_i and v_j are adjacent,

$$d(\hat{V}_1) = \delta(n_1) - d_e(V_1(i)) - d_e(V_2(j)) + d_i(V_1(i)) + d_i(V_2(j)) + 2 \tag{7.10.23}$$

and we can prove (7.10.13).

Theorem 7.10.1 *Let G be a connected homogeneous graph of degree α with $\alpha \leq n/2$ and $n > 2$ and let $n_1 \geq n_2 = n - n_1$. Then*

$$\alpha \geq [\delta(n_1)/n_1] + [\delta(n_1)/n_2]^* \tag{7.10.24}$$

where $[x]^*$ denotes the smallest integer greater than or equal to x.

Proof. Consider a set of n_1 vertices V_1 such that

$$d(V_1) = \delta(V_1)$$

and let $V_2 = \bar{V}_1$. The average of the external degrees of vertices in V_2 is $\delta(n_1)/n_2$. Thus, there is a v_j in V_2 with

$$d_e(V_2(j)) \geq [\delta(n_1)/n_2]^*. \tag{7.10.25}$$

CASE 1. Suppose $v_i \in V_1$ such that

$$d_e(V_1(i)) \geq [\delta(n_1)/n_1] + 1. \tag{7.10.26}$$

If v_i and v_j are adjacent, (7.10.25), (7.10.26), and Lemma 7.10.2 yield

$$1 + \alpha \geq [\delta(n_1)/n_1] + 1 + [\delta(n_1)/n_2]^*. \tag{7.10.27}$$

If v_i and v_j are not adjacent,

$$\alpha \geq [\delta(n_1)/n_1] + 1 + [\delta(n_1)/n_2]^* > [\delta(n_1)/n_1] + [\delta(n_1)/n_2]^*. \tag{7.10.28}$$

CASE 2. Assume for all $v_i \in V_1$

$$d_e(V_1(i)) = [\delta(n_1)/n_1] = \delta(n_1)/n_1. \tag{7.10.29}$$

Hence, $\delta(n_1)/n_1$ is an integer. If $v_k \in V_1$ such that $[k,j] \notin \Gamma$,

$$d_e(V_1(k)) = [\delta(n_1)/n_1] \tag{7.10.30}$$

and

$$d_e(V_2(j)) \geq [\delta(n_1)/n_2]^*. \tag{7.10.31}$$

Combining (7.10.30) and (7.10.31) with Lemma 7.10.2 again gives the theorem. Suppose v_j is adjacent to each vertex of V_1. Then

$$d_e(V_2(j)) = n_1. \tag{7.10.32}$$

By definition,

$$n_1 \geq n/2 \geq \alpha \geq d_e(V_2(j)). \tag{7.10.33}$$

From (7.10.32) and (7.10.33),

$$n_1 = n/2 = \alpha = d_e(V_2(j)). \tag{7.10.34}$$

Furthermore, since v_i and v_j are adjacent,

$$1 + n_1 = 1 + \alpha \geq d_e(V_1(i)) + d_e(V_2(j)) = \frac{\delta(n_1)}{n_1} + n_1 \tag{7.10.35}$$

or

$$\frac{\delta(n_1)}{n_1} \leq 1. \tag{7.10.36}$$

Since the graph is connected, $\delta(n_1) > 0$ and since $\delta(n_1)/n_1$ is an integer,

$$\delta(n_1) = n_1. \tag{7.10.37}$$

Now consider any $v_l \in V_2$ with $l \neq j$. Since $d_e(V_2(j)) = n_1$, $d_e(V_2(l)) = 0$. Furthermore, since $n_1 = m$, $d_i(V_2(j)) = 0$. Hence, any other vertex $v_l \in V_2$ is not adjacent to v_j. Since the graph is connected, $V_2 - \{v_j\}$ is empty and $n_2 = n_1 = 1$ or $n = 2$. This contradicts the hypothesis and the proof is completed.

Example 7.10.2 The statement of Theorem 7.10.1 excludes $n = 2$, since for $\alpha = m = 1 = n_1$ we have $\delta(n_1) = 1$. For G in Fig. 7.10.1, with $n_1 = 3$, (7.10.24) yields

$$4 \geq \left[\frac{\delta(3)}{3}\right] + \left[\frac{\delta(3)}{3}\right]^*. \tag{7.10.38}$$

Trying several values of $\delta(3)$ in (7.10.38), we see that $\delta(3) = 5$ yields the inequality $4 \geq 3$, $\delta(3) = 6$ yields $4 \geq 4$, and $\delta(3) = 7$ is impossible. Hence,

$$\delta(3) \leq 6. \tag{7.10.39}$$

Combining (7.10.11c) with (7.10.39), for the graph in Fig. 7.10.1, we get

$$\delta(3) = 6. \tag{7.10.40}$$

For the special case $\alpha = n/2$, we now derive a stronger upper bound on $\delta(n_1)$ and show that for this case we can always construct a graph for which this upper bound is satisfied with equality.

***Theorem* 7.10.2** *For a homogeneous graph of degree* $\alpha = n/2$

$$\delta(n_1) \leq \left[\frac{n_1 n_2}{2}\right]^*. \tag{7.10.41}$$

Proof. Since n is even, n_1 and n_2 are either both even or both odd.

CASE 1. Suppose for some positive integers q and l, $n_1 = 2q$ and $n_2 = 2l$. Then

$$\left[\frac{n_1 n_2}{2}\right]^* = 2ql \tag{7.10.42}$$

and

$$\alpha = n/2 = q + l. \tag{7.10.43}$$

Let

$$\delta(n_1) = 2ql + \beta. \tag{7.10.44}$$

Assume $\beta \geq 1$. The average external degree of a vertex in V_1 is

$$l + (\beta/2)q \tag{7.10.45}$$

and the average external degree of a vertex in V_2 is

$$q + (\beta/2)l. \tag{7.10.46}$$

Hence there exists a vertex $v_i \in V_1$ such that

$$d_e(V_1(i)) \geq l + 1 \tag{7.10.47}$$

and there exists a vertex $v_j \in V_2$ such that

$$d_e(V_2(j)) \geq q + 1. \tag{7.10.48}$$

If v_i and v_j are adjacent, combining (7.10.43), (7.10.47), and (7.10.48) with Lemma 7.10.2 gives

$$1 + q + l \geq 2 + q + l. \tag{7.10.49}$$

If v_i and v_j are not adjacent,

$$q + l \geq q + l + 2, \tag{7.10.50}$$

Hence $\beta < 1$. Since β is an integer, $\beta = 0$ and (7.10.44) yields

$$\delta(n_1) \leq 2ql = \left[\frac{n_1 n_2}{2}\right]^*. \tag{7.10.51}$$

CASE 2. Suppose for some integers q and l, $n_1 = 2q + 1$ and $n_2 = 2l + 1$. Then

$$\alpha = n/2 = q + l + 1 \tag{7.10.52}$$

and

$$\left[\frac{n_1 n_2}{2}\right]^* = 2ql + q + l + 1. \tag{7.10.53}$$

Let

$$\delta(n_1) = 2ql + q + l + k \tag{7.10.54}$$

and assume $k \geq 2$. The average of the external degrees of the vertices in V_1 is

$$l + (q + k)/(2q + 1) \tag{7.10.55}$$

and the average of the external degrees of the vertices in V_2 is

$$q + \frac{l + k}{2l + 1}. \tag{7.10.56}$$

Since $k \geq 2$,

$$1/2 < \frac{q + k}{2q + 1} \leq k. \tag{7.10.57}$$

Thus there exists $v_j \in V_2$ such that

$$d_e(V_2(j)) \geq q + 1. \tag{7.10.58}$$

Suppose there exists $v_i \in V_1$ such that

$$d_e(V_1(i)) \geq l + 2. \tag{7.10.59}$$

If v_i and v_j are adjacent, using (7.10.52), (7.10.58), and (7.10.59) and Lemma 7.10.2 yields

$$q + l + 2 \geq q + l + 3. \tag{7.10.60}$$

If v_i and v_j are not adjacent, Lemma 7.10.2 yields

$$q + l + 1 \geq q + l + 3. \tag{7.10.61}$$

The only remaining possibility is that for all $v_i \in V_1$

$$d_e(V_1(i)) \leq l + 1. \tag{7.10.62}$$

Similarly, we can show for all $v_j \in V_2$

$$d_e(V_2(j)) \leq q + 1. \tag{7.10.63}$$

Suppose there are r vertices in V_1 of external degree $l + 1$ and suppose there are x vertices in V_2 of external degree $q + 1$. Each of these r vertices in V_1 must be adjacent to each of these x vertices in V_2 since, otherwise,

$$q + l + 1 \geq q + l + 2. \tag{7.10.64}$$

Therefore,

$$d(V_1) = q + l + 2ql + k = \sum_{i=1}^{2q+1} d_e(V_1(i)) = r(l + 1) + \sum_{j:d(V_1(j)) \leq l} d_e(V_1(j)). \tag{7.10.65}$$

But

$$\sum_{j:d_e(V_1(j))\leq l} d_e(V_1(j)) \leq (2q + 1 - r)l. \tag{7.10.66}$$

Equations (7.10.65) and (7.10.66) give

$$q + l + 2ql + k \leq r(l + 1) + (2q + 1 - r)l \tag{7.10.67}$$

or

$$r \geq q + k \geq q + 2. \tag{7.10.68}$$

Similarly, we can show that

$$x \geq l + k \geq l + 2. \tag{7.10.69}$$

However, each of the vertices of V_1 is adjacent to each of the x vertices of V_2. Hence

$$l + 1 \geq x. \tag{7.10.70}$$

But

$$x \geq l + 2. \tag{7.10.71}$$

Hence, there is a contradiction and $k \leq 1$.

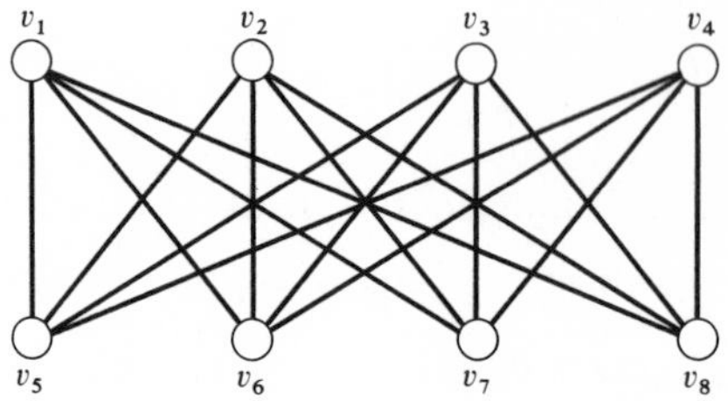

Fig. 7.10.2 Graph for Examples 7.10.3, 7.10.4, and 7.10.5.

Example 7.10.3 For the graph G in Fig. 7.10.2, Theorem 7.10.2 yields

$$\text{for } n_1 = 1, \qquad \delta(1) \leq 4, \tag{7.10.72a}$$

$$\text{for } n_1 = 2, \qquad \delta(2) \leq 6, \tag{7.10.72b}$$

$$\text{for } n_1 = 3, \qquad \delta(3) \leq 8, \tag{7.10.72c}$$

$$\text{for } n_1 = 4, \qquad \delta(4) \leq 8. \tag{7.10.72d}$$

We now show that for a class of complete bipartite graphs, (7.10.41) is satisfied with equality. Consider a complete bipartite graph $G_0 = (V,\Gamma)$ with $V = A \cup B$, $A \cap B = \phi$, $|A| = |B|$, and $\Gamma = [A,B]$. Let V_1 be an arbitrary set of n_1 vertices and let $A' = A \cap V_1$, $B' = B \cap V_1$, $a = |A'|$, and $b = |B'|$. Hence

$$a + b = |V_1| = n_1. \tag{7.10.73}$$

If every vertex has degree α then the maximum value of $d_e(V_1)$ is an αn_1, which occurs only if every vertex in V_1 is adjacent to every vertex in $\bar{V}_1$. For each branch with both endpoints in V_1 this number is reduced by two. In the complete bipartite graph there are ab branches between vertices in A' and vertices in B'. Hence

$$d_e(V_1) = \alpha n_1 - 2ab. \tag{7.10.74}$$

Thus

$$\delta(n_1) = \min_{a+b=n_1} [\alpha n_1 - 2ab]. \tag{7.10.75}$$

The minimum of $\delta(n_1)$ occurs where the product ab is maximum. If n_1 is even this occurs when $a = b = n_1/2$. In this case setting $\alpha = n/2$ gives

$$\delta(n_1) = \frac{n}{2}(n_1) - 2\left(\frac{n_1}{2}\right)^2 = \frac{n_1}{2}(n - n_1) = \frac{n_1 n_2}{2}$$

where $n_2 = n - n_1$.

In general for any value of n_1

$$\delta(n_1) = \left[\frac{n_1 n_2}{2}\right]^*. \tag{7.10.76}$$

Example 7.10.4 For the graph in Fig. 7.10.2,

$$\delta(1) = 4 = \left[\frac{1 \times 7}{2}\right]^*,$$

$$\delta(2) = 6 = \left[\frac{2 \times 6}{2}\right]^*,$$

$$\delta(3) = 8 = \left[\frac{3 \times 5}{2}\right]^*,$$

$$\delta(4) = 8 = \left[\frac{4 \times 4}{2}\right]^*.$$

Example 7.10.5 We finally show by means of a counter-example that (7.10.24) and (7.10.41) are independent, i.e., that

$$\frac{n}{2} \geq \left[\frac{\delta(n_1)}{n_1}\right] + \left[\frac{\delta(n_1)}{n_2}\right]^*$$

does not imply

$$\delta(n_1) < \left[\frac{n_1 n_2}{2}\right]^*$$

Consider $n = 10$, $\alpha = 5$, and $n_1 = n_2 = 5$. Then (7.10.24) yields

$$5 \geq [\delta(5)/5] + [\delta(5)/5]^* \text{ or } \delta(5) \leq 14.$$

However,

$$\left[\frac{n_1 n_2}{2}\right]^* = 13.$$

Hence $\delta(n_1)$ may satisfy (7.10.24) and not (7.10.41).

7.11 FURTHER REMARKS

The results of Sections 2, 3, and 4 form the basis of a computer program to determine v for a graph [FR4, FR6]. The result that $v = \omega$ is due to Boesch and Frisch [BO1]. Theorem 7.3.1 can be found in Berge [BE1]. The $\omega_{s,t}$ Labeling Algorithm and the Flow Variation Algorithm are due to Frisch [FR2, FR3]. Theorem 7.4.1, which reduces the number of pairs of v_s and v_t for which $\omega_{s,t}$ must be found is due to Steiglitz *et al.* [ST1]. Theorems 7.4.2 to 7.4.4 are due to Kleitman [KL1].

The necessary and sufficient conditions for the realizability of a set of integers as the degrees of a graph have been obtained by several authors with varying degrees of generality [BE1, FU2, HA1, HA2]. The necessary and sufficient conditions in the form we have given them were derived by Hakimi [HA1, HA2]. The Branch Exchange Algorithm was derived by Edmonds [ED1] although a similar technique has been used by Berge [BE1] and Hakimi [HA2] in realizing the degrees of a graph. The fact that Min $[d(i)] > [n/2]$ guarantees $\theta =$ Min $[d(i)]$ was proved by Chartrand [CH1]. Fulkerson and Shapley [FU1] have given an alternative method of constructing a graph with a maximum value of θ. Their process is iterative with the number of vertices and branches being increased at each step. Hence the procedure is not applicable to improving partially constructed systems unless one is willing to increase the number of vertices as well as the number of branches. In the case in which parallel branches are permitted, Frisch [FR5] has given a method for maximizing θ.

The results in Section 6 are due to Frank and Chou [FR1]. If parallel branches are permitted, the problem is considerably simpler and can be reduced to realizing terminal capacity matrices [CH2]. The results in Section 7 were first stated by Rosenbaum and Friedman [RO1] and by Harary [HA4]. A proof enumerating the vertex disjoint paths was first given for even v by Boesch and Thomas [BO2] and for odd v by Hakimi [HA3]. The proof given here is a combination of the proofs in references BO2, HA3, and HA4. The results on bipartite graphs are due to Boesch and Thomas [BO2], as are the results in Section 10 on the minimum n_1 degree. The proof of Theorem 7.8.1 is due to Felzer and Boesch [FE1]. The results on bipartite graphs have been extended to "k-partite" graphs by Felzer and Boesch [FE2].

The idea of adding shortest Hamilton circuits was introduced by Frisch [FR6] using the solution to the Traveling Salesman Problem by Lin [LI1]. The approximate scheme to realize a graph with $\omega_{i,j} \geq r_{i,j}$ at minimum cost and Theorem 7.9.1 are due to Steiglitz *et al.* [ST1]. In reference ST1 there is a detailed description of a computer program to generate random starting graphs and find approximately minimum cost realizations by means of branch exchanges.

PROBLEMS

1. Using the $\omega_{s,t}$ Labeling Algorithm, find $\omega_{s,t}$ for the graph in Fig. P7.1.

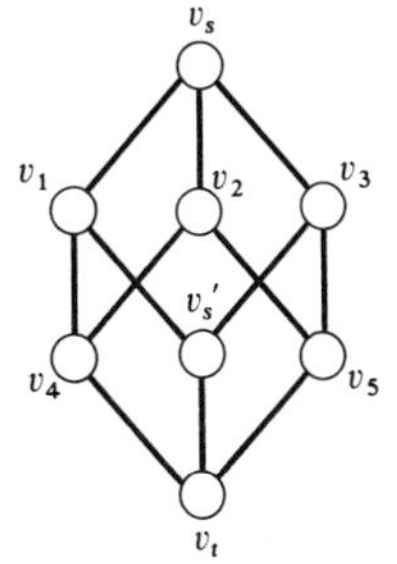

Figure P7.1

2. Using the Flow Variation Algorithm, find $\omega_{s',t}$ for the graph shown in Fig. P7.1.
3. For the directed graph below, use the $\omega_{s,t}$ Labeling Algorithm to find $\omega_{s,t}$. Begin with the initial flow pattern indicated by the branch weights.

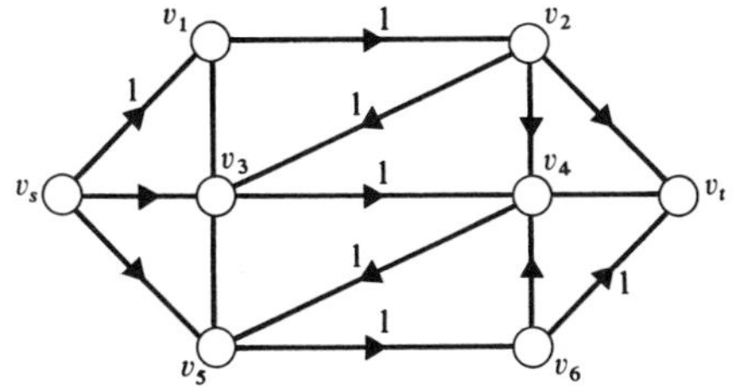

Figure P7.2

4. In the directed graph in Fig. P7.3 the first weight on each branch represents capacity. The second weight represents the branch flow which maximizes $f_{s,t}$. Use flow variation to maximize $f_{s',t}$.

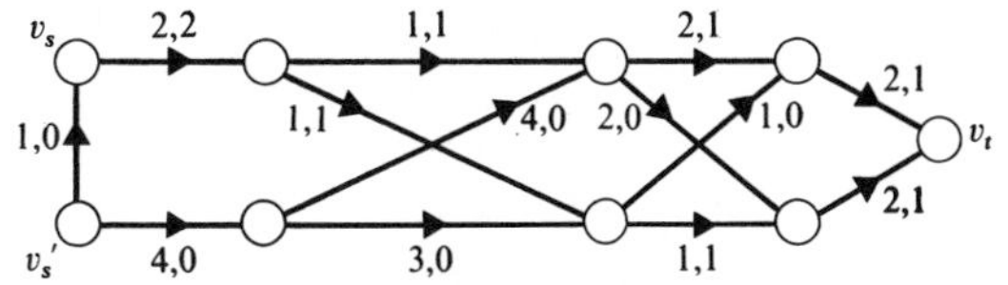

Figure P7.3

5. Consider a graph with 500 vertices. Indicate a procedure for determining whether $\omega \geq 30$. This procedure should use at most 1965 applications of the $\omega_{s,t}$ Labeling Algorithm rather than the maximum of $\dfrac{500 \times 499}{2}$ applications.

6. Determine whether the following sets of numbers are realizable as the degrees of the vertices of a graph. If they are, construct a graph which realizes them.

   a) $\{6,3,5,4,5,4,4,3\}$
   b) $\{7,3,6,5,8,4,3,3,2\}$

7. Synthesize a graph with $\theta = 3$ and vertex degrees specified by $\{5,5,5,4,4,4,4,3\}$.

8. Starting with the disconnected graph in Fig. P7.4 use branch exchanges to obtain a new graph with the same degrees as the original graph, but with θ equal to 6.

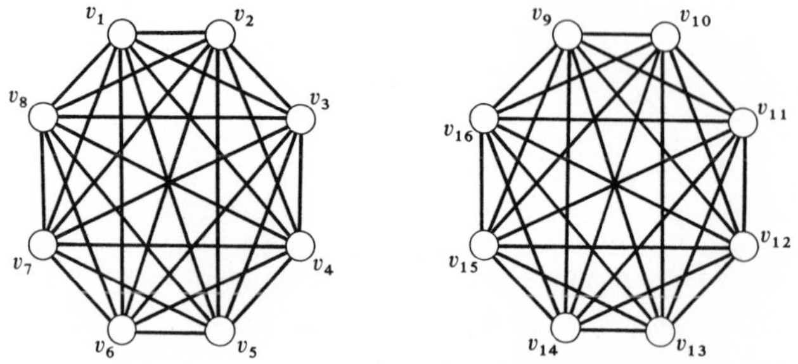

Figure P7.4

9. Consider two undirected graphs G_1 and G_2. Form a new graph G by connecting each vertex in G_1 to at least one vertex in G_2 and each vertex in G_2 to at least one vertex in G_1. If the smallest vertex cut-set in G_1 and in G_2 has ω vertices, then prove that the smallest vertex cut-set in G has at least $(\omega + 1)$ vertices [BU1].

10. Prove that if each vertex of an undirected graph G has degree at least k, then any branch of G is contained in a set of $[(k + 1)/2]$ branches, such that no two branches in the set are incident with a common vertex.

11. Using the methods of Section 7, construct graphs with $\omega = [2m/n]$ where m is the number of branches and n is the number of vertices, for

    a) $m = 18$ and $n = 9$
    b) $m = 17$ and $n = 10$
    c) $m = 17$ and $n = 9$
    d) $m = 18$ and $n = 10$

12. For the graphs formed in the last problem enumerate $\omega = [2m/n]$ vertex-disjoint paths between two arbitrary nonadjacent vertices.

13. Construct a bipartite graph with 10 vertices, a minimum number of branches, and $\omega = 4$.

14. Construct a bipartite graph with 11 vertices, a minimum number of branches, and $\omega = 5$.

15. Prove that Theorem 7.5.3 is a consequence of Lemma 7.10.1.

16. Prove Theorem 7.7.1 without enumerating vertex-disjoint paths as in (7.7.4) and (7.7.5). [*Hint:* Use Theorem 7.7.2.]

17. Use Lemma 7.10.1 to calculate lower bounds on $\delta(i)$ for $i = 1,\ldots,7$ for the graph below.

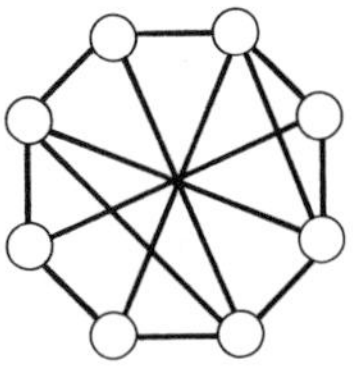

Figure P7.5

Find the actual value of $\delta(i)$ for $i = 1,\ldots,7$.

18. Repeat the previous problem using Theorem 7.10.1 rather than Lemma 7.10.1.

19. Calculate an upper bound on $\delta(i)$ for $i = 1,2,10,50,100$ for a homogeneous graph of degree 100 with 200 vertices.

20. Let v be the minimum number of elements in a directed graph whose removal breaks all directed paths between at least one remaining pair of vertices. Prove

$$v \leq \left[\frac{m}{n}\right].$$

21. Let v be as defined in Problem 20. Prove that, if for a directed graph $v = m/n$, then for every vertex v_i, $d^+(i) = d^-(i) = m/n$.

22. Let v be as defined in Problem 20. Given n vertices and a specified integer $v = k \leq [(n-1)/2]$ prove that the graph G resulting from the following algorithm realizes v and n with $v = e/n$.
   Let $V = \{v_1,\ldots,v_n\}$ be the set of vertices. Let $t = [(n-1)/2]$. Let $G = (V,\Gamma)$ be the directed graph constructed as follows.
   For all i add branches

$$(i,m_j)$$

where

$$m_j = (i + j) \bmod n, j = 1,\ldots,k, \text{ and } i = 1,\ldots,n.$$

[AY1]

REFERENCES

AY1 J. N. Ayoub and I. T. Frisch, "Optimally Invulnerable Directed Communication Networks," *IEEE Trans. Commun. Technol.* **COM-18,** 484–489 (1970).

BE1 C. Berge, *The Theory of Graphs and Its Applications,* Chapter 9, Wiley, New York, 1962.

BO1 F. T. Boesch and I. T. Frisch, "On the Smallest Disconnecting Set in a Graph," *IEEE Trans. Circuit Theory* **CT-15,** 286–288 (1968).

BO2 F. T. Boesch and R. E. Thomas, "Optimal Damage Resistant Communication Nets," *IEEE Trans. Circuit Theory* **CT-17,** 183–192 (1970).

BU1 D. E. Butler, "Communication Nets with Specified Survivability," Coordinated Science Laboratory Report R-359, University of Illinois, Urbana, July 1967.

CH1 Chartrand, "A Graph Theoretic Approach to the Communication Problem," *SIAM J. Appl. Math.* **14,** 778–781 (1966).

CH2 W. Chou and H. Frank, "Survivable Communication Networks and the Terminal Capacity Matrix," *IEEE Trans. Circuit Theory,* **CT-17,** 192–197 (1970).

ED1 J. Edmonds, "Existence of k-Edge Connected Ordinary Graphs with Prescribed Degrees," *J. Res. Natl Bur. Std B* **68B,** 73–74 (1964).

FE1 W. Felzer and F. T. Boesch, "A New Class of Optimally Invulnerable Graphs," *SIAM J. Appl. Math.,* in press (1971).

FE2 W. Felzer and F. T. Boesch, "Optimally Invulnerable k-Partite Graphs," to be published.

FR1 H. Frank and W. Chou, "Connectivity Considerations in the Design of Survivable Networks," *IEEE Trans. Circuit Theory* **CT-17**, 486–490 (1970).

FR2 I. T. Frisch, "Flow Variation in Multiterminal Min-Cut Calculations," *J. Franklin Inst.* **287,** 61–72 (1969).

FR3 I. T. Frisch, "An Algorithm for Vertex Pair Connectivity," *Intern. J. Control* **6,** 579–593 (1967).

FR4 I. T. Frisch, "Analysis of the Vulnerability of Communication Nets," in Proceedings of the First Annual Princeton Conference on Systems Science, Princeton, 1967, pp. 188–192.

FR5 I. T. Frisch, "Optimization of Communication Nets with Switching," *J. Franklin Inst.* **275,** 405–430 (1963).

FR6 I. T. Frisch, "A Computer Technique for Design and Analysis of Hard Communication Systems," *Digest of the 1967 IEEE International Conference on Communications, Minneapolis,* p. 96 (1967).

FU1 D. R. Fulkerson and L. S. Shapley, "Minimal k-Arc Connected Graphs," *RAND Report* **P-2371** (1961).

FU2 D. R. Fulkerson, A. J. Hoffman, and H. McAndrew, "Some Properties of Graphs with Multiple Edges," *Can. J. Math.* **17,** 166–177 (1965).

HA1 S. I. Hakimi, "On the Realizability of Integers as the Degrees of the Vertices of a Linear Graph—I," *J. Soc. Ind. Appl. Math.* **10,** 496–506 (1962).

HA2 S. L. Hakimi, "On the Realizability of a Set of Integers as the Degrees of the Vertices of a Linear Graph—II Uniqueness," *J. Soc. Ind. Appl. Math.* **11,** 135–147 (1963).

HA3 S. L. Hakimi, "An Algorithm for Construction of the Least Vulnerable Communication Network," *IEEE Trans. Circuit Theory* **CT-16,** 229–230 (1969).

HA4 F. Harary, "The Maximum Connectivity of a Graph," *Proc. Natl Acad. Sci. U.S.* **48,** 1142–1146 (1962).

KL1 D. J. Kleitman, "Methods for Investigating Connectivity of Large Graphs," *IEEE Trans. Circuit Theory* **CT-16,** 232–233 (1969).

LI1 S. Lin, "Computer Solutions of the Travelling Salesman Problem," *Bell System Tech. J.* **44,** 2245–69 (1965).

RO1 D. M. Rosenbaum and J. B. Friedman, "Redundant Networks," MITRE Technical Memorandum TM-3195, October 27, 1961.

ST1 K. Steiglitz, P. Weiner, and D. J. Kleitman, "The Design of Minimum-Cost Survivable Networks," *IEEE Trans. Circuit Theory* **CT-16,** 455–460 (1969).

VI1 I. M. Vinigradov, *Elements of Number Theory,* Dover Publications, New York, 1954.

CHAPTER 8

CONNECTIVITY AND VULNERABILITY OF PROBABILISTIC GRAPHS

8.1 RANDOM MODELS OF NETWORKS

In any network, random factors may affect performance. One example is the problem of survival of a network under enemy attack. The network may represent a system consisting of railroads, highways, bridges, and terminals; a system composed of power-generating stations, transformer relay points, and power transmission lines; or a radio communication system of transmitters and receivers. An enemy attack could be a bombing raid on the stations and links of the system or a "jamming" of communication between selected pairs of stations. Consequently, if the network is represented by a graph G, an attack could result in the removal of a subset of branches and vertices from G.

We would like to determine the probability that the network "survives" such an attack. As we have seen, "survivability" may be defined in a number of ways. We could consider the network to have survived if, after attack, a key pair of stations are still able to communicate; i.e., at least one path still exists between these points. Alternatively, we might say that the network has survived if, after attack, *all* pairs of stations may communicate with each other. This means that the network survives if the corresponding graph remains connected. With these definitions of survivability, we would want to know either the probability that there is a path in G between a pair of given points or the probability that G is connected.

If the network is disconnected, it may be important to know into how many sections it has been divided. Hence, the expected number of components or, more generally, the probability distribution of the number of components must be determined. Another quantity of interest could be the probability distribution or the expected number of stations which can communicate after an attack.

It is not necessary for the system to be subjected to a hostile environment to exhibit the above uncertainties. If the vertices of the graph represent central telephone offices, some of the interconnecting lines may be busy. Suppose a call can be routed over any idle line in the network. Then, given probabilities of various lines being idle, we may want to find the probability that a specified office can call another specified office. Furthermore, the probability that an appropriately defined graph is connected gives the probability that *every* office can call *every other* office.

Random graphs can model problems in a wide variety of disciplines. Consider the following problem in the theory of epidemics. A number of individuals in a closed environment contract a contagious disease which will either kill them or make them immune within a finite time. The disease may be transmitted to other members of the group with some probability. It is then desirable to know the expected number of individuals who will contract the disease within a specified time and the expected number of individuals who will eventually contract the disease. Other examples of random graphs may be taken from the theory of neural nets or from certain sociological problems concerning the relations between individuals in a group.

The basic structure of a random graph is described by specifying a set of vertices $\{v_1,v_2,\ldots,v_n\}$ and a set of elementary events $\{\mathscr{E}_{i,j};\ i,j = 1,\ldots,n\}$; $\mathscr{E}_{i,j}$ is the event that v_i is connected to v_j by a directed branch from v_i to v_j. Hence, there are n^2 elementary events or, if the random graph under consideration is undirected, there are $n(n+1)/2$ elementary events. The random nature of the graph is exhibited by specifying Prob $\{\mathscr{E}_{i,j}\}$ for each event, or joint probability distributions if the events are not independent. If the events are independent and equiprobable, then we will say that the graph is *unbiased*; otherwise, the graph is said to be *biased*.

In this chapter we are primarily concerned with problems of connectivity for random graphs. The connectivity constant ρ of a random graph G is a random number with values ranging between 1 and n. If $\rho = k$, then there are k components in G. Often we will be interested in finding $E\{\rho\}$, the expected value of ρ; Prob $\{\rho = 1\}$, the probability that G is connected; or the probability distribution of ρ. Other quantities of interest are related to the number of vertices which can be reached from a given point. Let $\delta(v_i)$ be a random number corresponding to the number of descendants of vertex v_i. In many cases we would like to find $E\{\delta(v_i)\}$ and the probability distribution of $\delta(v_i)$. Related to these numbers is the weak connectivity constant $\gamma(v_i)$ of G; $\gamma(v_i)$ is defined to be the expected fraction of vertices of G which are descendants of v_i. Hence $\gamma(v_i) = E\{\delta(v_i)/n\}$ and, clearly, if the graph is unbiased, $\gamma(v_i) = \gamma(v_j) = \gamma$ for all i and j.

In Section 2, we investigate the behavior of the above quantities for a restricted class of directed graphs defined by random mapping functions. These graphs have exactly one outwardly directed branch at each vertex. In Section 3 the class of directed s-graphs with equal outward demi-degrees is studied. Expressions for $\delta(v_i)$ and other quantities of interest are derived. In Section 4, several classes of homogeneous graphs are considered, including the complete graph in which every branch exists with probability p. Some interesting asymptotic properties of "evolving" graphs are also given. Section 5 discusses similar problems for biased graphs and examines several kinds of bias. In Section 6, "percolation" processes on *infinite* graphs are considered, as well as the problem of determining the nature of the spread of an excitation through a graph whose branches are randomly blocked. "Critical" probabilities are shown to exist and lower and upper bounds to these probabilities are given. In Section 7, the results developed in the preceding sections are applied to the problem of determining the vulnerability of a network to an attack and designing networks which can "survive" attack.

8.2 RANDOM MAPPING FUNCTIONS

To begin our treatment of random graphs, we shall examine a special case which is, in itself, of considerable interest. A directed graph $G = (V,\Gamma)$ consists of a set of n vertices such that every vertex v_i in V is adjacent to each of the vertices of the set $\Gamma(v_i)$. In general, Γ is a multivalued relation; in this section we will assume that Γ is a single-valued mapping function. Thus, the *outward demi-degree* of each vertex will be unity. Graphs constructed from *random* single-valued mapping functions will be called *outwardly homogeneous degree-one* graphs. Our reason for examining these graphs, aside from their inherent interest, is to demonstrate the nature of connectivity problems and to illustrate a number of techniques for solving these problems.

An example of an outwardly homogeneous graph of degree one is G in Fig. 8.2.1. Such graphs play an important role in the theory of relations, where, for example, each of a group of n people might be asked to name his best friend. In Fig. 8.2.1, the subgraphs G_1 and G_2 would then represent two "cliques." As another example, consider an n-state deterministic sequential machine. Each input of the machine defines an n-vertex state transition diagram such that the outward demi-degree of each vertex is unity. A third example consists of a network routing scheme in which information can arrive at a vertex from a variety of locations but all outgoing messages are routed over a single line.

We now assume that the existence of directed branches is uncertain. We will assume that the $\mathscr{E}_{i,j}$ are equiprobable and independent and the random graph under consideration is unbiased. Clearly, Prob $\{\mathscr{E}_{i,j}\} = 1/n$ for all i and j. Note that in this model we allow *self-loops*; i.e., the event $\mathscr{E}_{i,i}$ has nonzero probability. As an example, suppose the random graph represents a sequential machine. Then, if the machine is in any state and any input is applied, the machine will transmit to any other state with probability $1/n$. In the problem of individual relations mentioned above, the assumptions are equivalent to the hypothesis that individual preferences are equiprobable.

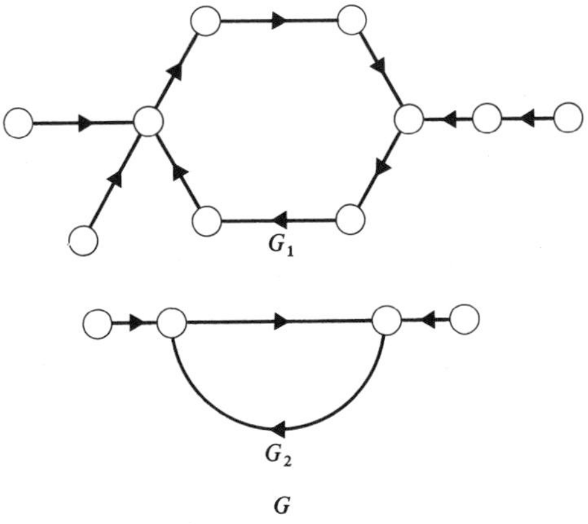

Fig. 8.2.1 Outwardly homogeneous graph of degree one.

We note that since Γ is a single-valued function, each component of G must contain exactly one circuit and this circuit must be directed.[1] This establishes the important fact that *the number of circuits in G is equal to ρ*. The directed circuit in each component of G can be easily found by enumerating the descendants of an arbitrary vertex, say v_1. If these descendants are labeled v_2, v_3, and so on, eventually we must reach a vertex v_t such that $\Gamma(v_t) = v_k$, $k \leq t$, and therefore, $v_k, v_{k+1}, \ldots, v_t$ are the vertices on this circuit of G. If $t < n$, another vertex v_{t+1} may be picked, and the process of enumeration begun again. If v_{t+1} is the same component as v_1, then eventually we will find that a descendant of v_{t+1} is adjacent to a vertex in the set $v_1, \ldots, v_t$. If not, a new circuit will be generated. This process may be continued until all of the vertices of G have been examined. For simplicity, assume that the ith vertex examined is labeled v_i for $i = 1, \ldots, n$.

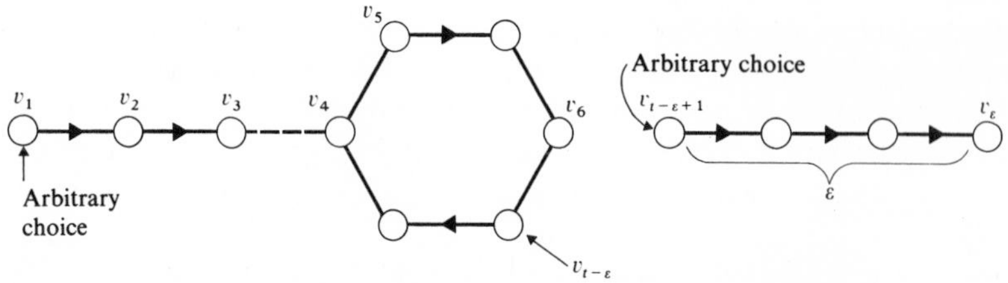

Fig. 8.2.2 Illustration of vertex-counting procedure.

From the above process, we see that each time we contact a vertex which has already been previously contacted, we arbitrarily select a new vertex. Suppose that vertices $v_1, v_2, \ldots, v_t$ have already been contacted. Then the last arbitrarily selected vertex may have been any of $v_1, v_2, \ldots, v_t$. If v_1 is the last arbitrary choice, a circuit has not yet been formed; if v_2 is the last arbitrary choice, a circuit (self-loop) was formed at the first step, and no vertex other than v_1 has been contacted more than once; finally, if v_t is the last arbitrary choice, a circuit was formed at the next to the last step or a previously contacted vertex was again contacted at that step. Let ε be the number of vertices counted since the last arbitrary vertex was selected, and let $p_t(\varepsilon)$ be the probability that, after t vertices are examined, exactly ε vertices have been counted since the last arbitrary vertex was selected. These numbers are illustrated in Fig. 8.2.2. Clearly,

$$p_1(1) = 1 \tag{8.2.1a}$$

$$p_1(\varepsilon) = 0 \qquad \text{for } 2 \leq \varepsilon \leq n, \tag{8.2.1b}$$

$$p_t(1) = \frac{t-1}{n} \qquad \text{for } 2 \leq t \leq n, \tag{8.2.1c}$$

$$p_t(\varepsilon) = 0 \qquad \text{for } t < \varepsilon. \tag{8.2.1d}$$

[1] See Problem 1.

We can derive a recurrence relation for $p_t(\varepsilon)$. If we have counted t elements, the last ε of which have been counted since the last arbitrary choice, after counting $t - 1$ elements, the last $\varepsilon - 1$ vertices must have been contacted since the last arbitrary choice and $\Gamma(v_{t-1}) \notin \{v_1, \ldots, v_{t-1}\}$. Hence

$$p_t(\varepsilon) = \frac{n - t + 1}{n} p_{t-1}(\varepsilon - 1) \qquad \text{for } 2 \le t \le n \text{ and } 2 \le \varepsilon \le n. \tag{8.2.2}$$

If exactly ε elements have been counted since the last arbitrary vertex was selected, the probability of forming a circuit on the next step is ε/n. Let Z_t be the probability of forming a circuit after t vertices have been examined:

$$Z_t = \sum_{\varepsilon=1}^{n} \frac{\varepsilon}{n} p_t(\varepsilon). \tag{8.2.3}$$

Using the recursion relation for $p_t(\varepsilon)$, we can derive one for Z_t. Thus

$$Z_t = \frac{1}{n} p_t(1) + \sum_{\varepsilon=2}^{n} \frac{\varepsilon}{n} p_t(\varepsilon) = \frac{1}{n}\frac{t-1}{n} + \sum_{\varepsilon=2}^{n} \frac{\varepsilon}{n} \frac{n-t+1}{n} p_{t-1}(\varepsilon - 1) \qquad \text{for } 2 \le t \le n \tag{8.2.4a}$$

and, since $p_{t-1}(n) = 0$,

$$Z_t = \frac{t-1}{n^2} + \frac{n-t+1}{n} \sum_{\varepsilon=1}^{n} \frac{\varepsilon+1}{n} p_{t-1}(\varepsilon)$$

$$= \frac{t-1}{n^2} + \frac{n-t+1}{n} \left[\sum_{\varepsilon=1}^{n} \frac{\varepsilon}{n} p_{t-1}(\varepsilon) + \frac{1}{n} \sum_{\varepsilon=1}^{n} p_{t-1}(\varepsilon) \right] \tag{8.2.4b}$$

Moreover,

$$\sum_{\varepsilon=1}^{n} p_{t-1}(\varepsilon) = 1, \qquad Z_t = \frac{t-1}{n^2} + \frac{n-t+1}{n} \left(Z_{t-1} + \frac{1}{n} \right) \tag{8.2.5}$$

with

$$Z_1 = \frac{1}{n}.$$

Therefore,

$$Z_t = \frac{n-t+1}{n} Z_{t-1} + \frac{1}{n} \tag{8.2.6}$$

with

$$Z_1 = \frac{1}{n}.$$

Lemma 8.2.1 *The expected number of components of G is given by*

$$E\{\rho\} = \sum_{t=1}^{n} Z_t. \tag{8.2.7}$$

Proof. Since every component has exactly one directed circuit, the expected number of components is equal to the expected number of circuits. Let $Z_t^{(r)}$ be the probability that the rth circuit is formed after t vertices have been counted (but only $t - 1$ outwardly directed branches have been counted). Then,

$$Z_t = Z_t^{(1)} + Z_t^{(2)} + \cdots + Z_t^{(t-1)}, \tag{8.2.8a}$$

where

$$Z_t^{(i)} = 0 \qquad \text{for } i \geq t, \tag{8.2.8b}$$

and $\sum_{t=1}^{n} Z_t(r)$ is the probability that G has at least r circuits.

Now

$$E\{\rho\} = \sum_{k=0}^{n} k \text{ Prob } \{\rho = k\} = \sum_{k=1}^{n} \text{Prob } \{\rho > k\} = \sum_{k=0}^{n} \sum_{t=0}^{n} Z_t^{(k+1)}$$

$$= \sum_{t=1}^{n} \sum_{k=1}^{n-1} Z_t^{(k)} = \sum_{t=1}^{n} Z_t. \tag{8.2.9}$$

Theorem 8.2.1 *The expected number of components of G is*

$$E\{\rho\} = \sum_{k=1}^{n} \frac{n!}{(n-k)!kn^k}. \tag{8.2.10}$$

Proof. From Lemma 8.2.1, the expected number of components of G is

$$E\{\rho\} = \sum_{t=1}^{n} Z_t. \tag{8.2.11}$$

Define the generating function [FE1] $g(y)$ by

$$g(y) = \sum_{t=1}^{n} Z_t y^t. \tag{8.2.12}$$

Then, by the recursion relation (8.2.6)

$$g(y) - \frac{y}{n} = \sum_{t=2}^{n} Z_t y^t = \sum_{t=1}^{n-1} \frac{n-t}{n} Z_t y^{t+1} + \frac{y}{n} \sum_{t=1}^{n-1} y^t. \tag{8.2.13}$$

Furthermore,

$$g(y) - \frac{y}{n} = \sum_{t=1}^{n} \frac{n-t}{n} Z_t y^{t+1} + \left(\frac{y}{n}\sum_{t=1}^{n} y^t - \frac{1}{n} y^{n+1}\right) \tag{8.2.14}$$

or

$$g(y) - \frac{y}{n} + \frac{1}{n} y^{n+1} = \sum_{t=1}^{n} \frac{n-t}{n} Z_t y^{t+1} + \frac{y}{n}\left[\frac{y - y^{n+1}}{1-y}\right]. \tag{8.2.15}$$

In (8.2.15) we have utilized the fact that

$$\sum_{t=1}^{n} y^t = \frac{y - y^{n+1}}{1-y} \qquad \text{for } y \neq 1. \tag{8.2.16}$$

The right-hand side of (8.2.15) may be written as

$$\sum_{t=1}^{n} Z_t y^{t+1} - \frac{1}{n}\sum_{t=1}^{n} t Z_t y^{t+1} + \frac{y^2 - y^{n+2}}{n(1-y)}, \tag{8.2.17}$$

so that

$$g(y) - \frac{y}{n} + \frac{1}{n} y^{n+1} = y g(y) - \frac{1}{n} y^2 \frac{d}{dy}[g(y)] + \frac{y^2 - y^{n+2}}{n(1-y)}. \tag{8.2.18}$$

Hence, we have obtained the differential equation

$$\frac{d}{dy}[g(y)] + n\frac{1-y}{y^2} g(y) = \frac{1-y^n}{y(1-y)}. \tag{8.2.19}$$

The solution of this differential equation can be written as the integral equation

$$g(y) = y^n e^{n/y} \int_0^y u^{-n} e^{-n/u} \frac{1-u^n}{u(1-u)}\, du, \tag{8.2.20}$$

where the constant of integration is determined by the condition that $g(y)$ is regular[2] at $y = 0$.

Since

$$E\{\rho\} = \sum_{t=1}^{n} Z_t, \tag{8.2.21}$$

we have

$$E\{\rho\} = g(1) = \int_0^1 \frac{u^{-n} - 1}{u(1-u)} e^{n - n/u}\, du \tag{8.2.22}$$

[2] A point y_0 is a regular point of $g(y)$ if $\lim_{h \to 0} \frac{1}{h} \int_{y_0}^{y_0 + h} [g(y_0) - g(y)]\, dy = 0.$

and, if we make the transformation $u = n/(n + z)$,

$$E\{\rho\} = \int_0^\infty \left[\left(1 + \frac{z}{n}\right)^n - 1\right] e^{-z} \frac{dz}{z}. \tag{8.2.23}$$

From the binomial theorem,

$$\left(1 + \frac{z}{n}\right)^n = \sum_{k=0}^{n} \left(\frac{z}{n}\right)^k \frac{n!}{k!(n-k)!}. \tag{8.2.24}$$

Consequently, (8.2.23) becomes

$$E\{\rho\} = \int_0^\infty \sum_{k=1}^{n} \left(\frac{z}{n}\right)^k \frac{e^{-z}}{z} \frac{n!}{k!(n-k)!} dz. \tag{8.2.25}$$

Now, since

$$\int_0^\infty z^k e^{-z} dz = k! \tag{8.2.26}$$

we obtain

$$E\{\rho\} = \sum_{k=1}^{n} \frac{1}{n^k} \frac{n!}{k!(n-k)!}(k-1)! = \sum_{k=1}^{n} \frac{n!}{(n-k)!kn^k}, \tag{8.2.27}$$

which is the desired expression.

Approximation 8.2.1. The asymptotic properties of $E\{\rho\}$ for large n may be examined by using the expression for $E\{\rho\}$ given by (8.2.23) as a starting point. As $n \to \infty$, $(1 + z/n)^n \to e^z$, and since the integral of $(1 - e^{-z})/z$ diverges, $E\{\rho\} \to \infty$. Furthermore, for large n, Kruskal [KR1] has shown that

$$E\{\rho\} = \tfrac{1}{2}(\log 2n + K) + o(1), \tag{8.2.28}$$

where K is Euler's constant given by

$$K = -\int_0^\infty e^{-x} (\log x)\, dx = 0.5772\ldots. \tag{8.2.29}$$

$E\{\rho\} \to \infty$ as $n \to \infty$, and hence Prob $\{\rho = 1\} \to 0$ as $n \to \infty$. We would like to compute this probability as a function of n. The condition $\rho = 1$ is equivalent to the condition that the graph contains exactly one circuit. Let $\{v_{i_1}, v_{i_2}, \ldots, v_{i_q}\} = V_c$ be the vertices of G on this circuit, and let n_j be the number of vertices not in V_c, which are at distance j from the circuit, as shown in Fig. 8.2.3. That is, there are n_j vertices which are connected by directed paths, containing $j - 1$ other vertices not in V_c, to an element of V_c. We will say that the distance between such a vertex and the circuit is j for $j = 1,2,\ldots,n - 1$. Moreover, if π is a path containing k branches, we will say that the length of π is k.

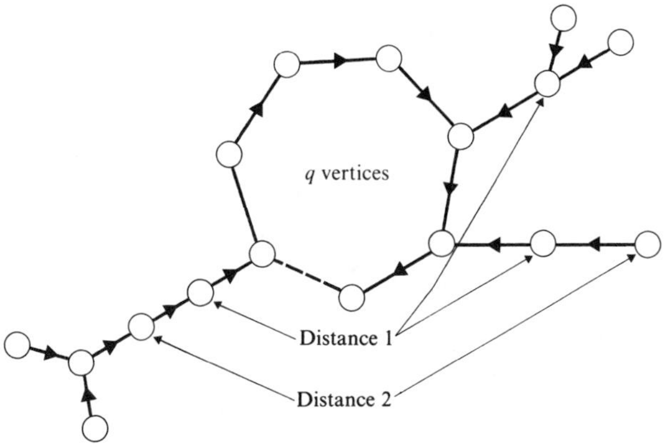

Fig. 8.2.3 Vertices at distance 1 and 2 from a circuit.

Let $[M]_k$ represent the collection of nonempty k-part partitions of a set of M vertices. Then we have:

Lemma 8.2.2 *The probability that G is connected is*

$$\text{Prob}\,\{\rho = 1\} = \sum_{q=1}^{n} \frac{n!}{n^n} \left\{ \sum_{k} \sum_{[M]_k} \frac{1}{q} \frac{q^{n_1} n_1^{n_2} \ldots n_{k-1}^{n_k}}{n_1! n_2! \ldots n_k!} \right\} \tag{8.2.30}$$

where $M = n - q$ and n_i is the number of vertices in the ith partition.

Proof. We first note that there are

$$\frac{n!}{q! n_1! \ldots n_k!} = \binom{n}{q, n_1, \ldots, n_k} \tag{8.2.31}$$

ways of selecting $k + 1$ sets of $q, n_1, \ldots, n_k$ vertices from a total of n vertices, where

$$q + \sum_{i=1}^{k} n_i = n.$$

We want to form a graph which contains a circuit of q vertices and sets of $n_1, n_2, \ldots, n_k$ vertices at respective distance $1, 2, \ldots, k$ from the circuit. Starting with a set of q vertices, there are $(q - 1)!$ ways of interconnecting these vertices to form a directed circuit. The total number of ways of connecting these vertices to any vertex of G is n^q. Hence for a given set of q vertices, the probability of forming a circuit is $(q - 1)!/n^q$.

Any vertex at distance 1 from the circuit is connected directly to the circuit. Any vertex at distance 2 is connected to a vertex at distance 1. Any vertex at distance k is connected to a vertex at distance $k - 1$. Hence, the probability of the event that the n_j vertices selected to be at distance j are actually at that distance is

$$\left(\frac{n_{j-1}}{n}\right)^{n_j} \qquad \text{if } 2 \leq j \leq k \tag{8.2.32a}$$

or

$$\left(\frac{q}{n}\right)^{n_1} \qquad \text{if } j = 1. \tag{8.2.32b}$$

Combining the above expressions, we find that the probability of forming a connected graph with a circuit of q vertices and n_j vertices at distance j from this circuit for $j = 1,2,\ldots,k$ is

$$\binom{n}{q,n_1,\ldots,n_k}\frac{(q-1)!}{n^q}\left(\frac{q}{n}\right)^{n_1}\left(\frac{n_1}{n}\right)^{n_2}\cdots\left(\frac{n_{k-1}}{n}\right)^{n_k}. \tag{8.2.33}$$

Summing this quantity over all possible partitions and over all possible values of q, we obtain the lemma.

Lemma 8.2.3

$$\sum_k \sum_{[M]_k} \frac{1}{q}\,\frac{q^{n_1}n_1^{n_2}\ldots n_{k-1}^{n_k}}{n_1!n_2!\ldots n_k!} = \frac{n^{q-1}}{(n-q)!} \tag{8.2.34}$$

where $M = n - q$.

Proof. From the binomial theorem

$$\frac{n^{q-1}}{(n-q)!} = \frac{(M+q)^{q-1}}{M!} = \sum_{n_1=1}^{M}\binom{M-1}{n_1-1}\frac{1}{M!}q^{n_1-1}M^{M-n_1}. \tag{8.2.35}$$

If $M_1 = M - n_1$ and $M_i = M_{i-1} - n_i$ for $i = 2,3,\ldots,$ then

$$\frac{(M+q)^{M-1}}{M!} = \sum_{n_1=1}^{M}\frac{q^{n_1-1}}{(n_1-1)!}\,\frac{(M_1+n_1)^{M_1-1}}{M_1!}. \tag{8.2.36}$$

Since

$$\frac{(M_i+n_i)^{M_i-1}}{M_i!} = \sum_{n_i=1}^{M}\binom{M_i-1}{n_i-1}\frac{1}{M_i!}n_i^{n_{i+1}-1}M_i^{M_i-n_i}$$

$$= \sum_{n_i=1}^{M_i}\frac{n_i^{n_{i+1}-1}}{(n_i-1)!}\,\frac{(M+n_i)^{M_i-1}}{M_i!} \tag{8.2.37}$$

we obtain

$$\frac{(M+q)^{M-1}}{M!} = \sum_{n_1=1}^{M}\frac{q^{n_1-1}}{(n_1-1)!}\sum_{n_2=1}^{M_1}\frac{n_1^{n_2-1}}{(n_2-1)!}\cdots\sum_{n_k=1}^{M_{k-1}}\frac{n_{k-1}^{n_k-1}}{(n_k-1)!}n_k^{-1} \tag{8.2.38}$$

where k is arbitrary. However, the summations on the right-hand side of the last equation are equivalent to summations over all k and all nonempty k-part partitions.

Combining Lemmas 8.2.2 and 8.2.3, we obtain

Theorem 8.2.2 *The probability that the random graph G is connected is*

$$\text{Prob}\,\{\rho = 1\} = \frac{(n-1)!}{n^n} \sum_{M=0}^{n-1} \frac{n^M}{M!}. \tag{8.2.39}$$

Approximation 8.2.2. The cumulative distribution function of a Poisson variable X with parameter n is

$$\text{Prob}\,\{X < n\} = \text{Prob}\,\{X \le n-1\} = \sum_{M=0}^{n-1} \frac{n^M}{M!} e^{-n}. \tag{8.2.40}$$

Thus,

$$\text{Prob}\,\{\rho = 1\} = \frac{(n-1)!}{n^n} e^n \,\text{Prob}\,\{X \le n-1\} = \frac{n!}{n^{n+1}} e^n \,\text{Prob}\,\{X \le n-1\}. \tag{8.2.41}$$

In addition, if n is large, we can use Stirling's approximation

$$n! \sim (2\pi)^{1/2}\, n^{n+1/2}\, e^{-n} \tag{8.2.42}$$

for the factorial, and the fact that

$$P(X \le n-1) \doteq \tfrac{1}{2} \tag{8.2.43}$$

to obtain

$$\text{Prob}\,\{\rho = 1\} \doteq \frac{n^{n+1/2}\, e^n}{2n^{n+1}} (2\pi)^{1/2} = \left(\frac{\pi}{2n}\right)^{1/2} \quad \text{for } n \text{ large.} \tag{8.2.44}$$

Now that we have found the probability that G is connected, we would like to find the probability that there are exactly k connected components. This problem has been solved by several authors including Folkert [FO1], Rubin and Sitgreaves [RU1], and Harris [HA6]. Since the derivation is long and complicated, we state only the result.

Theorem 8.2.3 *The probability that there are exactly k components in G is*

$$\text{Prob}\,\{\rho = k\} = \frac{n!}{n^k k!} \sum_{\mu=k}^{n} s(k,\mu) \sum_{(k_1,\ldots,k_n)} \frac{1}{k_1!\ldots k_n!} \left(\frac{1}{1!}\right)^{k_1} \left(\frac{2^2}{2!}\right)^{k_2} \cdots \left(\frac{n^n}{n!}\right)^{k_n} \tag{8.2.45}$$

where $s(k,\mu)$ is Stirling's number of the first kind, defined by

$$s(0,0) = 1 \tag{8.2.46a}$$

$$t(t-1)\ldots(t-k+1) = \sum_{\mu=0}^{k} s(k,\mu) t^{\mu} \quad \text{for } k > 0 \tag{8.2.46b}$$

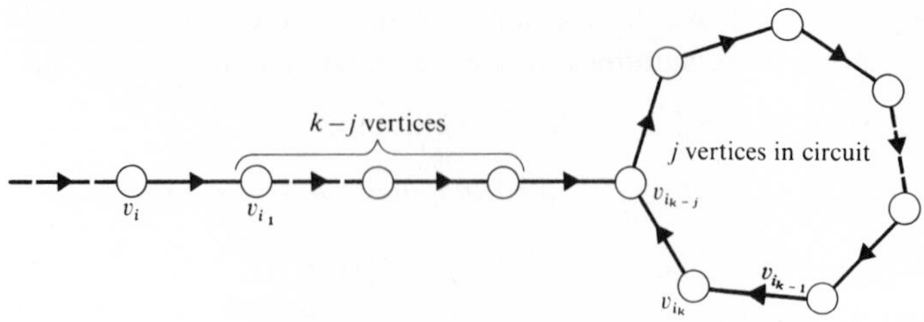

Fig. 8.2.4 Vertex v_i has exactly k descendants, exactly j of which are on the circuit.

and the summation over $(k_1, \ldots, k_n)$ *is over all n-part partitions such that* $k_i \geq 0$ *and*

$$\sum_{i=1}^{n} ik_i = n.$$

Theorems 8.2.1, 8.2.2, and 8.2.3 are each concerned with the number of connected subgraphs of G. As we have seen, an outwardly homogeneous degree-one random graph can be expected to be disconnected if the number of vertices is large enough. If the graph is disconnected, the size of each component is a factor of interest. Closely related to this quantity are the number of descendants, δ, of an arbitrary vertex v_i and the random number, β, of vertices in a circuit of G.

We first compute the probability that an arbitrary vertex v_i has exactly k descendants, exactly j of which are in the circuit. This event is illustrated in Fig. 8.2.4. The event $\delta = k$ and $\beta = j$ for $j \leq k$ is realized if v_{i_1} is connected to v_{i_2}, where v_{i_1} is any one of $(n - 1)$ vertices; if v_{i_2} is connected to v_{i_3}, where v_{i_2} is any one of $(n - 2)$ vertices; and in general if $v_{i_{l-1}}$ is connected to v_{i_l}, where v_{i_l} is any one of $(n - l + 1)$ vertices. However, v_{i_k} *must* be connected to vertex $v_{i_{k-j}}$ and therefore there is only *one* possible connection for v_{i_k}. The total number of ways of connecting the k vertices is n^k and we thus have:

Theorem 8.2.4 *The probability that an arbitrary vertex has k descendants, j of which are on a circuit of G is*

$$\text{Prob}\,\{\delta = k, \beta = j\} = \frac{(n-1)(n-2)\ldots(n-k+1)}{n^k}$$

$$= \frac{(n-1)!}{(n-k)!}\,\frac{1}{n^k} \qquad \text{for } 1 \leq j \leq k \leq n. \qquad (8.2.47)$$

The marginal probabilities that v_i has exactly k descendants or that a component of G has a circuit of j elements are now easily determined by summing the right-hand side of (8.2.47) over j and k, respectively. We now have

Corollary 8.2.1

$$\text{Prob}\,\{\delta = k\} = \frac{(n-1)!k}{(n-k)!n^k}. \qquad (8.2.48)$$

Corollary 8.2.2

$$\text{Prob}\,\{\beta = j\} = \sum_{k=j}^{n} \frac{(n-1)!}{(n-k)!n^k}\,. \tag{8.2.49}$$

Approximation 8.2.3. We can obtain asymptotic densities for the random numbers whose probability distributions are given by Theorem 8.2.4 and its corollaries. Let $k = \sqrt{n}\,x$, and $j = \sqrt{n}\,y$. Then, replacing the factorials in (8.2.47) by Stirling's approximations, we get

$$\frac{(n-1)!}{(n-k)!n^k} = \frac{(n-1)!}{(n-\sqrt{n}\,x)!n^{\sqrt{n}x}} \sim \frac{n^{n-\sqrt{n}x-1/2}e^{-\sqrt{n}x}}{(n-\sqrt{n}\,x)^{n-\sqrt{n}x+1/2}}$$

$$= \frac{n^{n-\sqrt{n}x-1/2}e^{-\sqrt{n}x}}{n^{n-\sqrt{n}x+1/2}\left(1-\dfrac{x}{\sqrt{n}}\right)^{n-\sqrt{n}x+1/2}}\,. \tag{8.2.50}$$

For large n,

$$\left(1-\frac{x}{\sqrt{n}}\right)^{n-\sqrt{n}x+1/2}$$

can be approximated by $\exp\left[(n-\sqrt{n}\,x+1/2)\log(1-x/\sqrt{n})\right]$. Expanding log $(1-x/\sqrt{n})$ in a power series gives

$$\text{Prob}\,\{\delta = k, \beta = j\} = \frac{(n-1)!}{(n-k)!n^k} \sim \frac{1}{n}e^{-1/2\,x^2} = \frac{1}{n}e^{-1/2\,k^2/n}. \tag{8.2.51}$$

From (8.2.51),

$$\text{Prob}\,\{\delta = k\} = \frac{(n-1)!k}{(n-k)!n^k} \sim \frac{k}{n}e^{-1/2\,x^2} = \frac{k}{n}e^{-1/2\,k^2/n} \tag{8.2.52}$$

and

$$\text{Prob}\,\{\beta = j\} = \sum_{k=j}^{n}\frac{(n-1)!}{(n-k)n^k} \sim \int_{\sqrt{n}}^{\infty}\frac{1}{n}e^{-1/2\,x^2}\,dx$$

$$= \frac{\sqrt{2\pi}}{\sqrt{n}}\int_{j}^{\infty}\frac{1}{\sqrt{2\pi n}}e^{-1/2\,k^2/n}\,dk. \tag{8.2.53}$$

Moreover, (8.2.53) can be written as

$$\text{Prob}\,\{\beta = j\} = \sqrt{\frac{2\pi}{n}}\left[1-\Phi\left(\frac{j}{\sqrt{n}}\right)\right], \tag{8.2.54}$$

where $\Phi(\cdot)$ is the cumulative distribution function of the standard normal variable.

Finally, we turn our attention to finding the weak connectivity constant, $\gamma = E\{\delta\}/n$, of G. First, we note that by definition

$$E\{\delta\} = \sum_{k=1}^{n} k \frac{(n-1)!k}{(n-k)!n^k} \tag{8.2.55}$$

and

$$E\{\beta\} = \sum_{j=1}^{n} j \sum_{k=j}^{n} \frac{(n-1)!}{(n-k)!n^k}. \tag{8.2.56}$$

Furthermore,

$$\begin{aligned} E\{\beta\} &= \sum_{j=1}^{n} \sum_{k=j}^{n} j \frac{(n-1)!}{(n-k)!n^k} \\ &= \sum_{k=1}^{n} \sum_{j=1}^{k} j \frac{(n-1)!}{(n-k)!n^k} \\ &= \sum_{k=1}^{n} \frac{k(k+1)}{2} \frac{(n-1)!}{(n-k)!n^k} \\ &= \frac{1}{2} \sum_{k=1}^{n} \frac{k^2(n-1)!}{(n-k)!n^k} + \frac{1}{2} \sum_{k=1}^{n} \frac{k(n-1)!}{(n-k)!n^k} \\ &= \frac{1}{2} \sum_{k=1}^{n} \frac{k^2(n-1)!}{(n-k)!n^k} + \frac{1}{2}, \end{aligned} \tag{8.2.57}$$

since $\sum_{k=1}^{n} k(n-1)!/(n-k)!n^k = 1$. Therefore,

$$E\{\beta\} = \tfrac{1}{2}E\{\delta\} + \tfrac{1}{2}. \tag{8.2.58}$$

From the asymptotic expressions obtained above, it is easily shown that

$$E\{\beta\} \sim \tfrac{1}{4}\sqrt{2\pi n}. \tag{8.2.59}$$

The weak connectivity constant γ is exactly

$$\gamma = \frac{1}{n} \sum_{k=1}^{n} \frac{k^2(n-1)!}{n^k(n-k)!} \tag{8.2.60}$$

and is asymptotically given by

$$\gamma = \frac{\sqrt{\pi}}{\sqrt{2n}}. \tag{8.2.61}$$

Note that the asymptotic values of γ and Prob $\{\rho = 1\}$ are identical.

8.3 OUTWARDLY HOMOGENEOUS GRAPHS

In the last section, we examined the class of random graphs whose vertices have outward demi-degrees constrained to be unity. These graphs correspond to single-valued random mapping functions. In this section, we generalize the preceding model to the class of graphs which contain only vertices with identical outward demi-degrees.

An n-vertex random graph $G = (V,\Gamma)$ is generated in the following manner. If v_i is an arbitrary element of V, then let $\Gamma(v_i) = \{v_{i_1}, \ldots, v_{i_\alpha}\}$ where, if v_j and v_k are *any* elements in V, Prob $\{v_j \in \Gamma(v_i)\}$ = Prob $\{v_k \in \Gamma(v_i)\}$. Thus, for each vertex $v_i \in V$, we choose $\alpha \leq n$ vertices $v_{i_1}, \ldots, v_{i_\alpha}$ from the set $v_1, \ldots, v_n$ such that all possible choices are equiprobable. Directed branches are then drawn from v_i to the α vertices so chosen.

Given a deterministic graph G, we could easily determine the number of descendants of a vertex v_i. Because G is random, we must find the expected number of descendants of v_i, and from this number the weak connectivity constant. We note that the random graphs generated by the above scheme are unbiased; consequently, $\gamma(v_i) = \gamma$ for $i = 1, \ldots, n$.

If $\alpha > 1$, there will be more than one branch emanating from each vertex. In this case, there is no correspondence between the circuits and the components of G. Each component may have several directed circuits, and, if we follow the procedure of generating vertex sequences suggested in the last section, a given vertex may be encountered a number of times. Hence, the vertex-counting procedure of the last section must be generalized.

Suppose we are given an urn with n balls, each ball corresponding to a vertex of G. The balls in the urn are either black or white. If, starting at a vertex v_i, we have already counted t of the descendants of v_i, the urn will contain t black balls and $n - t = w$ white balls. Assume a player has s_0 tickets; each ticket entitles him to draw one ball from the urn. If he withdraws a black ball from the urn, he must replace it and draw again; he is then left with $s_0 - 1$ tickets. This event corresponds to *tracing* a branch from a known vertex to an unknown vertex, only to find that the unknown vertex has been contacted and counted before. If the player draws a white ball from the urn, he replaces it with a black ball and is given α additional tickets for a total of $s = s_0 + \alpha - 1$ tickets to continue the game. The α additional tickets represent the α new branches which may be traced when a new vertex is encountered. The player continues drawing balls until he has no more tickets left, i.e., $s = 0$.

The game is initiated by placing n white balls in the urn and giving the player one ticket. At each stage, the number of tickets in the possession of the player represents the number of untraced branches emanating from vertices which have been already contacted. The first ball drawn corresponds to the initial selection of an arbitrary starting vertex. Let w_0 be the number of white balls in the urn at the start. Define $E(w_0,s_0)$ as the expected number of white balls in the urn at the end of the game, if the game started with w_0 white balls and s_0 tickets. Then,

$$E\{\delta\} = n - E(n,1) \tag{8.3.1}$$

and

$$\gamma = \frac{E\{\delta\}}{n} = 1 - \frac{E(n,1)}{n}. \tag{8.3.2}$$

Let $p(w_0,s_0,u)$ represent the probability that the game ends with u white balls in the urn, given that it started with w_0 white balls and s_0 tickets. The probability that G is connected is $p(n,1,0)$ and, in terms of the $p(w,s_0,u)$, $E(w_0,s_0)$ may be written as

$$E(w_0,s_0) = \sum_{u=0}^{w_0} up(w_0,s_0,u). \tag{8.3.3}$$

Lemma 8.3.1 *$p(w_0,s_0,u)$ and $E(w_0,s_0)$ are solutions of the difference equations*

$$p(w_0,s_0,u) = \left(1 - \frac{w_0}{n}\right)p(w_0,s_0 - 1,u) + \left(\frac{w_0}{n}\right)p(w_0 - 1,s_0 - 1 + \alpha,u) \tag{8.3.4}$$

and

$$E(w_0,s_0) = \left(1 - \left(\frac{w_0}{n}\right)\right)E(w_0,s_0 - 1) + \left(\frac{w_0}{n}\right)E(w_0 - 1,s_0 - 1 + \alpha) \tag{8.3.5}$$

for $s_0 > 0$, and the initial conditions,

$$p(w_0,0,u) = \begin{cases} 0 & \text{for } u \neq w_0, \\ 1 & \text{for } u = w_0, \end{cases} \tag{8.3.6a}$$

$$E(w_0,0) = w_0, \tag{8.3.6b}$$

$$E(0,s_0) = 0. \tag{8.3.6c}$$

Proof. $(1 - w_0/n)$ is the probability of drawing a black ball. If there is no change in the number of white balls, $u = w_0$ and

$$p(w_0,s_0,w_0) = \left(1 - \frac{w_0}{n}\right)p(w_0,s_0 - 1,w_0). \tag{8.3.7}$$

For $u < w_0$,

$$p(w_0,s_0,u) = \left(1 - \frac{w_0}{n}\right)p(w_0,s_0 - 1,u) + \frac{w_0}{n}p(w_0 - 1,s_0 - 1 + \alpha,u). \tag{8.3.8}$$

Thus, the first part of the lemma is proved. The initial conditions for $w_0 = 0$ and

$s_0 = 0$ are obvious. Equation (8.3.5) is easily shown by substituting (8.3.8) into (8.3.3).

Theorem 8.3.1

$$E(w_0,s_0) = \sum_{i=1}^{w_0} \beta_i \binom{w_0}{i} \left(1 - \frac{i}{n}\right)^{s_0 + (w_0 - i)\alpha}, \tag{8.3.9}$$

where the β_i are constants, independent of w_0 and s_0, which are obtained from the equations

$$k = \sum_{i=1}^{k} \beta_i \binom{k}{i} \left(1 - \frac{i}{n}\right)^{(k-i)\alpha} \qquad \textit{for } k = 1,2,\ldots,n-1. \tag{8.3.10}$$

Proof. From Lemma 8.3.1, if $w_0 = 1$,

$$E(1,s_0) = \left(1 - \frac{1}{n}\right) E(1,s - 1); \qquad E(1,0) = 1. \tag{8.3.11}$$

Hence,

$$E(1,s_0) = \left(1 - \frac{1}{n}\right)^{s_0}. \tag{8.3.12}$$

Substituting this value into (8.3.5) gives

$$E(2,s_0) = \left(1 - \frac{2}{n}\right) E(2,s_0 - 1) + \frac{2}{n}\left(1 - \frac{1}{n}\right)^{s_0 - 1 + \alpha}. \tag{8.3.13}$$

The solution of (8.3.13) is

$$E(2,s_0) = K_2\left(1 - \frac{2}{n}\right)^{s_0} + 2\left(1 - \frac{1}{n}\right)^{s_0 + \alpha} \tag{8.3.14}$$

where K_2 is found from the equation $E(2,0) = 2$ so that

$$E(2,s_0) = 2\left(1 - \frac{1}{n}\right)^{s_0 + \alpha} + 2\left[1 - \left(1 - \frac{1}{n}\right)^{\alpha}\right]\left(1 - \frac{2}{n}\right)^{s_0}. \tag{8.3.15}$$

In the same way,

$$\begin{aligned} E(3,s_0) = {} & 3\left(1 - \frac{1}{n}\right)^{s_0 + 2\alpha} + \left[1 - \left(1 - \frac{1}{n}\right)^{\alpha}\right]\left(1 - \frac{2}{n}\right)^{s_0 + \alpha} \\ & + 3\left\{1 - \left(1 - \frac{1}{n}\right)^{2\alpha} - 2\left[1 - \left(1 - \frac{1}{n}\right)^{\alpha}\right]\left(1 - \frac{2}{n}\right)^{\alpha}\right\}\left(1 - \frac{3}{n}\right)^{s_0}. \end{aligned} \tag{8.3.16}$$

Continuing in this manner, the general form may be written as

$$E(w_0,s_0) = \sum_{i=1}^{w_0} \xi_i \left(1 - \frac{i}{n}\right)^{s_0 + (w_0 - i)\alpha}. \tag{8.3.17}$$

Now, from Lemma 8.3.1, $E(w_0,0) = w_0$ and hence

$$w_0 = \sum_{i=1}^{w_0} \xi_i \left(1 - \frac{i}{n}\right)^{(w_0 - i)\alpha}. \qquad (8.3.18)$$

Furthermore, we claim that

$$\xi_i = \beta_i \binom{w_0}{i}$$

where the β_i are constants independent of w_0 and s_0. Equivalently,

$$E(w_0,s_0) = \sum_{i=1}^{w_0} \beta_i \binom{w_0}{i} \left(1 - \frac{i}{n}\right)^{s_0 + (w_0 - i)\alpha}. \qquad (8.3.19)$$

From (8.3.18) the β_i may be calculated as the solutions of the equations

$$k = \sum_{i=1}^{k} \beta_i \binom{k}{i} \left(1 - \frac{i}{n}\right)^{(k - i)\alpha}. \qquad (8.3.20)$$

To prove (8.3.19), we will show that $E(w_0,s_0)$ as given by (8.3.19) satisfies the difference equation (8.3.5) given in Lemma 8.3.1. From (8.3.19), $E(w_0,s_0 - 1)$ and $E(w_0 - 1, s_0 - 1 + \alpha)$ are

$$E(w_0,s_0 - 1) = \sum_{i=1}^{w_0} \beta_i \binom{w_0}{i} \left(1 - \frac{i}{n}\right)^{s_0 - 1 + (w_0 - i)\alpha}; \qquad (8.3.21a)$$

$$E(w_0 - 1,s_0 - 1 + \alpha) = \sum_{i=1}^{w_0 - 1} \beta_i \binom{w_0 - 1}{i} \left(1 - \frac{i}{n}\right)^{s_0 - 1 + (w_0 - i)\alpha}. \qquad (8.3.21b)$$

Hence, the right-hand side of (8.3.5) is

$$\left(1 - \frac{w_0}{n}\right) \sum_{i=1}^{w_0} \beta_i \binom{w_0}{i} \left(1 - \frac{i}{n}\right)^{s_0 - 1 + (w_0 - i)\alpha}$$

$$+ \frac{w_0}{n} \sum_{i=1}^{w_0 - 1} \beta_i \binom{w_0 - 1}{i} \left(1 - \frac{i}{n}\right)^{s_0 - 1 + (w_0 - i)\alpha} \qquad (8.3.22)$$

which may be rewritten as

$$\beta_{w_0} \left(1 - \frac{w_0}{n}\right)^{s_0 - 1} \left(1 - \frac{w_0}{n}\right) + \sum_{i=1}^{w_0 - 1} \beta_i \left(1 - \frac{i}{n}\right)^{s_0 - 1 + (w_0 - i)\alpha}$$

$$\cdot \left[\left(1 - \frac{w_0}{n}\right)\binom{w_0}{i} + \frac{w_0}{n}\binom{w_0 - 1}{i}\right]. \qquad (8.3.23)$$

However,

$$\left[\left(1 - \frac{w_0}{n}\right)\binom{w_0}{i} + \frac{w_0}{n}\binom{w_0 - 1}{i}\right]$$

$$= \left[\left(1 - \frac{w_0}{n}\right)\binom{w_0}{i} + \frac{w_0(w_0 - 1)!}{ni!(w_0 - i - 1)!}\right] \tag{8.3.24}$$

$$= \left[\left(1 - \frac{w_0}{n}\right)\binom{w_0}{i} + \frac{w_0!}{ni!(w_0 - i - 1)!}\,\frac{(w_0 - i)!}{(w_0 - i)!}\right]$$

$$= \left[\left(1 - \frac{w_0}{n}\right) + \frac{1}{n}(w_0 - i)\right]\binom{w_0}{i}$$

$$= \left(1 - \frac{i}{n}\right)\binom{w_0}{i}. \tag{8.3.25}$$

Therefore, (8.3.23) becomes

$$\sum_{i=1}^{w_0} \beta_i \binom{w_0}{i}\left(1 - \frac{i}{n}\right)^{s_0 + (w_0 - i)\alpha}, \tag{8.3.26}$$

which is $E(w_0,s_0)$ as specified by (8.3.19), and the theorem is proved.

Corollary 8.3.1 *The expected number of descendants of an arbitrary vertex v_i is given by*

$$\begin{aligned} E\{\delta\} &= n - E(n,1) \\ &= n - \sum_{i=1}^{n-1} \beta_i \binom{n}{i}\left(1 - \frac{i}{n}\right)^{1 + (n-i)\alpha} \\ &= n - \sum_{i=1}^{n-1} \beta_i \binom{n-1}{i}\left(1 - \frac{i}{n}\right)^{(n-i)\alpha}. \end{aligned} \tag{8.3.27}$$

Example 8.3.1 Consider a graph with four vertices in which the outward demi-degree of each vertex is two. We will find the expected number of vertices which *cannot* be reached from a vertex selected at random, and hence the weak connectivity γ of the graph.

From Theorem 8.3.1, for $n = 4$ and $\alpha = 2$,

$$E(4,1) = \sum_{i=1}^{3} \beta_i \binom{3}{i}\left(1 - \frac{i}{4}\right)^{(4-i)2} \tag{8.3.28}$$

where the β_i satisfy the equations

$$k = \sum_{i=1}^{k} \beta_i \binom{k}{i}\left(1 - \frac{i}{4}\right)^{(k-i)2} \qquad \text{for } k = 1,2,3. \tag{8.3.29}$$

We first find $\beta_1, \beta_2, \beta_3$.

$$\beta_1 = 1, \tag{8.3.30a}$$

$$2 = \beta_1 \binom{2}{1}\left(1 - \frac{1}{4}\right)^2 + \beta_2 \binom{2}{2}, \tag{8.3.30b}$$

$$\beta_2 = \tfrac{7}{8}, \tag{8.3.30c}$$

$$3 = \beta_1 \binom{3}{1}\left(1 - \frac{1}{4}\right)^4 + \beta_2 \binom{3}{2}\left(1 - \frac{2}{4}\right)^2 + \beta_3 \binom{3}{3}, \tag{8.3.30d}$$

$$\beta_3 = \tfrac{357}{256}. \tag{8.3.30e}$$

Therefore, from (8.3.29), $E(4,1)$ is

$$E(4,1) = \binom{3}{1}\left(1 - \frac{1}{4}\right)^6 + \frac{7}{8}\binom{3}{2}\left(1 - \frac{2}{4}\right)^4 + \frac{357}{256}\binom{3}{3}\left(1 - \frac{3}{4}\right)^2 \tag{8.3.31a}$$

or

$$E(4,1) = 3\left(\frac{3}{4}\right)^6 + \frac{7}{8}(3)\left(\frac{1}{2}\right)^4 + \frac{357}{256}\left(\frac{1}{4}\right)^2 = 0.7851. \tag{8.3.31b}$$

Consequently, from (8.3.2)

$$\gamma = 1 - \frac{E(4,1)}{4} = 1 - \frac{0.7851}{4} = 0.8037. \tag{8.3.32}$$

The probabilities $p(w_0,s_0,u)$ may be calculated in the same manner by solving the difference equation (8.3.4) for $p(w_0,s_0, u)$. If this is done, it can be shown that

$$\text{Prob}\{\rho = 1\} = 1 - \sum_{i=1}^{n-1} k_i \binom{n-1}{i}\left(1 - \frac{i}{n}\right)^{1+(n-i)\alpha} \tag{8.3.33a}$$

$$\sum_{i=1}^{k} k_i \binom{k}{i}\left(1 - \frac{i}{n}\right)^{(k-i)\alpha} = 1 \qquad \text{for } k = 1,2,\ldots,n \tag{8.3.33b}$$

and

$$\lim_{n\to\infty} k_i = (-1)^{i-1}. \tag{8.3.33c}$$

It is possible to show that $\text{Prob}\{\rho = 1\} \to 0$ as $n \to \infty$.

Approximation 8.3.1. If n is large, we can obtain asymptotic expressions for δ and γ. For example, Landau [LA1] has found that in the expression

$$\gamma = 1 - \sum_{i=1}^{n-1} \frac{\beta_i}{n}\binom{n-1}{i}\left(1 - \frac{i}{n}\right)^{\alpha(n-i)}, \tag{8.3.34}$$

as $n \to \infty$,

$$\frac{\beta_i}{n} \sim \frac{(i\alpha)^{i-1}}{n^i} \tag{8.3.35a}$$

$$\binom{n-1}{i} \sim \frac{n^i}{i!} \tag{8.3.35b}$$

and

$$\left(1 - \frac{i}{n}\right)^{\alpha(n-i)} \to e^{-i\alpha}. \tag{8.3.35c}$$

This indicates that, as $n \to \infty$,

$$\gamma = 1 - e^{-\alpha} \sum_{i=1}^{\infty} \frac{(i\alpha e^{-\alpha})^{i-1}}{i!}. \tag{8.3.36}$$

If $\alpha > 1$, it is easily shown by comparing consecutive terms that the series on the right-hand side of (8.3.36) converges. The proof of (8.3.36) is not complete since it remains to be shown that the limit of the sum in (8.3.34) is equal to the sum of the limiting forms shown in (8.3.36). A direct proof of the equivalence is complicated. Instead, Landau shows that (8.3.36) is equivalent to an alternate expression for γ first derived by Solomonoff and Rapoport [SO2].

Approximation 8.3.2. Solomonoff and Rapoport found the following transcendental equation for γ for infinite n

$$\gamma = 1 - e^{-\alpha\gamma}. \tag{8.3.37}$$

If $\gamma = 0$, every α is a solution of (8.3.37) and, if $\gamma \neq 0$, we can solve for α as a function of γ:

$$\alpha = \frac{-\ln(1-\gamma)}{\gamma}. \tag{8.3.38}$$

Thus, α may be plotted against γ, and, if the physically meaningless points are discarded, we obtain the graph shown in Fig. 8.3.1. If $\alpha = 2$, we see that $\gamma = 0.8$, so

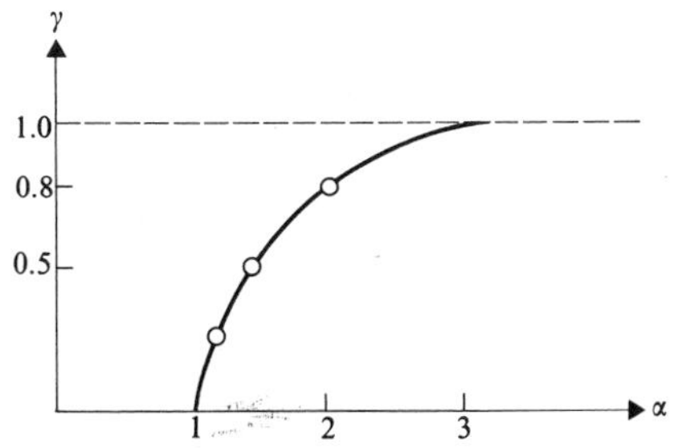

Fig. 8.3.1 Plot of weak connectivity versus outward demi-degree.

that approximately 80% of the vertices of G can be expected to be descendants of an arbitrary vertex. This approximation is extremely accurate. To see this, we can compare the values of γ predicted by (8.3.37) with the exact values of γ found from (8.3.2) and (8.3.27). Table 8.3.1 shows the exact values of γ for $n = 2,3,\ldots,50$ and $\alpha = 2,3,4,5$. Clearly, even for relatively small n, the exact value γ is very close to the values of γ given by (8.3.37), which are shown in the last row of Table 8.3.1.

Since the techniques used to find the asymptotic relations are of interest, we will derive (8.3.37). First define the sets $S_0 = \{v_i\}$, $S_1 = \Gamma(v_i)$, $S_2 = \Gamma(S_1),\ldots, S_k = \Gamma(S_{k-1})$, where by $S_j = \Gamma(S_{j-1})$ we mean the union of the sets $\Gamma(v_j)$ for v_j any vertex in S_{j-1}. Clearly, any vertex in S_j can be at no distance greater than j from v_i. Next use the following vertex-counting procedure. First, trace all branches emanating from v_i and count the new vertices encountered; i.e., first examine the vertices in $\Gamma(v_i)$. At the jth step, trace the branches emanating from S_{j-1} and examine the vertices in S_j. During the jth step, if a vertex is encountered which has already been counted, do not retrace the branches emanating from this vertex. In the derivation to follow, we compute the expected number of new vertices to be contacted at stage $t + 1$, *based on the assumption* that the expected number of vertices was encountered at stage t.

Let $p(t)$ be the probability that a given vertex is counted at the tth stage and let $q(t) = 1 - p(t)$. The probability that a vertex is counted *for the first time* at the tth stage is

$$p(t)\prod_{i=0}^{t-1} q(i) = [1 - q(t)]\prod_{i=0}^{t-1} q(i). \tag{8.3.39}$$

At the $t + 1$st stage, we will examine all vertices in S_{t+1} by tracing the branches emanating from S_t. However, if a vertex in S_t is also in S_j for $j < t$, we will not trace the branches emanating from this vertex. Since the outward demi-degree of each vertex is α, the expected number of branches to be traced at the $t + 1$st stage is

$$\alpha n(1 - q(t))\prod_{i=0}^{t-1} q(i) \tag{8.3.40}$$

and the probability that any given vertex in the graph is *not* contacted at the $t + 1$st stage is

$$q(t + 1) = \left(1 - \frac{1}{n}\right)^{\alpha n[1 - q(t)]\prod_{i=0}^{t-1} q(i)} \tag{8.3.41}$$

with

$$q(0) = \frac{n - 1}{n}.$$

For large n, this may be written as

$$q(t + 1) \sim \exp\left\{-\alpha[1 - q(t)]\prod_{i=0}^{t-1} q(i)\right\} = \exp\left\{-\alpha\left[\prod_{i=0}^{t-1} q(i) - \prod_{i=0}^{t} q(i)\right]\right\}, \tag{8.3.42}$$

Table 8.3.1

n	$\alpha = 2$	$\alpha = 3$	$\alpha = 4$	$\alpha = 5$
2	0.8750	0.9375	0.9688	0.9844
3	0.8271	0.9241	0.9674	0.9861
4	0.8037	0.9222	0.9699	0.9882
5	0.7910	0.9236	0.9722	0.9896
6	0.7840	0.9258	0.9740	0.9904
7	0.7802	0.9279	0.9753	0.9909
8	0.7782	0.9298	0.9761	0.9913
9	0.7774	0.9313	0.9767	0.9915
10	0.7774	0.9326	0.9772	0.9917
11	0.7777	0.9336	0.9775	0.9918
12	0.7784	0.9344	0.9778	0.9920
13	0.7792	0.9351	0.9780	0.9920
14	0.7801	0.9356	0.9782	0.9921
15	0.7810	0.9361	0.9784	0.9922
16	0.7810	0.9364	0.9785	0.9922
17	0.7828	0.9368	0.9786	0.9923
18	0.7836	0.9370	0.9787	0.9923
19	0.7844	0.9373	0.9788	0.9924
20	0.7852	0.9375	0.9789	0.9924
21	0.7859	0.9376	0.9789	0.9924
22	0.7865	0.9378	0.9790	0.9925
23	0.7871	0.9379	0.9790	0.9925
24	0.7877	0.9380	0.9791	0.9925
25	0.7882	0.9382	0.9791	0.9925
26	0.7887	0.9383	0.9792	0.9926
27	0.7891	0.9384	0.9792	0.9926
28	0.7895	0.9384	0.9793	0.9926
29	0.7899	0.9385	0.9793	0.9926
30	0.7902	0.9386	0.9793	0.9926
31	0.7906	0.9387	0.9794	0.9926
32	0.7908	0.9387	0.9794	0.9927
33	0.7911	0.9388	0.9794	0.9927
34	0.7914	0.9388	0.9794	0.9927
35	0.7916	0.9389	0.9794	0.9927
36	0.7918	0.9389	0.9795	0.9927
37	0.7920	0.9390	0.9795	0.9927
38	0.7922	0.9390	0.9795	0.9927
39	0.7924	0.9391	0.9795	0.9927
40	0.7925	0.9391	0.9795	0.9927
41	0.7927	0.9391	0.9796	0.9927
42	0.7928	0.9392	0.9796	0.9927
43	0.7929	0.9392	0.9796	0.9928
44	0.7931	0.9392	0.9796	0.9928
45	0.7932	0.9393	0.9796	0.9928
46	0.7933	0.9393	0.9796	0.9928
47	0.7934	0.9393	0.9796	0.9928
48	0.7935	0.9393	0.9797	0.9928
49	0.7936	0.9394	0.9797	0.9928
50	0.7937	0.9394	0.9797	0.9928
∞	0.7968	0.9405	0.9802	0.9930

Thus,

$$q(1)q(2)\ldots q(t+1) \sim \prod_{j=1}^{t} \exp\left\{-\alpha\left[\prod_{i=0}^{j-1} q(i) - \prod_{i=0}^{j} q(i)\right]\right\}$$

$$= \exp\left\{-\alpha \sum_{j=1}^{t}\left[\prod_{i=0}^{j-1} q(i) - \prod_{i=0}^{j} q(i)\right]\right\}. \qquad (8.3.43)$$

But

$$\sum_{j=0}^{t}\left[\prod_{i=0}^{j-1} q(i) - \prod_{i=0}^{j} q(i)\right] = q(0) - q(0)q(1) + q(0)q(1) - q(0)q(1)q(2)$$
$$+ q(0)q(1)q(2) + \cdots - \prod_{i=0}^{t-1} q(i) + \prod_{i=0}^{t-1} q(i) - \prod_{i=0}^{t} q(i).$$
$$(8.3.44a)$$

Hence

$$\prod_{j=1}^{t+1} q(j) \sim \exp\left\{-\alpha\left[q(0) - \prod_{i=1}^{t} q(i)\right]\right\}. \qquad (8.3.44b)$$

However, $q(0) = (n-1)/n \sim 1$ if n is large. Also, for large n, $1 - \gamma$ is the probability that a given vertex is never contacted. Therefore

$$\lim_{t\to\infty} \prod_{j=1}^{t+1} q(j) = \lim_{t\to\infty} \prod_{j=1}^{t} q(j) = 1 - \gamma. \qquad (8.3.45)$$

It follows that

$$1 - \gamma = \lim_{t\to\infty} \prod_{j=1}^{t+1} q(j) = \lim_{t\to\infty} \exp\left\{-\alpha\left[\frac{n-1}{n} - \prod_{i=1}^{t} q(i)\right]\right\} = e^{-\alpha\gamma} \qquad (8.3.46)$$

which is the desired asymptotic expression.

Equation (8.3.37) holds for graphs in which any vertex may be connected to any other vertex by up to α parallel directed branches. These graphs are constructed by sampling from the n possible vertices α times to determine the location of the branches which are directed out of any given vertex. Since the same vertex may be selected several times, we may have parallel branches. This sampling scheme is equivalent to sampling from a population of n points *with replacement.*

An alternative to sampling with replacement is to establish branches sequentially so that any vertex can be selected at most once. Such a scheme is equivalent to sampling *without replacement*. Thus, we can generate a random graph by the following method. For each v_i, select directed branches $(v_i,v_{i_1}), \ldots, (v_i,v_{i_\alpha})$ so that: v_{i_1} is selected equiprobably out of the set of vertices $V = \{v_1, \ldots, v_n\}$; v_{i_2} is selected equiprobably out of the set of vertices $V - \{v_{i_1}\}$; and in general v_{i_k} is selected equiprobably out of the set of vertices $V - \{v_{i_1}, v_{i_2}, \ldots, v_{i_{k-1}}\}$, for $k = 2,3,\ldots$. The weak connectivity constant, γ, can be found for a graph constructed in this manner. In fact, the same development can be used to show that the expression $\gamma = 1 - e^{-\alpha\gamma}$ is still valid for large graphs constructed by sampling without replacement. This result is not surprising since it is well known that for large population-to-sample ratios, the effect of replacement is negligible (see reference FE1, p. 57).

In the remainder of this section, we investigate the "time response" of an outwardly homogeneous degree-α graph. An excitation (e.g., information, disease, inventory, etc.) is applied to one or more vertices in the graph at time $t = 0$. At time $t = 1$, the vertices adjacent to these vertices are excited. At time $t = 2$, the vertices adjacent to the newly excited vertices are excited and the excitation is propagated in this manner.

Each vertex has α outwardly directed branches, and each of these branches is incident at any other vertex of G with probability 1/n. Under this assumption, the inward demi-degree of each vertex v_i is a binomial random variable $D^-(i)$ such that

$$\text{Prob}\,\{D^-(i) = k\} = \binom{\alpha n}{k}\left(\frac{1}{n}\right)^k\left(1 - \frac{1}{n}\right)^{\alpha n - k}.$$

Moreover, if n is large, $D^-(i)$ is approximately a Poisson random variable with parameter α. Hence Prob $\{D^-(i) = k\} \approx \alpha^k/k!e^{-\alpha}$. With each vertex v_i, we associate a threshold θ_i such that v_i must be adjacent to at least θ_i excited vertices before it can itself become excited. Most often $\theta_1 = \theta_2 = \cdots = \theta_n = \theta$. We might also associate a recovery time r_i with vertex v_i, so that if v_i is excited at time $t = k$, it cannot be excited again until time $t = k + r_i$, no matter what number of adjacent vertices are excited in the interim.

To examine the "expected" course of events in G, we assume that the expected number of events actually occur at each instant of time. Let $x(t)$ be the number of vertices excited at time t. For large n, the probability of an arbitrary vertex receiving k stimuli, at time $t + 1$, is

$$p(k) = \frac{\alpha^k x(t)^k}{k!n^k}\exp\left(\frac{-\alpha x(t)}{n}\right) \qquad \text{for } k = 0,1,2,\ldots. \tag{8.3.47}$$

Since the sum of these probabilities is unity, the probability of receiving *at least* k stimuli at time $t + 1$ is

$$P(k) = 1 - \exp\left(\frac{-\alpha x}{n}\right)\mathscr{E}_{k-1}\left(\frac{\alpha x}{n}\right) \tag{8.3.48}$$

where

$$\mathscr{E}_k(z) = \sum_{j=0}^{k}\frac{z^j}{j!}.$$

Suppose the threshold of each vertex is θ and the recovery time is unity. Then the probability of receiving at least θ stimuli at time $t + 1$ when $x(t)$ vertices are excited at time t is

$$P(\theta) = 1 - \exp\left(\frac{-\alpha x(t)}{n}\right)\mathscr{E}_{\theta-1}\left(\frac{\alpha x(t)}{n}\right). \tag{8.3.49}$$

Thus, the *average* fraction of vertices to be stimulated at time $t + 1$ is

$$1 - \exp\left(\frac{-\alpha x(t)}{n}\right)\mathscr{E}_{\theta-1}\left(\frac{\alpha x(t)}{n}\right) \sim \frac{x(t+1)}{n}. \tag{8.3.50}$$

Suppose we express (8.3.50) as a differential equation which holds approximately for $t = 0,1,2,3,\ldots$

$$\frac{dx(t)}{dt} = n\left(1 - \exp\left(\frac{-\alpha x(t)}{n}\right)\mathscr{E}_{\theta-1}\left(\frac{\alpha x(t)}{n}\right)\right) - x(t). \tag{8.3.51}$$

The right-hand side of (8.3.51) represents the difference between the expected number of vertices excited at time $t + 1$ and the number of vertices excited at time t. A steady state exists if $dx/dt = 0$ or

$$x(t) = n\left(1 - \exp\left(\frac{-\alpha x(t)}{n}\right)\mathscr{E}_{\theta-1}\left(\frac{\alpha x(t)}{n}\right)\right). \tag{8.3.52}$$

If the steady state exists, let $\hat{\gamma} = \lim_{t\to\infty} x(t)/n$. Then

$$\hat{\gamma} = 1 - e^{-\alpha\hat{\gamma}}\mathscr{E}_{\theta-1}(\alpha\hat{\gamma}). \tag{8.3.53}$$

Theorem 8.3.2 *If the threshold of each vertex is unity, then a steady state always exists and the fraction of vertices excited at any given time is equal to the weak connectivity constant of the graph.*

Proof. If $\theta = 1$, (8.3.51) becomes

$$\frac{dx}{dt} = n\left(1 - \exp\left(\frac{-\alpha x(t)}{n}\right)\right) - x(t). \tag{8.3.54}$$

Let $\hat{\gamma} = x(t)/n$; then the positive solution of the equation

$$\hat{\gamma}(t) = 1 - \exp(-\alpha\hat{\gamma}(t)) \tag{8.3.55}$$

always exists for $1 \leq \alpha < \infty$, and for fixed α is unique and independent of t. For an outwardly homogeneous graph of degree α, γ given by the transcendental equation

$$\gamma = 1 - \exp(-\alpha\gamma) \tag{8.3.56}$$

is the weak connectivity constant of the graph. This completes the proof.

If each vertex has threshold θ, the steady-state values of $x(t)$ is given by the roots of

$$1 - \frac{x(t)}{n} = \exp\left(\frac{-\alpha x(t)}{n}\right)\mathscr{E}_{\theta-1}\left(\frac{\alpha x(t)}{n}\right) \tag{8.3.57}$$

or equivalently, in terms of $\hat{\gamma}(t)$, by

$$1 - \hat{\gamma}(t) = e^{-\alpha\hat{\gamma}(t)}\mathscr{E}_{\theta-1}(\alpha\hat{\gamma}(t)). \tag{8.3.58}$$

Theorem 8.3.3 *If the threshold of each vertex is $\theta > 1$, there are two nonzero steady-state values of $\hat{\gamma}$, say $\hat{\gamma}_1$ and $\hat{\gamma}_2$, provided α exceeds a certain value dependent on θ.*

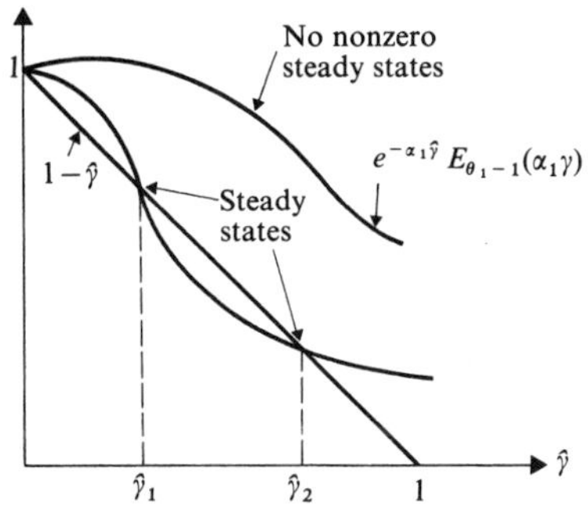

Fig. 8.3.2 Graphical representation of Eq. (8.3.58).

Proof. Holding α and θ fixed, plot the left- and right-hand sides of (8.3.58) against $\hat{\gamma}$, as shown in Fig. 8.3.2. We now show that if $\theta > 1$, for sufficiently small values of α, no nonzero intersections of the curves in Fig. 8.3.2 occur and for sufficiently large values of α, there are exactly two such intersections.

If $\theta > 1$,

$$\frac{d}{d\hat{\gamma}}e^{-\alpha\hat{\gamma}}\mathscr{E}_{\theta-1}(\alpha\hat{\gamma}) = -\alpha e^{-\alpha\hat{\gamma}}\mathscr{E}_{\theta-1}(\alpha\hat{\gamma}) + \alpha e^{-\alpha\hat{\gamma}}\mathscr{E}_{\theta-2}(\alpha\hat{\gamma}). \tag{8.3.59}$$

At $\hat{\gamma} = 0$,

$$\frac{d}{d\hat{\gamma}}e^{-\alpha\hat{\gamma}}\mathscr{E}_{\theta-1}(\alpha\hat{\gamma}) = -\alpha + \alpha = 0. \tag{8.3.60}$$

Furthermore,

$$\frac{d^2}{d\hat{\gamma}^2}e^{-\alpha\hat{\gamma}}\mathscr{E}_{\theta-1}(\alpha\hat{\gamma}) = \alpha^2 e^{-\alpha\hat{\gamma}}\mathscr{E}_{\theta-1}(\alpha\hat{\gamma}) - 2\alpha^2 e^{-\alpha\hat{\gamma}}\mathscr{E}_{\theta-2}(\alpha\hat{\gamma}) + \alpha^2 e^{-\alpha\hat{\gamma}}\mathscr{E}_{\theta-3}(\alpha\hat{\gamma})$$

$$\tag{8.3.61}$$

where $\mathscr{E}_{-1} = 0$. If

$$\frac{d^2}{d\hat{\gamma}^2}e^{-\alpha\hat{\gamma}}\mathscr{E}_{\theta-1}(\alpha\hat{\gamma}) = 0, \tag{8.3.62}$$

then, since $\alpha^2 e^{-\alpha\hat{y}} \neq 0$, we have from (8.3.61)

$$\mathscr{E}_{\theta-1}(\alpha\hat{y}) - 2\mathscr{E}_{\theta-2}(\alpha\hat{y}) + \mathscr{E}_{\theta-3}(\alpha\hat{y}) = 0. \tag{8.3.63}$$

Equation (8.3.63) can be rewritten as

$$(\mathscr{E}_{\theta-1}(\alpha\hat{y}) - \mathscr{E}_{\theta-2}(\alpha\hat{y})) - \mathscr{E}_{\theta-2}(\alpha\hat{y}) - \mathscr{E}_{\theta-3}(\alpha\hat{y})) = \frac{\alpha^{\theta-1}\hat{y}^{\theta-1}}{(\theta-1)!} - \frac{\alpha^{\theta-2}\hat{y}^{\theta-2}}{(\theta-2)!} = 0. \tag{8.3.64}$$

The roots of (8.3.64) are $\hat{y} = 0$ and $\hat{y} = (\theta - 1)/\alpha$. Hence, the graph of $e^{-\alpha\hat{y}}\mathscr{E}(\alpha\hat{y})$ has at most two inflection points, namely at $\hat{y} = 0$ and $\hat{y} = (\theta - 1)/\alpha$. Furthermore, if we substitute the value $\hat{y} = (\theta - 1)/\alpha$ into (8.3.59), we obtain

$$\begin{aligned}\frac{d}{d\hat{y}} e^{-\alpha\hat{y}}\mathscr{E}_{\theta-1}(\alpha\hat{y})\bigg|_{\hat{y}=\theta-1/\alpha} &= -\alpha e^{-(\theta-1)}\mathscr{E}_{\theta-1}(\theta-1) + \alpha e^{-(\theta-1)}\mathscr{E}_{\theta-2}(\theta-1) \\ &= -\alpha e^{-(\theta-1)}\frac{(\theta-1)^{\theta-1}}{(\theta-1)!} \sim \frac{-\alpha}{\sqrt{2\pi(\theta-1)}}\end{aligned} \tag{8.3.65}$$

where $\theta > 1$. The above results are summarized in Fig. 8.3.3.

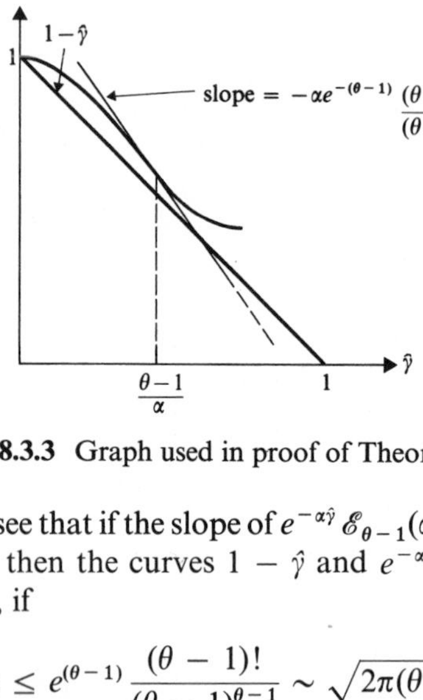

Fig. 8.3.3 Graph used in proof of Theorem 8.3.3.

From Fig. 8.3.3, we see that if the slope of $e^{-\alpha\hat{y}}\mathscr{E}_{\theta-1}(\alpha\hat{y})$ evaluated at $\hat{y} = (\theta - 1)/\alpha$ is not smaller than -1, then the curves $1 - \hat{y}$ and $e^{-\alpha\hat{y}}\mathscr{E}_{\theta-1}(\alpha\hat{y})$ will not intersect except at $\hat{y} = 0$. Hence, if

$$\alpha \le e^{(\theta-1)}\frac{(\theta-1)!}{(\theta-1)^{\theta-1}} \sim \sqrt{2\pi(\theta-1)} \tag{8.3.66}$$

then no positive steady state exists. Furthermore, if α is sufficiently large, the magnitude of the slope of $e^{-\alpha\hat{y}}\mathscr{E}_{\theta-1}(\alpha\hat{y})$ at $\hat{y} = (\theta - 1)/\alpha$ can be made large enough so that the curve $e^{-\alpha\hat{y}}\mathscr{E}_{\theta-1}(\alpha\hat{y})$ monotonically approaches zero from above. If the curve has crossed the line $1 - \hat{y}$ at some point where $\hat{y} < 1$, it must cross it again at some point where $\hat{y} < 1$ (neglecting the special case where the two curves are tangent). Hence the theorem is proved.

Corollary 8.3.2 *Let the positive steady states, if they exist, occur at $\hat{\gamma}_1$ and $\hat{\gamma}_2$ with $\hat{\gamma}_1 < \hat{\gamma}_2$. Then $x/n = \hat{\gamma}_1$ gives an unstable steady state while $x/n = \hat{\gamma}_2$ gives a stable steady state.*

Proof. From (8.3.51) we note that the sign of dx/dt is identical to the sign of $1 - e^{-\alpha\hat{\gamma}}\mathscr{E}_{\theta-1}(\alpha\hat{\gamma}) - \hat{\gamma}$. Referring to Fig. 8.3.2, we find

$$\frac{dx}{dt} < 0 \qquad \text{for } \frac{x}{n} < \hat{\gamma}_1; \tag{8.3.67a}$$

$$\frac{dx}{dt} > 0 \qquad \text{for } \hat{\gamma}_1 < \frac{x}{n} < \hat{\gamma}_2; \tag{8.3.67b}$$

$$\frac{dx}{dt} < 0 \qquad \text{for } \frac{x}{n} > \hat{\gamma}_2. \tag{8.3.67c}$$

Hence, if the initial number of excited vertices is $x(0)$, then if $x(0) < n\hat{\gamma}_1$, the activity will die out. If $n\hat{\gamma}_1 < x(0) < n\hat{\gamma}_2$, the initial activity will increase until the steady state $n\hat{\gamma}_2$ is reached. If $x(0) > n\hat{\gamma}_2$ the initial activity will decrease until the steady state $n\hat{\gamma}_2$ is reached.

Corollary 8.3.2 may be interpreted in the following way. The graph has two steady states, but the steady state $n\hat{\gamma}_1$ can only be excited with an initial excitation $x(0) = n\hat{\gamma}_1$. If any other initial excitation is applied, either the excitation will eventually disappear from the graph or the steady state $n\hat{\gamma}_2$ will be reached. Thus, we have a type of "ignition" phenomenon. Given values of θ and α such that two steady states exist, the entire graph has a threshold $n\hat{\gamma}_1$ such that an initial excitation in excess of this number will "ignite" the graph, which will remain active at an average level of $n\hat{\gamma}_2$ excited vertices per second. Thus, we might consider $n\hat{\gamma}_1$ to be the threshold of the graph and $n\hat{\gamma}_2$ to be its steady state. Moreover, if $\theta < \theta'$, then $e^{-\alpha\hat{\gamma}}\mathscr{E}_{\theta-1}(\alpha\hat{\gamma}) < e^{-\alpha\hat{\gamma}}\mathscr{E}_{\theta'-1}(\alpha\hat{\gamma})$ for all values of $\hat{\gamma} > 0$. Therefore, for fixed α, the curve $e^{-\alpha\hat{\gamma}}\mathscr{E}_{\theta'-1}(\alpha\hat{\gamma})$ lies completely above the curve $e^{-\alpha\hat{\gamma}}\mathscr{E}_{\theta-1}(\alpha\hat{\gamma})$ and so, if $\hat{\gamma}_1'$ and $\hat{\gamma}_2'$ are the threshold and steady-state parameters associated with θ', then $\hat{\gamma}_1' > \hat{\gamma}_1$ and $\hat{\gamma}_2' < \hat{\gamma}_2$. Thus, if the thresholds of the individual vertices are uniformly raised, the overall threshold of the graph will be raised and the steady-state activity lowered.

8.4 GENERALIZATIONS

There are several ways to modify the structure of directed graphs with identical outward demi-degrees while still retaining many of their essential features. For example, in each of the preceding models, Prob $\{v_i \in \Gamma(v_i)\}$ is nonzero and the graph may thus have self-loops. In many cases the graph models a physical situation in which a self-loop is meaningless or impossible. We must therefore adjust the results to reflect this new constraint. The assumption that a graph is outwardly homogeneous could be unwarranted since the demi-degree of an arbitrary vertex may be a random variable. The asymptotic results contained in the last section are valid for graphs whose vertices have demi-degrees with equal expectations. Finally, the graph of a network is often undirected; consequently, we would like to determine the effect on the connectivity properties if the branch directions are removed. The effect will be shown to be significant.

We first consider the class of outwardly homogeneous graphs which contain no self-loops. Fortunately, this constraint does not change the techniques necessary to compute the quantities of interest. For example, let us find the probability that an arbitrary vertex, of a degree-one outwardly homogeneous graph, has k descendants, exactly j of which are on a circuit. There are still $(n-1)(n-2)\dots(n-k+1)$ ways of realizing this event. However, there are now only $(n-1)^n$ total possible events. Hence the probability is

$$\frac{(n-1)!}{(n-k)!(n-1)^k} = \frac{(n-2)!}{(n-k)!(n-1)^{k-1}} \tag{8.4.1}$$

and, proceeding as in Section 2, we find

$$\text{Prob}\,\{\delta = k\} = \frac{(n-1)!(k-1)}{(n-k)^k(n-k)!} = \frac{(n-2)!(k-1)}{(n-k)!(n-1)^{k-1}}. \tag{8.4.2}$$

If we have an $n+1$-vertex graph $\tilde{G}$, with no self-loops, then the probability that a vertex of G has $k+1$ descendants is identical to the probability that a vertex in an n-vertex graph G, *with self-loops*, has k descendants. Thus, if γ_n and γ_n^* are weak connectivity constants for n-vertex graphs with and without self-loops, respectively

$$\gamma_{n+1}^* = \frac{n\gamma_n + 1}{n+1}. \tag{8.4.3}$$

Prob $\{\rho = 1\}$ is easily seen to be

$$\text{Prob}\,\{\rho = 1\} = \frac{(n-1)!}{(n-1)^n}\sum_{m=0}^{n-2}\frac{n^m}{m!} \tag{8.4.4a}$$

and

$$\lim_{n\to\infty} \text{Prob}\,\{\rho = 1\} = \exp\sqrt{\frac{\pi}{2(n-1)}}. \tag{8.4.4b}$$

In Fig. 8.4.1 Prob $\{\rho = 1\}$ is plotted for graphs with and without self-loops. The probability that the graph without self-loops is connected is substantially higher than for the graph in which self-loops are allowed.

For outwardly homogeneous graphs in which $\alpha > 1$ and no self-loops are allowed, the method given in Section 3 is also easily modified. Thus, in the urn model, when the first ball is withdrawn, it is not replaced. There are then $n-1$ balls in the urn, and the missing one corresponds to the vertex from which branches are being traced. Then $\gamma = 1 - E(n-1,1)/n$. The asymptotic result, $\gamma = 1 - e^{-\alpha\gamma}$, obtained in the last section is valid if α is the *average number* (not necessarily an integer) of directed branches emanating from an arbitrary vertex. This follows since the derivation of (8.3.37) holds this more general assumption.

Consider the n-vertex graph without self-loops generated according to the following scheme. There are $n(n-1)/2$ pairs of distinct vertices, which may be connected

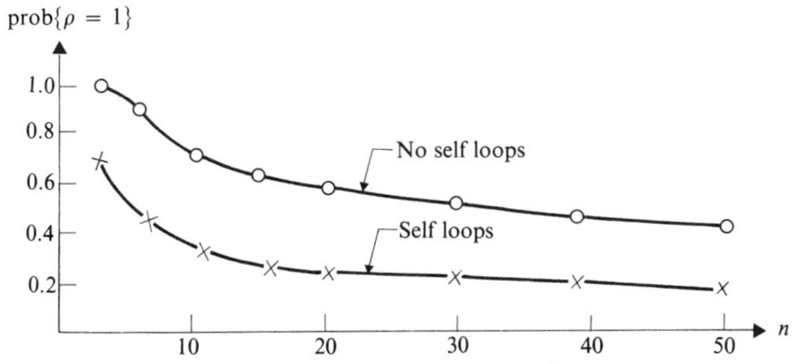

Fig. 8.4.1 Plot of Prob $\{\rho = 1\}$ versus number of vertices for graphs with and without self-loops.

by $n(n-1)$ directed branches. If a subset of the set of possible branches is drawn, there are $2^{n(n-1)}$ distinct possible graphs. For each i and j we let Prob $\{v_j \in \Gamma(v_i)\} = p$. Then, we generate a random graph by assigning a branch from v_i to v_j with probability p for $i, j = 1, \ldots, n$. The expected outward demi-degree of each vertex is $\alpha = np$. Thus (8.3.37) is valid for this class of graphs, and, for large n, the effect of varying p on the weak connectivity constant may be easily studied. If G is undirected there are $n(n-1)/2$ possible branches each of which is in Γ with probability p. We now find the probability that such a graph is connected.

Let $g(n,b)$ denote the number of all possible connected subgraphs with n vertices and b branches. The probability that G has exactly b branches is clearly

$$p^b(1-p)^{(n(n-1)/2)-b}. \tag{8.4.5}$$

Hence, we can express Prob $\{\rho = 1\}$ as

$$\text{Prob}\,\{\rho = 1\} = \sum_{b=n-1}^{n(n-1)/2} g(n,b)p^b(1-p)^{(n(n-1)/2)-b}. \tag{8.4.6}$$

Unfortunately, there are no known *simple* formulae for the $g(n,b)$. Gilbert [GI2] and Riddell and Uhlenbeck [RI1] have given a generating series of the form

$$\sum_{n=1}^{\infty}\sum_{b=1}^{\infty} \frac{g(n,b)x^n y^b}{n!} = \log\left(1 + \sum_{k=1}^{\infty} \frac{(1+y)^{k(k-1)/2}x^k}{k!}\right). \tag{8.4.7}$$

By formally expanding the logarithm into a power series and equating coefficients of $x^n y^b$, the $g(n,b)$ may be found. The following recursion formula is also available:

$$\binom{n(n+1)/2}{b} = \sum_{k=0}^{n}\binom{n}{k}\sum_{t=k}^{k(k+1)/2}\binom{(n-k)/2}{n-t} g(k+1,t). \tag{8.4.8}$$

In general, it is extremely tedious to determine $g(n,b)$. Using the expansion given by (8.4.7), it can be shown that

$$\text{Prob}\,\{\rho = 1\} = \sum_{r_1,\ldots,r_n} \frac{(-1)^k(k-1)!n!(1-p)^{(n^2-1^2r_1-\cdots-n^2r_n)/2}}{r_1!\ldots r_n!(1!)^{r_1}\ldots(n!)^{r_n}} \tag{8.4.9}$$

where the sum is over all partitions of n vertices and where $k = r_1 + \cdots + r_n$. Prob $\{\rho = 1\}$ can be computed as a function of n. For this purpose let Prob $\{\rho = 1|k\}$ be the probability that a k-vertex graph, generated by the above procedure, is connected. The probability that a given vertex v_i, in the n-vertex graph, has exactly $k - 1$ descendants (excluding itself) is then

$$\binom{n-1}{k-1} \text{Prob}\{\rho = 1|k\}(1-p)^{k(n-k)}. \tag{8.4.10}$$

Prob $\{\rho = 1\}$ = Prob $\{\rho = 1|n\}$ is the probability that a given vertex has exactly $n - 1$ descendants. Therefore,

$$\text{Prob}\{\rho = 1|n\} = 1 - \sum_{k=0}^{n-1} \binom{n-1}{k-1} \text{Prob}\{\rho = 1|k\}\, q^{k(n-k)} \tag{8.4.11}$$

where $q = 1 - p$.

From (8.4.10), the weak connectivity constant, γ, is

$$\gamma = \frac{1}{n} \sum_{k-1}^{n-1} k \binom{n-1}{k-1} \text{Prob}\{\rho = 1|k\}(1-p)^{k(n-k)}. \tag{8.4.12}$$

Approximation 8.4.1. Gilbert [GI2] has found the following upper and lower bounds for Prob $\{\rho = 1|n\}$:

$$\left\{1 - \frac{n-1}{2} q^{n-1}\right\} nq^{n-1} \leq 1 - \text{Prob}\{\rho = 1|n\} \tag{8.4.13}$$

and

$$1 - \text{Prob}\{\rho = 1|n\} \leq q^{n-1}\{(1 + q^{(n-2)/2(n-1)} - q^{(n-2)(n-1)/2}\}. \tag{8.4.14}$$

When n becomes large, both the upper and lower bounds in (8.4.13) and (8.4.14) approach nq^{n-1}. Hence

$$\text{Prob}\{\rho = 1\} \sim 1 - nq^{n-1}. \tag{8.4.15}$$

This probability is plotted in Fig. 8.4.2 as a function of p for $n = 6$.

The bounds found by Gilbert are obtained in the following way. A lower bound on 1 − Prob $\{\rho = 1|n\}$ is the probability Q that at least one vertex is connected to no other vertex. Q may be bounded from below by using the inequality[3]

$$\sum_{i=1}^{n} \text{Prob}\{Q_i\} - \sum_{i<j} \text{Prob}\{Q_i Q_j\} < Q, \tag{8.4.16}$$

where Q_i is the event that v_i is not connected to any other vertex. However,

$$\text{Prob}\{Q_i\} = (1-p)^{n-1} \tag{8.4.17a}$$

[3] This inequality is known as Bonferroni's Inequality; see reference [FE1], pp. 100–101.

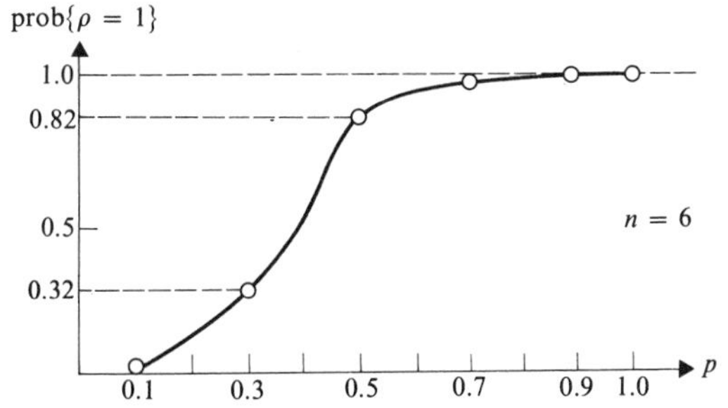

Fig. 8.4.2 Plot of Prob $\{\rho = 1\}$ versus probability of branch existence.

and

$$\text{Prob}\ \{Q_i Q_j\} = (1 - p)^{2n-3}. \tag{8.4.17b}$$

The lower bound given by (8.4.13) now follows. Since Prob $\{\rho = 1|k\} \le 1$,

$$1 - \text{Prob}\ \{\rho = 1|n\} \le \sum_{k=1}^{n-1} \binom{n-1}{k-1} (1 - p)^{k(n-k)}. \tag{8.4.18}$$

The function $k(n - k)$ is a convex function of k and

$$k(n - k) \ge \frac{(n - 2)k + n}{n} \qquad \text{if } 1 \le k \le \frac{n}{2}, \tag{8.4.19a}$$

$$k(n - k) \ge \frac{(n - 2)(n - k) + n}{2} \qquad \text{if } \frac{n}{2} \le k \le n - 1. \tag{8.4.19b}$$

Therefore, since $q \le 1$,

$$q^{k(n-k)} \le q^{n/2}\{q^{(n-2)k/2} + q^{(n-2)(n-k)/2}\}. \tag{8.4.20}$$

Substituting this inequality into (8.4.18) and simplifying gives the upper bound on $1 - \text{Prob}\ \{\rho = 1|n\}$ of (8.4.15).

The above model may be generalized by associating with each vertex a probability $\hat{p}$ that the vertex is in G. For this model Jacobs [JA1] has shown that

$$\text{Prob}\ \{\rho = 1\} \sim \sum_{i=0}^{n} \binom{n}{i} \hat{p}^i (1 - \hat{p})^{n-i} [1 - iq^{i-1}]. \tag{8.4.21}$$

Performing the summation in (8.4.21) gives

$$\text{Prob}\ \{\rho = 1\} \sim 1 - (\hat{p}n)q^{n-1}. \tag{8.4.22}$$

We now consider a simplified model of the growth of a graph. Given an existing graph, from time to time new branches and vertices will be added to meet the changing demands on the system. It is desirable to investigate the evolving structural properties of the graph, and, in particular, we consider an n-vertex, m-branch random graph $G_{n,m}$ as m varies as a function of n. The random graph $G_{n,m}$ is generated by selecting m branches from the $\binom{n}{2}$ possible branches so that all $\binom{\binom{n}{2}}{m}$ possible choices are equiprobable. The following results are contained in a monumental paper by Erdös and Rényi [ER1]. Because of space limitations we only present some of its highlights and direct the reader to the original for proofs.

Let the number of branches of G be a function $m(n)$. Erdös and Rényi attempt to find the "typical" structure of the graph at a given stage of evolution. A structure is "typical" if its probability of existence tends to 1 as $n \to \infty$. Let $P_{n,m(n)}(\mathscr{E})$ represent the probability that the event "$\mathscr{E}$" occurs for $G_{n,m(n)}$. Then, we will say that "almost all" graphs $G_{n,m(n)}$ have property $\mathscr{E}$ if $\lim_{n\to\infty} P_{n,m(n)}(\mathscr{E}) = 1$. Surprising results can be obtained for a number of fundamental structural properties $\mathscr{E}$. There exists a function $\mathscr{E}(n)$, called a *threshold function* of the property $\mathscr{E}$, such that $\mathscr{E}(n)$ monotonically increases to infinity and

$$\lim_{n\to\infty} P_{n,m(n)}(\mathscr{E}) = \begin{cases} 0 & \text{if } \lim\limits_{n\to\infty} \dfrac{m(n)}{\mathscr{E}(n)} = 0, \\[2ex] 1 & \text{if } \lim\limits_{n\to\infty} \dfrac{m(n)}{\mathscr{E}(n)} = \infty. \end{cases} \tag{8.4.23}$$

In other words, if the number of branches of $G_{n,m(n)}$ increase at a "faster" rate than $\mathscr{E}(n)$, almost all graphs $G_{n,m(n)}$ have property $\mathscr{E}$, whereas, if the number of branches of $G_{n,m(n)}$ increases at a "slower" rate than $\mathscr{E}(n)$, almost no graphs $G_{n,m(n)}$ have property $\mathscr{E}$.

If a graph G has n vertices and m branches, we will call the number $2m/n$ the "density" of G. If a graph G has the property that no subgraph of G has a larger density we will say that G is a *balanced* graph. We can find a threshold function for balanced graphs. This result is contained in:

Theorem 8.4.1 *Let k and l be positive integers such that $k \geq 2$ and $k - 1 \leq l \leq \binom{k}{2}$.*

Let $\beta_{k,l}$ denote an arbitrary nonempty set of connected balanced graphs consisting of k vertices and l branches. The threshold function, for the property that the random graph $G_{n,m(n)}$ should contain at least one subgraph isomorphic with some element of $\beta_{k,l}$, is $n^2 - k/l$.

Theorem 8.4.1 has several nontrivial corollaries which follow from the fact that trees and complete graphs are balanced graphs.

Corollary 8.4.1 *The threshold function, for the property that the random graph contains a subgraph which is a k-vertex tree, is $n^{(k-2)/(k-1)}$ for $k = 3,4,\ldots$.*

Corollary 8.4.2 *The threshold function, for the property that the random graph contains a complete k-vertex subgraph for* $k \geq 3$, *is* $n^{2(1-(1/(k-1)))}$.

A threshold function $\mathscr{E}(n)$, for the property $\mathscr{E}$, may have the additional property that there exists a probability distribution function $P(x)$ such that, if $0 < x < \infty$ and x is a point of continuity of $P(x)$, then

$$\lim_{n\to\infty} P_{n,m(n)}(\mathscr{E}) = P(x) \qquad \text{if } \lim_{n\to\infty} \frac{m(n)}{\mathscr{E}(n)} = x. \tag{8.4.24}$$

If (8.4.24) holds, then $\mathscr{E}(n)$ is said to be a *regular threshold function* for the property $\mathscr{E}$ and $P(x)$ is called the threshold distribution function of the property $\mathscr{E}$. The existence of this type of threshold function is demonstrated by:

Theorem 8.4.2 *If* $\lim_{n\to\infty}\left[\frac{m(n)}{n^{(k-2)/(k-1)}}\right] = l > 0$ *and* t_k *is the number of k-vertex trees which are components of* $G_{n,m(n)}$, *then*

$$\lim_{n\to\infty} P_{n,m(n)}(t_k = j) = \frac{\lambda^j e^{-\lambda}}{j!} \tag{8.4.25}$$

for $j = 0,1,\ldots,$ *where*

$$\lambda = \frac{(2l)^{k-1}k^{k-2}}{k!}.$$

Theorem 8.4.2 states that if $m(n) \sim ln^{(k-2)/(k-1)}$, then the number of k-vertex trees in $G_{n,m(n)}$ has, in the limit, a Poisson distribution with mean value λ. This implies that the threshold distribution function for k-vertex trees is $1 - e^{-\lambda}$. In the proof of Theorem 8.4.2, another interesting result is obtained. The average number of k-vertex trees is asymptotically $nm_k(2m/n)$ where

$$m_k(t) = \frac{k^{k-2}t^{k-1}e^{-kt}}{k!} \qquad \text{for } k = 1,2,\ldots. \tag{8.4.26}$$

For a fixed value of k, considered as a function of t, the value of $m_k(t)$ increases for $t < (k-1)/k$ and decreases for $t > (k-1)/k$; consequently, the average number of isolated k-vertex trees is maximum for $m \sim (n/2)(1 - 1/k)$ with

$$nm_k\left(\frac{2^{n/2}(1-1/k)}{n}\right) = \frac{(1-1/k)^{k-1}e^{-(k-1)}k^{k-2}}{k!} \tag{8.4.27}$$

and the average number of k-vertex isolated trees for large k is asymptotically

$$\frac{n}{\sqrt{2\pi k^5}}.$$

Several results concerning the number of circuits of a graph have also been obtained. Typical of these results is

Theorem 8.4.3 *Suppose $m(n) \sim dn/2$ for $d > 0$. Let β_k denote the number of k-vertex circuits contained in $G_{n,m(n)}$ for $k = 3,4,\ldots$. Then,*

$$\lim_{n\to\infty} P_{n,m(n)}(\beta_k = j) = \frac{\lambda^j e^{-\lambda}}{j!} \tag{8.4.28}$$

where

$$\lambda = \frac{(d)^k}{2k}.$$

Thus, the threshold distribution corresponding to the threshold function $\mathscr{E}(n) = n$, for the property that $G_{n,m(n)}$ contain a k-vertex circuit, is $1 - \exp\{-(1/2k)(d)^k\}$.

Numerous other theorems of this type can be given. However, we now consider "global" properties of $G_{n,m(n)}$. Among the many possible results are those concerning the number of vertices contained in trees or circuits. For example, if $V_{n,m}$ denotes the number of vertices of $G_{n,m}$ which belong to a component which is a tree, we can show:

Theorem 8.4.4 *If $\lim_{n\to\infty} \frac{m(n)}{n} = \frac{d}{2} > 0$, then*

$$\lim_{n\to\infty} \frac{E\{V_{n,m(n)}\}}{n} = \begin{cases} 1 & \text{for } d \le 1, \\ \dfrac{x(d)}{2d} & \text{for } d > 1, \end{cases} \tag{8.4.29}$$

where $x = x(d)$ is the only root satisfying $0 < x < 1$ of the equation $xe^{-x} = de^{-d}$.

Other interesting global properties are those which concern the number of components of $G_{n,m(n)}$.

Theorem 8.4.5 *If $\rho_{n,m}$ denotes the number of components of $G_{n,m}$, then if $m \sim dn/2$ with $0 < d < 1$ we have*

$$E\{\rho_{n,m(n)}\} = n - n(m) + 0(1) \tag{8.4.30a}$$

where the bound of the 0-term depends only on d. If $m(n) \sim n/2$, we have

$$E\{\rho_{n,m(n)}\} = n - m(n) + 0(\log n). \tag{8.4.30b}$$

If $m(n) \sim dn/2$ with $d > 1$, we have

$$\lim_{n\to\infty} \frac{E\{\rho_{n,m(n)}\}}{n} = \frac{1}{d}\left\{x(d) - \frac{x^2(d)}{2}\right\} \tag{8.4.30c}$$

where $x = x(d)$ is the only solution of the equation $xe^{-x} = de^{-d}$ satisfying $0 < x < 1$.

Note that for $d \ge 4$, $x(d) \doteq de^{-d}$ so that

$$\lim_{n\to\infty} \frac{E\{\rho_{n,m(n)}\}}{n} \doteq e^{-d}. \tag{8.4.31}$$

The *size* of the *largest* component of $G_{n,m(n)}$ is also of interest.

Theorem 8.4.6 *Let $\zeta_{n,m(n)}$ denote the number of vertices in the component of $G_{n,m(n)}$ with the greatest number of vertices. If $m(n) \sim dn$ where $d > 1$, we have for any $\varepsilon > 0$,*

$$\lim_{n\to\infty} \text{Prob}\left\{\left|\frac{\zeta_{n,m(n)}}{n} - G(d)\right| < \varepsilon\right\} = 1 \tag{8.4.32}$$

where $G(d) = 1 - x(d)/d$ and $x = x(d)$ is the only solution of the equation $xe^{-x} = de^{-d}$ satisfying $0 < x < 1$.

Again, if $d \geq 4$, then $x(d) \doteq de^{-d}$, so that

$$\lim_{n\to\infty} \text{Prob}\left\{\left|\frac{\zeta_{n,m(n)}}{n} - 1 + e^{-d}\right| < \varepsilon\right\} = 1. \tag{8.4.33}$$

Theorem 8.4.7 *Suppose $m(n) \sim dn/2$. If $d < 1$, the probability that the graph $G_{n,m(n)}$ is planar*[4] *tends to 1, whereas if $d > 1$ this probability tends to 0.*

We conclude this section with an additional theorem due to Erdös and Rényi [ER3]. This theorem concerns the minimum vertex and branch cut-sets of G. As in Chapter 7, let $\omega(G)$ and $\theta(G)$ be the minimum number of vertices and branches, respectively, which must be removed from G in order to disconnect it.

Theorem 8.4.8 *If $m(n) = (1/2)\log n + (r/2)n \log\log n + \alpha n + o(n)$, where α is a real constant and r a nonnegative integer, then*

$$\lim_{n\to\infty} \text{Prob}\,\{\omega(G_{n,m(n)}) = r\} = \lim_{n\to\infty} \text{Prob}\,\{\theta(G_{n,m(n)}) = r\}$$
$$= 1 - \exp\left(-\frac{e^{-2\alpha}}{r!}\right). \tag{8.4.34}$$

Moreover, if $d_r(G)$ is the number of vertices in G with degree r, then

$$\lim_{n\to\infty} \{d_r(G_{n,m(n)}) = k\} = \frac{\lambda^k e^{-\lambda}}{k!} \qquad \textit{for } k = 0,1,2,\ldots \tag{8.4.35}$$

where $\lambda = e^{-2\alpha}/r!$. In other words, the distribution of $d_r(G_{n,m(n)})$ tends to a Poisson distribution.

*8.5 BIASED GRAPHS

The probability of a direct connection between pairs of vertices in a random graph may vary for different pairs of vertices. For example, if the vertices of G represent the stations of a radio communication network, the connection reliability can be expected to be affected by the distance between stations. Connections between pairs of stations that are large distances apart can be expected to be more vulnerable than connections between pairs of stations that are close together. Any random graph such that the probability of connection between two points is a function of the

[4] A graph is planar if it can be drawn on a plane such that no branches intersect except at vertices.

distance between these points will be called a graph with *distance bias*. The epidemic model previously mentioned is a graph of this type. For a directed graph, if the occurrence of the event $\mathscr{E}_{i,j}$ increases the probability of the event $\mathscr{E}_{j,i}$, we will say that the event has *reciprocity bias*. Suppose the members of a group are asked to name their best friends. If individual v_i names v_j, an observer could expect that the *a priori* probability that individual v_j will name v_i is higher than the probability that v_j will name any other individual.

All the problems we have already discussed for unbiased graphs can be posed for biased graphs. However, the biased problems are considerably more difficult. Therefore, in this section we sometimes deal with expectations rather than with probability distributions. As we shall see, exact solutions results for biased graphs are complicated and computationally infeasible for large graphs.

Consider n points distributed through a region in some specified manner. A distance metric is defined on this region so that the distance between any two points may be calculated. The probability that any two points are adjacent is a function of the distance between them. Let $p(x)$ be a probability density such that $p(x)ds$ is the probability that a given vertex is adjacent to a vertex *in the vicinity* of a point at distance x from the given vertex. The differential ds may represent length, area, or volume. Assume the graph is outwardly homogeneous of degree α, and let $\gamma(x)$ represent the probability that a point at distance x from a given vertex v_1 is connected to v_1 by means of a directed path from v_1. We would like to find $\gamma(x)$. The approach we will use is an extension of the technique used to derive the equation $\gamma = 1 - e^{-\alpha\gamma}$. First we will reconsider the unbiased graph.

We again use the branch-tracing procedure discussed in Section 3 and let $p(t)$ denote the probability that a vertex is contacted at the tth stage of the tracing procedure. The difficulty with the approach used previously is that $p(1), p(2), \ldots, p(t)$ are no longer independent. To avoid this difficulty, let $P(t)$ be the probability that a vertex is contacted *for the first time* at the tth stage. Clearly, if $q(t) = 1 - p(t)$

$$\left[1 - \sum_{j=0}^{t} P(j)\right] q(t+1) = 1 - \sum_{j=0}^{t+1} P(j) \tag{8.5.1}$$

or

$$q(t+1) = \left[1 - \sum_{j=0}^{t+1} P(j)\right]\left[1 - \sum_{j=0}^{t} P(j)\right]^{-1}$$

$$= \frac{\left[1 - \sum_{j=0}^{t} P(j)\right] - P(t+1)}{\left[1 - \sum_{j=1}^{t} P(j)\right]}$$

$$= 1 - P(t+1)\left[1 - \sum_{j=0}^{t} P(j)\right]^{-1}. \tag{8.5.2}$$

Therefore,

$$P(t+1) = \left[1 - \sum_{j=0}^{t} P(j)\right](1 - q(t+1)). \tag{8.5.3}$$

We have seen before that for large n,

$$q(t+1) = e^{-\alpha P(t)}. \tag{8.5.4}$$

Therefore, we obtain the recurrence relation

$$P(t+1) = \left[1 - \sum_{j=0}^{t} P(j)\right][1 - e^{-\alpha P(t)}] \tag{8.5.5}$$

with $P(0) = 1 - (1 - 1/n)^{\alpha}$.

In deriving (8.5.5), we assume that n is large. We now make the additional assumption that the region occupied by the graph may be subdivided into small regions Δx, each of which also contains a large number of points. Furthermore, for simplicity, we will assume that the region is one-dimensional with the origin $x = 0$ at v_i. Then, $p(|x - x_0|)\,dx$ is the probability that a branch emanating from a point at x_0 terminates at a point in the vicinity of a vertex at distance $|x - x_0|$ from x_0. Let ν be the number of vertices per unit length in the graph.

Because of the distance bias, $P(t)$ is also a function of x, which we will write as $P(x,t)$. The expected number of vertices to be contacted for the first time at the tth stage in the region Δx is $\nu P(x,t)\,\Delta x$, and the expected number of branches to be traced from this region is $\alpha\nu P(x,t)\,\Delta x$. Here we have assumed that Δx is sufficiently small so that $P(x,t)$ is constant over the interval.

The probability that the vertex at x_0 is *not* contacted at the t + 1st stage, by tracing a branch emanating from any region, is

$$\prod_{x} [1 - p(|x - x_0|)\,\Delta x]^{\alpha\nu P(x,t)\,\Delta x} \tag{8.5.6}$$

where a representative point x is taken from each region Δx and the product is taken over all Δx. Therefore, the probability that the vertex at x_0 is contacted by at least one branch at the t + 1st stage is

$$1 - \prod_{x} [1 - p(|x - x_0|)\,\Delta x]^{\alpha\nu P(x,t)\,\Delta x}. \tag{8.5.7}$$

The probability that the vertex at x_0 has not been contacted in the first t stages is

$$1 - \sum_{j=0}^{t} P(x_0,j). \tag{8.5.8}$$

Since the dependence on distance has been included in finding the probability of at least one contact at the t + 1st stage, it follows that the probability of first contacting the vertex at x_0 at the $t + 1$st stage is

$$P(x_0,t+1) = \left[1 - \sum_{j=0}^{t} P(x_0,j)\right]\left\{1 - \prod_{x} [1 - p(|x - x_0|)\,\Delta x]^{\alpha v P(x,t)\,\Delta x}\right\}. \tag{8.5.9}$$

In general,

$$P(x,t+1) = \left[1 - \sum_{j=0}^{t} P(x,j)\right]\left\{1 - \prod_{y} [1 - p(|y - x|)\,\Delta y]^{\alpha v P(y,t)\,\Delta y}\right\} \tag{8.5.10}$$

with the initial condition

$$P(y,1) = p(|y - x|)\,\Delta y.$$

This recursion relation can be used to compute successive values of $P(x,t)$, given $P(x,1)$. The probability that the vertex at x will eventually be contacted is then

$$\gamma(x) = \sum_{t=0}^{\infty} P(x,t). \tag{8.5.11}$$

It should be noted that if the density v of vertices is too small, or the distance bias is too strong, the assumption that $P(x,t)$ is constant over each interval Δx is invalid. Rapoport [RA1] has suggested that this approach represents more accurately a graph which may be subdivided into a number of subgraphs, such that, within each subgraph, interconnections are equiprobable and the distance bias is a factor between the various subgraphs. In the contagion model, the subgraphs might represent "generalized households" and the graph a community. For the communication network problem, it seems reasonable to assume that distance bias would not be a factor for stations which are all reasonably close together.

In addition to distance bias, many communication systems exhibit reciprocity bias since establishing a link in one direction often aids in establishing a link in the other. For an unbiased graph, the probability that a directed branch will be reciprocated is $1 - (1 - 1/n)^{\alpha}$ and for large n and fixed α this probability is infinitesimal. A reciprocity bias is imposed on the graph by assigning a nonzero probability φ that a branch is reciprocated for infinite n. Clearly, if $\varphi = 1$, the graph is undirected.

For an unbiased graph, the probability distribution of the inward demi-degree of an arbitrary vertex is easy to find. There are $n\alpha$ directed branches, each of which is incident at a given vertex with probability $1/n$. Therefore, the inward demi-degree is binomially distributed. Let $D^-(i)$ be the random variable corresponding to the inward demi-degree of an arbitrary vertex v_i. For large n, $D^-(i)$ has a probability distribution which is approximately Poisson and is given by

$$\text{Prob}\,\{D^-(i) = k\} = \frac{\alpha^k}{k!}e^{-\alpha}. \tag{8.5.12}$$

For a graph with reciprocity bias, this is no longer true.

If the outward demi-degree of a given vertex is α, the probability that j of the branches incident out of this vertex are reciprocated is

$$\binom{\alpha}{j} \varphi^j (1 - \varphi)^{\alpha - j} \qquad \text{for } j = 0,1,2,\ldots,\alpha. \tag{8.5.13}$$

There are essentially $\alpha(1 - \varphi)$ "uncommitted" branches in the graph. This follows since for large n, φ is the fraction of the total number of branches emanating from a vertex which will be reciprocated. Therefore, the probability of the incidence of $(k - j)$ additional unreciprocated branches is

$$\frac{[\alpha(1 - \varphi)]^{k-j}}{(k - j)!} e^{-\alpha(1-\varphi)}. \tag{8.5.14}$$

Hence, the probability that the inward demi-degree of a given vertex is k is

$$\text{Prob}\,\{D^-(i) = k\} = \left[\sum_{j=0}^{\text{Min}(\alpha,k)} \binom{\alpha}{j} \varphi^j (1 - \varphi)^{\alpha+k-2j} \frac{\alpha^{k-j}}{(k - j)!}\right] e^{-\alpha(1-\varphi)}. \tag{8.5.15}$$

The behavior of Prob $\{D^-(i) = k\}$ for small values of bias can be examined by finding the derivatives with respect to φ of Prob $\{D^-(i) = k\}$ and evaluating at $\varphi = 0$. If this is done, we find

$$\frac{\partial}{\partial \varphi} \text{Prob}\,\{D^-(i) = k\}\big|_{\varphi=0} = 0, \tag{8.5.16a}$$

$$\frac{\partial^2}{\partial \varphi^2} \text{Prob}\,\{D^-(i) = k\}\big|_{\varphi=0} = \frac{e^{-\alpha k-1}}{k!}[k - (k - \alpha)^2]. \tag{8.5.16b}$$

Since the sign of the second derivative is the same as the sign of $[k - (k - \alpha)^2]$, we can conclude that, as the reciprocity bias begins to increase from zero, Prob $\{D^-(i) = k\}$ begins to increase for $k \leq \alpha + \sqrt{\alpha}$ and $k \geq \alpha - \sqrt{\alpha}$ and decreases for the remaining values of k.

We know that if $\varphi = 1$, then Prob $\{D^-(i) = k\} = 0$ for all $k \neq \alpha$ and Prob $\{D^-(i) = \alpha\} = 1$. Therefore, for φ sufficiently close to one, all terms except Prob $\{D^-(i) = \alpha\}$ can be shown to decrease monotonically as the bias increases from zero to one. To see the behavior of Prob $\{D^-(i) = \alpha\}$, consider

$$\begin{aligned}\frac{\partial}{\partial \varphi} \text{Prob}\,\{D^-(i) = \alpha\} &= \sum_{j=0}^{\alpha} \frac{\alpha^{\alpha-j+1}}{(\alpha - j)!} \binom{\alpha}{j} \varphi^j (1 - x)^{2(\alpha-j)} \\ &+ \sum_{j=1}^{\alpha} \frac{\alpha^{\alpha-1}}{(\alpha - j)!} \binom{\alpha}{j} j\varphi^{j-1} (1 - \varphi)^{2(\alpha-j)} \\ &- \sum_{j=0}^{\alpha-1} \frac{\alpha^{\alpha-j}}{(\alpha - j)!} \binom{\alpha}{j} \varphi^j 2(\alpha - j)(1 - \varphi)^{2(\alpha-j)-1}.\end{aligned} \tag{8.5.17}$$

We may treat $\partial/\partial \varphi$ Prob $\{D^-(i) = \alpha\}$ given in (8.5.17) as a polynomial in φ^j, for $j \leq \alpha$, whose coefficients involve powers of $(1 - \varphi)$. The coefficient of φ^j, after some manipulation, may be written as

$$\frac{\alpha^{\alpha-j-1}}{(\alpha-j)!}\binom{\alpha}{j}(1-\varphi)^{2(\alpha-j)-2}[\alpha(1-\varphi)-(\alpha-j)]^2 \tag{8.5.18}$$

which is always nonnegative. Consequently, Prob $\{D^-(i) = \alpha\}$ is a monotone increasing function of φ.

Only partial results are available for the behavior of the other probability terms. For example, if $\alpha = 1$, all the terms decrease as φ increases except Prob $\{D^-(i) = 1\}$ and Prob $\{D^-(i) = 2\}$, and the term Prob $\{D^-(i) = 2\}$ has a single maximum. Other results are: for all $\alpha > 2$, Prob $\{D^-(i) = 1\}$ decreases monotonically as φ increases; for $\alpha = 2$, Prob $\{D^-(i) = 1\}$ has a single maximum at $\varphi = \frac{1}{2}$ and Prob $\{D^-(i) = 0\}$ decreases monotonically.

The random graph with reciprocity bias is difficult to treat analytically because the elementary events $\mathscr{E}_{i,j}$ and $\mathscr{E}_{j,i}$ are not independent. Assume we are given a graph such that the probability of the existence of each possible branch b_k is specified to be p_k. Furthermore, suppose for all i and j, $\mathscr{E}_{i,j}$ and $\mathscr{E}_{j,i}$ are independent. Given two vertices, v_1 and v_2, we would like to find the probability that v_1 and v_2 are connected by at least one path from v_1 to v_2. For undirected graphs, this event is equivalent to the event that v_1 and v_2 are in the same component of G. The procedure to be discussed is applicable to both directed or undirected graphs. For simplicity, we will assume G is undirected.

The set of all 1-2 paths $\{\pi_k : k = 1, \ldots, q\}$ may be enumerated. Let $\lambda = [\lambda_{i,j}]$ be a $q \times m$ path matrix of zeros and ones, such that $\lambda_{i,j} = 1$ if and only if branch b_j is in path π_i.

Define the operation $*$ between 0 and 1 as

$$\begin{aligned} 1 * 0 = 0 * 1 = 1 * 1 = 1 \\ 0 * 0 = 0 \end{aligned} \tag{8.5.19}$$

and let $\mathscr{E}_i$ be the event that all of the branches in π_i are present in G. Clearly

$$\text{Prob}\,\{\mathscr{E}_i\} = \sum_{j=1}^{m} p_j \lambda_{i,j}. \tag{8.5.20}$$

The probability that there is *at least one* path between v_1 and v_2 is

$$\text{Prob}\,\{\mathscr{E}_1 \cup \mathscr{E}_2 \cup \cdots \cup \mathscr{E}_q\} = \sum_{i=1}^{q} \text{Prob}\,\{\mathscr{E}_i\} - \sum_{\substack{i,j \\ i \neq j}} \text{Prob}\,\{\mathscr{E}_i \cap \mathscr{E}_j\} + \cdots + (-1)^{q-1}\,\text{Prob}\,\{\mathscr{E}_1 \cap \mathscr{E}_2 \cap \cdots \cap \mathscr{E}_q\} \tag{8.5.21}$$

and, if the events $\mathscr{E}_{i,j}$ are independent, this probability may be written as

$$\text{Prob}\,\{\mathscr{E}_1 \cup \cdots \cup \mathscr{E}_q\} = \sum_{i=1}^{q} \prod_{k=1}^{m} p_k^{\lambda_{i,k}} - \sum_{\substack{i,j \\ i \neq j}} \prod_{k=1}^{m} p_k^{\lambda_{i,k} * \lambda_{j,k}} + \cdots$$

$$+ (-1)^{q-1} \prod_{k=1}^{m} p_k^{\lambda_{1,k} * \lambda_{2,k} * \cdots * \lambda_{q,k}}. \tag{8.5.22}$$

Thus, the probability that v_1 and v_2 are in the same component could be found by performing the set of tedious operations described by (8.5.22).

We can find an exact solution even if the probabilities of the existence of branches in G are not independent. To do this, we must have the joint probability distribution of the events $\mathscr{E}_{i,j}$. Let X_i be a random variable such that $X_i = 1$, if branch b_i is in G, and $X_i = 0$ otherwise and let $u_1, u_2, \ldots, u_m$ be a set of real variables. If the joint probability distribution of $X_1, \ldots, X_m$ is known, we find the generating function $g(u_1, \ldots, u_m)$ defined by

$$g(u_1, \ldots, u_m) = E\{u_1^{X_1} u_2^{X_2} \cdots u_m^{X_m}\}. \tag{8.5.23}$$

For example, if $X_1, \ldots, X_m$ are independent,

$$g(u_1, \ldots, u_m) = \prod_{i=1}^{m} E\{u_i^{X_i}\} = \prod_{i=1}^{m} (q_i + p_i u). \tag{8.5.24}$$

Define an index set I_i such that

$$I_i = \{j | \lambda_{j,i} = 1\} \qquad \text{for } i = 1, 2, \ldots, m. \tag{8.5.25}$$

I_i is simply the set of numbers of the paths in which branch b_i appears.

We must find the joint probability distribution of the random variables $Y_1, Y_2, \ldots, Y_q$ where Y_j is the number of branches that appear in path π_j. The Y_j's are related to the X_i's by

$$\begin{bmatrix} Y_1 \\ Y_2 \\ \vdots \\ \cdot \\ \cdot \\ Y_q \end{bmatrix} = \begin{bmatrix} \lambda_{1,1} & \cdots & \lambda_{1,m} \\ \cdot & & \cdot \\ \cdot & & \cdot \\ \cdot & & \cdot \\ \lambda_{q,1} & \cdots & \lambda_{q,m} \end{bmatrix} \begin{bmatrix} X_1 \\ X_2 \\ \vdots \\ \cdot \\ \cdot \\ X_m \end{bmatrix}. \tag{8.5.26}$$

Let $h(s_1, \ldots, s_q)$ be the generating function of the Y_j's and define the functions $S_1, S_2, \ldots, S_m$ by

$$S_i = \prod_{j \in I_i} s_j \qquad \text{for } i = 1, \ldots, m. \tag{8.5.27}$$

Then, $h(s_1, \ldots, s_q)$ is easily found by noting that

$$h(s_1, \ldots, s_q) = g(S_1, \ldots, S_m). \tag{8.5.28}$$

The generating function $h(s_1, \ldots, s_q)$ contains more information than is actually needed. We are interested in whether the path π_i is present in G. This is equivalent to the random variable Y_i attaining its maximum value, which we will denote as m_i. Hence, we define the random variable Z_i such that $Z_i = 1$ if $Y_i = m_i$ and $Z_i = 0$ if $Y_i < m_i$ for $i = 1,2,\ldots,m$. Clearly, the event $Z_1 = Z_2 = \cdots = Z_q = 0$ corresponds to the event that v_1 and v_2 are in different components of G.

Let

$$h^*(w_1, \ldots, w_q) = E\left\{\prod_{i=1}^{q} w_i^{Z_i}\right\} \tag{8.5.29}$$

be the generating function of $(Z_1, \ldots, Z_q)$. This function may be written as a sum of products of the form

$$h^*(w_1, \ldots, w_q) = \sum_{k_1,\ldots,k_q} \text{Prob}\,\{Z_1 = k_1, \ldots, Z_q = k_q\} \prod_{i=1}^{q} w_i^{k_i} \tag{8.5.30}$$

where $k_i = 0$ or $k_i = 1$. We also have

$$h(s_1, \ldots, s_q) = \sum_{k'_1,\ldots,k'_q} \text{Prob}\,\{Y_1 = k'_1, \ldots, Y_q = k'_q\} \prod_{i=1}^{q} s_i^{k'_i} \tag{8.5.31}$$

where $k_i' = 0,1,2,\ldots,m_i$. The relationship between $h(s_1, \ldots, s_q)$ and $h^*(w_1, \ldots, w_q)$ is simple. If in $h(s_1,\ldots,s_q)$, we replace any term $s_k^{m_i}$ by w_i and $s_i^{m_i - j}$ by unity for $j = 1,2,\ldots,m_i$, we obtain $h^*(w_1, \ldots, w_q)$. The probability that v_1 and v_2 are in the same component is then found to be $1 - h^*(0, \ldots, 0)$. The number of computations required for the above procedure could be enormous since $g(u_1, \ldots, u_m)$ has 2^m terms, and hence the method is not directly useful for graphs with more than about 25 branches.

*8.6 PERCOLATION PROCESSES

The major difficulty with all probabilistic approaches to random graph problems is that exact results for graphs with a large number of vertices are nearly always too cumbersome to yield useful results. Thus we usually make the assumption that we are dealing with an infinite random graph and search for asymptotic properties. This approach often gives useful expressions which are close to the exact solution.

In this section, we will examine "percolation" on infinite graphs. Conceptually, percolation processes are quite similar to the threshold and connectivity problems that we have already discussed. Roughly speaking, if we "excite" a single vertex of the graph, we would like to determine whether the number of vertices which are eventually excited is finite or infinite. If we excite a subset of vertices of G, which in some sense form a boundary around a large but finite subset of vertices, we would also like to determine the proportion of the vertices in this bounded set which will eventually be excited.

Since we begin our study with the assumption that the graph G has an infinite number of vertices, it is appropriate to limit the structure of our graphs. Hence, we will make several assumptions which, in the light of the physical motivation of our study, are entirely reasonable. However, first we must introduce several terms.

Let $\mu_i(k)$ denote the number of distinct k-branch paths starting at vertex v_i. Two vertices v_i and v_j are *outlike* if for each k the number of distinct k-branch paths starting from v_i is equal to the number of distinct k-branch paths starting from v_j. In other words, v_i and v_j are outlike if $\mu_i(k) = \mu_j(k)$ for $k = 1,2,3,\ldots$. An *outlike class* is a set of vertices any pair of which are outlike.

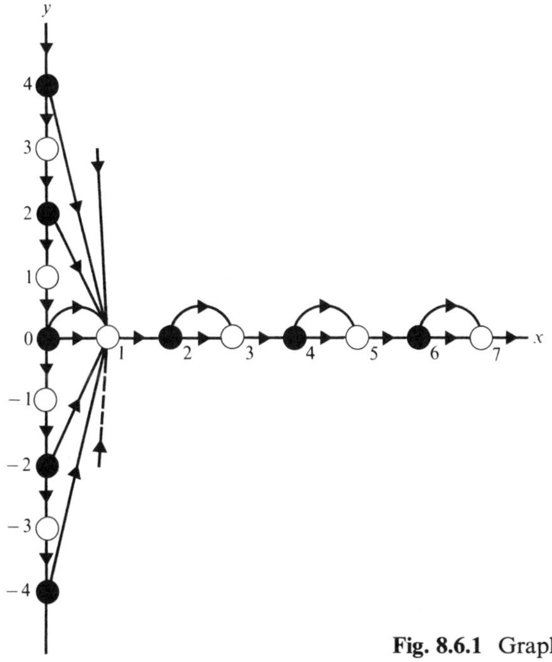

Fig. 8.6.1 Graph for Example 8.6.1.

Example 8.6.1 Vertices are located at integral points of the x- and y-axis of the Euclidian plane and connected by directed branches, as shown in Fig. 8.6.1. On the x-axis for each i the vertex located at $(2i - 1,0)$ is connected to $(2i,0)$ by a single outwardly directed branch and the vertex at $(2i,0)$ is connected to the vertex at $(2i + 1,0)$ by two outwardly directed branches; on the y-axis, for all integer j, the vertex at $(0,2j)$ is connected to the vertices at $(0,2j - 1)$ and $(1,0)$ by single outwardly directed branches and the vertex at $(0,2j + 1)$ is connected to $(0,2j)$ by a single outwardly directed branch.

Vertices with even coordinates, $(2i,0)$ or $(0,2j)$, have the same number of distinct k-branch paths; similarly vertices with odd coordinates, $(2i + 1,0)$ or $(0,2j + 1)$, have the same number of distinct k-branch paths. Hence there are two outlike classes; the sets of vertices $U_1 = \{(2i,0),(0,2j) \text{ for } j = \cdots -3,-2,-1,0,1,2,\ldots\}$ and $U_2 = \{(2i + 1,0),(0,2j + 1), \text{ for } i = 0,1,2,\ldots\}$.

We now restrict the structure of the graphs to be considered.

Assumption 8.6.1 *Each vertex of G belongs to one of a finite number of outlike classes, denoted by* $U_1, U_2, \ldots, U_S$.

Assumption 8.6.2 *The outward demi-degree of each vertex of G is finite.*

Assumption 8.6.3(a) *Any finite subset of vertices of G contains a vertex with a descendant not in the subset.*

Assumption 8.6.3(b) *If a subset of vertices does not contain at least one vertex from every outlike class, then it contains a vertex with a descendant not in that subset.*

Assumption 8.6.3(a) eliminates graphs which are composed of an infinite number of finite unconnected components. Without 8.6.3(a), we could construct graphs where each such component has a different asymptotic behavior. Clearly, general results would not be meaningful in such a case. Assumption 8.6.3(b) eliminates graphs composed of two or more infinite graphs each of which satisfies Assumptions 8.6.1, 8.6.2, and 8.6.3(a) but whose outlike classes are different.

If G satisfies Assumptions 8.6.1 to 8.6.3(b), it is possible to define a constant k which gives information concerning the number of distinct "long" paths in the graph. Let $\mu_i(k,r(k))$ denote the number of distinct k-branch directed *chains*[5] starting from v_i which can be broken into $r(k)$ or fewer paths. If $r(k) = 1$, then each of these chains must itself be a path and so $\mu_i(k,1)$ represents the number of distinct directed paths, with k branches, which originate at v_i. For any two vertices v_i and v_t in an outlike class U_j, we have that $\mu_i(k,1) = \mu_t(k,1) = \mu_t(k)$. Define the numbers $n_j(k)$ by

$$n_j(k) = \mu_i(k) \qquad \text{for } v_i \in U_j, \text{ and } j = 1,2,\ldots,S. \tag{8.6.1}$$

Then, if we define $\psi(k) = k^{-1} \log \left(\operatorname*{Max}_{1 \le j \le S} [n_j(k)] \right)$, there exists a constant K defined by

$$K = \operatorname*{Inf}_{k \ge 1} \psi(k) \tag{8.6.2}$$

and we can show:

Theorem 8.6.1 *For any vertex v_i of a graph G satisfying Assumptions 8.6.1 to 8.6.3(b),*

$$0 \le K = \operatorname*{Inf}_{k \ge 1} \psi(k) = \lim_{k \to \infty} \left(\frac{1}{k}\right) \log \mu_i(k,r(k)) < \infty \tag{8.6.3}$$

provided that

$$\lim_{k \to \infty} \frac{r(k)}{k} = 0. \tag{8.6.4}$$

The proof of this theorem is complicated [HA2] and is therefore omitted. If $r(k) = 1$ for all k, Theorem 8.5.1 may be restated as Corollary 8.6.1.

---

[5] A k-branch directed *chain* is a sequence of branches $b_1, b_2, \ldots, b_k$, in which each branch b_j has its initial vertex in common with b_{j-1} and its terminal vertex in common with b_{j+1} for $j = 2, \ldots, (k-1)$. A branch may appear more than once in a chain.

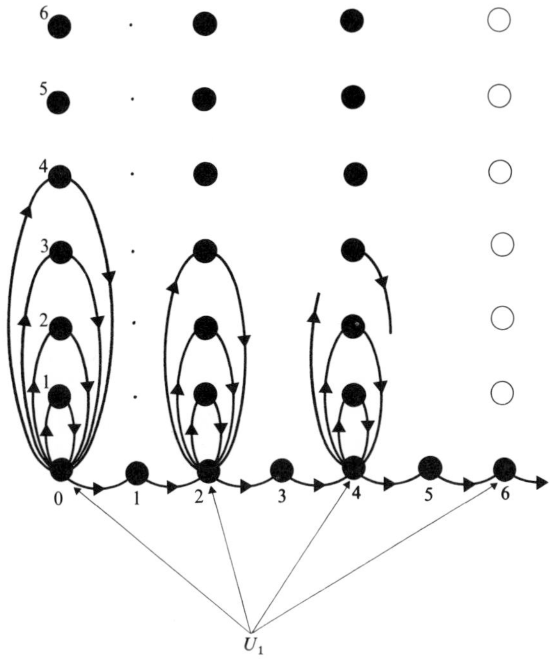

Fig. 8.6.2 Graph used to show the necessity of Assumptions 8.6.1 to 8.6.3(b) for Theorem 8.6.1 to hold.

Corollary 8.6.1 *The number of distinct k-branch directed paths starting from any vertex of a graph G with constant K defined by (8.6.2) is* exp $(kK + 0(k))$ *as* $k \to \infty$.

It can be shown by counterexamples that each of the Assumptions 8.6.1 to 8.6.3(b) is necessary for the conclusions of Theorem 8.6.1 to be true. For example, 8.6.3(a) or 8.6.3(b) must be satisfied since, otherwise, we can take two graphs with different constants K and consider these graphs as a single graph. As another example, consider the graph shown in Fig. 8.6.2. The vertices are the points $(2i,j)$ and $(2i - 1,0)$ in the Euclidian plane for $i,j = 0,1,2,\ldots$. Vertex $(i,0)$ is connected to $(i + 1,0)$ by a directed branch incident into the point $(i + 1,0)$ for $i = 0,1,2,\ldots,$ and the point $(2i,j)$ where $j \neq 0$ is connected to the point $(2i,0)$ by two oppositely directed branches. Assumption 8.6.2 is violated at every vertex on the x-axis. Clearly, there are two outlike classes U_1 and U_2; U_1 consists of every vertex on the x-axis of the form $(2i,0)$ while U_2 consists of the remaining vertices. Then, obviously $n_1(1) = \infty$ and since any vertex $(2i,0)$ in U_1 can reach a vertex in U_1 of the form $(2i + 2j,0)$ by a path containing $2j$ branches, it follows that $n_1(3) = n_1(5) = \cdots n_1(1 + 2j) = \infty$ for $j = 0,1,2,3,\ldots$. For any other k, $n_1(k) = 1$, since for each even k there is only one distinct path of k branches emanating from any element in U_1. For any vertex in U_2, the situation is reversed. There is only one way to leave a vertex in U_2. Hence $n_2(2j) = \infty$ for $j = 1,2,3,\ldots,$ and $n_2(1 + 2j) = 1$ for $j = 1,2,\ldots$. Thus, $\lim_{k\to\infty} (1/k)n_i(k)$ does not exist for either $i = 1$ or $i = 2$.

We may weaken the restrictions on the graph G. Suppose that only Assumptions 8.6.2 and 8.6.3(a) hold. The *connective bound* λ of the graph is defined to be

$$\lambda = \operatorname*{Sup}_i \left[\lim_{k\to\infty} \operatorname{Sup} \{k^{-1} \log n_i(k)\}\right]. \tag{8.6.5}$$

Clearly, if the constant K exists, then $K = \lambda$. It can be shown that Corollary 8.6.1 remains valid if K is replaced by λ. Suppose now that some of the branches of the infinite graph G are broken at random. Furthermore, suppose the probability that any branch in G *is not* broken is p. This model would seem to coincide with the unbiased models discussed in earlier sections. However, it is more general since we specifically allow parallel branches in G. Hence, by suitable choice of p and the number of branches between a given pair of vertices, we can arbitrarily closely approximate the biased model discussed in the latter part of Section 5.

A further restriction on the class of graphs we are treating is necessary. A graph G which satisfies Assumptions 8.6.1 to 8.6.3(b) is said to be *reversible* if G', the graph obtained from G by reversing all branch directions, also satisfies Assumptions 8.6.1 to 8.6.3(b). A random graph G which satisfies Assumptions 8.6.4 to 8.6.5 below is said to be a *random maze*.

Assumption 8.6.4 *G is reversible.*

Assumption 8.6.5 *Each branch of G has, independent of all other branches, a fixed probability $q = 1 - p$ of being broken.*

If G is a random maze, we obtain a *reverse maze* G' by reversing the direction of each branch of G. Vertices in G' are denoted by primes; that is, if v_i is a vertex of G, then v_i' is the corresponding vertex of G'. Suppose that an excitation is applied to vertex v_i. We would like to find the probability that only a finite number of vertices are eventually excited. Let $\hat{p}(v_i,p)$ denote this probability. Clearly $\hat{p}(v_i,0) = 1$ and by Assumptions 8.6.3(a) and 8.6.3(b), $\hat{p}(v_i,1) = 0$; moreover, $\hat{p}(v_i,p)$ is a monotone decreasing function of p. Hence we can define the *critical probability* $p_c(v_i)$ by

$$p_c(v_i) = \operatorname*{Sup}_{\hat{p}(v_i,p)=1} \{p\}. \tag{8.6.6}$$

Theorem 8.6.2 *If there is a finite path from v_i to v_j and another, not necessarily distinct, path from v_j to v_i in the graph G, then $p_c(v_i) = p_c(v_j)$.*

Proof. Suppose the shortest i-j path has k branches. The probability that this path exists is p^k. If this path exists and an infinite number of vertices are eventually excited when v_j is excited, then an infinite number of vertices will be excited when v_i is excited. Hence

$$1 - \hat{p}(v_i,p) \geq p^k\{1 - \hat{p}(v_j,p)\}. \tag{8.6.7}$$

Clearly $\hat{p}(v_i,p) = 1$ implies $\hat{p}(v_j,p) = 1$. Thus,

$$p_c(v_j) = \operatorname*{Sup}_{\hat{p}(v_j,p)=1} \{p\} = \operatorname{Max}\left[p_c(v_i), \operatorname*{Sup}_{\substack{\hat{p}(v_j,p)=1\\ \hat{p}(v_i,p)=0}} \{p\}\right] \geq p_c(v_i). \tag{8.6.8}$$

A similar argument shows that $p_c(v_j) \leq p_c(v_i)$ and the theorem follows.

Corollary 8.6.2 *If G is a strongly connected graph, then $p_c(v_i)$ is independent of v_i. (G is strongly connected if there is at least one directed path between every pair of vertices.)*

We can define another type of critical probability as follows. Let $S_k(v_i)$ be the set of *ancestors* of v_i such that the shortest path from an element in this set to v_i has exactly k branches. Suppose we excite every vertex in this set and no other vertices. We now seek the probability that v_i is eventually excited. We denote this probability by $w(v_i,k,p)$ and note that this probability is a monotone decreasing function of k. The set $S_k(v_i)$ cannot be empty since the reverse graph G' must have at least one path of every length starting at v_i. A bounded sequence of monotone functions must have a limit. Hence, let $w(v_i,p) = \lim_{k\to\infty} w(v_i,k,p)$. Clearly, $w(v_i,0) = 0$ and $w(v_i,p)$ is a monotone *increasing* function of p. Then, we define a critical probability

$$p_w(v_i) = \operatorname*{Sup}_{w(v_i,p)=0} \{p\}. \tag{8.6.9}$$

We will show that the two critical probabilities $p_c(v_i)$ and $p_w(v_i)$ are closely related; in fact, if v_i' is the vertex of the reverse graph G' which corresponds to v_i of G, then $p_w(v_i) = p_c(v_i')$. This result is derived in the next theorem.

Theorem 8.6.3 $w(v_i,p) + \hat{p}(v_i',p) = 1.$

Proof. Let $\mathscr{E}_k'$ denote the event that in the reverse maze v_i' excites at least k vertices when it is the only initially excited vertex. Similarly, let $\mathscr{E}_\infty'$ denote the event that v_i' excites infinitely many vertices. Clearly,

$$\lim_{k\to\infty} \text{Prob}\,\{\mathscr{E}_k'\} = \text{Prob}\,\{\mathscr{E}_\infty'\} \tag{8.6.10}$$

and, by definition of $\hat{p}(v_i',p)$,

$$\text{Prob}\,\{\mathscr{E}_\infty'\} = 1 - \hat{p}(v_i',p). \tag{8.6.11}$$

Suppose that v_i is excited when the vertices of $S_k(v_i)$ are the only excited vertices. Let this event be denoted by $\mathscr{E}_k$; by definition Prob $\{\mathscr{E}_k\} = w(v_i,k,p)$. The occurrence of $\mathscr{E}_k$ implies there exists at least one j-i path containing k branches. Hence, in the reverse maze there exists a k-branch path from v_i' to v_j'. Therefore, the k vertices on this path, not including v_i', are excited and $\mathscr{E}_k$ implies $\mathscr{E}_k'$. Consequently

$$w(v_i,k,p) = \text{Prob}\,\{\mathscr{E}_k\} \le \text{Prob}\,\{\mathscr{E}_k'\} \qquad \text{for all } k. \tag{8.6.12}$$

Since $w(v_i,k,p)$ is a nonincreasing function of k,

$$w(v_i,k,p) \le \text{Prob}\,\{\mathscr{E}_k'\} \qquad \text{for all } k. \tag{8.6.13}$$

Letting $k \to \infty$, we obtain

$$w(v_i,p) \le 1 - \hat{p}(v_i,p). \tag{8.6.14}$$

We must now show that the reverse inequality is also true. Suppose that the event $\mathscr{E}_\infty'$ occurs (i.e., in G' an infinite number of vertices are excited). This means that in random maze G', either an infinite length path from v_i exists or there are an infinite number of finite length paths. The latter possibility cannot occur because G' must obey Assumption 8.6.2. Thus, there must be an infinite length path originating at v_i'. But, this means that in G the event $\mathscr{E}_k$ occurs for every k. If we excite every vertex in $S_k(v_i)$, we must excite all vertices on the infinite path *to* v_i. Therefore,

$$1 - \hat{p}(v_i',p) = \text{Prob}\,\{\mathscr{E}'_\infty\} \leq \text{Prob}\,\{\mathscr{E}_k\} = w(v_i,k,p) \tag{8.6.15}$$

and, if we let $k \to \infty$,

$$1 - \hat{p}(v_i',p) \leq w(v_i,p). \tag{8.6.16}$$

Hence, the theorem is proved:

Corollary 8.6.3 *If there is a finite path from v_i to v_j and another, not necessarily distinct, finite path from v_j to v_i, then $p_w(v_i) = p_w(v_j)$.*

Corollary 8.6.4 *$p_w(v_i) = p_c(v_i')$ and hence we need consider only one critical probability, say $p_c(v_i)$.*

Corollary 8.6.5 *If the random maze G is strongly connected when $p = 1$, then $p_c(v_i)$ is independent of v_i.*

The constant defined in (8.6.2) and critical probability are related by the following theorem.

Theorem 8.6.4 *If G is a random maze, $p_c(v_i) \geq e^{-K}$.*

Proof. Let $p(j,k)$ be the probability that in the maze G, exactly j of the $\mu_i(k+1)$ k-branch paths exist. If infinitely many vertices are excited when v_i is excited, at least one k-branch path must exist for every k. Thus

$$0 \leq 1 - \hat{p}(v_i,p) \leq \sum_{j=1}^{\infty} p(j,k) \leq \sum_{j=0}^{\infty} jp(j,k). \tag{8.6.17}$$

The expected number of distinct k-branch paths in the maze G' is $\sum_{j=0}^{\infty} jp(j,k)$, and

$$\sum_{j=0}^{\infty} jp(j,k) = p^k n_l(k+1) \qquad \text{where } v_i \in U_l. \tag{8.6.18}$$

Therefore, for $n_l(k+1)$ defined in (8.6.1),

$$0 \leq 1 - \hat{p}(v_i,p) \leq p^k n_l(k+1). \tag{8.6.19}$$

By Corollary 8.6.1, $n_l(k+1) = e^{-Kk+0(k)}$ as $k \to \infty$ and thus, if $p < e^{-K}$, as $k \to \infty$

$$0 \leq 1 - \hat{p}(v_i,p) \leq e^{-kK}\, e^{-Kk+0(k)} \to 0 \qquad \text{as } k \to \infty. \tag{8.6.20}$$

Therefore, $p < e^{-K}$ implies $\hat{p}(v_i,p) = 1$. Since $p_c(v_i) = \sup_{p(v_i,p)=1} \{p\}$, it follows that $p_c(v_i) \geq e^{-K}$. (We have previously defined the connective bound λ, making only the

Assumptions 8.6.2 and 8.6.3(a). It is also possible to show that under these assumptions $p_c(v_i) \geq e^{-\lambda}$.)

We can also give upper bounds for the critical probability. First we must introduce several terms. Given a set of vertices S, any branch which is incident out of an element of S but incident into a vertex not in S will be called an *out branch*. An out branch is an *external out branch* if it is the first branch of an infinite path whose vertices are not in S, except for the initial vertex. A *neighborhood* of v_i is the union of all vertices which appear in a finite set of finite paths starting with v_i. Two *neighborhoods* are distinct if they do not contain the same vertices. The *principal neighborhood of order* k of v_i is the union of all vertices in the set of k-branch paths starting with v_i. Finally, an *assembly* is a neighborhood in which all out branches are external.

From Assumption 8.6.2, each neighborhood of v_i has a finite number of out branches. Also, the number of distinct neighborhoods of v_i which have a given finite number of vertices is finite; therefore, the set of distinct *neighborhoods* is countable.

Let $\xi(v_i)$ be the set of assemblies of the vertex v_i. Partition this set into not necessarily disjoint sets $\xi_1(v_i), \xi_2(v_i), \ldots$, such that each element of $\xi_k(v_i)$ has at least k out branches. Let $\xi_n^*(v_i)$ be the set of assemblies of v_i which have exactly k out branches. These $\xi_k^*(v_i)$ are mutually disjoint for different values of k and clearly

$$\xi(v_i) = \bigcup_{k=1}^{\infty} \xi_k(v_i) = \bigcup_{k=1}^{\infty} \xi_k^*(v_i). \tag{8.6.21}$$

Now, let $g_k(v_i)$ be the number of distinct assemblies contained in $\xi_k(v_i)$ and let $g_k^*(v_i)$ be the number of distinct assemblies contained in $\xi_k^*(v_i)$. Furthermore, define the series $\mathscr{g}(z)$ and $\mathscr{g}^*(z)$ in the complex variable z by

$$\mathscr{g}(z) = \sum_{k=1}^{\infty} g_k(v_i) z^k \tag{8.6.22a}$$

$$\mathscr{g}^*(z) = \sum_{k=1}^{\infty} g_n^*(v_i) z^k. \tag{8.6.22b}$$

Let $\nu(v_i)$ and $\nu^*(v_i)$ be the radii of convergence of $\mathscr{g}(z)$ and $\mathscr{g}^*(z)$, respectively. The remainder of this section is devoted to the proof of the following theorem.

Theorem 8.6.5 $p_c(v_i) \leq 1 - \nu^*(v_i) \leq 1 - \nu(v_i)$.

Proof. First, examine the trivial case in which $g_k^* = \infty$ for some k. From the definition of $\xi_k(v_i)$ and $\xi_k^*(v_i)$, there exists an $l \leq k$ such that $g_l(v_i) = \infty$. Therefore $\nu(v_i) = \nu^*(v_i) = 0$ and the upper bound given in the statement of the theorem reduces to $p_c(v_i) \leq 1$.

Now, suppose $g_k(v_i)$ and $g_k^*(v_i)$ are finite for all finite k. Let z be a real number with $0 \leq z \leq 1$. Let N be a given assembly of v_i. Since the set of assemblies of v_i

are countable, we can write $g(z)$ and $g^*(z)$ as sums over the assemblies:

$$g(z) = \sum_N z^{r(N)} \tag{8.6.23a}$$

and

$$g^*(z) = \sum_N z^{r^*(N)} \tag{8.6.23b}$$

where the functions $r(N)$ and $r^*(N)$ are defined by

$$N \in \xi_{r(N)}(v_i) \tag{8.6.24a}$$

and

$$N \in \xi^*_{r^*(N)}(v_i). \tag{8.6.24b}$$

From the definitions of $\xi_k(v_i)$ and $\xi^*_k(v_i)$, we can see that N has exactly $r^*(N)$ out branches and at least $r(N)$ out branches. This implies

$$r^*(N) \geq r(N). \tag{8.6.25}$$

Hence, (8.6.23) to (8.6.25) give

$$g(z) \geq g^*(z) \qquad \text{for } 0 \leq z \leq 1.$$

Both $g(z)$ and $g^*(z)$ are infinite when $z = 1$. Therefore,

$$0 \leq v(v_i) \leq v^*(v_i) \leq 1. \tag{8.6.26}$$

Then, by virtue of (8.6.26), we have only to show that

$$p_c(v_i) \leq 1 - v^*(v_i). \tag{8.6.27}$$

Let q be a real number such that $0 \leq q < v^*$. Since $g^*(q)$ is finite, if $q = 1 - p$ is the probability that an arbitrary branch is broken, then q^k is the probability that for $N \in \xi^*_k(v_i)$, all out branches are broken (if the branches have independent probabilities of being broken). Let $\mathscr{E}_k$ be the event that at least one of the assemblies in $\xi^*_k(v_i)$ has all of its out branches broken. Then

$$\text{Prob}\,\{\mathscr{E}_k\} \leq g^*_k q^k \tag{8.6.28}$$

and, consequently,

$$\sum_{k=1}^{\infty} \text{Prob}\,\{\mathscr{E}_k\} \leq \sum_{k=1}^{\infty} g^*_k(v_i) q^k = g(q) < \infty; \tag{8.6.29}$$

$g^*(q)$ is a convergent series. Therefore, there exists a number k_0 such that

$$\sum_{k > k_0} \text{Prob}\,\{\mathscr{E}_k\} \leq \tfrac{1}{2}. \tag{8.6.30}$$

Let $\hat{\mathscr{E}}_0$ be the event that each assembly of v_i which possesses more than k_0 out branches has at least one unbroken out branch. We then have

$$\text{Prob}\,\{\hat{\mathscr{E}}_0\} \geq \tfrac{1}{2}. \tag{8.6.31}$$

Let $\hat{\mathscr{E}}_1$ be the event that each assembly of v_i with no more than k_0 out branches has at least one unbroken out branch. Since the number of such assemblies and the number of their out branches are finite

$$\text{Prob}\,\{\hat{\mathscr{E}}_1\} > 0. \tag{8.6.32}$$

Finally, if $\hat{\mathscr{E}}$ is the event that each assembly of v_i contains at least one unbroken out branch, we can write

$$\text{Prob}\,\{\hat{\mathscr{E}}\} = \text{Prob}\,\{\hat{\mathscr{E}}_1|\hat{\mathscr{E}}_0\}\ \text{Prob}\,\{\hat{\mathscr{E}}_0\} \geq \tfrac{1}{2}\,\text{Prob}\,\{\hat{\mathscr{E}}_1|\hat{\mathscr{E}}_0\} \geq \tfrac{1}{2}\,\text{Prob}\,\{\hat{\mathscr{E}}_1\} > 0, \tag{8.6.33}$$

because the conditional probability of $\hat{\mathscr{E}}_1$, given that $\hat{\mathscr{E}}_0$ is realized, is at least as large as the *a priori* probability of $\hat{\mathscr{E}}_1$.

Vertex v_i must be connected to an infinite number of vertices with nonzero probability. If v_i has a finite set S of descendant vertices, all out branches (and therefore all external out branches) must be broken. If the set S is not an assembly, consider any arbitrary out branch which is not an external out branch and add to it the largest possible neighborhood which does not have vertices in S. This neighborhood must be finite; if not, the out branch in question would be external. Repeating this process for each nonexternal out branch of S, we obtain from S an assembly S_0 such that all out branches of S_0 are external out branches of S. If all the out branches of S_0 are broken, the event $\hat{\mathscr{E}}$ cannot be realized when v_i has only a finite number of descendants. Since Prob $\{\hat{\mathscr{E}}\} > 0$, there exists a nonzero probability that v_i has an infinite number of descendants. Thus, $p = 1 - q \geq p_c$, for all $q < v^*$, which is the desired result. This completes the proof of the theorem.

8.7 SOME PROBLEMS RELATED TO THE SYNTHESIS OF INVULNERABLE NETWORKS

As we have seen, the evaluation of exact expressions for Prob $\{\rho = 1\}$ is complicated by the enormous number of terms which must usually be computed. Thus, the exact results we have given are usable only for small graphs or graphs which possess a high degree of symmetry. The designer of a network is faced with two alternatives. (1) If the desired network is small, he may use exact methods to compute the reliabilities of a number of proposed systems. Then, basing his decision on the calculations and all other pertinent factors, he can choose the most suitable network. (2) If the network is large, he can attempt to use asymptotic results to obtain qualitative information concerning the structure of "good" networks. For example, such information might indicate the structure of graphs which achieve given reliabilities with minimum number of branches. In this section, we will examine several variations of the second alternative.

We first consider a biased random undirected graph $G = (V,\Gamma)$. Initially we allow the graph to contain parallel branches. With each branch we associate the reliability p. Thus if a branch $[i,j] \in \Gamma$, Prob $\{\mathscr{E}_{i,j}\} = p$ and, if $[i,j] \notin \Gamma$, Prob $\{\mathscr{E}_{i,j}\} = 0$. For this model, we first give an upper bound on Prob $\{\rho = 1\}$. If G is to be connected, there can be no isolated vertices. This means that at least one branch must be incident at every vertex. As in Section 4, let Q_i be the event that there are no intact branches incident at v_i. The union of the events Q_i for $i = 1, \ldots, n$ is the event that at least one vertex has no intact branches and the complementary event is therefore the probability that every vertex has at least one existing incident branch. Thus,

$$\text{Prob}\,\{\rho = 1\} \leq 1 - \text{Prob}\left\{\bigcup_{i=1}^{n} Q_i\right\} \tag{8.7.1}$$

and by applying Bonferroni's Inequality (see Section 4), we obtain

$$\text{Prob}\,\{\rho = 1\} \leq 1 - \sum_{i=1}^{n} \text{Prob}\,\{Q_i\} + \sum_{\substack{i,j \\ i<j}} \text{Prob}\,\{Q_i \cap Q_j\}. \tag{8.7.2}$$

Let $d(i)$ again denote the degree of v_i, and let $q = 1 - p$ be the probability of branch failure. Clearly,

$$\text{Prob}\,\{Q_i\} = q^{d(i)} \qquad \text{for } i = 1,2,\ldots,n. \tag{8.7.3}$$

Furthermore, if $a_{i,j}$ denotes the number of parallel branches between vertices v_i and v_j, Prob $\{Q_i \cap Q_j\}$ can be written as

$$\text{Prob}\,\{Q_i \cap Q_j\} = q^{d(i)+d(j)-a_{i,j}}. \tag{8.7.4}$$

This means that (8.7.2) has the form

$$\text{Prob}\,\{\rho = 1\} \leq 1 - \sum_{i=1}^{n} q^{d(i)} + \sum_{i=1}^{n-1} \sum_{j=i+1}^{n} q^{d(i)+d(j)-a_{i,j}}. \tag{8.7.5}$$

The right-hand side of (8.7.5) is an upper bound on Prob $\{\rho = 1\}$. If, for a given graph, we obtain an upper bound close to one, the bound conveys little information. On the other hand, if the bound is low, we know that G is unreliable. Thus, we conclude that graphs with

$$\sum_{i=1}^{n} q^{d(i)} - \sum_{i=1}^{n-1} \sum_{j=i+1}^{n} q^{d(i)+d(j)-a_{i,j}} \tag{8.7.6}$$

large are worse than graphs for which this number is small. With this assumption, we would like to maximize the right-hand side of (8.7.5) by varying the $d(i)$'s and $a_{i,j}$'s. To guarantee that the numbers we obtain are realizable as a graph, we must impose the constraints

$$\sum_{i=1}^{n} d(i) = 2m; \tag{8.7.7a}$$

$$\sum_{i=1}^{n-1} \sum_{j=i+1}^{n} a_{i,j} = m; \tag{8.7.7b}$$

$$a_{i,j} \leq \text{Min}\,[d(i),d(j)]. \tag{8.7.7c}$$

The constrained optimization problem of maximizing the right-hand side of (8.7.5) subject to the constraints (8.7.7a) to (8.7.7b) can be solved by using Lagrange Multipliers [HA1]. The solution of this problem can be shown to satisfy (8.7.7c) and hence is a solution to our problem. In fact, for $n > 2$, it can be shown that

$$d(i) = \frac{2m}{n} = v \tag{8.7.8a}$$

and

$$a_{i,j} = \frac{2m}{n(n-1)} = \frac{v}{n-1}. \tag{8.7.8b}$$

Hence, we can bound Prob $\{\rho = 1\}$ by a function of v and investigate the effect on this bound of varying v:

$$\text{Prob}\,\{\rho = 1\} \le 1 - nq^{v}\left[1 - \frac{n-1}{2}\,q^{v(1-1/n)}\right]. \tag{8.7.9}$$

Note the relation of (8.7.8a) and (8.7.8b) to the discussion in Section 7.5. In most cases of interest, Prob $\{\rho = 1\} \ge \frac{3}{4}$ and $n \gg v$. Therefore the term in the brackets of (8.7.9) is greater than 1/2 and

$$\text{Prob}\,\{\rho = 1\} \le 1 - \frac{n}{2}\,q^{v}. \tag{8.7.10}$$

Another form of the above equation is

$$v \ge \frac{1}{\log q}\left[\log\,(1 - \text{Prob}\,\{\rho = 1\}) - \log\frac{n}{2}\right]. \tag{8.7.11}$$

Therefore, if a probability of connectivity is specified, we can solve for the vertex degree needed to ensure that the upper bound attains this specified value.[6]

Upper bounds on v can also be found for specified graphs to guarantee a given connectivity probability. For example, consider the graph G shown in Fig. 8.7.1. Each subgraph G_i is complete and, if c is the number of vertices in each complete subgraph G_i, the degree of each vertex is $d = c + 1$. For d large, $d \doteq c$. For this graph, it is easy to see that $v = d$. G will be connected if each G_i is connected and $G_1, \ldots, G_{n/c}$ are connected to each other. In Section 4 (see 8.4.15), we showed that for the completely connected c-vertex graph,

$$\text{Prob}\,\{\rho = 1\} \sim 1 - cq^{c-1} \quad \text{for } c \text{ large.} \tag{8.7.12}$$

[6] In [JA1], Jacobs derives an upper bound on Prob $\{\rho = 1\}$ for the graphs with unreliable vertices. If the reliability of each vertex is t, then it can be shown that

$$\text{Prob}\,\{\rho = 1\} \le 1 - \frac{tn}{2}\,q^{v}.$$

Here, $\rho = 1$ if all of the intact vertices are in one component.

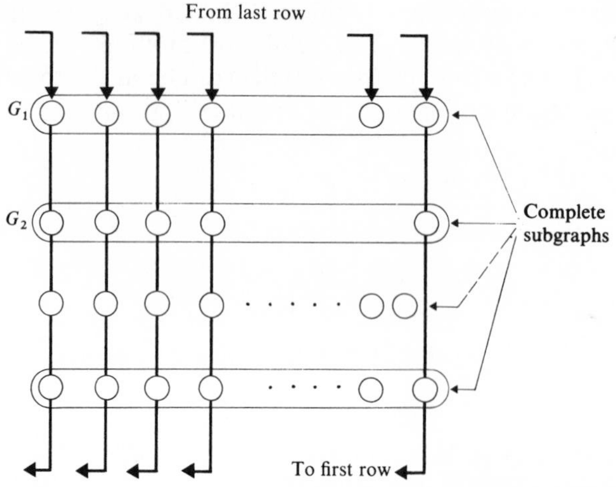

Fig. 8.7.1 Graph used to find lower bound on Prob $\{\rho = 1\}$.

Since there are approximately n/d subgraphs the probability that every subgraph G_i is connected is asymptotically

$$[1 - dq^d]^{n/d}. \tag{8.7.13}$$

We can compute a lower bound to Prob $\{\rho = 1\}$ as the product of the conditional probability that G is connected given that each G_i is connected and the probability that G_i is connected. The conditional probability that G is connected given that each G_i is connected is simply the probability that the graph G^* shown in Fig. 8.7.2 is connected. It can be shown that this probability is equal to

$$(1 - q^d)^{n/d} + \frac{n}{d}(1 - q^d)^{(n/d)-1}q^d. \tag{8.7.14}$$

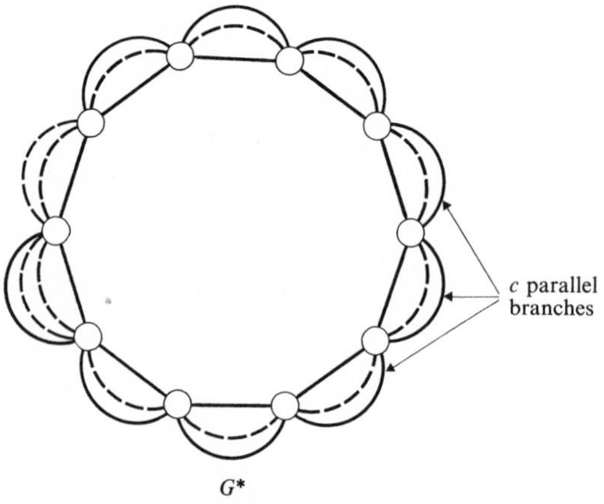

Fig. 8.7.2 Graph G^* with c parallel branches between vertices.

Therefore,

$$\text{Prob}\,\{\rho = 1\} \geq [1 - dq^d]^{n/d}\,[(1 - q^d)]^{n/d} + \frac{n}{d}(1 - q^d)^{(n/d)-1}q^d \tag{8.7.15a}$$

and for d large

$$\text{Prob}\,\{\rho = 1\} \geq [1 - (d + 1)q^d + q^{2d}]\left[1 + \frac{n}{d}\,\frac{q^d}{1 - q^d}\right]. \tag{8.7.15b}$$

If q is small, $dq^{2d} \doteq 0$ and $\frac{n}{d}\,\frac{q^d}{1 - q^d} \doteq 0$. Therefore

$$\text{Prob}\,\{\rho = 1\} \geq [1 - (d + 1)q^d]^{n/d}. \tag{8.7.16}$$

Given a graph with enough regularity or symmetry, it may be possible to find Prob $\{\rho = 1\}$ in terms of a few parameters. Then we can solve for these parameters in order to guarantee a given probability of connectivity. For example, given p_0 such that we desire Prob $\{\rho = 1\} \geq p_0$, (8.7.16) becomes

$$[1 - (d + 1)q^d] \geq p_0^{d/n} = [p_0^{1/n}]^d. \tag{8.7.17}$$

We can easily solve this equation for the smallest value of d which satisfies the equation.

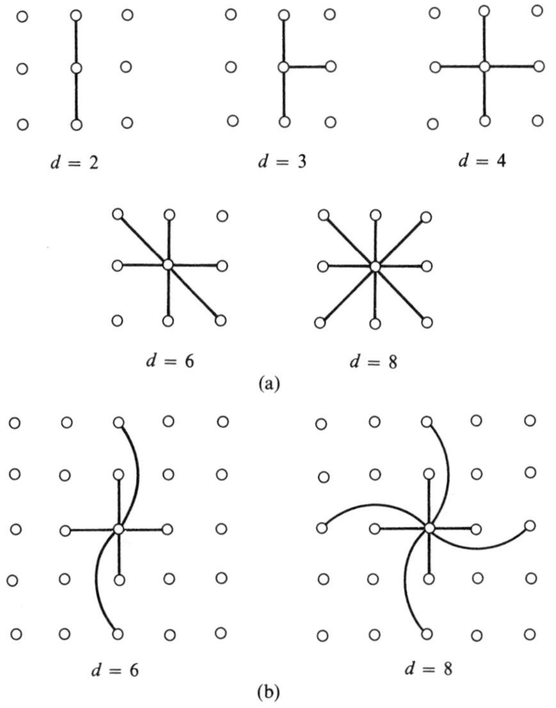

Fig. 8.7.3 Two connection schemes for the graphs considered by Baran.

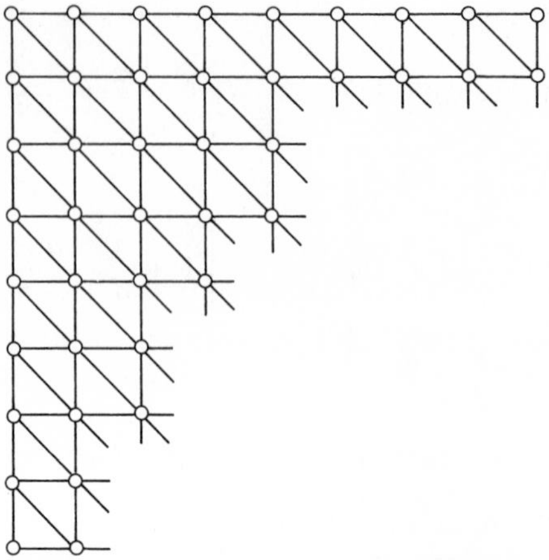

Fig. 8.7.4 A section of graph used in the Monte Carlo simulation.

Since it is usually quite complicated to obtain analytic results, we could use Monte Carlo programming. Baran [BA1] has taken this approach for homogeneous graphs using vertex degree as a measure of the complexity of the graph. The graphs considered by Baran are in the form of a grid in which each vertex is connected only to its nearest neighbors. The graph is then specified by the vertex degree d and the rule for connecting a single vertex to its neighbors. Two possible connection schemes are shown in Fig. 8.7.3 and a typical graph in Fig. 8.7.4.

Reliability indices such as Prob $\{\rho = 1\}$, or γ (the weak connectivity constant), are both reasonable. In [BA1] Baran uses a parameter closely related to γ. In addition to considering the average fraction of vertices in an arbitrary component, he finds the average fraction of vertices in the *largest* component, with varying probability of branch reliability and vertex degree. Typical results are given in Figure 8.7.5. To obtain this figure, an 18×18 array of 324 vertices was simulated for different branch probabilities and vertex degrees.

Graphs of the above type seem to be highly invulnerable for a reasonable size of d. However, it is difficult to make quantitative judgments because detailed analytic information is not available. In the remainder of this section we develop asymptotic and recursion formulas for computing the vulnerability of the classes of graphs that we studied in the preceding sections. At all times, we use the simplest formulation of a model which illustrates a particular point. We assume that the graphs are subject to an enemy attack aimed at isolating stations from one another. As a parameter we will use the weak connectivity constant γ, which is the average fraction of vertices which can be reached after the attack from a vertex picked at random.

Consider a directed graph G modeling a communication network. The structure of G may be fixed, as in the case of a microwave relay system, or it may be time-varying, as in a nonsynchronous network of satellites. Alternatively, the structure may be either deterministic or random. In fact, the same graph may simultaneously exhibit both qualities; for example, the builders of the network may know the exact

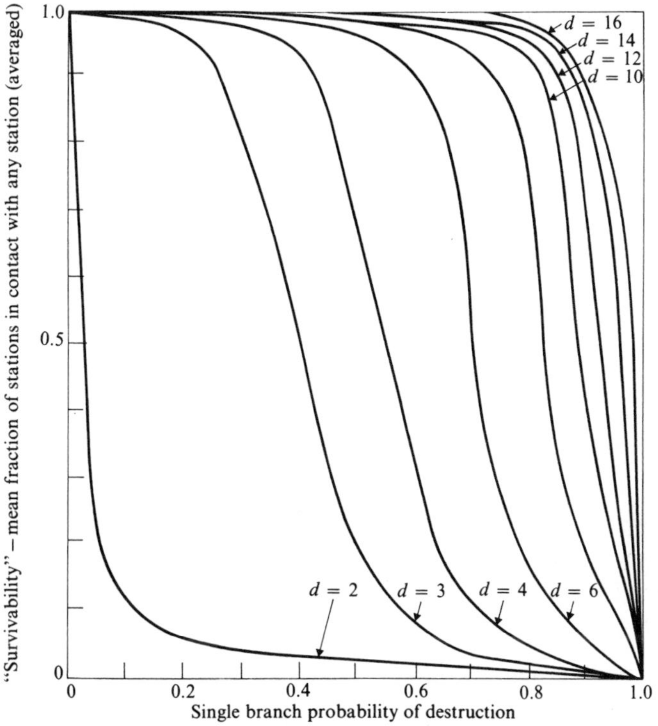

Fig. 8.7.5 Results of Baran's Monte Carlo analysis.

location of every channel and station while the enemy may know only the probable location of some of the channels and stations. In addition, atmospheric conditions, reliability problems, and other random factors may contribute to a state of uncertainty about the exact structure of the network.

The design of an invulnerable network may not be the only consideration. A reasonable objective is to generate a class of networks, with the same survivability features, from which a specific network can be selected. We will consider a simple class of directed graphs generated by the *random* process discussed in Section 3. This class is described by the assumptions:

Assumption 8.7.1a *Each vertex has on the average d outwardly directed branches. There are no self-loops.*

Assumption 8.7.1b *Each branch incident at v_i is also incident at vertex v_j with probability $1/(n-1)$ for $i, j = 1, \ldots, n$ and $i \neq j$.*

Assumption 8.7.1c *All vertices and branches are identical.*

We use the following model for the enemy attack strategy:

Assumption 8.7.2a *Weapons are directed at random into a region of area A. The probability that any given weapon is directed at a region of area Δ, for $\Delta < A$, is Δ/A.*

Assumption 8.7.2b *The density of weapons is $\eta(t)$ weapons per unit area.*

Assumption 8.7.2c *All weapons are identical.*

Assumption 8.7.2d *All weapons are fired at the same instant of time; that is, $\eta(t)$ is an impulse function.*

In our model, a weapon could represent a ballistic missile. In this case, the assumption that the missiles are falling *uniformly* at random into the area A is not unrealistic. If the weapons are bombs, Assumption 8.7.2a is most appropriate when the enemy is uncertain about the physical location of his targets or when he is aiming not directly at the network but at other targets in the same area.

Finally, we must specify the interaction between the weapons and the network. We will say that a branch or vertex experiences a *hit* if a weapon is directed to within a fixed distance from that branch or vertex. Naturally, this distance is a function of the power of the weapon and the hardness of the element. The vulnerability of the graph is defined by the following assumptions:

Assumption 8.7.3a *Vertices are destroyed if they are hit by at least K_s weapons.*

Assumption 8.7.3b *Branches are destroyed if they are hit by at least K_l weapons.*

Assumption 8.7.3c *There is no repair; i.e., once a vertex or branch is hit, it remains hit.*

Our assumptions describe a class of graphs whose behavior is, in a sense, uniform both before and after attack. That is, the probability of any vertex or branch surviving the attack is identical to the probability of any other vertex or branch surviving the attack. If n is large, the Law of Large Numbers may be applied. Consequently, the expected fraction of vertices which are destroyed is approximately equal to the probability that any given vertex is destroyed, and the expected fraction of branches which are destroyed is approximately equal to the probability that any given branch is destroyed.

Let $g_k(\eta)$ and $\hat{g}_k(\eta)$ denote the expected fraction of vertices and branches, respectively, which experience exactly k hits from an attack of density η. From the discussion in the last paragraph, the probability that any given vertex survives is approximately

$$\sum_{k=0}^{K_s-1} g_k(\eta) \tag{8.7.18a}$$

and the probability that any given branch survives is approximately

$$\sum_{k=0}^{K_l-1} \hat{g}_k(\eta). \tag{8.7.18b}$$

Let t_1 and $\hat{t}_1$ be the probability that any given weapon hits a given vertex and branch, respectively. These probabilities depend on the area of vulnerability of the vertex and branch. Given t_1 or $\hat{t}_1$, it is simple to compute $g_k(\eta)$ or $\hat{g}_k(\eta)$. In fact, it is well known that $g_k(\eta)$ or $\hat{g}_k(\eta)$ is the probability that a Poisson variable with parameter t_1 or $\hat{t}_1$ has k successes. In other words,

$$g_k(\eta) = e^{-t_1\eta}\frac{(t_1\eta)^k}{k!} \tag{8.7.19a}$$

and

$$\hat{g}_k(\eta) = e^{-\hat{t}_1\eta}\frac{(\hat{t}_1\eta)^k}{k!}. \tag{8.7.19b}$$

We derive (8.7.19a) using a method that is readily extended to more general situations which do not result in Poisson statistics.

On average, $\eta g_k(\eta)$ vertices experience exactly k hits in an attack of density η. Suppose that η is increased by the infinitesimal amount $d\eta$. Then $g_k(\eta)$ will change by the amount $dg_k(\eta)$. The change in $g_k(\eta)$ will result when vertices which have already been hit $k - 1$ or k times are hit again. The higher-order contributions, caused by $j \geq 2$ hits on targets with $k - j$ prior hits, can be ignored. Thus,

$$dg_k(\eta) = -t_1 d\eta g_k(\eta) + t_1 d\eta g_{k-1}(\eta) \qquad \text{for } k = 0,1,2,\ldots, \tag{8.7.20}$$

where $t_1 d\eta g_j(\eta)$ is the expected fraction of vertices which experience j hits from the attack of density η and an additional hit caused by the increase of $d\eta$ for $j = k - 1,k$.

The system of equations (8.7.20) may be rewritten as:

$$dg_0(\eta)/d\eta + t_1 g_0(\eta) = 0, \tag{8.7.21a}$$

$$dg_k(\eta)/d\eta + t_1 g_k(\eta) = t_1 g_{k-1}(\eta) \qquad \text{for } k = 1,2,\ldots \tag{8.7.21b}$$

Equation (8.7.21a) can be routinely solved to yield

$$g_0(\eta) = e^{-t_1\eta}. \tag{8.7.22}$$

All of the other $g_k(\eta)$ can be successively computed. For example,

$$dg_1(\eta)/d\eta + t_1 g_1(\eta) = t_1 e^{-t_1\eta} \tag{8.7.23}$$

or

$$g_1(\eta) = t_1\eta e^{-t_1\eta}, \tag{8.7.24}$$

and it is easily shown that, in general,

$$g_k(\eta) = e^{-t_1\eta}\frac{(t_1\eta)^k}{k!}. \tag{8.7.25}$$

This is the desired expression.

Now, suppose that branches $b_{i_1}, b_{i_2}, \ldots, b_{i_d}$ are directed away from station v_i. After the attack, suppose there is an average of α intact branches directed away from v_i, assuming that v_i survives. We have already shown in Section 3 that graphs described by Assumptions 8.7.1a to 8.7.1c satisfy

$$\gamma = 1 - e^{-\alpha\gamma}. \tag{8.7.26}$$

If we include Assumptions 8.7.2 and 8.7.3 in the model it is not difficult to show that

$$\gamma = 1 - \exp\left\{-d\left[\sum_{k=0}^{K_s-1} g_k(\eta)\right]\left[\sum_{k=0}^{K_l-1} \hat{g}_k(\eta)\right]\gamma\right\}. \tag{8.7.27}$$

The proof is a simple extension of the proof of (8.3.37). γ is now the average fraction of vertices which *both* survive the attack *and* can be reached from a vertex picked at random. In other words,

$$\gamma \sum_{k=0}^{K_s-1} g_k(\eta)$$

is the average fraction of vertices in the original graph which can communicate after the attack with the station picked at random. Rather than multiply γ by

$$\sum_{k=0}^{K_s-1} g_k(\eta),$$

for simplicity we will use γ as the vulnerability index.

A reasonable goal in the design of a communication system is to find d, K_s, and K_l, the "redundancy" and "hardness" of the network, so that on the average at least $\gamma_0 \times 100\%$ of the surviving vertices of the network can communicate after the attack, where γ_0 is a prescribed constant. Clearly, if d, K_s, and K_l are large enough, then $\gamma \geq \gamma_0$. Therefore, we seek a set of values which guarantee γ_0 with minimum cost.

If cost of the system can be expressed as a function of d, K_s, and K_l, say $H(d,K_s,K_l)$, then the design problem is: find d, K_s, and K_l such that $H(d,K_s,K_l)$ is minimized and $\gamma \geq \gamma_0$.

We solve for the exponent in the right-hand side of (8.7.27). Thus

$$d\left[\sum_{k=0}^{K_s-1} g_k(\eta)\right]\left[\sum_{k=0}^{K_l-1} \hat{g}_k(\eta)\right] = \frac{-\ln(1-\gamma)}{\gamma} \tag{8.7.28}$$

and the constraint $\gamma \geq \gamma_0$ is equivalent to

$$d\left[\sum_{k=0}^{K_s-1} g_k(\eta)\right]\left[\sum_{k=0}^{K_l-1} \hat{g}_k(\eta)\right] \geq \frac{-\ln(1-\gamma_0)}{\gamma_0}. \tag{8.7.29}$$

The design problem is now: find d, K_s, and K_l such that $H(d,K_s,K_l)$ is minimized and

$$d\left[\sum_{k=0}^{K_s-1} \frac{(t_1\eta)^k}{k!}\right]\left[\sum_{k=0}^{K_l-1} \frac{(\hat{t}_1\eta)^k}{k!}\right] \geq e^{(t_1+\hat{t}_1)\eta}\left(\frac{-\ln(1-\gamma_0)}{\gamma_0}\right). \tag{8.7.30}$$

The solution of the above is routine; K_s and K_l must be integers and usually will be bounded by some number, beyond which it is not feasible to increase the hardness of a branch or vertex. If we substitute all pairs of feasible K_s and K_l into (8.7.30) and then solve for the minimum d, we can evaluate the cost functions at each K_s, K_l, and d which satisfy this equation.

Example 8.7.1 We now design a graph with a large number of vertices so that 90% of the surviving vertices can be reached from a vertex picked at random after an attack. Suppose that the branches are invulnerable, that is, $K_l = \infty$, but the vertices are not, and let the cost function $H(d,K_s) = d\,(\exp 3K_s/2)$. Assume the enemy has been successful in locating targets to within one square mile, but that each target has a zone of vulnerability of only 0.05 square miles. Then, for a weapon directed at random into the target area, the probability of a hit is $t_1 = 0.05$. Let the density of weapons be 100 weapons per square mile. From (8.7.27),

$$\gamma = 1 - \exp\left\{-d \sum_{k=0}^{K_s-1} \frac{5^k}{k!} e^{-5}\gamma\right\} \tag{8.7.31}$$

and from (8.7.30) our constraint is

$$d \sum_{k=0}^{K_s-1} \frac{5^k}{k!} \geq e^5 \frac{(-\ln 0.1)}{0.9} = 374.8. \tag{8.7.32}$$

Suppose it is not feasible to harden vertices beyond survivability of five hits. Then, from (8.7.32), if $K_s = 1$, $d = 375$; if $K_s = 2$, $d = 63$; if $K_s = 3$, $d = 23$; if $K_s = 4$, $d = 10$; and if $K_s = 5$, $d = 5$. Evaluating the cost function we obtain $H(375,1) = 1680$; $H(63,2) = 1265$; $H(23,3) = 2070$; $H(10,4) = 4030$; and $H(5,5) = 5400$. The minimum of the cost function occurs at $d = 63$ and $K_s = 2$. Thus, each station of the system should be built to survive one direct hit and each station should have 63 outwardly directed links. Note that these numbers may be unrealistic because only 4% of the total vertices will survive since

$$\sum_{k=1}^{K_s-1} g_k(\eta) = 0.0404.$$

However, assume the vertices are invulnerable ($t_1 = 0$) and the branches are vulnerable with $\hat{t}_1 = 0.05$ and $K_l = 2$. If we let $d = 63$, then even though only 4% of the branches survive, 90% of the vertices can be reached from the vertex picked at random.

The graphs generated by the process defined by Assumption (8.7.1) can be modified in several ways. First, we note that these graphs may have parallel branches. The reason is that the process of selecting vertices adjacent to a given vertex is equivalent to sampling a population of $n - 1$ points *with replacement*. Consequently, the same point may be selected more than once. A more reasonable method of selection is to establish branches sequentially. The first vertex adjacent to a given vertex is selected equiprobably out of the $n - 1$ possible vertices; and the kth vertex is selected equiprobably out of the $n - k$ remaining vertices, where $k = 2, \ldots, d$. This process is equivalent to sampling a population of $n - 1$ points *without replacement*. We will call this scheme Assumption 8.7.1b′. In other words, we have:

Assumption 8.7.1b′ *The branches incident out of vertex v_i are determined by sampling vertices $v_1, \ldots, v_{i-1}, v_{i+1}, \ldots, v_n$ without replacement a total of d times for $i = 1, 2, \ldots, n$.*

We can include this assumption in our model without difficulty; in fact, it is easily shown that (8.7.27) is still valid. However, even though both schemes of sampling are equivalent for infinite populations, we can expect some difference for finite populations. In the finite case the number calculated by means of (8.7.27) is a lower bound for γ. We can investigate the quantitative difference between sampling with and without replacement for finite populations and simultaneously introduce another important factor into the study.

Most networks have some processing time associated with the branches and vertices. This processing time may be the time necessary to transmit information through the branches or the time needed at a vertex to decode, recode, and retransmit the information. In any event, it is usually desirable to limit the total time a message remains in the network. In many cases, limiting the total time is equivalent to limiting the number of branches in the paths traversed by the messages. Thus, instead of determining the total fraction of vertices that can be reached from a given vertex, it is reasonable to determine the fraction of vertices that can be reached from a given vertex by a path of no more than b branches.

Select a vertex v_x at random. Since the graph contains a large number of vertices, the probability that any vertex is a descendant of v_x is approximately equal to the average fraction of descendants of v_x. Similarly, the probability that any vertex is connected to v_x by a path of no more than b branches is approximately equal to the average fraction of vertices at distance b or less from v_x. We can compute this average by a recursion formula. For the system satisfying Assumptions 8.7.1, we have from (8.3.41)

$$q(b) = \left[1 - \frac{1}{n-1}\right]^{\alpha n[1 - q(b-1)]} \prod_{i=0}^{b-2} q(i) \tag{8.7.33}$$

where $q(i)$ is the probability that a vertex is more than i branches removed from v_x and $q(0) = (n-1)/n$. If we include Assumptions 8.7.2 and 8.7.3, we must then solve the recursion formula

$$q(b) = \left[1 - \frac{1}{n-1}\right]^{d\left\{\sum_{k=0}^{K_s-1} g_k(\eta) \sum_{k=0}^{K_l-1} \hat{g}_k(\eta)[1 - q(b-1)] \prod_{i=0}^{b-2} q(i)\right\} n} \tag{8.7.34}$$

where

$$q(0) = 1 - \frac{1}{n} \sum_{k=0}^{K_s-1} g_k(\eta).$$

Then, if $p(i)$ is the probability that a given vertex is *no more* than i branches removed from v_x, $p(i) = 1 - q(i)$ and, in particular $p(b) = 1 - q(b)$.

A similar recursion formula can be derived for the scheme of sampling without replacement. Let $q^*(b)$ represent the probability that any vertex is more than b branches removed from a point picked at random, when the system is described by Assumptions 8.7.1a, 8.7.1b′, 8.7.1c, 8.7.2, and 8.7.3. Then, it can be shown that $q^*(b)$ satisfies the recursion formula

$$q^*(b) = \left\{ 1 - \frac{d \sum_{k=0}^{K_s - 1} g_k(\eta) \sum_{k=0}^{K_l - 1} \hat{g}_k(\eta)}{n - 1} \right\}^{n[1 - q(b-1)] \prod_{i=0}^{b-2} q(i)} \tag{8.7.35}$$

where $q^*(0) = 1 - \frac{1}{n} \sum_{k=0}^{K_s - 1} g_k(\eta)$.

Let $P(b)$ and $P^*(b)$ represent the probabilities that any vertex is *exactly* b branches removed from v_x, under Assumptions 8.7.1b and 8.7.1b′, respectively. $P(b)$ and $P^*(b)$ are approximately equal to the expected fraction of vertices which are connected by at least one path of b branches and no path with fewer than b branches to v_x. By (8.5.3), $P(b)$ and $P^*(b)$ are given by

$$P(b) = \left[1 - \sum_{j=0}^{b-1} P(j)\right][1 - q(b)] \tag{8.7.36}$$

and

$$P^*(b) = \left[1 - \sum_{j=0}^{b-1} P^*(j)\right][1 - q^*(b)]. \tag{8.7.37}$$

If n is large, (8.5.5) implies that both $P(b)$ and $P^*(b)$ satisfy the recurrence relation

$$P(b) = \left[1 - \sum_{j=0}^{b-1} P(j)\right]\left[1 - \exp\left\{-P(b-1)\, d \sum_{k=0}^{K_s - 1} g_k(\eta) \sum_{k=0}^{K_l - 1} \hat{g}_k(\eta)\right\}\right] \tag{8.7.38}$$

with

$$P(0) = P^*(0) = \frac{1}{n} \sum_{k=0}^{K_s - 1} g_k(\eta). \tag{8.7.39}$$

We can now define an alternative vulnerability criterion. As we have already indicated, if the only available path between a pair of vertices is "too long," we may consider that the enemy has effectively disconnected the two vertices. The vulnerability index γ does not take this factor into account. In fact,

$$\gamma = \sum_{k=0}^{\infty} P(k). \tag{8.7.40}$$

Define $\gamma(b)$ by

$$\gamma(b) = \sum_{k=0}^{b} P(k); \tag{8.7.41}$$

$\gamma(b)$ is approximately equal to the average number of vertices connected by a path of length b or less to v_x. A reasonable vulnerability constraint is now $\gamma(b) \geq \gamma_0$. Again, the objective is to find d, K_s, and K_l so that the constraint is satisfied with minimum cost. In general, the recurrence relations given above must be solved for $P(b)$ or $P^*(b)$, given values of d, K_s, and K_l. The least-cost solution is then selected from those which satisfy the constraint.

Example 8.7.2 Consider a graph with 100 vertices and an average of 20 branches per vertex. Suppose the branches are invulnerable but the vertices are not. If the enemy has been successful at locating targets to within one square mile and each target has an area of vulnerability of 0.05 square miles, the probability of a hit is 0.05. Let the density of weapons be 50 weapons per square mile. We now find the average fraction of surviving vertices that can be reached after the attack from a vertex, v_x, picked at random and the average fraction of surviving vertices that can be reached by a path of *no more than* three branches. Assume that K_s is either 2 or 3.

If $K_s = 2$, the probability that a given vertex survives is

$$\sum_{k=0}^{1} g_k(\eta) = (1 + 2.5)e^{-2.5} = 0.287. \tag{8.7.42}$$

Then, from (8.7.27),

$$\gamma = 1 - e^{-20(0.287)\gamma} = 1 - e^{-5.74\gamma} \tag{8.7.43}$$

and γ is very close to unity. However, from (8.7.36), $P(b)$ is

$$P(0) = (1/100)(0.287) = 0.00284; \tag{8.7.44a}$$

$$P(1) = (1 - 0.00284)(1 - e^{-20(0.284)(0.00284)}) = 0.016; \tag{8.7.44b}$$

$$P(2) = (1 - 0.0189)(1 - e^{-20(0.284)(0.016)}) = 0.087; \tag{8.7.44c}$$

$$P(3) = (1 - 0.106)(1 - e^{-20(0.284)(0.086)}) = 0.352. \tag{8.7.44d}$$

Therefore, although nearly 100% of the surviving vertices can be reached from v_x,

$$\gamma(3) = 0.00287 + 0.016 + 0.087 + 0.352 = 0.45 \tag{8.7.45}$$

and only about 45% of the vertices can be reached with paths of 3 or fewer branches.

If $K_s = 3$, the situation changes drastically. Again it is easy to see that γ is very close to unity. The probability that a vertex survives is now $(1 + 2.5 + 6.25/2)e^{-2.5} = 0.544$. From (8.7.36),

$$P(0) = 0.00544; \tag{8.7.46a}$$

$$P(1) = 0.0665; \tag{8.7.46b}$$

$$P(2) = 0.431; \tag{8.7.46c}$$

$$P(3) = 0.502. \tag{8.7.46d}$$

Therefore, $\gamma(3) = 0.993$, or in other words, about 99% of the surviving vertices in the graph can be reached from v_x by a path of no more than 3 branches. We also note that if we compute fractions of the total number, rather than of the surviving number, of vertices in the original graph, $K_s = 2$ implies that only 28% of the vertices will survive. Hence, only about 12% of the original vertices can be reached after the attack, while if $K_s = 3$, over 50% of the original vertices can be reached from v_x, via a path of 3 or fewer branches.

In our previous models, we considered only graphs with directed branches. Suppose we are given a system with undirected channels whose stations have an average of d receivers and transmitters. Once a branch is established between a pair of stations, both stations can converse with each other. The parameter γ then becomes the average fraction of vertices in an arbitrary connected component of the graph. The random process used to generate the graph is described by the assumptions:

Assumption 8.7.1a″ *Each vertex has on the average d undirected branches. The total number of branches is $m = m(n) \sim dn/2$. There are no self-loops or parallel branches.*

Assumption 8.7.1b″ *There is a total of $n(n-1)/2$ possible branches. The graph is constructed by choosing the first branch equiprobably from among the $n(n-1)/2$ possible branches, and the kth branch is chosen equiprobably from among the $n(n-1)/2 - k$ remaining branches for $k = 2, \ldots, m$.*

We again find the average fraction of vertices that can be reached from a vertex chosen at random. Although this problem can be treated by the techniques already discussed, we will use an alternate approach based upon the work of Erdös and Rényi outlined in Section 4.

As a first step, we restate a theorem which gives the expected number of components of the system.

Theorem 8.7.1 *Let $B(n) \sim dn/2$ with $d > 1$. For large n, the expected number of components is approximately*

$$\frac{n}{d}\left[x(d) - \frac{x^2(d)}{2}\right] \tag{8.7.47}$$

where $x(d)$ is the only solution of the equation $x(d)e^{-x(d)} = de^{-d}$ that satisfies $0 < x(d) < 1$.

We can use this theorem to find the average fraction of vertices in an arbitrary component. The size of a component is a random variable which is identically distributed for each component. Therefore, the average number of vertices in a component is simply n divided by the average number of components. Hence, for large n, the average fraction of vertices in a component is

$$\gamma = \frac{d/n}{x(d) - x^2(d)/2} \tag{8.7.48}$$

which, for $d \geq 4$, is closely approximated by

$$\gamma = \frac{1}{n} e^d. \tag{8.7.49}$$

Equation (8.7.49) shows that for large n, γ becomes small. This is because almost all vertices of the network belong to some small component or to a "giant" component.[7] The size of the "giant" component can be found.

Theorem 8.7.2 *If $\bar{\gamma}$ denotes the fraction of vertices in the largest component of the graph and if $m(n) \sim dn/2$ with $d > 1$, then for any $\varepsilon > 0$,*

$$\lim_{n \to \infty} \text{Prob} \left\{ \left| \bar{\gamma} - 1 + \frac{x(d)}{d/2} \right| < \varepsilon \right\} = 1 \tag{8.7.50}$$

where $x(d)$ is the solution of the equation $x(d)e^{-x(d)} = de^{-d}$ that satisfies $0 < x(d) < 1$.

Theorem 8.7.2 states that the expected fraction of vertices in the largest component is $1 - x(d)/d$. For $d \geq 4$ this number is closely approximated by

$$\bar{\gamma} = 1 - 2e^{-d}. \tag{8.7.51}$$

The effect of an enemy attack can now be included in the model. An attack such as the one described by Assumption 8.7.2 has the effect of reducing the average degree of each vertex uniformly. Consequently, d becomes a random variable. Since all the necessary probabilities are known, we could easily find the expected value of $\bar{\gamma}$.

Assumption 8.7.3 describes the interaction between the enemy attack and the communication system. A valid objection to this hypothesis is the limitation of not allowing repair. We noted that the derivation of (8.7.19) could be extended to more general situations that did not involve Poisson statistics. We now replace Assumption 8.7.3c by

Assumption 8.7.3c *If a vertex has experienced k hits at time t, the probability that it will be completely repaired in the time interval $[t,t + dt]$ is $r_k^s(t)\,dt$, for $k = 1,2,\ldots$. If a branch has experienced k hits at time t, the probability that it will be completely repaired in the time interval $[t,t + dt]$ is $r_k^l(t)\,dt$, for $k = 1,2,\ldots$.*

The functions $r_k^s(t)$ and $r_k^l(t)$ are known as repair-rate functions [BA10]. Consider the system of differential equations which now describes the number of hits per station:

$$dg_0(\eta,t) = -t_1\, d\eta\, g_0(\eta,t) = \sum_k r_k^s(t)\, dt\, g_k(\eta,t); \tag{8.7.52a}$$

$$dg_k(\eta,t) = -t_1\, d\eta\, g_k(\eta,t) + t_1\, d\eta\, g_{k-1}(\eta,t) - r_k^s(t)\, dt\, g_k(\eta,t), \qquad \text{for } k = 1,2,\ldots. \tag{8.7.52b}$$

[7] These facts suggest that we should use Baran's vulnerability criterion for this system.

We no longer restrict η to be an impulse function. We then have

$$\frac{dg_0(t)}{dt} = -t_1 \frac{d\eta(t)}{dt} g_0(t) + \sum_k r_k^s(t) g_k(t); \tag{8.7.53a}$$

$$\frac{dg_k(t)}{dt} = -t_1 \frac{d\eta(t)}{dt} g_k(t) + t_1 \frac{d\eta(t)}{dt} g_{k-1}(t) - r_k^s(t) g_k(t), \qquad \text{for } k = 1,2,\ldots. \tag{8.7.53b}$$

This system of differential equations can be expressed as the following matrix equation.

$$\begin{bmatrix} g_0'(t) \\ g_1'(t) \\ g_2'(t) \\ \cdot \\ \cdot \\ \cdot \end{bmatrix} = \begin{bmatrix} -t_1\eta'(t) & r_1^s(t) & r_2^s(t) & r_3^s(t) & \cdots \\ t_1\eta'(t) & -r_1^s(t) - t_1\eta'(t) & 0 & 0 & \cdots \\ 0 & t_1\eta'(t) & -r_2^s(t) - t_1\eta'(t) & 0 & \cdots \\ \cdot & & & & \cdot \\ \cdot & & & & \cdot \\ \cdot & \cdots & \cdots & \cdots & \cdot \end{bmatrix} \begin{bmatrix} g_0(t) \\ g_1(t) \\ g_2(t) \\ \cdot \\ \cdot \\ \cdot \end{bmatrix} \tag{8.7.54}$$

where $'$ indicates differentiation with respect to t and the initial conditions are $g_0(0^-) = 1$, $g_1(0^-) = g_2(0^-) = \cdots = 0$.

The difficulty in numerically solving this equation depends on the functions $\eta'(t)$ and $r_k^s(t)$. However, in some cases, the computational problems are relatively easy. One such case occurs when $r_k^s(t) = r_k$ for all k (exponential repair) and $\eta(t)$ is a periodic function. Another routine case occurs if $\eta'(t)$ can be approximated by a piecewise constant function and $r_k^s(t) = r_k$ for all k.

Let us consider one special case. Suppose each vertex cannot survive any direct hits but will be repaired in time dt with probability $r_1\, dt$ if it has suffered exactly one hit. The probability that a vertex is operating is thus $g_0(t)$ and the differential equation which describes the behavior of $g_0(t)$ is

$$\begin{bmatrix} g_0'(t) \\ g_1'(t) \end{bmatrix} = \begin{bmatrix} -t_1\eta'(t) & r_1 \\ t_1\eta'(t) & -r_1 - t_1\eta'(t) \end{bmatrix} \begin{bmatrix} g_0(t) \\ g_1(t) \end{bmatrix} \tag{8.7.55}$$

with initial conditions $g_0(0^-) = 1$ and $g_1(0^-) = 0$.

Example 8.7.3 A network with a large number of stations is attacked periodically with a density of weapons shown in Fig. 8.7.6. Suppose stations cannot survive direct hits but that stations which have experienced exactly one direct hit will be repaired with probability 0.1 dt in time dt (i.e., $r_1 = 0.1$). Assume that channels are invulnerable and as before suppose that the probability that a given weapon hits a given station is $t_1 = 0.05$. We compute γ as a function of time.

From (8.7.55), the equation describing the system is

$$\begin{bmatrix} g_0'(t) \\ g_1'(t) \end{bmatrix} = \begin{bmatrix} -0.05\,\eta'(t) & 0.1 \\ 0.05\,\eta'(t) & -0.1 - 0.05\,\eta'(t) \end{bmatrix} \begin{bmatrix} g_0(t) \\ g_1(t) \end{bmatrix} \tag{8.7.56}$$

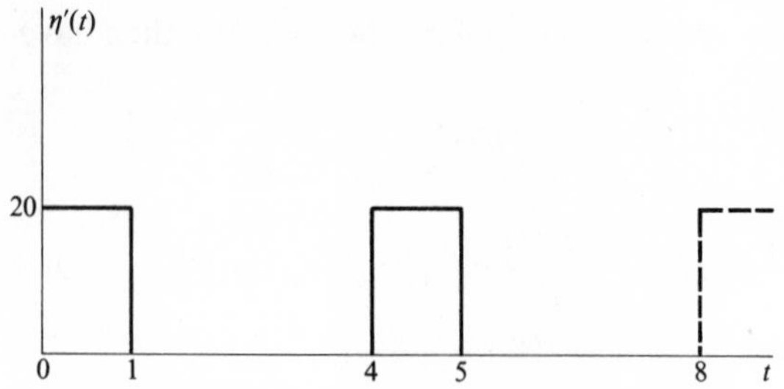

Fig. 8.7.6 Weapon density for Example 8.7.3.

with $g_0(0^-) = 1$ and $g_1(0^-) = 0$. From time $t = 0$ to time $t = 1$, $\eta'(t) = 20$ and (8.7.56) is

$$\begin{bmatrix} g_0'(t) \\ g_1'(t) \end{bmatrix} = \begin{bmatrix} -1 & 0.1 \\ 1 & -1.1 \end{bmatrix} \begin{bmatrix} g_0 \\ g_1 \end{bmatrix}. \tag{8.7.57}$$

On this time interval, the solution is

$$\begin{bmatrix} g_0(1) \\ g_1(t) \end{bmatrix} = \begin{bmatrix} \alpha_0(t) - \alpha_1(t) & 0.1\,\alpha_1(t) \\ \alpha_1(t) & -1.1\,\alpha_1(t) \end{bmatrix} \begin{bmatrix} 1 \\ 0 \end{bmatrix} \tag{8.7.58}$$

where $\alpha_0(t) = 2.14\,e^{-0.724}$ and $\alpha_1(t) = 1.14\,e^{-1.37t}$. Therefore, at $t = 1$

$$\begin{bmatrix} g_0(1) \\ g_1(1) \end{bmatrix} = \begin{bmatrix} 0.383 \\ 0.36 \end{bmatrix}. \tag{8.7.59}$$

In the time interval $1 \le t \le 4$, the system is described by the equation

$$\begin{bmatrix} g_0'(t) \\ g_1'(t) \end{bmatrix} = \begin{bmatrix} 0 & 0.1 \\ 0 & -0.1 \end{bmatrix} \begin{bmatrix} g_0(t) \\ g_1(t) \end{bmatrix} \tag{8.7.60}$$

and the solution is

$$\begin{bmatrix} g_0(t) \\ g_1(t) \end{bmatrix} = \begin{bmatrix} 1 & 1 - e^{-0.1(t-1)} \\ 0 & e^{-0.1(t-1)} \end{bmatrix} \begin{bmatrix} g_0(1) \\ g_1(1) \end{bmatrix}. \tag{8.7.61}$$

Therefore,

$$\begin{bmatrix} g_0(4) \\ g_1(4) \end{bmatrix} = \begin{bmatrix} 0.47 \\ 0.266 \end{bmatrix}. \tag{8.7.62}$$

During the time interval $4 \le t \le 5$, (8.7.57) is again applicable. Thus

$$\begin{bmatrix} g_0(5) \\ g_1(5) \end{bmatrix} = \begin{bmatrix} 0.27 \\ 0.25 \end{bmatrix} \tag{8.7.63}$$

and on the interval $5 \le t \le 8$ we can use (8.7.60) to show that

$$\begin{bmatrix} g_0(8) \\ g_1(8) \end{bmatrix} = \begin{bmatrix} 0.332 \\ 0.19 \end{bmatrix}. \tag{8.7.64}$$

Without repair, an instantaneous attack of density $\eta' = 20$ gives

$$g_0(\eta) = e^{-t_1} = 0.368. \tag{8.7.65}$$

However, applying the same attack scheme as shown in Fig. 8.7.6, without repair (i.e., $r_1 = 0$), we find

$$g_0(1) = 0.368, \tag{8.7.66a}$$

$$g_0(5) = 0.135, \tag{8.7.66b}$$

$$g_0(8) = 0.135. \tag{8.7.66c}$$

The weak connectivity γ can be computed as a function of time for a given d. Thus

$$\gamma = 1 - e^{-dg_0(t)\gamma}. \tag{8.7.67}$$

For example, if $d = 6$, for the system with repair rate $r_1 = 0.1$, $\gamma \doteq 0.8$ at $t = 8$ while for the system without repair, $\gamma \doteq 0.03$.

Once the $g_k(t)$ and $\hat{g}_k(t)$ are found, they may be substituted into the vulnerability equations given in the preceding sections. We are now dealing with time-varying vulnerability expressions and thus the synthesis goals must be modified. The functions $r_k^s(t)$ and $r_k^l(t)$ could be included among the unknowns. The cost of repair must also be included in the objective function which is to be minimized. A network can then be designed by repetitive analysis.

Another generalization can be introduced if we consider the fact that the kth hit on a target may not be independent of the number k. That is, the probability of a given vertex or branch being hit after it already has suffered k hits may vary with k. This situation corresponds to a system with memory. Furthermore, each weapon may be capable of delivering more than a single hit. For example, the weapons may vary in power, a hit may cause secondary damage, or each weapon may actually be composed of a number of smaller weapons.

Both generalizations can be included in the model without difficulty. These extensions are due to Biermann [BI1, BI2], who considered similar problems in a different context. We will briefly formulate the appropriate equations and state the solutions. First, we examine the possibility of multiple hits per weapon. Let t_i be the probability that a given vertex receives exactly i hits from a given weapon; let $\hat{t}_i$ be the probability of this event for the branches. Then, if η changes by the amount $d\eta$, $g_k(\eta)$ will change by the amount

$$dg_0(\eta) = -(1 - t_0)g_0(\eta); \tag{8.7.68a}$$

$$dg_k(\eta) = -(1 - t_0)g_k(\eta)\,d\eta + \sum_{j=1}^{k} t_j g_{k-j}(\eta)\,d\eta, \qquad \text{for } k = 1,2,\ldots. \tag{8.7.68b}$$

We can write (8.7.68) as the differential equation

$$\frac{dg_k(\eta)}{d\eta} + (1 - t_0)g_0 = 0; \tag{8.7.69a}$$

$$\frac{dg_k(\eta)}{d\eta} + (1 - t_0)g_k(\eta) = \sum_{j=1}^{k} t_j g_{k-j}(\eta), \qquad \text{for } k = 1,2,\ldots. \tag{8.7.69b}$$

Biermann has shown that the solution of this system of equations is given by

$$g_0(\eta) = \exp\{-(1 - t_0)\eta\}, \tag{8.7.70a}$$

$$g_k(\eta) = \exp\{-(1 - t_0)\eta\}\left\{\sum_{r=1}^{k} \frac{\eta^r}{r!} \sum_i \psi_i(k;r)\right\}, \qquad \text{for } k = 1,2,\ldots. \tag{8.7.70b}$$

The term $\sum_i \psi_i(k;r)$ is found as follows. A composition of k into r parts ($r \le k$) is a set of r positive integers whose sum is k. The same r integers arranged in a different order are considered to be a different composition. $\psi_i(k;r)$ is defined to be the product

$$\prod_{j=1}^{r} t_{i_j}$$

such that the indices j are compositions of k into r parts. The expression $\sum_i \psi_i(k;r)$ represents the sum over all such compositions of k into r parts.

Example 8.7.4 Suppose two types of weapon are directed at the network such that the first type has twice the power of the second type. Furthermore, suppose that the density of weapons of the first type is η_1 while the density of weapons of the second type is η_2. The weapons are assumed to be randomly mixed. Let $p_0^{(1)}$ and $p_0^{(2)}$ be the probabilities that a given vertex receives no hits from weapons of the first and second type, respectively.

The probability that a given weapon is type i is $\eta_i/(\eta_1 + \eta_2)$ for $i = 1,2$. Hence t_0, t_1, and t_2 may be computed in terms of conditional probabilities

$$t_k = \sum_{i=1}^{2} \text{Prob}\,\{k \text{ hits} \mid \text{type } i \text{ weapon}\}\ \text{Prob}\,\{\text{type } i \text{ weapon}\}. \tag{8.7.71}$$

Hence, if $\eta = \eta_1 + \eta_2$,

$$t_0 = p_0^{(1)}\frac{\eta_1}{\eta_1 + \eta_2} + p_0^{(2)}\frac{\eta_2}{\eta_1 + \eta_2} = \frac{p_0^{(1)}\eta_1}{\eta} + \frac{p_0^{(2)}\eta_2}{\eta}; \tag{8.7.72a}$$

$$t_1 = (1 - p_0^{(2)})\frac{\eta_2}{\eta_1 + \eta_2} = \frac{(1 - p_0^{(2)})\eta_2}{\eta}; \tag{8.7.72b}$$

$$t_2 = \frac{(1 - p_0^{(1)})\eta_1}{\eta_1 + \eta_2} = (1 - p_0^{(1)})\frac{\eta_1}{\eta}. \tag{8.7.72c}$$

Suppose each vertex has been hardened to survive at most four hits. Then, the average fraction of surviving vertices is

$$\sum_{i=1}^{4} g_k(\eta).$$

Table 8.7.1

(k,r)	Compositions of k into r parts	$\sum_i \psi_i(k,r)$
(1,1)	(1)	t_1
(2,1)	(2)	t_2
(2,2)	(1,1)	t_1^2
(3,1)	(3)	$t_3 = 0$
(3,2)	(2,1),(1,2)	$t_2t_1 + t_1t_2$
(3,3)	(1,1,1)	t_1^3
(4,1)	(4)	$t_4 = 0$
(4,2)	(2,2),(3,1),(1,3)	$t_2^2 + t_3t_1 + t_1t_3 = t_2^2$
(4,3)	(1,1,2),(1,2,1),(2,1,1)	$t_1^2t_2 + t_1t_2t_1 + t_2t_1^2$
(4,4)	(1,1,1,1)	t_1^4

We must therefore find $\sum_i \psi_i(1,1)$, $\sum_i \psi_i(2,1)$, $\sum_i \psi_i(2,2)$, $\sum_i \psi_i(3,1)$, $\sum_i \psi_i(3,2)$, $\sum_i \psi_i(3,3)$, $\sum_i \psi_i(4,1)$, $\sum_i \psi_i(4,2)$, $\sum_i \psi_i(4,3)$, and $\sum_i \psi_i(4,4)$. All compositions of k into r parts for the above quantities are shown in the second column of Table 8.7.1. In the third column, the corresponding $\sum_i \psi_i(k,r)$ is given.

From this table and (8.7.70), we have

$$g_0(\eta) = \exp\{-(1 - p_0^{(1)}\eta_1 - p_0^{(2)}\eta_2)\}; \tag{8.7.73a}$$

$$g_1(\eta) = \exp\{-(1 - p_0^{(1)}\eta_1 - p_0^{(2)}\eta_2)\}\{\eta t_1\}; \tag{8.7.73b}$$

$$g_2(\eta) = \exp\{-(1 - p_0^{(1)}\eta_1 - p_0^{(2)}\eta_2)\}\left\{\eta t_2 + \frac{\eta^2}{2} t_1^2\right\}; \tag{8.7.73c}$$

$$g_3(\eta) = \exp\{-(1 - p_0^{(1)}\eta_1 - p_0^{(2)}\eta_2)\}\left\{\eta^2 t_2 t_1 + \frac{\eta^3}{6} t_1^3\right\}; \tag{8.7.73d}$$

$$g_4(\eta) = \exp\{-1 - p_0^{(1)}\eta_1 - p_0^{(2)}\eta_2)\}\left\{\frac{\eta^2}{2} t_2^2 + \frac{\eta^3}{2} t_1^2 t_2 + \frac{\eta^4}{4!} t_1^4\right\}. \tag{8.7.73e}$$

The system with memory is then easily described by a minor modification of (8.7.69). If the system has memory, the probability t_i depends on the number of hits already taken by the target. Therefore we must consider probabilities of the form $t_{i,j}$, which represents the probability that a given vertex which has already been hit j times will suffer i additional hits from a single weapon. The differential equations of the system are

$$\frac{dg_0(\eta)}{d\eta} + (1 - t_{0,0})g_0(\eta) = 0; \tag{8.7.74a}$$

$$\frac{dg_k(\eta)}{d\eta} + (1 - t_{0,k})g_k(\eta) = \sum_{j=1}^{k} t_{j,k-j}g_{k-j}(\eta), \qquad \text{for } k = 1,2,\ldots. \tag{8.7.74b}$$

The solution of this system of equations is

$$g_0(\eta) = \exp\{-(1 - t_{0,0})\eta\}; \tag{8.7.75a}$$

$$g_k(\eta) = \sum_{j=1}^{k} c_{j,k} \exp\{-(1 - t_{0,0})\eta\} \qquad \text{for } k = 1,2,\ldots; \tag{8.7.75b}$$

$$c_{k,k} = -\sum_{j=0}^{k-1} c_{j,k} \qquad \text{for } k = 1,2,\ldots, \tag{8.7.75c}$$

$$c_{j,k} = [(1 - t_{0,k}) - (1 - t_{0,j})]^{-1} \sum_{i=j}^{k-1} c_{j,i} t_{k-i,i}$$

$$\text{for } k = 1,2,\ldots, \text{ and } j = 0,1,2,\ldots. \tag{8.7.75d}$$

An alternative approach to using the above equations for finding the $g_k(\eta)$ is to solve (8.7.69) or (8.7.74) successively for g_k in terms of $g_{k-1}, \ldots, g_0$.

8.8 FURTHER REMARKS

The material presented in this chapter is due to a number of researchers. Metropolis and Ulam [ME1] defined the random mapping function and posed the problem of finding the expected number of components. Kruskal [KR1] solved this problem and subsequently Katz [KA1] solved the problem of finding the probability that the random graph is connected. Rubin and Sitgreaves [RU1], Folkert [FO1], and Harris [HA6] derived a number of important results, many of which are discussed in Section 2.

The material in Section 3 is due mainly to Rapoport, Solomonoff, and Landau. Solomonoff and Rapoport [SO2] examined the problem of finding the weak connectivity constant of an outwardly homogeneous graph of degree α and they developed an asymptotic formula for this constant. Solomonoff [SO1] subsequently found an exact method for the computation of the connectivity of random graphs. This method, by means of example, demonstrated the accuracy of the approximate asymptotic expression given by Solomonoff and Rapoport. However, the required computations were very involved and did not lead to an explicit expression for the weak connectivity constant. Landau [LA1] then developed a procedure, based on an urn model, to find both the weak connectivity constant and the probability that the graph is connected. This procedure is complicated but could be used for graphs with several hundred vertices. For graphs with a large number of vertices, the asymptotic expressions given by Landau [LA1] and Solomonoff and Rapoport [SO2] can be used with great accuracy. The results on ignition phenomena are based on a paper by Rapoport [RA4]. Similar results have been reported by Allanson [AL1].

Some of the extensions given in Section 4 are contained in the papers mentioned above. Gilbert [GI2] derived the results for the random graph in which each branch exists with probability p. He showed that the probability that the graph is connected is asymptotically $1 - n(1 - p)^{n-1}$. In the same paper, Gilbert derived an expression for the probability that two given vertices are in the same component; this probability is asymptotically $1 - 2(1 - p)^{n-1}$ as $n \to \infty$. Related results appear in a paper by Austin *et al.* [AU1]. Here a different random process is used to generate a random graph and the authors find a generating function for the number of components. The properties of the random graphs $G_{n,m(n)}$ as $n \to \infty$ were derived by Erdös and Rényi [ER1]. Other results were obtained by these authors in references [ER2] and [ER3].

Rapoport discussed the theory of biased graphs [RA1-RA3] and introduced the concepts of distance and reciprocity bias. A number of different problems are formulated in these papers and partial solutions of several are given. Fu and Yau [FU1] examined the problem of finding the probability that two points are in the same component when branches have independent reliabilities. The procedure for calculating this probability, when the events that branches exist are not independent, is a modification of a procedure given in Section 4.4.

Section 6 is derived from a paper by Broadbent and Hammersley [BR1] and from several papers by Hammersley [HA2-HA4]. Additional results on percolation processes are contained in other papers by Hammersley *et al.* [HA5, FR4]. Harris [HA7] has considered percolation processes defined on *crystals* and has shown that the critical probability for such structures is bounded from below by 1/2. Hammersley has shown that this critical probability is also bounded from above by 0.646790. In reference [HA5], Hammersley and Welsh consider the problem of finding "first passage time" probabilities (the time that the excitation first reaches a given vertex) for percolation processes. This problem is extremely important but not many concrete results are available. In reference [GI3], Gilbert relates percolation processes to graphs constructed by drawing lines in the plane at random.

Jacobs [JA1] considered the problem of finding upper and lower bounds on the probability of connectivity for graphs whose branches exist with probability p. He used branch density as the parameter in order to compare the relative "efficiencies" of graphs. Another method of comparing the efficiencies of graphs is based on Monte Carlo programming. Well-documented programs are available for this purpose [DE1]. Baran [BA1] examined a class of networks he termed distributed networks. In a series of Rand Corporation Memoranda [BA2-BA9, BO1, SM1] he and his colleagues studied, in great detail, the routing, vulnerability, and cost problems of these networks. In another paper, Frank [FR1] studied the analysis and design of invulnerable networks. Applying the methods of Biermann [BI1, BI2] he gave asymptotic and recurrence equations for the vulnerability of networks under enemy attack. These equations, which are developed earlier in this chapter, are discussed in Section 7.

PROBLEMS

1. Show that each component of an outwardly homogeneous graph of degree one contains exactly one circuit and this circuit is directed.
2. Verify that

$$g(y) = y^n e^{n/y} \int_0^y u^{-n} e^{-n/y} \frac{1 - u^n}{u(1 - u)} du$$

is the solution of the differential equation

$$\frac{d}{dy}[g(y)] + n\frac{(1 - y)}{y^2} g(y) = \frac{1 - y^n}{y(1 - y)}.$$

3. Prove that for an outwardly homogeneous graph of degree α, constructed by sampling equiprobably with replacement from an n-vertex population (as in Section 3),

$$\text{Prob}\,\{\rho = 1\} = 1 - \sum_{i=1}^{n-1} k_i \binom{n-1}{i} \left(1 - \frac{i}{n}\right)^{i+(n-i)\alpha}$$

where the k_i satisfy

$$\sum_{i=1}^{k} k_i \binom{k}{i} (1 - i)^{(k-i)\alpha} \qquad \text{for } k = 1,2,\ldots,n.$$

4. Find the probability that the d-graph shown in Fig. P8.1 is connected.
5. a) Find schemes of connecting undirected graphs so that γ computed from the equation $\gamma = 1 - \exp(-\alpha\gamma)$ is an upper bound to the weak connectivity. In particular, how can directed graphs be used to model the undirected networks considered by Baran, as discussed in Section 7?
   b) Compare the Monte Carlo results shown in Fig. 8.7.5 with the solution of the equation of $\gamma = 1 - \exp(-\alpha\gamma)$.

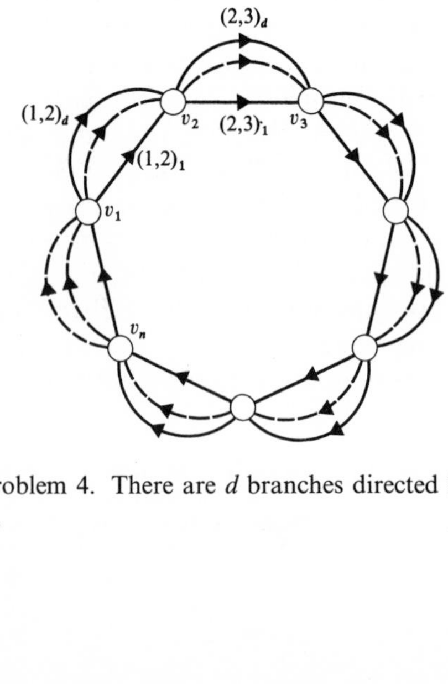

Fig. P8.1 d-graph for Problem 4. There are d branches directed from v_n to v_1 and from v_i to v_{i+1} for $i = 1,\ldots,n - 1$.

6. Fill in the missing steps in the computation of bounds on Prob $\{\rho = 1\}$ for the complete unbiased random graph discussed in Section 8.4 (each branch exists with probability p).

7. Consider a complete n-vertex graph in which each branch has probability p of existing and each vertex has probability t of existing. Prove that

$$\text{Prob}\,\{\rho = 1\} \sim 1 - \exp\{(n-1)\log 1/q + \log tn\}$$

where $q = 1 - p$ (see Jacobs [JA1]).

8. Let G be a 100-vertex random graph with vulnerable vertices and invulnerable branches. Suppose $t_1 = 0.05$, and the cost function is $H(d,K_s) = d\exp\{3K_s/2\}$. Find d and K_s so that the average fraction of surviving vertices reachable by a directed path of three or less branches from a vertex picked at random is at least 0.9. Assume

   a) The random graph is generated by a process satisfying Assumption 8.7.1b.
   b) The random graph is generated by a process satisfying Assumption 8.7.1b′.

   Compare the results of (a) and (b).

9. Let G be the graph discussed in Example 8.7.3. Plot γ as a function of time as the repair rate r_1 varies from 0 to 1.

10. Let G be an n-vertex, m-branch graph. Let each branch of G have capacity c and probability p of existence.

    a) Show that for variable n, but fixed m,

$$\text{Max } E\{\tau_{s,t}\} = \begin{cases} p + \frac{1}{2}(m-1)p^2 & \text{if } m \text{ is odd,} \\ \frac{1}{2}mp^2 & \text{if } m \text{ is even.} \end{cases}$$

    Under what conditions is the maximum value of $E\{\tau_{s,t}\}$ attained?

    b) What is the maximum value of $E\{\tau_{s,t}\}$ when the number of vertices is fixed? Under what conditions is this maximum attained?

    c) For the case in which the number of vertices of G is fixed, use the results of (b) to show that

$$\sum_t \sum_{\substack{s \\ s \neq t}} E\{\tau_{s,t}\}$$

    is maximized when G is complete.

11. Let T given in Fig. P8.2 be the terminal capacity matrix of a graph G with unreliable branches. Let each branch of G exist with probability p. Shown in Fig. P8.2 are three realizations of T.

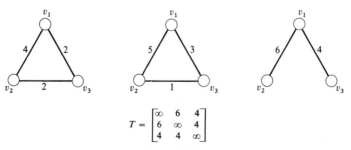

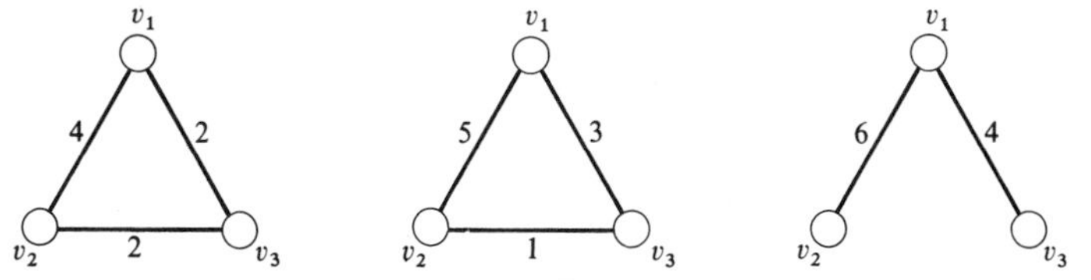

$$T = \begin{bmatrix} \infty & 6 & 4 \\ 6 & \infty & 4 \\ 4 & 4 & \infty \end{bmatrix}$$

Figure P8.2

Find the realization of T for which

$$\sum_{t}\sum_{\substack{s \\ s\neq t}} E\{\tau_{s,t}\}$$

is maximum. Generalize this result to the realization of arbitrary symmetric terminal capacity matrices.

12. Let $G = (V,\Gamma)$ be an n-vertex, m-branch undirected graph. Define an n-vertex, m^*-branch graph $G^* = (V,\Gamma^*)$ such that $\Gamma^* \subset \Gamma$ and

$$\text{Prob}\,\{[i,j] \in \Gamma^*\} = p \qquad \text{for all } [i,j] \in \Gamma,$$

G^* is said to be induced by G. Let ρ^* be the number of components of G^*.

a) Show that

$$\text{Prob}\,\{\rho^* = 1\} = \sum_{k=n-1}^{m} N_k p^k (1-p)^{m-k}$$

and

$$\text{Prob}\,\{\rho^* = 1\} = 1 - \sum_{k=\theta}^{m} \hat{N}_k p^{m-k} (1-p)^k$$

where N_k is the number of connected $n - k$-branch, n-vertex subgraphs of G, $\hat{N}_k$ is the number of $m - k$-branch, n-vertex unconnected subgraphs of G, and θ is the number of elements in the smallest branch cut-set of G.

b) Let G be any undirected n-vertex, m-branch graph and let $\tilde{G}^*$ be a subgraph of G induced by the above process. Let $\tilde{\rho}^*$ be the number of components of $\tilde{G}^*$ and $\tilde{\theta}$, the size of the minimum branch cut-set of $\tilde{G}$. Show

i) if G has more trees than $\tilde{G}$, then there exists a $p_0 > 0$ such that Prob $\{\rho^* = 1\} >$ Prob $\{\tilde{\rho}^* = 1\}$ for all $p \leq p_0$.

ii) if $\theta > \tilde{\theta}$, there exists a $p_1 < 1$ such that Prob $\{\rho^* = 1\} >$ Prob $\{\tilde{\rho}^* = 1\}$ for all $p \geq p_1$.

13. Let $G = (V,\Gamma)$ be an n-vertex, m-branch undirected graph. Define an n^*-vertex, m^*-branch graph $G^* = (V^*,\Gamma^*)$ such that $V^* \subset V$, $\Gamma^* \subset \Gamma$, and

$$\text{Prob}\,\{v_i \in V^*\} = p,$$

and $[i,j] \in \Gamma^*$ if and only if $v_i \in V^*$, $v_j \in V^*$, and $[i,j] \in \Gamma$. Let ρ^* be the number of components of G^*.

a) Show that

$$\text{Prob}\,\{\rho^* = 1\} = 1 - \sum_{k=\omega}^{n-1} N_k p^{n-k} (1-p)^k$$

where N_k is the number of unconnected $n - k$-vertex subgraphs of G and ω is the number of elements in the minimum vertex cut-set of G.

b) Let $\bar{G}$ be any n-vertex, m-branch undirected graph and $\bar{G}^*$ a subgraph induced by $\bar{G}$ by the above process. Let $\bar{\rho}^*$ be the number of components of $\bar{G}^*$ and $\bar{\omega}$ the size of the minimum vertex cut-set of $\bar{G}$. Show that, if $\omega > \bar{\omega}$, then there exists a $p_1 < 1$ such that

$$\text{Prob}\{\rho^* = 1\} > \text{Prob}\{\bar{\rho}^* = 1\} \qquad \text{for all } p \geq p_1.$$

c) Suppose p is close to unity. Among what class of n-vertex, m-branch graphs can the graph with maximum probability of connectivity be found?
[FR2]

14. Let G be an n-vertex, m-branch undirected graph with $\omega = [2m/n]$, and $[2(m-1)/n] < [2m/n]$. Suppose that G contains only one vertex cut-set with exactly $[2m/n]$ elements. Prove that G is bipartite [FR2].

15. Let $G = (V,\Gamma)$ be a complete undirected bipartite graph with $V = A \cup B$, $A \cap B = \phi$, and $\Gamma = [A,B]$. Let $|A| = k_1$, $|B| = k_2 = n - k_1$, $k_1 \leq k_2$, and $m \leq k_1 k_2$.
   a) Let $G^* = (V^*,\Gamma^*)$ be induced by G as in Problem 13. Let $\tilde{G}$ be any undirected n-vertex graph with $k_1 k_2$ branches whose minimum vertex cut-set has k_1 elements and let $\tilde{G}^*$ be the graph induced by $\tilde{G}$. Show that there exists a $p_1 < 1$ such that for all $p \geq p_1$, Prob $\{\rho^* = 1\} \geq$ Prob $\{\tilde{\rho}^* = 1\}$ [FR2].
   b) Suppose that $k_1 \geq k_2 - 1$. Show that for any n-vertex, m-branch graph $\tilde{G}$, there exists a $p_1 < 1$ such that for all $p \geq p_1$, Prob $\{\rho^* = 1\} >$ Prob $\{\tilde{\rho}^* = 1\}$.
   c) Can the result in (b) be extended to $k_1 < k_2 - 1$? If so, for what values of k_1 and n is this result true?

16. Let $G = (V,\Gamma)$ be an n-vertex, m-branch undirected graph and let $G^* = (V,\Gamma^*)$ be induced by G such that $\Gamma^* \subset \Gamma$ and Prob $\{[i,j] \in \Gamma^*\} = p$. Assign to each branch of G a unit capacity and let T be the terminal capacity matrix of G. The following heuristic procedure is suggested for approximating Prob $\{\rho^* = 1\}$ where ρ^* is the number of components of G^* [FR3].
   i) Let $G_t = (V,\Gamma_t)$ be the tree resulting from the Multiterminal Analysis Algorithm of Section 5.3; G_t realizes T.
   ii) Let c_k be the capacity of b_k in G_t for all k.
   iii) Approximate Prob $\{\rho^* = 1\}$ by Prob $\{\rho_t^* = 1\}$ where ρ_t^* is the number of components in the graph G_t^* defined by

$$\text{Prob}\{b_k \in \Gamma_t^*\} = \begin{cases} 0 & \text{if } b_k \notin \Gamma_t, \\ 1 - (1-p)^{c_k} & \text{if } b_k \in \Gamma_t. \end{cases}$$

   a) Give an expression for Prob $\{\rho_t^* = 1\}$.
   b) Compute an exact expression for Prob $\{\rho^* = 1\}$ for the graph induced by G shown in Fig. P8.3.
   c) Find G_t for G shown in Fig. P8.3.
   d) Compare Prob $\{\rho^* = 1\}$ and Prob $\{\rho_t^* = 1\}$ for $p = 0.4, 0.5, 0.6, 0.8$.

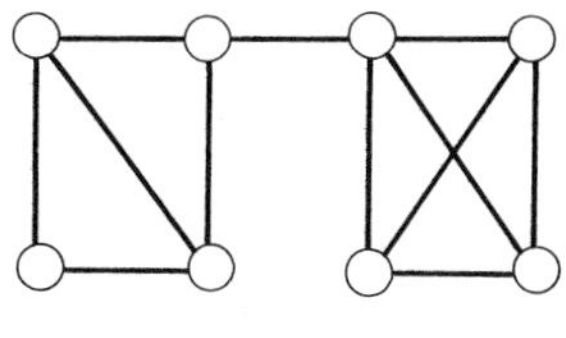

Figure P8.3

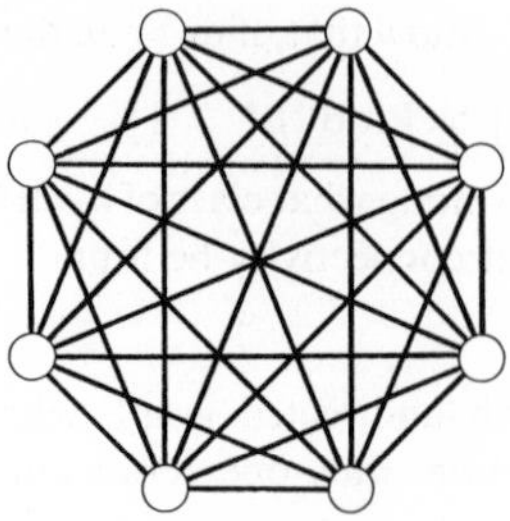

Figure P8.4

e) Repeat (b) to (d) for the complete graph shown in Fig. P8.4.

f) Generalize the above heuristic method to the case where $G = (V,\Gamma)$ induces a graph $G^* = (V,\Gamma^*)$ such that $\Gamma^* \subset \Gamma$ and

$$\text{Prob}\,\{b_k \in \Gamma^*\} = p_k \qquad \text{for } k = 1,\ldots,m.$$

Let G be the graph shown in Fig. P8.5. The branch weights on G are probabilities. Compute Prob $\{\rho^* = 1\}$ and Prob $\{\rho_t^* = 1\}$ where the branch probabilities on G_t are appropriately chosen. Compare these probabilities for $p_1 = 0.8$ and $p_2 = 0.6$.

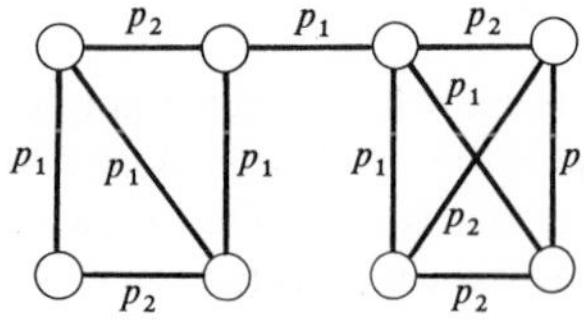

Figure P8.5

REFERENCES

AL1 J. T. Allanson, "Some Properties of a Randomly Connected Neural Network," *Symposium on Information Theory*, ed. C. Cherry, Wiley, New York, 1955, pp. 303–313.

AU1 T. L. Austin, R. E. Fagen, W. F. Penny, and J. Riordan, "The Number of Components in Random Linear Graphs," *Ann. Math. Stat.* **30,** 747–754 (1959).

BA1 P. Baran, "On Distributed Communication Networks," *IEEE Trans. Commun. Systems* **COM-12,** 1–9 (1964).

BA2 P. Baran, "Priority, Precedence, and Overload," *Rand Corporation Memorandum* (August 1964).

BA3 P. Baran, "History, Alternative Approaches and Comparisons," *Rand Corporation Memorandum* (August 1964).

BA4 P. Baran, "Mini-Cost Microwave," *Rand Corporation Memorandum* (August 1964).

BA5 P. Baran, "Tentative Engineering Specifications and Preliminary Design for a High Data Rate Distributed Network Switching Node," *Rand Corporation Memorandum* (August 1964).

BA6 P. Baran, "The Multiplexing Station," *Rand Corporation Memorandum* (August 1964).

BA7 P. Baran, "Security, Secrecy, and Tamper-Free Considerations," *Rand Corporation Memorandum* (August 1964).

BA8 P. Baran, "Cost Analysis," *Rand Corporation Memorandum* (August 1964).

BA9 P. Baran, "Summary Overview," *Rand Corporation Memorandum* (August 1964).

BA10 R. E. Barlow and F. Proschan, *Mathematical Theory of Reliability,* Wiley, New York, 1965.

BI1 A. Biermann, "A General Target Theory of Radiobiological Action," *Bull. Math. Bio-Phys.* **25,** 273–296 (1963).

BI2 A. Biermann, "A General Target Theory of Radiobiological Action: II," *Bull. Math. Biophys.* **25,** 367–385 (1963).

BO1 S. Boehm and P. Baran, "Digital Simulation of Hot-Potato Routing in a Broadband Distributed Communication Network," *Rand Corporation Memorandum* (August 1964).

BR1 S. R. Broadbent and J. M. Hammersley, "Percolation Processes, Crystals, and Mazes," *Proc. Cambridge Phil. Soc.* **53,** 629–641 (1957).

DE1 P. Demetriou, "Probabilistic Network Programs," Ph.D. Dissertation, Columbia University, New York, 1963.

ER1 P. Erdös and A. Rényi, "On the Evolution of Random Graphs," *Publ. Math. Inst. Hung. Acad. Sci.* **5,** 17–61 (1960).

ER2 P. Erdös and A Rényi, "On Random Graphs I," *Publ. Math. Inst. Hung. Acad. Sci.* **4,** 290–297 (1959).

ER3 P. Erdös and A. Rényi, "On the Strength of Connectedness of a Random Graph," *Acta Math.* 261–267 (1961).

FE1 W. Feller, *An Introduction to Probability Theory and Its Application*, Vol. 1, Wiley, New York, 1957.

FO1 J. E. Folkert, "The Distribution of the Number of Components of a Random Mapping Function," Ph.D. Dissertation, Michigan State University, 1955.

FR1 H. Frank, "Vulnerability of Communication Networks," *IEEE Trans. Commun. Technol.* **COM-15,** 778–789 (1967).

FR2 H. Frank, "Maximally Reliable Vertex Weighted Graphs," Proceedings of Third Princeton Conference on Information Sciences and Systems, 1969, pp. 1–6.

FR3 H. Frank, "Some New Results in the Design of Survivable Networks," in Proceedings of the 12th Midwest Symposium on Circuit Theory, April 1969, I.3.1–I.3.8.

FR4 H. L. Frisch and J. M. Hammersley, "Percolation Processes and Related Topics," *J. Soc. Ind. Appl. Math.* **11,** 894–918 (1963).

FU1 Y. Fu and S. S. Yau, "A Note on the Reliability of Communication Networks," *J. Soc. Ind. Appl. Math.* **10,** 469–474 (1962).

GI1 E. N. Gilbert, "Enumeration of Labeled Graphs," *Can. J. Math.* **8,** 405–411 (1956).

GI2 E. N. Gilbert, "Random Graphs," *Ann. Math. Statistics* **30,** 1141–1144 (1959).

GI3 E. N. Gilbert, "Random Plane Networks," *J. Soc. Ind. Appl. Math.* **9,** 543–553 (1961).

HA1 G. Hadley, *Nonlinear and Dynamic Programming,* Addison-Wesley, Reading, Mass., 1964.

HA2 J. M. Hammersley, "Percolation Process—the Connective Constant," *Proc. Cambridge Phil. Soc.* **53,** 642–647 (1957).

HA3 J. M. Hammersley, "Percolation Processes—Lower Bounds on the Critical Probability," *Ann. Math. Statistics* **28,** 790–795 (1957).

HA4 J. M. Hammersley, "Bornes Supérieures de la Probabilité Critique dans un Processus de Filtration," *Le Calcul des Probabilités et ses Applications,* Centre National de la Recherche Scientifique, Paris, 1959, pp. 17–37.

HA5 J. M. Hammersley and D. J. A. Welsh, "First-Passage Percolation, Sub-additive Processes, Stochastic Networks and Generalized Renewal Theory," in *Bernoulli, Bayes, Laplace,* ed. by J. Neyman and L. M. LeCam, Springer-Verlag, New York, 1965, pp. 61–110.

HA6 B. Harris, "Probability Distributions Related to Random Mappings," *Ann. Math. Statistics* **31,** 1045–1062 (1960).

HA7 T. E. Harris, "A Lower Bound for the Critical Probability in a Certain Percolation Process," *Proc. Cambridge Phil. Soc.* **56,** 13–20 (1960).

HO1 F. E. Hohn and L. R. Schissler, "Boolean Matrices and the Design of Combinatorial Relay Switching Circuits," *Bell System Tech. J.* **34,** 177–202 (1955).

JA1 I. Jacobs, "Connectivity of Probabilistic Graphs," MIT Research Laboratory of Electronics, Technical Report no. 356, Cambridge, Mass., September 15, 1959.

KA1 I. Katz, "Probability of Indecomposibility of a Random Mapping Function," *Ann. Math. Statistics* **26,** 512–517 (1955).

KR1 M. D. Kruskal, "The Expected Number of Components under a Random Mapping Function," *Am. Math. Monthly* **61,** 392–397 (1954).

LA1 H. G. Landau, "On Some Problems of Random Nets," *Bull. Math. Biophys.* **14,** 203–212 (1952).

ME1 N. Metropolis and S. Ulam, "A Property of Randomness of an Arithmetical Function," *Am. Math. Monthly* **60,** 252–253 (1953).

RA1 A. Rapoport, "Nets with Distance Bias," *Bull. Math. Biophys.* **13,** 85–91 (1951).

RA2 A. Rapoport, "Contribution to the Theory of Random and Biased Nets," *Bull. Math. Biophys.* **19,** 257–277 (1957).

RA3 A. Rapoport, "Nets with Reciprocity Bias," *Bull. Math. Biophys.* **20,** 191–210 (1958).

RA4 A. Rapoport, "Ignition Phenomena in Random Nets," *Bull. Math. Biophys.* **14,** 35–44 (1952).

RE1 A. Rényi, "On Connected Graphs, I," *Publ. Math. Inst. Hung. Acad. Sci.* **4,** 385–388 (1959).

RI1 R. J. Riddell, Jr., and G. E. Uhlenbeck, "On the Virial Development of the Equation of State of Monotonic Gases," *J. Chem. Phys.* **21,** 2056–2064 (1953).

RU1 H. Rubin and R. Sitgreaves, "Probability Distributions Related to Random Transformations on a Finite Set," Technical Report no. 19A, Applied Mathematics and Statistics Laboratory, Stanford University, California, 1954.

SM1 J. W. Smith, "Determination of Path Lengths in a Distributed Network," *Rand Corporation Memorandum* (August 1965).

SO1 R. Solomonoff, "An Exact Method for the Computation of the Connectivity of Random Nets," *Bull. Math. Biophys.* **14,** 153–157 (1952).

SO2 R. Solomonoff and A. Rapoport, "Connectivity of Random Nets," *Bull. Math. Biophys.* **13,** 107–117 (1951).

CHAPTER 9

TIME DELAY IN NETWORKS

9.1 INTRODUCTION

In many flow systems it is the element of time rather than maximum flow which is of major importance. In our previous discussion, we usually assumed that, for a flow between two stations over a given path, all of the branches in that path were simultaneously in use. Such an assumption accurately describes most telephone networks as well as many other communication systems. However, there are many networks which cannot be reasonably represented by this model. For example, a fixed time may be necessary to code and decode messages at stations along the transmission path or a fixed transit time may be associated with each branch. The transmission path is then established sequentially and at any instant of time the branches along any path may accommodate various commodities where each commodity is distinguished by its origin-destination pair. Such networks are usually called store-and-forward networks.

In a store-and-forward network, a set of flows originate at certain stations and have various destinations. The flows are stored at their sources until they can be transmitted. For example, a number of messages may arrive at a given station during a short interval of time. These messages form a queue and are then transmitted on a "first come, first served" basis to adjacent stations along the message routes. When a message arrives at an adjacent station it must again be stored until it can be transmitted to the next station on its route. In this manner the message finally reaches its destination.

In networks with delay, flow and routing, as well as delays associated with channels and stations, may be either deterministic or random. Hence, the time to transmit a given flow from its origin to its destination could be either a fixed number or a random variable. For these networks, the formulation of a reasonable performance measure is necessary. Among the factors which may affect such a performance measure are the variation of transit time as a function of branch capacity and the size of memory units storing flow at the stations.

In this chapter, we discuss some of the many aspects of the above problems. First, investigating deterministic graphs, we determine the equations governing flow for graphs with certain routing procedures. Attention is focused on the time required to transport flow from a set of sources to a set of terminals. We then consider the problem of determining optimal memory unit sizes and branch capacities to satisfy flow rate requirements in deterministic store-and-forward networks. The treatment leads to a linear program which indicates the constraints and the difficulties encountered in problems of this nature. In Section 4 and 5 we examine networks with stochastic flows. The properties of single exponential service stages are first developed. These properties are then applied to find the average message delay in networks with random messages arrivals. Finally, the results of this analysis are used to find values of branch capacities which minimize the average time delay.

9.2 GRAPHS WITH DETERMINISTIC BRANCH DELAYS

Let G be a directed graph. With each branch b_i, associate a branch delay δ_i and infinite branch capacity. To simplify the computations, we assume that the delay δ_i is rational. Therefore, without loss of generality, we can take $\delta_1 = \delta_2 = \cdots = \delta_m = \delta$ by introducing additional nodes as necessary.[1] If b_i is directed from v_j to v_k, a flow in b_i leaving v_j at time t_0 will arrive at v_k at time $t_0 + \delta$.

Let $x_i(t)$ represent the flow in b_i at time t and let $\mathbf{x}(t) = (x_1(t), \ldots, x_m(t))'$ be an $m \times 1$ *flow vector*. At each vertex it may be possible to insert or extract flow; that is, each vertex may behave as a source, a terminal, or both. Let $u_i(t)$ be the external input to v_i at time t and let $y_j(t)$ be the output at v_j at time t for $i,j = 1, \ldots, n$. Furthermore, define $\mathbf{u}(t)$ and $\mathbf{y}(t)$ as $\mathbf{u}(t) = (u_1(t), \ldots, u_n(t))'$, and $\mathbf{y}(t) = (y_1(t), \ldots, y_n(t))'$. If it is not possible to insert or remove flow at v_k, then $u_k(t) \equiv 0$ or $y_k(t) \equiv 0$, respectively. We call $\mathbf{u}(t)$ and $\mathbf{y}(t)$, the input and output flow vectors, respectively. Figure 9.2.1 illustrates these definitions.

If at time t_0 a flow a is initiated in b_i, we assume $x_i(t) \equiv a$ for all t such that $t_0 \leq t \leq t_0 + \delta$. Hence, $\mathbf{x}(t)$ is completely specified over $t_0 \leq t \leq t_0 + k\delta$ by specifying $\mathbf{x}(t_0)$, $\mathbf{x}(t_0 + \delta), \ldots, \mathbf{x}(t_0 + k\delta)$. Since all branch flows were initially established at t_0, we also assume that inputs and outputs are discrete and that flow can only be added to, or withdrawn from, the graph at time t_0, $t_0 + \delta, \ldots$. In most systems, the input vector may be considered as an independent variable, while the output vector is a dependent variable. In this section, we assume that the flow and

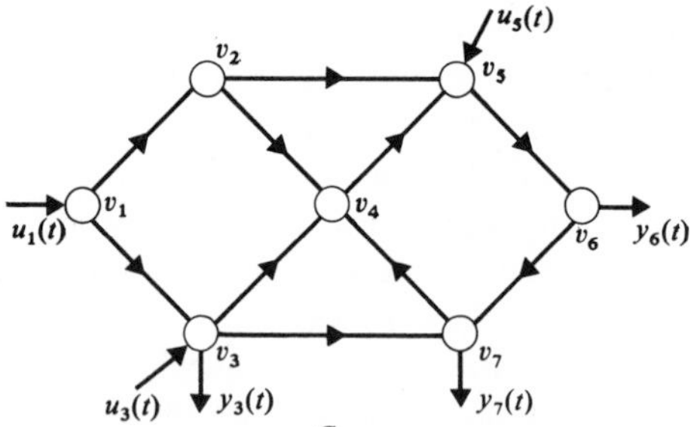

$\mathbf{u}(t) = (u_1(t), 0, u_3(t), 0, u_5(t), 0, 0)'$

$\mathbf{y}(t) = (0, 0, y_3(t), 0, 0, y_6(t), y_7(t))'$

Fig. 9.2.1 Graph with several sources and terminals.

[1] If $\delta_1 \neq \delta_2 \neq \cdots \neq \delta_m$, we can construct a new graph $\hat{G}$, by replacing the ith branch by a directed path of δ_i/δ series branches each with time delay δ, where δ is the greatest common divisor of $\delta_1, \delta_2, \ldots, \delta_m$.

output vectors at time $t_0 + k\delta$ can be expressed as a function of the flow and input vectors at $t_0 + (k-1)\delta$.

With the above assumptions, we may write

$$\mathbf{x}(t_0 + k\delta) = \mathbf{f}[\mathbf{x}(t_0 + (k-1)\delta),\mathbf{u}(t_0 + (k-1)\delta)] \tag{9.2.1a}$$

and

$$\mathbf{y}(t_0 + k\delta) = \hat{\mathbf{f}}[\mathbf{x}(t_0 + (k-1)\delta),\mathbf{u}(t_0 + (k-1)\delta)] \tag{9.2.1b}$$

where $\mathbf{f}[\cdot] = (f_1(\cdot),\ldots,f_m(\cdot))'$ and $\hat{\mathbf{f}}[\cdot] = (\hat{f}_1(\cdot),\ldots,\hat{f}_m(\cdot))'$ are vector-valued vector functions. For simplicity, let $t_0 = 0$ and let $\mathbf{x}(k\delta),\mathbf{u}(k\delta),\mathbf{y}(k\delta)$ be denoted by $\mathbf{x}(k),\mathbf{u}(k),\mathbf{y}(k)$. Equations (9.2.1a) and (9.2.1b) then become

$$\mathbf{x}(k) = \mathbf{f}[\mathbf{x}(k-1),\mathbf{u}(k-1)] \tag{9.2.2a}$$

and

$$\mathbf{y}(k) = \hat{\mathbf{f}}[\mathbf{x}(k-1),\mathbf{u}(k-1)]. \tag{9.2.2b}$$

We refer to the ordered pair of functions $(\mathbf{f}[\cdot],\hat{\mathbf{f}}[\cdot])$ as the flow strategy.

The techniques used in this section are those ordinarily associated with linear system theory. For example, the reader may recognize that (9.2.2) is the input-output state equation of the system and the vector $\mathbf{x}(t)$ is the state vector of the system [ZA1]. Consequently, the reader unfamiliar with these techniques may wish to omit the proofs of the theorems. In general, both $\mathbf{f}$ and $\hat{\mathbf{f}}$ are explicit functions of time (that is, $\mathbf{f} = \mathbf{f}[\cdot,t],\hat{\mathbf{f}} = \hat{\mathbf{f}}[\cdot,t]$) and since negative flow is meaningless in a directed graph, the vectors, $\mathbf{x}$, $\mathbf{u}$, and $\mathbf{y}$ have nonnegative components for all time.

Definition 9.2.1 *Let $\mathcal{A}^+$ be the matrix formed by setting to zero all negative entries in the incidence matrix $\mathcal{U}$ and let $-\mathcal{A}^-$ be the matrix formed by setting to zero all positive entries in the incidence matrix. Then, the flow strategy* $(\mathbf{f},\hat{\mathbf{f}})$ *is conservative if*

$$\mathcal{A}^+\mathbf{x}(k) + \mathbf{u}(k) = \mathcal{A}^-\mathbf{x}(k+1) + \mathbf{y}(k+1) \textit{ for all } k. \tag{9.2.3}$$

An alternative expression of (9.2.3) is

$$\mathcal{A}^+\mathbf{x}(k) + \mathbf{u}(k) = \mathcal{A}^-\mathbf{f}[\mathbf{x}(k),\mathbf{u}(k)] + \hat{\mathbf{f}}[\mathbf{x}(k),\mathbf{u}(k)]. \tag{9.2.4}$$

Thus, a flow strategy is conservative if, for all i, the sums of the flows entering v_i at one instant of time are equal to the sums of the flows leaving v_i at the next instant of time.

$\mathcal{A}^-$ has exactly one "1" and $n - 1$ "0's" in each column. Suppose every vertex has at least one outwardly directed branch. Then, there is at least one nonzero entry in every row of $\mathcal{A}^-$. Hence, the vertices and branches can be reordered so that, in the resulting matrix $\mathcal{A}^-$, the first n rows and columns form an identity matrix. That is, there is a reordering such that if $\underset{\sim}{\mathcal{A}}^-$, $\underset{\sim}{\mathcal{A}}^+$, $\underset{\sim}{\mathbf{x}}$, $\underset{\sim}{\mathbf{u}}$, $\underset{\sim}{\mathbf{f}}$, and $\underset{\sim}{\hat{\mathbf{f}}}$ represent $\mathcal{A}^-$, $\mathcal{A}^+$, $\mathbf{x}$, $\mathbf{u}$, $\mathbf{f}$, and $\hat{\mathbf{f}}$ after the reordering, and

$$\mathcal{A}^+\underset{\sim}{\mathbf{x}}(k) + \underset{\sim}{\mathbf{u}}(k) = \underset{\sim}{\mathcal{A}}^-\begin{bmatrix}\underset{\sim}{\mathbf{x}}_a(k+1)\\ \underset{\sim}{\mathbf{x}}_b(k+1)\end{bmatrix} + \underset{\sim}{\hat{\mathbf{f}}}[\underset{\sim}{\mathbf{x}}(k),\mathbf{u}(k)] \tag{9.2.5}$$

where $\underset{\sim}{\mathcal{A}}^- = [I\ \underset{\sim}{B}^-]$, $\underset{\sim}{\mathbf{x}}_a(k+1) = (x_{i_1}(k+1), \ldots, x_{i_n}(k+1))'$ and $\underset{\sim}{\mathbf{x}}_b(k+1) = (x_{i_{n+1}}(k+1), \ldots, x_{i_{n+m}}(k+1)$. Replacing $\underset{\sim}{\mathcal{A}}^-$ by $[I\ \underset{\sim}{B}^-]$ in (9.2.5) and solving for $\underset{\sim}{x}_a$ gives

$$\underset{\sim}{\mathbf{x}}(k+1) = \begin{bmatrix} \mathcal{A}^+ \\ \underset{\sim}{0} \end{bmatrix} \underset{\sim}{\mathbf{x}}(k) + \underset{\sim}{\mathbf{u}}(k) + \begin{bmatrix} -B^- \\ I \end{bmatrix} \mathbf{f}_b[\underset{\sim}{\mathbf{x}}(k),\underset{\sim}{\mathbf{u}}(k)] - \underset{\sim}{\hat{\mathbf{f}}}[\underset{\sim}{\mathbf{x}}(k),\mathbf{u}(k)], \tag{9.2.6}$$

where $\underset{\sim}{\mathbf{f}}_b[\cdot]$ consists of the last $m - n$ components of $\underset{\sim}{\mathbf{f}}[\cdot]$.

This result may be considerably strengthened in the case where there are at least d^+ branches incident out of each vertex. In this case, there is a reordering such that

$$\underset{\sim}{\mathcal{A}}^- = [I_1, I_2, \ldots, I_{d^+}, B^*] \tag{9.2.7}$$

where the I_i are identity matrices. For example, if the outward demi-degree of each vertex is exactly d^+, then B^* is empty and

$$[I, I, \ldots, I]\underset{\sim}{\mathbf{f}} = \underset{\sim}{\mathcal{A}}^+ \underset{\sim}{\mathbf{x}}(k) + \underset{\sim}{\mathbf{u}}(k) - \underset{\sim}{\mathbf{f}}[\underset{\sim}{\mathbf{x}}(k),\underset{\sim}{\mathbf{u}}(k)] \tag{9.2.8}$$

or $\underset{\sim}{\mathbf{f}}$ may be partitioned into d^+ subvectors $\underset{\sim}{\mathbf{f}}_a, \underset{\sim}{\mathbf{f}}_b, \ldots, \underset{\sim}{\mathbf{f}}_d$ such that

$$\underset{\sim}{\mathbf{f}}_a + \underset{\sim}{\mathbf{f}}_b + \cdots + \underset{\sim}{\mathbf{f}}_{d^+} = \underset{\sim}{\mathcal{A}}^+ \underset{\sim}{\mathbf{x}}(k) + \underset{\sim}{\mathbf{u}}(k) - \underset{\sim}{\mathbf{f}}[\underset{\sim}{\mathbf{x}}(k),\underset{\sim}{\mathbf{u}}(k)]. \tag{9.2.9}$$

Let $S^{\mathcal{H}} = [s^{\mathcal{H}}_{i,j}]$, $S^{\mathcal{J}} = [s^{\mathcal{J}}_{i,j}]$, $S^{\mathcal{L}} = [s^{\mathcal{L}}_{i,j}]$, and $S^{\mathcal{M}} = [s^{\mathcal{M}}_{i,j}]$ be $m \times m$, $m \times n$, $n \times m$, and $n \times n$ matrices of zeros and ones, respectively, such that, for any v_x,

$$s^{\mathcal{H}}_{i,j} = 1 \text{ if and only if } b_i \text{ is directed into } v_x \text{ and } b_j \text{ is directed out of } v_x \text{ for some } v_x; \tag{9.2.10a}$$

$$s^{\mathcal{J}}_{i,j} = 1 \text{ if and only if } u_j \not\equiv 0 \text{ and } b_i \text{ is directed out of } v_j; \tag{9.2.10b}$$

$$s^{\mathcal{L}}_{i,j} = 1 \text{ if and only if } y_i \not\equiv 0 \text{ and } b_j \text{ is directed into } v_i; \tag{9.2.10c}$$

$$s^{\mathcal{M}}_{i,j} = 1 \text{ if and only if } y_i \not\equiv 0 \text{ and } u_j \not\equiv 0. \tag{9.2.10d}$$

It is impossible to route flow from branch b_j to b_i without using intermediate branches unless $s^{\mathcal{H}}_{i,j} = 1$. Similarly, it is impossible to route an input flow directly from v_j to b_i unless $s^{\mathcal{J}}_{i,j} = 1$, and an output flow cannot be removed at v_i from b_j unless $s^{\mathcal{L}}_{i,j} = 1$.

We now consider "linear" routing strategies. Such strategies specify that the flow in b_i at time $k + 1$ is the sum of percentages of flow. This sum consists of flows from the branches b_j at time k for all j such that $s^{\mathcal{H}}_{i,j} = 1$ and a percentage of the input at vertex v_x where $(x,i) \in \Gamma$. A precise definition of this concept is now given.

***Definition* 9.2.2** *The flow strategy* $(\mathbf{f},\hat{\mathbf{f}})$ *in a graph G with infinite branch capacities is said to be linear if*

$$\mathbf{x}(k+1) = \mathbf{f}[\mathbf{x}(k),\mathbf{u}(k)] = \mathcal{H}\mathbf{x}(k) + \mathcal{J}\mathbf{u}(k), \tag{9.2.11a}$$

$$\mathbf{y}(k+1) = \hat{\mathbf{f}}[\mathbf{x}(k),\mathbf{u}(k)] = \mathcal{L}\mathbf{x}(k) + \mathcal{M}\mathbf{u}(k), \tag{9.2.11b}$$

where $\mathscr{H}$, $\mathscr{J}$, $\mathscr{L}$, and $\mathscr{M}$ are $m \times m$, $m \times n$, $n \times m$, and $n \times n$ matrices, respectively,[2] *such that $\mathscr{H} \leq S^{\mathscr{H}}$, $\mathscr{J} \leq S^{\mathscr{J}}$, $\mathscr{L} \leq S^{\mathscr{L}}$, and $\mathscr{M} \leq S^{\mathscr{M}}$.*

If $(\mathbf{f},\hat{\mathbf{f}})$ is linear, (9.2.4) becomes

$$\mathcal{C}^{+}\mathbf{x}(k) + \mathbf{u}(k) = \mathcal{C}^{-}\mathscr{H}\mathbf{x}(k) + \mathcal{C}^{-}\mathscr{J}\mathbf{u}(k) + \mathscr{L}\mathbf{x}(k) + \mathscr{M}\mathbf{u}(k) \qquad (9.2.12a)$$

or

$$[\mathcal{C}^{+} - \mathscr{L} - \mathcal{C}^{-}\mathscr{H}]\mathbf{x}(k) = [\mathcal{C}^{-}\mathscr{J} + \mathscr{M} - I]\mathbf{u}(k), \qquad (9.2.12b)$$

and since $\mathbf{x}(k)$ and $\mathbf{u}(k)$ can be arbitrary, for strategy $(\mathbf{f},\hat{\mathbf{f}})$ to be conservative, we must have

$$\mathcal{C}^{+} = \mathscr{L} + \mathcal{C}^{-}\mathscr{H} \qquad (9.2.13a)$$

and

$$\mathcal{C}^{-}\mathscr{J} + \mathscr{M} = I. \qquad (9.2.13b)$$

Let $(\mathbf{f},\hat{\mathbf{f}})$ be a conservative linear flow strategy. Let there be at time $t = 0$, an initial flow $\mathbf{x}(0)$. We wish to examine the behavior of the branch flows as a function of time when no additional flow is inserted into the graph. Intuitively, it would seem that if $\mathscr{L} \neq [0]$, a reasonable strategy should eventually result in all zero branch flows; that is, all of the initial flow should eventually appear as outputs of the system.

An upper bound for the flow in each branch as time approaches infinity is given in Theorem 9.2.2.

***Theorem* 9.2.2** *Let $(\mathbf{f},\hat{\mathbf{f}})$ be a conservative linear strategy with flow equations given by (9.2.11a) and (9.2.11b). If $\mathscr{L} \neq [0]$, there exists an $m \times m$ matrix $\hat{M} = [\hat{m}_{i,j}]$ satisfying*

$$0 \leq \hat{m}_{i,j} \leq 1, \qquad \text{for all } i,j \qquad (9.2.14a)$$

$$\sum_{i=1}^{m} \hat{m}_{i,j} = 1, \qquad \text{for } j = 1,2,\ldots,m \qquad (9.2.14b)$$

$$\hat{m}_{i,j} = \hat{m}_{i,k}, \qquad \text{for } j, k = 1,\ldots,m \qquad (9.2.14c)$$

such that, if $\mathbf{u}(t) \equiv \mathbf{0}$,

$$\mathbf{x}(k) \leq \hat{M}\mathbf{x}(0) + \boldsymbol{\varepsilon}(k) \qquad (9.2.14d)$$

where $\boldsymbol{\varepsilon}(k) \geq \mathbf{0}$ is a vector whose components approach zero as $k \to \infty$.

Proof. If $\mathbf{u}(t) \equiv 0$, $\mathbf{x}(k) = \mathscr{H}^{k}\mathbf{x}(0)$. Since $(\mathbf{f},\hat{\mathbf{f}})$ is conservative, by (9.2.13a),

$$\sum_{i=1}^{m} h_{i,j} \leq 1 \qquad (9.2.15)$$

---

[2] Since negative flows are prohibited, all entries of $\mathscr{H}$, $\mathscr{J}$, $\mathscr{L}$, and $\mathscr{M}$ must be nonnegative. If $u_i(t) \equiv 0$ for all t, the ith columns of $\mathscr{J}$ and $\mathscr{M}$ are identically zero while, if $y_j(t) \equiv 0$ for all t, the jth rows of $\mathscr{L}$ and $\mathscr{M}$ are identically zero.

where $\mathscr{H} = [h_{i,j}]$. By assumption, $\mathscr{L} \neq [0]$; hence, there exists at least one j such that the inequality in (9.2.15) is strict. Consider the matrix $\mathscr{H}'$, where the prime indicates transpose. There exists an $m \times m$ matrix $M = [m_{i,j}]$ such that $0 \leq m_{i,j} \leq 1$, for all i,j, $h_{i,j} \leq m_{i,j}$, and M has at least one row with no zero entries.[3] Then, obviously,

$$\mathscr{H}' \leq M \tag{9.2.16a}$$

and

$$(\mathscr{H}')^k \leq M^k \qquad \text{for } k = 0,1,2,\ldots. \tag{9.2.16b}$$

For a matrix M with these properties [FI1], there exists a matrix $\hat{M}' = [\hat{m}_{j,i}]$ such that

$$\lim_{k\to\infty} M^k = \hat{M}', \tag{9.2.17a}$$

$$0 \leq \hat{m}_{j,i} \leq 1, \qquad \text{for all } i,j \tag{9.2.17b}$$

$$\sum_{i=1}^{m} \hat{m}_{j,i} = 1, \qquad \text{for } j = 1,2,\ldots,m, \tag{9.2.17c}$$

$$\hat{m}_{j,i} = \hat{m}_{k,i}, \qquad \text{for } j,k = 1,\ldots,m. \tag{9.2.17d}$$

Hence, there exists a vector $\boldsymbol{\varepsilon}(k) \geq \mathbf{0}$ such that

$$\mathscr{H}^k \leq \lim_{k\to\infty} (M^k)' + \boldsymbol{\varepsilon}(k) = \hat{M} + \boldsymbol{\varepsilon}(k), \tag{9.2.18}$$

where $\boldsymbol{\varepsilon}(k) \to \mathbf{0}$ as $k \to \infty$. This completes the proof of the theorem.

There exists a *finite* k such that $\mathbf{x}(k) = \mathbf{0}$ if and only if $\mathbf{x}(0)$ is in the null space of the operator $\mathscr{H}^k$. If $\mathscr{H}$ is nilpotent,[4] there always exists such a $k \leq r$, where r is the degree of the minimum polynomial of $\mathscr{H}$. Furthermore, it can be shown that $\lim_{k\to\infty} \mathbf{x}(k) = \mathbf{0}$ if and only if the eigenvalues of $\mathscr{H}$ are within the unit circle.

If $\mathbf{u}(t) \equiv 0$, a desirable property of the network might be that $\mathbf{x}(k+1) < \mathbf{x}(k)$ or $\underset{1\leq i\leq m}{\text{Max}} [x_i(k+1)] < \underset{1\leq i\leq m}{\text{Max}} [x_i(k)]$ for all k. In the first case we see that the value of flow in each branch decreases as k increases, while in the second case the

---

[3] If the i-jth element $h_{j,i}$ of $\mathscr{H}'$ is nonzero, then let $m_{i,j} = h_{i,j}$. $\mathscr{H}'$ has at least one row, say row α, such that

$$\sum_{j=1}^{m} h_{j,\alpha} < 1.$$

If for any j, $h_{j,\alpha} = 0$, then let $m_{\alpha,j}$ be any element such that $0 < m_{\alpha,j} < 1$ and

$$\sum_{j=1}^{m} m_{\alpha,j} = 1.$$

M is known as a Markov matrix.

[4] $\mathscr{H}$ is nilpotent if there exists a finite integer k such that $\mathscr{H}^k = 0$.

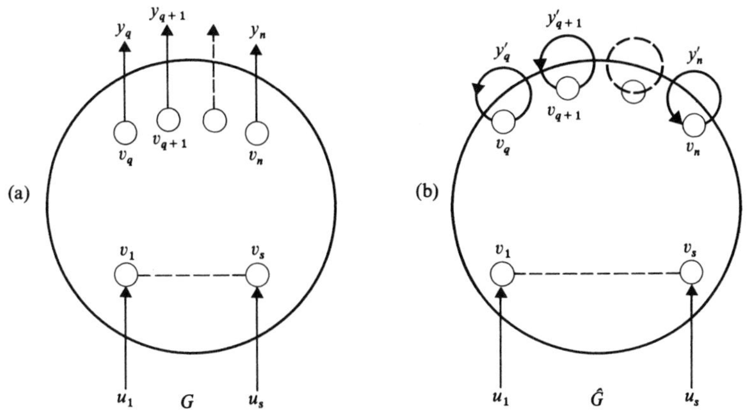

Fig. 9.2.2 (a) $y_{q+j}(k)$ is the output at time $t = t_0 + k\delta$ for $j = 1,\ldots,n - q$; (b) $y'_{q+j}(K) = \sum_{k=0}^{K} y_{q+j}(k)$ for $j = 1,\ldots,n - q$.

maximum branch flow decreases. The following theorem gives the necessary and sufficient condition for the second of these properties. The proof is left as an exercise.

Theorem 9.2.3 *Let* $(\mathbf{f},\hat{\mathbf{f}})$ *be a linear flow strategy defined by (9.2.11). Then*

$$\operatorname*{Max}_{1\le i\le m}\,[x_i(k+1)] < \operatorname*{Max}_{1\le i\le m}\,[x_i(k)], \tag{9.2.19a}$$

for all k and $\mathbf{x}(0)$, *if and only if*

$$\sum_{j=1}^{m} h_{i,j} < 1 \qquad \text{for } i = 1,\ldots,m. \tag{9.2.19b}$$

We next consider graphs with vertex demands. For the graph G shown in Fig. 9.2.2(a), vertices $v_1,\ldots,v_s$ represent the input locations and vertices $v_q,\ldots,v_n$ represent the output locations. Suppose the total output at v_i from time t_0 to time $t_0 + K\delta$ must be at least r_i units of flow for $i = q,\ldots,m$. This requirement can be alternatively specified by considering the graph $\hat{G}$ of Fig. 9.2.2(b). If, for $\hat{G}$, $h_{m+j,m+j} = 1$, where the $x_{m+j}(k)$ is the flow in self-loop b_{m+j} at time k, then $x_{m+j}(K)$ is equivalent to the sum of the output at v_{q+j} from time $t = t_0$ to time $t = t_0 + K\delta$. Hence, demands $r_q,\ldots,r_n$ are satisfied if

$$x_{m+j}(K) \ge r_{q+j} \qquad \text{for } j = 1,\ldots,n - q. \tag{9.2.20a}$$

or, equivalently, if $\hat{\mathbf{x}}$ is the flow vector for $\hat{G}$,

$$\hat{\mathbf{x}}(K) \ge \mathbf{r}, \tag{9.2.20b}$$

where $\mathbf{r} = (0,\ldots,0,r_q,\ldots,r_n)'$. These ideas lead to the following definition.

Definition 9.2.3 *The flow* $\hat{\mathbf{x}}_f$ *is reachable from the zero flow under the flow strategy* $(\mathbf{f},\hat{\mathbf{f}})$, *if there exists an integer K and nonnegative vectors* $\mathbf{u}(0),\mathbf{u}(1),\ldots,\mathbf{u}(K-1)$ *such* *that the zero flow* $\mathbf{x}(0) = \mathbf{0}$ *is transformed into the flow* $\mathbf{x}(K)$ *by* $\mathbf{u}(0), \ldots, \mathbf{u}(K-1)$ *where* $\mathbf{x}(K) \geq \mathbf{x}_f$.

Theorem 9.2.4 *Let* $(\mathbf{f},\hat{\mathbf{f}})$ *be a conservative linear strategy with flow equations given by (9.2.11). for a graph G with infinite branch capacities. Define* $Q(K)$ *by*

$$Q(K) = [\mathscr{J}, \mathscr{H}\mathscr{J}, \mathscr{H}^2\mathscr{J}, \ldots, \mathscr{H}^{K-1}\mathscr{J}]. \tag{9.2.21}$$

Then, flow $\mathbf{x}_f$ *is reachable from the zero flow in time K if and only if there exists a vector* $\mathbf{x}^*$ *in the subspace spanned by the columns of* $Q(K)$ *such that* $\mathbf{x}^* \geq \mathbf{x}_f$. *Furthermore,* $\mathbf{x}_f$ *is reachable from the zero flow in a finite time if and only if it is reachable from the zero flow in time* θ, *where* θ *is the degree of the minimal polynomial of* $\mathscr{H}$.

Proof. From (9.2.11), if $\mathbf{u}(0), \mathbf{u}(1), \ldots, \mathbf{u}(K-1)$ is an arbitrary sequence of inputs, the flow at time K is [ZA1]

$$\mathbf{x}(K) = \mathscr{H}^K\mathbf{x}(0) + \sum_{i=0}^{K-1} \mathscr{H}^i\mathscr{J}\mathbf{u}(K-i-1), \tag{9.2.22a}$$

and

$$\mathbf{x}(K) = \sum_{i=0}^{K-1} \mathscr{H}^i\mathscr{J}\mathbf{u}(K-i-1), \qquad \text{if } \mathbf{x}(0) = \mathbf{0}. \tag{9.2.22b}$$

Let $\mathscr{H}^i\mathscr{J} = [\boldsymbol{\mu}_{i,1}, \ldots, \boldsymbol{\mu}_{i,n}]$ where the $\boldsymbol{\mu}_{i,j}$ are column vectors and $i = 0,1,\ldots,K-1$. Then

$$\mathbf{x}(K) = \sum_{i=0}^{K-1} \sum_{j=1}^{n} \boldsymbol{\mu}_{i,j} u_j(K-i-1), \tag{9.2.23}$$

where $u_j(K-i-1)$ is the jth component of $\mathbf{u}_j(K-i-1)$. Clearly, if there is no $\mathbf{x}^* \geq \mathbf{x}_f$ in the space spanned by the columns of $Q(k)$ (i.e., the $\boldsymbol{\mu}_{i,j}$) then there does not exist any input sequence such that $\mathbf{x}_f$ can be reached in time K. From the Cayley-Hamilton theorem [GA2] the spaces spanned by the $Q(K)$ for $K \geq \theta$ are identical to the space spanned for $K = \theta$. Hence, if $\mathbf{x}_f$ is not reachable in time θ, it is not reachable. Thus, we have established the necessity of both conditions.

Suppose $\mathbf{x}^* \geq \mathbf{x}_f$ is in the space of the columns of $Q(K)$. This implies that there exist numbers $\xi_{i,j}$ for $i = 0, \ldots, K-1$ and for $j = 1, \ldots, n$ such that

$$\mathbf{x}^* = \sum_{i=0}^{K-1} \sum_{j=1}^{n} \boldsymbol{\mu}_{i,j}\xi_{i,j}. \tag{9.2.24}$$

Let $\xi_{i,j}^+ = \text{Max}\,[\xi_{i,j},0]$. Then, if we pick $\mathbf{u}^*(K-i-1) = (\xi_{i,l}^+, \ldots, \xi_{i,n}^+)$, we have

$$\mathbf{x}_f \leq \mathbf{x}^* = \sum_{i=0}^{K-1} \sum_{j=1}^{n} \boldsymbol{\mu}_{i,j}\xi_{i,j} \leq \sum_{i=0}^{K-1} \mathscr{H}'\mathscr{J}\mathbf{u}^*(K-i-1). \tag{9.2.25}$$

Hence, the sufficiency follows.

Theorem 9.2.4 is strongly dependent on the linear flow strategy $(\mathbf{f},\hat{\mathbf{f}})$. Therefore, given a graph, it is desirable to know whether there exists some linear flow strategy under which a given set of terminal flows are reachable. The proof of the following theorem is left as Problem 1.

***Theorem* 9.2.5** *Let* $\mathbf{x}_f$ *be a specified terminal state of the graph* C. *Statements* (1) *and* (2) *are equivalent.*

1) *There exists a conservative linear strategy, under which* $\mathbf{x}_f$ *is reachable from the zero flow, with equation*

$$\mathbf{x}(k+1) = \mathscr{H}\mathbf{x}(k) + \mathscr{J}\mathbf{u}(k) \tag{9.2.26}$$

such that $\mathscr{H} = [h_{i,j}]$, $\mathscr{J} = [j_{i,k}]$, and $s_{i,j}^{\mathscr{H}} = 1$ *implies* $h_{i,j} \geq \varepsilon > 0$, *and* $s_{i,k}^{\mathscr{J}} = 1$ *implies* $j_{i,k} \geq \varepsilon > 0$.

2) *There exists an* $\mathbf{x}^* \geq \mathbf{x}_f$ *in the space spanned by the columns of*

$$\bar{Q} = [S^{\mathscr{J}}, S^{\mathscr{H}}S^{\mathscr{J}}, \ldots, (S^{\mathscr{H}})^{\theta}S^{\mathscr{J}}] \tag{9.2.27}$$

where θ *is the degree of the minimal polynomial of* $S^{\mathscr{H}}$.

Often, the maximum input at a given vertex is limited. If this is the case, flows reachable in an unconstrained system may no longer be reachable. Furthermore, we can no longer set a uniform upper limit on the number of time intervals necessary to reach a given flow vector.

Suppose $(\mathbf{f},\hat{\mathbf{f}})$ is a linear conservative flow strategy. Let the input $\mathbf{u}(k)$ be bounded by some vector $\mathbf{U}$ such that for all k, $\mathbf{u}(k) \leq \mathbf{U}$. Let $\mathbf{x}(0) = \mathbf{0}$ and let $\mathbf{x}_f$ be a specified terminal flow. Let $\lambda_1, \lambda_2, \ldots, \lambda_m$ be the eigenvalues of the matrix $\mathscr{H}$. We assume that the λ_i are distinct. For the case of nondistinct eigenvalues see [FR1]. From the fundamental formula for a function of a matrix [GA2],

$$\mathscr{H}^k = \sum_{i=1}^{m} \lambda_i^k Z_i, \tag{9.2.28}$$

where the Z_i are $m \times m$ matrices called the *components of* $\mathscr{H}$ and are independent of k. Then

$$\mathbf{x}(K) = \sum_{k=0}^{K-1} \mathscr{H}^k \mathscr{J}\mathbf{u}(K-k-1) = \sum_{k=0}^{K-1} \sum_{i=1}^{m} \lambda_i^k Z_i \mathscr{J}\mathbf{u}(K-k-1), \tag{9.2.29}$$

and

$$\mathbf{x}(K) \leq \sum_{k=0}^{K-1} \sum_{i=1}^{m} \lambda_i^k \boldsymbol{\mu}_i, \tag{9.2.30}$$

where

$$\boldsymbol{\mu}_i = Z_i \mathscr{J}\mathbf{U}, \qquad \text{for } i = 1, \ldots, m. \tag{9.2.31}$$

Furthermore,

$$\sum_{k=0}^{K-1}\sum_{i=1}^{m}\lambda_i^k\boldsymbol{\mu}_i = \sum_{i=1}^{m}\boldsymbol{\mu}_i\left(\sum_{k=0}^{K-1}\lambda_i^k\right). \tag{9.2.32}$$

For $i = 1,\ldots,m$ define β_i as

$$\beta_i = \sum_{k=0}^{\infty}\lambda_i^k = \begin{cases}\dfrac{1}{1-\lambda_i} & \text{for } |\lambda_i| < 1,\\ \infty & \text{otherwise.}\end{cases} \tag{9.2.33}$$

Hence,

$$\mathbf{x}(K) \le \sum_{i=1}^{m}\beta_i\boldsymbol{\mu}_i = \mathbf{x}_{\max}(\infty), \tag{9.2.34}$$

where $\mathbf{x}(K) \to \mathbf{x}_{\max}(\infty)$ if $\mathbf{u}(k) = \mathbf{U}$ for all k and $K \to \infty$. Since for any given K,

$$\sum_{k=0}^{K-1}\lambda_i^k = \frac{1-\lambda_i^K}{1-\lambda_i} \qquad \text{for } \lambda_i \ne 1, \tag{9.2.35}$$

the set of reachable flows can easily be found.

We conclude this section with a brief statement of the flow reachability problem for graphs with capacity constrained branches. For two arbitrary real vectors $\mathbf{W} = (W_1,\ldots,W_m)'$ and $\mathbf{Z} = (Z_1,\ldots,Z_m)'$, $\boldsymbol{\xi} = \text{Min}\,[\mathbf{W},\mathbf{Z}]$, is defined as $\boldsymbol{\xi} = (\xi_1,\ldots,\xi_m)'$ such that $\xi_i = \text{Min}\,[W_i,Z_i]$ for $i = 1,2,\ldots,m$. With this convention, we now define linear flow strategies for graphs with capacity constraints. We assume zero storage capability; that is, if more flow arrives at a vertex than can be transmitted, the excess flow is lost.

Definition 9.2.4 *The flow strategy* $(\mathbf{f},\hat{\mathbf{f}})$ *in a graph G with branch capacity vector* $\mathbf{c} = (c_1,c_2,\ldots,c_m)'$ *is said to be a memoryless linear strategy if*

$$\mathbf{x}(k+1) = \text{Min}\,[\mathscr{H}\mathbf{x}(k) + \mathscr{J}\mathbf{u}(k),\mathbf{c}] \tag{9.2.36a}$$

and

$$\mathbf{y}(k+1) = \mathscr{L}\mathbf{x}(k) + \mathscr{M}\mathbf{u}(k) \tag{9.2.36b}$$

where $\mathscr{H} = [h_{i,k}]$, $\mathscr{J} = [j_{i,k}]$, $\mathscr{L} = [l_{i,k}]$, *and* $\mathscr{M} = [m_{i,k}]$ *are* $m \times m, m \times n, n \times m$, *and* $n \times n$ *matrices, respectively.*

The following theorem can be proved [FR2].

Theorem 9.2.6 *Let* $(\mathbf{f},\hat{\mathbf{f}})$ *be a memoryless linear flow strategy of the graph G with branch capacity constraints and let the flow equations for* $(\mathbf{f},\hat{\mathbf{f}})$ *be given by (9.2.36a) and*

(9.2.36b). Let $\hat{Q}(K)$ be the set of vectors

$$\hat{Q}(K) = \left\{\zeta \mid \zeta \le \sum_{i=1}^{K} \zeta^{(i)}\right\} \tag{9.2.37}$$

where

$$\zeta^{(1)} = (c_1\beta_1, c_2\beta_2, \ldots, c_m\beta_m)'$$

$$\beta_l = \begin{cases} 1 & \text{if } j_{l,k} \neq 0 \quad \text{for any } k = 1,2,\ldots,n \\ 0 & \text{otherwise} \end{cases}$$

and

$$\zeta^{(k)} = \mathrm{Min}\left[\mathscr{H}\zeta^{(k-1)}, \mathbf{c} - \sum_{i=1}^{k-1} \zeta^{(i)}\right] \qquad \text{for } k = 2,3,\ldots,K.$$

Then, the flow $\mathbf{x}_f$ is reachable from the zero flow in time K if and only if $x_f \in \hat{Q}(K)$.

Several corollaries of the above theorem allow us to obtain similar results with time varying capacities $c_1(k), c_2(k), \ldots, c_m(k)$ and inputs bounded by time-varying functions [FR2]. In addition if all branch capacities are equal, the form of $\hat{Q}(K)$ is greatly simplified.

9.3 GRAPHS WITH MEMORY

The capacity of a branch often models the maximum allowable rate at which flow can be transmitted through that branch. If more flow arrives at some vertex than can be simultaneously transmitted over the outwardly directed branches, the excess flow must be stored at the vertex or the flow is lost. Hence, the vertex must have a memory system, which might be a computer, a warehouse, or other storage device. In this section, we present a general formulation for analysis and synthesis of deterministic networks with memory. This leads to a linear program with an unmanageable set of constraints. However, the formulation points out the critical features of networks with memory.

Let G be a weighted directed graph. Let c_i be the flow capacity of branch b_i and m_i the memory capacity of vertex v_i. From G, we construct a directed graph G^* by splitting each vertex v_j into two vertices, v_{j1} and v_{j2}. In G^*, all branches incident into v_j in G are incident into v_{j1} and all branches incident out of v_j in G are incident out of v_{j2} for each v_j. In G^*, there is a branch directed from v_{j1} to v_{j2} for all j. This construction is shown in Fig. 9.3.1.

Let the flow rate at time t due to messages originating from v_{i1} in branch b_j be $z_{i,j}(t)$. Let $x_{i,j}(t)$ be the sum of the corresponding flows entering v_j at time t and let $y_{i,j}(t)$ be the sum of these flows leaving v_j. If there is no memory at v_j (that is, $m_j = 0$), then $x_{i,j} = y_{i,j}$ for all t. If $m_j > 0$, then it is possible to have either $x_{i,j}(t) > y_{i,j}(t)$ or $x_{i,j}(t) < y_{i,j}(t)$.

We first consider the case when all flow in G^* originates at one source vertex, say $v_{1,1}$. Let $r_1(t)$ be the rate at which messages are produced at $v_{1,1}$ and let $r_{1,1}(t)$, $r_{1,2}(t), \ldots, r_{1,n}(t)$ be the demands at $v_{1,2}, v_{2,2}, \ldots, v_{n,2}$, respectively. At each vertex

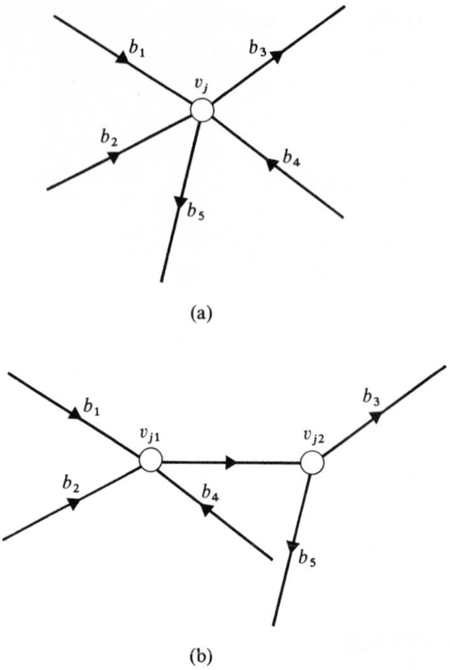

Fig. 9.3.1 (a) Vertex before splitting; (b) vertex after splitting.

of G^* we can write a linear algebraic equation expressing a relationship between the above quantities. Define the following column vectors:

$$\mathbf{Z}_1(t) = (z_{1,1}(t), z_{1,2}(t), \ldots, z_{1,n}(t))'; \tag{9.3.1a}$$

$$\mathbf{X}_1(t) = (x_{1,1}(t), x_{1,2}(t), \ldots, x_{1,n}(t))'; \tag{9.3.1b}$$

$$\mathbf{Y}_1(t) = (y_{1,1}(t), y_{1,2}(t), \ldots, y_{1,n}(t))'; \tag{9.3.1c}$$

$$\mathbf{R}_1(t) = (r_{1,1}(t), r_{1,2}(t), \ldots, r_{1,n}(t))'. \tag{9.3.1d}$$

Then

$$W_1 \begin{bmatrix} \mathbf{Z}_1(t) \\ r_1(t) \end{bmatrix} = \mathbf{X}_1(t) \tag{9.3.2a}$$

and

$$W_2 \begin{bmatrix} \mathbf{Z}_1(t) \\ \mathbf{R}_1(t) \end{bmatrix} = \mathbf{Y}_1(t), \tag{9.3.2b}$$

where W_1 and W_2 are appropriately defined matrices of zeros and ones.

In addition to the above equations, we have two sets of inequality constraints. The first set is simply $\mathbf{0} \le \mathbf{Z}_1(t) \le \mathbf{c}$ where $\mathbf{c} = (c_1, c_2, \ldots, c_m)'$. The second set of constraints expresses limitation on memory storage. Let $\mathbf{S}_1(t) = (s_{1,1}(t), s_{1,2}(t), \ldots, s_{1,n}(t))'$ be a column vector whose ith component represents the amount stored in the memory at v_i at time t. Then, for each t, we have $\mathbf{0} \le \mathbf{S}_1(t) \le \mathbf{M}$ where $\mathbf{M} = (m_1, m_2, \ldots, m_n)'$. In other words,

$$0 \le \mathbf{S}_1(t_0) + \int_{t_0}^{t} [\mathbf{X}_1(\tau) - \mathbf{Y}_1(\tau)]\, d\tau \le \mathbf{M}. \tag{9.3.3}$$

Now, suppose that there are sources at many vertices in G. Let $r_i(t)$ be the flow at time t from source v_i and let $\mathbf{R}_i(t) = (r_{i,1}, r_{i,2}, \ldots, r_{i,n})'$ be a vector whose component $r_{i,j}(t)$ represents the demand at v_j of a flow originating at v_i. Let $\mathbf{S}_i(t)$ be a vector whose components represent the amount stored in the various memories due to the source at v_i. The above equations can now be generalized as

$$W_1 \begin{bmatrix} \mathbf{Z}_i(t) \\ r_i(t) \end{bmatrix} = \mathbf{X}_i(t), \qquad \text{for } i = 1,2,\ldots,n; \tag{9.3.4a}$$

$$W_2 \begin{bmatrix} \mathbf{Z}_i(t) \\ \mathbf{R}_i(t) \end{bmatrix} = \mathbf{Y}_i(t), \qquad \text{for } i = 1,2,\ldots,n; \tag{9.3.4b}$$

$$\sum_{i=1}^{n} \mathbf{Z}_i(t) \le \mathbf{c}; \tag{9.3.4c}$$

$$\mathbf{Z}_i(t) \ge 0, \qquad \text{for } i = 1,2,\ldots,n; \tag{9.3.4d}$$

$$\sum_{i=1}^{n} \mathbf{S}_i(t_0) + \int_{t_0}^{t} [\mathbf{X}_i(\tau) - \mathbf{Y}_i(\tau)]\, d\tau \le \mathbf{M}; \tag{9.3.4e}$$

$$\mathbf{S}_i(t_0) + \int_{t_0}^{t} [\mathbf{X}_i(\tau) - \mathbf{Y}_i(\tau)]\, d\tau \ge \mathbf{0} \qquad \text{for } i = 1,2,\ldots,n, \tag{9.3.4f}$$

for all $t \ge t_0$. In the above equations $\mathbf{Z}_i(t)$, $\mathbf{X}_i(t)$, and $\mathbf{Y}_i(t)$ are column vectors defined by

$$\begin{aligned} \mathbf{Z}_i(t) &= (z_{i,1}, z_{i,2}, \ldots, z_{i,m})', \\ \mathbf{X}_i(t) &= (x_{i,1}, x_{i,2}, \ldots, x_{i,m})', \\ \mathbf{Y}_i(t) &= (y_{i,1}, y_{i,2}, \ldots, y_{i,m})'. \end{aligned}$$

We now formulate two problems that in principle can be solved by linear programming. In the formulation of the linear programs the expressions in (9.3.4) are constraints.

Problem 9.3.1 Optimum Design Let G, $r_i(t)$, and $\mathbf{R}_i(t)$ be specified for $i = 1,2,\ldots,n$. Suppose that the initial condition vectors $\mathbf{S}_1(0), \ldots, \mathbf{S}_n(0)$ are given. Also assume linear cost coefficients h_i and k_j such that $h_i c_i$ and $k_j m_j$ are the cost of c_i units of capacity for branch b_i and m_j units of memory for the memory at vertex v_j, respectively. We would like to find $\mathbf{c}$ and $\mathbf{M}$ such that the flow requirements are satisfied on the finite time interval $[0,T]$. This problem can be formulated as: find $\mathbf{c} \geq \mathbf{0}$ and $\mathbf{M} \geq \mathbf{0}$ such that

$$\mathbf{h}'\mathbf{c} + \mathbf{k}'\mathbf{m} \text{ is minimized} \tag{9.3.5}$$

subject to (9.3.4a) through (9.3.4f).

Problem 9.3.2 Evaluation of Traffic Handling Capability We would like to determine the maximum flow capabilities of a given graph. Thus, assume we are given G, $\mathbf{c}$, $\mathbf{M}$, and $\mathbf{S}_1(0), \mathbf{S}_2(0), \ldots, \mathbf{S}_n(0)$. We are also given $r_1(t), r_2(t), \ldots, r_n(t)$ but the $\mathbf{R}_i(t)$ are not specified. Subject to (9.3.4a) through (9.3.4f), we maximize

$$\sum_{i=1}^{n} \left[\delta_i \sum_{j=1}^{n} \hat{\delta}_j \int_0^T r_{i,j}(t)\, dt \right] \tag{9.3.6}$$

where the δ_i and $\hat{\delta}_j$ are zero or one depending upon the choice of the desired subsets of flows to be maximized.

To solve either one of Problems 9.3.1 or 9.3.2, we can quantize time and then replace all of the integrals by their equivalent summations. Then, all constraints and objective functions become linear algebraic equations. Hence, the problems can be formulated as linear programs although the size of the programs could be extremely large. For example, if the interval $[0,T]$ is quantized into T^* subintervals, the number of constraints in (9.3.4a) through (9.3.4f) is $3n^2T^* + mT^* + nT^* + nmT^*$. Because of the large number of constraints, we consider a suboptimal approach.

Suppose that $\gamma_{i,j}(t)$ is the flow generated at v_i at time t which has destination v_j. Moreover, suppose that $\gamma_{i,j}(t)$ is a *periodic* function with period T, and let $\bar{\gamma}_{i,j}$ be its mean. That is,

$$\bar{\gamma}_{i,j} = \frac{1}{T} \int_0^T \gamma_{i,j}(t)\, dt \qquad \text{for } i,j = 1, \ldots, n. \tag{9.3.7}$$

Flow will be routed from v_i to v_j over the *shortest path* between these vertices. Moreover, the output flow at time t from v_i will be taken to be the average flow $\bar{\gamma}_{i,j}$. Hence, if $s_i^j(t)$ is the amount of flow which is destined for v_j stored at time t at v_i, then $s_i^j(0) = s_i^j(T)$. Once a flow leaves the memory at v_i it is routed directly from v_i to v_j without further delay.

We consider the source producing flow at v_i for v_j independently of the other flow sources in the system. Let the cost of providing c units of capacity on the shortest path from v_i to v_j be $h(i,j)c$. The cost of providing sufficient capacity to transmit all flow produced at v_i with destination v_j is

$$H(i,j) = h(i,j) \operatorname*{Sup}_{0 \leq t \leq T} \{\gamma_{i,j}(t)\}. \tag{9.3.8}$$

On the other hand, the cost of providing sufficient capacity to transmit at rate $\bar{\gamma}_{i,j}$ while storing excess flow is

$$K(i,j) = h(i,j)\bar{\gamma}_{i,j} + k_i \left\{ s_i^j(0) + \operatorname*{Sup}_{0 \leq t \leq T} \int_0^t [\gamma_{i,j}(\tau) - \bar{\gamma}_{i,j}]\, d\tau \right\}. \tag{9.3.9}$$

A decision to use direct routing without delay over store-and-forward traffic is made if

$$\text{Min}\,[H(i,j),K(i,j)] = H(i,j)$$

and, if the minimum cost decision is made for all i and j, the cost of the system is

$$\sum_{i=1}^{n}\sum_{j=1}^{n} \text{Min}\,[H(i,j),K(i,j)]. \tag{9.3.10}$$

The cost of storing flow destined from v_i to v_j in the memory unit at v_i is

$$s_i^j(0) + \sup_{0 \le t \le T} \int_0^t [\gamma_{i,j}(\tau) - \bar{\gamma}_{i,j}]\, d\tau \tag{9.3.11}$$

and hence the smaller $s_i^j(0)$, the smaller the cost $K(i,j)$. In addition, $s_i^j(0)$ must satisfy the inequality

$$s_i^j(0) + \int_0^t [\gamma_{i,j}(\tau) - \bar{\gamma}_{i,j}]\, d\tau \ge 0 \tag{9.3.12}$$

for all t in $[0,T]$ since otherwise we will be transmitting more than 100% of the available flow. In order to reduce (9.3.11) to its smallest value without violating (9.3.12), we select $s_i^j(0)$ to be

$$s_i^j(0) = -\inf_{0 \le t \le T} \int_0^t [\gamma_{i,j}(\tau) - \bar{\gamma}_{i,j}]\, d\tau. \tag{9.3.13}$$

With this value of $s_i^j(0)$, we can maintain a constant transmission rate $\bar{\gamma}_{i,j}$ from v_i to v_j with minimum memory requirement at v_i.

9.4 ELEMENTARY QUEUING THEORY

In this section we consider some elementary concepts of queuing theory which are prerequisite to the study of stochastic message delay. We will examine a system with a single input terminal and a single output terminal. Inputs arrive at random intervals of time and are processed according to a fixed strategy and then appear at the output terminal. This system is called a single-service stage and is illustrated in Fig. 9.4.1. The input arrival times are characterized by a probability distribution called the input process. An input to the system is a request for service. The strategy with which inputs are processed is called the service process. The goal of queuing theory is to

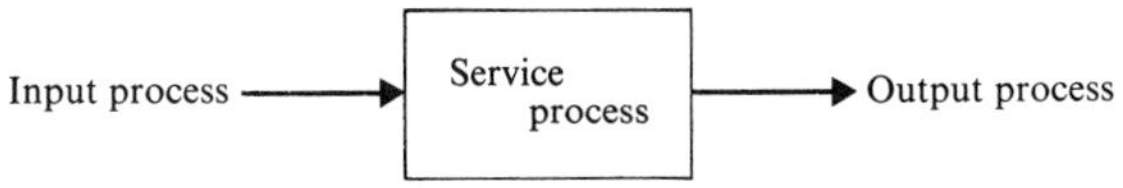

Fig. 9.4.1 A single-service stage.

compute the probability distribution of the output, or, in other words, to find the output process.

The simplest of input processes is the Poisson process, which is specified as follows. An input to the system is a request for service. At any time t, the probability that a service request occurs in the small interval of time $(t,t + \Delta t)$ is assumed to be $\lambda\Delta t + o(\Delta t)$ where λ is a constant and $o(\Delta t)$ is a quantity of higher than linear order in Δt. The two aspects of the Poisson process of special interest are: (1) the distribution function for the interarrival intervals for consecutive service requests; and (2) the probability distribution for the number of arrivals in a fixed interval of time.

Let $\eta(t)$ be the cumulative probability distribution for the interarrival intervals. That is, $\eta(t)$ is the probability that two service requests arrive at the input terminal no more than t seconds apart. The probability $1 - \eta(t)$ is therefore the probability that the time separating the arrivals of two consecutive service requests is greater than t seconds. Now, consider the time interval $(0,t + \Delta t)$; $1 - \eta(t + \Delta t)$ is the probability that the interarrival time is greater than $t + \Delta t$ seconds. This probability may be written as the product of the probability that the interarrival time is greater than t seconds and the probability of no arrivals in the interval $(t,t + \Delta t)$. Thus,

$$1 - \eta(t + \Delta t) = (1 - \eta(t))(1 - \lambda\Delta t - o(\Delta t)) \tag{9.4.1a}$$

or

$$\frac{(1 - \eta(t + \Delta t)) - (1 - \eta(t))}{\Delta t} = (1 - \eta(t))\lambda - \frac{(1 - \eta(t))o(\Delta t)}{\Delta t}. \tag{9.4.1b}$$

Taking the limit of (9.4.1b) as $\Delta t \to 0$ gives

$$\frac{d\eta(t)}{dt} = (1 - \eta(t))\lambda. \tag{9.4.2}$$

Since arrivals prior to time $t = 0$ are impossible, we have the initial condition $\eta(0) = 0$. The solution of (9.4.2) is therefore

$$\eta(t) = \begin{cases} 1 - e^{-\lambda t} & \text{for } t \geq 0, \\ 0 & \text{for } t < 0. \end{cases} \tag{9.4.3}$$

The probability distribution of the number of service requests in a fixed interval of time $(0,t)$ can be determined in a similar fashion. Let $P_k(t)$ be the probability that there are k arrivals in the interval $(0,t)$ for any t. If there have been k service requests up to time $t + \Delta t$, there must have been either k requests in $(0,t)$ and then no requests in $(t,t + \Delta t)$ or $k - 1$ requests in $(0,t)$ and one request in $(t,t + \Delta t)$. Since the probability that there will be more than one request in $(t,t + \Delta t)$ is not significant as $\Delta t \to 0$

$$P_k(t + \Delta t) = P_k(t)(1 - \lambda\Delta t) + P_{k-1}(t)\lambda\Delta t + o(\Delta t). \tag{9.4.4}$$

This equation is true for all k including $k = 0$ if $P_k(t) = 0$ for $k < 0$. Hence, the appropriate differential recurrence relation is

$$\frac{dP_k(t)}{dt} = -\lambda P_k(t) + \lambda P_{k-1}(t) \qquad \text{for } k = 0,1,\ldots. \tag{9.4.5}$$

The solution of this equation can be verified by substitution to be

$$P_k(t) = e^{-\lambda t}\frac{(\lambda t)^k}{k!} \qquad \text{for } k = 0,1,2,\ldots. \tag{9.4.6}$$

The expected number of arrivals in time t is λt, and the number of requests for service in a time interval $(0,t)$ has a Poisson distribution.

We now specify the service process. The service stage will be assumed to serve arrivals on a "first come, first served" basis. The service distribution will be assumed to be an *exponential* distribution. That is, if service begins at time 0, it will be completed in less than t seconds with probability $\xi(t)$ given by

$$\xi(t) = \begin{cases} 1 - e^{-vt} & \text{for } t \geq 0, \\ 0 & \text{for } t < 0. \end{cases} \tag{9.4.7}$$

The probability density of the service process is

$$\frac{d\xi(t)}{dt} = \begin{cases} ve^{-vt} & \text{for } t \geq 0, \\ 0 & \text{for } t < 0, \end{cases} \tag{9.4.8}$$

and the average service rate is v where $1/v$ is the average service time.

Now that we have specified the input and service processes for the service stage, we will consider their interaction to determine such quantities as the expected number of customers in the system and the expected time for service for an arbitrary service request. Our primary goal will be to find steady-state quantities.

Consider a single-service stage with Poisson input and exponential service. Let $Q_k(t)$ be the probability that there are a total of k requests for service in the system at time t. We can easily derive a set of differential equations for the $Q_k(t)$. In fact, for Δt sufficiently small,

$$\begin{aligned} Q_k(t + \Delta t) = {} & Q_k(t)(1 - \lambda\Delta t - o(\Delta t))(1 - v\Delta t - o(\Delta t)) \\ & + Q_{k+1}(v\Delta t)(1 - \lambda\Delta t - o(\Delta t)) \\ & + Q_{k-1}(\lambda\Delta t)(1 - v\Delta t - o(\Delta t)) \qquad \text{for } k \geq 1. \end{aligned} \tag{9.4.9}$$

In (9.4.9), $Q_k(t)(1 - \lambda\Delta t - o(\Delta t))(1 - v\Delta t - o(\Delta t))$ is the probability that there are k requests in the system at time t and no service requests arrive or are completed in the interval $(t,t + \Delta t)$; $Q_{k+1}(t)(v\Delta t)(1 - \lambda\Delta t - o(\Delta t))$ is the probability that there are $k + 1$ service requests in the system at time t, one service completion in the interval $(t,t + \Delta t)$, and no new requests in that interval; and $Q_{k-1}(t)(\lambda\Delta t)(1 - v\Delta t - o(\Delta t))$ is the probability that there are $k - 1$ requests in the system at time t, an additional arrival within the interval of time $(t,t + \Delta t)$, and no completions of service within that interval.

If we let $\Delta t \to 0$ in (9.4.9), we obtain the differential equations

$$\frac{dQ_k(t)}{dt} = -(\lambda + v)Q_k(t) + \lambda Q_{k-1}(t) + vQ_{k+1}(t) \qquad \text{for } k \geq 1 \tag{9.4.10a}$$

and

$$\frac{dQ_0(t)}{dt} = -\lambda Q_0(t) + \nu Q_1(t) \qquad \text{for } k = 0. \tag{9.4.10b}$$

The time-dependent solution of (9.4.10) is not of interest here. Instead, we seek the steady-state solution for $Q_k(t)$ defined by

$$\frac{dQ_k(t)}{dt} = 0.$$

If $\lambda \geq \nu$, the system will be unstable and the number of requests in the system will grow without bound. Therefore we consider the case $\lambda < \nu$. The set of equations (9.4.10) has a solution of the form $Q_k(t) = q_k$ for $k = 0,1,2,\ldots$, where the q_k satisfy the equations

$$0 = -\lambda q_0 + \nu q_1, \tag{9.4.11a}$$

$$0 = -(\lambda + \nu)q_k + \lambda q_{k-1} + \nu q_{k+1} \qquad \text{for } k \geq 1. \tag{9.4.11b}$$

Let K be an arbitrary positive integer and add the first $K - 1$ equations of this set to obtain

$$(-\lambda q_0 + \nu q_1) + (\lambda q_0 - \lambda q_1 - \nu q_1 + \nu q_2) + (\lambda q_1 - \lambda q_2 - \nu q_2 + \nu q_3) + \cdots + (\lambda q_{K-2} - \lambda q_{K-1} - \nu q_{K-1} + \nu q_K) = 0. \tag{9.4.12}$$

After cancelling the appropriate terms, we get

$$-\lambda q_{K-1} + \nu q_K = 0 \tag{9.4.13a}$$

or

$$q_K = \frac{\lambda}{\nu} q_{K-1} \qquad \text{for } K \geq 1. \tag{9.4.13b}$$

Hence

$$q_k = \rho^k q_0 \qquad \text{for } k = 0,1,2,\ldots \tag{9.4.14}$$

where

$$\rho = \frac{\lambda}{\nu} < 1.$$

The sum of the q_k's must equal unity. Therefore, for $\rho < 1$,

$$\sum_{k=0}^{\infty} q_k = q_0 \sum_{k=0}^{\infty} \rho^k = q_0 \frac{1}{1 - \rho}. \tag{9.4.15}$$

Consequently,

$$q_0 = 1 - \rho \tag{9.4.16a}$$

and

$$q_k = \rho^k(1 - \rho) \qquad \text{for } k = 0,1,2,\ldots. \tag{9.4.16b}$$

The expected number of service requests in the system is

$$E\{\text{number in system}\} = \sum_{k=0}^{\infty} kq_k = (1-\rho)\sum_{k=0}^{\infty} k\rho^k$$

$$= (1-\rho)\rho \sum_{k=1}^{\infty} k\rho^{k-1} = (1-\rho)\frac{d}{d\rho}\left\{\frac{1}{1-\rho}\right\}$$

$$= \frac{\rho}{1-\rho}. \tag{9.4.17}$$

It is easily shown that the variance of the number of service requests in the system is

$$\text{Var}\{\text{number in system}\} = \frac{\rho}{1-\rho} + \frac{\rho^2}{(1-\rho)^2}. \tag{9.4.18}$$

The next quantity of interest is the average time, $\hat{T}$, that a request for service spends in the system. This time is the sum of the average time spent in the queue waiting for service and the expected time for service. The average time $\hat{T}$ can be heuristically justified as follows. Since the system is in steady state, on the average λT requests for service will arrive in $\hat{T}$ seconds. Since we are following a first-come-first-served policy, if a request for service arrives at time t and is satisfied at time $t + \hat{T}$, the only requests at time $t + \hat{T}$ in the system will be those that arrived in the interval $(t, t + \hat{T})$. The expected number of requests in the system will therefore be λ times the expected time in the system. Symbolically,

$$\lambda\hat{T} = \frac{\rho}{1-\rho} \qquad \text{for } \rho < 1, \tag{9.4.19a}$$

or

$$\hat{T} = \frac{1}{\nu(1-\rho)} \qquad \text{for } \rho < 1. \tag{9.4.19b}$$

A rigorous derivation of (9.4.19b) can be found in any standard text on queuing theory [SA1, RI1].

Finally, we develop the analogy between a single exponential service stage and the two-vertex, one-branch graph shown in Fig. 9.4.2. Suppose that branch b_1 has capacity c_1 in bits per second. That is, messages can be transmitted from v_1 to v_2 at c_1 bits per second. Suppose that messages arrive at v_1 at the average rate of λ_1 messages per second with a Poisson distribution. Moreover, assume that the length of each message is an independent random variable from an exponential distribution with mean $1/\mu_1$ bits per message. Then, the average *message* transmission rate from

Fig. 9.4.2 A two-vertex, one-branch graph equivalent to single-service stage.

v_1 to v_2 is $\nu_1 = \mu_1 c_1$ messages per second. If we assume that the transmission of any message is completed when the last bit of that message arrives at v_2, the transmission process is similar to an exponential service stage with Poisson input. The average time, T_1, for a message to be totally received at v_2 once it has arrived at v_1 is equal to $\hat{T}$ given by (9.4.19b) and is

$$T_1 = \frac{1}{\mu_1 c_1 - \lambda_1} \tag{9.4.20}$$

where c_1 is the branch capacity in bits per second, $1/\mu_1$ is the average message length in bits, and λ_1 is the message arrival rate at v_1 in messages per second.

9.5 STOCHASTIC MESSAGE FLOW AND DELAY

We now examine networks in which probabilistic time delays are associated with the branches. The flows are assumed to be random variables and branch capacities are assumed to be transmission rates. The inability of the branches to transmit flows instantaneously causes queues of messages to form at the vertices.

The messages in these queues must be stored until they can be transmitted, on a first-come-first-served basis. We choose as a performance criterion the average time required for a message to reach its destination. In this section, we will summarize some results due to Kleinrock [KL1] who has treated networks of this type in great depth.

The expected message delay will be determined for graphs with specified stochastic processes governing message length and arrivals. The message routing will be assumed to be deterministic such that, if a message has its origin v_i and destination v_j, there is a unique directed *i-j* path over which the message must travel. Once the expected delay is found, the capacities of the branches which minimize this delay, subject to a fixed budget, will be calculated. To begin our study the appropriate assumptions and notation must be introduced.

Each message has a single origin and a single destination. A message is said to be received at a given vertex when the last bit of the message arrives at that vertex. Associated with each message are two random variables, its *length* and its *arrival time*. Message length will be assumed to be exponentially distributed. This assumption will be presently discussed in more detail. The input process at each source vertex is assumed to be a Poisson process. Message lengths and arrival times are assumed to be independent of one another and to have probability distributions that do not vary with time.

The following symbols will be used:[5]

$\gamma_{j,k}$ = average number of messages per second entering the graph at v_j with destination v_k

λ_i = average number of messages per second entering branch b_i

[5] We use Kleinrock's notation to facilitate reference to his book.

$\dfrac{1}{\mu_{j,k}} =$ average length (in bits) of messages entering the graph at v_j with destination v_k

$$\gamma = \sum_{j=1}^{n} \sum_{k=1}^{n} \gamma_{j,k} = \text{total arrival rate of messages to graph}$$

$$\lambda = \sum_{i=1}^{m} \lambda_i = \text{total rate of messages entering branches}$$

$$C = \sum_{i=1}^{m} c_i = \text{total branch capacity in graph (in bits per second)}$$

$T_i =$ average delay for messages passing through branch b_i

$T =$ average delay for messages arriving at vertices

In terms of the above notation, the average time delay for all messages can be written as the sum over i of the average delays of the messages passing through b_i divided by the total average number of messages entering the graph. Hence,

$$T = \sum_{i=1}^{m} \frac{\lambda_i}{\gamma} T_i = \text{average message delay.} \tag{9.5.1}$$

Assume we are given a fixed amount of capital Q and that c_i units of capacity on branch b_i costs $h_i c_i$. We will determine $c_1, \ldots, c_m$ in order to minimize

$$T = \sum_{i=1}^{m} \frac{\lambda_i}{\gamma} T_i, \tag{9.5.2a}$$

subject to a fixed deterministic routing procedure on a graph G and the cost constraint

$$Q = \sum_{i=1}^{m} h_i c_i \tag{9.5.2b}$$

where the h_i's are constants.

To solve the above synthesis problem, we must first solve the analysis problem. That is, we must calculate T given $c_1, \ldots, c_m$. This problem is totally intractable. The reason for this difficulty can be pointed out by considering a simple two-vertex, one-branch graph. Suppose that all messages enter the graph at v_1 with destination v_2. Each message entering the graph is assigned a length according to the exponential probability distribution. Now let us examine two successive messages which arrive at v_2. The second message may arrive at v_1 after the first message has been received at v_2, or the second message may arrive at v_1 while the first message is being transmitted. In the latter case, the interarrival time at v_2 between the two messages is equal to the length of the second message. Hence, the interarrival time at v_2 between a successive pair of messages is *not* independent of the length of the second message. Be-

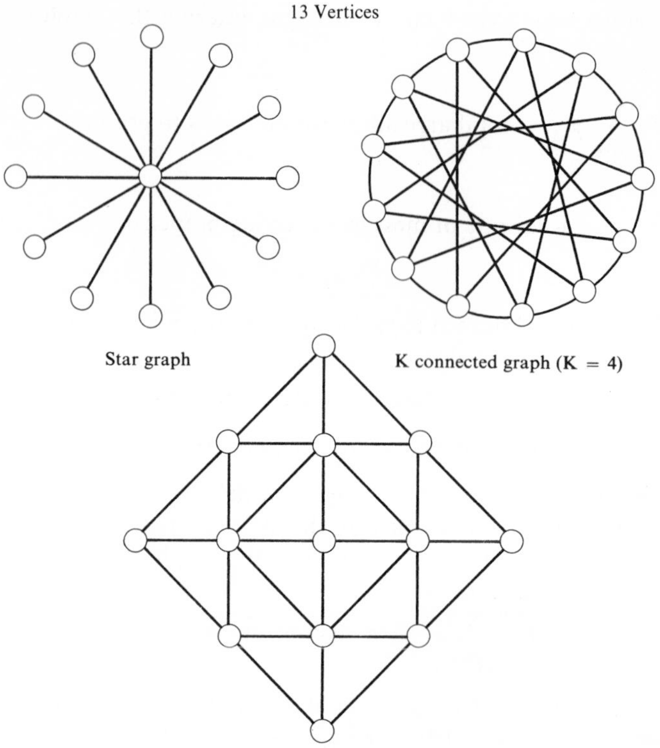

Figure 9.5.1(a)

cause of this dependence the problem of finding analytical solutions for average message delay is intractable.

Even if arrivals at the source vertices are independent of the length of the messages, arrivals at intermediate vertices in the graph are not independent of message length.

The mathematical difficulties created by message length and arrival time dependence can be eliminated if we alter the assumption that *permanent* lengths are assigned to each message. The following assumption will be made:

Independence Assumption *Each time a message is received at a vertex, a new length is chosen for this message from the probability density function* $p(l) = \mu e^{-\mu l}$.

The Independence Assumption does not correspond to usual schemes of information transmission. However, there are normally several outwardly directed and inwardly directed branches at any vertex. Hence, there are a multiplicity of paths in and out of each vertex. This multiplicity of paths tends to reduce the dependency between message arrivals and message lengths. If the Independence Assumption is made, a complete mathematical analysis of the problem becomes feasible. The validity of the assumption can then be tested by comparing the results of the mathematical analysis with results obtained by Monte Carlo programming. Kleinrock performed this com-

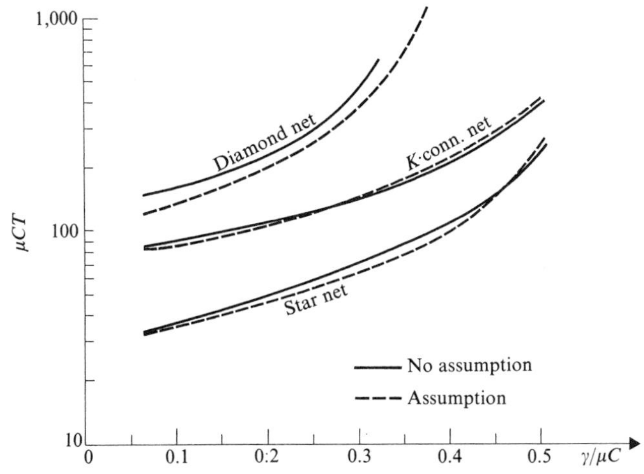

Fig. 9.5.1(b) Effect of Independence Assumption. (Taken from Fig. 3.5 of L. Kleinrock, *Communication Nets—Stochastic Message Flow and Delay*, McGraw-Hill, New York, 1964.)

parison for several graphs, three of which are shown in Fig. 9.5.1(a). The curves plotted in Fig. 9.5.1(b) indicate the accuracy of the Independence Assumption.

On the basis of the Independence Assumption, we will investigate the assignment of branch capacities to minimize the average delay. First, consider the simple graph of the form shown in Fig. 9.5.2. There are m branches emanating from v_1, each of which has capacity C/m. Messages arrive according to a Poisson process with average arrival rate λ messages per second. Moreover, all message lengths are identically and independently distributed with an exponential distribution of mean $1/\mu$ bits. These messages are transmitted on a first-come-first-served basis such that each message is transmitted over the first available branch. If several branches are available, the branch used is chosen from among the available branches uniformly at random.

The average time delay and the number m which minimizes this delay is given by the following theorem, proved in reference [KL1]. The conclusion of the theorem is that, to minimize T, we should minimize the number of paths out of v_1 and concentrate all traffic on a single branch.

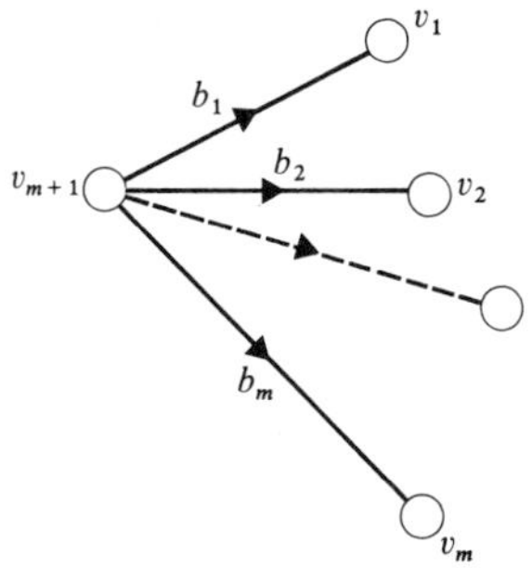

Fig. 9.5.2 System with several branches at input.

Theorem 9.5.1

a)

$$T = \frac{m}{\mu C}\left[1 + \frac{\dfrac{1}{m(1-\rho)}}{S_m(1-\rho)+1}\right] \tag{9.5.3}$$

where

$$S_m = \sum_{i=0}^{m-1} \frac{(m\rho)^{i-m} m!}{i!} > 0$$

and

$$\rho = \frac{\lambda}{\mu C}.$$

b) The value of m which minimizes T for all ρ, with $0 \le \rho < 1$, is $m = 1$. For $m = 1$,

$$T = \frac{1}{\mu C(1-\rho)}. \tag{9.5.4}$$

We next consider a graph for which there are m unconnected branches (Fig. 9.5.3). Vertex v_{i1} has Poisson arrivals at rate λ_i messages per second for $i = 1,2,\ldots,m$ and each message arriving at v_{i1} has an independent exponential length with mean $1/\mu_i$ bits. The capacity of b_i is c_i bits per second. The total branch capacity in the graph is fixed by the constraint

$$C = \sum_{i=1}^{m} c_i. \tag{9.5.5}$$

The total arrival rate at the input vertices is

$$\lambda = \sum_{i=1}^{m} \lambda_i. \tag{9.5.6}$$

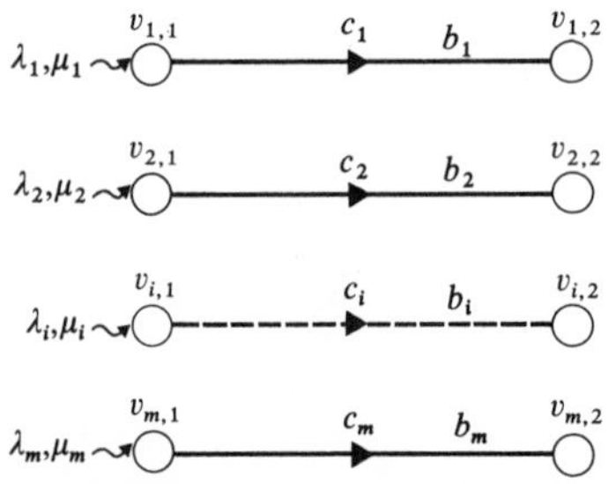

Fig. 9.5.3 System consisting of unconnected branches.

The average message time delay can then be defined as

$$T = \sum_{i=1}^{m} \frac{\lambda_i}{\lambda} T_i \tag{9.5.7}$$

where T_i is the sum of the average waiting time at v_{i1} and the average transmission time from v_{i1} to v_{i2}.

Each subgraph consisting of vertices v_{i1} and v_{i2}, and branch b_i behaves like the single exponential service stage discussed in Section 9.4. For such a system the average time delay is

$$T_i = \frac{1}{\mu_i c_i (1 - \rho_i)} \qquad \text{for } i = 1,2,\ldots,m \tag{9.5.8}$$

where

$$\rho_i = \frac{\lambda_i}{\mu_i c_i}.$$

Hence the total time delay is

$$T = \sum_{i=1}^{m} \frac{\lambda_i}{\lambda} \frac{1}{\mu_i c_i (1 - \rho_i)}. \tag{9.5.9}$$

We must now find the values of $c_1, c_2, \ldots, c_m$ constrained by (9.5.5) which minimizes T as given by (9.5.9). By straightforward use of Lagrange multipliers, we can obtain

Theorem 9.5.2 *The choice of $c_1, c_2, \ldots, c_m$ which minimizes the average time delay T subject to*

$$\sum_{i=1}^{m} c_i = C$$

is

$$c_i = \frac{\lambda_i}{\mu_i} + C(1 - \rho) \frac{\sqrt{\lambda_i/\mu_i}}{\sum_{j=1}^{m} \sqrt{\lambda_j/\mu_j}} \qquad \text{for } i = 1,2,\ldots,m \tag{9.5.10}$$

provided that

$$C > \sum_{i=1}^{m} \frac{\lambda_i}{\mu_i}, \tag{9.5.11a}$$

$$\rho = \frac{\lambda}{\mu C}, \tag{9.5.11b}$$

and

$$\frac{1}{\mu} = \sum_{i=1}^{m} \frac{\lambda_i}{\lambda} \frac{1}{\mu_i}. \tag{9.5.11c}$$

With this optimum assignment,

$$T_i = \frac{\sum_{j=1}^{m} \sqrt{\lambda_j/\mu_j}}{C(1-\rho)\sqrt{\lambda_i/\mu_i}} \quad \text{for } i = 1,2,\ldots,m \tag{9.5.12a}$$

and

$$T = \frac{\left[\sum_{i=1}^{m} \sqrt{\lambda_i/\mu_i}\right]^2}{\lambda C(1-\rho)}. \tag{9.5.12b}$$

In the optimum assignment, each branch is first given a capacity equal to the average flow λ_i/μ_i in that branch. The excess capacity is then assigned in proportion to the square roots of the average flows λ_i/μ_i.

In order to simplify the general synthesis problem, we assume that any message originating at v_i with destination v_j must follow a single unique path (for example, a shortest *i-j* path) through the graph. Such routing is called fixed routing since there are no alternate paths for messages to take in order to avoid saturated branches.

The following theorem is due to Burke [BU1].

Theorem 9.5.3 *The steady-state output of a system with m parallel servers, with Poisson arrival statistics, and with service times chosen independently from an exponential distribution is itself Poisson-distributed.*

The externally applied traffic to the graph is Poisson-distributed with exponential lengths. Then, if $\mu_i = \mu$ for all i, by Theorem 9.5.3, all interarrival times for messages anywhere in the graph are Poisson. The average message delay can be rewritten as

$$T = \sum_{j=1}^{n} \sum_{k=1}^{n} \frac{\gamma_{j,k}}{\gamma} z_{j,k} \tag{9.5.13}$$

where $\gamma_{j,k}$ is the average number of messages entering the graph at v_j with destination v_k, $z_{j,k}$ is the average delay for such messages, and

$$\gamma = \sum_{j=1}^{n} \sum_{k=1}^{n} \gamma_{j,k}.$$

The weighting factor $\gamma_{j,k}/\gamma$ for $z_{j,k}$ in (9.5.13) is the average number of messages which experience the delay $z_{j,k}$.

The quantity $z_{j,k}$ is the sum of the average delays encountered at the vertices and branches of the fixed path, say $\pi_{j,k}$, over which all messages with origin v_j and destination v_k are routed. Let λ_i be the sum of the $\gamma_{j,k}$'s such that b_i is in path $\pi_{j,k}$ and let λ be the sum of the λ_i's. That is,

$$\lambda_i = \sum_{\substack{j,k \\ b_i \in \pi_{j,k}}} \gamma_{j,k} \tag{9.5.14a}$$

and

$$\lambda = \sum_{i=1}^{m} \sum_{\substack{j,k \\ b_i \in \pi_{j,k}}} \gamma_{j,k}. \tag{9.5.14b}$$

The quantity λ_i corresponds to the average number of messages per second passing through b_i and hence λ is the total arrival rate for all branches in the graph. The delay, T, as given by (9.5.13), can also be written as

$$T = \sum_{i=1}^{m} \frac{\lambda_i}{\gamma} T_i. \tag{9.5.15}$$

The delay T_i is the service time for a single exponential service stage with Poisson input. Hence, $T_i = 1/\mu_i c_i(1 - \rho_i)$, and

$$T = \sum_{i=1}^{m} \frac{\lambda_i}{\gamma} \frac{1}{\mu c_i(1 - \rho_i)}. \tag{9.5.16}$$

A generalization of Theorem 9.5.2 is

Theorem 9.5.4 *The choice of $c_1, c_2, \ldots, c_m$ for a graph G with fixed routing which minimizes T subject to the constraint*

$$Q = \sum_{i=1}^{m} h_i c_i \tag{9.5.17}$$

is

$$c_i = \frac{\lambda_i}{\mu} + \frac{Q_e}{h_i} \frac{\sqrt{\lambda_i h_i}}{\sum_{j=1}^{m} \sqrt{\lambda_j h_j}}. \tag{9.5.18}$$

With this optimum assignment,

$$T_i = \frac{\sum_{j=1}^{m} \sqrt{\lambda_j h_j}}{\mu Q_e \sqrt{\lambda_i / h_i}} \tag{9.5.19a}$$

and

$$T = \frac{\bar{n}\left[\sum_{i=1}^{m} \sqrt{\lambda_i h_i/\lambda}\right]^2}{\mu Q_e} \tag{9.5.19b}$$

where

$$\bar{n} = \frac{\lambda}{\gamma}$$

provided that

$$Q_e = Q - \sum_{j=1}^{m} \lambda_j h_j/\mu > 0. \tag{9.5.20}$$

Proof. From (9.5.16),

$$T = \sum_{i=1}^{m} \frac{\lambda_i}{\gamma} \frac{1}{\mu c_i - \lambda_i}. \tag{9.5.21}$$

Using Lagrange multipliers, we define

$$H = T + \sigma\left(\sum_{i=1}^{m} c_i h_i - Q\right). \tag{9.5.22}$$

Differentiating H with respect to the c_i and setting these derivatives equal to zero gives

$$0 = -\frac{\lambda_i}{\gamma} \frac{\mu}{(\mu c_i - \lambda_i)^2} + \sigma h_i \tag{9.5.23a}$$

or

$$c_i = \frac{\lambda_i}{\mu} + \frac{1}{\sqrt{\sigma\gamma}} \sqrt{\frac{\lambda_i}{\mu h_i}}. \tag{9.5.23b}$$

Now, if we multiply (9.5.23b) by h_i and then sum over i, we obtain

$$\sum_{i=1}^{m} c_i h_i = Q = \sum_{i=1}^{m} \frac{\lambda_i h_i}{\mu} + \frac{1}{\sqrt{\sigma\gamma}} \sum_{i=1}^{m} \sqrt{\frac{\lambda_i}{\mu h_i}}. \tag{9.5.24}$$

Therefore, we can solve for $1/\sqrt{\sigma\gamma}$ as

$$\frac{1}{\sqrt{\sigma\gamma}} = \frac{Q - \sum_{i=1}^{m} \lambda_i h_i/\mu}{\sum_{i=1}^{m} \sqrt{\lambda_i h_i/\mu}}. \tag{9.5.25}$$

We now substitute the above expression for $1/\sqrt{\sigma\gamma}$ into (9.5.23b) and obtain (9.5.18). To establish the validity of (9.5.19a) and (9.5.19b), we substitute c_i, as given by (9.5.18), into (9.5.21).

Corollary 9.5.1 *$\bar{n} = \lambda/\gamma$ is the average number of branches in the path traversed by an arbitrary message.*

Proof. The average path length can be written as

$$\sum_{j=1}^{n} \sum_{k=1}^{n} \frac{\gamma_{j,k}}{\gamma} |\pi_{j,k}| \tag{9.5.26}$$

where $|\pi_{j,k}|$ is the number of branches in the path $\pi_{j,k}$ over which messages with origin v_j and destination v_k must be routed. Moreover, by definition,

$$\lambda = \sum_{i=1}^{m} \lambda_i = \sum_{i=1}^{m} \sum_{\substack{j,k \\ b_i \in \pi_{j,k}}} \gamma_{j,k}. \tag{9.5.27}$$

Since for each fixed j and k there are $|\pi_{j,k}|$ branches in $\pi_{j,k}$, the last sum can be rewritten as

$$\lambda = \sum_{\substack{j,k \\ b_i \in \pi_{j,k}}} \sum_{i=1}^{m} \gamma_{j,k} = \sum_{j,k} |\pi_{j,k}| \gamma_{j,k}. \tag{9.5.28}$$

Consequently, the average path length is λ/γ, which proves the corollary.

If $h_1 = h_2 = \cdots = h_m = h$ and $Q/h = C$, the fixed capital constraint is

$$C = \sum_{i=1}^{m} c_i.$$

The optimum branch capacities and the average time delay for these capacities are then

$$c_i = \frac{\lambda_i}{\mu} + C(1 - \bar{n}\rho) \frac{\sqrt{\lambda_i}}{\sum_{j=1}^{m} \sqrt{\lambda_j}} \qquad \text{for } i = 1,2,\ldots,m \tag{9.5.29a}$$

and

$$T = \frac{\bar{n}\left[\sum_{i=1}^{m} \sqrt{\frac{\lambda_i}{\lambda}}\right]^2}{\mu C(1 - \bar{n}\rho)} \tag{9.5.29b}$$

where $\rho = \lambda/\mu C$.

From (9.5.29b) we can make several observations. First, T can be made small when $\bar{n}$ is small. This means that the routing procedure and structure of the graph should be adjusted so that the average path length is small. For example, for a given G we can route messages over only a shortest path from origin to destination. *Both* the average time delay and the optimum branch capacities are functions of the message load $\rho = \lambda/\mu C$, which is the ratio of the arrival rate, in bits per second, to the total branch capacity. Hence, a graph designed optimally for low-traffic situations will not necessarily be optimum for high-traffic situations and vice versa.

The optimum capacities are functions of the λ_j's. However, the λ_j's are branch arrival rates and hence are not predetermined quantities. These arrival rates are functions of the traffic and are also dependent on the selection of the paths used in the fixed message routing. From (9.5.29) we can see that T is directly proportional to

$$\sum_{i=1}^{m} \lambda_i \tag{9.5.30}$$

and therefore we can reduce T by reducing this number. Using Lagrange Multipliers, it is not difficult to show that the λ_i's which minimize (9.5.29b) subject to

$$\sum_{i=1}^{m} \lambda_i = \lambda \tag{9.5.31}$$

are given by

$$\lambda_1 = \lambda \tag{9.5.32a}$$

$$\lambda_2 = \lambda_3 = \cdots = \lambda_m = 0. \tag{9.5.32b}$$

Furthermore, if the λ_i's are additionally constrained so that each branch must carry some minimum amount of traffic we can also compute the λ_i's which minimize (9.5.29b). That is, if we are given the constraints

$$\lambda_i \geq k_i > 0 \qquad \text{for } i = 1,2,\ldots,m \tag{9.5.33}$$

the λ_i's which minimize (9.5.29) subject to (9.5.31) and (9.5.33) are

$$\lambda_1 = \lambda - \sum_{i=2}^{m} k_i \tag{9.5.34a}$$

$$\lambda_i = k_i \qquad \text{for } i = 2,3,\ldots,m. \tag{9.5.34b}$$

Hence the optimum routing should *concentrate* the traffic into as few branches as possible.

9.6 FURTHER REMARKS

The state-space analysis of networks with branch time delays was given by Frank [FR1] and extended by Frank and El Bardai [FR2] to networks with capacity constraints. The model discussed here is easily adopted to networks with vertex processing times and can often be applied to study nonlinear flow strategies as well as continuous input systems. Other studies of networks with fixed branch delays are given by Gale [GA1] and by Ford [FO1]. In [FO1], Ford finds the maximum dynamic flow that can be established between a specified pair of terminals. This work is also discussed by Ford and Fulkerson [FO2].

Hakimi [HA1, HA2] developed the linear programming approach to the analysis and design of networks with memory. He also considered these problems for networks with messages of different priorities. It is interesting to observe that Hakimi's formulation can also be used to derive an approximate model of street or highway traffic.

The elementary queuing theory discussed in Section 9.4 can be found in any standard text [SA1, RI1]. The problem of stochastic message flow and delay is developed in great depth by Kleinrock [KL1]. Among the other problems he considers is the effect of message routing on time delay and waiting times for priority-weighted queue disciplines. Kleinrock, as well as Prosser [PR1], Shapiro [SH1], and Benes [BE1, BE2], also considers the effect of random network routing. The conclusion is that random routing is highly inefficient in terms of time delay but is relatively unaffected by small perturbations in traffic or network structure. Many additional results may be found on the analysis and design of store and forward computer-communication networks in papers by Kleinrock [KL2, KL3] and by Frank *et al.* [FR3, FR4].

PROBLEMS

1. Prove Theorem 9.2.5. Investigate the relationship between the conclusions of this theorem and the existence of directed paths in the graph.
2. Exhibit the explicit form of W_1 and W_2 in (9.3.2a) and (9.3.2b).
3. Generalize the linear programming formulation in Section 9.3 to store-and-forward schemes in which there are different priority classes for messages.
4. Derive input- output-state equations for networks with linear routings and deterministic time delays at the vertices.
5. Prove Theorem 9.2.6.
6. Prove Theorem 9.5.1.
7. Suppose that in the graph shown in Fig. P9.1 each branch (i,j) has a capacity $c(i,j)$ and a transit time $\delta(i,j)$, each of which varies with time but is constant over an interval Δt. We wish to find the maximum total s-t flow that can be sent in a given time $k\Delta t$. Formulate this problem as a maximum flow problem in a graph with fixed capacities and no transit times.

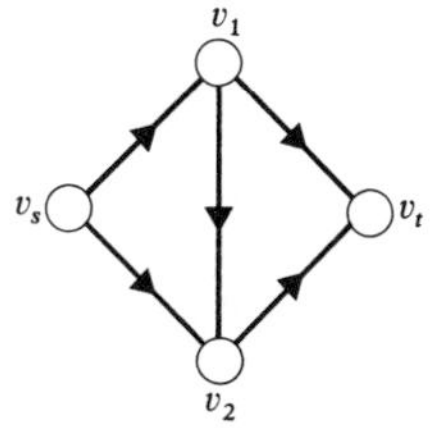

Figure P9.1

REFERENCES

BE1 V. E. Benes, *Mathematical Theory of Connecting Networks and Telephone Traffic,* Academic Press, New York, 1965.

BE2 V. E. Benes, "Programming and Control Problems Arising from Optimal Routing in Telephone Networks," *Bell System Tech. J.* **45,** 1373–1438 (1966).

BU1 P. J. Burke, "The Output of a Queueing System," *Operations Res.* **4,** 699–704 (1956).

FI1 M. Fisz, *Probability Theory and Mathematical Statistics,* Wiley, New York, 1963.

FO1 L. R. Ford, "Constructing Maximal Dynamic Flows from Static Flows," *Operations Res.* **6,** 419–433 (1958).

FO2 L. R. Ford, Jr., and D. R. Fulkerson, *Flows in Networks,* Princeton University Press, Princeton, N.J., 1962.

FR1 H. Frank, "Dynamic Communication Networks," *IEEE Trans. Commun. Tech.* **COM-15,** 156–163 (1967).

FR2 H. Frank and M. El Bardai, "Dynamic Communication Networks with Capacity Constraints," *IEEE Trans. Commun. Tech.* **COM-17,** 432–437 (1969).

FR3 H. Frank, I. T. Frisch, and W. Chou, "Topological Considerations in the Design of the ARPA Computer Network," *AFIPS Conf. Proc.* **36,** 581–587 (1970).

FR4 H. Frank, I. T. Frisch, W. Chou, and R. Van Slyke, "Optimal Design of Centralized Computer Networks," in Proceedings of the IEEE International Conference on Communications, June 1970, pp. 19-1–19-10.

GA1 D. Gale, "Transient Flows in Networks," *Mich. Math. J.* **6,** 59–63 (1958).

GA2 F. R. Gantmacher, *The Theory of Matrices,* Vol. 1, Chelsea, New York, 1960.

HA1 S. L. Hakimi, "A Linear Programming Formulation of Communication Networks with Memory," in Proceedings of First Annual Princeton Conference on Information Sciences and Systems, 1967, pp. 206–210.

HA2 S. L. Hakimi, "Analysis and Design of Communication Networks with Memory," *J. Franklin Inst.* **287,** 1–7 (1969).

KL1 L. Kleinrock, *Communication Nets—Stochastic Message Flow and Delay,* McGraw-Hill, New York, 1964.

KL2 L. Kleinrock, "Models for Computer Networks," in Proceedings of the IEEE International Conference on Communications, June 1969, pp. 21.9–21.16.

KL3 L. Kleinrock, "Analytic and Simulation Methods in Computer Network Design," *AFIPS Conf. Proc.* **36,** 569–579 (1970).

PR1 R. T. Prosser, "Routing Procedures in Communication Networks, Part I: Random Procedures," *IRE Trans. Commun. Systems* **CS-10,** 322–329 (1962).

RI1 J. Riordan, *Stochastic Service Systems,* Wiley, New York, 1962.

SA1 T. L. Saaty, *Elements of Queueing Theory,* McGraw-Hill, New York, 1961.

SH1 S. D. Shapiro, "Random Store and Forward Communication Networks," in Proceedings of the Polytechnic Institute of Brooklyn Symposium on Generalized Networks, Polytechnic Press, Brooklyn, N.Y., 1966, pp. 721–733.

ZA1 L. A. Zadeh and C. A. Desoer, *Linear System Theory,* McGraw-Hill, New York, 1963.

AUTHOR INDEX

SUBJECT INDEX

ABCDE7987654321